Microbiology Monographs

Volume 16

Series Editor
Alexander Steinbüchel
Münster, Germany

More information about this series at http://www.springer.com/series/7171

Héctor M. Alvarez
Editor

Biology of Rhodococcus

Second Edition

Springer

Editor
Héctor M. Alvarez
INBIOP, Universidad Nacional de la
Patagonia San Juan Bosco (UNPSJB) and
CONICET
Comodoro Rivadavia, Chubut, Argentina

ISSN 1862-5576 ISSN 1862-5584 (electronic)
Microbiology Monographs
ISBN 978-3-030-11460-2 ISBN 978-3-030-11461-9 (eBook)
https://doi.org/10.1007/978-3-030-11461-9

Library of Congress Control Number: 2019934756

This Springer imprint is published by the registered company Springer Nature Switzerland AG.
The registered company address is: Gewerbestrasse 11, 6330 Cham, Switzerland

Preface

Rhodococci are metabolically versatile actinobacteria frequently found in the environment with potential applications in bioremediation, biotransformations and biocatalysis, among other biotechnological processes. These microorganisms are currently the subject of research in many countries of the world. The number of publications and patents on rhodococci has increased significantly during the last several years. In this context, the knowledge acquired during the last decade about basic aspects of *Rhodococcus* biology is significant and reveals promising future prospects. Several public and private genomic projects involving *Rhodococcus* members are now in progress due to the increasing interest in their biotechnological applications. The large *Rhodococcus* genomes, which contain a multiplicity of catabolic genes, a high genetic redundancy of biosynthetic pathways and a sophisticated regulatory network, reflect the complexity of *Rhodococcus* biology. The combination of functional genomic studies with biochemical and physiological knowledge is providing new insights that will make it possible to put rhodococci to biotechnological use.

This *Microbiology Monographs* volume provides a thorough review of many aspects of biochemistry, physiology and genetics of *Rhodococcus* in the context of new genomic information. Expert international scientists have contributed reviews on the extraordinary capabilities of the *Rhodococcus* genus with regard to the biodegradation and biotransformation of diverse compounds, biosynthesis of lipids and biosurfactants, adaptation and tolerance to solvents and the interaction with metals and biotechnological applications. Chapters dealing with its taxonomy, genomes and plasmids and oligotrophic and central metabolism are also included in this volume. Moreover, the book examines the basic aspects of the phytopathogenic member *R. fascians*.

I would like to express my thanks to all of the authors, who contributed with high-quality reviews of each topic, to the series editor, Alexander Steinbüchel, and to the staff at Springer Nature, especially Bibhuti Sharma, for supporting the second edition of this book.

I hope that this volume will serve as a guidebook for researchers and students and will open new avenues for future research.

Comodoro Rivadavia, Argentina Héctor M. Alvarez

Contents

List of Contributors

Héctor M. Alvarez Institute of Bioscience of Patagonia-National Scientific and Technical Research Council (INBIOP-CONICET) and Faculty of Natural Sciences and Health Sciences, University of Patagonia San Juan Bosco, Comodoro Rivadavia, Chubut, Argentina

Martina Cappelletti Unit of General and Applied Microbiology, Department of Pharmacy and Biotechnology, University of Bologna, Bologna, Italy

Carla C. C. R. de Carvalho Department of Bioengineering, iBB-Institute for Bioengineering and Biosciences, Instituto Superior Técnico, Universidade de Lisboa, Lisbon, Portugal

Stefano Fedi Department of Pharmacy and Biotechnology, University of Bologna, Bologna, Italy

Isolde M. Francis Department of Biology, California State University Bakersfield, Bakersfield, CA, USA

Patrizia Di Gennaro Department of Biotechnology and Biosciences, University of Milano-Bicocca, Milan, Italy

Michael Goodfellow School of Biology, University of Newcastle, Newcastle upon Tyne, UK

Martín A. Hernández Institute of Bioscience of Patagonia-National Scientific and Technical Research Council (INBIOP-CONICET) and Faculty of Natural Sciences and Health Sciences, University of Patagonia San Juan Bosco, Comodoro Rivadavia, Chubut, Argentina

O. Marisa Herrero Institute of Bioscience of Patagonia-National Scientific and Technical Research Council (INBIOP-CONICET) and Faculty of Natural Sciences and Health Sciences, University of Patagonia San Juan Bosco, Comodoro Rivadavia, Chubut, Argentina

Irena B. Ivshina Institute of Ecology and Genetics of Microorganisms, Perm Scientific Centre of the Ural Branch of the Russian Academy of Sciences, Perm State University, Perm, Russia

Song Jiao Department of Chemical Engineering, Tsinghua University, Beijing, China

Key Laboratory of Industrial Biocatalysis, The Ministry of Education, Tsinghua University, Beijing, China

Amanda L. Jones Faculty of Health and Life Sciences, Department of Applied Sciences, Northumbria University, Newcastle upon Tyne, UK

Maria S. Kuyukina Institute of Ecology and Genetics of Microorganisms, Perm Scientific Centre of the Ural Branch of the Russian Academy of Sciences, Perm State University, Perm, Russia

Mariana P. Lanfranconi Institute of Bioscience of Patagonia-National Scientific and Technical Research Council (INBIOP-CONICET) and Faculty of Natural Sciences and Health Sciences, University of Patagonia San Juan Bosco, Comodoro Rivadavia, Chubut, Argentina

Shanshan Li School of Energy and Power Engineering, Xi'an Jiaotong University, Xi'an, China

Youxiang Liang Department of Chemical Engineering, Tsinghua University, Beijing, China

Key Laboratory of Industrial Biocatalysis, The Ministry of Education, Tsinghua University, Beijing, China

Ting Ma College of Life Sciences, Nankai University, Tianjin, China

Elena Piacenza Department of Biological Sciences, University of Calgary, Calgary, AB, Canada

Alessandro Presentato Environmental Microbiology Laboratory, Department of Biotechnology, University of Verona, Verona, Italy

Vartul Sangal Faculty of Health and Life Sciences, Department of Applied Sciences, Northumbria University, Newcastle upon Tyne, UK

Robert J. Seviour Department of Microbiology, La Trobe University, Bundoora, VIC, Australia

Roxana A. Silva Institute of Bioscience of Patagonia-National Scientific and Technical Research Council (INBIOP-CONICET) and Faculty of Natural Sciences and Health Sciences, University of Patagonia San Juan Bosco, Comodoro Rivadavia, Chubut, Argentina

Alexander Steinbüchel Institut für Molekulare Mikrobiologie und Biotechnologie der Westfälischen Wilhelms-Universität, Münster, Germany

Environmental Sciences Department, King Abdulaziz University, Jeddah, Saudi Arabia

Iain C. Sutcliffe Faculty of Health and Life Sciences, Department of Applied Sciences, Northumbria University, Newcastle upon Tyne, UK

Lingjun Tang Department of Chemical Engineering, Tsinghua University, Beijing, China

Key Laboratory of Industrial Biocatalysis, The Ministry of Education, Tsinghua University, Beijing, China

Raymond J. Turner Department of Biological Sciences, University of Calgary, Calgary, AB, Canada

Danny Vereecke Department of Plant and Crop, Faculty of Bioscience Engineering, Ghent University, Ghent, Belgium

María Soledad Villalba Institute of Bioscience of Patagonia-National Scientific and Technical Research Council (INBIOP-CONICET) and Faculty of Natural Sciences and Health Sciences, University of Patagonia San Juan Bosco, Comodoro Rivadavia, Chubut, Argentina

Miaomiao Wang Department of Chemical Engineering, Tsinghua University, Beijing, China

Key Laboratory of Industrial Biocatalysis, The Ministry of Education, Tsinghua University, Beijing, China

Nobuyuki Yoshida Department of Applied Chemistry and Biochemical Engineering, Graduate School of Engineering, Shizuoka University, Hamamatsu, Japan

Huimin Yu Department of Chemical Engineering, Tsinghua University, Beijing, China

Key Laboratory of Industrial Biocatalysis, The Ministry of Education, Tsinghua University, Beijing, China

Center for Synthetic and Systems Biology, Tsinghua University, Beijing, China

Jessica Zampolli Department of Biotechnology and Biosciences, University of Milano-Bicocca, Milan, Italy

Davide Zannoni Department of Pharmacy and Biotechnology, University of Bologna, Bologna, Italy

Refined Systematics of the Genus *Rhodococcus* Based on Whole Genome Analyses

Vartul Sangal, Michael Goodfellow, Amanda L. Jones, Robert J. Seviour, and Iain C. Sutcliffe

Contents

Abstract The genus *Rhodococcus* has had a long and complicated taxonomic history. With the advent of 16S rRNA gene sequence analyses and, subsequently, whole genome-based phylogenomic studies, the heterogeneity within the genus has become progressively more apparent and better resolved. We here review recent advances in the systematics of the genus, which currently contains more than 50 described species. Important developments include the proposed reclassifications of *Rhodococcus equi* and *Rhodococcus kunmingensis* into novel genera as *Prescottella equi* and *Aldersonia kunmingensis*, respectively. Notably, phylogenomic studies consistently resolve the genus *Rhodococcus* into multiple species-groups of coherent composition which thus likely merit separate genus status alongside *Rhodococcus* sensu stricto. In particular, a well-sampled group of plant-associated species we here define as the "*Rhodococcus fascians* assemblage" likely represents a novel genus.

V. Sangal · A. L. Jones · I. C. Sutcliffe (✉)
Faculty of Health and Life Sciences, Department of Applied Sciences, Northumbria University, Newcastle upon Tyne, UK
e-mail: iain.sutcliffe@northumbria.ac.uk

M. Goodfellow
School of Natural and Environmental Science, University of Newcastle, Newcastle upon Tyne, UK

R. J. Seviour
Department of Microbiology, La Trobe University, Bundoora, VIC, Australia

H. M. Alvarez (ed.), *Biology of Rhodococcus*, Microbiology Monographs 16,
https://doi.org/10.1007/978-3-030-11461-9_1

The rhodococci are remarkable organisms: interest in their ecology and biological activities (and the pathogenicity of some strains) has sustained thousands of scientific studies, including what is now two editions of this monograph (Alvarez 2010, 2018). Consequently, it is all the more remarkable that we haven't yet answered definitively the question "are they all actually rhodococci"? This chapter will review progress in clarifying the systematics of the genus *Rhodococcus* in the light of new knowledge and will provide a convincing answer to that question.

1 Historical Perspective and Recent Developments in the Traditional Classification of Rhodococci

The long and intricate taxonomic history of the genus *Rhodococcus* has been the subject of several comprehensive reviews (Bousfield and Goodfellow 1976; Goodfellow and Wayne 1982; Goodfellow et al. 1998; Gürtler and Seviour 2010) and hence will only be considered briefly here. The genus was proposed by Zopf (1891) for two species of red-coloured bacteria that had been classified in the genus *Micrococcus* by Overbeck (1891). Early classifications based upon form and function are difficult to navigate as "rhodococci" were assigned to 11 genera, including *Bacillus, Micrococcus*, *Mycobacterium* and *Nocardia*. The reintroduction of the epithet "rhodochrous" for members of this extensive taxonomic diaspora by Gordon and Mihm (1959) represented an important milestone in rhodococcal systematics as the strains belonging to this taxon were considered to represent a definitive species, albeit one that was tentatively classified in the genus *Mycobacterium*. However, it soon became apparent from a succession of studies based on chemotaxonomic and numerical phenetic methods that this taxon merited generic status (Bousfield and Goodfellow 1976). Following an extensive numerical taxonomic study by Goodfellow and Alderson (1977), the genus *Rhodococcus* was reintroduced for "rhodochrous" strains with *Rhodococcus rhodochrous* (Zopf 1891) Tsukamura 1974 as the type species, along with nine additional species. Several of the latter have since been assigned to the genus *Gordonia* (http://www.bacterio.net/rhodococcus.html).

The taxonomic status of most of the 32 validly named *Rhodococcus* species considered by Gürtler and Seviour (2010) and in the current edition of *Bergey's Manual of Systematic Bacteriology* (Jones and Goodfellow 2012) is underscored by a combination of genotypic and phenotypic data, although it is clear from associated 16S rRNA gene sequence studies that the genus is heterogeneous. Indeed, Jones and Goodfellow (2012) assigned the type strains of 30 rhodococcal species to three lineages, the *Rhodococcus equi*, *Rhodococcus erythropolis* and *R. rhodochrous* subclades. Similar intra-generic structure was also evident in other studies, although the position of *R. equi* strains tended to be unstable (Rainey et al. 1995; Goodfellow et al. 1998; McMinn et al. 2000; Gürtler et al. 2004; Gürtler and Seviour 2010). At the time of publication of Gürtler and Seviour (2010), the genus contained 32 validly named species. Since then, an additional 22 *Rhodococcus* species with either validly or effectively published names have been described (Table 1). In addition to the description of these new species, it has also been demonstrated that the biotechnologically important strain *Rhodococcus* sp. RHA1 (Yam et al. 2011; Ceniceros et al. 2017) belongs to the species *Rhodococcus jostii* (Jones et al. 2013a), whereas *Rhodococcus jialingiae* is a later synonym of *Rhodococcus qingshengii* (Táncsics et al. 2014). Furthermore, recent genome sequence data have been used to emend the descriptions of 23 rhodococcal species with more precise measures of genomic DNA G + C content and estimated genome sizes (Nouioui et al. 2018).

It is evident from 16S rRNA gene sequence analyses (as exemplified in Fig. 1) that the type strains of most of these species can be assigned to several multi-membered clades that correspond to a sister clade that encompasses representative *Nocardia* type strains. Genotypic data acquired from the analyses of nucleic acids and associated phenotypic data from chemotaxonomic and other expressed features (polyphasic taxonomy) has led to considerable improvements in the classification of rhodococci and other mycolic acid containing Actinobacteria classified in the order *Corynebacteriales* (Goodfellow and Jones 2012). Nevertheless, problems remain, notably the difficulty of distinguishing between genera within this taxon and between closely related species within individual genera (Goodfellow and Jones 2012; Baek et al. 2018). 16S rRNA gene sequence analysis in particular has played an important role in improving the classification of the genus although it is equally clear that this marker lacks sufficient resolution for distinguishing between closely related taxa. Classifications based on whole genome sequences and associated bioinformatic methods are proving to be effective ways of addressing such challenges, as they are based upon massively expanded amounts of sequence data, thereby providing a step change not only in information but also in reliability (Sangal et al. 2014b; Whitman 2014; Chun et al. 2018; Baek et al. 2018). However, it is important that whole genome sequence-based classifications (phylogenomics; see below) adhere to both minimal standards for sequencing quality (Chun et al. 2018) and sound taxonomic practices, such as the need to follow the nomenclatural type concept and the requirement to deposit type strains in a minimum of two public collections in different countries (Parker et al. 2019).

In addition to the application of whole genome sequence analysis to systematics, future improvements to the classification and identification of rhodococci will likely also derive from the refinement and implementation of matrix-assisted laser desorption ionisation-time of flight (MALDI-TOF) mass spectrometry methods for rapidly phenotyping bacteria (Hsueh et al. 2014; Schumann and Maier 2014; Pasciak et al. 2015; de Alegria Puig et al. 2017). The integration of phylogenomic and high-throughput MALDI-TOF-based approaches has much potential for microbial systematics generally (Ramasamy et al. 2014) and in application to the rhodococci and relatives specifically.

Table 1 *Rhodococcus* species described since 2010

Name	Type strain	16S rRNA accession number	Source	Reference
Rhodococcus artemisiae	YIM 65754^T	GU367155	Endophyte of *Artemisia annua* L.	Zhao et al. (2012)
Rhodococcus nanhaiensis	SCSIO 10187^T	JN582175	South China Sea sediment	Li et al. (2012)
Rhodococcus canchipurensis	MBRL 353^T	JN164649	Limestone quarry, India	Nimaichand et al. (2013)
Rhodococcus cerastii	C5^T	FR714842	Phyllosphere of *Cerastium holosteoides*	Kämpfer et al. (2013)
Rhodococcus trifolii	T8^T	FR714843	Phyllosphere of *Trifolium repens*	
Rhodococcus defluvii	Ca11^T	KC788572	Bioreactor wastewater, Germany	Kämpfer et al. (2014)
Rhodococcus enclensis	NIO-1009^T	HQ858009	Marine sediment sample, Goa	Dastager et al. (2014)
Rhodococcus kronopolitis[a]	NEAU-ML12^T	KF887492	*Kronopolites svenhedind* Verhoeff	Liu et al. (2014)
Rhodococcus biphenylivorans	TG9^T	KJ546454	Polychlorinated biphenyl-contaminated sediment, China	Su et al. (2015)
Rhodococcus soli	DSD51W^T	KJ939314	Park soil, Japan	Li et al. (2015)
Rhodococcus aerolatus	PAMC 27367^T	KM044053	Rainwater from above the Bering Sea	Hwang et al. (2015)
Rhodococcus agglutinans	CFH S0262^T	KP232908	Soil sample, Vietnam	Guo et al. (2015)
Rhodococcus lactis	DW151B^T	KP342300	Dairy waste sludge, India	Singh et al. (2015)
Rhodococcus antrifimi	D7-21^T	LN867321	Dried bat dung, South Korea	Ko et al. (2015)
Rhodococcus degradans	CCM 4446^T	JQ776649	Polluted soil, Switzerland	Švec et al. (2015)
Rhodococcus pedocola	UC12^T	KT301938	Forest topsoil, South Korea	Nguyen and Kim (2016)
Rhodococcus humicola	UC33^T	KT301939	Forest topsoil, South Korea	
Rhodococcus sovatensis	H004^T	KU189221	Hypersaline and heliothermal lake water, Romania	Táncsics et al. (2017)
Rhodococcus gannanensis	M1^T	NR_152643	Root endophyte of *Helianthus annuus* L.	Ma et al. (2017)
Rhodococcus olei	Ktm-20^T	MF405107	Oil-contaminated soil, Nepal	Chaudhary and Kim (2018)

(continued)

Table 1 (continued)

Name	Type strain	16S rRNA accession number	Source	Reference
Rhodococcus psychrotolerans[a]	CMAA 1533^T	KY317932	Rhizosphere of *Deschampsia antarctica*	Silva et al. (2018)
Rhodococcus electrodiphilus	JC435^T	LT630357	Marine coral reef	Ramaprasad et al. (2018)
Rhodococcus daqingensis[a]	Z1^T	MH205096	Petroleum-contaminated soil	Wang et al. (2018)

[a]Effectively published name at the time of writing

2 Reclassification of *R. equi* as *Prescottella equi*

Based on 16S rRNA gene trees, it is now clear that *R. equi* represents a distinct phylogenetic lineage, although as noted above its position is unstable with regard to other members of the genus *Rhodococcus*. Indeed, Gürtler and Seviour (2010) noted that "whether *R. equi* should continue to be placed in the genus *Rhodococcus* or moved to a separate or alternative existing genus still awaits resolution", while Jones and Goodfellow (2012) raised the question as to "whether it should be recognized as a genus in its own right". This matter was addressed with the proposed reclassification of *R. equi* as the type species of the genus *Prescottia* (Jones et al. 2013b), although the rules of nomenclate required that it was necessary subsequently to propose the replacement of the illegitimate genus name *Prescottia* with *Prescottella* (Jones et al. 2013c), with the type species *Prescottella equi*. The justification for the reclassification was based upon extensive phylogenetic, molecular fingerprinting and numerical taxonomic analyses (Jones et al. 2013b), which have since been supported by phylogenomic data (Sangal et al. 2016; see below). However, the question of the correct nomenclature has been complicated by the formal reclassification of *R. equi* into the species *Rhodococcus hoagii* (Kämpfer et al. 2014). This move has been challenged in a "Request for an Opinion" from the Judicial Commission of the International Committee on Systematics of Prokaryotes, which requested the retention of the species epithet *equi* (Garrity 2014). Until such an Opinion is issued, the correct name of the taxon remains *R. hoagii*, and the species name *P. equi* cannot be validated. As reviewed in Goodfellow et al. (2015), a pragmatic solution is to refer to the organism as "*Rhodococcus equi* (*Rhodococcus hoagii*/*Prescottella equi*)" at first usage in scientific communications, while the matter is under consideration by the Judicial Commission. We note that the community can use any of these names although currently *R. hoagii* is the "correct" one (Tindall 2014a).

While 16S rRNA gene sequence analysis recovered 26 *R. equi* strains in a single cluster, the molecular fingerprinting and numerical taxonomic analyses of Jones et al. (2013b) suggested that the species could be resolved into two subclusters. However, subsequent genomic analysis has indicated a high degree of relatedness

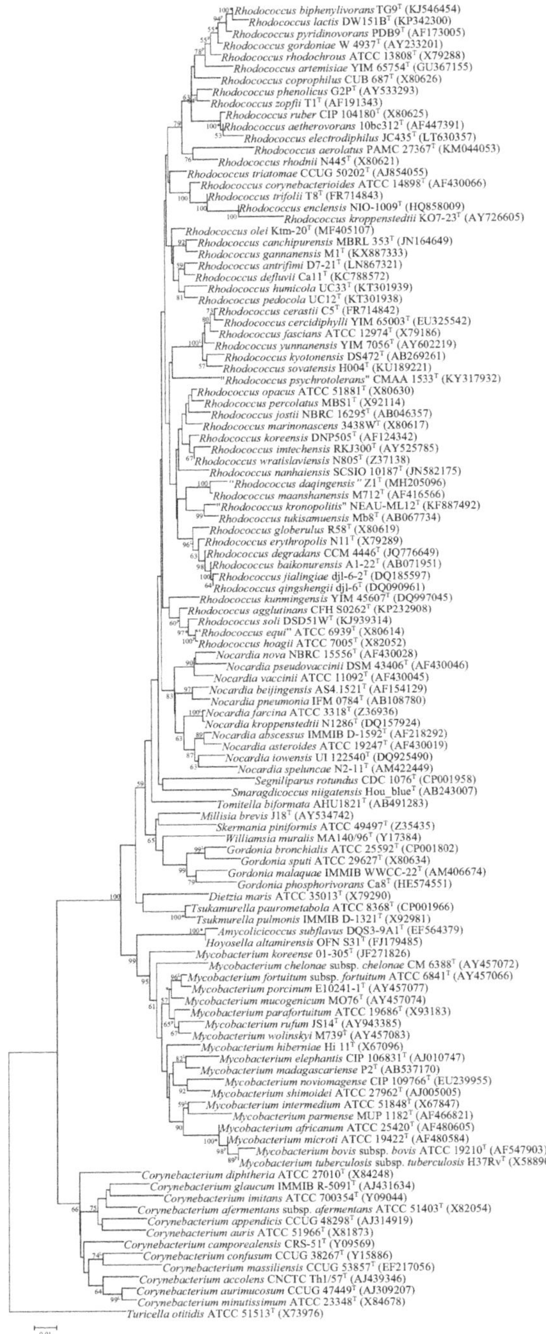

Fig. 1 16S rRNA gene sequence analysis of members of the genus *Rhodococcus* and representatives of related genera in the order *Corynebacteriales*. The tree was constructed using the neighbour-joining algorithm in MEGA. Nodes supported by greater than 50% of 1000 bootstrap replications are indicated. Nodes labelled L or P were also supported in the maximum likelihood or maximum parsimony trees, respectively. Nodes labelled * were supported by all three tree-making algorithms

between strains of *R. equi* (Sangal et al. 2014a, 2016; Anastasi et al. 2016), indicative of a clonal population, although six sub-clades were evident (Anastasi et al. 2016). Similarly, a multilocus sequence typing scheme for *R. equi* based on seven housekeeping loci (https://pubmlst.org/rhodococcus/) supported a clonal population structure and identified 37 sequence types within the species, 16 of which could be grouped into 6 clonal complexes (Duquesne et al. 2017).

3 Is the Name *Rhodococcus* Zopf 1891 "Legitimate"?

Apart from the above complications regarding the nomenclature of *R. equi*, Tindall (2014b) has reported that the genus name *Rhodococcus* Zopf 1891 is illegitimate under the rules of nomenclatural priority in the International Code of Nomenclature of Prokaryotes (Parker et al. 2019), as it post-dates the establishment of the algal genus *Rhodococcus* Hansgirg 1884. However, the algal genus *Rhodococcus* Hansgirg 1884 is currently regarded as synonymous with the earlier cyanobacterial genus *Chroococcus* Nägeli 1849 (http://www.algaebase.org/search/genus/detail/?genus_id=42893). Thus, the legitimacy (or otherwise) of the bacterial name *Rhodococcus* Zopf 1891 will ultimately depend on both confirming the assignment of *Rhodococcus* Hansgirg 1884 to *Chroococcus* Nägeli 1849 and also on complex revisions to both the International Code of Nomenclature of Algae, Fungi and Plants (ICNafp) and International Code of Nomenclature of Prokaryotes regarding which rules should apply to cyanobacteria (Tindall 2014b; Oren and Ventura 2017). Such revisions will undoubtedly be time-consuming, and so, in the meantime, it seems prudent that researchers studying members of the current genus *Rhodococcus* maintain the use of this name until these matters are resolved (e.g. by the International Committee on Systematics of Prokaryotes and its sister committees governing nomenclature). Ironically, the reclassification of various rhodococcal species-groups into new genera (see below) will help to provide these new taxa with legitimate names, so long as the rules of the International Code of Nomenclature of Prokaryotes are correctly followed.

4 Insights from Phylogenomic Analyses

Gürtler et al. (2004) posed the question "Can whole genome analysis refine the taxonomy of the genus *Rhodococcus*?" It is now clear that this question can be answered in the affirmative. Four phylogenomic studies have defined several robust species-groups within the genus. In a study focussed on plant pathogenic rhodococci and the intraspecific structure of *Rhodococcus fascians*, Creason et al. (2014) generated a phylogenetic tree based on average nucleotide identity (ANI) divergence for 59 genome-sequenced rhodococci, which was resolved into 7 clades with intragroup ANI values >70–75% (a threshold they considered relevant to the delineation of bacterial genera). The 25 plant-associated isolates included were recovered in two related clades. Eighteen of these isolates (including the type strain of *R. fascians*) were recovered in a clade with four recognisable subclades (1i–1iv). The remaining 34 rhodococci analysed were recovered in five clades, including Clade 5 containing *R. equi* strains. However, the clustering method and absence of outgroup taxa prevented relationships between these clades and other mycolic acid-containing bacteria from being inferred.

Subsequently, three further phylogenomic studies have confirmed the heterogeneity within *Rhodococcus*. In a study focussed on understanding relationships between 29 *R. equi* strains, Anastasi et al. (2016) used PhyloPhlAn to analyse the genomes of these strains, along with those of 46 rhodococci (discounting a sequence misattributed to "*Rhodococcus rhodnii*" NRRL B-16535^T; see below) and 57 other genomes representing strains belonging to 11 genera of the order *Corynebacteriales*. The genome of *Streptomyces albus* NBRC 1304^T was used as the out-group. This analysis recovered all the *Rhodococcus* strains in a single lineage but also identified five well-supported species-groups within this phyletic line, four of which were each resolved into two subclusters. Contemporaneously, we used PhyloPhlAn to analyse 100 rhodococcal genomes and 15 genomes representing seven genera of the order *Corynebacteriales* (Sangal et al. 2016). In this study, seven rhodococcal species-groups were identified, several of which were separated by lineages containing representatives of other genera, i.e. indicating *Rhodococcus* to be paraphyletic. Similar to Creason et al. (2014), we also found the ANIb threshold for separating potential genera to be ~74.8%. To explore this further, we have extended our PhyloPhlAn analysis to include 116 rhodococcal genomes and 15 genomes representing 7 genera of the order *Corynebacteriales* (Fig. 2; Table 2). Notably, this tree now includes genomes for type strains of 29 of the 52 currently validly named rhodococcal species. This analysis generated a tree with the same overall topology as that in Sangal et al. (2016), with the *Rhodococcus* strains recovered in the same seven species-groups (Table 3). In addition, *R. rhodnii* strain LMG 5362, which was considered to be a possibly misclassified singleton by Sangal et al. (2016), is now shown to belong to a newly defined species-group (H), along with *R. rhodnii* NBRC 100604^T and the type strain of *Rhodococcus triatomae* (Fig. 2). As noted by Anastasi et al. (2016), a genome deposited as "*R. rhodnii* NRRL B-16535^T" (Genbank GCA_000720375.1) is clearly misclassified, as it forms a lineage distant from the other two *R. rhodnii* strains (Fig. 2). As in the Sangal et al. (2016) analysis, *Rhodococcus kunmingensis* DSM 45001^T was recovered as a singleton lineage in Fig. 2, whereas it is typically loosely affiliated with *R. equi* (*Prescottella*) in 16S rRNA gene trees (e.g. Fig. 1).

Finally, a recent large-scale phylogenomic analysis of the phylum *Actinobacteria*, based on 1142 actinobacterial type strain genomes, has also

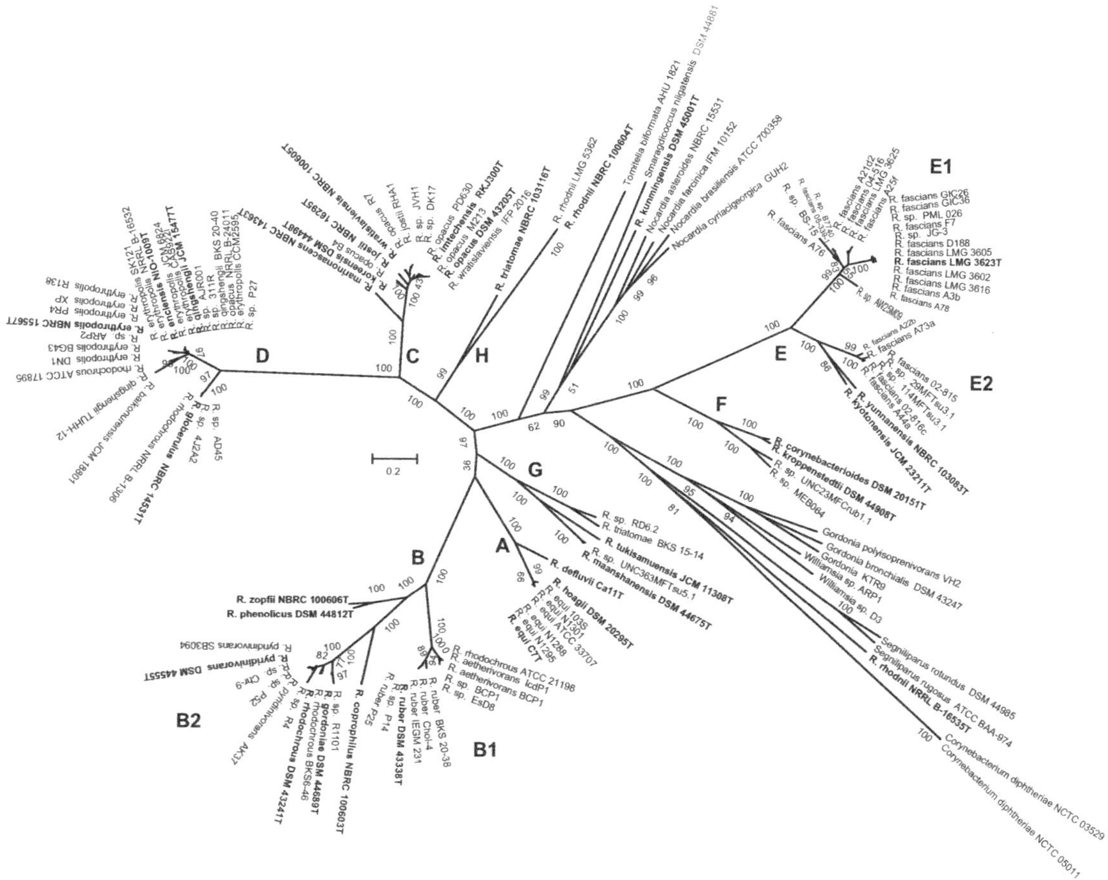

Fig. 2 Phylogenomic analysis of members of the genus *Rhodococcus* and representatives of related genera in the order *Corynebacteriales*. The genome sequence data were obtained from the GenBank and re-annotated using RAST server (Aziz et al. 2008). The phylogenetic tree was constructed from the protein sequences using PhyloPhlAn (Segata et al. 2013). PhyloPhlAn extracts a subset of amino acids from 400 universal proteins and constructs a maximum likelihood tree from the concatenated amino acid sequence alignment. The scale bar shows a normalised fraction of the total branch lengths as described by Segata et al. (2013)

Table 2 Data on rhodococcal and outgroup genomes analysed in this study

Taxon no.	Organism	Strain	Accession no.	Genome size (Mb)	GC%	Group
1	*Corynebacterium diphtheriae*	NCTC 03529	AJGI00000000	2.48	53.6	–
2	*Corynebacterium diphtheriae*	NCTC 05011	AJVH00000000	2.38	53.6	–
3	*Gordonia* sp.	KTR9	NC_018581	5.44	67.8	–
4	*Gordonia bronchialis*	DSM 43247	NC_013441	5.21	67.1	–
5	*Gordonia polyisoprenivorans*	VH2	NC_016906	5.67	67.0	–
6	*Nocardia asteroides*	NBRC 15531	BAFO02000000	6.96	69.9	–
7	*Nocardia brasiliensis*	ATCC 700358	NC_018681	9.44	68.1	–
8	*Nocardia cyriacigeorgica*	GUH2	NC_016887	6.19	68.4	–
9	*Nocardia farcinica*	IFM 10152	NC_006361	6.02	70.8	–
10	*Rhodococcus aetherivorans*	IcdP1	NZ_CP011341.1	5.92	70.6	B (B1)
11	***Rhodococcus aetherivorans***	**BCP1**	**AVAE00000000**	**6.23**	**70.4**	**B (B1)**
12	***Rhodococcus baikonurensis***	**JCM 18801**	**BBBO00000000**	**6.82**	**62.4**	**D**
13	***Rhodococcus coprophilus***	**NBRC 100603^T**	**BDAM00000000**	**4.55**	**66.9**	**B (B2)**
14	*Rhodococcus corynebacterioides*	DSM 20151^T	LPZL00000000	3.9	70.3	F
15	*Rhodococcus defluvii*	Ca11^T	JPOC00000000	5.14	68.7	A
16	***Rhodococcus enclensis***	**NIO-1009^T**	**FMBB00000000**	**7.48**	**62.3**	**D**
17	*Rhodococcus equi*	103S	NC_014659	5.04	68.8	A
18	*Rhodococcus equi*	N1288	LRQY00000000	5.17	68.8	A
19	*Rhodococcus equi*	N1295	LRQZ00000000	5.31	68.7	A
20	*Rhodococcus equi*	N1301	LRRA00000000	5.65	68.5	A
21	*Rhodococcus equi*	ATCC 33707	NZ_CM001149	5.26	68.8	A
22	*Rhodococcus equi*	C7^T	APJC00000000	5.2	68.8	A
23	*Rhodococcus erythropolis*	SK121	ACNO00000000	6.79	62.5	D
24	*Rhodococcus erythropolis*	CAS922i	LDPN01000000	7.21	62.3	D
25	*Rhodococcus erythropolis*	PR4	NC_012490	6.52	62.3	D
26	*Rhodococcus erythropolis*	CCM2595	NC_022115	6.28	62.5	D

27	*Rhodococcus erythropolis*	BG43	NZ_CP011295.1	6.33	62.4	D
28	***Rhodococcus erythropolis***	**NBRC 15567^{T}**	**BCRM00000000**	**6.59**	**62.4**	**D**
29	*Rhodococcus erythropolis*	JCM 6824	BBLL01000000	7.02	62.3	D
30	*Rhodococcus erythropolis*	NRRL B-16532	JOIL01000000	6.94	62.4	D
31	*Rhodococcus erythropolis*	XP	AGCF00000000	7.23	62.3	D
32	*Rhodococcus erythropolis*	DN1	AUZK00000000	6.55	62.4	D
33	*Rhodococcus erythropolis*	R138	ASKF01000000	6.77	62.3	D
34	*Rhodococcus fascians*	A22b	JOKB01000000	5.91	64.3	E (E1)
35	*Rhodococcus fascians*	A44a	JMEX01000000	5.95	64.5	E (E1)
36	*Rhodococcus fascians*	A73a	JMEW01000000	5.93	64.4	E (E1)
37	*Rhodococcus fascians*	02-815	JMFF01000000	6.24	64.3	E (E1)
38	*Rhodococcus fascians*	02-816c	JMFE01000000	6.08	64.5	E (E1)
39	*Rhodococcus fascians*	A76	JMEV01000000	6.03	64.6	E (E2)
40	*Rhodococcus fascians*	05-339-1	JMFC01000000	5.73	64.7	E (E2)
41	*Rhodococcus fascians*	04-516	JMFD01000000	5.82	64.2	E (E2)
42	*Rhodococcus fascians*	A21d2	JMFA01000000	5.98	64.1	E (E2)
43	*Rhodococcus fascians*	A25f	JMEZ01000000	5.87	64.1	E (E2)
44	*Rhodococcus fascians*	LMG 3625	JMEM01000000	5.94	64.1	E (E2)
45	*Rhodococcus fascians*	F7	LFDS01000000	5.25	64.7	E (E2)
46	*Rhodococcus fascians*	A3b	JMEY01000000	6.03	64.2	E (E2)
47	*Rhodococcus fascians*	A78	JMEU01000000	6	64.3	E (E2)
48	*Rhodococcus fascians*	GIC26	JMES01000000	5.34	64.5	E (E2)
49	*Rhodococcus fascians*	GIC36	JMER01000000	5.56	64.5	E (E2)
50	*Rhodococcus fascians*	D188	JMET01000000	5.44	64.6	E (E2)
51	*Rhodococcus fascians*	LMG 3602	JMEQ01000000	5.36	64.5	E (E2)
52	*Rhodococcus fascians*	LMG 3605	JMEP01000000	5.44	64.5	E (E2)
53	*Rhodococcus fascians*	LMG 3616	JMEO01000000	5.76	64.3	E (E2)

(continued)

Table 2 (continued)

Taxon no.	Organism	Strain	Accession no.	Genome size (Mb)	GC%	Group
54	*Rhodococcus fascians*	LMG 3623^T	JMEN01000000	5.78	64.4	E (E2)
55	***Rhodococcus globerulus***	**NBRC 14531^T**	**BCWX00000000**	**6.74**	**61.7**	**D**
56	*Rhodococcus gordoniae*	DSM 44689^T	LPZN00000000	4.82	67.9	B (B2)
57	*Rhodococcus hoagii*	DSM 20295^T	LRRF00000000	4.97	68.8	A
58	*Rhodococcus imtechensis*	RKJ300^T	AJJH00000000	8.23	67.2	C
59	*Rhodococcus jostii*	RHA1	NC_008268	7.8	67.5	C
60	***Rhodococcus jostii***	**NBRC 16295^T**	**BCWY00000000**	**9.73**	**66.9**	**C**
61	***Rhodococcus koreensis***	**DSM 44498^T**	**FNSV00000000**	**10.31**	**67.4**	**C**
62	*Rhodococcus kroppenstedtii*	DSM 44908^T	LPZO00000000	3.91	70.1	F
63	*Rhodococcus kunmingensis*	DSM 45001^T	LRRB00000000	5.62	66.2	*Aldersonia*
64	***Rhodococcus kyotonensis***	**JCM 23211^T**	**FZOW00000000**	**6.31**	**64.2**	**E2**
65	***Rhodococcus maanshanensis***	**DSM 44675^T**	**FOAW00000000**	**5.67**	**69.2**	**G**
66	***Rhodococcus marinonascens***	**NBRC 14363^T**	**BCXB00000000**	**4.92**	**64.4**	**C**
67	*Rhodococcus opacus*	DSM 43205^T	LRRG00000000	8.53	67.3	C
68	*Rhodococcus opacus*	B4	NC_012522	7.91	67.9	C
69	*Rhodococcus opacus*	M213	AJYC00000000	9.2	67.0	C
70	*Rhodococcus opacus*	R7	CP008947.1	8.47	67.2	C
71	*Rhodococcus opacus*	PD630	AGVD00000000	9.16	67.0	C
72	*Rhodococcus opacus*	NRRL B-24011	JOIM01000000	6.38	62.4	D
73	*Rhodococcus phenolicus*	DSM 44812^T	LRRH00000000	6.28	68.4	B
74	*Rhodococcus pyridinivorans*	SB3094	NC_023150	5.23	68.0	B (B2)
75	*Rhodococcus pyridinivorans*	AK37	AHBW00000000	5.25	67.9	B (B2)
76	*Rhodococcus pyridinivorans*	DSM 44555^T	LRRI00000000	5.18	67.9	B (B2)
77	*Rhodococcus qingshengii*	BKS 20-40	AODN00000000	6.6	62.4	D
78	*Rhodococcus qingshengii*	TUHH-12	JNCU01000000	7.43	61.7	D

79	*Rhodococcus qingshengii*	JCM 15477^{T}	LRRJ00000000	7.26	62.4	D
80	*Rhodococcus rhodnii*	LMG 5362	APMY00000000	4.39	69.7	H
81	*Rhodococcus rhodnii*	NRRL B-16535^{T}	JOAA01000000	12.42	66.3	See main text
82	***Rhodococcus rhodnii***	**NBRC 100604**	**BCXD00000000**	**4.46**	**69.7**	**H**
83	*Rhodococcus rhodochrous*	ATCC 21198	AZHI00000000	6.48	70.2	B (B1)
84	*Rhodococcus rhodochrous*	DSM 43241^{T}	LRRK00000000	5.18	68.2	B (B2)
85	*Rhodococcus rhodochrous*	BKS6-46	AGVW00000000	6.22	67.4	B (B2)
86	*Rhodococcus rhodochrous*	ATCC 17895	ASJJ00000000	6.87	62.3	D
87	*Rhodococcus rhodochrous*	NRRL B-1306	JNWS01000000	6.78	61.7	D
88	*Rhodococcus ruber*	DSM 43338^{T}	LRRL00000000	5.3	70.7	B (B1)
89	*Rhodococcus ruber*	IEGM 231	CCSD01000000	5.99	70.2	B (B1)
90	*Rhodococcus ruber*	P25	LDUF01000000	5.73	70.5	B (B1)
91	*Rhodococcus ruber*	Chol-4	ANGC00000000	5.4	70.6	B (B1)
92	*Rhodococcus ruber*	BKS 20-38	AOEX00000000	6.13	69.7	B (B1)
93	*Rhodococcus* sp.	BCP1	AVAE00000000	6.21	70.4	B (B1)
94	*Rhodococcus* sp.	EsD8	CAVJ000000000	6.63	70.1	B (B1)
95	*Rhodococcus* sp.	P14	AJFC00000000	5.68	70.4	B (B1)
96	*Rhodococcus* sp.	R1101	AJVB00000000	4.66	67.9	B (B2)
97	*Rhodococcus* sp.	Chr-9	JTIZ01000000	5.35	67.7	B (B2)
98	*Rhodococcus* sp.	R4	AFAQ00000000	8.86	69.5	B (B2)
99	*Rhodococcus* sp.	P52	JPJJ01000000	5.29	67.9	B (B2)
100	*Rhodococcus* sp.	JVH1	AKKP00000000	9.18	67.0	C
101	*Rhodococcus* sp.	DK17	AJLQ00000000	9.11	67.1	C
102	*Rhodococcus* sp.	311R	CFHW01000000	6.35	62.6	D
103	*Rhodococcus* sp.	AJR001	LPZM00000000	7.26	62.4	D
104	*Rhodococcus* sp.	ARP2	LEKE01000000	6.3	62.4	D
105	*Rhodococcus* sp.	P27	AVCO00000000	6.26	62.4	D

(continued)

Table 2 (continued)

Taxon no.	Organism	Strain	Accession no.	Genome size (Mb)	GC%	Group
106	*Rhodococcus* sp.	AD45	JYOP01000000	6.79	61.7	D
107	*Rhodococcus* sp.	4J2A2	CEDU01000000	6.45	61.8	D
108	*Rhodococcus* sp.	114MFTsu3.1	ARTN00000000	5.55	64.7	E (E1)
109	*Rhodococcus* sp.	29MFTsu3.1	ARND00000000	5.58	64.6	E (E1)
110	*Rhodococcus* sp.	BS-15	BAOX01000000	5.51	64.4	E (E2)
111	*Rhodococcus* sp.	B7740	NZ_CP010797.1	5.34	64.9	E (E2)
112	*Rhodococcus* sp.	PML 026	JZIS01000000	5.17	64.7	E (E2)
113	*Rhodococcus* sp.	JG-3	AXVF00000000	5.27	64.6	E (E2)
114	*Rhodococcus* sp.	AW25M09	CAPS00000000	5.64	64.1	E (E2)
115	*Rhodococcus* sp.	UNC23MFCrub1.1	JMLQ01000000	4.7	68.5	F
116	*Rhodococcus* sp.	MEB064	JXQS01000000	4.67	67.0	F
117	*Rhodococcus* sp.	UNC363MFTsu5.1	JMLO01000000	5.67	69.3	G
118	*Rhodococcus* sp.	RD6.2	CVQP01000000	5.57	68.4	G
119	*Rhodococcus triatomae*	BKS 15-14	AODO00000000	5.83	69.0	G
120	***Rhodococcus triatomae***	**NBRC 103116^T**	**BCXF00000000**	**4.74**	**68.7**	**H**
121	***Rhodococcus tukisamuensis***	**JCM 11308^T**	**FNAB00000000**	**5.49**	**69.8**	**G**
122	*Rhodococcus wratislaviensis*	IFP 2016	ANIU00000000	9.7	66.9	C
123	*Rhodococcus wratislaviensis*	NBRC 100605^T	BAWF01000000	10.41	66.8	C
124	***Rhodococcus yunnanensis***	**NBRC 103083^T**	**BCXH00000000**	**6.37**	**63.9**	**E2**
125	***Rhodococcus zopfii***	**NBRC 100606^T**	**BCXI00000000**	**6.3**	**68.2**	**B**
126	*Segniliparus rotundus*	DSM 44985	NC_014168	3.16	66.8	–
127	*Segniliparus rugosus*	ATCC BAA-974	ACZI00000000	3.57	68.1	–
128	*Smaragdicoccus niigatensis*	DSM 44881	AQXZ00000000	5.32	64.3	–
129	*Tomitella biformata*	AHU 1821	BAVQ00000000	4.71	68.1	–
130	*Williamsia* sp.	ARP1	JXYP00000000	4.75	68.6	–
131	*Williamsia* sp.	D3	AYTE00000000	5.62	64.6	–

Type strains are labelled ^{T}. Genomes in bold were not included in the analysis presented in Sangal et al. (2016)

Table 3 Species-groups evident within the current genus *Rhodococcus* as determined by phylogenomic analyses

Representative species[a]	Species-group[b] (Fig. 2)	Anastasi et al. (2016) clade	Creason et al. (2014) clade[b]	Affiliated species from 16S rRNA analysis[c]
"R. equi" *R. defluvii*	A	3a	5	*R. agglutinans* *R. soli*
R. ruber	B1	1b	6	*R. aetherivorans*
R. coprophilus *R. gordoniae* ***R. rhodochrous*** *R. phenolicus* *R. pyridinivorans* *R. zopfii*	B2	1a	7	*R. artemisiae* *R. biphenylivorans* *R. lactis*
R. imtechensis *R. jostii* *R. koreensis* *R. opacus* ***R. marinonascens*** *R. wratislaviensis*	C	2b	4	*R. percolatus*
R. enclensis ***R. erythropolis*** *R. qingshengii* *R. globerulus*	D	2a	3	*R. baikonurensis* *R. degradans*
R. fascians *R. kyotoensis* *R. yunnanensis*	E	5a	1 and 2	*R. cerastii* *R. cercidiphylli* *R. sovatensis*
R. corynebacterioides *R. kroppenstedtii*	F	5b	Not included	*R. trifolii*
R. maanshanensis *R. tukisamuensis*	G	3b	Strain BKS-51-14 associated with clade 5	*R. kronopolitis* *R. daqingensis*
R. rhodnii *R. triatome*	H	4 (*R. rhodnii* LMG 5362)	*R. rhodnii* LMG 5362 associated with clade 5	
R. kunmingensis[d]	Singleton		Not included	

[a]The earliest described species is shown in bold
[b]All these species-groups were also recovered as well-separated clusters in a seven protein multilocus sequence analysis (maximum likelihood tree) of 199 members of *Rhodococcus* (Savory et al. 2017)
[c]Species likely to belong to the same clusters/clades as judged from a consistent affiliation with related species in 16S rRNA gene trees
[d]Now reclassified as *Aldersonia kunmingensis* (Nouioui et al. 2018)

confirmed the species-groups indicated in Table 3 (Nouioui et al. 2018). Whole proteome analysis using the Genome BLAST Distance Phylogeny (GBDP) was used to determine both intra- and supra-generic relationships within the phylum. With the exception of *R. kunmingensis*, the 29 rhodococcal type strains included were recovered as a lineage within the family *Nocardiaceae*. Importantly, within this "core"

Rhodococcus lineage, the type strains were resolved into eight lineages in exact agreement with groups A–H in Table 3. However, similar to Fig. 2, *R. kunmingensis* DSM 45001^T was recovered as a singleton well separated from any of these species-groups, showing a poorly supported relationship with *Skermania piniformis*. Consequently, Nouioui et al. (2018) have proposed the reclassification of *R. kunmingensis* DSM 45001^T as the type species representing a novel genus, *Aldersonia kunmingensis* gen. nov., comb. nov.

It is notable that all of the above phylogenomic analyses recover well-supported species-groups of very similar composition (Table 3). *R. equi* strains, along with the type strain of *R. defluvii*, are recovered in a well-separated group corresponding to the proposed genus *Prescottella* (Jones et al. 2013b, c; Sangal et al. 2015, 2016). Group B (Fig. 2) contains the type strain of *R. rhodochrous*, the type species of the genus, and can therefore be defined as *Rhodococcus* sensu stricto. The formal proposal of novel genera in addition to *Aldersonia* and *Prescottella* may require further analyses (notably of genomes of as yet unsequenced rhodococcal type strains). However, it is already evident that Group E (Fig. 2) containing soil and plant-associated strains related to the *R. fascians* (Clades 1 and 2 sensu Creason et al. 2014) should be reclassified into at least one novel genus containing *R. fascians*, *Rhodococcus kyotonensis*, *Rhodococcus yunnanensis* and multiple novel species (Creason et al. 2014; Sangal et al. 2016). Indeed, multilocus sequence analysis combined with ANI analyses of a larger collection of plant-associated strains that map onto Group E indicated that Clades 1 and 2 may contain at least 13 novel species in addition to *R. fascians* (Savory et al. 2017). For now, we refer to these strains as the *R. fascians* assemblage. It is notable that Savory et al. (2017) also concluded that this grouping contains *R. kyotonensis* and is well separated from other rhodococci (*R. yunnanensis* was not included in their analysis). Their multilocus sequence analysis of 199 *Rhodococcus* strains (based on translated sequences for *ftsY*, *infB*, *rpoB*, *rsmA*, *secY*, *tsaD* and *ychF*) yielded a maximum likelihood tree with major groupings strikingly concordant with Groups A–H in Table 3 and also highlighted the distant relationship of *R. kunmingensis* DSM 45001^T to the rhodococci.

It is evident from the above, and Table 3, that several independent research groups using differing phylogenomic methodologies have all delineated at least eight groups of species currently classified in the genus *Rhodococcus*, aside from the reclassification of *R. kunmingensis* into the genus *Aldersonia* (Nouioui et al. 2018). As expanding our analysis from 100 to 116 strains, mostly with genome data for type strains, has not added to the number of clusters identified within the genus, it is likely that as further strains are sequenced and analysed they will fit mainly into the eight species-groups defined in Table 3. Indeed, the consistent association in 16S rRNA gene sequence trees (e.g. Fig. 1) of some species that have yet to have representative genomes sequenced with taxa already assigned to species-groups A–H strongly predicts that these species will eventually be assigned to the corresponding groups (Table 3). For example, 16S rRNA gene analyses (e.g. Guo et al. 2015; Li et al. 2015; Chaudhary and Kim 2018; Fig. 1) consistently indicate that *Rhodococcus agglutinans* and *Rhodococcus soli* are probably members of the genus *Prescottella*.

Similarly, 16S rRNA gene analyses (e.g. Silva et al. 2018; Táncsics et al. 2017) consistently associate *Rhodococcus cerastii*, *Rhodococcus cercidiphylli* and *Rhodococcus sovatensis* with *R. fascians*, *R. kyotonensis* and *R. yunnanensis*, as in Fig. 1.

5 Concluding Comments

Clearly, remarkable progress has been made recently in resolving the complex systematics of the genus *Rhodococcus*, in particular through the application of phylogenomic approaches. The heterogeneity within the genus is confirmed and the separation of two pathogenic species into the genus *Prescottella* (associated with animal hosts) and the *R. fascians* assemblage (plant-associated) is especially notable. A wealth of genomic data are now available for rhodococcal strains, including more than 50% of the currently recognised type strains. Completion of genome sequencing data for the remaining rhodococcal type strains will no doubt bring further insights into the groupings described herein (Table 3), along with important insights into their biology and defining characteristics. The division of *Rhodococcus* into multiple genera may be considered controversial as one persons' phylogenomically defined genus is another's intra-genus "species-group". Nevertheless, the taxonomic status of *Rhodococcus* seems to be currently poised like a game of Jenga: once blocks such as *Aldersonia*, *Prescottella* and the *R. fascians* assemblage have been moved, the tower will tumble into multiple genera, as was the case with the aerobic endospore-forming bacilli and, more recently, *Mycobacterium* (Gupta et al. 2018). Thus, it seems likely that a future edition of this monograph will need to be called "Biology of *Rhodococcus*, *Prescottella* and closely related genera".

References

Alvarez HM (2010) Biology of *Rhodococcus*. Springer, Heidelberg

Alvarez HM (2018) Biology of *Rhodococcus*. Springer, Heidelberg

Anastasi E, Macarthur I, Scortti M, Alvarez S, Giguere S, Vazquez-Boland JA (2016) Pangenome and phylogenomic analysis of the pathogenic actinobacterium *Rhodococcus equi*. Genome Biol Evol 8:3140–3148

Aziz RK, Bartels D, Best AA, Dejongh M, Disz T, Edwards RA, Formsma K, Gerdes S, Glass EM, Kubal M, Meyer F, Olsen GJ, Olson R, Osterman AL, Overbeek RA, Mcneil LK, Paarmann D, Paczian T, Parrello B, Pusch GD, Reich C, Stevens R, Vassieva O, Vonstein V, Wilke A, Zagnitko O (2008) The RAST Server: rapid annotations using subsystems technology. BMC Genomics 9:75

Baek I, Kim M, Lee I, Na SI, Goodfellow M, Chun J (2018) Phylogeny trumps chemotaxonomy: a case study involving *Turicella otitidis*. Front Microbiol 9:834

Bousfield IJ, Goodfellow M (1976) The '*rhodochrous*' complex and its relationships with allied taxa. In: Goodfellow M, Brownell GH, Serrano JA (eds) The biology of the nocardiae. Academic Press, London

Ceniceros A, Dijkhuizen L, Petrusma M, Medema M (2017) Genome-based exploration of the specialised metabolic capacities of the genus *Rhodococcus*. BMC Genomics 18:593

Chaudhary DK, Kim J (2018) *Rhodococcus olei* sp. nov., with the ability to degrade petroleum oil, isolated from oil-contaminated soil. Int J Syst Evol Microbiol 68:1749–1756

Chun J, Oren A, Ventosa A, Christensen H, Arahal DR, Da Costa MS, Rooney AP, Yi H, Xu XW, De Meyer S, Trujillo ME (2018) Proposed minimal standards for the use of genome data for the taxonomy of prokaryotes. Int J Syst Evol Microbiol 68:461–466

Creason AL, Davis EW 2nd, Putnam ML, Vandeputte OM, Chang JH (2014) Use of whole genome sequences to develop a molecular phylogenetic framework for *Rhodococcus fascians* and the *Rhodococcus* genus. Front Plant Sci 5:406

Dastager SG, Mawlankar R, Tang SK, Krishnamurthi S, Ramana VV, Joseph N, Shouche YS (2014) *Rhodococcus enclensis* sp. nov., a novel member of the genus *Rhodococcus*. Int J Syst Evol Microbiol 64:2693–2699

De Alegria Puig CR, Pilares L, Marco F, Vila J, Martinez-Martinez L, Navas J (2017) Comparison of the Vitek MS and Bruker matrix-assisted laser desorption ionization-time of flight mass spectrometry systems for Identification of *Rhodococcus equi* and *Dietzia* spp. J Clin Microbiol 55:2255–2260

Duquesne F, Houssin E, Sevin C, Duytschaever L, Tapprest J, Fretin D, Hebert L, Laugier C, Petry S (2017) Development of a multilocus sequence typing scheme for *Rhodococcus equi*. Vet Microbiol 210:64–70

Garrity GM (2014) Conservation of *Rhodococcus equi* (Magnusson 1923) Goodfellow and Alderson 1977 and rejection of *Corynebacterium hoagii* (Morse 1912) Eberson 1918. Int J Syst Evol Microbiol 64:311–312

Goodfellow M, Alderson G (1977) The actinomycete-genus *Rhodococcus*: a home for the "rhodochrous" complex. J Gen Microbiol 100:99–122

Goodfellow M, Alderson G, Chun J (1998) Rhodococcal systematics: problems and developments. Antonie Van Leeuwenhoek 74:3–20

Goodfellow M, Jones AL (2012) Order V. *Corynebacteriales* ord. nov. In: Goodfellow M, Kämpfer P, Busse H-J, Trujillo ME, Suzuki K-I, Ludwig W, Whitman WB (eds) Bergey's manual of systematic bacteriology. Springer, New York

Goodfellow M, Sangal V, Jones AL, Sutcliffe IC (2015) Charting stormy waters: a commentary on the nomenclature of the equine pathogen variously named *Prescottella equi*, *Rhodococcus equi* and *Rhodococcus hoagii*. Equine Vet J 47:508–509

Goodfellow M, Wayne LG (1982) Taxonomy and nomenclature. In: Ratledge C, Stanford JL (eds) Biology of the mycobacteria. Academic Press, London

Gordon RE, Mihm JM (1959) A comparison of four species of mycobacteria. J Gen Microbiol 21:736–748

Guo QQ, Ming H, Meng XL, Duan YY, Gao R, Zhang JX, Huang JR, Li WJ, Nie GX (2015) *Rhodococcus agglutinans* sp. nov., an actinobacterium isolated from a soil sample. Antonie Van Leeuwenhoek 107:1271–1280

Gupta RS, Lo B, Son J (2018) Phylogenomics and comparative genomic studies robustly support division of the genus *Mycobacterium* into an emended genus *Mycobacterium* and four novel genera. Front Microbiol 9:67

Gürtler V, Mayall BC, Seviour R (2004) Can whole genome analysis refine the taxonomy of the genus *Rhodococcus*? FEMS Microbiol Rev 28:377–403

Gürtler V, Seviour R (2010) Systematics of members of the genus *Rhodococcus* (Zopf 1891) emend Goodfellow et al. 1998. In: Alvarez HM (ed) Biology of *Rhodococcus*. Springer, Heidelberg

Hsueh PR, Lee TF, Du SH, Teng SH, Liao CH, Sheng WH, Teng LJ (2014) Bruker biotyper matrix-assisted laser desorption ionization-time of flight mass spectrometry system for identification of *Nocardia*, *Rhodococcus*, *Kocuria*, *Gordonia*, *Tsukamurella*, and *Listeria* species. J Clin Microbiol 52:2371–2379

Hwang CY, Lee I, Cho Y, Lee YM, Baek K, Jung YJ, Yang YY, Lee T, Rhee TS, Lee HK (2015) *Rhodococcus aerolatus* sp. nov., isolated from subarctic rainwater. Int J Syst Evol Microbiol 65:465–471

Jones AL, Davies J, Fukuda M, Brown R, Lim J, Goodfellow M (2013a) *Rhodococcus jostii*: a home for *Rhodococcus* strain RHA1. Antonie Van Leeuwenhoek 104:435–440

Jones AL, Goodfellow M (2012) Genus IV *Rhodococcus* (Zopf 1891) emended. Goodfellow, Alderson and Chun 1998a. In: Goodfellow M, Kämpfer P, Busse H-J, Trujillo ME, Suzuki K-I, Ludwig W, Whitman WB (eds) Bergey's manual of systematic bacteriology, The Actinobacteria, vol 5, 2nd edn. Springer, New York, pp 437–463

Jones AL, Sutcliffe IC, Goodfellow M (2013b) *Prescottia equi* gen. nov., comb. nov.: a new home for an old pathogen. Antonie Van Leeuwenhoek 103:655–671

Jones AL, Sutcliffe IC, Goodfellow M (2013c) Proposal to replace the illegitimate genus name *Prescottia* Jones et al. 2013 with the genus name *Prescottella* gen. nov. and to replace the illegitimate combination *Prescottia equi* Jones et al. 2013 with *Prescottella equi* comb. nov. Antonie Van Leeuwenhoek 103:1405–1407

Kämpfer P, Dott W, Martin K, Glaeser SP (2014) *Rhodococcus defluvii* sp. nov., isolated from wastewater of a bioreactor and formal proposal to reclassify [*Corynebacterium hoagii*] and *Rhodococcus equi* as *Rhodococcus hoagii* comb. nov. Int J Syst Evol Microbiol 64:755–761

Kämpfer P, Wellner S, Lohse K, Lodders N, Martin K (2013) *Rhodococcus cerastii* sp. nov. and *Rhodococcus trifolii* sp. nov., two novel species isolated from leaf surfaces. Int J Syst Evol Microbiol 63:1024–1029

Ko KS, Kim Y, Seong CN, Lee SD (2015) *Rhodococcus antrifimi* sp. nov., isolated from dried bat dung of a cave. Int J Syst Evol Microbiol 65:4043–4048

Li SH, Yu XY, Park DJ, Hozzein WN, Kim CJ, Shu WS, Wadaan MA, Ding LX, Li WJ (2015) *Rhodococcus soli* sp. nov., an actinobacterium isolated from soil using a resuscitative technique. Antonie Van Leeuwenhoek 107:357–366

Li J, Zhao GZ, Long LJ, Wang FZ, Tian XP, Zhang S, Li WJ (2012) *Rhodococcus nanhaiensis* sp. nov., an actinobacterium isolated from marine sediment. Int J Syst Evol Microbiol 62: 2517–2521

Liu H, Zhang Y, Liu C, Fang B, Li C, Guan X, Li L, Wang X, Xiang W (2014) *Rhodococcus kronopolitis* sp. nov., a novel actinobacterium isolated from a millipede (*Kronopolites svenhedind Verhoeff*). Antonie Van Leeuwenhoek 106:1207–1214

Ma J, Zhang L, Wang G, Zhang S, Zhang X, Wang Y, Shi C, Si L, Zhao H, Liu F, Zhang B, Huang X (2017) *Rhodococcus gannanensis* sp. nov., a novel endophytic actinobacterium isolated from root of sunflower (*Helianthus annuus L.*). Antonie Van Leeuwenhoek 110:1113–1120

Mcminn EJ, Alderson G, Dodson HI, Goodfellow M, Ward AC (2000) Genomic and phenomic differentiation of *Rhodococcus equi* and related strains. Antonie Van Leeuwenhoek 78:331–340

Nguyen TM, Kim J (2016) *Rhodococcus pedocola* sp. nov. and *Rhodococcus humicola* sp. nov., two antibiotic-producing actinomycetes isolated from soil. Int J Syst Evol Microbiol 66: 2362–2369

Nimaichand S, Sanasam S, Zheng LQ, Zhu WY, Yang LL, Tang SK, Ningthoujam DS, Li WJ (2013) *Rhodococcus canchipurensis* sp. nov., an actinomycete isolated from a limestone deposit site. Int J Syst Evol Microbiol 63:114–118

Nouioui I, Carro L, García-López M, Meier-Kolthoff JP, Woyke T, Kyrpides N, Pukall R, Klenk H-P, Goodfellow M, Göker M (2018) Genome-based taxonomic classification of the phylum *Actinobacteria*. Front Microbiol 9:2007

Oren A, Ventura S (2017) The current status of cyanobacterial nomenclature under the "prokaryotic" and the "botanical" code. Antonie Van Leeuwenhoek 110:1257–1269

Overbeck A (1891) Zur Kenntnis der Fettfarbstoff-Production bei Spaltpilzen. Nova Acta Leopoldina 55:399

Parker CT, Tindall BJ, Garrity GM (eds) (2019) International code of nomenclature of prokaryotes. Int J Syst Evol Microbiol 69:S1–S111

Pasciak M, Dacko W, Sikora J, Gurlaga D, Pawlik K, Miekisiak G, Gamian A (2015) Creation of an in-house matrix-assisted laser desorption ionization-time of flight mass spectrometry *Corynebacterineae* database overcomes difficulties in identification of *Nocardia farcinica* clinical isolates. J Clin Microbiol 53:2611–2621

Rainey FA, Burghardt J, Kroppenstedt RM, Klatte S, Stackebrandt E (1995) Phylogenetic analysis of the genera *Rhodococcus* and *Nocardia* and evidence for the evolutionary origin of the genus *Nocardia* from within the radiation of *Rhodococcus* species. Microbiology 141:523–528

Ramaprasad EVV, Mahidhara G, Sasikala C, Ramana CV (2018) *Rhodococcus electrodiphilus* sp. nov., a marine electro active actinobacterium isolated from coral reef. Int J Syst Evol Microbiol 68:2644–2649

Ramasamy D, Mishra AK, Lagier JC, Padhmanabhan R, Rossi M, Sentausa E, Raoult D, Fournier PE (2014) A polyphasic strategy incorporating genomic data for the taxonomic description of novel bacterial species. Int J Syst Evol Microbiol 64:384–391

Sangal V, Goodfellow M, Jones AL, Schwalbe EC, Blom J, Hoskisson PA, Sutcliffe IC (2016) Next-generation systematics: an innovative approach to resolve the structure of complex prokaryotic taxa. Sci Rep 6:38392. https://doi.org/10.1038/srep38392

Sangal V, Jones AL, Goodfellow M, Hoskisson PA, Kämpfer P, Sutcliffe IC (2015) Genomic analyses confirm close relatedness between *Rhodococcus defluvii* and *Rhodococcus equi* (*Rhodococcus hoagii*). Arch Microbiol 197:113–116

Sangal V, Jones AL, Goodfellow M, Sutcliffe IC, Hoskisson PA (2014a) Comparative genomic analyses reveal a lack of a substantial signature of host adaptation in *Rhodococcus equi* ("*Prescottella equi*"). Pathog Dis 71:352–356

Sangal V, Nieminen L, Tucker NP, Hoskisson PA (2014b) Revolutionising systematics through next-generation sequencing. In: Goodfellow M, Sutcliffe IC, Chun J (eds) Methods in microbiology: bacterial taxonomy. Elsevier, Oxford, UK

Savory EA, Fuller SL, Weisberg AJ, Thomas WJ, Gordon MI, Stevens DM, Creason AL, Belcher MS, Serdani M, Wiseman MS, Grunwald NJ, Putnam ML, Chang JH (2017) Evolutionary transitions between beneficial and phytopathogenic *Rhodococcus* challenge disease management. Elife 6. https://doi.org/10.7554/eLife.30925

Schumann P, Maier T (2014) MALDI-TOF mass spectrometry applied to classification and identification of bacteria. In: Goodfellow M, Sutcliffe IC, Chun J (eds) Methods in microbiology. Elsevier, Oxford, UK

Segata N, Bornigen D, Morgan XC, Huttenhower C (2013) PhyloPhlAn is a new method for improved phylogenetic and taxonomic placement of microbes. Nat Commun 4:2304

Silva LJ, Souza DT, Genuario DB, Hoyos HAV, Santos SN, Rosa LH, Zucchi TD, Melo IS (2018) *Rhodococcus psychrotolerans* sp. nov., isolated from rhizosphere of *Deschampsia antarctica*. Antonie Van Leeuwenhoek 111:629–636

Singh PK, Kumari A, Chawla N, Pinnaka AK, Korpole S (2015) *Rhodococcus lactis* sp. nov., an actinobacterium isolated from sludge of a dairy waste treatment plant. Int J Syst Evol Microbiol 65:4215–4220

Su X, Liu Y, Hashmi MZ, Hu J, Ding L, Wu M, Shen C (2015) *Rhodococcus biphenylivorans* sp. nov., a polychlorinated biphenyl-degrading bacterium. Antonie Van Leeuwenhoek 107: 55–63

Švec P, Černohlávková J, Busse HJ, Vojtková H, Pantůček R, Cnockaert M, Mašlaňová I, Králová S, Vandamme P, Sedláček I (2015) Classification of strain CCM 4446T as *Rhodococcus degradans* sp. nov. Int J Syst Evol Microbiol 65:4381–4387

Táncsics A, Benedek T, Farkas M, Mathe I, Marialigeti K, Szoboszlay S, Kukolya J, Kriszt B (2014) Sequence analysis of 16S rRNA, *gyrB* and *catA* genes and DNA-DNA hybridization reveal that *Rhodococcus jialingiae* is a later synonym of *Rhodococcus qingshengii*. Int J Syst Evol Microbiol 64:298–301

Táncsics A, Mathe I, Benedek T, Toth EM, Atasayar E, Sproer C, Marialigeti K, Felfoldi T, Kriszt B (2017) *Rhodococcus sovatensis* sp. nov., an actinomycete isolated from the hypersaline and heliothermal Lake Ursu. Int J Syst Evol Microbiol 67:190–196

Tindall BJ (2014a) The correct name of the taxon that contains the type strain of *Rhodococcus equi*. Int J Syst Evol Microbiol 64:302–308

Tindall BJ (2014b) A note on the genus name *Rhodococcus* Zopf 1891 and its homonyms. Int J Syst Evol Microbiol 64:1062–1064

Wang L, Zhang L, Zhang X, Zhang S, Yang L, Yuan H, Chen J, Liang C, Huang W, Liu J, Zhao Y, Yang Q (2018) *Rhodococcus daqingensis* sp. nov., isolated from petroleum-contaminated soil. Antonie Van Leeuwenhoek. https://doi.org/10.1007/s10482-018-1201-y

Whitman WB (2014) The need for change: embracing the genome. In: Goodfellow M, Sutcliffe IC, Chun J (eds) Methods in microbiology: bacterial taxonomy. Elsevier, Oxford, UK

Yam KC, Okamoto S, Roberts JN, Eltis LD (2011) Adventures in *Rhodococcus* – from steroids to explosives. Can J Microbiol 57:155–168

Zhao GZ, Li J, Zhu WY, Tian SZ, Zhao LX, Yang LL, Xu LH, Li WJ (2012) *Rhodococcus artemisiae* sp. nov., an endophytic actinobacterium isolated from the pharmaceutical plant *Artemisia annua L.* Int J Syst Evol Microbiol 62:900–905

Zopf W (1891) Ueber Ausscheidung von Fettfarbstoffen (Lipochromen) seitens gewisser Spaltpilze. Ber Dtsch Bot Ges 9:22–28

Genomics of *Rhodococcus*

Martina Cappelletti, Jessica Zampolli, Patrizia Di Gennaro, and Davide Zannoni

Contents

Abstract Members of the genus *Rhodococcus* have metabolic versatility and unique adaptation capacities to fluctuating environmental conditions, enabling the colonization of a wide variety of environments; they also play an important role in nutrient cycling and have potential applications in bioremediation, biotransformations and biocatalysis. *Rhodococcus* spp. are mainly distributed in soil, water and marine sediments, although some of them are also pathogens for humans, animals and plants. Consistent with the wide catabolic diversity, *Rhodococcus* spp. possess large and complex genomes (up to 10.1 Mbp), which contain a multiplicity of catabolic genes, high genetic redundancy of biosynthetic pathways and large catabolic plasmids, the latter encoding peculiar metabolic and physiological traits. Recently, the progress in sequencing technology led to a dramatic increase in the

M. Cappelletti (✉) · D. Zannoni
Department of Pharmacy and Biotechnology, University of Bologna, Bologna, Italy
e-mail: martina.cappelletti2@unibo.it; davide.zannoni@unibo.it

J. Zampolli · P. Di Gennaro
Department of Biotechnology and Biosciences, University of Milano-Bicocca, Milan, Italy
e-mail: j.zampolli@campus.unimib.it; patrizia.digennaro@unimib.it

H. M. Alvarez (ed.), *Biology of Rhodococcus*, Microbiology Monographs 16,
https://doi.org/10.1007/978-3-030-11461-9_2

number of sequenced *Rhodococcus* genomes, which have been investigated through diverse bioinformatic approaches. In particular, whole-genome comparative and genome-based functional studies were associated to *omic* technologies for the study of the global *Rhodococcus* cell response with the aim of providing insight into the genetic basis of specific catabolic capacities and phenotypic traits. Lastly, genome-based advances in *Rhodococcus* engineering led to the first design of molecular toolkits for tunable and targeted genome editing. Besides this, genome-based metabolic models were developed to make metabolic predictions of the *Rhodococcus* cell response to specific growth conditions. Both the synthetic and system approaches offered new opportunities for genome-scale rational design of *Rhodococcus* cell for environmental and industrial applications.

1 Introduction

The *Rhodococcus* genus comprises of Gram-positive, nonmotile, nonsporulating, aerobic bacteria, with a high G+C content and a mycolic acid-containing cell wall. Members of *Rhodoccocus* genus are genetically and physiologically diverse bacteria, widely distributed in soil, water and marine sediments; some *Rhodococcus* spp. are also pathogens for plants (*R. fascians*), animals and humans (*R. equi*). Members of *Rhodococcus* have also been recovered from extreme environments such as the deep sea, oil-contaminated soils and freeze-thaw tundra on glacial margins (Sheng et al. 2011; Shevtsov et al. 2013; Konishi et al. 2014).

Rhodococcus genus is featured by a broad metabolic versatility and environmental persistence supporting its clinical, industrial and environmental significance. In particular, *Rhodococcus* strains have peculiar degradative capacities towards a variety of organic compounds, including toxic and recalcitrant molecules like chlorinated hydrocarbons (Gröning et al. 2014; Cappelletti et al. 2017), herbicides (Fang et al. 2016), 4,4′-dithiodibutyric acid (DTDB) (Khairy et al. 2016) and dibenzothiophene (DBT) (Tao et al. 2011) as well as the ability to resist to various stress conditions (desiccation, radiation, heavy metals) (LeBlanc et al. 2008; Taketani et al. 2013; Cappelletti et al. 2016). They are also able to mediate a broad range of biotransformation, including enantioselective syntheses, and to produce biosurfactants, which facilitate the cell contact with hydrophobic substrates (Martìnkovà et al. 2009). Due to their metabolic flexibility and their tolerance to various stresses, they play an important role in nutrient cycling and have potential applications in bioremediation, biotransformations and biocatalysis.

Because of their biotechnological applications, the massive application of high-throughput genomic technologies has dramatically increased the number of sequenced *Rhodococcus* genomes, and a great effort has been directed towards the computational analysis of genomic data. Since the first complete genome (of *R. jostii* RHA1) published in 2006, an exponentially increasing number of genomes of *Rhodococcus* strains have been sequenced due to the progress of the sequencing technologies and to the reduction of the DNA sequencing costs. In July 2018, 218 complete genomes of *Rhodococcus* are available in NCBI (www.ncbi.nlm.nih.gov/genomes/GenomesGroup.cgi?taxid=1827), most of them being available as draft genomes. Collectively, the analysis of genome sequences has provided insights into the genetic basis for diversity, plasticity and adaptation characteristic of *Rhodococcus* spp.

2 The *Rhodococcus* Genome: General Features and Structure

Rhodococcus genomes show high GC amount (61–71%) and range in size from around 4 Mb to over 10 Mb. In particular, the genome size seems to be at least partially dependent on the *Rhodococcus* strain lifestyle and niche complexity; indeed, the pathogenic *R. equi* has a substantially smaller genome than the soil-associated versatile biodegrader *R. jostii* RHA1 (9.7 Mb) and the other two environmental rhodococci, *Rhodococcus erythropolis* PR4 (6.9 Mb) and *Rhodococcus opacus* B4 (8.2 Mb) (Letek et al. 2010). Indeed, obligate pathogens live inside a host, reducing the need to adapt to sudden environmental changes and presence of competitors, which may contribute to genome minimization (Moran 2002).

Among *Rhodococcus* strains of the same species, the amount of GC in the genome was quite constant, showing the highest values (>70% GC-content) in *R. aetherivorans* and *R. ruber* species (Table 1). On the other hand, a remarkable variability was observed in the rhodococcal genome structure and plasmid presence.

Chromosomal and plasmid DNA of *Rhodococcus* strains is either linear or circular. Both *R. jostii* RHA1 and *R. opacus* B4 have linear chromosomes, while *R. erythropolis* PR4 and *R. equi* chromosomes are circular (Table 1). Interestingly, chromosome topology does not seem to correlate with taxonomy or phylogenetic relationship, as *R. equi* and *R. erythropolis* belong to different subclades, and the latter is considered the prototype of the "*erythropolis* subclade", which also includes *R. opacus* (Letek et al. 2010). Further, not all actinomycetales genera present linear chromosomes. *Streptomycetes* typically have linear chromosomes of very large size (>8.5 Mb), so linearization appears to have occurred independently in different actinobacterial lineages during evolution, apparently in association with increasing genome size (Bentley et al. 2002).

The co-existence of large linear and circular plasmids of different sizes within the same cell is also a characteristic of several actinomycetes genera; however, linear plasmids are mainly described in *Streptomyces* and *Rhodococcus* strains. The linearity of DNA molecules in *Rhodococcus* strains has been mainly associated to the presence telomere-like structures comprising of terminal inverted repeats (TIRs) and proteins bound to their 5′ ends (Kikuchi et al. 1985; Dib et al. 2015). This structure, named as invertron, has been detected in both chromosomes and plasmids of *Rhodococcus* and has been extensively studied in *Streptomyces* strains (Ventura

Table 1 Genomic features of *Rhodococcus* strains selected on the basis of the completeness of the genomes sequence and/or on the availability of functional studies[a]

Strain	WGS accession no. (NCBI)	Genome size (Mb)	Presence of plasmid DNA (count)	G+C content (%)	Total no. of genes (CDS)[c]	Main features
R. jostii RHA1	CP000431	9.7	Yes (3)	67.5	9256	PCB degrader, lipid accumulation
R. opacus B4	AP011115	8.83	Yes (5)	68	8227	Lipid accumulation
R. opacus PD630	CP003949	9.17	Yes (9)	67	8402	Hydrocarbon degrader, lipid accumulation
R. opacus R7	CP008947	10.12	Yes (5)	67	8621	Hydrocarbon degrader
R. opacus 1CP	CP009111	8.64	Yes (2)	67	7973	Enantioselective oxygenation of styrene
R. opacus M213	AJYC00000000	9.19	Yes (2)	67	8680	PAH degrader
R. erythropolis PR4	AP008957	6.89	Yes (3)	62	6491	Hydrocarbon degrader
R. erythropolis CCM2595	CP003761	6.37	Yes (1)	62.5	5895	Phenol degrader
R. erythropolis R138	CP007255–7	6.8	Yes (2)	62	6301	Quorum-sensing signal degrader and plant protection capacities
R. erythropolis DN1	AUZK00000000	6.55	–[b]	62	6331	Crude oil degrader
R. erythropolis BG43	CP011295	6.86	Yes (3)	62	6396	Quorum quencher
R. erythropolis XP	AGCF00000000	7.23	–	62	6826	Oil desulfurizer
R. erythropolis JCM 6824	BBLL00000000	7.02	–	62	6635	Antibiotic producer
R. erythropolis VSD3	MLKO00000000	6.55	–	62	6026	Hydrocarbon degrader and plant growth-promoting bacterium
R. aetherivorans BCP1	AVAE00000000	6.23	Yes (2)	70.3	5713	Hydrocarbon and chlorinated alkane degrader, metal-nanoparticle producer
R. aetherivorans IcdP1	CP011341	5.92	No	70.6	5388	High-molecular-weight and chlorinated hydrocarbon degrader
R. ruber P14	CP024315	5.52	No	70.5	5053	High-molecular-weight PAH degrader
R. ruber YYL	CP024890	5.59	Yes (2)	70.5	5200	Tetrahydrofuran degrader

R. ruber SD3	CP029146	5.37	No	70.6	4905	Phenol and chlorobenzene degrader
R. ruber Chol-4	ANGC00000000	5.46	–	70.4	4970	Steroid degrader
R. qingshengii djl-6-2	CP025959	6.70	Yes (2)	62	6332	Carbendazim degrader
R. qingshengii TUHH-12	JNCU00000000	7.43	–	62	6903	Crude oil degrader
R. wratislaviensis IFP 2016	ANIU00000000	9.69	–	67	9589	Diesel and gasoline oil additive degrader
R. wratislaviensis NBRC 100605	BAWF00000000	10.4	–	67	9528	–
R. pyridinivorans SB3094	CP006996	5.59	Yes (2)	68	5165	Methyl ethyl ketone degrader
R. pyridinivorans GF3	CP022915	5.3	No	68	4966	Anthraquinone dye degrader
R. rhodochrous ATCC 17895	ASJJ00000000	6.9	No	62	6609	Biocatalysis
R. rhodochrous NCTC 10210	LT906450	5.27	No	68	4668	–
R. rhodochrous J3	FXAV00000000	6.11	–	68	5639	–
R. rhodochrous DSM 43241	LRRK00000000	5.18	–	68	4790	Carotenoid producer
R. imtechensis RKJ300	AJJH01	8.23		67	7962	Nitrophenol degrader
R. equi 103S	FN563149	5.04	Ye (1)	69	4649	Pathogen for animals
R. equi ATCC 33707	CM001149	5.26	No	69	4899	Pathogen for humans
R. equi DSSKP-R-001	CP027793	5.25	Yes (2)	69	5118	Pathogen for animals
R. fascians D188	CP015235	5.5	Yes (2)	65	5149	Phytopathogen
R. fascians NBRC 12155 = LMG 3623	JMEN00000000	5.77	–	64	5433	Phytopathogen
R. fascians A44A	JMEX00000000	5.95	–	64.5	5566	Phytopathogen

(continued)

Table 1 (continued)

Strain	WGS accession no. (NCBI)	Genome size (Mb)	Presence of plasmid DNA (count)	G+C content (%)	Total no. of genes (CDS)[c]	Main features
Rhodococcus sp. AD45	CM003191	6.79	Yes (1)	62	6252	Isoprene degrader
Rhodococcus sp. DK17	AJLQ00000000	9.11	Yes (3)	67	8615	Aromatic hydrocarbon degrader

[a]Genome data are collected from NCBI genome database
[b]"–" stands for not determined
[c]CDS = coding sequence

et al. 2007). Most commonly, both ends of linear plasmids are the same or featured by only few mismatches, giving recognizable terminal inverted repeats (TIRs). For instance, plasmid pHG207 of *R. opacus* MR22 carries an imperfect terminal inverted repeat (TIR) of 583/560 bp (Kalkus et al. 1993). The RHA1 replicons have been described as typical actinomycete invertrons, containing two sets of inverted repeats flanking the GCTXCGC central motif with covalently associated proteins (Warren et al. 2004). The chromosomal inverted repeats (of around 10 Kp in RHA1) are much longer than those of the plasmids (e.g. the pRHL1 telomeres are 500 bp). The RHA1 telomeric inverted repeats are similar to those of *Streptomycetes*, particularly over the first 300 bp. The replication of linear *Streptomyces* chromosomes and plasmids is initiated from a fairly centrally located replication origin rich in DnaA box sequences and proceeds bidirectionally towards the telomeres. The telomeres themselves are replicated by a mechanism that includes priming from the terminal proteins covalently bound to the 5′ ends. It has been proposed that linear plasmids evolved from bacteriophages and that linear chromosomes originated from single-crossover recombination between an initially circular chromosome and a linear plasmid (McLeod et al. 2006; Ventura et al. 2007).

2.1 *Genomic and Metabolic Traits of Reference* Rhodococcus *Strains*

The analysis of single bacterial genomes and its annotation confer a wide range of knowledge in relation to the general genome features and chromosomal structure but also in relation to identification of specific genes in a genome and the detection of regions containing functionally connected genes. Although most of them are still in a draft version, the sequencing of individual *Rhodococcus* genomes has often provided important inputs for wet-lab experiments aimed at investigating the relevant metabolic capacities shown by the isolates.

Table 1 reports the genome features of representative *Rhodococcus* strains belonging to different species featured by specific metabolic and physiological traits.

Members of *Rhodococcus jostii*, *R. opacus* and *R. wratislaviensis* species possess genomes of very large size, ranging from 8.5 to over 10 Mb (Table 1). *R. jostii* RHA1 is considered the model of the genus as its genome was the first completely sequenced from a *Rhodococcus* strain. *R. jostii* RHA1 was isolated from lindane-contaminated soil (McLeod et al. 2006) and was widely investigated for the strong ability to degrade a wide variety of aromatic compounds and polychlorinated biphenyls (PCBs), for lipid accumulation and for stress condition resistance (Gonçalves et al. 2006; LeBlanc et al. 2008; Patrauchan et al. 2012; Costa et al. 2015).

R. opacus R7 and *R. wratislaviensis* NBRC 100605 have the largest *Rhodococcus* genomes described up to date (10.1 and 10.4 Mbp, respectively) and two of the largest bacterial genome reported in the literature. While NBRC 100605 metabolic features have been scarcely investigated (Guo and Wu 2017), *R. opacus* R7 has been thoroughly studied considering both the broad degradative abilities towards aromatic and aliphatic hydrocarbons and the associated genome features (Di Gennaro et al. 2001, 2010; Zampolli et al. 2014; Orro et al. 2015; Di Canito et al. 2018). Based on the terminal sequence signatures defining the possible linearity of replicons, R7 possesses one linear chromosome and additional five linear replicons, ranging in size from 20 Kb to 430 Kb (Tables 1 and 2) (Di Gennaro et al. 2014). Other *Rhodococcus opacus* strains described from a genomic point of view have broad organic compound degradative abilities and capacity to produce and accumulate lipids for possible biodiesel application. In particular, *R. opacus* PD630 is the most studied bacterium for its ability to produce and accumulate lipids (mostly triacylglycerols, TAGs) using different carbon sources (Alvarez et al. 1996; Castro et al. 2016). Indeed, it was shown to accumulate significant amounts of lipids, namely, 76 and 87%, of its cellular dry weight, when grown on gluconate and olive oil, respectively (Alvarez et al. 1996; Voss and Steinbuchel 2001). *R. opacus* B4 was the first *R. opacus* with a completely sequenced genome; it was described to be able to transform and degrade several types of hydrocarbons, to stabilize water-oil phases and to accumulate TAG (Na et al. 2005; Honda et al. 2008; Sameshima et al. 2008). The analysis of *R. opacus* B4 genome revealed an 8.8-Mb-long genome composed by one linear chromosome, two linear plasmids of pROB series of size range of 2–4 Kb and three circular plasmids of pKNR series of size range of 110–550 Kb (Table 2).

Members of *R. erythropolis* are largely distributed in the environment and exhibit a remarkable metabolic versatility related to their capacity to degrade complex compounds, such as quorum-sensing signals N-acylhomoserine lactones, phenols, sterols and fuel derivatives (Table 1). Moreover, many *R. erythropolis* strains have been isolated from petroleum-contaminated sites and have been described for their ability to degrade various hydrocarbons using multiple catabolic pathways (de Carvalho and da Fonseca 2005). All the *R. erythropolis* strains whose complete genome has been sequenced showed the presence of plasmids ranging in length from 3 Kb to 261 Kb (Table 2).

The genomes of *R. aetherivorans* and *R. ruber* species showed the highest amount of GC, being around 70% in all the strains described up to date. Members of these two species have been described for different biosynthetic and biodegradative capacities. *R. aetherivorans* BCP1 (formerly *Rhodococcus* sp. BCP1) is able to degrade a wide range of alkanes, naphthenic acids and chlorinated alkanes (Frascari et al. 2006; Cappelletti et al. 2011, 2015; Presentato et al. 2018a; Ciavarelli et al. 2012) and to produce metal-based nanostructures (Presentato et al. 2016, 2018b, c). BCP1 has a genome of 6.2 Mbp composed by one chromosome and two plasmids (of 100–120 Kb size) (Tables 1 and 2). *R. aetherivorans* IcdP1 was isolated from an abandoned coking plant; it was proven to be able to degrade numerous high-molecular-weight polycyclic aromatic hydrocarbons and organochlorine pesticides (Qu et al. 2015). Although their genome data are not available, *R. aetherivorans* IAR1 was described to accumulate polyhydroxybutyrate PHB, while *R. aetherivorans* IG24 was employed by

Table 2 Details on the plasmid content of reference *Rhodococcus* genomes from Table 1[a]

Strain	WGS accession no. (NCBI)	Replicon	Size (Mbp)	Topology	G+C content (%)	Total no. of genes (CDS)[c]
R. jostii RHA1	CP000431	Chromosome	7.8	L	67.5	7262
	CP000432	pRHL1	1.12	L	65	1186
	CP000433	pRHL2	0.44	L	64	474
	CP000434	pRHL3	0.33	L	65	334
R. opacus B4	AP011115	Chromosome	7.91	L	68	7287
	AP011118	pKNR	0.004	C	65	7
	AP011119	pKNR01	0.002	C	64	2
	AP011120	pKNR02	0.55	C	64	598
	AP011116	pROB01	0.24	L	66	232
	AP011117	pROB02	0.11	L	64	101
R. opacus PD630	CP003949	Chromosome	9.17	C	67	7715
	CP003950	Plasmid 1	0.17	C	65	166
	CP003951	Plasmid 2	0.09	C	64	95
	CP003952	Plasmid 3	0.08	L	65	75
	CP003953	Plasmid 4	0.06	L	65	63
	CP003954	Plasmid 5	0.09	L	66	91
	CP003955	Plasmid 6	0.04	L	66	35
	CP003956	Plasmid 7	0.04	L	66	36
	CP003957	Plasmid 8	0.16	L	64	142
	CP003958	Plasmid 9	0.04	L	65	30
R. opacus R7	CP008947	Chromosome	10.1	L	67	7353
	CP008948	pPDG1	0.65	L	65	486
	CP008949	pPDG2	0.43	L	64	310
	CP008950	pPDG3	0.35	L	66	307
	CP008951	pPDG4	0.19	L	65	148
	CP008952	pPDG5	0.02	L	63	17
R. opacus 1CP	CP009111	Chromosome	8.63	C	67	7101
	CP009112	pR1CP1	0.88	L	64	815
	CP009113	pR1CP2	0.06	L	64	57
R. erythropolis PR4	AP008957	Chromosome	6.89	C	62	6092
	AP008932	pREC1	0.10	C	63	104
	AP008933	pREC2	0.003	C	62	5
	AP008931	pREL1	0.27	L	62	290
R. erythropolis CCM2595	CP003761	Chromosome	6.37	C	62.5	5793
	CP003762	pRECF1	0.09	C	62	102
R. erythropolis R138	CP007255	Chromosome	6.8	C	62	5759
	AP008932	pCRE138	0.09	C	61	7
	AP008933	pLRE138	0.48	L	60	33
R. erythropolis BG43	CP011295	Chromosome	6.33	C	62	5875

(continued)

Table 2 (continued)

Strain	WGS accession no. (NCBI)	Replicon	Size (Mbp)	Topology	G+C content (%)	Total no. of genes (CDS)[c]
	CP011296 CP011297 CP011298	pRLCBG43 pRLLBG43 pRSLBG43	0.24 0.26 0.03	C C C	60 61 59	225 273 23
R. aetherivorans BCP1	CM002177	Chromosome	6.2	L	70.3	5506
	CM002178 CM002179	pBMC1 pBMC2	0.12 0.10	L L	69 65	108 99
R. ruber YYL	CP024890	Chromosome	5.59	C	70	5200
	CP024892 CP024891	pYYL1.1 pYYL1.2	0.24 0.09	C L	67 66	206 91
R. qingshengii djl-6-2	CP025959	Chromosome	6.52	C	62	6154
	CP025960 CP025961 CP025962	pDJL1 pDJL2 pDJL3	0.08 0.08 0.02	C C C	62 62 61	87 73 18
R. pyridinivorans SB3094	CP006996	Chromosome	5.23	C	68	4818
	CP006997 CP006998	Unnamed Unnamed2	0.36 0	C C	65 66	344 3
R. equi 103S	FN563149	Chromosome	5.04	C	69	4649
	FN563149	pVAPA1037	0.08	C	65	73
R. equi DSSKP-R-001	CP027793	Chromosome	5.25	C	69	4872
	CP027794 CP027795	Plasmid 1 Plasmid 2	0.09 0.1	C C	69 66	151 95
R. fascians D188	CP015235	Chromosome	5.14	C	65	4808
	CP015236 CP015237	pFiD188 Unnamed	0.2 0.16	L C	62 64	176 165
R. sp. AD45	JYOP	Chromosome	6.45	C	62	5949
	JYOP	Unnamed	0.34	C	61	303
R. sp. DK17	AJLQ01	Chromosome	9.1	–[b]	67	–
	AJLQ01	pDK1 pDK2 pDK3	0.38 0.33 0.75	– –	– –	– –

[a]Genome data are collected from NCBI genome database
[b]"–" stands for not determined
[c]CDS = coding sequence

Merck Corporation for the production of the anti-HIV drug Crixivan™ (Treadway et al. 1999; Hori et al. 2009).

Rhodococcus pathogenic strains belong to *R. equi* and *R. fascians* species, which cause disease to animals and plants, respectively. In line with their lifestyle inside a host, their genomes are generally smaller (<6 Mb) compared to the environmental *Rhodococcus* strains (Table 1). Recently, numerous genomes of both these two species have been sequenced to conduct phylogenomic analyses and to investigate on the evolutionary mechanisms of virulence traits (Letek et al. 2010; Creason et al. 2014a, b). In particular, virulence genes have been associated to plasmids. Several *R. equi* strains have been described to possess circular plasmids, whereas *R. fascians* members are featured by linear plasmids. The presence of linear plasmids in *R. fascians* has been correlated with phytopathogenicity, although it was described not to be strictly necessary (Creason et al. 2014a).

Compared to the other species, only few genomic studies are available for *R. qingshengii*, *R. pyridinivorans* and *R. rhodochrous*. Nevertheless, members of these species have been reported to possess peculiar metabolic activities such as degradative capacity towards rubber, fungicide and other hydrocarbon-related complex molecules. In relation to this, additional genome sequencing efforts promise to expand the knowledge on the genetic and metabolic diversity of *Rhodococcus* bacterial strains (Table 1) (Xu et al. 2007; Dueholm et al. 2014; Watcharakul et al. 2016).

3 Comparative Genomics of *Rhodococcus*

Comparative genomics allow providing genomic basis for the differences observed among microorganisms in terms of phenotypes, metabolic capacities and cellular response to specific environmental conditions. Recent studies have described genome-wide comparative analyses of *Rhodococcus* strains with the aim of defining the phylogeny and evolutionary relationship, determining the *Rhodococcus* genus core genome, analysing their catabolic potentials and stress response (Orro et al. 2015; Cappelletti et al. 2016). Genome comparative analysis is performed through pairwise or multiple (>3 genomes) whole-genome alignments. Among the applicable programs, Artemis Comparison Tool (ACT), available within the Integrated Microbial Genomes (IMG) system, was used for genome comparison of strain *R. opacus* M213 with different *Rhodococcus* strains (Pathak et al. 2016). Mauve (http://darlinglab.org/mauve/mauve.html) performs multiple sequence alignment and has been used for comparative genomic analyses of different *Rhodococcus* strains, e.g. *R. opacus* R7 and *R. aetherivorans* BCP1 (Orro et al. 2015), *R. opacus* M213 (Pathak et al. 2016) and *Rhodococcus fascians* strains (Creason et al. 2014b) (Fig. 1).

Comparative genome analysis allows to detect genomic regions that are conserved or syntenic (Pathak et al. 2016), representing conserved genome sections that have not undergone internal rearrangements and inversions. This conservation typically reflects phylogenetic correlation and represents the indication of functional evolutionary constraints. In comparative studies concerning *Rhodococcus*, the conservation of collinear blocks in different *R. fascians* strains was shown to be in accordance with their phylogenetic relationship (Creason et al. 2014b; Pathak et al.

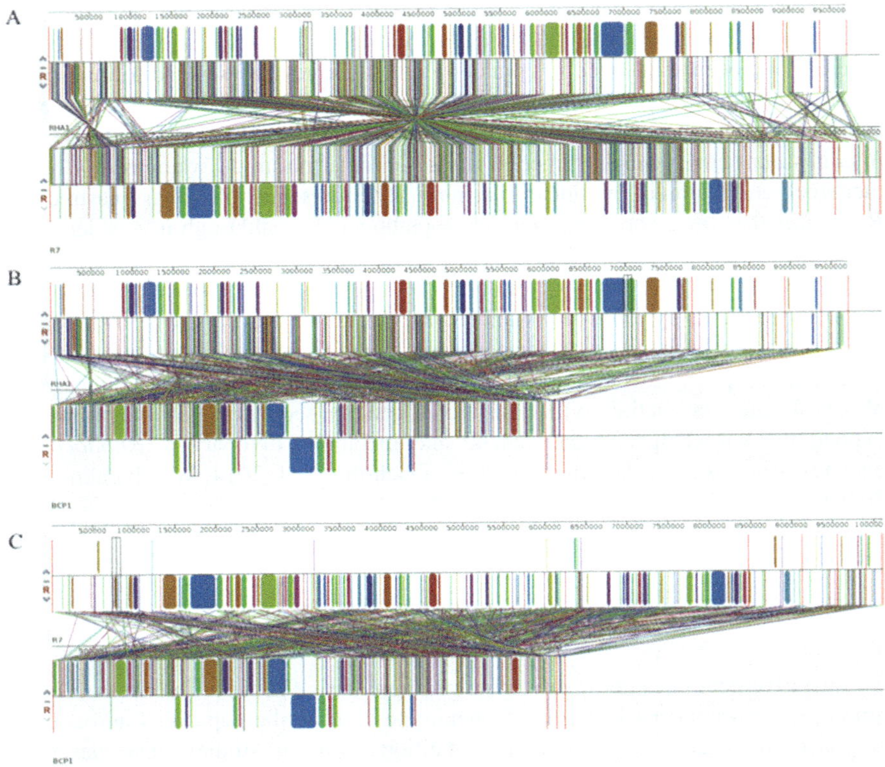

Fig. 1 Whole-genome sequence alignment with program Mauve diagram. The alignment of three pairs of genomes is shown: *R. jostii* RHA1 and *R. opacus* R7 (panel A), *R. jostii* RHA1 and *R. aetherivorans* BCP1 (panel B), *R. opacus* R7 and *R. aetherivorans* BCP1 (panel C). The coloured region represents a block of sequences that is collinear to a corresponding block of sequences in the other genome sequence. [Modified from Orro et al. (2015)]

2016). The 16S rRNA-based affiliation of *Rhodococcus* sp. M213 to *opacus* species was challenged by the higher M213 syntenic affiliation with *R. jostii* RHA1 compared to *R. opacus* strains B4 and PD630 (Pathak et al. 2016).

Another useful way to compare whole genomes is by computing a value that summarizes their similarity or distance. The average nucleotide identity (ANI) value calculated from whole-genome alignments was described to provide a tool for taxonomy and evolutionary studies and species definition (Creason et al. 2014b; Anastasi et al. 2016). Whole-genome alignment analysis of three *Rhodococcus equi* strains allowed to detect single nucleotide polymorphisms (SNPs) in specific coding regions that were suggested to possibly reflect differences in the lifestyle of these three isolates (Sangal et al. 2014).

In addition to gene/protein sequence-based alignment, gene content comparison can be done in terms of function using, for instance, the concept of functional categories (available in RAST subsystem) or COG (cluster of orthologous groups of proteins). Typically, statistical tools analyse gene functions (or annotations) as categorized by COGs and determine whether certain functions or functional categories are statistically enriched in certain genomes compared to others. In this respect, the *R. opacus* M213 genome analysis at functional (COG) level further reinforced the closer relationship of M213 with *R. jostii* RHA1 and other *Rhodococcus* strains (e.g. *R. wratislaviensis* IFP 2016, *R. imtechensis* RKJ300, *Rhodococcus* sp. strain DK17) compared to *R. opacus* strains (Pathak et al. 2016). On the other hand, by comparing the RAST-based functional categories, *R. opacus* R7 genome resulted to be enriched in genes belonging to all the functional categories, compared to *R. aetherivorans* BCP1, except for those involved in some central metabolic process (e.g. DNA, iron, potassium and phosphorous metabolism, dormancy and sporulation, motility and chemotaxis) (Orro et al. 2015).

3.1 Phylogenomics

The use of the only 16S rRNA sequence as taxonomic marker was proved to limit the resolution of *Rhodococcus* phylogenetic studies (Gürtler et al. 2004; Pathak et al. 2016; Creason et al. 2014b) and more in general to lead to artefacts in the bacterial phylogenetic tree construction (Ventura et al. 2007). More recent phylogenetic investigations have been carried out using alternate gene/protein sequences (e.g. functional genes or molecular chaperone-encoding genes) as phylogenetic markers, which often evolve faster than the rRNA operon. This provides a larger phylogenetic resolution that can help to distinguish closely related bacterial species. For this purpose, the alkane monooxygenase (*alkB*) gene was applicable as phylogenetic marker of *Rhodococcus* strains (Táncsics et al. 2015).

The availability of genome sequences has provided new opportunities to get information on applicable phylogenetic markers, and new approaches arose to examine rhodococcal phylogeny based upon multiple gene or protein sequences. Among these approaches, the construction of phylogenetic trees based on concatenated sequences for large numbers of proteins (also called multilocus sequence analysis—MLSA) has proven particularly useful for evolutionary relationship studies (Gao and Gupta 2012). Phylogenetic tree based upon multiple conserved (or slow-evolving) genes/proteins was demonstrated to properly assign environmental isolates to existing *Rhodococcus* species (Orro et al. 2015) and to resolve evolutionary relationships of *Rhodococcus* strains belonging to the same species (Creason et al. 2014b; Letek et al. 2010). MLSA mainly employed sequence concatenamers of housekeeping genes including 16S rRNA, *recA*, *gyrB*, *rpoB*, *rpoC* and *secY* (Orro et al. 2015; Kwasiborski et al. 2015). Orro et al. (2015) assigned the species to two *Rhodococcus* strains on the basis of MLSA with four of these marker genes (Fig. 2). MLSA provided a framework for defining *Rhodococcus* genus within the *Actinobacteria* phylum using seven marker genes (*ftsY*, *infB*, *rpoB*, *rsmA*, *secY*, *tsaD* and *ychF*) (Creason et al. 2014b). In line with the whole-genome based studies

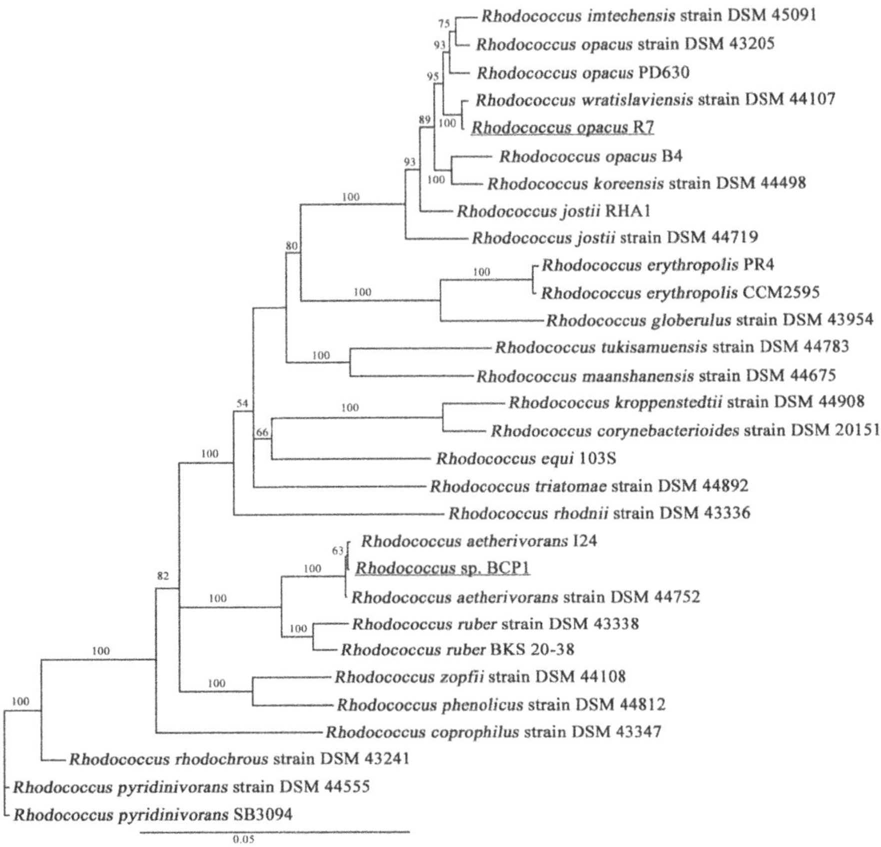

Fig. 2 Multilocus sequence analysis-based tree of 28 *Rhodococcus* strains. Phylogenetic tree of 28 *Rhodococcus* strains based on MLSA using the sequence of the four marker genes, 16S rRNA, *secY*, *rpoC* and *rpsA*. [Modified from Orro et al. (2015)]

by Letek et al. (2010), this study placed the *R. equi* species within the *Rhodococcus* genus, which was in contrast with other inconclusive phylogenetic studies based on 16S rDNA sequences (Goodfellow et al. 1998). More genome sequences and informative marker sequences might help in improving MLSA-mediated taxonomy and discern the finer details within the suborder. In this respect, available genome sequences were used to discover novel molecular characteristics that could be used for evolutionary and systematic studies of *Rhodococcus* genus. In this regard, novel specific conserved signature indels (CSIs) and conserved signature proteins (CSPs) have been proposed as possible molecular markers on the basis of their conservation among different *Rhodococcus* species (Gao and Gupta 2012).

In addition to phylogenomic analysis based on the concatenation of different taxonomic markers, *Rhodococcus* genome-based trees have been constructed on the basis of average sequence similarity. The genome similarity is at the basis of the wet-lab DNA-DNA hybridization (DDH) technique, which represents the gold standard for bacterial taxonomy classification (Richter and Rosselló-Móra 2009). In the era of genomics, different attempts have been made to replace the traditional wet-lab DDH with in silico genome-to-genome comparison (Goris et al. 2007; Richter and Rosselló-Móra 2009). Among these, the average nucleotide identity (ANI) estimates the average nucleotide identity between two genomic datasets, and an ANI value in the range of 94–96% has been proposed to be the criterion for bacterial species delineation (corresponding to a wet-lab DDH value of 70%), while ANI values of 70–75% identified members of the same genus (Konstantinidis and Tiedje 2005; Goris et al. 2007; Richter and Rosselló-Móra 2009). Creason et al. (2014b) conducted phylogeny study of *Rhodococcus* genus by using the average nucleotide identity (ANI) approach and found that the genus *Rhodococcus* can be represented by as many as 20 distinct species (Fig. 3).

Using 27 de novo sequenced *R. equi* genomes, the ANI value was 99.13%, well above the consensus 95–96% threshold for prokaryotic species definition (Anastasi et al. 2016). This result highlighted the high genetic homogeneity of *R. equi* group and was found to correspond to a 16S rRNA sequence identity of 100%.

3.2 Core and Accessory Genome

Whole-genome comparative analysis can be used to obtain information on the pan genome, core genome and accessory genome of a group of bacterial strains belonging to a single species or different species belonging a single genus. The pan genome comprises all the genes present in a given bacterial group (the collection of all genetic material). The core genome is defined as the set of all genes shared by the genomes as orthologs and therefore considered as conserved by all the bacterial strains under analysis (Ventura et al. 2007). The genes included within the core genome are supposed to have strong functional significance for cell physiology and replication. Moreover, as the core genome can be considered to define the members of the bacterial group under analysis, the core genes are suitable as molecular targets to infer phylogeny (Ventura et al. 2007). On the other hand, the accessory or variable genome is represented by the genes, which are confined to a single member of the bacterial group and are therefore specific for a single strain (Ali 2013). The accessory genome therefore represents those gene families that can be associated with phenotypic traits that differentiate each member within a given bacterial population under analysis.

Comparative genomic studies on *Rhodococcus* strains have assessed the core genome shared by either strains belonging to the same species or strains belonging to different species. The resulting core-genome size and identity reported in each work was dependent on the *Rhodococcus* strains under analysis. Within the *Rhodococcus* genome set composed by *R. opacus* R7, *R. aetherivorans* BCP1, *R. jostii* RHA1, *R. opacus* PD630, *R. opacus* B4 and *R. pyridinivorans* SB3094, the core genome corresponded to around 50% of all the identified CDSs in each strain (Orro et al.

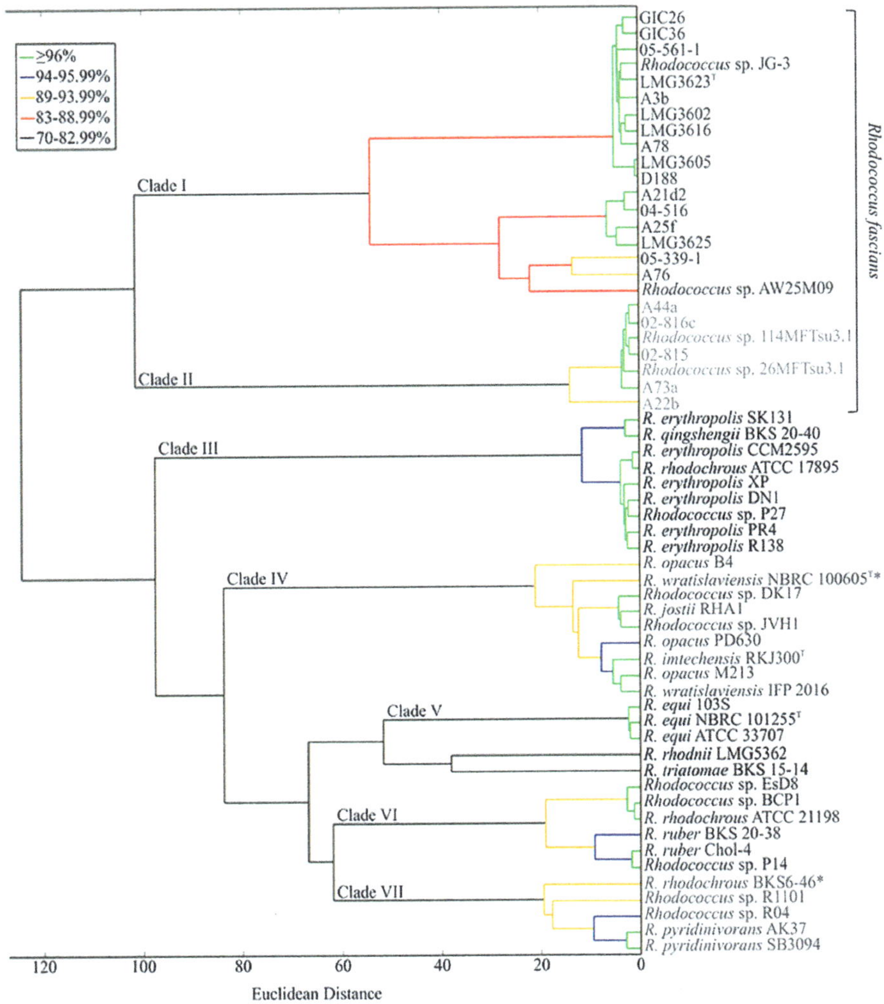

Fig. 3 Average nucleotide identity (ANI) dendrogram for 59 isolates of *Rhodococcus*. Complete genome sequences for 59 isolates of *Rhodococcus* were used to generate an ANI matrix. The matrix was used to calculate an ANI divergence dendrogram. Branches are coloured based on the ANI threshold values. [Modified from Creason et al. (2014b)]

2015) (Fig. 4a). This indicated a large genetic variability among these *Rhodococcus* strains, some of them belonging to different species. The portion of core genes increased by considering *Rhodococcus* strains belonging to the same species, such as *R. opacus* R7, B4 and PD630 (Fig. 4a) (Orro et al. 2015), *R. erythropolis* (Kwasiborski et al. 2015) and *R. equi* (Anastasi et al. 2016). In particular, the genome of *R. erythropolis* R138 was compared to the genomes of other ten

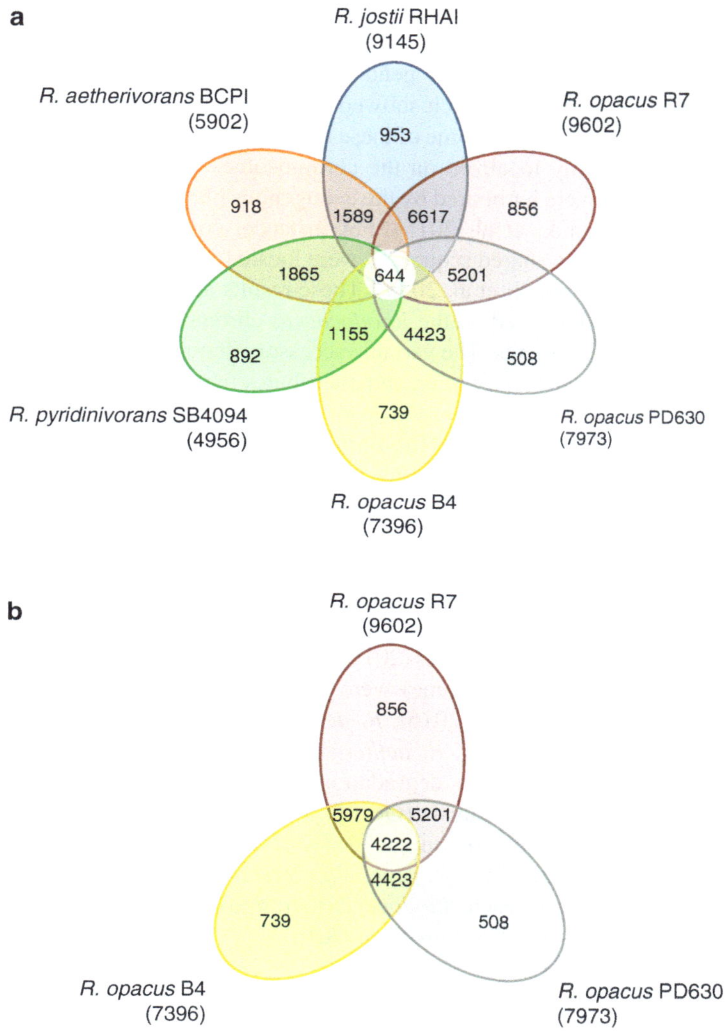

Fig. 4 Venn diagram displaying the core-genome and the unique genes (accessory genome) specific to each strain. The number of genes shared by all strains included within each genome set (i.e. the core genome) is in the centre. Numbers in nonoverlapping portions of each oval show the number of genes unique to each strain. The total number of protein coding genes found in each genome is listed below the strain name. Panel (A), core-genome and unique genes resulting from the comparison of the genomes from *R. opacus* R7, *R. aetherivorans* BCP1, *R. jostii* RHA1, *R. opacus* PD630, *R. opacus* B4, *R. pyridinivorans* SB4094; panel (B), core-genome and unique genes resulting from the comparison of the genomes from *R. opacus* R7, *R. opacus* PD630 and *R. opacus* B4

R. erythropolis strains, either available in the database or newly sequenced. The core-genome genes represented up to 87% of all the CDSs identified in *R. erythropolis* R138. A large core genome was also identified in a comparative study with 29 *R. equi* strains, which showed to have ~80% of shared genes in their genome (Anastasi et al. 2016). Some of these comparative studies also found that the core genes were mainly localized on the chromosome, whereas the unique genes (accessory genome) were harboured by the endogenous plasmids in each strain (Orro et al. 2015; Kwasiborski et al. 2015). For instance, considering *R. erythropolis* strains, up to 97% of the shared sequences were located on the circular chromosome of R138 strain (Kwasiborski et al. 2015). These results are general indications of a core set of functions localized on the *Rhodococcus* chromosome, whereas plasmids have variable genetic contents. The variable accessory genome underlies the peculiar capacities of each *Rhodococcus* strain and contributes functions for niche adaptation (Kwasiborski et al. 2015). In line with this, considering *R. equi* strains, the core genome included traits necessary for bacterial physiology, virulence and niche adaptation, including tolerance to desiccation and oxidative stress (Anastasi et al. 2016). Similarly, *R. erythropolis* core genome was highly represented by gene categorized within the primary metabolism of amino acids, carbohydrates, cofactors and proteins (Kwasiborski et al. 2015). Interestingly, genes encoding enzymes involved in the degradation of complex molecules (i.e. N-acylhomoserine lactones, phenol, catechol and sterol derivatives) were also shared by the 11 *R. erythropolis* under analysis by Kwasiborski et al. (2015). On the other hand, functions encoded by unique genes (accessory genome) were associated to degradative enzymes in *R. opacus* M213 (Pathak et al. 2016), *R. aetherivorans* BCP1 and *R. opacus* R7 (Orro et al. 2015). In particular, *R. aetherivorans* BCP1 possessed unique genes involved in short-chain *n*-alkanes degradation which were not present in the other *Rhodococcus* strains under analysis (Orro et al. 2015) Conversely, *R. opacus* R7 genome had a high number of unique genes involved in aromatic degradation metabolism. These genetic findings generally correlated with the different types of contaminants persisting in each *Rhodococcus* strain isolation source. (Di Gennaro et al. 2001; Frascari et al. 2006; Orro et al. 2015).

3.3 Functional Genomics

Genome sequence-based analysis relies on homology-based gene annotation and protein domain identification to annotate open reading frames (ORFs) and to perform functional predictions. Bioinformatic tools have been applied for the functional analysis of *Rhodococcus* genomes to determine the genetic basis possibly involved in relevant metabolic activities and phenotypic traits, e.g. degradation abilities of specific xenobiotics, plant growth-promoting effect and stress tolerance (Francis et al. 2016; Cappelletti et al. 2016; Pathak et al. 2016). For instance, genome-based analyses assessed the occurrence and distribution of key genes and pathways involved in the synthesis and accumulation of triacylglycerols (TAG), wax esters, polyhydroxyalkanoates, glycogen and polyphosphate (Hernández et al. 2008; Villalba et al. 2013). In line with the extensive experimental work demonstrating the ability of *Rhodococcus* strains to synthesize and accumulate neutral lipids, numerous genes/enzymes predicted to be involved in TAG biosynthesis and degradation and fatty acid β-oxidation were identified in the genomes of *R. jostii* RHA1, *R. opacus* PD630, *R. opacus* B4, *R. erythropolis* PR4, *R. equi* 103S and *R. fascians* F7 (Villalba et al. 2013). The comparative genome analysis of psychrophilic *Rhodococcus* sp. JG3 and other mesophilic rhodococci allowed to detect the cold adaptive traits which confer JG3 with the ability to survive the extremely arid, cold and oligotrophic conditions of permafrost (Goordial et al. 2016). A bioinformatic analysis of the genomes of 20 *Rhodococcus* strains allowed identifying numerous biosynthetic gene clusters encoding pathways involved in the possible production of novel secondary metabolites (Ceniceros et al. 2017).

Several works have assessed a large range of metabolic abilities of strains using Phenotype Microarray (PM) technologies. PM results were interpreted using genomic analysis to predict the genes involved in central metabolic pathways, xenobiotic degradation metabolisms and stress response (Letek et al. 2010; Holder et al. 2011; Orro et al. 2015; Cappelletti et al. 2016).

The large-scale functional analysis of *Rhodococcus* genomes has been further integrated with the results obtained through *omic* technologies. In particular, high-throughput experiments were conducted to obtain global datasets on the expression of genes (transcriptomic analyses through RNA-seq or microarray) and on the protein profiles induced under specific growth conditions or after specific cell treatments. Transcriptomic studies were performed to define the genes involved in the degradation of aromatic compounds and steroids (chlorate and cholesterol) (Gonçalves et al. 2006; van der Geize et al. 2007; Swain et al. 2012), of PCBs and diesel oil (Puglisi et al. 2010, Laczi et al. 2015) and of synthetic polymers (Gravouil et al. 2017) but also in lipid metabolism and accumulation (Chen et al. 2014; Amara et al. 2016) and isoprene metabolism (Crombie et al. 2015). Proteome studies were combined with genome analysis for the study of the degradation of xenobiotic 4,4-dithiodibutyric acid (DTDB) by *R. erythropolis* MI2 (Khairy et al. 2016) and the metabolism of short-chain alkanes in *R. aetherivorans* BCP1 (Cappelletti et al. 2015) and to study the global response to stress conditions such as desiccation and carbon starvation in *R. jostii* RHA1 (LeBlanc et al. 2008; Patrauchan et al. 2012). In many of these studies, the involvement of multiple homologous genes was identified in specific metabolic pathways, providing clues to the redundancy of the different catabolic pathways in *Rhodococcus* (Gonçalves et al. 2006; Swain et al. 2012). Further, some omic studies gave indications on the involvement of possible regulators in the specific catabolic pathways under analysis (Crombie et al. 2015). In *R. ruber* C208, transcriptomic results indicated the involvement of alkane degradation pathway and β-oxidation of fatty acids as the main catabolic route for polyethylene degradation; further, it also indicated metabolic limiting steps which could represent molecular target to optimize the biodegradation process (Gravouil et al. 2017). In addition to catabolic genes/proteins, many genes/proteins involved in stress response (e.g. chaperone-like proteins, superoxide dismutase, catalase/

peroxidase) were reported to be induced during the degradation of organic compounds (Tomás-Gallardo et al. 2006; Puglisi et al. 2010; Khairy et al. 2016). The microarray analysis of *R. aetherivorans* I24 cells exposed to PCBs reported the transcriptional response to be primarily directed towards reducing oxidative stress rather than catabolism (Puglisi et al. 2010).

4 The Genomic Basis of Metabolic Versatility in *Rhodococcus*

The extraordinary metabolic versatility of *Rhodococcus* strains is reflected in their genomic attributes, like (1) large genome sizes (up to 10 Mbp) encoding numerous catabolic pathways for a variety of chemical compounds; (2) a significant degree of gene redundancy that ensures functional robustness and free-to-evolve genetic reservoirs; and (3) the occurrence of circular and linear (mega)plasmids that generally evolve more rapidly than the chromosome and consist of an additional pool of DNA that can evolve and can be easily transferred (Redenbach and Altenbuchner 2002; Van der Geize and Dijkhuize 2004). In this context, the broad *Rhodococcus* adaptability is related to their genome "flexibility" which refers to the genomic rearrangements mainly occurring on the large linear plasmids and due to still mostly unknown molecular mechanisms promoting the frequent non-homologous illegitimate recombination. This last aspect has been discussed in several review articles by Larkin and collaborators (Larkin et al. 1998, 2010; Kulakov and Larkin 2002). Additionally, recent works have underlined how the equipment of genes for the transport of many different substrates (de Carvalho et al. 2014) and the numerous genes encoding oxygenases, typically catalysing the hydroxylation and cleavage of organic compounds, are at the basis of the wide metabolic capacities of *Rhodococcus* strains. Only a few portion of the oxygenase genes have been acquired through horizontal gene transfer events, while most of them were chromosomally located, suggesting their fundamental role in *Rhodococcus* physiology (Mcleod et al. 2006).

4.1 Catabolic Gene Redundancy

Since the first complete genome sequence obtained from a *Rhodococcus* strain, the presence of multiple catabolic genes encoding homologous enzymes, called isozymes, attracted research interest. Recently, the analysis of whole-genome sequences of several rhodococci confirmed the redundancy in catabolic pathways initially only hypothesized for this genus (McLeod et al. 2006). Catabolic gene redundancy is often considered at the basis of the *Rhodococcus* catabolic versatility, functional robustness, adaptation to polluted and extreme environments and high-performing environmental competition (Mcleod et al. 2006; Pérez-Pérez et al. 2009).

Several examples of genetic redundancy originated either from gene duplication or horizontal gene transfer events have been described in *Rhodococcus* strains, the latter being supported by the proximity of transposase and invertase sequences to duplicated gene clusters (Taguchi et al. 2007). In general, the duplicate genes have been described to evolve and to encode isoenzymes possessing similar sequences but different substrate specificities and induction patterns. On the other hand, some duplicates have been described to be, at some extent, functionally redundant (Zhang 2012), catalysing the same metabolic reaction and presenting overlapping substrate range and inducing profiles. This type of duplication has probably occurred in relative recent period of time or has not been subject to strong selective pressure. For instance, in *R. jostii* RHA1, it has been hypothesized that the multiple homologous genes involved in the central aromatic pathways have received low selective pressure for functional diversification or gene removal (McLeod et al. 2006).

In this framework, the presence of multiple homologous genes strongly contributes to the wide *Rhodococcus* versatility in terms of catabolic activities including aliphatic and aromatic hydrocarbon catabolism, chlorinated hydrocarbons transformation and steroid degradation. Further, the presence of multiple functional homologs was found in central metabolic pathways (e.g. tricarboxylic acid cycle) (Van der Geize and Dijkhuize 2004).

While the presence of multiple copies of specific genes was not a universal feature of *Rhodococcus*, genomic redundancy generally characterizes the members of this genus in relation to metabolic and physiological traits useful for the adaptation to the environmental niche from which each *Rhodococcus* strain was isolated. For instance, the genome analysis of the psychrophilic *Rhodococcus* sp. JG3 isolated from a permafrost soil with a possible alkane source showed genomic redundancy in relation to genes involved in the adaptation to cold temperatures such as those associated to osmotic stress and to genes encoding alkane 1-monooxygenases (Goordial et al. 2016).

Functional redundancy has referred to the multiple alkane hydroxylase genes found in *Rhodococcus* strains featured by versatile alkane degradation capacity and typically isolated from petroleum-contaminated (sediment or marine) sites (van Beilen et al. 2002) (Fig. 5a). In particular, *alkB* gene encodes the membrane alkane hydroxylase AlkB, which catalyses the initial oxidation of alkanes. Four *alkB* homologous genes have been detected in *R. erythropolis* NRRL B-16531 and *R. qingshengii* Q15 (Whyte et al. 2002); three to five *alkB* genes were found in different *R. erythropolis* strains from various contaminated soils and in *Rhodococcus* sp. strain TMP2 (van Beilen et al. 2002; Takei et al. 2008). The multiple *alkB* genes found in *Rhodococcus* strains were mainly localized on the chromosome, and some of them showed high sequence divergence and different flanking regions (Whyte et al. 2002). The different AlkBs often catalysed the oxidation of different ranges or classes of alkanes and were differently expressed depending on the substrate and growth condition (Takei et al. 2008; Laczi et al. 2015). Because of this, the presence of multiple AlkB hydroxylases has been associated to broad metabolic capacities of *Rhodococcus* strains towards medium- and long-chain alkane and also to branched alkanes. In *R. aetherivorans* BCP1, two homologous gene clusters encoding alkane

Fig. 5 Examples of genetic redundancy in *Rhodococcus*. Genetic organization of (**a**) alkB genes in *Rhodococcus* strain NRRLB-16531, (**b**) catabolic island including *tpa*, *pat* and *pad* gene that is duplicated in the two linear plasmids in *R. jostii* RHA1, (**c**) the three homologous gene clusters (*bphA*, *etb1A* and *etb2A* genes) encoding dioxygenase systems involved in the initial hydroxylation of substituted benzenes in *R. jostii* RHA1. The plasmidic or chromosomal localization is indicated in correspondence to each gene cluster. [Modified from Whyte et al. (2002), Hara et al. (2007), Gonçalves et al. (2006), respectively]

monooxygenases of SDIMO (soluble di-iron monooxygenases) family were involved in the oxidation of short-chain alkanes (Cappelletti et al. 2013, 2015). This functional redundancy was correlated with the strong capacity of BCP1 strain to utilize gaseous alkanes and to co-metabolize chlorinated alkanes (Frascari et al. 2006; Cappelletti et al. 2012). In some cases, alkane hydroxylation functional redundancy was also associated to the co-existence in a single *Rhodococcus* strains of genes encoding different alkane hydroxylases, AlkB, SDIMOs and cytochrome P450 belonging to the CYP153 family. For instance, the versatile and efficient degradation of alkanes by *R. erythropolis* PR4 was associated to the presence of four *alkB* genes, two CYP153 genes and other genes coding P450 on its genome, which are differently expressed on alkanes and hydrocarbon mixtures (Laczi et al. 2015).

Catabolic gene redundancy in some *Rhodococcus* strains has been extensively reported for aromatic compound degradation pathways. In particular, *R. jostii* RHA1, isolated from γ-hexachlorocyclohexane (lindane)-contaminated site, was described to be featured by high redundancy of catabolic pathways involved in aromatic hydrocarbon catabolism, which are also responsible for the co-metabolic transformation of polychlorinated biphenyls (PCBs). Several homologous genes were predicted to encode enzymes involved in the upper and lower degradation pathways of substituted benzenes, like ethylbenzene (ETB) and biphenyl (BPH). In RHA1, three homologous gene clusters (*bphA*, *etb1A* and *etb2A* genes) were found to encode dioxygenase systems, which are all induced by BPH and ETB by a possible common regulatory system (Fig. 5b). Nevertheless, these dioxygenase systems have shown distinct substrate specificity in terms of both aromatic hydrocarbons and PCB congeners (Iwasaki et al. 2006, 2007; Patrauchan et al. 2008). The gene cluster *bphA* is localized on the large linear plasmid pRHL1, while *etb1A* and *etb2A* are carried on the other linear plasmid pRHL2 and share high sequence similarity. Multiple copies of *bph* genes (*bphC-G*) were also predicted to be involved in the benzene ring cleavage downstream of the initial hydroxylation. Among these multiple homologs, only a few numbers of genes were transcriptionally induced during the growth on BPH and ETB, while the most part was expressed at constitutive levels (Gonçalves et al. 2006). In particular, only three out of the eight *bphEFG* clusters were up-regulated during RHA1 growth on BPH and ETB, independently on their genome localization (Gonçalves et al. 2006; Patrauchan et al. 2008). This indicated that out of the eight *bphEFG*, five homologous clusters produced paralog enzymes involved in distinct physiological processes (Irvine et al. 2000; Taguchi et al. 2004). Other *Rhodococcus* strains, isolated from the termite ecosystem typically exposed to plant-derived lignin and aromatics, showed multiple copies of *bph* gene clusters involved in PCB/biphenyl degradation; many of these were present on linear plasmids and in the proximity of transposase and invertase sequences (Taguchi et al. 2007). In addition to the catabolic enzymes, functional redundancy was found in the regulatory system responsible for the growth of RHA1 on aromatic compounds. In particular, two copies of the gene cluster *bphS* and *bphT* encode the two-component systems, BphS1T1 and BphS2T2, which showed high similarity in the amino acid sequence (>92%) and the same substrate spectrum, except for biphenyl (Takeda et al. 2010).

Additional genomic redundancy found in the genome of aromatic *Rhodococcus* degraders was related to genes involved in the catabolism of phthalate (*pad*), terephthalate (*tpa*), catechol (*cat*), protocatechuate (*pca*) and benzoate (*ben*) (Fig. 5c). In particular, two identical copies of a catabolic island including the *pad* and *tpa* clusters were found on the linear plasmids of *R. jostii* RHA1 and *Rhodococcus* sp. DK17, flanked by transposase-encoding genes (Patrauchan et al. 2005; Choi et al. 2005). The duplicated phthalate-degrading operons resulted to be simultaneously expressed and equally functional during the DK17 growth on phthalate, allowing this strain to achieve the maximal degradation of phthalate (Choi et al. 2007). Genes involved in naphthalene and phthalate were found duplicated on two separate genomic islands in *R. opacus* M213, and, in *Rhodococcus* sp. TFB, it was demonstrated that naphthalene degradation probably results from the activities of different isozymes (Tomás-Gallardo et al. 2006). Genetic redundancy in aromatic degradation pathways was observed in *R. opacus* R7 possibly in relation to both the size of the genome and the degradation abilities towards different aromatic classes (Di Gennaro et al. 2014; Orro et al. 2015) and in *R. opacus* M213.

The catabolic redundancy was also associated to different substrate specificities and to possible mechanisms of metabolic intermediate detoxification in the steroid metabolism (involving *kshAB* and *kshD* genes) by some *Rhodococcus* strains. Four homologous *ksh* gene clusters are present in *R. jostii* RHA1 genome; the cluster 1 was shown to encode enzymes involved in cholesterol catabolism (van der Geize et al. 2007), while the homologous cluster 3 supported the catabolism of cholate (Swain et al. 2012). In *R. erythropolis* SQ1, *ksh* cluster 1 was involved in steroid metabolism, while *ksh* cluster 2 was supposed to have a role in limiting the intracellular accumulation of toxic metabolic intermediates (van der Geize et al. 2008). The genome analysis of *R. rhodochrous* DSM43269A revealed five *kshA* homologous genes (*kshA1* to *kshA5*) which were phylogenetically distinct, and each one showed a unique steroid induction pattern and substrate range, ensuring a fine-tuned steroid catabolism (Petrusma et al. 2011).

The redundancy of genes encoding (chloro)phenol hydroxylases supported the ability of *R. opacus* 1CP to metabolize a large spectrum of phenolic and chlorophenolic compounds (Gröning et al. 2014). All the three homologous phenol hydroxylases were able to convert the tested phenolic substrates at significant rates. This was probably due to the broad substrate specificity of these phenol hydroxylases and, therefore, low specialization (Gröning et al. 2014). Multiple genes (from 3 to 5) encoding phenol hydroxylases (*pheA1/pheA2*) were identified in other *Rhodococcus* strains (Gröning et al. 2014) In particular, in *R. jostii* RHA1, differences in the transcriptional regulation of two *pheA1/pheA2* clusters suggested their activation under different growth conditions (Szőköl et al. 2014).

Interestingly, the functional redundancy found in *Rhodococcus* was strongly associated to evolutionary mechanisms related to niche adaptation. In this respect, the genome and microarray analysis of *R. aetherivorans* I24 study of this strain to PCB/biphenyl exposure indicated the involvement of the only core enzymes in the substituted benzene metabolism, while most of the isozymes found in RHA1 were missing (Puglisi et al. 2010). This was associated to the fact that, unlike RHA1, I24 strain was not isolated from a PCB-contaminated site. Therefore, as previously mentioned, despite the gene redundancy is recognized as a trait of *Rhodococcus* genus, the presence of functional isozymes involved in specific catabolic pathways reflects the selective pressure imposed by the environmental habitat and is also in part associated to the size of the genomes, as in the case of *R. jostii* and *R. opacus* strains featured by larger genomes compared to *R. aetherivorans* and *R. ruber* (Table 1).

4.2 Plasmids

Plasmids are circular or linear extrachromosomal DNA elements, which are capable of semi-autonomous or self-replication; they do not typically encode essential genes for the host but instead carry genes that may help the organism to adapt to novel environments or nutrient sources (Aminov 2011; Carroll and Wong 2018). In this respect, most of the *Rhodococcus* genes belonging to the variable accessory genome have plasmidic localization (see Par. 3.2) (Creason et al. 2014a; Orro et al. 2015; Pathak et al. 2016). Plasmids are considered a major driving force in prokaryotic evolution as they can be transferred between cells, as mobile genetic elements, mediating horizontal gene transfer events (Hülter et al. 2017). Accordingly, in bacteria, plasmids are thought to play critical roles in the evolution, propagation and assembly of catabolic pathways, antibiotic and metal resistance mechanisms, antimicrobial biosynthetic pathways and pathogenicity (Meinhardt et al. 1997; Shimizu et al. 2001).

Many *Rhodococcus* strains harbour plasmids, which can be linear or circular, of small (of few Kbp) or large size (of hundreds of Kbp), with cryptic or catabolic functions. Some *Rhodococcus* strains have been described to simultaneously possess several plasmids of different types and sizes, e.g. *R. erythropolis* PR4 contains one linear plasmid, pREL1 (~270 Kb), and two circular plasmids, pREC1 (~100 Kb) and pREC2 (~3.5 Kb) (Sekine et al. 2006); *R. opacus* B4 possesses two linear plasmids of pROB series (pROB01 of ~560 Kb, pROB01 of ~240 Kb) and three circular plasmids of pKNR series (pKNR of 111 Kb, pKNR01 of 4.4 Kb, pKNR02 of 2.8 Kb) (Na et al. 2005). Many *Rhodococcus* plasmids, both linear and circular, encode catabolic functions associated to xenobiotic degradation (Shimizu et al. 2001), chemolithoautotrophical growth on gaseous hydrogen and carbon dioxide, hydrogen production (Grzeszik et al. 1997), metal toxicity resistance (Cappelletti et al. 2016) and pathogenicity (Letek et al. 2010). In this respect, deletions in several *Rhodococcus* plasmids have resulted in the loss of degradative genes and specific growth deficiencies; e.g. deletions on pRHL1 and pRHL2 in *R. jostii* RHA1 and on pTA422 in *R. erythropolis* TA421 affected biphenyl degradation (Fukuda et al. 1998; Kosono et al. 1997). In some cases, catabolic plasmids were also found to harbour homologous genes, and their mutation was found to exert moderate or null phenotypic effects (Patrauchan et al. 2005). In this framework, catabolic plasmids are thought to play a key role in *Rhodococcus* catabolic versatility and efficiency by harbouring unique set of genes involved in specific catabolic pathways but also contributing to *Rhodococcus* multiple pathways and functional redundancy (Kim et al. 2018). Additionally, *Rhodococcus* plasmids have been described to have a much higher density of DNA mobilization genes (e.g. insertion sequences, transposase genes), pseudogenes, unique species-specific genes and niche-specific determinants (e.g. genes involved in pathogenesis in *R. equi* and in peripheral aromatic clusters in *R. jostii* RHA1 and *R. opacus* R7) than the corresponding chromosomes. Rhodococcal plasmids are therefore under less stringent selection and are key players in rhodococcal genome plasticity and niche adaptability (Sekine et al. 2006).

Linear plasmids, which are often very large, are widespread and diverse among *Rhodococcus* strains (Ventura et al. 2007). As they can reach lengths of several hundreds of kilobases, they are frequently referred to as megaplasmids. For instance, pRHL2 from *R. jostii* RHA1 is 443 kb (Shimizu et al. 2001), pPDG1 from *R. opacus* R7 is 656 kb (Di Gennaro et al. 2014), and pRHL1 from *R. jostii* RHA1 is 1.12 Mb (McLeod et al. 2006). The reason of extensive research attention on *Rhodococcus* large linear plasmids is connected with the fact that they are associated with catabolic genes and that they confer advantageous abilities on their hosts (Meinhardt et al. 1997). For instance, linear plasmids have been described to encode enzymes involved in the catabolism of naphthalene (Kulakov et al. 2005; Orro et al. 2015; Pathak et al. 2016), biphenyl (Taguchi et al. 2004), toluene (Priefert et al. 2004), alkylbenzene (Kim et al. 2002), isopropylbenzene and trichloroethylene (Meinhardt et al. 1997), phthalate (Patrauchan et al. 2005), gaseous *n*-alkanes (Cappelletti et al. 2015), chloroaromatic compounds (Konig et al. 2004), isoprene (Crombie et al. 2015) and triazine compounds (Dodge et al. 2011) and also in the desulphurization of organosulphur compounds (Denis-Larose et al. 1997), resistance to toxic metals (Meinhardt et al. 1997; Cappelletti et al. 2016), pathogenicity towards bovines (Valero-Rello et al. 2015) and phytopathogenicity (Francis et al. 2012). In many cases, catabolic genes present in linear plasmids contribute to degradation pathways together with genes located on the chromosome (Gonçalves et al. 2006).

As a distinctive feature, linear megaplasmids from *Rhodococcus* strains are capable of conjugal transfer, being responsible for genetic information sharing through HGT (Meinhardt et al. 1997; Dib et al. 2015). The importance of linear elements is also associated to their higher "flexibility" compared to circular plasmids. In particular, the telomeres are considered to be frequently subject to recombinational events (Volff and Altenbuchner 2000; Chen et al. 2002). In the case of HGT, intermolecular recombination events can take place between plasmids but also between host chromosomes of compatible species. Both the frequent observation of linear plasmids and illegitimate recombination in *Rhodococcus* have led to the hypothesis of the hyper-recombinational gene storage strategy. This is related to the function of the plasmids as storage of large number of catabolic genes that can represent recombination sources to respond and adapt to novel compounds in the native soil environments.

As for most bacteria, circular plasmids are also very common in *Rhodococcus* species. Circular plasmids in *Rhodococcus* have been generally shown to possess smaller size compared to the linear ones (Table 2). However, they often contribute to the catabolic capacities of the *Rhodococcus* host strain. Several circular plasmids harbour genes encoding part or complete xenobiotic degradation pathways in *Rhodococcus* strains such as pRTL1 (100 Kb) encoding haloalkane degradation enzymes in *R. rhodochrous* NCIMB13064 (Kulakova et al. 1995); pREC1 contains a complete set of genes for the β-oxidation of fatty acids (Sekine et al. 2006) and the large circular plasmids pKNR (111 kb) in *R. opacus* B4 (Honda et al. 2012).

Additionally, many *Rhodococcus* plasmids are self-transmissible but phenotypically cryptic. The cryptic plasmids described in *Rhodococcus* strains have principally small size and circular structure, e.g. pREC2 (3.6 Kb) in *R. erythropolis* PR4, pB264 (4.9 Kb) in *Rhodococcus* sp. B264-1. In some cases, these plasmids have been isolated to develop vector systems for DNA manipulation and protein expression in *Rhodococcus* strains. For instance, pKA22 (4.9 Kb) from *R. rhodochrous* NCIMB13064 (Kulakov et al. 1997), pFAJ2600 (5.9 Kb) from *R. erythropolis* NI86/21 (De Mot et al. 1997), pAN12 (6.3 Kb) from *R. erythropolis* AN12 and pKNR01 (4.4 Kb) from *R. opacus* B4 are some of the small circular cryptic plasmids, described in the literature, that have been used to construct *E. coli-Rhodococcus* shuttle plasmids (Kostichka et al. 2003). The first two plasmids require the activity of two replication proteins, RepA and RepB, to replicate. As these two replication proteins resemble replication proteins of the theta-type replicating *Mycobacterium* plasmid pAL5000 (Ventura et al. 2007), pKA22 and pFAJ2600 are categorized as replicons of pAL5000 family. The plasmid pAN12 is of the pIJ101/pJV1 family, which usually replicates thanks to a single replication. Detection of ssDNA intermediates in several of the pIJ101/pJV1 family of plasmids suggested that they replicate by rolling circle mechanism (Kostichka et al. 2003).

5 Towards Genome-Scale Modification of *Rhodococcus* Through System and Synthetic Biology

Genetic manipulation (or genetic engineering) methods have been applied to *Rhodococcus* strains in order to obtain new strains that express additional genetic properties to acquire new catabolic capacities (Xiong et al. 2012, 2016; Venkataraman et al. 2015; Rodrigues et al. 2001; Hirasawa et al. 2001), to perform promoter activity studies (van der Geize et al. 2008; Cappelletti et al. 2011) and to introduce genetic mutations or deletions to determine the phenotypic alterations and the function of a specific gene or gene cluster (van der Geize et al. 2008; Amara et al. 2016).

Genetic manipulation methods are based on the development of efficient *Rhodococcus* transformation procedure, as biochemical and genetic characteristics of this genus have for long time limited the genetic manipulation of this strain (Sallam et al. 2006; Cappelletti 2010). Although protoplast-mediated transformation methods have also been used for some *Rhodococcus* strains (Singer and Finnerty 1988; Duran 1998), the use of electroporation procedure showed the highest efficiency, and it is presently the method that is most widely used for *Rhodococcus* transformation (Shao et al. 1995; Sekizaki et al. 1998; Kalscheuer et al. 1999). Several genetic manipulation strategies during the last three decades led to the development of (1) *E. coli-Rhodococcus* shuttle vectors from cryptic plasmids of *Rhodococcus* strains (Kostichka et al. 2003; Na et al. 2005; Matsui et al. 2006), (2) expression vectors for heterologous gene expression and protein production

(Kalscheuer et al. 1999; Nakashima and Tamura 2004a, b), (3) random mutagenesis methods using transposons or spontaneous illegitimate recombination (Desomer et al. 1991; Sallam et al. 2006; Crespi et al. 1994) and (4) targeted gene disruption methods based on unmarked mutagenesis deletion systems with SacB as counter-selection (van der Geize et al. 2001).

Recently, advances in genomics and genome editing offered new opportunities to engineer *Rhodococcus* in a directed and combinatorial manner through a genomic-scale rational design of the cell through system and synthetic biology approaches.

With the aim of developing synthetic biology platforms for *Rhodococcus* engineering, a first BioBrick™-compatible plasmid system was designed for *R. opacus* PD630 by Ellinger and Schmidt-Dannert (2017). Generally, the BioBrick™ toolbox allows convenient and reproducible assembly of multigene pathways into series of vectors and the possibility to quickly mobilize any cloned gene into vectors with different features for gene expression and protein purification (Vick et al. 2011). Currently, the majority of the available BioBricks were designed for the Gram-negative model organism *Escherichia coli*. However, recently, novel BioBrick™ tools have been designed for the engineering of other bacteria, e.g. *Bacillus* (Radeck et al. 2013). The first BioBrick™-compatible vector designed for a *Rhodococcus* strain pSRKBB derived from the backbone of the *E. coli-Rhodococcus* shuttle vector pSRK21, which was modified by removing all BioBrick™-incompatible restriction, while it retained the capacity to replicate in *R. opacus* PD630 (Ellinger and Schmidt-Dannert 2017) (Fig. 6). This BioBrick-compatible plasmid was demonstrated to enable robust heterologous protein expression in *R. opacus* PD630 from the lac promoter, giving a first demonstration of the use of *Rhodococcus* as platform for synthetic biology.

Moreover, in order to expand the molecular toolkit for *Rhodococcus* genetic engineering, a recent work analysed the genetic tools to obtain gene expression control in *R. opacus* PD630 by characterizing a constitutive promoter library, optimizing antibiotic resistance markers and defining the copy numbers of different gene expression plasmids (DeLorenzo et al. 2018). The same work also reported methods for genome editing by using recombinase-based technique for site-specific genomic modifications and by describing the methodology to identify neutral integration sites for stable heterologous expression. Lastly, a tunable and targeted gene repression system was developed by utilizing for the first time in *Rhodococcus* the CRISPR interference method (DeLorenzo et al. 2018). Taken together, the heterologous gene expression tools along with the CRISPRi method open new prospective for the optimization of the bioconversion abilities of *R. opacus* and for the possible development of molecular tools for other *Rhodococcus* species with relevant environmental and/or industrial characteristics.

At the same time, genomic information have been used to develop genome-scale metabolic models, i.e. mathematical representations of the stoichiometry of the biochemical networks occurring in *Rhodococcus* cell. In this regard, an in silico genome-scale metabolic model (iMT1174) was developed to make metabolic predictions on the behaviour of *R. jostii* RHA1 in relation to the accumulation of three types of carbon storage (i.e. glycogen, polyhydroxyalkanoate and triacylglycerols),

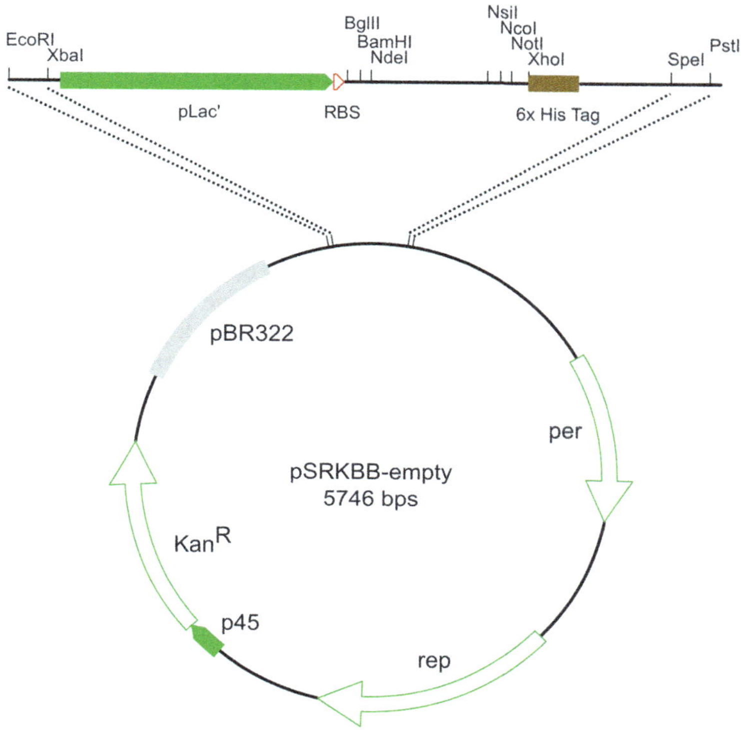

Fig. 6 Diagram of BioBrick™-compatible vector pSRKBB for *Rhodococcus*. The enlarged region includes the lac promoter and the ribosome binding site (RBS) upstream of the multi-cloning site where coding sequences can be cloned for constitutive expression. Further details are provided by Ellinger and Schmidt-Dannert (2017)

using different carbon sources (glucose or acetate) and under growth conditions typically occurring in activated sludge bioreactor systems for wastewater recovery (Tajparast et al. 2015, 2018). The simulations were compared with the experimentally measured metabolic fluxes through ^{13}C-labelling assays, and the predictive capacity of the model was established. These works represent the first steps towards systems biology approaches enabling to simulate and predict *Rhodococcus* metabolic fluxes leading to the production of industrially interesting metabolites. More efforts are needed to extend system and synthetic biology tools for genome-scale engineering of *Rhodococcus* species different from the model ones, i.e. *R. opacus* PD630 and *R. jostii* RHA1, in order to get deep into the extraordinary physiological and metabolic diversity featuring different *Rhodococcus* strains and to provide new opportunities for their utilization in targeted bioconversion and biodegradation processes.

6 Conclusions

Due to the biotechnological significance of this genus, a dramatic increase of sequenced *Rhodococcus* genomes has been published and become available in the database. The comparative and functional analyses of these genome data provided the opportunity to get insight into the genetic basis of the extraordinary metabolic diversity and versatility of *Rhodococcus* genus members. The high versatility and adaptability of members of *Rhodococcus* was found to be the reflection of the size and complexity of their genomes. The co-existence of linear and circular replicons (chromosome and plasmids) and the presence of large catabolic plasmids have strong implications in rhodococcal genome plasticity, fluidity and metabolic diversity. Further, the genetic and functional redundancy is a general characteristic of *Rhodococcus* genus, which supports metabolic robustness and versatility towards the degradation of different classes of substrates and organic compounds. The specific type of genomic redundancy was found to reflect the peculiar niche adaptation need of each *Rhodococcus* strain, and, at least in part, it was related to the size of the genome.

Lastly, advances in *Rhodococcus* genomics and genome editing provided new opportunities for the application of directed and combinatorial approaches to *Rhodococcus* engineering. Recent breakthroughs in genetic engineering of *Rhodococus* have included the use of synthetic biology platforms and new approaches for genome editing (CRISPR/Cas9 tool) of *R. opacus* PD630. Systems biology approaches have also been applied to *R. jostii* RHA1 to develop genome-scale metabolic models to predict cell metabolic fluxes for the production of relevant metabolites and storage compounds. Despite more efforts are needed to expand system and synthetic biology tools for genome-scale engineering of *Rhodococcus* species different from the model ones, i.e. *R. opacus* PD630 and *R. jostii* RHA1, the novel molecular toolkits along with the extended knowledge on genomic and functional genomics have largely contributed in making *Rhodococcus* strains relevant for future novel industrial applications.

References

Ali A (2013) Microbial comparative genomics: an overview of tools and insights into the genus *Corynebacterium*. J Bacteriol Parasitol 04:1–16. https://doi.org/10.4172/2155-9597.1000167

Alvarez HM, Mayer F, Fabritius D et al (1996) Formation of intracytoplasmic lipid inclusions by *Rhodococcus opacus* strain PD630. Arch Microbiol 165:377–386

Amara S, Seghezzi N, Otani H et al (2016) Characterization of key triacylglycerol biosynthesis processes in rhodococci. Sci Rep 6:24985. https://doi.org/10.1038/srep24985

Aminov RI (2011) Horizontal gene exchange in environmental microbiota. Front Microbiol 2:158. https://doi.org/10.3389/fmicb.2011.00158

Anastasi E, MacArthur I, Scortti M (2016) Pangenome and phylogenomic analysis of the pathogenic actinobacterium *Rhodococcus equi*. Genome Biol Evol 8(10):3140–3148

Bentley SD, Chater KF, Cerdeño-Tárraga AM et al (2002) Complete genome sequence of the model actinomycete *Streptomyces coelicolor* A3(2). Nature 417(6885):141–147
Cappelletti M (2010) Monooxygenases involved in the *n*-alkanes metabolism by *Rhodococcus* sp. BCP1. PhD Thesis Dissertation, University of Bologna
Cappelletti M, Fedi S, Frascari D et al (2011) Analyses of both the *alkB* gene transcriptional start site and *alkB* promoter-inducing properties of *Rhodococcus* sp. strain BCP1 grown on *n*-alkanes. Appl Environ Microbiol 77:1619–1627
Cappelletti M, Frascari ZD et al (2012) Microbial degradation of chloroform. Appl Microbiol Biotechnol 96:1395–1409
Cappelletti M, Di Gennaro P, D'Ursi P et al (2013) Genome sequence of *Rhodococcus* sp. strain BCP1, a biodegrader of alkanes and chlorinated compounds. Genome Announc 1(5):e00657–e00613. https://doi.org/10.1128/genomeA.00657-13
Cappelletti M, Presentato A, Milazzo G et al (2015) Growth of *Rhodococcus* sp. strain BCP1 on gaseous *n*-alkanes: new metabolic insights and transcriptional analysis of two soluble di-iron monooxygenase genes. Front Microbiol 6:393. https://doi.org/10.3389/fmicb.2015.00393
Cappelletti M, Fedi S, Zampolli J et al (2016) Phenotype microarray analysis may unravel genetic determinants of the stress response by *Rhodococcus aetherivorans* BCP1 and *Rhodococcus opacus* R7. Res Microbiol 167:1–9
Cappelletti M, Pinelli D, Fedi S (2017) Aerobic co-metabolism of 1,1,2,2-tetrachloroethane by *Rhodococcus aetherivorans* TPA grown on propane: kinetic study and bioreactor configuration analysis. J Chem Technol Biotechnol 93:155–165
Carroll AC, Wong A (2018) Plasmid persistence: costs, benefits, and the plasmid paradox. Can J Microbiol 64:293–304
Castro AR, Rocha I, Alves MM et al (2016) *Rhodococcus opacus* B4: a promising bacterium for production of biofuels and biobased chemicals. AMB Express 6:35. https://doi.org/10.1186/s13568-016-0207-y
Ceniceros A, Dijkhuizen L, Petrusma M et al (2017) Genome-based exploration of the specialized metabolic capacities of the genus *Rhodococcus*. BMC Genomics 18:593. https://doi.org/10.1186/s12864-017-3966-1
Chen CW, Huang CH, Lee HH et al (2002) Once the circle has been broken: dynamics and evolution of *Streptomyces* chromosomes. Trends Genet 18:522–529
Chen Y, Ding Y, Yang L et al (2014) Integrated omics study delineates the dynamics of lipid droplets in *Rhodococcus opacus* PD630. Nucleic Acids Res 42(2):1052–1064
Choi KY, Kim D, Sul WJ et al (2005) Molecular and biochemical analysis of phthalate and terephthalate degradation by *Rhodococcus* sp. strain DK17. FEMS Microbiol Lett 252:207–213
Choi KY, Kim D, Chae JC et al (2007) Requirement of duplicated operons for maximal metabolism of phthalate by *Rhodococcus* sp. strain DK17. Biochem Biophys Res Commun 357:766–771
Ciavarelli R, Cappelletti M, Fedi S et al (2012) Chloroform aerobic cometabolism by butane-growing *Rhodococcus aetherovorans* BCP1 in continuous-flow biofilm reactors. Bioprocess Biosyst Eng 35:667–681
Costa JSD, Herrero OM, Alvarez HM et al (2015) Label-free and redox proteomic analyses of the triacylglycerol-accumulating *Rhodococcus jostii* RHA1. Microbiology 161:593–610
Creason AL, Vandeputte OM, Savory EA et al (2014a) Analysis of genome sequences from plant pathogenic *Rhodococcus* reveals genetic novelties in virulence loci. PLoS One 9(7):e101996. https://doi.org/10.1371/journal.pone.0101996
Creason AL, Davis EW II, Putnam ML et al (2014b) Use of whole genome sequences to develop a molecular phylogenetic framework for *Rhodococcus fascians* and the *Rhodococcus* genus. Front Plant Sci 5:406. https://doi.org/10.3389/fpls.2014.00406
Crespi M, Vereecke D, Temmerman W et al (1994) The *fas* operon of *Rhodococcus fascians* encodes new genes required for efficient fasciation of host plants. J Bacteriol 176:2492–2501
Crombie TA, Rhodius VA, Miller MC et al (2015) Regulation of plasmid-encoded isoprene metabolism in *Rhodococcus*, a representative of an important link in the global isoprene cycle. Environ Microbiol 17:3314–3329

de Carvalho CCCR, da Fonseca MMR (2005) The remarkable *Rhodococcus erythropolis*. Appl Microbiol Biotechnol 67:715–726

de Carvalho CCCR, Costa SS, Fernandes P et al (2014) Membrane transport systems and the biodegradation potential and pathogenicity of genus *Rhodococcus*. Front Physiol 5:133. https://doi.org/10.3389/fphys.2014.00133

De Mot R, Nagy I, De Schrijver A et al (1997) Structural analysis of the 6 kb cryptic plasmid pFAJ2600 from *Rhodococcus erythropolis* NI86/21 and construction of *Escherichia coli* – *Rhodococcus* shuttle vectors. Microbiology 143:3137–3147

DeLorenzo DM, Rottinghaus AG, Henson WR et al (2018) Molecular toolkit for gene expression control and genome modification in *Rhodococcus opacus* PD630. ACS Synth Biol 7 (2):727–738

Denis-Larose C, Labbe D, Bergeron H et al (1997) Conservation of plasmid-encoded dibenzothiophene desulfurization genes in several rhodococci. Appl Environ Microbiol 63:2915–2919

Desomer J, Crespi M, Van Montagu M (1991) Illegitimate integration of non-replicative vectors in the genome of *Rhodococcus fascians* upon electrotransformation as an insertional mutagenesis system. Mol Microbiol 5:2115–2124

Di Canito A, Zampolli J, Orro A et al (2018) Genome-based analysis for the identification of genes involved in *o*-xylene degradation in *Rhodococcus opacus* R7. BMC Genomics 19:587

Di Gennaro P, Rescalli E, Galli E et al (2001) Characterization of *Rhodococcus opacus* R7, a strain able to degrade naphthalene and *o*-xylene isolated from polycyclic aromatic hydrocarbon-contaminated soil. Res Microbiol 152:641–651

Di Gennaro P, Terreni P, Masi G et al (2010) Identification and characterization of genes involved in naphthalene degradation in *Rhodococcus opacus* R7. Appl Microbiol Biotechnol 87 (1):297–308

Di Gennaro P, Zampolli J, Presti I et al (2014) Genome sequence of *Rhodococcus opacus* strain R7, a biodegrader of mono- and polycyclic aromatic hydrocarbons. Genome Announc 2(4):e00827–e00814. https://doi.org/10.1128/genomeA.00827-14

Dib JR, Wagenknecht M, Farías ME et al (2015) Strategies and approaches in plasmidome studies—uncovering plasmid diversity disregarding of linear elements? Front Microbiol 6:463. https://doi.org/10.3389/fmicb.2015.00463

Dodge AG, Wackett LP, Sadowsky MJ (2011) Plasmid localization and organization of melamine degradation genes in *Rhodococcus* sp strain Mel. Appl Environ Microbiol 78(5):1397–1403. https://doi.org/10.1128/AEM.06468-11

Dueholm MS, Albertsen M, D'Imperio S (2014) Complete genome of *Rhodococcus pyridinivorans* SB3094, a methyl-ethyl-ketone-degrading bacterium used for bioaugmentation. Genome Announc 2(3):e00525-14. https://doi.org/10.1128/genomeA.00525-14

Duran R (1998) New shuttle vectors for *Rhodococcus* sp. R312 (formerly *Brevibacterium* sp. R312), a nitrile hydratase producing strain. J Basic Microbiol 38:101–106

Ellinger J, Schmidt-Dannert C (2017) Construction of a BioBrick™ compatible vector system for *Rhodococcus*. Plasmid 90:1–4. https://doi.org/10.1016/j.plasmid.2017.01.004

Fang H, Xu T, Cao D et al (2016) Characterization and genome functional analysis of a novel metamitron-degrading strain *Rhodococcus* sp. MET via both triazinone and phenyl rings cleavage. Sci Rep 6:32339. https://doi.org/10.1038/srep32339

Francis I, De Keyser A, De Backer P et al (2012) pFiD188, the linear virulence plasmid of *Rhodococcus fascians* D188. MPMI 25:637–647

Francis IM, Stes E, Zhang Y et al (2016) Mining the genome of *Rhodococcus fascians*, a plant growth-promoting bacterium gone astray. New Biotechnol 33:706–717

Frascari D, Pinelli D, Nocentini M et al (2006) Chloroform degradation by butane-grown cells of *Rhodococcus aetherovorans* BCP1. Appl Microbiol Biotechnol 73(2):421–428

Fukuda M, Shimizu S, Okita N et al (1998) Structural alteration of linear plasmids encoding the genes for polychlorinated biphenyl degradation in *Rhodococcus* strain RHA1. Antonie Leeuwenhoek 74:169–173

Gao B, Gupta RS (2012) Phylogenetic framework and molecular signatures for the main clades of the phylum *Actinobacteria*. Microbiol Mol Biol Rev 76(1):66–112. https://doi.org/10.1128/MMBR.05011-11

Gonçalves ER, Hara H, Miyazawa D et al (2006) Transcriptomic assessment of isozymes in the biphenyl pathway of *Rhodococcus* sp. strain RHA1. Appl Environ Microbiol 72:6183–6193

Goodfellow M, Alderson G, Chun J (1998) Rhodococcal systematics: problems and developments. Antonie Van Leeuwenhoek 74:3–20

Goordial J, Raymond-Bouchard I, Zolotarov Y et al (2016) Cold adaptive traits revealed by comparative genomic analysis of the eurypsychrophile *Rhodococcus* sp. JG3 isolated from high elevation McMurdo Dry Valley permafrost, Antarctica. FEMS Microbiol Ecol 92. https://doi.org/10.1093/femsec/fiv154

Goris J, Konstantinidis KT, Klappenbach JA (2007) DNA-DNA hybridization values and their relationship to whole-genome sequence similarities. Int J Syst Evol Microbiol 57:81–91

Gravouil K, Ferru-Clément R, Colas S et al (2017) Transcriptomics and lipidomics of the environmental strain *Rhodococcus ruber* point out consumption pathways and potential metabolic bottlenecks for polyethylene degradation. Environ Sci Technol 51:5172–5181. https://doi.org/10.1021/acs.est.7b00846

Gröning JAD, Eulberg D, Tischler D et al (2014) Gene redundancy of two-component (chloro) phenol hydroxylases in *Rhodococcus opacus* 1CP. FEMS Microbiol Lett 361:68–75

Grzeszik C, Lubbers M, Reh M et al (1997) Genes encoding the NAD-reducing hydrogenase of *Rhodococcus opacus* MR11. Microbiology 143:1271–1286

Guo C, Wu ZL (2017) Construction and functional analysis of a whole-cell biocatalyst based on CYP108N7. Enzym Microb Technol 106:28–34

Gürtler V, Mayall BC, Seviour R (2004) Can whole genome analysis refine the taxonomy of the genus *Rhodococcus*? FEMS Microbiol Rev 28:377–403

Hara H, Eltis LD, Davies JE et al (2007) Transcriptomic analysis reveals a bifurcated terephthalate degradation pathway in *Rhodococcus* sp. strain RHA1. J Bacteriol 189:1641–1647

Hernández MA, Mohn WW, Martínez E et al (2008) Biosynthesis of storage compounds by *Rhodococcus jostii* RHA1 and global identification of genes involved in their metabolism. BMC Genomics 9:600. https://doi.org/10.1186/1471-2164-9-600

Hirasawa K, Ishii Y, Kobayashi M et al (2001) Improvement of desulfurization activity in *Rhodococcus erythropolis* KA2-5-1 by genetic engineering. Biosci Biotechnol Biochem 65:239–246

Holder JW, Ulrich JC, DeBono AC et al (2011) Comparative and functional genomics of *Rhodococcus opacus* PD630 for biofuels development. PLoS Genet 7(9):e1002219. https://doi.org/10.1371/journal.pgen.1002219

Honda K, Yamashita S, Nakagawa H et al (2008) Stabilization of water-in-oil emulsion by *Rhodococcus opacus* B-4 and its application to biotransformation. Appl Microbiol Biotechnol 78:767–773

Honda K, Imura M, Okano K et al (2012) Identification of replication region of a 111-kb circular plasmid from *Rhodococcus opacus* B-4 by λ red recombination-based deletion analysis. Biosci Biotechnol Biochem 76:1758–1764

Hori K, Kobayashi A, Ikeda H et al (2009) *Rhodococcus aetherivorans* IAR1, a new bacterial strain synthesizing poly(3-hydroxybutyrate-co-3-hydroxyvalerate) from toluene. J Biosci Bioeng 107:145–150

Hülter N, Ilhan J, Wein T et al (2017) An evolutionary perspective on plasmid lifestyle modes. Curr Opin Microbiol 38:74–80

Irvine VA, Kulakov LA, Larkin MJ (2000) The diversity of extradiol dioxygenase (edo) genes in cresol degrading rhodococci from a creosote-contaminated site that express a wide range of degradative abilities. Antonie Van Leeuwenhoek 78:341–352

Iwasaki T, Miyauchi K, Masai E et al (2006) Multiple-subunit genes of the aromatic-ring-hydroxylating dioxygenase play an active role in biphenyl and polychlorinated biphenyl degradation in *Rhodococcus* sp. strain RHA1. Appl Environ Microbiol 72:5396–5402

Iwasaki T, Takeda H, Miyauchi K et al (2007) Characterization of two biphenyl dioxygenases for biphenyl/PCB degradation in a PCB degrader, *Rhodococcus* sp. strain RHA1. Biosci Biotechnol Biochem 71:993–1002

Kalkus J, Dorrie C, Fischer D et al (1993) The giant linear plasmid pHG207 from *Rhodococcus* sp. encoding hydrogen autotrophy: characterization of the plasmid and its termini. J Gen Microbiol 139:2055–2065

Kalscheuer R, Arenskötter M, Steinbüchel A (1999) Establishment of a gene transfer system for *Rhodococcus opacus* PD630 based on electroporation and its application for recombinant biosynthesis of poly(3-hydroxyalkanoic acids). Appl Microbiol Biotechnol 52:508–515

Khairy H, Meinert C, Wübbeler JH et al (2016) Genome and proteome analysis of *Rhodococcus erythropolis* MI2: elucidation of the 4,4′-dithiodibutyric acid catabolism. PLoS One 11(12): e0167539. https://doi.org/10.1371/journal.pone.0167539

Kikuchi Y, Hirai K, Gunge N et al (1985) Hairpin plasmid-a novel linear DNA of perfect hairpin structure. EMBO J 4:1881–1886

Kim D, Kim YS, Kim SK et al (2002) Monocyclic aromatic hydrocarbon degradation by *Rhodococcus* sp. strain DK17. Appl Environ Microbiol 6:3270–3278

Kim D, Choi KY, Yoo M (2018) Biotechnological potential of *Rhodococcus* biodegradative pathways. J Microbiol Biotechnol 28(7):1037–1051

Konig C, Eulberg D, Groning J et al (2004) A linear megaplasmid, p1CP, carrying the genes for chlorocatechol catabolism of *Rhodococcus opacus* 1CP. Microbiology 150:3075–3087

Konishi M, Nishi S, Fukuoka T et al (2014) Deep-sea *Rhodococcus* sp. BS-15, lacking the phytopathogenic *fas* genes, produces a novel glucotriose lipid biosurfactant. Mar Biotechnol 16:484–493

Konstantinidis KT, Tiedje JM (2005) Towards a genome-based taxonomy for prokaryotes. J Bacteriol 187:6258–6264. https://doi.org/10.1128/JB.187.18.6258-6264.2005

Kosono S, Maeda M, Fuji F et al (1997) Three of the seven *bphC* genes of *Rhodococcus erythropolis* TA421, isolated from a termite ecosystem, are located on an indigenous plasmid associated with biphenyl degradation. Appl Environ Microbiol 63:3282–3285

Kostichka K, Tao L, Bramucci M et al (2003) A small cryptic plasmid from *Rhodococcus erythropolis*: characterization and utility for gene expression. Appl Microbiol Biotechnol 62:61–68

Kulakov LA, Larkin MJ (2002) Genomic organization of *Rhodococcus*. In: Danchin A (ed) Genomics of GC-rich gram-positive bacteria. Caister Academic Press, Norfolk, pp 15–46

Kulakov LA, Larkin MJ, Kulakova AN (1997) Cryptic plasmid pKA22 isolated from the naphthalene degrading derivative of *Rhodococcus rhodochrous* NCIMB13064. Plasmid 38:61–69

Kulakov LA, Chen S, Allen CC et al (2005) Web-type evolution of *Rhodococcus* gene clusters associated with utilization of naphthalene. Appl Environ Microbiol 71:1754–1764

Kulakova AN, Stafford TM, Larkin MJ et al (1995) Plasmid pRTL1 controlling 1-chloroalkane degradation by *Rhodococcus rhodochrous* NCIMB13064. Plasmid 33:208–217

Kwasiborski A, Mondy S, Teik-Min C et al (2015) Core genome and plasmidome of the quorum-quenching bacterium *Rhodococcus erythropolis*. Genetica 143:253–261

Laczi K, Kis A, Horváth B et al (2015) Metabolic responses of *Rhodococcus erythropolis* PR4 grown on diesel oil and various hydrocarbons. Appl Microbiol Biotechnol 99:9745–9759

Larkin MJ, DeMot R, Kulakov LA et al (1998) Applied aspects of *Rhodococcus* genetics. Antonie Van Leeuwenhoek 74:133–153

Larkin MJ, Kulakov LA, Allen CCR (2010) Genomes and plasmids in *Rhodococcus*. In: Alvarez HM (ed) Biology of *Rhodococcus*. Springer, Berlin, pp 73–90

LeBlanc JC, Gonçalves ER, Mohn WW (2008) Global response to desiccation stress in the soil actinomycete *Rhodococcus jostii* RHA1. Appl Environ Microbiol 74:2627–2636

Letek M, González P, MacArthur I et al (2010) The genome of a pathogenic *Rhodococcus*: cooptive virulence underpinned by key gene acquisitions. PLoS Genet 6(9):e1001145. https://doi.org/10.1371/journal.pgen.1001145

Martìnkovà L, Uhnàkovà B, Pàtek M et al (2009) Biodegradation potential of the genus *Rhodococcus*. Environ Int 35:162–177
Matsui T, Saeki H, Shinzato N et al (2006) Characterization of *Rhodococcus* – *E. coli* shuttle vector pNC9501 constructed from the cryptic plasmid of a propene-degrading bacterium. Curr Microbiol 52:445–448
McLeod MP, Warren RL, Hsiao WW et al (2006) The complete genome of *Rhodococcus* sp. RHA1 provides insights into a catabolic powerhouse. Proc Natl Acad Sci U S A 103(42):15582–15587
Meinhardt F, Schaffrath R, Larsen M (1997) Microbial linear plasmids. Appl Microbiol Biotechnol 47:329–336
Moran NA (2002) Microbial minimalism: genome reduction in bacterial pathogens. Cell 108:583–586
Na K, Kuroda A, Takiguchia N (2005) Isolation and characterization of benzene-tolerant *Rhodococcus opacus* strains. J Biosci Bioeng 99:378–382
Nakashima N, Tamura T (2004a) A novel system for expressing recombinant proteins over a wide temperature range from 4 to 35°C. Biotechnol Bioeng 86:136–148
Nakashima N, Tamura T (2004b) Isolation and characterization of a rolling-circle-type plasmid from *Rhodococcus erythropolis* and application of the plasmid to multiple-recombinant-protein expression. Appl Environ Microbiol 70:5557–5568
National Center for Biotechnology Information (NCBI) (2018) U.S. National Library of Medicine, Rockville Pike. www.ncbi.nlm.nih.gov/genomes/GenomesGroup.cgi?taxid=1827 Accessed July 2018
Orro A, Cappelletti M, D'Ursi P et al (2015) Genome and phenotype microarray analyses of *Rhodococcus* sp. BCP1 and *Rhodococcus opacus* R7: genetic determinants and metabolic abilities with environmental relevance. PLoS One 10(10):e0139467. https://doi.org/10.1371/journal.pone.0139467
Pathak A, Chauhan A, Blom J et al (2016) Comparative genomics and metabolic analysis reveals peculiar characteristics of *Rhodococcus opacus* strain M213 particularly for naphthalene degradation. PLoS One 17:1–32
Patrauchan MA, Florizone C, Dosanjh M et al (2005) Catabolism of benzoate and phthalate in *Rhodococcus* sp. strain RHA1: redundancies and convergence. J Bacteriol 187:4050–4063
Patrauchan MA, Florizone C, Eapen S et al (2008) Roles of ring-hydroxylating dioxygenases in styrene and benzene catabolism in *Rhodococcus jostii* RHA1. J Bacteriol 190(1):37–47
Patrauchan MA, Miyazawa D, LeBlanc JC et al (2012) Proteomic analysis of survival of *Rhodococcus jostii* RHA1 during carbon starvation. Appl Environ Microbiol 78(18):6714–6725
Pérez-Pérez JM, Candela H, Micol JL (2009) Understanding synergy in genetic interactions. Trends Genet 8:368–376
Petrusma M, Hessels G, Dijkhuizen L et al (2011) Multiplicity of 3-ketosteroid-9α-hydroxylase enzymes in *Rhodococcus rhodochrous* DSM43269 for specific degradation of different classes of steroids. J Bacteriol 193(15):3931–3940. https://doi.org/10.1128/JB.00274-11
Presentato A, Piacenza E, Anikovskiy M et al (2016) *Rhodococcus aetherivorans* BCP1 as cell factory for the production of intracellular tellurium nanorods under aerobic conditions. Microb Cell Factories 15:204. https://doi.org/10.1186/s12934-016-0602-8
Presentato A, Cappelletti M, Sansone A et al (2018a) Aerobic growth of *Rhodococcus aetherivorans* BCP1 using selected naphthenic acids as the sole carbon and energy sources. Front Microbiol 9:672. https://doi.org/10.3389/fmicb.2018.00672
Presentato A, Piacenza E, Anikovskiy M et al (2018b) Biosynthesis of selenium-nanoparticles and -nanorods as a product of selenite bioconversion by the aerobic bacterium *Rhodococcus aetherivorans* BCP1. New Biotechnol 41:1–8
Presentato A, Piacenza E, Darbandi A et al (2018c) Assembly, growth and conductive properties of tellurium nanorods produced by *Rhodococcus aetherivorans* BCP1. Sci Rep 8:3923. https://doi.org/10.1038/s41598-018-22320-x
Priefert H, O'Brien XM, Lessard PA et al (2004) Indene bioconversion by a toluene inducible dioxygenase of *Rhodococcus* sp. I24. Appl Microbiol Biotechnol 65:168–176

Puglisi E, Cahill MJ, Lessard PA et al (2010) Transcriptional response of *Rhodococcus aetherivorans* I24 to polychlorinated biphenyl-contaminated sediments. Microb Ecol 60 (3):505–515. https://doi.org/10.1007/s00248-010-9650-5

Qu J, Miao L-L, Liu Y et al (2015) Complete genome sequence of *Rhodococcus* sp. Strain IcdP1 shows diverse catabolic potential. Genome Announc 3(4):e00711-15. https://doi.org/10.1128/genomeA.00711-15

Radeck J, Kraft K, Bartels J (2013) The *Bacillus* BioBrick Box: generation and evaluation of essential genetic building blocks for standardized work with *Bacillus subtilis*. J Biol Eng 7:29

Redenbach M, Altenbuchner J (2002) Why do some bacteria have linear chromosomes and plasmids. BIOspektrum 8:158–163

Richter M, Rosselló-Móra R (2009) Shifting the genomic gold standard for the prokaryotic species definition. Proc Natl Acad Sci U S A 106:19126–19131

Rodrigues JLM, Maltseva OV, Tsoi TV et al (2001) Development of a *Rhodococcus* recombinant strain for degradation of products from anaerobic dechlorination of PCBs. Environ Sci Technol 35:663–668

Sallam KI, Mitani Y, Tamura T (2006) Construction of random transposition mutagenesis system in *Rhodococcus erythropolis* using IS*1415*. J Biotechnol 121:13–22

Sameshima Y, Honda K, Kato J et al (2008) Expression of *Rhodococcus opacus alkB* genes in anhydrous organic solvents. J Biosci Bioeng 106:199–203

Sangal V, Jones AL, Goodfellow M et al (2014) Comparative genomic analyses reveal a lack of a substantial signature of host adaptation in *Rhodococcus equi* ("*Prescottella equi*"). Pathog Dis 71:352–356

Sekine M, Tanikawa S, Omata S et al (2006) Sequence analysis of three plasmids harboured in *Rhodococcus erythropolis* strain PR4. Environ Microbiol 8:334–346

Sekizaki T, Tanoue T, Osaki M (1998) Improved electroporation of *Rhodococcus equi*. J Vet Med Sci 60:277–279

Shao Z, Dick WA, Behki RM (1995) An improved *Escherichia coli* – *Rhodococcus* shuttle vector and plasmid transformation in *Rhodococcus* spp. using electroporation. Lett Appl Microbiol 21:261–266

Sheng HM, Gao HS, Xue LG et al (2011) Analysis of the composition and characteristics of culturable endophytic bacteria within subnival plants of the Tianshan Mountains, northwestern China. Curr Microbiol 62:923–932

Shevtsov A, Tarlykov P, Zholdybayeva E et al (2013) Draft genome sequence of *Rhodococcus erythropolis* DN1, a crude oil biodegrader. Genome Announc 1:e00846–e00813. https://doi.org/10.1128/genomeA.00846-13

Shimizu S, Kobayashi H, Masai E et al (2001) Characterization of the 450-kb linear plasmid in a polychlorinated biphenyl degrader, *Rhodococcus* sp. strain RHA1. Appl Environ Microbiol 67:2021–2028

Singer ME, Finnerty WR (1988) Construction of an *Escherichia coli–Rhodococcus* shuttle vector and plasmid transformation in *Rhodococcus* spp. J Bacteriol 170:638–645

Swain K, Casabon I, Eltis LD et al (2012) Two transporters essential for the reassimilation of novel cholate metabolites by *Rhodococcus jostii* RHA1. J Bacteriol 194(24):6720–6727

Szőköl J, Rucká L, Šimčíková M et al (2014) Induction and carbon catabolite repression of phenol degradation genes in *Rhodococcus erythropolis* and *Rhodococcus jostii*. Appl Microbiol Biotechnol 98:8267–8279

Taguchi K, Motoyama M, Kudo T (2004) Multiplicity of 2, 3-dihydroxybiphenyl dioxygenase genes in the Gram-positive polychlorinated biphenyl degrading bacterium *Rhodococcus rhodochrous* K37. Biosci Biotechnol Biochem 68:787–795

Taguchi K, Motoyama M, Iida T et al (2007) Polychlorinated biphenyl/biphenyl degrading gene clusters in *Rhodococcus* sp. K37, HA99, and TA431 are different from well-known *bph* gene clusters of rhodococci. Biosci Biotechnol Biochem 71:1136–1144

Tajparast M, Frigon D (2015) Genome-scale metabolic model of *Rhodococcus jostii* RHA1 (iMT1174) to study the accumulation of storage compounds during nitrogen-limited condition. BMC Syst Biol 9:43. https://doi.org/10.1186/s12918-015-0190-y

Tajparast M, Frigon D, Virolle MJ (2018) Predicting the accumulation of storage compounds by *Rhodococcus jostii* RHA1 in the feast-famine growth cycles using genome-scale flux balance analysis. PLoS One 13:e0191835. https://doi.org/10.1371/journal.pone.0191835

Takeda H, Shimodaira J, Yukawa K et al (2010) Dual two-component regulatory systems are involved in aromatic compound degradation in a polychlorinated-biphenyl degrader, *Rhodococcus jostii* RHA1. J Bacteriol 192:4741–4751

Takei D, Washio K, Morikawa M (2008) Identification of alkane hydroxylase genes in *Rhodococcus* sp. strain TMP2 that degrades a branched alkane. Biotechnol Lett 30:1447–1452

Taketani RG, Zucchi TD, Soares de Melo I et al (2013) Whole-genome shotgun sequencing of *Rhodococcus erythropolis* strain P27, a highly radiation-resistant actinomycete from Antarctica. Genome Announc 1(5):e00763–e00713. https://doi.org/10.1128/genomeA.00763-13

Táncsics A, Benedek T, Szoboszlay S et al (2015) The detection and phylogenetic analysis of the alkane 1-monooxygenase gene of members of the genus *Rhodococcus*. Syst Appl Microbiol 38:1–7

Tao F, Zhao P, Li Q et al (2011) Genome sequence of *Rhodococcus erythropolis* XP, a biodesulfurizing bacterium with industrial potential. J Bacteriol 193:6422–6423

Tomás-Gallardo L, Canosa I, Santero E et al (2006) Proteomic and transcriptional characterization of aromatic degradation pathways in *Rhodococcus* sp. strain TFB. Proteomics 6:S119–S132. https://doi.org/10.1002/pmic.200500422

Treadway SL, Yanagimachi KS, Lankenau E (1999) Isolation and characterization of indene bioconversion genes from *Rhodococcus* strain I24. Appl Microbiol Biotechnol 51:786–793

Valero-Rello A, Hapeshi A, Anastasi E et al (2015) An invertron-like linear plasmid mediates intracellular survival and virulence in bovine isolates of *Rhodococcus equi*. Infect Immun 83:2725–2737. https://doi.org/10.1128/IAI.00376-15

Van Beilen JB, Smits THM, Whyte LG et al (2002) Alkane hydroxylase homologues in gram-positive strains. Environ Microbiol 4:676–682

Van der Geize, Dijkhuize L (2004) Harnessing the catabolic diversity of rhodococci for environmental and biotechnological applications. Curr Opin Microbiol 7:255–261

Van der Geize R, Hessels GI, van Gerwen R (2001) Unmarked gene deletion mutagenesis of *kstD*, encoding 3-ketosteroid Δ1-dehydrogenase, in *Rhodococcus erythropolis* SQ1 using *sacB* as counters electable marker. FEMS Microbiol Lett 205:197–202

Van der Geize R, Yam K, Heuser T et al (2007) A gene cluster encoding cholesterol catabolism in a soil actinomycete provides insight into *Mycobacterium tuberculosis* survival in macrophages. Proc Natl Acad Sci U S A 104:1947–1952

Van der Geize R, de Jong W, Hessels GI et al (2008) A novel method to generate unmarked gene deletions in the intracellular pathogen *Rhodococcus equi* using 5-fluorocytosine conditional lethality. Nucleic Acids Res 36:e151

Venkataraman H, Evelien MP, Rosłoniec KZ et al (2015) Biosynthesis of a steroid metabolite by an engineered *Rhodococcus erythropolis* strain expressing a mutant cytochrome P450 BM3 enzyme. Appl Microbiol Biotechnol 99:4713–4721

Ventura M, Canchaya C, Tauch A et al (2007) Genomics of *Actinobacteria:* tracing the evolutionary history of an ancient phylum. Microbiol Mol Biol Rev 71:495–548

Vick J, Johnson E, Choudhary S et al (2011) Optimized compatible set of BioBrick™ vectors for metabolic pathway engineering. Appl Microbiol Biotechnol 92:1275–1286

Villalba MS, Hernandéz MA, Silva RA et al (2013) Genome sequences of triacylglycerol metabolism in *Rhodococcus* as a platform for comparative genomics. J Mol Biochem 2:94–105

Volff JN, Altenbuchner J (2000) A new beginning with new ends: linearization of circular chromosomes during bacterial evolution. FEMS Microbiol Lett 186:143–150. https://doi.org/10.1111/j.1574-6968.2000.tb09095.x

Voss I, Steinbuchel A (2001) High cell density cultivation of *Rhodococcus opacus* for lipid production at a pilot-plant scale. Appl Microbiol Biotechnol 55:547–555

Warren R, Hsiao WW, Kudo H et al (2004) Functional characterization of a catabolic plasmid from polychlorinated-biphenyl degrading *Rhodococcus* sp. strain RHA1. J Bacteriol 186:7783–7795
Watcharakul S, Röther W, Birke J (2016) Biochemical and spectroscopic characterization of purified Latex Clearing Protein (Lcp) from newly isolated rubber degrading *Rhodococcus rhodochrous* strain RPK1 reveals novel properties of Lcp. BMC Microbiol 16:92. https://doi.org/10.1186/s12866-016-0703-x
Whyte LG, Smits THM, Labbe' D et al (2002) Gene cloning and characterization of multiple alkane hydroxylase systems in *Rhodococcus* strains Q15 and NRRL B-16531. Appl Environ Microbiol 68:5933–5942
Xiong X, Wang X, Chen S (2012) Engineering of a xylose metabolic pathway in *Rhodococcus* strains. Appl Environ Microbiol 78(16):5483–5491. https://doi.org/10.1128/AEM.08022-11
Xiong X, Lian J, Yu X et al (2016) Engineering levoglucosan metabolic pathway in *Rhodococcus jostii* RHA1 for lipid production. J Ind Microbiol Biotechnol 43:1551–1560
Xu JL, He J, Wang ZC et al (2007) *Rhodococcus qingshengii* sp. nov., a carbendazim-degrading bacterium. Int J Syst Evol Microbiol 57:2754–2757
Zampolli J, Collina E, Lasagni M et al (2014) Biodegradation of variable-chain-length *n*-alkanes in *Rhodococcus opacus* R7 and the involvement of an alkane hydroxylase system in the metabolism. AMB Express 4:73
Zhang J (2012) Genetic redundancies and their evolutionary maintenance. Evol Syst Biol 751:279–300

Central Metabolism of Species of the Genus *Rhodococcus*

Martín A. Hernández, Héctor M. Alvarez, Mariana P. Lanfranconi, Roxana A. Silva, O. Marisa Herrero, and María Soledad Villalba

Contents

Abstract Metabolism of *Rhodococcus* has evolved for adapting to a wide range of nutritional conditions. This adaptation often involves the flexibility of the central metabolism, which usually provides energy and precursors for biosynthesis processes, either during growth or during non-replicative metabolically active periods. The pathways of central metabolism are almost identical across widely divergent organisms, which share essentially the same metabolic network. However, this network possesses species-specific components, which depends on the biology of rhodococci. The central metabolism of members of *Rhodococcus* genus in the context of their physiology is the main topic of this chapter. An overview of main pathways of the central metabolism and their link with other metabolic processes is given. Glycolytic pathways, gluconeogenesis, phosphoenolpyruvate-pyruvate-oxaloacetate node, tricarboxylic acid cycle (TCA), glyoxylate pathway, and some lithoautotrophic pathways are included.

M. A. Hernández (✉) · H. M. Alvarez · M. P. Lanfranconi · R. A. Silva · O. M. Herrero · M. S. Villalba
Institute of Bioscience of Patagonia-National Scientific and Technical Research Council (INBIOP-CONICET) and Faculty of Natural Sciences and Health Sciences, University of Patagonia San Juan Bosco, Comodoro Rivadavia, Chubut, Argentina
e-mail: mahernandez@unpata.edu.ar

H. M. Alvarez (ed.), *Biology of Rhodococcus*, Microbiology Monographs 16,
https://doi.org/10.1007/978-3-030-11461-9_3

1 Introduction

Members of the genus *Rhodococcus* are aerobic non-sporulating bacteria widely distributed in diverse natural environments. They have been detected in tropical, artic, and arid soils, as well as marine and very deep-sea sediments (Whyte et al. 1999; Heald et al. 2001; Peressutti et al. 2003; Alvarez et al. 2004; Luz et al. 2004; Peng et al. 2008). One striking feature of rhodococci is the metabolic versatility with capabilities for biodegradation and transformation of a wide range of pollutant compounds, such as hydrocarbons, pesticides, and xenobiotics (Warhurst and Fewson 1994; Larkin et al. 2005; Martínková et al. 2009). This feature, in addition to their extraordinary environmental persistence and tolerance to stress conditions, makes such microorganisms promising candidates for in situ bioremediation of contaminated soil environments. Moreover, members of *Rhodococcus* are able to synthesize diverse compounds, such as surfactants, wax esters, and oils, which are of interest for biotechnological purposes. For these reasons, the studies on rhodococci have shown a significant increase within the last years. In this context, most knowledge on metabolism acquired during the last years has been focused principally on catabolism of diverse compounds and on the biosynthesis of lipids. For detailed overviews on these topics, the readers should refer to recent reviews (Larkin et al. 2005; Alvarez and Steinbüchel 2002; Alvarez 2006; Martínková et al. 2009; Yam et al. 2011). Despite the importance of the central metabolism for understanding the biology of rhodococci, this area has received little attention in comparison with the other metabolic processes mentioned above. The pathways of central metabolism are highly conserved among organisms; however, each species adopts a metabolic configuration specific to its biology. In this context, the pathways of central metabolism in rhodococci must provide energy and metabolic intermediates not only during growth periods but also in those environmental conditions in which cells do not grow but are still metabolically active. Under stress conditions, the physiology of rhodococci seems to depend on the metabolism of diverse storage compounds. Members of *Rhodococcus* genus possess an extensive capacity to synthesize and metabolize diverse storage compounds, such as triacylglycerols, wax esters, polyhydroxyalkanoates, glycogen, and polyphosphate (Anderson et al. 1995; Alvarez et al. 1997; Alvarez 2003; Hernández et al. 2008, 2017).

During the last years, numerous complete genome sequences of several *Rhodococcus* species such as *R. opacus* (strains B4 and PD630), *R. rhodochrous*, *R. pyridinivorans*, *R. ruber*, *R. hoagie* (*equi*), *R. erythropolis*, *R. qingshengii*, and *R. fascians*, among others, have been uploaded in public database. This fact reflects the importance of this genus as resources for biotechnological purposes or in the case of *R. hoagie* and *R. fascians* for their association with some pathologies of animals, humans, and plants. Among them, the genome sequence of *Rhodococcus jostii* strain RHA1 was the first sequence publicly available for screening and identification of genes and metabolic pathways (http://www.ncbi.nlm.nih.gov/genomes/lproks.cgi). For this reason, *R. jostii* RHA1 became a good model for understanding the genetics, physiology, and metabolism of the *Rhodococcus* genus. Strain RHA1 possesses one of the largest bacterial genomes sequenced to date, containing 9.7 Mbp arranged in a linear chromosome (7,802,028 bp) and three linear plasmids: pRHL1 (1,123,075 bp), pRHL2 (442,536 bp), and pRHL3 (332,361 bp) (McLeod et al. 2006). RHA1 is a soil bacterium with the ability to degrade and transform polychlorinated biphenyls and other aromatic compounds (Masai et al. 1995; Van der Geize et al. 2007; Patrauchan et al. 2008). The large RHA1 genome contains a multiplicity of catabolic genes, a high genetic redundancy of biosynthetic pathways, and a sophisticated regulatory network. Based on peer-reviewed literature and a thorough genome examination of *R. jostii* RHA1 as model, this chapter summarizes some aspects of the central metabolism of species of the *Rhodococcus* genus, including glycolytic pathways and NADPH-generating systems, gluconeogenesis, and phosphoenolpyruvate-pyruvate-oxaloacetate node, the tricarboxylic acid cycle (TCA), and the glyoxylate shunt. In addition, some new information about the energy pathways is given.

2 Glycolytic Pathways and NADPH-Generating Systems

Glycolysis (**E**mbden-**M**eyerhof-**P**arnas pathway, EMP) is a common pathway for glucose oxidation used by eukaryotic cells and some aerobic and facultative anaerobic bacteria, but not by Archaea. In this pathway, glucose is split into two molecules of pyruvic acid with the formation of two ATP molecules. Many aerobic bacteria and archaea use an alternative glycolytic pathway called **E**ntner-**D**oudoroff pathway (ED pathway), which produces glyceraldehyde 3-P, pyruvate, and only one ATP molecule from each molecule of glucose. This pathway is found in prokaryotes such as *Pseudomonas*, *Azotobacter*, *Rhizobium*, and other Gram negatives. However, the occurrence of the ED pathway has been also described in actinobacteria (Gunnarsson et al. 2004; Borodina et al. 2005; Juarez et al. 2017).

The analysis of the *R. jostii* RHA1 genome revealed the occurrence of key genes involved in glycolytic pathways. RHA1 strain is able to use both EMP and ED glycolytic pathways, to catabolize carbohydrates. Figure 1 and Table 1 show an overview on the pathways and enzymes/genes implicated in the glucose oxidation by this strain.

The RHA1 genome also contains all necessary genes/enzymes for the **P**entose **P**hosphate **P**athway (PPP) (Fig. 1 and Table 1). This pathway produces the precursor ribose 5-phosphate used in nucleic acid biosynthesis as well as erythrose 4-phosphate for the synthesis of aromatic amino acids. In addition, the PPP has been linked with the production of NADPH used in anabolic processes (Spaans et al. 2015).

Whereas the glucose metabolism by EMP generates two reducing equivalents in the form of NADH, the ED and PPP pathways generate one and two reducing equivalents in the form of NADPH, respectively. The initial step of ED and PPP pathways involves the conversion of glucose 6-phosphate in 6-phosphogluconolactone catalyzed by the enzyme glucose-6-phosphate dehydrogenase (Zwf), in which NADPH is generated.

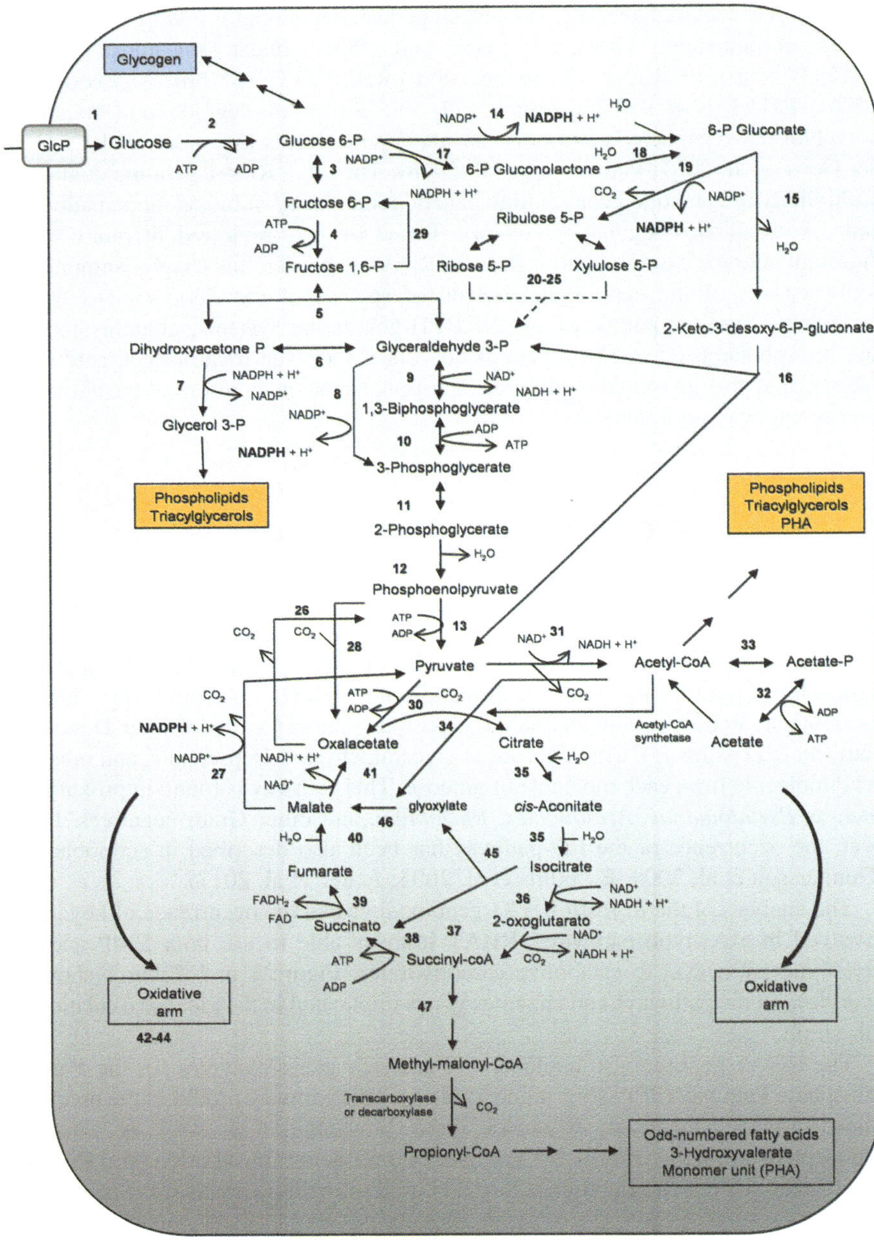

Fig. 1 Main pathways involved in central metabolism of *Rhodococcus jostii* RHA1 deduced from the genome database

Table 1 Main genes involved in the central metabolism of *R. jostii* RHA1

Reaction number[a]	Gene ID	Gene name	Predicted function based on gene annotation	Metabolic pathway
1	*RHA1_RS11555*	*glcP*	Glucose transporter major facilitator superfamily (MFS)	Glucose uptake
2	*RHA1_RS05385*		Glucokinase	Glycolysis (EMP-ED), glycogen degradation
2	*RHA1_RS06585*		Glucokinase	Glycolysis (EMP-ED), glycogen degradation
2	*RHA1_RS20805*		Glucokinase	Glycolysis (EMP-ED), glycogen degradation
3	*RHA1_RS43575*	*pgi*	Glucose-6-phosphate isomerase	Glycolysis (EMP), gluconeogenesis
3	*RHA1_RS43120*	*pgi*	Glucose-6-phosphate isomerase	Glycolysis (EMP), gluconeogenesis
3	*RHA1_RS44310*	*pgi*	Glucose-6-phosphate isomerase	Glycolysis (EMP), gluconeogenesis
3	*RHA1_RS27160*	*pgi*	Glucose-6-phosphate isomerase	Glycolysis (EMP), gluconeogenesis
4	*RHA1_RS00295*		Phosphofructokinase	Glycolysis (EMP)
4	*RHA1_RS31690*	*pfkA*	6-Phosphofructokinase	Glycolysis (EMP)
4	*RHA1_RS33150*		1-Phosphofructokinase	Glycolysis (EMP)
5	*RHA1_RS27000*	*fba*	Class II fructose-bisphosphate aldolase	Glycolysis (EMP), gluconeogenesis
5	*RHA1_RS26975*	*fba*	Fructose-bisphosphate aldolase	Glycolysis (EMP), gluconeogenesis
6	*RHA1_RS35040*	*tpiA*	Triose-phosphate isomerase	Glycolysis (EMP)
7	*RHA1_RS31815*	*gpsA*	Glycerol-3-phosphate dehydrogenase [NAD(P)+] 2	Glycerol 3-phosphate synthesis
7	*RHA1_RS11710*	*gpsA*	Glycerol-3-phosphate dehydrogenase [NAD(P)+] 1	Glycerol 3-phosphate synthesis
8	*RHA1_RS16630*	*tadD*	NADP+-dependent Glyceraldehyde-3-phosphate dehydrogenase	Glycolysis (EMP/ED)
9	*RHA1_RS35030*	*gap2*	Glyceraldehyde 3-phosphate dehydrogenase	Glycolysis (EMP), gluconeogenesis
10	*RHA1_RS35035*	*pgk*	Phosphoglycerate kinase	Glycolysis (EMP), gluconeogenesis
11	*RHA1_RS04205*		Phosphoglycerate mutase	Glycolysis (EMP), gluconeogenesis
11	*RHA1_RS06270*	*gpmA*	Phosphoglycerate mutase	Glycolysis (EMP), gluconeogenesis

(continued)

Table 1 (continued)

Reaction number[a]	Gene ID	Gene name	Predicted function based on gene annotation	Metabolic pathway
11	*RHA1_RS31045*		Phosphoglycerate mutase	Glycolysis (EMP), gluconeogenesis
12	*RHA1_RS28200*	*eno*	Enolase	Glycolysis (EMP), gluconeogenesis
13	*RHA1_RS04835*	*pyk*	Pyruvate kinase	Glycolysis (EMP-ED)
13	*RHA1_RS12545*	*pyk*	Pyruvate kinase	Glycolysis (EMP-ED)
13	*RHA1_RS15635*	*pyk*	Pyruvate kinase	Glycolysis (EMP-ED)
13	*RHA1_RS23435*	*pyk*	Pyruvate kinase	Glycolysis (EMP-ED)
14	*RHA1_RS12070*		PQQ-dependent sugar dehydrogenase	Glucose degradation (PPP-ED pathway)
14	*RHA1_RS31700*		PQQ-dependent sugar dehydrogenase	Glucose degradation (PPP-ED pathway)
15	*RHA1_RS11570*	*edd*	Phosphogluconate dehydratase	Glucose degradation (ED)
16	*RHA1_RS11565*	*eda*	KHG/KDPG aldolase	Glucose degradation (ED)
17	*RHA1_RS02710*	*zwf1*	Glucose-6-phosphate 1-dehydrogenase	PPP/ED pathway
17	*RHA1_RS11575*	*zwf2*	Glucose-6-phosphate 1-dehydrogenase	PPP/ED pathway
17	*RHA1_RS27530*	*zwf3*	Glucose-6-phosphate 1-dehydrogenase	PPP/ED pathway
17	*RHA1_RS49040*	*zwf4*	Glucose-6-phosphate 1-dehydrogenase (NADP+)	PPP/ED pathway
17	*RHA1_RS44335*	*zwf5*	Glucose-6-phosphate 1-dehydrogenase	PPP/ED pathway
17	*RHA1_RS43115*	*fgd*	Glucose-6-phosphate dehydrogenase coenzyme-F420	PPP/ED pathway
17	*RHA1_RS43570*	*fgd*	Glucose-6-phosphate dehydrogenase coenzyme-F420	PPP/ED pathway
18	*RHA1_RS35055*	*pgl*	6-Phosphogluconolactonase	PP pathway
19	*RHA1_RS02715*	*gnd*	Phosphogluconate dehydrogenase (decarboxylating)	PP pathway
19	*RHA1_RS17785*	*gnd*	Phosphogluconate dehydrogenase (decarboxylating)	PP pathway
19	*RHA1_RS35365*	*gnd*	Phosphogluconate dehydrogenase ((NADP(+)-dependent, decarboxylating))	PP pathway

(continued)

Table 1 (continued)

Reaction number[a]	Gene ID	Gene name	Predicted function based on gene annotation	Metabolic pathway
19	*RHA1_RS43110*	*gnd*	Phosphogluconate dehydrogenase (decarboxylating)	PP pathway
20	*RHA1_RS34980*	*rpe*	Ribulose-phosphate 3-epimerase	PP pathway
21	*RHA1_RS14120*		Ribose-5-phosphate isomerase	PP pathway
21	*RHA1_RS06675*		Ribose 5-phosphate isomerase	PP pathway
22	*RHA1_RS49045*	*tal*	Transaldolase	PP pathway
22	*RHA1_RS36885*	*tal*	Transaldolase	PP pathway
23	*RHA1_RS03880*		Transketolase, C-terminal subunit	PP pathway
23	*RHA1_RS03885*		Transketolase, N-terminal subunit	PP pathway
23	*RHA1_RS18640*		Transketolase, N-terminal subunit	PP pathway
23	*RHA1_RS18645*		Transketolase, C-terminal subunit	PP pathway
23	*RHA1_RS35075*		Transketolase	PP pathway
24	*RHA1_RS16375*		Sugar phosphate isomerase/ epimerase	PP pathway
24	*RHA1_RS25680*		Sugar phosphate isomerase/ epimerase	PP pathway
24	*RHA1_RS22195*		Sugar phosphate isomerase/ epimerase	PP pathway
24	*RHA1_RS14555*		Sugar phosphate isomerase/ epimerase	PP pathway
24	*RHA1_RS10065*		Sugar phosphate isomerase/ epimerase	PP pathway
25	*RHA1_RS22585*		Class II aldolase	PP pathway
25	*RHA1_RS34445*		Class II aldolase	PP pathway
26	*RHA1_RS25350*		Phosphoenolpyruvate carboxykinase	Gluconeogenesis/ anaplerotic pathway
27	*RHA1_RS02430*		NAD-dependent malic enzyme	Gluconeogenesis/ anaplerotic pathway
27	*RHA1_RS12570*		NAD-dependent malic enzyme	Gluconeogenesis/ anaplerotic pathway
27	*RHA1_RS29320*		NADP-dependent malic enzyme	Gluconeogenesis/ anaplerotic pathway
27	*RHA1_RS44255*		NADP-dependent malic enzyme	Gluconeogenesis/ anaplerotic pathway
27	*RHA1_RS39800*		NADP-dependent malic enzyme	Gluconeogenesis/ anaplerotic pathway
28	*RHA1_RS35050*	*ppc*	Phosphoenolpyruvate carboxylase	Anaplerotic pathway

(continued)

Table 1 (continued)

Reaction number[a]	Gene ID	Gene name	Predicted function based on gene annotation	Metabolic pathway
29	*RHA1_RS28640*	*glpX*	Fructosa 1,6-bisphosphatase class II	Gluconeogenesis
30	*RHA1_RS31870*	*pycA*	Pyruvate carboxylase	Anaplerotic pathway
31	*RHA1_RS02650*		Pyruvate dehydrogenase E1 component	Acetyl-CoA synthesis
31	*RHA1_RS02655*		Dihydrolipoyllysine-residue succinyltransferase (PDH)	Acetyl-CoA synthesis
31	*RHA1_RS05135*		Pyruvate oxidase	Acetyl-CoA synthesis
31	*RHA1_RS05795*	*aceE1*	Pyruvate dehydrogenase (acetyl-transferring), homodimeric type	Acetyl-CoA synthesis
31	*RHA1_RS07665*	*pdhA*	Pyruvate dehydrogenase E1 component alpha subunit	Acetyl-CoA synthesis
31	*RHA1_RS07670*		Alpha-ketoacid dehydrogenase subunit beta	Acetyl-CoA synthesis
31	*RHA1_RS07675*		2-Oxo acid dehydrogenase subunit E2	Acetyl-CoA synthesis
31	*RHA1_RS10445*	*lpdA*	Dihydrolipoyl dehydrogenase	Acetyl-CoA synthesis
31	*RHA1_RS12550*	*aceE*	Pseudo	Acetyl-CoA synthesis
31	*RHA1_RS14255*	*lpdA*	Dihydrolipoyl dehydrogenase	Acetyl-CoA synthesis
31	*RHA1_RS16000*		MFS transporter	Acetyl-CoA synthesis
31	*RHA1_RS16020*		Pyruvate dehydrogenase	Acetyl-CoA synthesis
31	*RHA1_RS16085*	*lpdA*	Dihydrolipoyl dehydrogenase	Acetyl-CoA synthesis
31	*RHA1_RS16090*		Alpha-ketoacid dehydrogenase subunit beta	Acetyl-CoA synthesis
31	*RHA1_RS16095*	*pdhB1*	Pyruvate dehydrogenase E1 component beta subunit	Acetyl-CoA synthesis
31	*RHA1_RS16100*	*pdhA*	Pyruvate dehydrogenase (acetyl-transferring) E1 component subunit alpha	Acetyl-CoA synthesis
31	*RHA1_RS16380*	*lpdA*	Dihydrolipoyl dehydrogenase	Acetyl-CoA synthesis
31	*RHA1_RS16385*		2-Oxo acid dehydrogenase subunit E2	Acetyl-CoA synthesis
31	*RHA1_RS16390*		Alpha-ketoacid dehydrogenase subunit beta	Acetyl-CoA synthesis

(continued)

Table 1 (continued)

Reaction number[a]	Gene ID	Gene name	Predicted function based on gene annotation	Metabolic pathway
31	*RHA1_RS16395*	*pdhA*	Pyruvate dehydrogenase (acetyl-transferring) E1 component subunit alpha	Acetyl-CoA synthesis
31	*RHA1_RS27135*		NAD(P)/FAD-dependent oxidoreductase	Acetyl-CoA synthesis
31	*RHA1_RS30615*		NAD(P)H-quinone dehydrogenase	Acetyl-CoA synthesis
31	*RHA1_RS37395*		Hypothetical protein	Acetyl-CoA synthesis
31	*RHA1_RS42945*	*lpdA*	Dihydrolipoyl dehydrogenase	Acetyl-CoA synthesis
31	*RHA1_RS42970*		2-oxo acid dehydrogenase subunit E2	Acetyl-CoA synthesis
31	*RHA1_RS42975*		Alpha-ketoacid dehydrogenase subunit beta	Acetyl-CoA synthesis
31	*RHA1_RS42980*		Thiamine pyrophosphate-dependent dehydrogenase E1 component subunit alpha	Acetyl-CoA synthesis
32	*RHA1_RS10720*		Acetate kinase	AK-PTA pathway
33	*RHA1_RS10725*		Phosphate acetyltransferase	AK-PTA pathway
34	*RHA1_RS01765*	*citA1*	Citrate (Si)-synthase	TCA cycle (oxidative)
34	*RHA1_RS07835*		Possible citrate synthase, C-terminal	TCA cycle (oxidative)
34	*RHA1_RS24420*	*citA2*	Citrate (Si)-synthase	TCA cycle (oxidative)
34	*RHA1_RS24440*		Citrate (Si)-synthase	TCA cycle (oxidative)
34	*RHA1_RS34010*		Citrate synthase	TCA cycle (oxidative)
35	*RHA1_RS11725*	*acnA1*	Aconitate hydratase	TCA cycle (oxidative/reductive)
35	*RHA1_RS35180*	*acnA2*	Aconitate hydratase	TCA cycle (oxidative/reductive)
35	*RHA1_RS36835*	*acnA3*	Aconitate hydratase	TCA cycle (oxidative/reductive)
36	*RHA1_RS02995*		Isocitrate dehydrogenase (NADP +)	TCA cycle (oxidative/reductive)
36	*RHA1_RS30490*		Isocitrate dehydrogenase (NADP +)	TCA cycle (oxidative/reductive)
37	*RHA1_RS29380*	*odhA*	2-Oxoglutarate dehydrogenase, E1 and E2 components	TCA cycle (oxidative)
38	*RHA1_RS27190*	*sucC*	Succinate CoA ligase β-subunit	TCA cycle (oxidative)

(continued)

Table 1 (continued)

Reaction number[a]	Gene ID	Gene name	Predicted function based on gene annotation	Metabolic pathway
38	*RHA1_RS27195*	*sucD*	Succinate CoA ligase α-subunit	TCA cycle (oxidative)
38	*RHA1_RS17525*		CoA binding protein (previously annotated as probable succinate CoA ligase)	TCA cycle (oxidative)
39	*RHA1_RS05045*	*sdhB1*	Succinate dehydrogenase iron-sulfur protein	TCA cycle (oxidative)
39	*RHA1_RS05050*	*sdhA1*	Succinate dehydrogenase flavoprotein subunit	TCA cycle (oxidative)
39	*RHA1_RS05055*		Possible succinate dehydrogenase	TCA cycle (oxidative)
39	*RHA1_RS23305*		Hypothetical protein	TCA cycle (oxidative)
39	*RHA1_RS27810*		FAD-binding protein	TCA cycle (oxidative)
39	*RHA1_RS30525*	*sdhB2*	FAD-binding protein	TCA cycle (oxidative)
39	*RHA1_RS30530*	*sdhA2*	Succinate dehydrogenase	TCA cycle (oxidative)
39	*RHA1_RS30535*		Probable succinate dehydrogenase hydrophobic membrane anchor protein	TCA cycle (oxidative)
39	*RHA1_RS30540*		Probable succinate dehydrogenase (cytochrome)	TCA cycle (oxidative)
39	*RHA1_RS36845*		Ferredoxin	TCA cycle (oxidative)
39	*RHA1_RS39385*		Probable succinate dehydrogenase hydrophobic membrane anchor protein	TCA cycle (oxidative)
39	*RHA1_RS47620*	*sdhC*	Succinate dehydrogenase cytochrome β-subunit	TCA cycle (oxidative)
39	*RHA1_RS39395*		Ferredoxin	TCA cycle (oxidative)
39	*RHA1_RS39400*	*sdhB3*	Succinate dehydrogenase Fe-S protein subunit	TCA cycle (oxidative)
40	*RHA1_RS28635*	*fumC*	Fumarate hydratase class II	TCA cycle (oxidative/reductive)
40	*RHA1_RS28815*		Fumarate hydratase class I	TCA cycle (oxidative/reductive)
40	*RHA1_RS39380*		Fumarate hydratase, class I	TCA cycle (oxidative/reductive)
41	*RHA1_RS30520*	*mdh*	Malate dehydrogenase	TCA cycle (oxidative/reductive)
41	*RHA1_RS32320*		Malate dehydrogenase (acceptor)	TCA cycle (oxidative/reductive)

(continued)

Table 1 (continued)

Reaction number[a]	Gene ID	Gene name	Predicted function based on gene annotation	Metabolic pathway
42	*RHA1_RS00955*		Citrate lyase	TCA cycle (reductive)
42	*RHA1_RS01540*		CoA ester lyase	TCA cycle (reductive)
42	*RHA1_RS03190*		Aldolase	TCA cycle (reductive)
42	*RHA1_RS03695*		CoA ester lyase	TCA cycle (reductive)
42	*RHA1_RS03710*		CoA ester lyase	TCA cycle (reductive)
42	*RHA1_RS11960*		Possible citrate lyase beta subunit	TCA cycle (reductive)
42	*RHA1_RS13995*		CoA ester lyase (previously annotated as citrate (pro-3S)-lyase)	TCA cycle (reductive)
42	*RHA1_RS14330*		Citrate lyase beta subunit	TCA cycle (reductive)
42	*RHA1_RS29250*		CoA ester lyase	TCA cycle (reductive)
42	*RHA1_RS29815*		CoA ester lyase	TCA cycle (reductive)
42	*RHA1_RS32175*		CoA ester lyase	TCA cycle (reductive)
42	*RHA1_RS38600*		Possible citrate lyase beta subunit (C-terminal)	TCA cycle (reductive)
42	*RHA1_RS39235*		Aldolase	TCA cycle (reductive)
43	*RHA1_RS11675*		2-Oxo acid ferredoxin oxidoreductase subunit beta	TCA cycle (reductive)
44	*RHA1_RS11825*		FAD-binding dehydrogenase	TCA cycle (reductive)
44	*RHA1_RS40180*		FAD-binding dehydrogenase	TCA cycle (reductive)
44	*RHA1_RS40490*		FAD-binding dehydrogenase	TCA cycle (reductive)
45	*RHA1_RS10355*		Isocitrate lyase	Glyoxylate pathway
46	*RHA1_RS04300*	*glcB*	Malate synthase G	Glyoxylate pathway
46	*RHA1_RS12540*		Malate synthase G	Glyoxylate pathway
47	*RHA1_RS35295*	*mutA*	Methylmalonyl-CoA mutase small subunit	Methylmalonyl-CoA pathway
47	*RHA1_RS35300*	*mutB*	Methylmalonyl-CoA mutase	Methylmalonyl-CoA pathway

[a]Numbers refer to reactions shown in Fig. 1.

Nguyen et al. (2017) have characterized an alternative enzyme that catalyzes this step in *R. jostii* RHA1 called F420-dependent glucose-6-phosphate dehydrogenase (FGD). This enzyme uses low redox potential cofactor that could play an important role in the central metabolism and xenobiotics degradation.

The NADPH is an essential reducing equivalent in most biosynthetic pathways such as the fatty acid biosynthesis (Spaans et al. 2015). Recent proteomic and transcriptomic analyses (Dávila Costa et al. 2015; Juarez et al. 2017) of *R. jostii* RHA1 under lipid accumulation conditions (low nitrogen levels) showed a significant induction of genes/proteins that participate in ED and PPP pathways. Likewise, studies conducted in *R. opacus* PD630 using ^{13}C also show the activation of these pathways during lipid accumulation (Hollinshead et al. 2015). In addition, MacEachran and Sinskey (2013) identified and characterized an enzyme that contributes to the formation of NADPH, a non-phosphorylative glyceraldehyde dehydrogenase enzyme, which catalyzes the direct conversion of the glyceraldehyde 3-P to 3-phosphoglycerate (Fig. 1). This enzyme was also induced specifically during lipid accumulation. On the other hand, the occurrence of $NADP^+$-dependent malic enzymes (reaction 27 in Fig. 1 and Table 1) and their effect on lipid accumulation have been described recently in the RHA1 strain (Hernández and Alvarez 2018). Then, the NADPH necessary for lipid biosynthesis seems to come from several pathways in these bacteria.

Glycolytic pathway also contributes with key precursors for lipid biosynthesis such as triacylglycerols (TAG) in oleaginous *Rhodococcus* species. For example, the glycolytic intermediate dihydroxyacetone phosphate is converted to glycerol 3-phosphate by the NADP-dependent glycerol 3-phosphate dehydrogenase. This enzyme is upregulated in *R. jostii* RHA1 under TAG-accumulating condition (Dávila Costa et al. 2015).

All genes involved in the carbohydrate metabolism are widely distributed throughout the RHA1 genome. However, some genes such as the *tpiA*, *gap2*, and *pgk* involved in the glycolytic pathways are clustered in the genome. Similarly, the genes *RHA1_RS11570* coding for a phosphogluconate dehydratase (Edd) and *RHA1_RS11565* coding for a KHG/KDPG aldolase (Edda), both involved specifically in the ED pathway, are clustered in RHA1 genome with the gene coding for glucose-6-phosphate dehydrogenase (Zwf). In this same cluster, it is located in the gene *glcp* (*RHA1_RS11555*), involved in glucose uptake components in *R. jostii* RHA1 (Araki et al. 2011).

3 Glycogen Synthesis and the Link with the Central Metabolism

The ability to synthesize and accumulate glycogen has been studied based on bioinformatics and/or experimental approaches in different members of the *Rhodococcus* genus such as *R. jostii* (Hernández et al. 2008; Tajparast and Frigon

2015), *R. opacus* (Hernández and Alvarez 2010; Hernández et al. 2013), *R. erythropolis*, *R. ruber*, *R. equi*, and *R. fascians* (Hernández and Alvarez 2010; Yano et al. 2016). In general, the total content of glycogen in those microorganisms reached up to 2–6% of cellular dry weight. The studied strains accumulated glycogen during exponential growth phase, and the content tends to decrease during stationary growth phase. The occurrence of glycogen accumulation during exponential growth phase has been reported a long time ago in other actinomycetes such as *Mycobacterium smegmatis* (Belanger and Hatfull 1999) and *Corynebacterium glutamicum* (Seibold et al. 2007).

Similarly to other microorganisms, genes of glycogen synthesis and degradation in *Rhodococcus* species are usually arranged in clusters and seem to be non-redundant (*glgC*, *glgA*, *glgB*, *glgX*) (Hernández et al. 2008; Hernández and Alvarez 2010). In addition, as occur in other actinobacteria such as *Mycobacterium* and *Streptomyces* (Kalscheuer et al. 2010; Elbein et al. 2010), *Rhodococcus* species also contain the alternative GlgE α-glucan pathway mediated by the *glgE* gene (Hernández et al. 2008). Glycogen may have a role as metabolic intermediate since it is accumulated mainly during the exponential growth phase by cells and is mobilized later in the stationary phase; thus, glycogen has been proposed as a carbon capacitor for glycolysis during exponential growth (Belanger and Hatfull 1999). Glycogen may be a part of a mechanism for controlling sugar excess in *Rhodococcus* or may act as part of a sensing/signaling mechanism. Persson et al. (2007) proposed that the expression of some genes in *E. coli*, such as the universal stress protein *uspA* involved in the carbon starvation conditions or stationary phase, is regulated by glycolytic intermediates such as fructose-6-phosphate. Alteration in the pool size of phosphorylated sugars of the upper glycolytic pathway may ensure expression of stress proteins preceding the complete depletion of the external carbon source and growth arrest (Persson et al. 2007). Thus, glycogen formation may act to attenuate phosphorylated sugar signals and to protect cells from sudden increases in fluxes of sugars.

Transcriptome and proteome analyses in RHA1 and PD630 strains have shown that gene expression and/or protein abundance associated with glycogen synthesis remains constant between culture conditions with high or low nitrogen levels, which suggest that glycogen synthesis is not essentially regulated at the transcriptional level in those conditions (Dávila Costa et al. 2015; Chen et al. 2014). On the contrary, proteins associated with glycogen synthesis seem to be regulated at a posttranslational level. In this context, Cereijo et al. (2016) reported an allosteric regulation of ADP-glucose pyrophosphorylase of RHA1 strain, a key regulatory step in bacterial glycogen metabolism, being the glucose-6P, mannose-6P, fructose-6P, ribose-5P, and phosphoenolpyruvate, the major activators, whereas the NADPH and 6P-gluconate behaved as main inhibitors of this enzyme. The authors also suggested a potential role of glycogen as a temporal reserve that provides a pool of carbon able to be re-routed to produce long-term storage of lipids under certain conditions in oleaginous species of *Rhodococcus*. In contrast, Dávila Costa et al. (2015) reported a higher abundance of enzymes involved in glycogen degradation, such as glycogen phosphorylase (GlgP, EC 2.4.1.1), glycogen debranching enzyme (GlgX,

EC3.2.1.33), and phosphoglucomutase (PGM,EC5.4.2.2) during cell cultivation under conditions leading to lipid accumulation (low-nitrogen conditions), which suggest that glycogen degradation can be regulated at a transcriptional level. In conclusion, glycogen metabolism seems to respond to a complex regulation at both transcriptional and posttranslational levels and probably in orchestration with the synthesis of other compounds such as exopolysaccharides and lipids. In this context, recent studies have described the potential role of a capsular α-glucan-like compound exported from internal glycogen in the pathogenic strain of *M. tuberculosis* in virulence by evading the host immune response (Koliwer-Brandl et al. 2016). Although a capsular α-glucan-like compound have not been specifically described in *Rhodococcus*, it is known that several species of this genus are able to produce exopolysaccharides of varied nature (Perry et al. 2007; Hernández et al. 2017).

4 Gluconeogenesis and the Phosphoenolpyruvate-Pyruvate-Oxaloacetate Node

The metabolic link between glycolysis/gluconeogenesis and the TCA cycle is represented by the PEP-pyruvate-oxaloacetate node, also called the anaplerotic node (Sauer and Eikmanns 2005) (Fig. 1). This node comprises a set of reactions that direct the carbon flux into appropriate directions in a flexible manner. The node represents a relevant switch point for carbon flux distribution within the central metabolism (Sauer and Eikmanns 2005).

During growth on acetate, fatty acids or ethanol, which enters to the central metabolism via the acetyl-CoA or as TCA cycle intermediates, the intermediates malate and oxaloacetate must be converted to pyruvate and phosphoenolpyruvate (PEP) for the synthesis of sugars. Once PEP is formed, the synthesis of sugar phosphates is accomplished by reversible reactions of glycolysis (gluconeogenesis). Under gluconeogenic conditions, the TCA cycle intermediates oxaloacetate and malate must be decarboxylated (C4-decarboxylation) to form pyruvate and PEP, which serve as precursors for sugar phosphate synthesis. The formation of PEP can be achieved either by the PEP carboxykinase enzyme (Pck) directly or oxaloacetate decarboxylase and/or malic enzyme in combination with PEP synthetase. As in *C. glutamicum* (Sauer and Eikmanns 2005), PEP carboxykinase (RHA1_RS25350) is probably the only enzyme responsible for PEP synthesis from TCA cycle intermediates in *R. jostii* RHA1, since the gene coding a PEP synthetase is lacking in its genome. Nevertheless, some PEP syntheses are present in other *Rhodococcus* species such us *R. equi* (REQ_RS07445), *R. pyridinivorans* (Y013_RS04990), and *R. aetherivorans* (AAT18_RS13085). In *M. tuberculosis*, the role of Pck enzyme in gluconeogenesis has been demonstrated in presence of lipid substrates, being essential for intracellular growth and survival in macrophages during infection (Basu et al. 2018). On the other hand, pyruvate can be provided from malate by the action of some malic enzymes (ME). The ME enzymes have been associated also with the production of reducing equivalents for anabolic processes in several organisms (Spaans et al. 2015). In RHA1 strain as well as other *Rhodococcus* species, there are several genes coding for these kinds of enzymes, and some of them seem to be associated also with the generation of NADPH for lipid biosynthesis in oleaginous species (Hernández and Alvarez, unpublished results) (Fig. 1).

Under glycolytic conditions, the final products of glycolysis, PEP and pyruvate through the pyruvate kinase and pyruvate dehydrogenase complex, feed acetyl-CoA into the TCA cycle (Fig. 1 and Table 1). Anaplerotic reactions (C3-carboxylation) must replenish TCA cycle intermediates that were bled off for anabolic processes. This function is accomplished in most bacteria by PEP carboxylase and/or pyruvate carboxylase, which convert PEP and pyruvate, respectively, to oxaloacetate (Sauer and Eikmanns 2005). In *M. tuberculosis* H37Rv, it has been demonstrated that these anaplerotic functions involve not only pyruvate carboxylase (PCA) but also Pck and ME enzymes. The anaplerotic node was analyzed through different variant mutants of these genes, showing they are also essential for intracellular growth of *M. tuberculosis* strain. Particularly it was demonstrated that PCA is required for growth on pyruvate and that the phenotype could be complemented by addition of aspartate (as proxy for oxaloacetate) or by growing cultures in presence of 5% CO_2. These results showed the anaplerotic role of PCA replenishing oxaloacetate into TCA cycle and that this function could be complemented by CO_2 addition probably by action of either ME or Pck (Basu et al. 2018).

R. jostii RHA1 seems to possess PEP carboxylase and pyruvate carboxylase in its genome as anaplerotic enzymes (Fig. 1 and Table 1). However, as occurs in *M. tuberculosis H37Rv*, Pck enzyme of *R. jostii* RHA1 would be operating in the carboxylating direction (anaplerotic) or in gluconeogenic direction (Basu et al. 2018). It is known that pyruvate carboxylase plays a major anaplerotic role in mammals, yeast, and in a few prokaryotes such as *Rhodopseudomonas spheroids*, *Arthrobacter globiformis*, and *Mycobacterium smegmatis*. Many other bacteria such as *Pseudomonas citronellolis*, *Azotobacter vinelandii*, or *C. glutamicum* are able to use both PEP carboxylase and pyruvate carboxylase as anaplerotic enzymes (Sauer and Eikmanns 2005).

As occurs in *C. glutamicum* (Sauer and Eikmanns 2005), putative PEP carboxylase gene (*ppc*) of RHA1 strain is clustered with the genes glyceraldehyde-3-phosphate dehydrogenase (*gap*2), phosphoglycerate kinase (*pgk*), and triose-phosphate isomerase (*tpi*A) of the glycolytic pathway.

Feisthauer et al. (2008) reported the dependence of externally provided CO_2 for growth in *R. opacus* 1CP in comparison with *Pseudomonas knackmussii* B13, which was able to grow in the absence of external CO_2 under similar culture conditions. Using $^{13}CO_2$, the authors demonstrated that during growth on glucose, *R. opacus* 1CP showed lower C yield than *P. knackmussii* B13. In addition, fatty acids (principally the odd-numbered fraction) and amino acids (principally the aspartate family) contained in *R. opacus* 1CP were much higher enriched in ^{13}C than those in strain B13 (Feisthauer et al. 2008). The authors concluded that *R. opacus* 1CP possesses an essential dependence on heterotrophic CO_2 fixation by anaplerotic reactions. The odd-numbered fatty acids are usually produced by *Rhodococcus* species using propionyl-CoA as precursor for biosynthesis, as has been reported previously (Alvarez et al. 1997). Propionyl-CoA is produced through the methyl malonyl-CoA pathway using TCA cycle intermediates as precursors (Fig. 1). These intermediates can be formed by direct carboxylation of PEP or pyruvate through anaplerotic reactions, as has been reported for *R. ruber* (Anderson et al. 1995), *R. opacus* PD630 (Alvarez et al. 1997), and *R. opacus* 1CP (Feisthauer et al. 2008). The use of TCA cycle intermediates for the biosynthesis of odd-numbered fatty acids, which may account up to 20–30% of the total fatty acids in many *Rhodococcus* strains, is probably in detriment of the energy gain by cells and may explain the lower C yields of *Rhodococcus* in comparison with other Gram-negative bacteria (Feisthauer et al. 2008). Altogether, these results emphasize the role of the PEP carboxylase and pyruvate carboxylase enzymes in *R. jostii* RHA1 metabolism, a very close species to *R. opacus*. The occurrence of additional anaplerotic enzymes as Pck enzyme (C3 carboxylation) in *R. jostii* and *R. opacus* genomes should be investigated in the future.

Pyruvate kinase and the pyruvate dehydrogenase complex, which are responsible for the production of acetyl-CoA from PEP, are also implicated in the PEP-pyruvate-oxaloacetate node (Fig. 1 and Table 1). The acetyl-CoA fuels the TCA cycle to produce energy and anabolic precursors. The pyruvate:quinine oxidoreductase enzyme, which is present in *C. glutamicum*, seems not to occur in RHA1 genome; thus, the bypass of the pyruvate dehydrogenase reaction via pyruvate:quinine oxidoreductase, in combination with AskA and Pta enzymes, may be not functional in *R. jostii*. However, the *askA* and *pta* genes encoding a putative acetate kinase and a phosphotransacetylase, respectively, are present in the RHA1 genome. When acetyl-CoA is in excess and the C flux exceeds the amphibolic capability of the central metabolism, acetyl-CoA can be converted to acetate and ATP by the AK-PTA pathway (Yasuda et al. 2007). The acetyl-CoA synthetase enzymes can reactivate back acetate to acetyl-CoA when the metabolic situation changes (Fig. 1). The occurrence of the AK-PTA pathway in *R. jostii* RHA1 and probably in other members of the genus may contribute with the control of C fluxes and the maintenance of the intracellular acetyl-CoA pools during fluctuating nutritional conditions, as found in the environment. In this context, *askA* and *pta* genes were downregulated by strain RHA1 during cultivation of cells under nitrogen-limiting conditions, which promote a high flux of acetyl-CoA toward the fatty acid biosynthesis pathway for the accumulation of triacylglycerols. The *askA* and *pta* genes are also present in *R. opacus* B4 and *R. erythropolis* PR4 genomes.

In general, the demand of acetyl-CoA in rhodococci is probably high, considering the high content of different lipid species found in their cellular structures, such as the cell envelope and the storage lipids as triacylglycerols. As *Mycobacterium tuberculosis*, the genome of *R. jostii* RHA1 possesses a multiplicity of genes involved in lipid metabolism (Cole et al. 1998; Hernández et al. 2008). The synthesis of acetyl-CoA from PEP via pyruvate kinase and pyruvate dehydrogenase complex may be the main source for acetyl-CoA production in rhodococci. Otherwise, other reactions could contribute with the intracellular acetyl-CoA pool, such as the reaction catalyzed by citrate lyase enzyme, which converts citric acid into acetyl-CoA and oxaloacetate and the eventual activation of acetate to acetyl-CoA by an acetyl-CoA synthetase. Transcriptomic analysis of *R. jostii* RHA1 under conditions that lead to TAG accumulation revealed a partial repression of pyruvate dehydrogenase enzyme to generate acetyl-CoA; however, it was determined the induction of several enzymes acetyl-CoA synthetases (Juarez et al. 2017).

The significant induction of a gene coding for a putative citrate lyase enzyme by cells of strain RHA1 has been observed during their cultivation under nitrogen starvation conditions.

5 The Tricarboxylic Acid Cycle (TCA)

In aerobic bacteria like *Rhodococcus*, the TCA cycle in the oxidative direction oxidizes acetate to CO_2 and provides reducing power (NADH and $FADH_2$) for energy production and metabolic intermediates for use in biosynthesis pathways. *R. jostii* RHA1 seems to be able to drive the entire TCA cycle oxidatively, since it possesses all necessary genes/enzymes (Table 1 and Fig. 1). The reducing power generated under aerobic conditions via the TCA cycle, in addition to the glycolysis or β-oxidation pathway, could be reoxidized through the respiratory electron transport chain. As mentioned above, the TCA cycle in *Rhodococcus* members serves as a source for the production of propionyl-CoA, via methyl malonyl-CoA pathway, which is used for the biosynthesis of odd-numbered fatty acids (Anderson et al. 1995; Alvarez et al. 1997; Feisthauer et al. 2008) (Fig. 1).

Interestingly, genomic data revealed the presence of the key enzymes of the reductive TCA cycle in the RHA1 genome, including genes encoding citrate lyase and 2-oxoglutarate synthase, in addition to the rest of the TCA cycle enzymes, which may act reversible (Fig. 1 and Table 1). The occurrence of those enzymes in the RHA1 genome suggests that this strain should be able to drive the entire reductive TCA cycle. Citrate lyase converts citric acid into acetyl-CoA and oxaloacetate, whereas the 2-oxoglutarate synthase produces 2-oxoglutarate from succinyl-CoA. This allows the metabolism to incorporate CO_2 and to serve as an engine for synthesis instead of energy production (Srinivasan and Morowitz 2006). CO_2 fixation also occurs in *R. erythropolis* N9T-4 which possess an alternative TCA oxidative cycle. In Yano et al. (2015), the authors propose a bypass of oxoglutarate to succinate through succinic semialdehyde instead of the classic succinyl-CoA in this strain. They also demonstrate a shortcut to prevent CO_2 losses that occur within the TCA cycle. The possible autotrophic metabolism in *Rhodococcus* will be exposed deeper in a specific chapter of this book by Yoshida et al. The switch from oxidative to reductive TCA cycle may both facilitate carbon fixation and restore the balance of oxidative and reductive reactions during environmental fluctuations (Srinivasan and Morowitz 2006). The ability to drive the TCA cycle in both directions, either oxidative or reductive, has been also reported for other actinomycetes, such as *Mycobacterium tuberculosis* and *Streptomyces coelicolor* (Srinivasan and Morowitz 2006). In addition, the genes coding for key enzymes of the reductive TCA cycle seem also to be present in *R. opacus* B4 and *R. erythropolis* PR4 genomes. A partial or fully functional oxidative or reductive TCA cycle may be used by these microorganisms to balance metabolism and to adapt to diverse environmental conditions, for example, (a) starvation as demonstrated for RHA1 that showed a decrease in seven enzymes involved in the oxidative and reductive TCA cycle due to the shutdown of energy production or carbon-containing compounds (Patrauchan et al. 2012) or (b) pollutant stress as shown in a rhodococcal strain isolated from lithium-mining areas that overexpress TCA enzymes involved in the synthesis of intermediates of the TCA cycle (Belfiore et al. 2017). The authors postulate that these metabolites act as scavengers of ROS. A similar behavior was observed during degradation of tetrahydrofuran by *Rhodococcus* sp.YYL (He et al. 2014).

6 The Glyoxylate Pathway

Strain RHA1 possesses both enzymes involved in the glyoxylate pathway: isocitrate lyase and malate synthase (Table 1). This pathway serves as a mechanism to replenish oxaloacetate during growth on acetate and fatty acids (Fig. 1). The glyoxylate pathway is relevant not only for *R. jostii* RHA1 but also for other triacylglycerol-accumulating *Rhodococcus* strains, such as *R. opacus* PD630, because it links the gluconeogenesis with the oxidation of acetyl residues obtained by the β-oxidation of fatty acids during mobilization of the stored triacylglycerols. Members of the genus *Rhodococcus* are able to accumulate variable amounts of triacylglycerols, which are degraded during carbon starvation (Alvarez et al. 2000; Alvarez 2006). The upregulation of isocitrate lyase by cells of strain RHA1 suggested the activation of the glyoxylate shunt under C-starvation, which correlated with the use of intracellular fatty acids as carbon and energy source (Patrauchan et al. 2012). The synthesis of triacylglycerols seems to be an important metabolic pathway in some species of rhodococci for the maintenance of energy homeostasis. Triacylglycerols are synthesized in these bacteria when a carbon source is available and then degraded to provide carbon and energy during C-starvation via the successive operation of β-oxidation, the glyoxylate cycle, the partial TCA, and gluconeogenesis. Thus, the glyoxylate cycle may be very active in rhodococci due to their dependence on the degradation of stored triacylglycerols under nutritional fluctuating conditions. In this context, the glyoxylate cycle plays a pivotal role in the persistence of *Mycobacterium tuberculosis* in mice by sustaining intracellular infection in inflammatory macrophages (Eisenreich et al. 2010). This pathway enables mycobacteria to utilize carbon sources when TCA cycle is shut down during O_2 and nutrient limitation (Boshoff and Barry 2005; Tang et al. 2009). Furthermore, it was recently described the importance of malate synthase for the *M. tuberculosis* growth and survival on even fatty acids by detoxification of glyoxylate, arising from acetyl-CoA metabolism, by assimilation of a second molecule of acetyl-CoA (Puckett et al.

2017). This function is also important in *R. fascians*, the only phytopathogen described so far, to avoid the toxicity of the glyoxylate produced when bacteria catabolize compounds from plants as nutrient factors during issues of infection (Vereecke et al. 2002). On the other hand, Yano et al. (2015) described the relevance of gluconeogenesis and glyoxylate cycle to oligotrophic growth and CO_2 fixation in strain *R. erythropolis* N9T-4; this pathway avoids the loss of carbon as CO_2 molecules and preserves CoA units in the TCA cycle.

7 Litoautotrophic Processes in *Rhodococcus*

Rhodococcus bacteria are usually considered chemoheterotrophic microorganisms, which use organic compounds as sources of carbon and energy. However, there is bioinformatic and experimental evidence that members of this genus are rather facultative chemolithoautotrophs. Lithotrophy is the use of an inorganic compound as a source of energy. Some aerobic bacteria are able to remove electrons from a substrate and put them through an electron transport system that will produce ATP by electron transport phosphorylation. When lithotrophs take their carbon from CO_2, they are autotrophs. In this context, Grzeszik et al. (1997) reported the capability of *R. opacus* strain MR11, which is very close to strain RHA1, for growing on CO_2 and gaseous H_2 as the sole carbon and energy sources. Aragno and Schlegel (1992) previously identified a soluble hydrogenase system in strain MR11, which was localized in the cytoplasm and catalyzed the transfer of electrons directly to NAD. On the other hand, *R. equi* was able to consume H_2 in the late exponential and stationary phase (Meredith et al. 2014). The examination of genomic data revealed that *R. jostii* RHA1 possesses a gene cluster (*RHA1_RS22440* to *RHA1_RS22555*) encoding a putative hydrogenase system. Hydrogenase genes also occur in the genome of *R. opacus* B4 and *R. erythropolis* PR4. Altogether, these results indicated that *Rhodococcus* members are not only able to use a wide range of organic compounds as carbon and energy sources but also to oxidize H_2 as an additional energy source. Moreover, analysis of the genome database of *R. jostii* RHA1 revealed the occurrence of additional litoautotrophic pathways, including genes coding for putative carbon monoxide dehydrogenases (CODH) and a thiocyanate hydrolase enzyme. Thiocyanate hydrolase, which usually occurs in obligate as well as in facultative chemolithotrophs, is a cobalt(III)-containing enzyme catalyzing the degradation of thiocyanate to carbonyl sulfide and ammonia (Yamasaki et al. 2002). The genes coding for the three subunits of the enzyme (α, β, and γ) are localized into an operon (*RHA1_RS21555-RHA1_RS21565*) in the RHA1 genome. The homologous of these RHA1 genes can be identified in *R. opacus* B4 genome with the same locus organization. Curiously, thiocyanate hydrolase genes are missing in *R. erythropolis* PR4 genome. On the other hand, the genomes of *R. jostii* RHA1 and *R. opacus* B4 contain three clusters including genes coding for putative CO dehydrogenases (CODH). Patrauchan et al. (2012) reported a 250-fold increase in CO dehydrogenase concurrent with a 130-fold increase in CODH activity during cultivation of RHA1 strain under carbon starvation conditions, suggesting a switch from heterotrophic to autotrophic metabolism. The authors suggested that CODH enables assimilation of CO_2 or CO into cellular building blocks under these environmental conditions. Autotrophic growth has also been demonstrated in *R. erythropolis* N9T-4, a bacterial strain isolated from stored crude oil (Ohhata et al. 2007). This strain was able to grow on a basal medium without any additional carbon, nitrogen, sulfur, and energy sources but required CO_2 for oligotrophic growth. The authors suggested that N9T-4 possesses a novel CO_2 fixation pathway linked to the formaldehyde metabolism (Yoshida et al. 2011). Strain N9T-4 can utilize carbon, nitrogen, and sulfur compounds, directly from the atmosphere. It has been suggested that N9T-4 can utilize traces amounts of atmospheric ammonia as a nitrogen source through the strong expression of an ammonium transporter gene (*amtB*) (Yoshida et al. 2014).

Altogether, these studies confirmed the existence of diverse litoautotrophic pathways in rhodococci that allow them to survive and thrive in oligotrophic environments. CODH, hydrogenase, thiocyanate hydrolase, and other still unknown metabolic systems may avoid the release of carbon as CO_2 by rhodococcal metabolism and may serve as auxiliary mechanisms for energy metabolism during nutritional starvation conditions.

8 Concluding Remarks

Rhodococcus bacteria are endowed with a robust, flexible, and versatile central metabolism, which is necessary for metabolically adapting to environments with fluctuating nutritional conditions. The central metabolism must provide all necessary intermediates for the biosynthesis of a wide diversity of molecules and complex macromolecules for assembling cellular structures like the cellular envelope or secondary metabolites. These bacteria exert a very efficient management of their nutritional resources, based on the flexibility of their metabolism and the diversity of metabolic reactions. The partitioning of carbon through the central metabolism of the substrates to either energy production or synthesis of compounds depends on the metabolic demands of cells. Rhodococci seem to possess the ability to conserve useful energy during catabolism, distributing the carbon flux of the substrate between the energy production and the biosynthesis of essential molecules, reserving a part of the carbon into diverse storage compounds, such as triacylglycerols, polyhydroxyalkanoates, and glycogen. Thus, cells may resign energetic productivity and growth yield, which may be compensated to some extent with additional processes providing energy and carbon intermediates from inorganic compounds, such as H_2, CO_2, or thiocyanate, among others. This feature may be environmentally favorable for growth and survival in environments with low-energy fluxes, such as in soil or marine sediments. The flexibility of the central metabolism is another essential feature of rhodococci. These microorganisms seem to have the potential to use alternatively different glycolytic pathways, such as EMP or ED pathways, the

PPP pathway, and a partial or fully functional TCA in both, oxidative or reductive directions, according to the conditions. Moreover, the central metabolism of rhodococci possesses some mechanisms that probably permit cells to respond rapidly to changes in nutritional state and to balance metabolism. One of them may be the glycogen biosynthesis, which may deal with the sugar excess during exponential growth phase, and probably serve as a pool of sugars for using when necessary. Other point may be the AK-PTA pathway, which may contribute with the control of C fluxes and the maintenance of the intracellular acetyl-CoA pools during fluctuating conditions. The management of acetyl-CoA pool may be a key point for rhodococci metabolism, since they usually synthesize many different lipid species, which perform important function in their interactions with the environment. In this context, the PEP-pyruvate-oxaloacetate node may play a key role in the C flux distribution within the overall cell metabolism.

Despite the numerous *Rhodococcus* genomes now available, the biochemistry and the molecular biology of the central metabolism of the *Rhodococcus* genus are still poorly known. More studies involving the regulation of metabolic genes and enzymes, the kinetic characterization of enzymes, or the analysis of carbon fluxes through the metabolism should help to define a clearer picture of the functionality and regulation of the central metabolism within the cellular context in rhodococci.

References

Alvarez HM (2003) Relationship between β-oxidation pathway and the hydrocarbon-degrading profile in actinomycetes bacteria. Int Biodeterior Biodegrad 52:35–42

Alvarez HM (2006) Chapter 6: Bacterial triacylglycerols. In: Welson LT (ed) Triglycerides and cholesterol research. Nova Science Publishers, Hauppauge, NY, pp 159–176

Alvarez HM, Steinbüchel A (2002) Triacylglycerols in prokaryotic microorganisms. Appl Microbiol Biotechnol 60:367–376

Alvarez HM, Kalscheuer R, Steinbüchel A (1997) Accumulation of storage lipids in species of *Rhodococcus* and *Nocardia* and effect of inhibitors and polyethylene glycol. Fett-Lipid 99:239–246

Alvarez HM, Kalscheuer R, Steinbüchel A (2000) Accumulation and mobilization of storage lipids by *Rhodococcus opacus* PD630 and *Rhodococcus ruber* NCIMB 40126. Appl Microbiol Biotechnol 54:218–223

Alvarez HM, Silva RA, Cesari AC, Zamit AL, Peressutti SR, Reichelt R, Keller U, Malkus U, Rasch C, Maskow T, Mayer F, Steinbüchel A (2004) Physiological and morphological responses of the soil bacterium *Rhodococcus opacus* strain PD630 to water stress. FEMS Microbiol Ecol 50:75–86

Anderson AJ, Williams D, Dawes EA, Ewing D (1995) Biosynthesis of poly(3-hydroxybutyrate-co-3-hydroxyvalerate) in *Rhodococcus ruber*. Can J Microbiol 41:4–13

Aragno M, Schlegel HG (1992) The mesophilic hydrogen-oxidizing (Knallgas) bacteria. In: Balows A, Trüper HG, Dworkin M, Harder W, Schleifer KH (eds) The prokaryotes. Springer, New York, pp 344–384

Araki N, Suzuki Miyauchi K, Kasai D, Masai E, Fukuda M (2011) Identification and characterization of uptake systems for glucose and fructose in *Rhodococcus jostii* RHA1. Mol Microbiol Biotechnol 20:125–136

Basu P, Snadhu N, Bhatt A, Singh A, Balhana R, Gobe I, Crowhurst NA, Mendum TA, Gao L, Ward JL, Beale MH, McFadden J, Beste DJV (2018) The anaplerotic node is essential for the intracellular survival of *Mycobacterium tuberculosis*. J Biol Chem 15:5695–5704

Belanger AE, Hatfull GF (1999) Exponential-phase glycogen recycling is essential for growth of *Mycobacterium smegmatis*. J Bacteriol 181:6670–6678

Belfiore C, Curia MV, Farias ME (2017) Characterization of *Rhodococcus* sp. $A5_{wh}$ isolated from a high altitude Andean lake to unravel the survival strategy under lithium stress. Rev Argent Microbiol 50(3):311–322

Borodina I, Schöller C, Eliasson A, Nielsen J (2005) Metabolic network analysis of *Streptomyces tenebrarius*, a *Streptomyces* species with an active Entner-Doudoroff pathway. Appl Environ Microbiol 71:2294–2302

Boshoff HI, Barry CE 3rd (2005) Tuberculosis–metabolism and respiration in the absence of growth. Nat Rev Microbiol 3(1):70–80

Cereijo AE, Asencion Diez MD, Dávila Costa JS, Alvarez HM, Iglesias AA (2016) On the kinetic and allosteric regulatory properties of the ADP-glucose pyrophosphorylase from *Rhodococcus jostii*: an approach to evaluate glycogen metabolism in oleaginous bacteria. Front Microbiol 7:830

Chen Y, Ding Y, Yang L, Yu J, Liu G, Wang X, Zhang S, Yu D, Song L, Zhang H, Zhang C, Huo L, Huo C, Wang Y, Du Y, Zhang H, Zhang P, Na H, Xu S, Zhu Y, Xie Z, He T, Zhang Y, Wang G, Fan Z, Yang F, Liu H, Wang X, Zhang X, Zhang MQ, Li Y, Steinbüchel A, Fujimoto T, Cichello S, Yu J, Liu P (2014) Integrated omics study delineates the dynamics of lipid droplets in *Rhodococcus opacus* PD630. Nucleic Acids Res 42(2):1052–1064

Cole ST, Brosch R, Parkhill J, Garnier T, Churcher C, Harris D, Gordon SV, Eiglmeier K, Gas S, Barry CE 3rd, Tekaia F, Badcock K, Basham D, Brown D, Chillingworth T, Connor R, Davies R, Devlin K, Feltwell T, Gentles S, Hamlin N, Holroyd S, Hornsby T, Jagels K, Krogh A, McLean J, Moule S, Murphy L, Oliver K, Osborne J, Quail MA, Rajandream MA, Rogers J, Rutter S, Seeger K, Skelton J, Squares R, Squares S, Sulston JE, Taylor K, Whitehead S, Barrell BG (1998) Deciphering the biology of *Mycobacterium tuberculosis* from the complete genome sequence. Nature 393:537–544

Dávila Costa JS, Herrero OM, Alvarez HM, Leichert L (2015) Label-free and redox proteomic analyses of the triacylglycerol-accumulating Rhodococcus jostii RHA1. Microbiology 161(Pt 3):593–610

Eisenreich W, Dandekar T, Heesemann J, Goebel W (2010) Carbon metabolism of intracellular bacterial pathogens and possible links to virulence. Nat Rev Microbiol 8:401–412

Elbein AD, Pastuszak I, Tackett AJ, Wilson T, Pan YT (2010) Last step in the conversion of trehalose to glycogen: a mycobacterial enzyme that transfers maltose from maltose 1-phosphate to glycogen. J Biol Chem 285:9803–9812

Feisthauer S, Wick LY, Kästner M, Kaschabek SR, Schlömann M, Richnow HH (2008) Differences of heterotrophic $^{13}CO_2$ assimilation by *Pseudomonas knackmussii* strain B13 and *Rhodococcus opacus* 1CP and potential impact on biomarker stable isotope probing. Environ Microbiol 10:1641–1651

Grzeszik C, Lübbers M, Reh M, Schlegel HG (1997) Genes encoding the NAD-reducing hydrogenase of *Rhodococcus opacus* MR11. Microbiology 143:1271–1286

Gunnarsson N, Mortensen UH, Sosio M, Nielsen J (2004) Identification of the Entner-Doudoroff pathway in an antibiotic-producing actinomycete species. Mol Microbiol 52:895–902

He Z, Yao Y, Lu Z, Ye Y (2014) Dynamic metabolic and transcriptional profiling of *Rhodococcus* sp. strain YYL during the degradation of tetrahydrofuran. Appl Environ Microbiol 80:2656–2664

Heald SC, Brandão PF, Hardicre R, Bull AT (2001) Physiology, biochemistry and taxonomy of deep-sea nitrile metabolising *Rhodococcus* strains. Antonie Van Leeuwenhoek 80:169–183

Hernández MA, Alvarez HM (2010) Glycogen formation by *Rhodococcus* species and the effect of inhibition of lipid biosynthesis on glycogen accumulation in *Rhodococcus opacus* PD630. FEMS Microbiol Lett 312:93–99

Hernández MA, Alvarez HM (2018) Increasing lipid production using an NADP⁺-dependent malic enzyme from *Rhodococcus jostii*. Microbiology 165(1):4–14

Hernández MA, Mohn WW, Martínez E, Rost E, Alvarez AF, Alvarez HM (2008) Biosynthesis of storage compounds by *Rhodococcus jostii* RHA1 and global identification of genes involved in their metabolism. BMC Genomics 9:600

Hernández MA, Arabolaza A, Rodríguez E, Gramajo H, Alvarez HM (2013) The *atf2* gene is involved in triacylglycerol biosynthesis and accumulation in the oleaginous *Rhodococcus opacus* PD630. Appl Microbiol Biotechnol 97:2119–2130

Hernández MA, Gleixner G, Sachse D, Alvarez HM (2017) Carbon allocation in *Rhodococcus jostii* RHA1 in response to disruption and overexpression of *nlpR* regulatory gene, based on 13C-labeling analysis. Front Microbiol 8:1992

Hollinshead WD, Henson WR, Abernathy M, Moon TS, Tang YJ (2015) Rapid metabolic analysis of *Rhodococcus opacus* PD630 via parallel 13C-metabolite fingerprinting. Biotechnol Bioeng 113:91–100

Juarez A, Villa JA, Lanza VF, Lázaro B, de la Cruz F, Alvarez HM, Moncalián G (2017) Nutrient starvation leading to triglyceride accumulation activates the Entner Doudoroff pathway in *Rhodococcus jostii* RHA1. Microb Cell Factories 16:35

Kalscheuer R, Syson K, Veeraraghavan U, Weinrick B, Biermann KE, Liu Z, Sacchettini JC, Besra G, Bornemann S, Jacobs WR (2010) Self-poisoning of *Mycobacterium tuberculosis* by targeting GlgE in an α-glucan pathway. Nat Chem Biol 6:376–384

Koliwer-Brandl H, Syson K, van de Weerd R, Chandra G, Appelmelk B, Alber M et al (2016) Metabolic network for the biosynthesis of intra- and extracellular α-Glucans required for virulence of *Mycobacterium tuberculosis*. PLoS Pathog 12(8):e1005768

Larkin MJ, Kulakov LA, Allen CC (2005) Biodegradation and *Rhodococcus*-masters of catabolic versatility. Curr Opin Biotechnol 16:282–290

Luz AP, Pellizari VH, Whyte LG, Greer CW (2004) A survey of indigenous microbial hydrocarbon degradation genes in soils from Antarctica and Brazil. Can J Microbiol 50:323–333

MacEachran DP, Sinskey AJ (2013) The *Rhodococcus opacus* TadD protein mediates triacylglycerol metabolism by regulating intracellular NAD(P)H pools. Microb Cell Fact 12:104

Martínková L, Uhnáková B, Pátek M, Nésvera J, Krén V (2009) Biodegradation potential of the genus *Rhodococcus*. Environ Int 35:162–177

Masai E, Yamada A, Healy JM, Hatta T, Kimbara K, Fukuda M, Yano K (1995) Characterization of biphenyl catabolic genes of gram-positive polychlorinated biphenyl degrader *Rhodococcus* sp. strain RHA1. Appl Environ Microbiol 61:2079–2085

McLeod MP, Warren RL, Hsiao WWL, Araki N, Myhre M, Fernandes C, Miyazawa D, Wong W, Lillquist AL, Wang D, Dosanjh M, Hara H, Petrescu A, Morin RD, Yang G, Stott JM, Schein JE, Shin H, Smailus D, Siddiqui AS, Marra MA, Jones SJM, Holt R, Brinkman FSL, Miyauchi K, Fukuda F, Davies JE, Mohn WW, Eltis LD (2006) The complete genome of *Rhodococcus* sp. RHA1 provides insights into a catabolic powerhouse. PNAS 103:15582–15587

Meredith LK, Rao D, Bosak T, Klepac-Ceraj V, Tada KR, Hansel CM, Ono S, Prinn RG (2014) Consumption of atmospheric hydrogen during the life cycle of soil-dwelling Actinobacteria. Environ Microbiol 6:226–238

Nguyen QT, Trinco G, Binda C, Mattevi A, Fraaije MW (2017) Discovery and characterization of an F420-dependent glucose-6-phosphate dehydrogenase (Rh-FGD1) from *Rhodococcus jostii* RHA1. Appl Microbiol Biotechnol 101:2831–2842

Ohhata N, Yoshida N, Egami H, Katsuragi T, Tani Y, Takagi H (2007) An extremely oligotrophic bacterium, *Rhodococcus erythropolis* N9T-4, isolated from crude oil. J Bacteriol 189 (19):6824–6831

Patrauchan MA, Florizone C, Eapen S, Gómez-Gil L, Sethuraman B, Fukuda M, Davies J, Mohn WW, Eltis LD (2008) Roles of ring-hydroxylating dioxygenases in styrene and benzene catabolism in *Rhodococcus jostii* RHA1. J Bacteriol 190:37–47

Patrauchan MA, Miyazawa D, LeBlanc JC, Aiga C, Florizone C, Dosanjh M, Davies J, Eltis LD, Mohn WW (2012) Proteomic analysis of survival of *Rhodococcus jostii* RHA1 during carbon starvation. Appl Environ Microbiol 78:6714–6725

Peng F, Wang Y, Sun F, Liu Z, Lai Q, Shao Z (2008) A novel lipopeptide produced by a Pacific Ocean deep-sea bacterium, *Rhodococcus* sp. TW53. J Appl Microbiol 105:698–705

Peressutti SR, Alvarez HM, Pucci OH (2003) Dynamic of hydrocarbon-degrading bacteriocenosis of an experimental oil pollution on patagonic soil. Int Biodeterior Biodegrad 52:21–30

Perry MB, MacLean LL, Patrauchan MA, Vinogradov E (2007) The structure of the exocellular polysaccharide produced by *Rhodococcus* sp. RHA1. Carbohydr Res 342(15):2223–2229

Persson O, Valadi A, Nyström T, Farewell A (2007) Metabolic control of the *Escherichia coli* universal stress protein response through fructose-6-phosphate. Mol Microbiol 65:968–978

Puckett S, Trujillo C, Wang Z, Eoh H, Ioerger TR, Krieger I, Sacchettini J, Schnappinger DA, Rhee KY, Ehrt S (2017) Glyoxylate detoxification is an essential function of malate synthase required for carbon assimilation in *Mycobacterium tuberculosis*. Proc Natl Acad Sci 114(11):E2225–E2232

Sauer U, Eikmanns BJ (2005) The PEP-pyruvate-oxaloacetate node as the switch point for carbon flux distribution in bacteria. FEMS Microbiol Rev 29:765–794

Seibold G, Dempf S, Schreiner J, Eikmanns BJ (2007) Glycogen formation in *Corynebacterium glutamicum* and role of ADP-glucose pyrophosphorylase. Microbiology 153:1275–1285

Spaans SK, Weusthuis RA, van der Oost J, Kengen SWM (2015) NADPH-generating systems in bacteria and archaea. Front Microbiol 6:742

Srinivasan V, Morowitz HJ (2006) Ancient genes in contemporary persistent microbial pathogens. Biol Bull 210:1–9

Tajparast M, Frigon D (2015) Genome-scale metabolic model of *Rhodococcus jostii* RHA1 (iMT1174) to study the accumulation of storage compounds during nitrogen-limited condition. BMC Syst Biol 9:43

Tang YJ, Shui W, Myers S, Feng X, Bertozzi C, Keasling JD (2009) Central metabolism in *Mycobacterium smegmatis* during the transition from O(2)-rich to O (2)-poor conditions as studied by isotopomer-assisted metabolite analysis. Biotechnol Lett 31:1233–1240

Van der Geize R, Yam K, Heuser T, Wilbrink MH, Hara H, Anderton MC, Sim E, Dijkhuizen L, Davies JE, Mohn WW, Eltis LD (2007) A gene cluster encoding cholesterol catabolism in a soil actinomycete provides insight into *Mycobacterium tuberculosis* survival in macrophages. PNAS 104:1947–1952

Vereecke D, Cornelis K, Temmerman W, Jaziri M, Van Montagu M, Holsters M, Goethals K (2002) Chromosomal locus that affects pathogenicity of *Rhodococcus fascians*. J Bacteriol 184 (4):1112–1120

Warhurst AM, Fewson CA (1994) Biotransformation catalyzed by the genus *Rhodococcus*. Crit Rev Biotechnol 14:29–73

Whyte LG, Slagman SJ, Pietrantonio F, Bourbonnière L, Koval SF, Lawrence JR, Inniss WE, Greer CW (1999) Physiological adaptations involved in alkane assimilation at a low temperature by *Rhodococcus* sp. strain Q15. Appl Environ Microbiol 65:2961–2968

Yam KC, Okamoto S, Roberts JN, Eltis LD (2011) Adventures in *Rhodococcus* – from steroids to explosives. Can J Microbiol 57(3):155–168

Yamasaki M, Matsushita Y, Namura M, Nyunoya H, Katayama Y (2002) Genetic and immunochemical characterization of thiocyanate-degrading bacteria in lake water. Appl Environ Microbiol 68:942–946

Yano T, Yoshida N, Yu F, Wakamatsu M, Takagi H (2015) The glyoxylate shunt is essential for CO_2-requiring oligotrophic growth of *Rhodococcus erythropolis* N9T-4. Appl Microbiol Biotechnol 99:5627–5637

Yano T, Funamizu Y, Yoshida N (2016) Intracellular accumulation of trehalose and glycogen in an extreme oligotroph, *Rhodococcus erythropolis* N9T-4. Biosci Biotechnol Biochem 80 (3):610–613

Yasuda K, Jojima T, Suda M (2007) Analyses of the acetate-producing pathways in *Corynebacterium glutamicum* under oxygen-deprived conditions. Appl Microbiol Biotechnol 77:853–860

Yoshida N, Hayasaki T, Takagi H (2011) Gene expression analysis of methylotrophic oxidoreductases involved in the oligotrophic growth of *Rhodococcus erythropolis* N9T-4. Biosci Biotechnol Biochem 75(1):123–127
Yoshida N, Inaba S, Takagi H (2014) Utilization of atmospheric ammonia by an extremely oligotrophic bacterium, *Rhodococcus erythropolis* N9T-4. J Biosci Bioeng 117(1):28–32

Oligotrophic Growth of *Rhodococcus*

Nobuyuki Yoshida

Contents

Abstract We can relatively easily isolate "super" oligotrophs from various environments, which can grow on minimal medium without additional carbon and energy sources. The super oligotrophs isolated to date have all belonged to the genera *Rhodococcus* and *Streptomyces*. This chapter primarily describes the genetics and biochemistry of one of the super oligotrophs, *R. erythropolis* N9T-4, which was isolated from crude oil. It is suggested that C2 metabolism starting from acetaldehyde is the key to oligotrophic carbon metabolism of N9T-4. Intriguingly, this bacterium is also oligotrophic for nitrogen and sulfur sources. The most unequivocal evidence for the oligotrophy of N9T-4 relates to the utilization of nitrogen, in which ammonium transporter is involved in the incorporation of atmospheric ammonia under oligotrophic conditions. A unique intracellular structure found in the oligotrophically grown super oligotroph is also introduced.

N. Yoshida (✉)
Department of Applied Chemistry and Biochemical Engineering, Graduate School of Engineering, Shizuoka University, Hamamatsu, Japan
e-mail: yoshida.nobuyuki@shizuoka.ac.jp; https://wwp.shizuoka.ac.jp/yoshida-cb-shizuoka/en

H. M. Alvarez (ed.), *Biology of Rhodococcus*, Microbiology Monographs 16,
https://doi.org/10.1007/978-3-030-11461-9_4

1 What Are Oligotrophs?

An oligotroph is a microorganism that can grow at very low concentrations of nutrients. Kuznetsov defined an oligotroph as one that develops at the first cultivation from nature on a medium with minimal organic matter, containing about 1–15 mg of carbon per liter (Kuznetsov et al. 1979). However, this definition is still nebulous, and there may be no sense in arguing whether concentration of nutrients is high or low for microbial growth in various environments, since low nutrient concentrations would be enough for the growth of oligotrophs. Recent metagenomic analyses have revealed that numerous microorganisms exist in various oligotrophic environments (Bonilla-Rosso et al. 2012; Siering et al. 2013). However, there are few research groups combining the biochemical, genetic, and molecular biological approaches required to investigate microbial oligotrophy. In general, autotrophs, which use CO_2 as their sole carbon source in conjunction with some energy or reducing power, such as sunlight, reducing metal, and molecular hydrogen, are not categorized as oligotrophs. However, it is possible that some oligotrophs fix atmospheric CO_2 to use low concentrations of in situ organic carbon sources effectively.

We have suggested that oligotrophs are useful as hosts for production of useful compounds, since the cost of their cultivation can be restrained. Oligotrophs would also be suitable for in situ bioremediation, because such microorganisms can grow on the contaminated site without additional nutrients. Thus, oligotrophs have a biotechnological potential, motivating us to attempt their isolation from various environments.

First, we used 1000 times-diluted nutrient broth to isolate numerous oligotrophs from various natural samples. However, during the screening, we surprisingly found that oligotrophs could be relatively easily isolated using an inorganic basal medium without carbon sources (BM), which was composed of 1% $NaNO_3$, 1% KH_2PO_4, 0.1% K_2HPO_4, 0.05% $MgSO_4{\cdot}7H_2O$, 0.01% $CaCl_2{\cdot}2H_2O$, and a vitamin mixture (Ohhata et al. 2007). Most of the oligotrophs isolated using BM did not grow under CO_2-limiting conditions, in which a CO_2 absorbent was placed with a petri dish containing BM inoculated by an isolated oligotroph in a plastic bag. However, even under the CO_2-limiting conditions, growth was observed on BM containing $NaHCO_3$ or restored by injection of CO_2 gas into the plastic bag; this indicated that oligotrophic growth depends on CO_2 in the atmosphere. It should be noted that we have not used the term "autotrophic" but rather "super oligotrophic" for such CO_2-requiring oligotrophs as described above, since the CO_2 fixation mechanism is still unknown and any additional energy sources to fix CO_2 are not required for the oligotrophic growth. Indeed, the reason why $NaNO_3$ has been used as the nitrogen source is to avoid isolation of ammonia-oxidizing autotrophs, which could be isolated when NH_4^+ was used as the nitrogen source.

Most organisms, from bacteria to man, have CO_2-dependent carboxylase reactions in their metabolic pathways: acetyl-CoA carboxylase and anaplerotic enzymes such as phosphoenolpyruvate carboxylase, pyruvate carboxylase, and malic enzyme.

In that sense, all organisms are CO_2-dependent, though the CO_2 does not necessarily come from the atmosphere. Meanwhile, microorganisms that require a higher concentration of CO_2 (ca. 5%) are known as capnophiles (Al-Haideri et al. 2016; Bury-Moné et al. 2008; Sahuquillo-Arce et al. 2017). Such capnophilic bacteria lack a carbonic anhydrase activity, which catalyzes the reversible hydration of CO_2 to bicarbonate (HCO_3^-). Capnophilic bacteria should be dependent on nonenzymatically produced HCO_3^- from CO_2 due to the shift of the chemical equilibration under the high concentrations of CO_2. Unlike the capnophiles, super oligotrophs can grow on BM under the ambient air. Is CO_2 used as the main carbon source or in the anaplerotic pathways for super oligotrophs? What is the energy source for the CO_2 fixation associated with oligotrophic growth? The process of solving this puzzle has begun and continues to date.

2 Super Oligotrophs in *Rhodococcus*

The first super oligotroph to be discovered was *R. erythropolis* N9T-4, isolated from crude oil stored in a Japanese oil stockpile (Ohhata et al. 2007). N9T-4 grew well on BM solidified with agar or silica gel without any additional carbon and energy sources. This bacterium did not show oligotrophic growth under the CO_2-limiting conditions described above but grew on a BM plate containing $NaHCO_3$ (Fig. 1). Intriguingly, the only organic compounds that could sustain the growth under the CO_2-limiting conditions were *n*-alkanes, such as *n*-tetradecane and *n*-hexadecane, whereas sugars (glucose, fructose, and galactose) and alcohols (ethanol and methanol) did not restore the growth. The reason why only *n*-alkanes reversed the growth defect under CO_2-limiting conditions is unclear but is suggested to be due to C2 metabolism in N9T-4 as mentioned in the next section. The draft genome of N9T-4 has been sequenced, and it reveals that there are no genes encoding the key enzymes involved in the six known microbial CO_2 fixation pathways and cycles (Berg 2011). Recently, a seventh CO_2 fixation pathway has been found in a phosphite-oxidizing chemolithoautotroph (Figueroa et al. 2018). The gene encoding a formate-tetrahydrofolate ligase, which is the key enzyme in the seventh pathway, is also not found in the genome of N9T-4.

It is interesting to consider whether such unique oligotrophy of N9T-4 is due to specific environmental conditions, such as crude oil, where N9T-4 was isolated, or is widely distributed across environments. After the discovery of N9T-4, further isolation of super oligotrophs was attempted extensively from various environments (Yoshida et al. 2007). As a result, three super oligotrophic *Streptomyces* spp. were isolated from soil samples, which were close to *S. viridobrunneus*, *S. bikiniensis*, and *S. exfoliatus*. Super oligotrophs were also isolated from some culture collections, and besides *R. erythropolis*, *R. rhodochrous*, *R. opacus*, *R. ruber*, *and R. wratislaviensis* showed similar oligotrophic growth to that of N9T-4 (Fig. 2). As for *Streptomyces* sp., *S. aureus* NBRC 100912, *S. griseus* NBRC 15421, *S. bikiniensis* NBRC 14598, and *S. venezuelae* NBRC 13096 could also grow on BM (Yoshida et al. 2007). Thus,

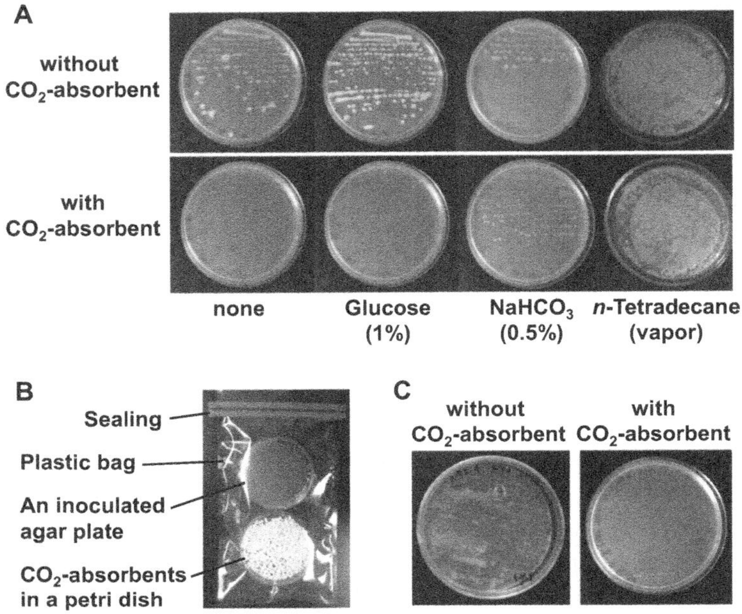

Fig. 1 Oligtorphic growth of *R. erythropolis* N9T-4. (**a**) N9T-4 formed its colonies on BM medium solidified by agar after 5-day cultivation at 30 °C. The oligotrophic growth was repressed when CO_2 was removed from the culture environment by a CO_2 absorbent, while $NaHCO_3$ and *n*-tetradecane, which was soaked in two filter papers on the lid of a glass petri dish and provided as the vapor, sustained the growth under the CO_2-limiting conditions. (**b**) CO_2-limiting conditions were constructed by using a CO_2 absorbent, which was sealed with a BM plate inoculated by N9T-4 in a plastic bag. (**c**) Inorganic growth of N9T-4 was verified using a BM solidified by silica gel

to date, super oligotrophs have been confined to the genera of *Rhodococcus* and *Streptomyces*. These results suggest that certain properties of actinomycetes are responsible for the super oligotrophy.

3 Genetics and Biochemistry of a Super Oligotroph

To elucidate the oligotrophic metabolism of N9T-4, differential two-dimensional electrophoresis (2-DE) was first performed using the cells grown on BM without carbon source and that with *n*-tetradecane as the sole carbon source (Ohhata et al. 2007). The protein profiles on the 2-DE showed that six proteins were highly induced in the oligotrophically grown cells. From analysis of their amino acid sequences, four of them were found to be similar to chaperones such as ClpB, DnaK, and DnaJ. The other two proteins with the apparent molecular masses of 45 kDa and 55 kDa were very close to methanol: *N,N*′-demethyl-4-nitrosoaniline (NDMA) oxidoreductase (MNO) from *Rhodococcus* sp. strain NI86/21 (Nagy et al.

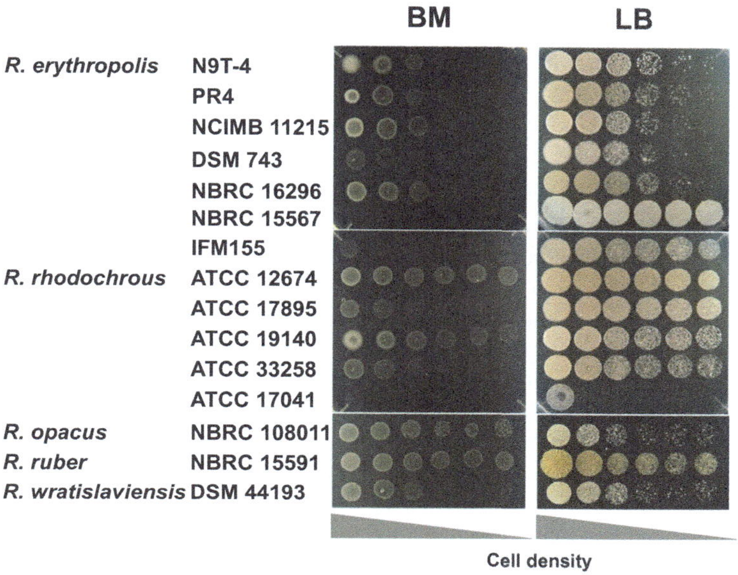

Fig. 2 Super oligotrophs in various *Rhodococcus* spp.

1995) and an NAD-dependent aldehyde dehydrogenase (ALDH) from *R. erythropolis* UPV-1, respectively. *R. erythropolis* UPV-1 could remove phenol and formaldehyde in industrial wastewater, and the ALDH had a broad substrate specificity for aliphatic aldehydes from formaldehyde to *n*-octanal (Jaureguibeitia et al. 2007). MNO is a nicotinoprotein, which was found in gram-positive methylotrophic bacteria, containing a tightly but noncovalently bound NADP (H) per subunit (Bystrykh et al. 1993). MNO catalyzes the oxidation of methanol by using NDMA as an artificial electron acceptor, whereas the enzyme has a remarkable dismutase activity for formaldehyde to produce methanol and formate without NDMA. ALDH and MNO from N9T-4 also had activities of NAD-dependent formaldehyde dehydrogenase and formaldehyde dismutase, respectively, and the enzyme activities and mRNA expression of these enzymes from the cells grown oligotrophically were higher than those from the cells grown on LB (Ohhata et al. 2007; Yoshida et al. 2011). From these results, it was suggested that CO_2 was converted to formaldehyde by an unknown reaction, which was followed by formaldehyde metabolism found in methylotrophs. However, currently, this hypothesis has been denied since the conversion of CO_2 to formaldehyde in the N9T-4 cells has not been detected and the genes involved in formaldehyde assimilation, such as the ribulose monophosphate pathway (Mitsui et al. 2000) and serine pathway (Vorholt 2002) in methylotrophic bacteria, were not found in the N9T-4 genome. These are in accordance with the fact that N9T-4 could not utilize methanol as the sole carbon source. However, intriguingly, N9T-4 has a methylotrophic formaldehyde dehydrogenase (FADH) besides oligotrophic ALDH, which is probably dependent on NAD and mycothiol as the cofactors. Mycothiol is a thiol compound found in *Mycobacterium* and *Rhodococcus* spp., and its functions correspond to glutathione in gram-negative bacteria, plants, and animals. In the formaldehyde-dissimilation pathway in gram-positive bacteria, *S*-hydroxymethyl-mycothiol is formed from formaldehyde and mycothiol, and then the NAD-dependent dehydrogenation occurs to produce *S*-formyl mycothiol. It is of interest that the gene encoding the methylotrophic FADH was not upregulated under the oligotrophic conditions but expressed remarkably when formaldehyde was added to the growth medium, whereas that oligotrophic *aldA* was regulated in the opposite manner (Yoshida et al. 2011).

Another approach to elucidate the oligotrophic metabolism in N9T-4 was random mutagenesis to obtain oligotrophy-deficient mutants. Random mutation library is still a powerful tool to clarify the unknown metabolisms of various microorganisms, and it was attempted to obtain mutants that could not grow on BM but form their colonies on LB (nutrient-rich conditions) from a random mutation library constructed using transposons (Yano et al. 2015). Intriguingly, transposase and its recognition sequences were not necessary, and only linear DNA fragments containing a selection marker were enough to construct the random mutation library for wild-type N9T-4. Furthermore, not insertion mutation but deletion mutation occurred in all of the mutants, suggesting that the mechanism of nonhomologous end joining is involved in the formation of the mutants. From nearly 10,000 independent mutants, 3 types of deletion mutants were obtained, in which *aceA*, *aceB*, and *pckG* encoding isocitrate lyase, malate synthase, and phosphoenolpyruvate, respectively, were deleted. The results indicated that glyoxylate shunt and gluconeogenesis play important roles in the oligotrophy of N9T-4. Although the deletion mutants of *aldA* and *mnoA* encoding ALDH and MNO described above, respectively, could not be obtained from the random mutation library, a Δ*aldA* strain constructed by homologous recombination did not grow under the oligotrophic conditions, whereas a Δ*mnoA* strain showed the same level of oligotrophic growth as the wild-type strain.

Since it was revealed that the glyoxylate shunt was essential for the oligotrophic growth, the analysis of the TCA cycle in N9T-4 is of interest. Most of the activities of the TCA cycle were clearly observed in the cell extract prepared from the oligotrophically grown N9T-4 cells, except for the α-ketoglutarate dehydrogenase (KGDH) (Yano et al. 2015). No KGDH activity was detected in the cell extracts from the cells grown on BM and LB. It is known that *Mycobacterium tuberculosis*, which is phylogenetically closed to the *Rhodococcus* sp., also lacks KGDH activity and has the remarkable α-ketoglutarate decarboxylase (KGD) activity to produce succinic semialdehyde and CO_2 from α-ketoglutarate (Tian et al. 2005a, b). It was elucidated that protein–protein interaction between E1 (SucA) and E2 (DlaT) components of KGDH is weak and SucA catalyzes the KGD reaction without E2 and E3 components. Subsequently, succinic semialdehyde is converted to succinate by succinic semialdehyde dehydrogenase in the TCA cycle in *M. tuberculosis*. Such a TCA bypass route via succinic semialdehyde was also found in cyanobacteria (Steinhauser et al. 2012) and links to the γ-aminobutyric acid shunt via glutamate.

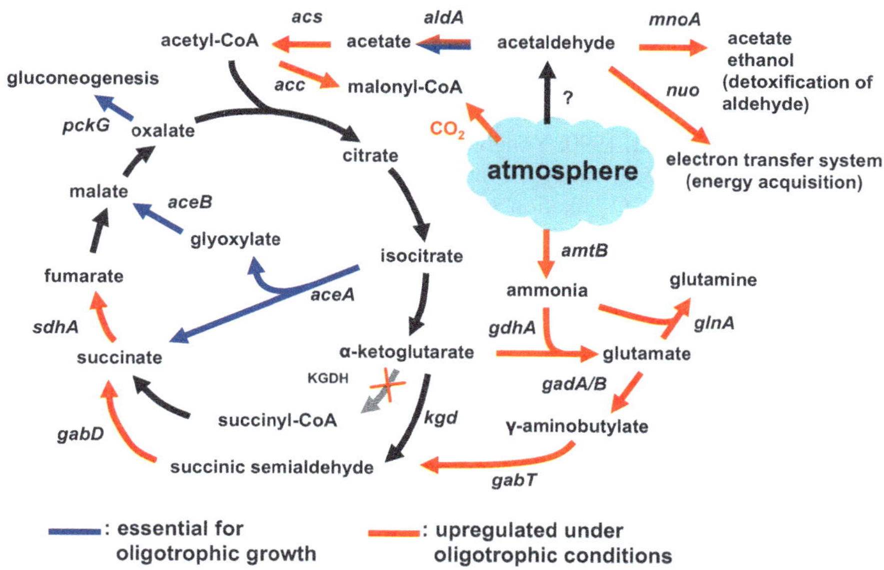

Fig. 3 Postulated oligotrophic carbon and nitrogen metabolisms in N9T-4

It is suggested that the TCA cycle in N9T-4 has similar features to those in *Mycobacterium* and cyanobacteria (Fig. 3). The physiological role of the TCA bypass routes is still unclear, but Steinhauser suggested that the bypass route is associated with low oxygen conditions, which lead to α-ketoglutarate and glutamate via the oxidative branch and succinate via the reductive branch in the TCA cycle in cyanobacteria. One presumed role of the variant TCA cycle in N9T-4 is that the bypass does not proceed via succinyl-CoA and may facilitate the CoA supply to help to maintain sufficient levels of acetyl-CoA.

Figure 3 shows carbon and nitrogen metabolisms in N9T-4 drawn based on the growth of the gene-deletion mutants and the transcriptome analysis. It should be noted that *acs* encoding acetyl-CoA synthetase was also highly expressed under the oligotrophic conditions. Since it was revealed that ALDH and MNO showed the remarkable activity toward acetaldehyde at the same levels toward formaldehyde in N9T-4, the following hypothesis can be built. Trace amounts of acetaldehyde, probably from the atmosphere, may convert to acetyl-CoA via acetate produced by ALDH and MNO, and then acetyl-CoA flows into the TCA cycle. Thus, C2 metabolism may be the key to the oligotrophic carbon metabolism in N9T-4, which is consistent with the fact that the glyoxylate shunt is essential for its oligotrophic growth. Furthermore, NADH produced from dehydrogenation of acetaldehyde by ALDH may be led to the electron transfer system to produce ATP that is used for oligotrophic growth, which was speculated based on upregulation of the genes encoding NADH oxidoreductase subunits (*nuo* genes). The function of MNO in this hypothesis is equivocal, but high concentrations of acetaldehyde are very toxic against cells and N9T-4 may have such different types of aldehyde-utilizing and aldehyde-detoxifying systems. As described above, MNO also catalyzes aldehyde dismutation reaction, and aldehyde dismutases were found in some bacteria that can remove high concentrations of formaldehyde in industrial wastewater (Kato et al. 1986; Adroer et al. 1990; Yonemitsu and Kikuchi 2018). In spite of extensive genetic and biochemical analyses, no remarkable expression of any genes or enzymes involved in CO_2 fixation was observed. It has thus far been concluded that CO_2 is not main carbon source for the super oligotrophs and the CO_2 absorbent that has been used to construct the CO_2-limiting conditions may also remove CO_2 molecules that are essential for carboxylase reactions in the general biological metabolic pathways. This may be one of the reasons why most carbohydrates could not restore growth under CO_2-limiting conditions. A large amount of acetyl-CoA, which should be produced from *n*-alkanes such as *n*-tetradecane and *n*-hexadecane via β-oxidation, may be enough to accelerate the C2 metabolism and overcome a deficit of CO_2 under CO_2-limiting conditions.

4 Mystery in the Super Oligotroph

Some researchers have leveled the criticism that oligotrophs only utilize various volatile compounds in the laboratory in which numerous other microorganisms are cultivated in the same room or the same incubators. Another has suggested that oligotrophs could use gaseous hydrocarbons produced from laboratory wares made of synthetic resin such as polypropylene. Even if these suggestions are possible, it is certain that super oligotrophs can utilize an extremely trace amount of organic compounds, that is, oligotrophy. Furthermore, N9T-4 could grow outside of the laboratory on BM prepared in glassware, suggesting that the oligotrophy of N9T-4 is not due to such artifacts. However, it is hard to consider that N9T-4 utilizes a trace amount of acetaldehyde in the atmosphere, since the concentrations of acetaldehyde and formaldehyde in the atmosphere are at ppb levels.

There is another interesting fact that may be important to infer the carbon oligotrophic metabolism. N9T-4 shows oligotrophy not only for carbon source but also for nitrogen source, and N9T-4 could grow on BM without $NaNO_3$ (BM-N) (Yoshida et al. 2014). When the BM-N plate inoculated with N9T-4 was sealed in a plastic bag, no colony was observed (Fig. 4). Diazotrophic growth was first expected to account for the nitrogen oligotrophy, but nitrogenase activity was not detected in the cells and the putative gene encoding nitrogenase was not found in the N9T-4 genome. Transcriptome analysis revealed that one of the ammonium transporter genes (*amtB*) was strongly upregulated under oligotrophic conditions, and Δ*amtB* could not grow under nitrogen-limiting conditions. There results indicated that N9T-4 could utilize ammonia in the atmosphere as the nitrogen source. Actually, injection of ammonia vapor enhanced the growth of N9T-4 on BM-N in the closed system prepared by using a 9-L desiccator. The minimum concentration of ammonia that sustained the growth under the nitrogen-limiting conditions was 4 ppb, which is

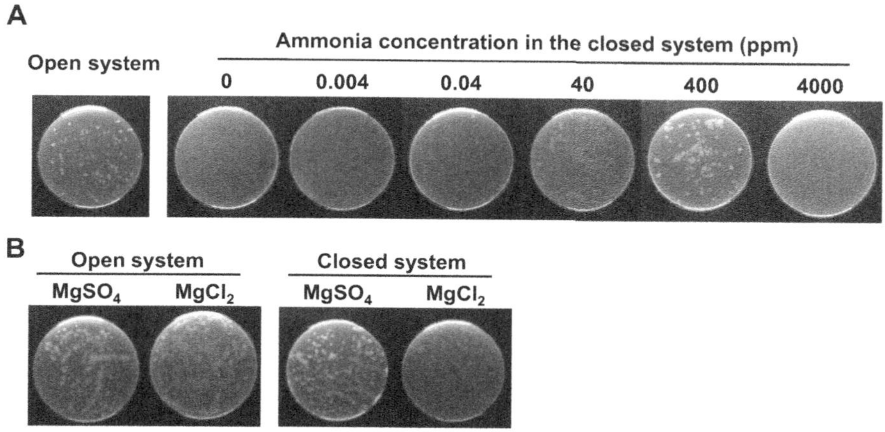

Fig. 4 Nitrogen and sulfur oligotrophy of N9T-4. (**a**) Injection of ammonium vapor restored the oligotrophic growth of N9T-4 on BM without nitrogen source in a closed culture system using 9-L desiccator. The concentrations of ammonia indicated in the figure were those estimated, if ammonia water added in the desiccator was completely volatilized. (**b**) N9T-4 could also grow on BM containing $MgCl_2$ instead of $MgSO_4$. The sulfur oligotrophy was also repressed in the closed system in a similar way as nitrogen oligotrophy

comparable to the atmospheric ammonia concentration in the place where the experiments were carried out (Matsumoto et al. 1998). In this experiment, N9T-4 cells could utilize only the amount of gaseous ammonia in 9 L of the air, whereas a constant concentration of gaseous ammonia would be supplied in the open culture system, suggesting that nitrogen oligotrophy involving incorporation of ammonia is a possible feature in nature. Recently, in environmental microbiology, attention has been drawn to the fact that some actinomycetes can uptake and oxidize atmospheric molecular hydrogen (0.53 ppm/0.40 nM) using their high-affinity [NiFe]-hydrogenase (Constant et al. 2010; Greening et al. 2014). These findings may help validate the utilization of atmospheric acetaldehyde by N9T-4.

Intriguingly, preliminary examination showed that N9T-4 could also grow on BM without a sulfur source ($MgSO_4$ was replaced with $MgCl_2$ in BM) (Yoshida et al. 2014). Similar to nitrogen oligotrophic growth, sulfur oligotrophic growth was repressed in the closed system, suggesting that gaseous sulfur molecules such as sulfur oxides and dimethylsulfide are candidates for the sulfur source for N9T-4. Furthermore, carbon monoxide (CO) could also be the carbon source for N9T-4, but further examination is necessary since a putative gene encoding CO dehydrogenase was not essential for CO utilization (Yano et al. 2012). Thus, N9T-4 can utilize the three biologically principal elements, carbon, nitrogen, and sulfur, from the atmosphere, lending it the name of "air-utilizing bacterium."

5 A Unique Intracellular Compartment in Super Oligotroph

Given that a super oligotroph, N9T-4, shows unique oligotrophic growth as described above, the intracellular structure of the cells grown under oligotrophic conditions is of interest. Transmission electron microscopic observation revealed that a relatively large and spherical compartment (named oligobody) was observed in N9T-4 cells grown on BM (Yoshida et al. 2017). Approximately 30% of the oligotrophic cells had the oligobodies, and each was around 150 nm in diameter (Fig. 5a, b). In most cases, only one oligobody was observed per cell, but in some cases they were localized at each pole of the cell with a septum at the center, suggesting that they separate at cell division (Fig. 5g). The oligobody was not observed or rarely observed (6–7% of total cells) in small sizes (approximately 70 nm in diameter) in the cells under nutrient-rich conditions (LB). The formation frequency and the size of the oligobody on BM containing glucose or *n*-tetradecane as the carbon source were at the same levels as those on BM without carbon source (Fig. 5e and f). The cells grown on BM containing *n*-tetradecane formed the other icicle-like inclusions, which were similar to wax bodies found in various alkane-degrading bacteria (Ishige et al. 2002).

To determine the elemental composition of the oligobody, energy dispersive X-ray analysis was performed. The results showed remarkable peaks corresponding phosphorous and potassium in an oligobody, whereas these peaks in the cytoplasm were very weak. The remarkable phosphorous content indicated accumulation of inorganic polyphosphate (polyP) in the oligobodies. It is known that polyphosphate kinases catalyze the reversible transfer of the terminal phosphate of ATP as the substrate to form a polyP chain. These kinases are divided into two categories, PPK1 and PPK2, which tend to catalyze the elongation and degradation of a polyP chain, respectively (Zhang et al. 2002). PolyP granules in microorganisms are well known, as they were formerly called volutin granules (Meyer 1904) and have recently been called acidocalcisomes (Docampo et al. 2005). Docampo said that acidocalcisomes are the only "organella" conserved from bacteria to man. The genes encoding PPK1 and PPK2 (*ppk1* and *ppk2*, respectively) were found in the N9T-4 genome, and as expected, Δ*ppk1* showed lower polyP content and oligobody formation than the wild type, whereas those of Δ*ppk2* were not affected. However, Δ*ppk1* and Δ*ppk2* showed the same levels of oligotrophic growth as the wild type, suggesting that polyP accumulation and oligobody formation are not involved in the oligotrophy of N9T-4 (Yoshida et al. 2017).

As described above, polyP synthesis requires ATP and was not essential for oligotrophic growth; nevertheless, why does N9T-4 accumulate polyP and form oligobodies under low carbon and energy conditions? This phenomenon is also one of the mysteries of N9T-4, which is interesting and should be clarified. It is known that *Escherichia coli* accumulates polyP under nutrient starvation conditions and polyP activates Lon protease, which helps to facilitate ribosomal protein degradation and supply amino acids to the cells (stringent response) (Kuroda et al. 2001). Indeed,

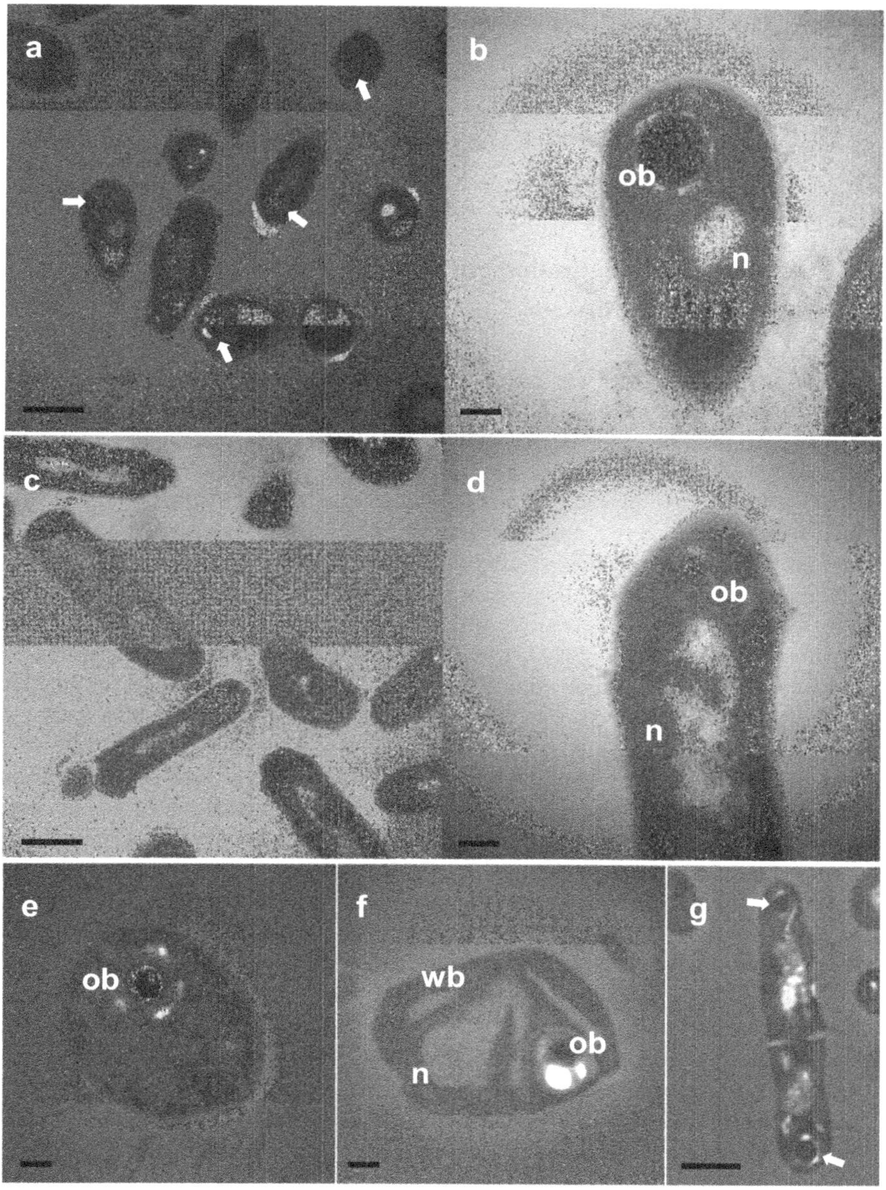

Fig. 5 Oligobodies found in N9T-4 cells. (**a**) and (**b**) BM-grown cells. (**c**) and (**d**) LB-grown cells. (**e**) BMG-grown cells. (**f**) BMT-grown cells. (**g**) a BM-grown cell during cell division. (**a/b**) and (**c/d**) shows images at two different magnifications focused on a single cell or multiple cells. All cultured cells were cultivated at 30 °C for 3 days (LB-grown cells) or 4 days (all others). Structures are labeled as follows: n, nucleoid; ob, oligobody; wb, lipid body. Arrows also indicate oligobody. Scale bars are 500 nm in **a, c,** and **g** and 100 nm in **b** and **d–f**

the conditions for polyP accumulation are quite different between *E. coli* and the super oligotrophy, and polyP was accumulated when cultivation was shifted to starvation from nutrient-rich conditions in *E. coli* and during oligotrophic conditions in N9T-4. However, it may be possible that oligotrophic conditions can be starvation or stress signals in N9T-4 to synthesize polyP in the cells.

6 Future and Biotechnological Application of Super Oligotrophs

As described above, there are many unsolved mysteries to the super oligotroph, N9T-4. In particular, the carbon metabolism should be clarified for the biotechnological application of super oligotrophs. Is it really possible that super oligotrophs utilize atmospheric acetaldehyde as carbon and energy sources? Cultivation in a closed system using a plastic bag was attempted to elucidate the carbon utilization, which was used to clarify ammonia utilization from the atmosphere, but N9T-4 could grow on BM in the closed system. A rapid concentration system for acetaldehyde may exist in N9T-4, considering that mRNAs for *aldA* and *mnoA* were expressed at levels of 20–100-fold after 30-min induction under the oligotrophic conditions (Yoshida et al. 2011). It was confirmed that ALDHs were also highly expressed in the other super oligotrophs isolated until now, so it is certain that aldehyde metabolism is the key in the carbon metabolism in the super oligotrophs in nature.

As the authors in the other chapters mention, *Rhodococcus* spp. bacteria have a great potential for biotechnological application through production of useful compounds (Yamada and Kobayashi 1996; Tajparast and Frigon 2015; Castro et al. 2016) and bioremediation (Soleimani et al. 2007; Kim et al. 2018). N9T-4, a super oligotrophic *Rhodococcus* sp., accumulates polyP under oligotrophic conditions, and recently, it was found that N9T-4 accumulated trehalose and glycogen when grown on BM, with the amount of trehalose being relatively high (0.28 g/g dry cell weight) while that of glycogen was very low (6.7 mg/g dry cell weight) (Yano et al. 2016). Thus, carbon metabolism in this super oligotrophy is still powerful even under oligotrophic conditions, so that it may be feasible to apply it to produce useful compounds at low cost by metabolic engineering.

One of the problems for the biotechnological application of the super oligotrophs is that super oligotrophs cannot grow well in a liquid culture medium ($OD_{660} < 0.1$). This growth defect in liquid culture may be due to their preference for air and/or necessity of cell scaffolding during cultivation. For its industrial application, an effective cultivation system should be established to obtain a large amount of the cells. Recently, submerged cultivation of N9T-4 was attempted using a polyurethane foam sponge with a supporting material to achieve approximately ten times the oligotrophic growth of the bacterium in the liquid culture medium (Matsuoka and Yoshida 2018). This method also facilitated the collection of bacterial cells, and a high concentration of bacterial cells can be obtained simply by squeezing the sponge. It is expected that the physiology of oligotrophy in N9T-4 will be elucidated completely using the sponge.

References

Adroer N, Casas C, de Mas C, Solà C (1990) Mechanism of formaldehyde biodegradation by *Pseudomonas putida*. Appl Microbiol Biotechnol 33:217–220

Al-Haideri H, White MA, Kelly DJ (2016) Major contribution of the type II beta carbonic anhydrase CanB (Cj0237) to the capnophilic growth phenotype of *Campylobacter jejuni*. Environ Microbiol 18:721–735

Berg IA (2011) Ecological aspects of the distribution of different autotrophic CO_2 fixation pathways. Appl Environ Microbiol 77:1925–1936

Bonilla-Rosso G, Peimbert M, Alcaraz LD, Hernández I, Eguiarte LE, Olmedo-Alvarez G, Souza V (2012) Comparative metagenomics of two microbial mats at Cuatro Ciénegas Basin II: community structure and composition in oligotrophic environments. Astrobiology 12:659–673

Bury-Moné S, Mendz GL, Ball GE, Thibonnier M, Stingl K, Ecobichon C, Avé P, Huerre M, Labigne A, Thiberge JM, De Reuse H (2008) Roles of alpha and beta carbonic anhydrases of Helicobacter pylori in the urease-dependent response to acidity and in colonization of the murine gastric mucosa. Infect Immun 76:497–509

Bystrykh LV, Vonck J, van Bruggen EF, van Beeumen J, Samyn B, Govorukhina NI, Arfman N, Duine JA, Dijkhuizen L (1993) Electron microscopic analysis and structural characterization of novel NADP(H)-containing methanol: N,N′-dimethyl-4-nitrosoaniline oxidoreductases from the gram-positive methylotrophic bacteria *Amycolatopsis methanolica* and *Mycobacterium gastri* MB19. J Bacteriol 175:1814–1822

Castro AR, Rocha I, Alves MM, Pereira MA (2016) *Rhodococcus opacus* B4: a promising bacterium for production of biofuels and biobased chemicals. AMB Express 6:35

Constant P, Chowdhury SP, Pratscher J, Conrad R (2010) Streptomycetes contributing to atmospheric molecular hydrogen soil uptake are widespread and encode a putative high-affinity [NiFe]-hydrogenase. Environ Microbiol 12:821–829

Docampo R, de Souza W, Miranda K, Rohloff P, Moreno SN (2005) Acidocalcisomes—conserved from bacteria to man. Nat Rev Microbiol 3:251–261

Figueroa IA, Barnum TP, Somasekhar PY, Carlström CI, Engelbrektson AL, Coates JD (2018) Metagenomics-guided analysis of microbial chemolithoautotrophic phosphite oxidation yields evidence of a seventh natural CO_2 fixation pathway. Proc Natl Acad Sci U S A 115:E92–E101

Greening C, Berney M, Hards K, Cook GM, Conrad R (2014) A soil actinobacterium scavenges atmospheric H2 using two membrane-associated, oxygen-dependent [NiFe] hydrogenases. Proc Natl Acad Sci U S A 111:4257–4261

Ishige T, Tani A, Takabe K, Kawasaki K, Sakai Y, Kato N (2002) Wax ester production of *n*-alkanes by *Acinetobacter* sp. strain M-1: ultrastructure of cellular inclusions and role of acyl coenzyme A reductase. Appl Environ Microbiol 68:1192–1195

Jaureguibeitia A, Saá L, Llama MJ, Serra JL (2007) Purification, characterization and cloning of aldehyde dehydrogenase from *Rhodococcus erythropolis* UPV-1. Appl Microbiol Biotechnol 73:1073–1086

Kato N, Yamagami T, Shimao M, Sakazawa C (1986) Formaldehyde dismutase, a novel NAD-binding oxidoreductase from *Pseudomonas putida* F61. Eur J Biochem 156:59–64

Kim D, Choi KY, Yoo M, Zylstra GJ, Kim E (2018) Biotechnological potential of *Rhodococcus* biodegradative pathways. J Microbiol Biotechnol 28:1037–1051

Kuroda A, Nomura K, Ohtomo R et al (2001) Role of inorganic polyphosphate in promoting ribosomal protein degradation by the lon protease in *E. coli*. Science 293:705–708

Kuznetsov S, Dubinina G, Lapteva N (1979) Biology of oligotrophic bacteria. Annu Rev Microbiol 33:377–387

Matsumoto M, Ueda N, Ono N (1998) Precise measurements of acidic and alkaline gaseous and particulate species in the atmosphere using annular denuder system. Annual Report of Nara Prefectural Institute of Public Health 32:39–46

Matsuoka T, Yoshida N (2018) Establishment of an effective oligotrophic cultivation system for *Rhodococcus erythropolis* N9T-4. Biosci Biotechnol Biochem 82:1652–1655

Meyer A (1904) Orientirende Untersuchungen uber Verbreitung, Morphologie und chemie des volutins. Botanische Zeitung 7:113–152

Mitsui R, Sakai Y, Yasueda H, Kato N (2000) A novel operon encoding formaldehyde fixation: the ribulose monophosphate pathway in the gram-positive facultative methylotrophic bacterium *Mycobacterium gastri* MB19. J Bacteriol 182:944–948

Nagy I, Verheijen S, De Schrijver A, Van Damme J, Proost P, Schoofs G, Vanderleyden J, De Mot R (1995) Characterization of the *Rhodococcus* sp. NI86/21 gene encoding alcohol: N,N-′-dimethyl-4-nitrosoaniline oxidoreductase inducible by atrazine and thiocarbamate herbicides. Arch Microbiol 163:439–446

Ohhata N, Yoshida N, Egami H, Katsuragi T, Tani Y, Takagi H (2007) An extremely oligotrophic bacterium, *Rhodococcus erythropolis* N9T-4, isolated from crude oil. J Bacteriol 189:6824–6831

Sahuquillo-Arce JM, Chouman-Arcas R, Molina-Moreno JM, Hernández-Cabezas A, Frasquet-Artés J, López-Hontangas JL (2017) Capnophilic Enterobacteriaceae. Diagn Microbiol Infect Dis 87:318–319

Siering P, Wolfe G, Wilson M, Yip A, Carey C, Wardman C, Shapiro R, Stedman K, Kyle J, Yuan T (2013) Microbial biogeochemistry of boiling springs lake: a physically dynamic, oligotrophic, low-pH geothermal ecosystem. Geobiology 11:356–376

Soleimani M, Bassi A, Margaritis A (2007) Biodesulfurization of refractory organic sulfur compounds in fossil fuels. Biotechnol Adv 25:570–596

Steinhauser D, Fernie AR, Araújo WL (2012) Unusual cyanobacterial TCA cycles: not broken just different. Trends Plant Sci 17:503–509

Tajparast M, Frigon D (2015) Genome-scale metabolic model of *Rhodococcus jostii* RHA1 (iMT1174) to study the accumulation of storage compounds during nitrogen-limited condition. BMC Syst Biol 9:43

Tian J, Bryk R, Itoh M, Suematsu M, Nathan C (2005a) Variant tricarboxylic acid cycle in *Mycobacterium tuberculosis*: identification of α-ketoglutarate decarboxylase. Proc Natl Acad Sci U S A 102:10670–10675

Tian J, Bryk R, Shi S, Erdjument-Bromage H, Tempst P, Nathan C (2005b) *Mycobacterium tuberculosis* appears to lack α-ketoglutarate dehydrogenase and encodes pyruvate dehydrogenase in widely separated genes. Mol Microbiol 57:859–868

Vorholt JA (2002) Cofactor-dependent pathways of formaldehyde oxidation in methylotrophic bacteria. Arch Microbiol 178:239–249

Yamada H, Kobayashi M (1996) Nitrile hydratase and its application to industrial production of acrylamide. Biosci Biotechnol Biochem 60:1391–1400

Yano T, Yoshida N, Takagi H (2012) Carbon monoxide utilization of an extremely oligotrophic bacterium, *Rhodococcus erythropolis* N9T-4. J Biosci Bioeng 114:53–55

Yano T, Yoshida N, Yu F, Wakamatsu M, Takagi H (2015) The glyoxylate shunt is essential for CO_2-requiring oligotrophic growth of *Rhodococcus erythropolis* N9T-4. Appl Microbiol Biotechnol 99:5627–5637

Yano T, Funamizu Y, Yoshida N (2016) Intracellular accumulation of trehalose and glycogen in an extreme oligotroph, *Rhodococcus erythropolis* N9T-4. Biosci Biotechnol Biochem 80:610–613

Yonemitsu H, Kikuchi Y (2018) Biodegradation of high concentrations of formaldehyde using *Escherichia coli* expressing the formaldehyde dismutase gene of *Methylobacterium* sp. FD1. Biosci Biotechnol Biochem 82:49–56

Yoshida N, Ohhata N, Yoshino Y, Katsuragi T, Tani Y, Takagi H (2007) Screening of carbon dioxide-requiring extreme oligotrophs from soil. Biosci Biotechnol Biochem 71:2830–2832

Yoshida N, Hayasaki T, Takagi H (2011) Gene expression analysis of methylotrophic oxidoreductases involved in the oligotrophic growth of *Rhodococcus erythropolis* N9T-4. Biosci Biotechnol Biochem 75:123–127
Yoshida N, Inaba S, Takagi H (2014) Utilization of atmospheric ammonia by an extremely oligotrophic bacterium, *Rhodococcus erythropolis* N9T-4. J Biosci Bioeng 117:28–32
Yoshida N, Yano T, Kedo K, Fujiyoshi T, Nagai R, Iwano M, Taguchi E, Nishida T, Takagi H (2017) A unique intracellular compartment formed during the oligotrophic growth of *Rhodococcus erythropolis* N9T-4. Appl Microbiol Biotechnol 101:331–340
Zhang H, Ishige K, Kornberg A (2002) A polyphosphate kinase (PPK2) widely conserved in bacteria. Proc Natl Acad Sci U S A 99:16678–16683

Adaptation of *Rhodococcus* to Organic Solvents

Carla C. C. R. de Carvalho

Contents

Abstract Most of the commercially interesting compounds and those affecting the environment are poor water soluble. Bacteria able to carry out the bioconversion or bioremediation of such compounds in systems using organic solvents as substrate and/or product reservoir are valuable. Strains of *Rhodococcus* have been reported to be particularly solvent tolerant whilst presenting a broad array of enzymes with potential for the production of industrially relevant compounds and/or for the metabolism of recalcitrant organic solvents. Under stressful conditions, these cells can adapt the cell wall and membrane compositions, as well as the physicochemical properties of the cell surface, can degrade or bioconvert toxic compounds such as benzene and toluene, and can aggregate and produce exopolymeric substances to protect the cell population. The adaptability and versatility of *Rhodococcus* cells can further broaden their application scope.

C. C. C. R. de Carvalho (✉)
Department of Bioengineering, iBB-Institute for Bioengineering and Biosciences, Instituto Superior Técnico, Universidade de Lisboa, Lisbon, Portugal
e-mail: ccarvalho@tecnico.ulisboa.pt

H. M. Alvarez (ed.), *Biology of Rhodococcus*, Microbiology Monographs 16,
https://doi.org/10.1007/978-3-030-11461-9_5

1 Introduction

Sound and successful biocatalytic and bioremediation processes require bacterial strains tolerant or adapted to organic solvents. These cells should work in the presence of organic solvents necessary to overcoming the low water solubility of hydrophobic substrates and/or products in biocatalytic processes (de Carvalho and da Fonseca 2002a; Daugulis 2001; Heipieper et al. 2007) or be able to biodegrade organic solvents released into the environment (Aislabie et al. 2006; de Carvalho and da Fonseca 2007; Leisinge 1996; Parales and Haddock 2004). The use of an organic phase in bioreactors enhances productivity and facilitates downstream processing: (1) by controlled delivery of a toxic substrate from the organic to the aqueous phase containing the biocatalyst, or (2) by continuous extraction of a byproduct or of an inhibitory product. The group of extremophilic bacteria able to tolerate organic solvents retains their enzymatic activity in the presence of these compounds (Inoue 2011; Sardessai and Bhosle 2004).

The first report of a bacterium tolerant to organic solvents was made by Inoue and Horikoshi (1989) who discovered that *Pseudomonas putida* IH-2000 was able to tolerate and grow in the presence of 50% (v/v) toluene. This strain was, nonetheless, unable to metabolize toluene as sole carbon source. Since then, several Gram-positive strains belonging mainly to the genus *Bacillus*, *Arthrobacter* and *Rhodococcus* have shown high tolerance to organic solvents including benzene, which is much more toxic than toluene. *Rhodococcus* strains have been reported as efficient catalysts in the presence of organic solvents and also as degraders of these compounds because of their ability to metabolize a wide range of organic compounds under a wide set of conditions (de Carvalho and da Fonseca 2005; Larkin et al. 2006). *Rhodococcus* cells are ubiquitous and are able to thrive in stressful environments, having been found in Antarctica (Bej et al. 2000; Pini et al. 2007), in the Artic (Thomassin-Lacroix et al. 2001;Whyte et al. 2002), at sea level (Bell et al. 1998), in the deep sea (Colquhoun et al. 1998), at high altitude (Margesin et al. 2003), and in semiarid soils (Pucci et al. 2000).

The high hydrophobicity of *Rhodococcus* cells allows them to adhere directly to organic solvents containing the substrate of interest, which improves mass transfer between reservoir and biocatalyst (de Carvalho et al. 2007, 2009). The production of surface-active compounds, which decrease the surface tension of water and increase the solubility of low water-soluble compounds, further enhances the biocatalytic/degradation abilities of *Rhodococcus* cells. However, the ability of these cells to adapt the lipid composition and surface properties of their cell envelop is probably the most remarkable feature observed during adaptation to organic solvents (de Carvalho et al. 2009, 2016; Heipieper et al. 2007).

1.1 Predicting Solvent Toxicity

Organic solvents are, in general, toxic to bacterial cells even at low concentrations. The toxicity of water-immiscible solvents may result from (1) phase toxicity by direct contact between solvents and cells, and/or (2) molecular toxicity caused by the molecules of solvent dissolved in the aqueous phase. These compounds partition to cell membranes, increasing their fluidity and disrupting the phospholipid bilayer and interfering with membrane-bound proteins and ultimately causing cell death (Heipieper et al. 1991, 1994; Sikkema et al. 1995; Vermuë et al. 1993). In fact, organic solvents such as alcohols and phenols contained within natural oils and balsams have been widely used as antimicrobial agents since antiquity, long before their mode of action and effect was known.

Several authors have tried to present methods to predict a solvent's toxicity based on its physicochemical properties. In 1985, Brink and Tramper proposed the use of the solvent polarity and its molecular size to predict the suitability of a solvent to be used in a multiphasic biocatalytic system. Solvent polarity could be determined by the Hildebrand solubility parameter, δ, whilst the latter could be expressed as molecular weight or molar volume. In this system, high biocatalytic rates should be obtained by choosing organic solvents with low polarity ($\delta < \sim 8$) and high molecular weight ($M > \sim 150$). A few years later, the hydrophobicity of a solvent measured as the logarithm of the octanol-water partition coefficient, $\log P_{\mathrm{O/W}}$, was proposed as a better predictor of the antimicrobial action of a solvent (Laane et al. 1987; Osborne et al. 1990; Sikkema et al. 1994). According to the $\log P_{\mathrm{O/W}}$ scale, enzymes and microorganisms present a minimum of activity with solvents with log P values of 0–2 and 2–4, respectively, after which, the use of solvents with increasing log P values will result in increased biocatalyst stability. However, the actual concentration of the solvent in the bacterial cell membrane will depend both on the solvent concentration in the water phase and on the partitioning of the solvent from the water phase to the membrane. In 1994, Sikkema and co-workers proposed the following equation to correlate the $\log P_{\mathrm{O/W}}$ value of a solvent and its partitioning value between the membrane and water, $\log P_{\mathrm{M/W}}$:

$$\log P_{\mathrm{M/W}} = 0.97 \times \log P_{\mathrm{O/W}} - 0.64 \quad (1)$$

Hydrophobic solvents, with $\log P_{\mathrm{O/W}} > 4$, accumulate in cellular membranes but will not reach a high membrane concentration and are not toxic because of their low water solubility. On the contrary, solvents with $\log P_{\mathrm{O/W}}$ between 1 and 4 present higher water solubility values whilst being also able to partition to biological membranes, resulting in relatively high concentrations of these solvents in the membranes and high toxicity to the cells (de Bont 1998). The fact that solvents with high partition coefficients to the membrane, such as *n*-dodecane and *n*-hexadecane, are not toxic to bacterial cells is, apparently, contradictory. However, a "cut-off" in toxic effect around log P values of 4–5 was observed for microorganisms (Laane et al. 1985; Vermuë et al. 1993), above which the solvents do not

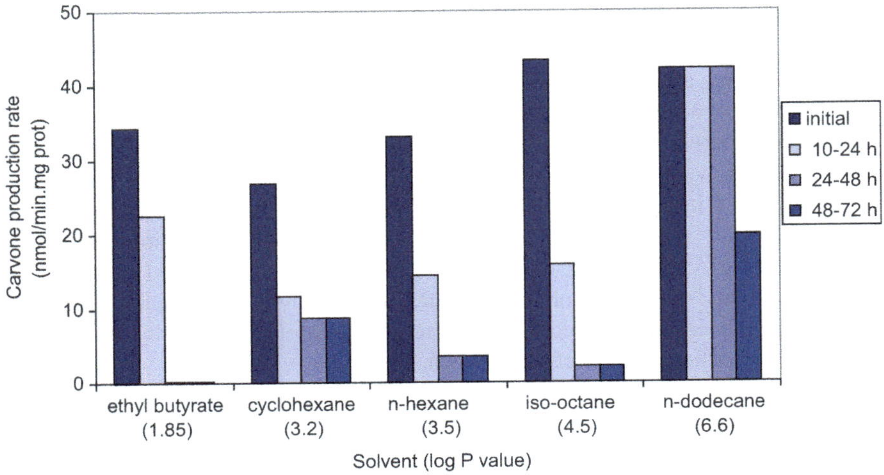

Fig. 1 Effect of the log *P* value of the solvent on *R. erythropolis* DCL14 cells carrying out the biotransformation of carveol into carvone in an organic-aqueous system (adapt. from de Carvalho and da Fonseca 2002a)

present toxicity. The reasons presented involve low solubility of the solvents in the membrane bilayer or absence of a membrane-disturbing effect, with significant responses being observed with different organisms and solvents (Sikkema et al. 1994). As pointed out by de Bont (1998), calculating the actual solvent concentration in membranes could be helpful. The author however noted that several parameters that can also affect the partitioning of solvents, such as the composition of the biological membrane, are disregarded. Furthermore, many of the potentially interesting substrates are not very soluble in apolar solvents (those with higher log *P* values), and the reaction rates in favourable solvents are often low (Cassells and Halling 1990). The transition between toxic and biocompatible solvents depends on the type of compound and also on the concentration of the solvent used in the test, and thus the concentration in the aqueous phase which causes loss of half of the metabolic activity of the bacterial cells should be determined (Vermuë et al. 1993). Nevertheless, the log *P* values could be used as an indication of the biocatalyst behaviour, even in processes requiring cofactor regeneration by viable cells (Fig. 1).

Hamada and co-workers (2008) compared several methods to predict bacterial predilection for organic solvents, namely, bacterial adhesion to hydrocarbon (BATH), contact angle measurement (CAM), hydrophobic interaction chromatography (HIC) and glass adhesion test (GAT). They concluded that CAM could be used to predict the dispersibility of bacteria in anhydrous organic solvents, whilst the BATH assays were better to predict the behaviour of bacterial cells in organic-aqueous two-phase systems. The differences in the obtained results using the two techniques were ascribed to the effect of electrostatic interactions between bacteria and oil droplets.

As shown later by de Carvalho et al., *R. erythropolis* cells presented apparent contradictory results in the BATH and CAM when grown in C5–C16 alkanes: cell surface hydrophobicity, estimated by BATH, increased with the chain length of the alkane, but decreasing contact angles were observed (de Carvalho et al. 2009). These results could be explained by the fact that these cells could increase their zeta potential with the number of carbons in the alkane molecule: from −35.8 to +4.7 mV when the cells grew on pentane and hexadecane, respectively. The surprising change of polarity observed in *R. erythropolis* cells grown on long-chain-length alkanes could explain the strong cell adhesion to hexadecane observed during the BATH assay (where both hydrophobic and electrostatic forces are involved) and the contact angle measurements (resulting from van der Waals forces).

1.2 Effect of Solvents on Bacterial Cells

In a review on the cellular toxicity of lipophilic compounds, Sikkema et al. (1994) discussed the results showing that the accumulation of these compounds occurs at varying depths in the membrane bilayer, depending on the presence of hydroxyl, carboxyl or phenyl groups. Hydrophobic cyclic hydrocarbons should accumulate in the acyl region of the membrane, whilst hydrophobic compounds should interact with hydrophobic end of acyl chains, and more hydrophilic compounds should affect the hydration of the head groups of the membrane lipids. Small lipophilic compounds should intercalate with the acyl chains, resulting in membrane swelling and in an increased fluidity of the membrane. Since larger hydrophobic molecules should affect both the inner and outer leaflet of the lipid bilayer, an increased ordering of the membrane lipids reduces the membrane permeability and fluidity but increases the bilayer width.

The cellular membrane of bacteria acts as a permeability barrier to solutes, regulating the intracellular environment, but is also responsible for the maintenance of the energy status of the cell, turgor pressure, signal transduction and other energy transduction processes (Sikkema et al. 1995). When solvents disrupt the membrane, loss of ions, metabolites, lipids and proteins and impairment of the proton-motive force across the membrane may occur, which stops the pH gradient and the electrical potential, and inhibits the function of membrane proteins. The microorganisms' ability to maintain their biological functions under stressful conditions, such as the presence of organic solvents, results from changes in protein, sterol, hopanoid and carotenoid content but mainly from changes in the membrane lipid composition (Heipieper et al. 1994; Weber and de Bont 1996). The cells try to modify the fatty acid composition of the cellular membrane in order to maintain the membrane fluidity. The membrane fluidity is kept constant by changes in the degree of saturation of the fatty acids of the membrane phospholipids through a mechanism called "homeoviscous adaptation" (Sinensky 1974). Gram-negative bacteria, such as those of the genera *Pseudomonas* and *Vibrio*, isomerize unsaturated fatty acids from conformation *cis* to *trans* (Heipieper et al. 1992; Weber et al. 1994), which seems to be a special mechanism of adaptation to high concentrations of toxic compounds when de novo synthesis of lipids is not possible (Diefenbach et al. 1992). The studies carried out suggested that *cis* to *trans* conversion increases membrane ordering, decreasing the membrane fluidity. The level of isomerization has been found to correlate with the concentration (Diefenbach et al. 1992; Heipieper et al. 1992) and hydrophobicity (Heipieper et al. 2003) of the toxic compounds. Changes in the saturation-unsaturated and in the long-chain-short-chain fatty acid ratios can be seen as long-term changes to regulate membrane fluidity since synthesis of fatty acids is required. Other mechanisms involved in the response of Gram-negative bacteria to organic solvents include changes in the phospholipids headgroups, in the outer membrane proteins and lipopolysaccharides, and the action of efflux pumps. These mechanisms have been reviewed by Segura et al. (1999). Several other reviews have been published on solvent-tolerant bacteria (de Bont 1998; Isken and de Bont 1998; Ramos et al. 2002; Sardessai and Bhosle 2002; Sikkema et al. 1994). Although these works report solvent-tolerant *Rhodococcus* and other Gram-positive strains, the authors initially agreed that without an outer membrane, Gram-positive bacteria should be less tolerant to organic solvents. However, recent studies indicate that Gram-positive cells share tolerance mechanisms with their Gram-negative counterparts, including efflux pumps systems, isomerization of unsaturated fatty acids and modifications of the membrane phospholipids headgroups, formation of vesicles where toxic compounds may be stored and repair processes involving increased rate of phospholipids biosynthesis (de Carvalho et al. 2014a; Pacífico et al. 2018; Torres et al. 2011).

The permeable cell wall of Gram-positive bacteria does not usually restrict the penetration of antimicrobial agents, but vancomycin-intermediate-resistant *Staphylococcus aureus* strains present a significant thickened cell wall (Lambert 2002). Bacteria containing mycolic acids, such as mycobacteria and *Rhodococcus*, have cells walls with a high lipid content which may act as a barrier to both hydrophobic and hydrophilic antimicrobials (Brennan and Nikaido 1995). In this chapter, the resistance and the mechanisms conferring adaptation capabilities to *Rhodococcus* strains will be presented and discussed.

2 Intrinsic Resistance to Organic Solvents

Some microorganisms have the property of being able to resist or tolerate concentrations of a certain compound that would kill or stress others. Intrinsic resistance is defined as an innate genetically controlled property of a bacterial cell that enables it to elude the action of a biocide (Russell 1995). Resistance genes may reside on the chromosome, on a plasmid or on a transposon. This natural resistance is the result of penetration barriers, metabolic pathways or effective efflux pumps, whilst acquired resistance results from an increased tolerance gained through a genetic change by which an organism and its progeny will be able to remain viable and/or multiply under the stress conditions. As long as the basal level of the repair systems is not surpassed by the stress damages, the cells will be able to survive.

Gram-negative bacteria are generally considered less susceptible to biocides than Gram-positive bacteria because their outer membrane acts as a permeability barrier. However, mycobacteria can present a high intrinsic tolerance because the cell is highly hydrophobic due to the mycolarabinogalactan-peptidoglycan skeleton (McDonnell and Russell 1999). In the case of staphylococci, the cell wall is mainly composed of peptidoglycan and teichoic acid, which are not effective as penetration barriers. However, the physiological state of the cells can influence the susceptibility of the cells to biocides as the growth rate or growth under limiting nutrient conditions may change the thickness and degree of cross-linking of the peptidoglycan (Gilbert and Brown 1995). Slime-producing *S. aureus* strains also present a higher resistance to antimicrobial agents than non-mucoid strains, indicating that exopolymeric substances may act as physical barrier or as absorbent of biocide molecules (McDonnell and Russell 1999). When the physical barriers allow the passage of the toxic compound, intrinsic resistance is given by efflux pumps that may extrude different types of compounds (Piddock 2006; Poole 2008).

In a study comparing the resistance of Gram-negative *P. putida* and the Gram-positive *Mycobacterium* sp. and *R. erythropolis* cells to various water-miscible and water-immiscible solvents, it was found that of the three strains only *R. erythropolis* was able to endure high concentrations of water-miscible solvents (de Carvalho et al. 2004). This result showed that *R. erythropolis* was able to maintain viability at higher solvent concentrations than *P. putida*, which was expected to be more tolerant to organic solvents than Gram-positive bacteria because of the protective effect of the outer membrane (Isken and de Bont 1998). Fang and co-workers (2007) also found the Gram-positive *Bacillus subtilis* to be more tolerant than *P. putida* towards fullerene-based manufactured nanomaterials. *Bacillus*, *Rhodococcus* and *Arthrobacter* have been found to be organic solvent tolerant even to benzene, one of the most toxic organic compounds (Sardessai and Bhosle 2002). The dominance of BTEX-degrading Gram-positive bacteria in nearshore surface water and in sediments from the Pacific Ocean also contradicts the general idea that Gram-negative bacteria are the main group of organic pollutant degraders (Wang et al. 2008). Apparently, the more extensively linked peptidoglycan in Gram-positive cells (Sikkema et al. 1995), and the enzymes/substances excreted by these cells to emulsify/solubilize the organic solvents (Abe et al. 1995), could play an important role in decreasing solvent toxicity.

Rhodococcus strains have been found to endure considerably high concentrations of known toxic compounds, e.g. phenol, benzene and toluene (Table 1). The work of Gutiérrez and co-workers (2003) with *Rhodococcus* sp. showed that although benzene caused an increase in membrane fluidity after 0.5 h, the cells did not change the fluidity of the membrane in the 6 h of the assay, during which the cells were alive. The authors ascribed this to a well-adapted inherent mechanism allowing the cells to survive benzene and other solvent "shocks" in the environment. Gutiérrez et al. (2009) studied the constitutive processes that confer resistance to benzene. Non-adapted *Rhodococcus* sp. 33 cells were able to endure shock concentrations

Table 1 Intrinsic resistance of several *Rhodococcus* strains to organic solvents

Strain	Compound	Tolerance	References
R. phenolicus G2P[T]	Phenol	0.75%[a]	Rehfuss and Urban (2005)
R. opacus B-4, B-9, B-10	Benzene	90%[b]	Na et al. (2005)
R. opacus B-4	*n*-Tetradecane	100%	Hamada et al. (2008)
	Oleyl alcohol, bis (2-ethylhexyl) phthalate	100%	Yamashita et al. (2007)
Rhodococcus sp. CN6	*p*-Nitrophenol	100 mg/L[a]	Zhang et al. (2009)
R. rhodochrous S-2	*n*-Hexadecane	10%	Iwabuchi et al. 2000
R. sp. NO14-1	*n*-Hexadecane Diesel oil Phenol	34%[c] 27%[c] 12.5 mM[c]	Margesin et al. (2005)
Rhodococcus sp. NO20-3	*n*-Hexadecane Diesel oil Phenol	54%[c] 37%[c] 12.5 mM[c]	Margesin et al. (2005)
Rhodococcus sp.	Benzene	200 mg/L	Gutiérrez et al. (2003)
R. equi 85F	Hydrogen peroxide	150 mM	Benoit et al. (2002)
R. erythropolis UPV-1	Phenol	1000 mg/L	Prieto et al. (2002)
	Ethanol Butanol Dimethylformamide Dodecane bis(2-Ethylhexyl) phthalate	40% 20% 50% 5% 5% 1%	de Carvalho et al. (2004)
	Toluene	20%[d]	de Carvalho et al. (2007)
R. erythropolis DCL14	Methanol Ethanol Butanol Cyclohexanol Dodecanol	15%[a] 20%[a] 2%[a] 1%[a] 5%[a]	de Carvalho et al. (2005)
	Iso-octane	99.99%	de Carvalho et al. (2000)
	C5–C16 *n*-alkanes Motor oils Fuel oil	0.25%[a] 2%[a] 1.6%[a]	de Carvalho and da Fonseca (2007)
R. erythropolis 17S	Phenol	100 mg/L	Korobov et al. (2017)

[a]Organic solvent used as sole carbon source
[b]Growth observed
[c]At 10 °C
[d]After 1 h incubation

of up to 1000 mg/L of pre-solubilized benzene, a concentration usually lethal to most microorganisms. The production of an extracellular polymer and the composition of the cell wall and cell plasma membrane may be responsible for this high tolerance. In some cases, tolerance is related to the capacity of the cells to degrade the toxic molecule and to use it as sole carbon and energy source (de Carvalho et al. 2005; de Carvalho and da Fonseca 2007). In the case of *R. rhodochrous*, the extracellular polysaccharide produced by the mucoidal strains was responsible for the tolerance of the cells to *n*-hexadecane (Iwabuchi et al. 2000). The tolerance of *R. erythropolis* IBB_{Po1}, isolated from Poeni crude oil-contaminated soil, to alkanes and aromatics was ascribed to the presence of catabolic (*alkB*, *alkB1*, *todC1*, *todM*, *xylM*), transporter (HAE1) and trehalose-6-phosphate synthase (*otsA1*, KF059973.1) genes in the large genome (Stancu 2014). Following 1 and 24 h of exposure to 1% to alkanes and aromatics, the cells presented changes in cell growth pattern, morphology, hydrophobicity and also on the metabolic fingerprinting and *otsA1* gene sequence. Trehalose-6-phosphate synthases, TPSs and OtsAs, are involved in the production of surface-active trehalose lipids during *Rhodococcus* growth on medium with long-chained *n*-alkanes (Kim et al. 1990; Rapp et al. 1979). In strain *R. opacus* 1CP, OtsA1 was found to be involved in the overproduction of trehalose lipids, whilst OstA2 presented a major role in trehalose metabolism and trehalose lipid formation for the cell wall (Tischler et al. 2013). The role of OtsA2 was further supported by the presence of ORFs with putative function in cell envelope biogenesis next to the *ostA2* gene in this strain.

In 1942, Withell observed an exponential relation between the duration of a stress episode and bacterial death, which could be explained by the presence of cells with different tolerance within the population. According to Booth (2002), survival to a stress agent largely occurs at the level of a single cell (as seen by the ability of a single cell to form a colony on an agar plate). After exposure to stressors that kill the majority of the cells, the heterogeneity of a population determines its survival as a small fraction of the cells may possess the necessary tools to endure the stress. The results obtained by de Carvalho and co-workers (2007), indicate that cells that are able to remain viable after the first seconds of exposure to high toluene concentrations will also be able to endure this toxic for longer periods. Two explanations could justify this result: (1) the existence of toluene-tolerant persistent cells within the population; and (2) the tolerance is induced by cell exposure to a toxic, and until tolerance is acquired, cell death occurs at a toluene-concentration-dependent rate.

3 Adaptation Mechanisms to Organic Solvents

Several reviews have been published on the tolerance of bacterial cells to organic solvents where the mechanisms of cell adaptation are discussed (de Bont 1998; Chapman 2003; Isken and de Bont 1998; Sardessai and Bhosle 2002; Segura et al. 1999; Sikkema et al. 1994; Torres et al. 2011). Among the most important mechanisms described are (1) changes in the cell membrane to modulate its fluidity, (2) the metabolism of the toxic compound or its inactivation, and (3) increased efflux of the toxic compound. Most of the papers published on this subject dealt with Gram-negative bacteria. Although the interest in using Gram-positive bacteria in biocatalysis and bioremediation processes is increasing, the number of studies on the adaptation of non-pathogenic strains to nonantibiotic compounds is still limited.

In a study carried out to assess the effects of organic solvents in organic-aqueous systems on *R. erythropolis*, *Xanthobacter* Py2, *Arthrobacter simplex* and *Mycobacterium* sp. NRRL B-3805, principal component analysis was used to interpret the data (de Carvalho and da Fonseca 2004). The variables used to construct the data matrix were cell viability, cell morphological parameters (e.g. size, elongation factor, circularity), number of cells in clusters and the conditions to which the cells were exposed (substrate concentration, carbon source used for growth, adaptation time to the solvent prior to substrate addition) and physical properties of the solvents such as density, molecular weight and log *P*. Over a third of the variability of the data related to *R. erythropolis* could be explained by solvent toxicity. When studying the effect of several solvents, present at different organic/aqueous ratios, on whole cells of *R. erythropolis* DCL14 carrying out the biotransformation of (−)-carveol to (−)-carvone, principal components analysis showed that 41.2% of the variance of the data responsible for the cell behaviour could also be ascribed to solvent toxicity (de Carvalho et al. 2003).

One basic idea observed in adaptation studies is that cells growing at a slow rate acquire general tolerance to the stress agent (Booth 2002; Sonnleitner 1998). Furthermore, by growing at slower growth rates, separate but overlapping pathways that confer tolerance to diverse stresses are induced, and the cells become simultaneously resistant to, e.g. acid, heat, alkali and hydrogen peroxide (Booth 2002). This mechanism of tolerance acquisition is particularly important in biofilms. The slow growth observed in the matrix-embedded cells and the limited transport of nutrients, metabolites and oxygen between the surface and the interior of the biofilm could be responsible for an increased antibiotic and disinfectant resistance of biofilm cells over planktonic cells (Stewart and Costerton 2001; Donlan and Costerton 2002). Isolates from equine samples of *R. equi*, a facultative intracellular pathogen, able to produce biofilms presented higher resistance to macrolide antibiotics (Gressler et al. 2015). The same could be true regarding organic solvents as cells inside aggregates and biofilms are less exposed to the deleterious action of these compounds. *Rhodococcus* adhesion to both metallic and nonmetallic surfaces is promoted by a modulation of the cell lipid composition in response to both the type of surface and nutrients available (Rodrigues and de Carvalho 2015). Although *Mycobacterium vaccae* cells adapted to organic solvents presented an increased tolerance towards efflux pump inhibitors but not towards antibiotics (Pacífico et al. 2018), *R. erythropolis* cells adapted to toluene presented an increased resistance against 50% ethanol and Betadine® and Micropur® tablets used for surface disinfection and water purification (de Carvalho et al. 2007). The ability of *Rhodococcus* cells to form biofilms on the surface of indwelling central venous catheters led to the suggestion of adding them to the list of biofilm-forming organisms in immunocompromised hosts such as cancer patients (Al Akhrass et al. 2012). Since several rhodococcal species, including *R. opacus*, *R. erythropolis* and *R. jostii*, present genes allowing them to transport many different substrates including solvents and drugs, not only their biodegradation potential may be increased but also their pathogenicity. The rationale to study the adaptation mechanisms to organic solvents thus presents a further motivation.

3.1 Adaptation of the Cell Wall and of the Cellular Membrane

When challenged by organic solvents, the most important adaptive response shown by tolerant bacterial cells is probably the mechanism that allows them to control the fluidity and viscosity of the membrane which is termed "homeoviscous adaptation" (Sinensky 1974). This mechanism, initially shown for cells responding to changes in the growth temperature, is based on the phase transition temperature which reflects the temperature that causes a change in the physical state of the phospholipid bilayer from an ordered gel phase to a disordered liquid-crystalline phase (Sinensky 1974; Eze 1991). By changing the length of the fatty acids and their configuration, the cells alter the phase transition temperature of lipids and thus membrane fluidity (de Carvalho and Caramujo 2018). For example, whilst the saturated fatty acid C16:0 (palmitic acid) presents a melting temperature of 63 °C, the monounsaturated fatty acid in the *trans* conformation (*trans*-C16:1, palmitelaidic acid) and *cis* conformation (*cis*-C16:1, palmitoleic acid) has a melting temperature of 33 and 0 °C, respectively (Heipieper et al. 2007). Parameters such as the degree of saturation, *cis*/*trans* ratio and *iso*/*anteiso* ratio thus allow the assessment of the changes in membrane fluidity promoted by the cells.

Several papers have reported that, when exposed to toxic organic compounds, tolerant bacterial strains change the fatty acid profile of their membrane (e.g. de Carvalho et al. 2009; Heipieper et al. 1994; Isken and de Bont 1998; Pacífico et al. 2018; Sardessai and Bhosle 2002; Sikkema et al. 1995; Torres et al. 2011). The existence of an outer membrane allows Gram-negative bacteria to quickly modify and adapt the lipopolysaccharides, efflux pumps and the fatty acid composition of the cellular membrane (Ramos et al. 2002). The *cis*/*trans* isomerization which results in a lower penetration of the solvents through the inner membrane (Cronan 2002) is a short-term response which takes place within 1 min after solvent exposure, whilst the remaining changes are long-term responses (Junker and Ramos 1999). As previously mentioned, the mechanisms involved in the adaptation of Gram-positive strains have been less studied, but it has been suggested that the mechanisms should be similar to those presented by Gram-negative bacteria (Fang et al. 2007; Nielsen et al. 2005; Ramos et al. 2002; Torres et al. 2011).

When the effect of organic solvents was studied in 29 strains belonging to 6 species of the *Rhodococcus* genus, namely, *R. erythropolis*, *R. fascians*, *R. longus*, *R. opacus*, *R. rhodochrous* and *R. ruber*, it was found that composition of the medium, solvent concentration and individual traits of each strain strongly affected bacterial resistance (Korshunova et al. 2016). The cells resistant to solvents increased the relative surface area and reduced the rigidity of the bonds between the cell wall components.

Responses in *R. erythropolis* DSM 1069 occurring in the few minutes following osmotic stress caused by sodium chloride were observed in the lipid composition and net surface charge of the cells (de Carvalho et al. 2014b). The doubling time of strain DSM 1069, under the most favourable conditions of the study, was 3.5 h, but salt-induced changes in the total lipid composition of the cells could be observed after only 6 min. The most surprising result was the production of polyunsaturated fatty acids (PUFAs), which reached 36.3% of the total lipids in bacteria exposed to 5.5% NaCL for 35 min. The production of PUFAs is unusual in bacteria, and until the 1990s, it was thought that bacteria, with the exception of certain cyanobacteria, could not produce them. In the case of strain *R. erythropolis* DSM 2016, the production of PUFAs was accompanied by a concomitant decrease in the percentage of monounsaturated fatty acids, suggesting the action of fatty acid desaturases which have been found in *R. erythropolis* strains. Production of PUFAs was also observed in *R. erythropolis* DCL14 during adaptation to extreme conditions (de Carvalho 2012). A stepwise adaptation strategy allowed the growth of these cells at 4–37 °C, pH 3–11 and in the presence of up to 7.5% sodium chloride and 1% copper sulphate. The adapted cells adjusted the relative proportion of straight, methyl and cyclopropyl saturated, unsaturated and hydroxyl-substituted fatty acids, produced PUFAs and regulated their net surface charge. Furthermore, the adapted *R. erythropolis* cells could degrade C6 to C16 *n*-alkanes and alcohols under conditions that did not allow the growth of non-adapted cells.

Metz et al. have also described a pathway for the synthesis of long PUFA which uses a polyketide synthase-like gene cluster instead of multiple desaturases and elongases (Metz et al. 2001). Seven polyketide synthase genes were identified in whole genome of *Rhodococcus* sp. RHA1 (McLeod et al. 2006). Polyketide synthases are also involved in the production of secondary metabolites (McLeod et al. 2006) and mycolic acid synthesis (Portevin et al. 2004).

Mycolic acids of rhodococci act as a permeability barrier for hydrophilic compounds, and water-filled channels are required for their entrance in the cell. Lichtinger et al. (2000) identified and purified a channel-forming protein in *R. erythropolis* with a molecular mass of just 8.4 kDa with no significant homology to known protein sequences. The authors suggested that 2.0 nm channels were formed by protein oligomers in the cell wall, being highly cation selective due to negative charges located at the channel mouths. Organic solvent extracts of *R. equi* cells allowed the identification of two channel-forming proteins: protein PorA(Req) is cation selective, has a single-channel conductance of 4 nS in 1 M KCl and is highly permeable to positively charged solutes and porB(Req) prefers the passage of anions, is not voltage gated and presents a conductance of 300 pS in 1 M KCl (Rieβ et al. 2003).

n-Alkane droplets have negative zeta potentials, e.g. the value for *n*-hexadecane droplets is -46.0 ± -3.4 mV (de Carvalho et al. 2009). The negative zeta potential of *n*-alkane droplets is the result of selective adsorption of OH^- ions, which causes gathering of the excessive negative charge at the oil-water interface (Stachurski and Michalek 1996). Since *n*-alkanes and the channels at the cell wall have both localized negative charges, the entrance of these compounds could be prevented in the cells.

However, several papers published showed that *Rhodococcus* cells are able to adhere directly to organic layers in organic-aqueous two-phase systems (de Carvalho and da Fonseca 2002b, 2003; de Carvalho et al. 2007). Bouchez-Naïtali and co-workers (2001) also observed a direct uptake of *n*-hexadecane by four *R. equi* strains, which did not produce biosurfactants. *R. erythropolis* S+14He is able to select and transport, by an energy-driven transport system, *n*-hexadecane from mixtures of *n*-alkanes into intracellular inclusions (Kim et al. 2002). *Rhodococcus* sp. BAP-1 transports fluoranthene across the cell membrane by a process requiring energy and concurrent catabolism with significant production of CO_2 (Li et al. 2014). *R. erythropolis* NTU degrades long-chain *n*-alkanes at concentrations up to 20%, and a correlation between the amount of H^+ ions in the culture medium and biodegradation of C14–C16 alkanes could be found (Liu et al. 2012). This may be used to follow alkane consumption by pH monitoring.

Yoneda et al. adaptively evolved *R. opacus* PD630 to phenol (to improve the tolerance of the strain to lignin-derived phenolic compounds) over 40 passages using phenol as sole carbon source (Yoneda et al. 2016). Two strains isolated from passages 33 and 40 showed increased phenol consumption rates, and a twofold increase in the production of lipids from phenol was observed when compared to the wild-type strain. Whole-genome sequencing and comparative transcriptomics indicated that both adapted strains presented highly upregulated degradation pathways and putative transporters for phenol. As pointed out by the authors, this suggests a new mechanism of bacterial aromatic tolerance related to the transport of toxic phenolic compounds into the cell which requires a proper balance between transport and degradation rates to prevent accumulation of toxic compounds inside the cell.

Kurosawa et al. studied the tolerance and adaptive evolution of an engineered triacylglycerol producing strain, *R. opacus* MITXM-61, to lignocellulose-derived inhibitors such as phenols, furans and organic acids (Kurosawa et al. 2015). Levulinic acid affected lipid biosynthesis and altered the principal fatty acid component of *Rhodococcus* TAGs from even-numbered C16–C18 fatty acids to odd-numbered C15–C17 fatty acids. An evolutionary adaptation strategy using short-term serial transfers of the cell cultures in inhibitor-containing liquid medium resulted in an evolved strain with improved tolerance to lignin, 4-hydroxybenzaldehyde and syringaldehyde.

When comparing the effect of low and high doses of buckminsterfullerene (*n*-C60) on *Bacillus subtilis* with the effect of hyperosmotic conditions, Fang et al. (2007) concluded that Gram-negative bacteria can present different responses to the same type of membrane-active compounds under different conditions, whereas Gram-positive bacteria can have the same response in terms of fatty acid composition under different stress conditions. *Rhodococcus* strains responded to the presence of organic solvents by changing the degree of saturation of the fatty acids of the cellular membrane, by changing the length of the fatty acids and mycolic acids according to the chain length of the carbon source and by altering the percentage of branched fatty acids (Table 2). Rodgers et al. (2000), by using ^{13}C-enriched C16 and C18 alkanes and electrospray ionization Fourier transform ion cyclotron resonance mass spectrometry, showed that complete mineralization was achieved by *R. rhodochrous*, with

Table 2 Mechanisms of solvent tolerance observed at the cellular membrane level in *Rhodococcus* strains

Strain	Compound	Reported mechanism	References
Rhodococcus sp. 33	Benzene	Increased degree of saturation	Gutiérrez et al. (1999)
R. opacus GM-14, GM-29, and 1CP	Benzene, phenol, 4-chlorophenol, chlorobenzene or toluene	Increase content of branched (10-methyl) fatty acids	Tsitko et al. (1999)
R. opacus PDW4	4-Chlorophenol	Adjustments in the average chain length and unsaturation index of mycolic acids	de Carvalho et al. (2016)
R. opacus MITXM-61	Phenols, furans and organic acids	Altered TAGs from even-numbered to odd-numbered fatty acids	Kurosawa et al. (2015)
R. erythropolis DCL14	Short-chain alcohols (methanol, ethanol, butanol) n-Alkanes and n-alkanols	Decrease of degree of saturation Increase in degree of saturation	de Carvalho et al. (2005)
R. erythropolis DCL14	C5–C16 *n*-alkanes	Saturated fatty acids with chain length corresponding to the substrate used; net surface charge increased with chain length	de Carvalho et al. (2009)
R. erythropolis DCL14	Carveol and carvone	Lower percentage of long-chain fatty acids; decrease of the unsaturation index	de Carvalho and da Fonseca (2007)
R. erythropolis DCL14	Toluene	C14:0 and C16:0 increased whilst C18:0 decreased; increased percentage of *iso*-branched fatty acids when compared to straight-chain	de Carvalho et al. (2007)
R. erythropolis CCM 2595	Phenol	Changes in the chain length and unsaturation level of mycolic acids	Kolouchová et al. (2012)
R. erythropolis E1	C2–C7 *n*-alkanoic acid salts C9–C15 *n*-alkanes	MA profile according to the even-odd nature of the carbon chain of substrate Changed cell wall permeability	Sokolovská et al. (2003)
R. erythropolis 17, *Rhodococcus* sp. 20, *R. opacus*	Pentadecane, hexadecane	Fatty acids related to the chain length of the substrate	Alvarez (2003)
Rhodococcus sp. Q15	Alkanes at low T	Decrease of degree of saturation	Whyte et al. (1999)

complete ^{13}C incorporation in the bacterial lipids. The incorporation of saturated fatty acids in the membrane phospholipids with chain length corresponding to the substrate used for cell growth has been found frequently. Studies carried out by Alvarez (2003) showed that most fatty acids in actinomycetes were related to the chain length of the substrate and also to β-oxidation-derived fatty acids. Only when *R. erythropolis* DCL14 cells were grown on *n*-tridecane and *n*-pentane were the fatty acids C13:0 and C15:0 found in the cellular membrane, respectively (de Carvalho et al. 2009). Sokolovská et al. (2003) also observed two types of responses in mycolic acid patterns of *R. erythropolis* E1: lack of odd-numbered carbon chains when the cells grew on linear alkanes with even number of carbon atoms and mycolic acids with both even and odd carbon chains in cells grown on branched alkanes or on mixtures of substrates. Furthermore, hydrocarbons can also be used in the biosynthesis of triacylglycerols and wax esters under nitrogen starvation (Alvarez et al. 1996; Voss and Steinbüchel 2001).

Kolouchová et al. (2012) studied the mycolic acids of two *R. erythropolis* strains able to degrade phenol. One of the strains, *R. erythropolis* CCM 2595, was adapted to phenol as sole carbon source for 6 months, whilst *R. erythropolis* CCM 2595 *pSKRK 21 phe* contains a plasmid allowing the cells to present higher phenol hydroxylase activity. Soluble humic acids were added to the growth media to help the transport of organic substances into the cells as result of their interaction with components of the cell surface envelop.

R. opacus PDW4 challenged with 4-chlorophenol, a standard monoaromatic chemical used to investigate cellular toxicity and adaptation, changed their mycolic acid content according to the xenobiotic concentration (de Carvalho et al. 2016). Both the average chain length and unsaturation index of the mycolic acids decreased up to a concentration of 4-chlorophenol of 150 mg/L but increased for higher concentrations. Curiously, Solyanikova and co-workers showed that *R. opacus* 1 cp, after storage in a dormant state, could degrade chlorinated phenols that had not been degradable before dormancy (Solyanikova et al. 2011). During the dormant period, the cells were incubated at 20 °C under static conditions for 4 months, and it was not clear if this process induced the appearance of a novel metabolic pathway or favoured an existing one. Energy and carbon starvation or nutrient-induced stationary phase of *R. erythropolis* SQ1 cells also promoted higher resistance to heat shock and oxidative damage than that observed in exponentially grown cells (Fanget and Foley 2011).

The traditional way of favouring the growth of variants more adapted within a population uses repeated passages of the culture under selective conditions. As mentioned previously, using a stepwise strategy to adapt *R. erythropolis* DCL14 cells to extreme conditions of temperature, pH, salt and copper concentrations, it was possible to degrade C6–C16 *n*-alkanes and alcohols under conditions that previously did not allow the growth of this bacterium (de Carvalho 2012). The cells made the necessary adjustments to maintain the polarization of the membrane, changing the fatty acid composition of the cellular membrane and the net surface charge. The same strategy resulted in *R. erythropolis* DCL14 cells adapted to 20–65% toluene concentrations (de Carvalho et al. 2007). The cells adapted by increasing the percentage of tetradecanoic and hexadecenoic acids whilst decreasing the percentage of octadecanoic acid. The proportion of saturated *iso*-branched fatty acids also increased during toluene adaptation, whilst the amount of straight-chain fatty acids decreased. By increasing the content of *iso*-branched fatty acids, the cells decreased the fluidity or flexibility of the cellular membrane.

The most interesting physicochemical surface properties adaptation of *Rhodococcus* cells to organic solvents was observed with *R. erythropolis* DCL14 growing on C5–C16 *n*-alkanes (de Carvalho et al. 2009). A strong correlation between the *n*-alkane chain length and the zeta potential of the bacterial cells was observed, the cells even becoming positive when the cells grew on C14–C16. Most known bacteria only exhibit negative surface charges at circum-neutral pH (Jucker et al. 1996), and the fact that *R. erythropolis* DCL14 presents a positive surface charge is quite remarkable. According to the extended Derjaguin, Landau, Verwey and Overbeek (DLVO) theory of colloidal stability (Van Oss 1995), in which electrostatic repulsion, van der Waals attraction and acid-base (hydrophobic) interactions are considered, the adhesion capacity of bacterial cells is inversely correlated with the (negative) surface charge of the cell. Since most natural surfaces are negative, adhesion will only take place when the electrostatic repulsion is overcome by attractive forces (e.g. van der Waals, hydrophobic interactions) between the bacterial cells and the surface. In the case of strain DCL14, the positive surface charge contributes to the attachment of the cells to negatively charged surfaces such as *n*-hexadecane droplets (zeta potential of -46.0 ± -3.4 mM).

Gutiérrez et al. (2003) compared a *Rhodococcus* sp. strain, able to tolerate and degrade high concentrations of benzene, with a benzene-sensitive mutant obtained by mutagenesis. The mutant was unable to increase the saturation degree of the fatty acids to the levels achieved by the wild type when exposed to benzene. The fluidity of the membranes increased after only 0.5 h of exposure to benzene, but the wild type was able to respond by changing the saturation:unsaturation ratio of the cellular membrane, especially by changing the proportion of myristic and oleic acids.

The content of 10-methyl branched fatty acids increased in *R. opacus* GM-14, GM-29, and 1CP, when the cells were grown on benzene, phenol, 4-chlorophenol, chlorobenzene or toluene as sole carbon sources, as compared to fructose-grown cells (Tsitko et al. 1999). A dose-related increase in the percentage of 10-methyl-branched fatty acids was also observed as a response to increasing concentrations of phenol and toluene in strain GM-14, which is unable to metabolize toluene. 10-Methyl-branched fatty acids are also present in *Rhodococcus koreensis* DNP505^T, which is able to degrade 2,4-dinitrophenol (Yoon et al. 2000). The role and position of 10-methyl-branched fatty acids in *Rhodococcus* are still unclear, but the results of Tsitko et al. (1999) suggest that the cell envelope lipids containing 10-methyl-branched fatty acids should be involved in the adaptation of *Rhodococcus* strains to compounds affecting the cellular membrane, such as aromatics. The melting temperature of methyl-branched fatty acids depends on the position of the methyl substitution which will divide the lipid chain into longer and shorter sections (de Carvalho and Caramujo 2018).

The cell surface hydrophobicity of *Rhodococcus* sp. Q15, which is able to mineralize alkanes, diesel and Bunker C crude oil at both 5 and 24 °C, was also higher after cell growth in diesel fuel and hexadecane than when the cells were grown on glucose-acetate (Whyte et al. 1999). The carbon source strongly influenced the fatty acid profile of the cells, with small amounts of C18 fatty acids and greater amounts of C16 and C14:0 fatty acids being observed in cells grown on *n*-hexadecane when compared to those grown on glucose-acetate. The degree of saturation of the fatty acids of the membrane of strain Q15 decreased as response to a lower growth temperature: the membrane contained relatively saturated fatty acids at 24 °C and relatively unsaturated fatty acids at 5 °C. This happened independently of the substrate used, although the decrease in the degree of saturation occurred at a lesser extent when the cells grew on hydrocarbons then on glucose-acetate. The cells were thus able to modulate the membrane fluidity to respond to both low temperature and hydrocarbon toxicity.

Changes in cell hydrophobicity may be promoted by using the cell response to organic compounds. By exposing *R. erythropolis* cells to the terpenes carveol and carvone in organic-aqueous systems, it was possible to demote biofilm formation and even to disrupt established biofilms (de Carvalho and da Fonseca 2007). Strain DCL14 responded to the presence of the solvents tested by decreasing the unsaturation index, which reflects the average number of double bonds per fatty acid chain, with increasing number of carbons in the alkane chain. In the presence of carveol or carvone, the cells increased the unsaturated index, thus counteracting the effect of the solvents. The presence of unsaturated fatty acids has been reported to allow the cells to change membrane fluidity (Melchior 1982) and permeability (Russell 1988). The major response to the presence of these two terpenes was, however, a decrease in the percentage of fatty acids with a number of carbons higher than 16 (de Carvalho and da Fonseca 2007). Since a direct relation between these fatty acids and cell hydrophobicity was found, by decreasing the percentage of fatty acids with more than 16 carbon atoms, cell hydrophobicity decreased which resulted in the dispersion of cells previously aggregated because of the presence of organic solvents.

3.2 Biocatalysis and Biodegradation of the Toxic Compound

Bioconversion or mineralization of a toxic compound has been presented as a mechanism for bacterial strains to thrive in its presence. However, whilst some studies indicate that tolerance to solvents derives from the capacity of the cells to metabolize them, other works suggest that conversion or metabolism of organic solvents is not essential to tolerance. The degradation may mediate the resistance of some bacterial strains to solvents, but it cannot be the main mechanism conferring tolerance to a broad number of solvents (Isken and de Bont 1998).

Rhodococci present a broad catabolic diversity and enzymatic capabilities, increasing their importance in environmental and biotechnological processes (Bell et al. 1998; de Carvalho 2016; de Carvalho and da Fonseca 2005; Martínková et al. 2009; Warhurst and Fewson 1994). Their exceptional ability to resist and degrade hydrophobic compounds and xenobiotics is related to the presence and mobilization of large linear and circular plasmids, whilst the presence of multiple pathways and

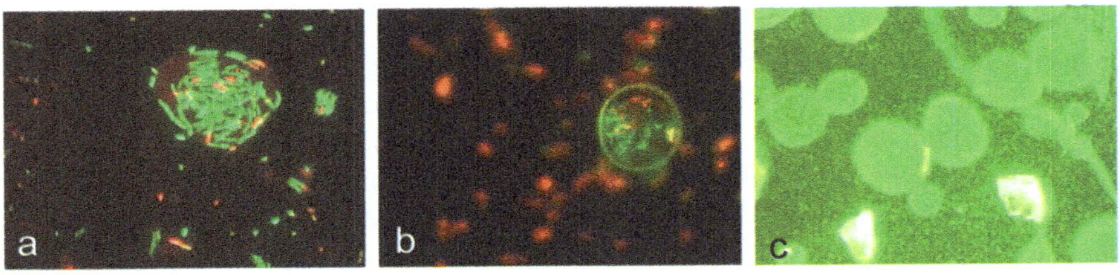

Fig. 2 *R. erythropolis* cells inside *n*-dodecane droplets. (**a**, **b**) Solvent droplets trapped between a slide and cover slip (horizontal and vertical field widths equal to 0.16 and 0.12 mm, respectively); (**c**) solvent droplets adsorbed on silica gel particles (horizontal and vertical field widths equal to 0.4 and 0.3 mm, respectively). Viable cells are stained green whilst non-viable cells are stained red

gene homologous enhance their versatility (Larkin et al. 2005, 2006; van der Geize and Dijkhuizen 2004). Studies reporting the ability of *Rhodococcus* strains to act as whole-cell biocatalyst even in anhydrous organic solvents have also been published (e.g. Sameshima et al. 2008; Yamashita et al. 2007).

Rhodococcus sp. N-774, found in 1980, was the first strain to produce acrylamide at industrial scale, but *R. rhodochrous* J1 was found to contain a much more active nitrile hydratase (Nagasawa et al. 1993). The latter strain produces 30,000 tons per year of acrylamide (Mitsubishi Rayon, pers. comm.) and 6000 tons per year of nicotinamide by Lonza Guangzhou Fine Chemicals (de Carvalho 2016).

In two-phase systems, *R. erythropolis* cells migrate towards the organic phase because of their high cell hydrophobicity (de Carvalho and da Fonseca 2002a, b, 2003). When emulsion samples were collected and droplets of solvent were observed by fluorescence microscopy, it was found that the cells partitioned between the organic and the aqueous phase. Part of the cell population was even inside the solvent droplets (Fig. 2a, b). The images with cells on the organic phase were not just the result of a superposition of different planes, as shown by a technique developed by de Carvalho and da Fonseca in 2003 (Fig. 2c), which allows the observation of 3-D solvent droplets adsorbed to solid particles, e.g. of silica gel. Cells were preferentially inside organic droplets in systems with solvents with high log *P* value, being the percentage of cells in droplets lower when the organic phase was more toxic (Table 3). However, in the presence of cyclohexane, which has a log *P* value of 3.2 and should be toxic, most of the cells were also positioned inside the solvent droplets. Cells directly positioned in the organic phase can access the dissolved hydrophobic substrates in biocatalytic processes (de Carvalho et al. 2000; de Carvalho and da Fonseca 2002b) and can degrade organic solvents by direct uptake in bioremediation systems (Bouchez-Naïtali et al. 2001; de Carvalho and da Fonseca 2007). In these cases, the growth and conversion rates obtained are independent of the interfacial area, due to the strong adsorption of the bacterial cells at the solvent-aqueous interface. The use of organic solvents for in situ product removal in fermentation and biocatalytic systems is a powerful technique to increase bioprocess productivity (Dafoe and Daugulis 2014). Strains able to tolerate different

Table 3 Percentage of *R. erythropolis* cells inside solvent droplets with different toxicities at different phase ratios

Phase ratio	0.0005				0.0025						0.025		
log *P* (solvent)	1.85	3.2	4.5	6.6	−1.35	1.85	3.2	3.5	9.6	12.88	−1.35	6.6	12.88
Cells in droplets (%)	64.9	93.5	72.5	97.7	73.4	58.5	99.0	57.2	73.0	99.3	68.0	91.7	97.6

The cells were carrying out the conversion of carveol into carvone in the organic-aqueous two-phase systems solvents may provide a broader choice during bioprocess development as the best solvents to extract product(s) may be used instead of only the most biocompatible.

R. erythropolis NTU-1 degrades *n*-alkanes and branched alkanes in mineral medium and also under saline conditions, but in the latter case, cells formed aggregates (Liu et al. 2009). The biofloccules trapped significant amount of the substrate, enhancing removal of the alkane in addition to bioremediation. During the bioremediation of contaminated sea water, the biofloccules may allow a physical separation and a relatively fast removal of hydrocarbon compounds.

Pospíšilová et al. reported an increased growth and degradation of phenol by *R. erythropolis* cells stimulated by a homogenous electromagnetic field (Pospíšilová et al. 2015). The electromagnetic field stimulated biofilm formation although cells presented a lower hydrophobicity, suggesting the possibility of using a biofilm reactor for phenol degradation. Enhanced aggregation and production of extracellular polymeric substances was also observed in *R. jostii* exposed to toluene and a mixture of perfluoroalkyl acids (Weathers et al. 2015). Furthermore, the results suggested that flocculation was a biological response rather than the result of chemical interactions of surfactant and perfluoroalkyl acids.

R. erythropolis IEGM 267 and *R. rhodochrous* IEGM 107 pre-grown in the presence of 0.1% (v/v) *n*-hexadecane were able to partially and completely degrade 500 mg/L of dehydroabietic acid, a toxic tricyclic diterpenoid (Cheremnykh et al. 2018). Pregrowth in the presence of the C16 alkane was essential for the degradation of the compound. In the presence of *n*-hexadecane, the cells formed 10–50 μm aggregates with more than ten viable cells, and when exposed to dehydroabietic acid, strain IEGM 107 promoted cell aggregates with up to 0.5 mm in size. Strain IEGM 265 formed aggregates with up to 5 mm of diameter. The cells also decreased the zeta potential during the biodegradation and increased their length and width. An increase in the ratio of the cell surface area to its volume promoted a higher cell-substrate contact.

In a paper regarding adaptation of microbes, Sonnleitner (1998) discussed the role of static effectors (such as concentrations of substrates and products) that affect the system instantaneously and independently of time (limiting or inhibiting the cells) and dynamic effectors which are time dependent and associated to changes in the physiological state, protein synthesis, or degradation. When in contact with a toxic compound, the cells may express an enzyme(s), or use alternative pathways to catabolize and decrease the concentration of the biocide in the microenvironment surrounding the cells. For economical reasons, the cells usually express only the required enzymes after being exposed to the toxic compound.

R. pyridinivorans NT2 was found able to degrade 4-nitrotoluene, recognized as priority pollutant, at a concentration of 400 mg/L, through partial reduction of the amino group (Kundu et al. 2013). The strain was isolated from an effluent sediment of a pesticide manufacturing facility in Gujarat, India. The cells produced trehalose-based biosurfactants which reduced the surface tension of the media to 29 mN/m. Strain NT2 was also able to degrade 2,6-dinitrotoluene which is a common isomer produced during the synthesis of 2,4,6-trinitrotoluene (TNT) and used in the production of herbicides, dyes and foams (Kundu et al. 2016). The cells were able to use

2,6-dinitrotoluene as sole carbon and nitrogen source. The major byproduct of TNT synthesis, 2,4-dinitrotoluene, which is used as waterproofing and plasticizing agent in explosives and propellants, is also degraded by strain NT2 (Kundu et al. 2015). Besides the ability to degrade the intermediates of the TNT biodegradation pathway, some *R. erythropolis* strains are able to degrade TNT through the formation of Meisenheimer complexes (Vorbeck et al. 1998). The complex is further reduced to a yellow metabolite identified as the protonated 3,5-dihydride complex of TNT. The 1000 tons of TNT produced annually and the numerous contaminated sites resulting from military and industrial activities worldwide make bioremediation of TNT and its derivatives of paramount importance (Serrano-González et al. 2018).

By slowly increasing the concentration of solvent, substrate and product, it was possible to adapt *R. erythropolis* DCL14 cells in an air-driven direct contact bioreactor (de Carvalho and da Fonseca 2002b). By recirculating *n*-dodecane through the column reactor at a rate slow enough so that the biotransformation could only take place at a small extent, the cells were able to adapt. When the biotransformation was performed after the adaptation period, the product carvone reached a concentration of 94 mM after 310 h of operation, thus overcoming carvone inhibition (observed at carvone concentrations of 50 mM). By increasing the adaptation period to 136 h, carvone accumulation reached 259 mM. Since the biomass was kept relatively constant during the experiments, the adapted cells were able to produce much more product than the non-adapted cells.

A similar strategy could be used to adapt strain DCL14 cells to toluene to allow its degradation at high concentrations in *n*-dodecane-aqueous systems (de Carvalho et al. 2007). Only 10.5% of the initially non-adapted cells remained viable after 1-h exposure to 20% (v/v) toluene. Cell adaptation was carried out by adding a toluene pulse, whenever its concentration reached ca. 50% of the initial value, to double the previous initial concentration (i.e. if the initial concentration was 1%, a pulse of toluene was added when toluene concentration reached 0.5% so that the concentration would be 2%). Curiously, toluene degradation rate increased with the increasing toluene concentrations added to the reactor. By using this strategy, the cells could endure a maximum toluene concentration of 4.9 M, which corresponds to 52.4% (v/v) in the organic phase, toluene being consumed at 10.7 mg/(h.mg protein). Once more, the amount of biomass was kept nearly constant throughout the 35 days of the experiment. The fatty acid composition of the adapted cells presented a higher amount of branched fatty acids and a ca. 40% decrease in the amount of straight-chain fatty acids. Interestingly, cells adapted to toluene presented a significant increased resistance to ethanol, silver ions and iodine when compared to non-adapted cells.

The results obtained with *R. erythropolis* DCL14 suggest that toluene metabolism should be related to toluene tolerance mechanisms, as a correlation between toluene consumption rate and initial toluene concentration was found. The presence and co-metabolism of *o*-, *m*- and *p*-xylene further enhanced the degradation of toluene (de Carvalho et al. 2007). Leneva and co-workers (2009) could also adapt *R. opacus* 412 and *R. rhodnii* 135 to phenanthrene and anthracene on solid medium. The cells accelerated the metabolism and became able to grow on phenanthrene as sole carbon and energy source in liquid medium. *R. erythropolis* CCM 2595 although able to use phenol, pyrocatechol, resorcinol, *p*-nitrophenol, *p*-chlorophenol, hydroquinone and hydroxybenzoate was strongly affected by the substrate and initial concentration, and whilst some of the monoaromatic compounds suppressed the ability of strain CCM 2595 to use phenol in binary mixtures, others were strong inducers of phenol 2-monooxygenase (Čejková et al. 2005). In the latter case, the mixtures were more rapidly metabolized if the cells were pre-exposed to the substrates. Cold-adapted *Rhodococcus* sp. strain NO14-1 and strain NO20-3 were able to fully degrade 12.5 mM phenol after 25 and 28 days, respectively, but when exposed to 15 mM phenol, only 4 mM were degraded after 10 days, and no further degradation was observed over 36 days (Margesin et al. 2005). Adaptation of *R. rhodochrous* 172 on agar mineral medium with fluorene for 6 months resulted in rapid growth without lag phase of the adapted cells: complete degradation of 12 mg/L fluorene was achieved in liquid medium within 5 days, whilst the non-adapted cells were unable to grow (Rubashko et al. 2006). The results presented are an indication that in these strains tolerance and degradation capacity may be strongly related. On the contrary, other studies done with *R. opacus* B-4 suggested that benzene tolerance is not related to the utilization or degradation of this organic compound (Na et al. 2005). A deletion-insertion mutant defective in the benzene dioxygenase genes, strain B-4KSN01, grew in 10% benzene as well as the wild strain B-4.

A *R. rhodochrous* isolated from a bacterial consortium enriched on toluene from a sample collected on a gasoline-contaminated aquifer was able to degrade benzene, toluene, ethylbenzene and xylene as primary carbon source (Deeb, Alvarez-Cohen 1999). When different combinations were tested, it was found that benzene and toluene degradation rates were enhanced in the consortium by the presence of *o*-xylene, whilst toluene, benzene or ethylbenzene inhibited xylene degradation rates. In the case of *R. rhodochrous*, ethylbenzene degradation decreased toluene degradation, benzene induced a slight inhibitory effect and *o*-xylene had no significant effect on toluene degradation.

Two mutants of *Rhodococcus* sp. 33 unable to degrade benzene were still tolerant to 500–800 mg/L of benzene (non-adapted strain 33 cells were able to tolerate ca. 1000 mg/L), also confirming that benzene degradation is of minor importance to the tolerance of these variants (Gutiérrez et al. 2009). Mosqueda et al. (1999) also suggested that toluene metabolism is not involved in toluene tolerance in *P. putida* DOT-T1. Independently of the mechanisms used by each strain, solvent tolerance is important as it allows bacterial growth at high organic solvent concentrations, which is of paramount importance in the bioremediation of sites contaminated with compounds such as benzene and toluene (Chen et al. 2009; Na et al. 2005).

Lately, organic solvent-tolerant strains have received a further notice because biodesulphurization of petroleum occurs in the presence of high concentrations of hydrocarbons. Many of the described competent bacteria to perform desulphurizations are *R. erythropolis* strains, e.g. IGTS8, N1-43, D-1 and KA2-5-1, with strain IGTS8 being the best characterized. The *dszA*, *B* and *C* genes primarily responsible for DBT metabolism are located in a single operon on a large plasmid in strain IGTS8 and in other related strains (Monticello 2000). Five strains, able to utilize dibenzothiophene (DBT) as sole sulphur source and to convert it to hydroxybiphenyl (HBP), isolated from coal storage sites in the north of France and from soils contaminated with heavy crude oil with high sulphur content belonged to the *Rhodococcus/Gordonia* cluster (Abbad-Andaloussi et al. 2003). All strains were able to use DBT in 95% *n*-hexadecane, used as model for diesel oil, although no activity was observed at *n*-hexadecane concentration of 99%. Some of strains used by Bouchez-Naïtali et al. (2004), namely, *Rhodococcus* sp. MK7C1 and MK2.4, exhibited good resistance to solvents, being even more tolerant than non-desulphurizing *Pseudomonas* strains, which could explain why transference of biodesulphurizing genes into Gram-negative strains did not promote biodesulphurization activity. An increased DBT desulphurization activity could be increased in biphasic systems, as reported by Ohshiro et al. (1995) with *R. erythropolis* H-2 in 70% *n*-tetradecane, Patel et al. (1997) with *R. erythropolis* IGTS8 in 50% *n*-hexadecane, and Abbad-Andaloussi et al. (2003) with *Rhodococcus* sp. strain I-2207 in the presence of 50% *n*-hexadecane.

R. erythropolis ATCC 4277 is able to carry out the biodesulphurization of heavy gas oil (HGO) at a rate of 5 g sulphur/(kg HGO.h) when grown in cheap media such as residues (cassava wastewater) and byproducts (trub) (Porto et al. 2017). Kawaguchi et al. made a recombinant strain from the organic-tolerant strain *R. opacus* B-4 to express the genes *dszA*, *dszB* and *dszC* from *R. erythropolis* IGTS8 to prevent inhibition by the sulphate end product of the 4S pathway of dibenzothiophene observed in this strain (Kawaguchi et al. 2012). Unlike the wild-type B-4 strain, the recombinant strain could grow in dibenzothiophene as sole sulphur source. Besides, the reaction could take place in biphasic systems with *n*-hexadecane as organic solvent, which allowed higher concentrations of substrate and thus higher desulphurization activities. *R. erythropolis* PD1 is also able to carry out the biodesulphurization of dibenzothiophene in both aqueous and biphasic systems (Derikvand et al. 2015). Nevertheless, total biodesulphurization of fossil fuel at industrial scale is not expected to occur in the near future (Soleimani et al. 2007). There are two main reasons that hamper the industrial application of biodesulphurization processes: failure to explore the full range of applications of biodesulphurization biocatalysts (e.g. to detoxify warfare agents or to produce fine chemicals) and failure to develop biocatalysts with activities high enough for a commercial petroleum desulphurization process (e.g. able to overcome inhibition caused by the product 2-hydroxybiphenyl (Kilbane 2017)).

3.3 Other Mechanisms of Protection

Iwabuchi and co-workers (2000) reported an association between colony morphotypes and oil tolerance in *R. rhodochrous*. The mucoidal strain was resistant to 10% (v/v) *n*-hexadecane, whilst the rough derivatives were sensitive to this concentration. Furthermore, when the extracellular polymeric substance (EPS) produced by the mucoidal strain was added to cultures of the rough strain, the latter acquired resistance to *n*-hexadecane. Rough strains are hydrophobic and mucoidal strains are hydrophilic. The EPS produced could confer tolerance to organic solvents by lowering the surface hydrophobicity, since, as Kobayashi et al. (1999) showed, low cell surface hydrophobicity could act as a defence mechanism against these compounds. The EPS produced by *R. rhodochrous* S-2 was even effective in stimulating the degradation of aromatic compounds in crude oil by native marine bacteria (Iwabuchi et al. 2002). Urai et al. (2007) also suggested that the large quantity of extracellular polysaccharides produced by *R. erythropolis* PR4 cells (able to degrade several hydrocarbons, including pristine) plays an important role in hydrocarbon tolerance.

Cells of *R. erythropolis* DCL14 present non-EPS and EPS producer variants. When exposed to the terpenes carveol and carvone and to organic solvents, the degree of saturation of the membrane phospholipids decreased, whilst the reverse was observed on the EPS producer counterpart (de Carvalho and da Fonseca 2007). The presence of EPS altered the level of cell exposure to solvents and terpenes. In the presence of organic solvents, part of the initially rough DCL14 population may start producing EPS (de Carvalho and da Fonseca 2002b, 2007). When these cells were under organic solvent stress-induced conditions, the small part of the population that remained viable produced colonies with a different phenotype: they were yellow or white, whilst the non-stressed cells produced pink colonies (de Carvalho et al. 2004). Furthermore, under conditions that allow high cell viability, cells that died presented no significant morphological changes when compared to viable cells. However, under aggressive conditions, nonviable cells were much larger, probably due to an increase in membrane fluidity, with the viable cells succeeding in decreasing their surface area to minimize the area of contact with the toxic agent (de Carvalho and da Fonseca 2004; de Carvalho et al. 2004).

Several rhodococci strains have been reported to produce biosurfactants, usually glycolipids, such as *R. erythropolis* (Lang and Philp 1998), *R. erythropolis* 51T7 (Marqués et al. 2009), and *R. equi* Ou2 (Bouchez-Naïtali and Vandecasteele 2008). A relation between the surface tension of the culture medium and the alkane chain length during C5–C16 growth of *R. erythropolis* DCL14 indicated the production of a biosurfactant in the presence of *n*-alkanes with longer carbon chains (de Carvalho et al. 2009). The significant reduction of the surface tension to values lower than 30 mN/m indicates the biosurfactant power of the compound produced. Kuyukina et al. proposed the use of methyl *tert*-butyl ether (MTBE) for the extraction of biosurfactants from *R. ruber*. The EPS produced contained ca. 10% polar lipids and 45% non-polar lipids and reduced the surface tension of water to 29.2 mN/m. Pen et al. studied the adhesion forces between *Rhodococcus* RC291 and a silicon oxide surface by force spectroscopy (Pen et al. 2015). They showed that cells in the late growth phase have a greater adhesion to the surface than those in the early growth stage. The EPS molecules from late exponential phase are less densely bound but have chains able to extend which could result in contraction and extension of the EPS and to changes in the density of the binding sites. On the other hand, the denser EPS of the late stationary phase provides shelter for the cells. Besides it could endure higher compression and retained higher amounts of water.

A common pattern for growth on long-chain alkanes in bacteria not producing biosurfactants is the formation of cellular flocs (Bouchez-Naïtali et al. 2001). In this case, the degradation rates during linear growth do not increase with interfacial area but with the efficiency of stirring, the interfacial uptake being limited by floc formation. Cell clustering was also visible with *R. erythropolis* DCL14 cells growing on *n*-dodecane, *n*-tetradecane and *n*-hexadecane (de Carvalho and da Fonseca 2005; de Carvalho et al. 2009). Biofilm formation can also been seen as a form of protection of cells to stress environments as cells inside the biofilm matrix are more protected (Heipieper et al. 1991). For example, no toxic effect of *n*-octane on biofilm growth was observed, whilst the cells in suspension were strongly inhibited by the accumulation of 1-octanol, a metabolite of the alkane monooxygenase during growth on *n*-octane (de Carvalho et al. 2009).

Coaggregation studies demonstrated interactions between *Rhodococcus* sp. strain MF3727, *Rhodococcus* sp. strain MF3803 and two strains of *Acinetobacter calcoaceticus* (MF3293, MF3627), all isolated from food processing surfaces (Møretrø et al. 2015). Coaggregation may contribute to biofilm formation whilst providing protection against cleaning and disinfection agents.

The action of efflux pumps responsible for the efflux of organic solvents from the inside of cells could also be responsible for adaptation of *Rhodococcus* cells to these compounds. Membrane transport systems are involved in several cellular processes, including uptake of essential nutrients, secretion of metabolites, extrusion of toxic byproducts and maintenance of cell homeostasis (Saier 2000). A recent review discussing membrane transport system in *Rhodococcus* shows that cells of this genus show a plastic genome coding for numerous efflux pumps (de Carvalho et al. 2014a). These cells specialized in transporting substrates such as solvents and drugs, and some species such as *R. opacus*, *R. erythropolis* and *R. jostii*, present genes for the transport of several different substrates. The association of efflux systems and adapted phenotypes clearly indicate that the cells have the equipment to respond to challenging compounds and conditions.

4 Application

A large set of enzymes from rhodococci cells have been reported and identified, allowing these cells to carry out a large array of bioconversions and degradations: from oxidations to dehydrogenations, epoxidations, hydrolysis, hydroxylations, dehalogenations, and desulphurizations. Since the cells are able to tolerate and adapt to organic solvents, bioreactions can be done in nonconventional media, using low water-soluble substrates, which makes this genus one of most promising in biotechnology.

The ability of *Rhodococcus* cells to metabolize hydrocarbons (even at significantly high concentrations and under a wide variety of environments), to produce biosurfactants (required to increase the bioavailability of low water-soluble organic solvents), and to change the physicochemical properties of the cellular surface (making these cells highly adaptive), makes these cells ideal candidates to in situ bioremediation of hydrocarbon-contaminated sites.

Acknowledgements The author acknowledges Fundação para a Ciência e a Tecnologia, I.P. (FCT), Portugal, for financial support under programme "FCT Investigator 2013" (IF/01203/2013/CP1163/CT0002).

References

Abbad-Andaloussi S, Lagnel C, Warzywoda M, Monot F (2003) Multi-criteria comparison of resting cell activities of bacterial strains selected for biodesulfurization of petroleum compounds. Enzym Microb Technol 32:446–454

Abe A, Inoue A, Usami R, Moriya K, Horikoshi K (1995) Properties of a newly isolated marine bacterium that can degrade polyaromatic hydrocarbons in the presence of organic solvents. J Mar Biotechnol 2:182–186

Aislabie J, Saul DJ, Foght JM (2006) Bioremediation of hydrocarbon-contaminated polar soils. Extremophiles 10:171–179

Al Akhrass F, Al Wohoush I, Chaftari A-M, Reitzel R, Jiang Y, Ghannoum M et al (2012) *Rhodococcus* bacteremia in cancer patients is mostly catheter related and associated with biofilm formation. PLoS One 7:e32945

Alvarez HM (2003) Relationship between beta-oxidation pathway and the hydrocarbon-degrading profile in actinomycetes bacteria. Int Biodeterior Biodegrad 52:35–42

Alvarez HM, Mayer F, Fabritius D, Steinbüchel A (1996) Formation of intracytoplasmic lipid inclusions by *Rhodococcus opacus* strain PD630. Arch Microbiol 165:377–386

Bej AK, Saul D, Aislabie J (2000) Cold-tolerant alkane-degrading *Rhodococcus* species from Antarctica. Polar Biol 23:100–105

Bell K, Philp J, Aw D, Christofi N (1998) The genus *Rhodococcus*. J Appl Microbiol 85:195–210

Benoit S, Benachour A, Taouji S, Auffray Y, Hartke A (2002) H2O2, which causes macrophage-related stress, triggers induction of expression of virulence-associated plasmid determinants in *Rhodococcus equi*. Infect Immun 70:3768–3776

Booth IR (2002) Stress and single cell: intrapopulation diversity is a mechanism to ensure survival upon exposure to stress. Int J Food Microbiol 78:19–30

Bouchez-Naïtali M, Vandecasteele JP (2008) Biosurfactants, an help in the biodegradation of hexadecane? The case of *Rhodococcus* and *Pseudomonas* strains. World J Microbiol Biotechnol 24:1901–1907

Bouchez-Naïtali M, Blanchet D, Bardin V, Vandecasteele JP (2001) Evidence for interfacial uptake in hexadecane degradation by *Rhodococcus equi*: the importance of cell flocculation. Microbiology 147:2537–2543

Bouchez-Naïtali M, Abbad-Andaloussi S, Warzywoda M, Monot F (2004) Relation between bacterial strain resistance to solvents and biodesulfurization activity in organic medium. Appl Microbiol Biotechnol 65:440–445

Brennan PJ, Nikaido H (1995) The envelope of mycobacteria. Annu Rev Biochem 64:29–63

Brink LES, Tramper J (1985) Optimization of organic solvent in multiphase biocatalysis. Biotechnol Bioeng 27:1258–1269

Cassells JM, Halling PJ (1990) Protease-catalyzed peptide-synthesis in aqueous-organic 2-phase systems: reactant precipitation and interfacial inactivation. Enzym Microb Technol 12:755759

Čejková A, Masak J, Jirku V, Vesely M, Patek M, Nesvera J (2005) Potential of *Rhodococcus erythropolis* as a bioremediation organism. World J Microbiol Biotechnol 21:317–321

Chapman JS (2003) Disinfectant resistance mechanisms, cross-resistance, and co-resistance. Int Biodeterior Biodegrad 51:271–276

Chen HL, Yao J, Wang L, Wang F, Bramanti E, Maskow T, Zaray G (2009) Evaluation of solvent tolerance of microorganisms by microcalorimetry. Chemosphere 74:1407–1411

Cheremnykh KM, Luchnikova NA, Grishko VV, Ivshina IB (2018) Bioconversion of ecotoxic dehydroabietic acid using *Rhodococcus* actinobacteria. J Hazard Mater 346:103–112
Colquhoun JA, Heald SC, Li L, Tamaoka J, Kato C, Horikoshi K, Bull AT (1998) Taxonomy and biotransformation activities of some deep-sea actinomycetes. Extremophiles 2:269–277
Cronan JE Jr (2002) Phospholipid modifications in bacteria. Curr Opin Microbiol 5:202–205
Dafoe JT, Daugulis AJ (2014) In situ product removal in fermentation systems: improved process performance and rational extractant selection. Biotechnol Lett 36:443–460
Daugulis AJ (2001) Two-phase partitioning bioreactors: a new technology platform for destroying xenobiotics. Trends Biotechnol 19:457–462
de Bont JAM (1998) Solvent-tolerant bacteria in biocatalysis. Tibtech 16:493–499
de Carvalho CCCR (2012) Adaptation of *Rhodococcus erythropolis* cells for growth and bioremediation under extreme conditions. Res Microbiol 163:125–136
de Carvalho CCCR (2016) Whole cell biocatalysts: essential workers from Nature to the industry. Microb Biotechnol 10:250–263
de Carvalho CCCR, Caramujo MJ (2018) The various roles of fatty acids. Molecules 23:2583
de Carvalho CCCR, da Fonseca MMR (2002a) Maintenance of cell viability in the biotransformation of (-)-carveol with whole cells of *Rhodococcus erythropolis*. J Mol Catal B Enzym 19:389–398
de Carvalho CCCR, da Fonseca MMR (2002b) Influence of reactor configuration on the production of carvone from carveol by whole cells of *Rhodococcus erythropolis* DCL14. J Mol Catal B Enzym 19:377–387
de Carvalho CCCR, da Fonseca MMR (2003) A simple method to observe organic solvent drops with a standard optical microscope. Microsc Res Tech 60:465–466
de Carvalho CCCR, da Fonseca MMR (2004) Solvent toxicity in organic–aqueous systems analysed by multivariate analysis. Bioprocess Biosyst Eng 26:361–375
de Carvalho CCCR, da Fonseca MMR (2005) The remarkable *Rhodococcus erythropolis*. Appl Microbiol Biotechnol 67:715–726
de Carvalho CCCR, da Fonseca MMR (2007) Preventing biofilm formation: promoting cell separation with terpenes. FEMS Microbiol Ecol 61:406–413
de Carvalho CCCR, van Keulen F, da Fonseca MMR (2000) Biotransformation of limonene-1,2-epoxide to limonene-1,2-diol by *Rhodococcus erythropolis* cells—an introductory approach to selective hydrolysis and product separation. Food Technol Biotechnol 38:181–185
de Carvalho CCCR, Pons MN, da Fonseca MMR (2003) Principal components analysis as a tool to summarise biotransformation data: influence on cells of solvent type and phase ratio. Biocatal Biotransform 21:305–314
de Carvalho CCCR, da Cruz AARL, Pons MN, Pinheiro HMRV, Cabral JMS, da Fonseca MMR, Ferreira BS, Fernandes P (2004) *Mycobacterium* sp., *Rhodococcus erythropolis*, and *Pseudomonas putida* behavior in the presence of organic solvents. Microsc Res Tech 64:215–222
de Carvalho CCCR, Parreno-Marchante B, Neumann G, da Fonseca MMR, Heipieper HJ (2005) Adaptation of *Rhodococcus erythropolis* DCL14 to growth on n-alkanes, alcohols and terpenes. Appl Microbiol Biotechnol 67:383–388
de Carvalho CCCR, Fatal V, Alves SS, da Fonseca MMR (2007) Adaptation of *Rhodococcus erythropolis* cells to high concentrations of toluene. Appl Microbiol Biotechnol 76:1423–1430
de Carvalho CCCR, Wick LY, Heipieper HJ (2009) Cell wall adaptations of planktonic and biofilm *Rhodococcus erythropolis* cells to growth on C5 to C16 n-alkane hydrocarbons. Appl Microbiol Biotechnol 82:311–320
de Carvalho CCCR, Costa SS, Fernandes P, Couto I, Viveiros M (2014a) Membrane transport systems and the biodegradation potential and pathogenicity of genus *Rhodococcus*. Front Physiol 5:133
de Carvalho CCCR, Marques MPC, Hachicho N, Heipieper HJ (2014b) Rapid adaptation of *Rhodococcus erythropolis* cells to salt stress by synthesizing polyunsaturated fatty acids. Appl Microbiol Biotechnol 98:5599–5606
de Carvalho CCCR, Fischer MA, Kirsten S, Würz B, Wick LY, Heipieper HJ (2016) Adaptive response of *Rhodococcus opacus* PWD4 to salt and phenolic stress on the level of mycolic acids. AMB Express 6:66

Deeb RA, Alvarez-Cohen L (1999) Temperature effects and substrate interactions during the aerobic biotransformation of BTEX mixtures by toluene-enriched consortia and *Rhodococcus rhodochrous*. Biotechnol Bioeng 62:526–536
Derikvand P, Etemadifar Z, Biria D (2015) RSM optimization of dibenzothiophene biodesulfurization by newly isolated strain of *Rhodococcus erythropolis* PD1 in aqueous and biphasic systems. Microbiology 84:65–72
Diefenbach R, Heipieper HJ, Keweloh H (1992) The conversion of cis- into trans- unsaturated fatty acids in *Pseudomonas putida* P8: evidence for a role in the regulation of membrane fluidity. Appl Environ Microbiol 38:382–387
Donlan RM, Costerton JW (2002) Biofilms: survival mechanisms of clinically relevant microorganisms. Clin Microbiol Rev 15:167–193
Eze MO (1991) Phase transitions in phospholipid bilayers: lateral phase separations play vital roles in biomembranes. Biochem Edu 19:204–208
Fang J, Lyon D, Wiesner M, Dong J, Alvarez PJJ (2007) Effect of a fullerene water suspension on bacterial phospholipids and membrane phase behavior. Environ Sci Technol 41:2636–2642
Fanget NVJ, Foley S (2011) Starvation/stationary-phase survival of *Rhodococcus erythropolis* SQ1: a physiological and genetic analysis. Arch Microbiol 193:1–13
Gilbert P, Brown MRW (1995) Some perspectives on preservation and disinfection in the present day. Int Biodeterior Biodegrad 36:219–226
Gressler LT, Vargas AC, Costa MM, Sutili FJ, Schwab M, Pereira DIB et al (2015) Biofilm formation by *Rhodococcus equi* and putative association with macrolide resistance. Pesqui Vet Bras 35:835–841
Gutierrez JA, Nichols P, Couperwhite I (1999) Changes in whole-cell derived fatty acids induced by benzene and occurrence of the unusual 16:1ω6c in *Rhodococcus* sp. 33. FEMS Microbiol Lett 176:213–218
Gutiérrez T, Learmonth RP, Nichols PD, Couperwhite I (2003) Comparative benzene-induced fatty acid changes in a *Rhodococcus* species and its benzene-sensitive mutant: possible role of myristic and oleic acids in tolerance. J Chem Ecol 29:2369–2378
Gutiérrez T, Learmonth RP, Couperwhite I (2009) Analysis of benzene-induced effects on *Rhodococcus* sp. 33 reveals that constitutive processes play a major role in conferring tolerance. Sci World J 9:209–223
Hamada T, Sameshima Y, Honda K, Omasa T, Kato J, Ohtake H (2008) A comparison of various methods to predict bacterial predilection for organic solvents used as reaction media. J Biosci Bioeng 106:357–362
Heipieper HJ, Keweloh H, Rehm HJ (1991) Influence of phenols on growth and membrane permeability of free and immobilized *Escherichia coli*. Appl Environ Microbiol 57:1213–1217
Heipieper HJ, Diefenbach R, Keweloh H (1992) Conversion of cis unsaturated fatty acids to trans, a possible mechanism for the protection of phenol-degrading *Pseudomonas putida* P8 from substrate toxicity. Appl Environ Microbiol 58:1847–1852
Heipieper HJ, Weber FJ, Sikkema J, Keweloh H, de Bont JAM (1994) Mechanisms behind resistance of whole cells to toxic organic solvents. Trends Biotechnol 12:409–415
Heipieper HJ, Meinhardt F, Segura A (2003) The *cis-trans* isomerase of unsaturated fatty acids in *Pseudomonas* and *Vibrio*: biochemistry, molecular biology and physiological function of a unique stress adaptive mechanism. FEMS Microbiol Lett 229:1–7
Heipieper HJ, Neumann G, Cornelissen S, Meinhardt F (2007) Solvent-tolerant bacteria for biotransformations in two-phase fermentation systems. Appl Microbiol Biotechnol 74:961–973
Inoue A (2011) Diversity and ecology of organic solvent tolerant microorganisms. In: Horikoshi K (ed) Extremophiles handbook. Springer Japan, Tokyo, pp 945–970
Inoue A, Horikoshi K (1989) A *Pseudomonas* thrives in high concentrations of toluene. Nature 338:264–266
Isken S, de Bont JAM (1998) Bacteria tolerant to organic solvents. Extremophiles 2:229–238

Iwabuchi N, Sunairi M, Anzai H, Nakajima M, Harayama S (2000) Relationships between colony morphotypes and oil tolerance in *Rhodococcus rhodochrous*. Appl Environ Microbiol 66:5073–5077
Iwabuchi N, Sunairi M, Urai M, Itoh C, Anzai H, Nakajima M, Harayama S (2002) Extracellular polysaccharides of *Rhodococcus rhodochrous* S-2 stimulate the degradation of aromatic components in crude oil by indigenous marine bacteria. Appl Environ Microbiol 68:2337–2343
Jucker BA, Harms H, Zehnder AJB (1996) Adhesion of the positively charged bacterium *Stenotrophomonas* (*Xanthomonas*) *maltophilia* 70401 to glass and ntarc. J Bacteriol 178:5472–5479
Junker F, Ramos J (1999) Involvement of the cis-trans isomerase Cti in the solvent resistance of *Pseudomonas putida* DOT-T1E. J Bacteriol 181:5693–5700
Kawaguchi H, Kobayashi H, Sato K (2012) Metabolic engineering of hydrophobic *Rhodococcus opacus* for biodesulfurization in oil-water biphasic reaction mixtures. J Biosci Bioeng 113:360–366
Kilbane JJ (2017) Biodesulfurization: how to make it work? AJSE 42:1–9
Kim J-S, Powalla M, Lang S, Wagner F, Lünsdorf H, Wray V (1990) Microbial glycolipid production under nitrogen limitation and resting cell conditions. J Biotechnol 13:257–266
Kim IS, Foght JM, Gray RM (2002) Selective transport and accumulation of alkanes by *Rhodococcus erythropolis* S+14He. Biotechnol Bioeng 80:650–659
Kobayashi H, Takami H, Hirayama H, Kobata K, Usami R, Horikoshi K (1999) Outer membrane changes in a toluene-sensitive mutant of toluene-tolerant *Pseudomonas putida* IH-2000. J Bacteriol 181:4493–4998
Kolouchová I, Schreiberová O, Masák J, Sigler K, Řezanka T (2012) Structural analysis of mycolic acids from phenol-degrading strain of *Rhodococcus erythropolis* by liquid chromatography-tandem mass spectrometry. Folia Microbiol 57:473–483
Korobov VV, Zhurenko EI, Zharikova NV, Iasakov TR, Markusheva TV (2017) Possibility of using phenol- and 2,4-dichlorophenol-degrading strain, *Rhodococcus erythropolis* 17S, for treatment of industrial wastewater. Mosc Univ Biol Sci Bull 72:201–205
Korshunova IO, Pistsova ON, Kuyukina MS, Ivshina IB (2016) The effect of organic solvents on the viability and morphofunctional properties of *Rhodococcus*. Appl Biochem Microbiol 52:43–50
Kundu D, Hazra C, Dandi N, Chaudhari A (2013) Biodegradation of 4-nitrotoluene with biosurfactant production by *Rhodococcus pyridinivorans* NT2: metabolic pathway, cell surface properties and toxicological characterization. Biodegradation 24:775–793
Kundu D, Hazra C, Chaudhari A (2015) Biodegradation of 2,4-dinitrotoluene with *Rhodococcus pyridinivorans* NT2: characteristics, kinetic modeling, physiological responses and metabolic pathway. RSC Adv 5:38818–38829
Kundu D, Hazra C, Chaudhari A (2016) Biodegradation of 2,6-dinitrotoluene and plant growth promoting traits by *Rhodococcus pyridinivorans* NT2: identification and toxicological analysis of metabolites and proteomic insights. Biocatal Agric Biotechnol 8:55–65
Kurosawa K, Laser J, Sinskey AJ (2015) Tolerance and adaptive evolution of triacylglycerol-producing *Rhodococcus opacus* to lignocellulose-derived inhibitors. Biotechnol Biofuels 8:76
Laane C, Boeren S, Vos K (1985) On optimizing organic solvents in multi-liquid-phase biocatalysis. *Trends Biotechnol* 3:251–252
Laane C, Boeren S, Vos K, Veeger C (1987) Rules for optimization of biocatalysis in organic solvents. Biotechnol Bioeng 30:81–87
Lambert PA (2002) Cellular impermeability and uptake of biocides and antibiotics in Gram-positive bacteria and mycobacteria. J Appl Microbiol 92:46S–54S
Lang S, Philp JC (1998) Surface active lipids in rhodococci. Antonie Van Leeuwenhoek 74:59–70
Larkin MJ, Kulakov LA, Allen CCR (2005) Biodegradation and *Rhodococcus*-masters of catabolic versatility. Curr Opin Biotechnol 16:282–290
Larkin MJ, Kulakov LA, Allen CCR (2006) Biodegradation by members of the genus *Rhodococcus*: biochemistry, physiology, and genetic adaptation. Adv Appl Microbiol 59:1–29

Leisinge T (1996) Biodegradation of chlorinated aliphatic compounds. Curr Opin Biotechnol 7:295–300

Leneva NA, Kolomytseva MP, Baskunov BP, Golovleva LA (2009) Phenanthrene and anthracene degradation by microorganisms of the genus *Rhodococcus*. Appl Biochem Microbiol 45:169–175

Li Y, Wang H, Hua F, Su M, Zhao Y (2014) Trans-membrane transport of fluoranthene by *Rhodococcus* sp. BAP-1 and optimization of uptake process. Bioresour Technol 155:213–219

Lichtinger T, Reiss G, Benz R (2000) Biochemical identification and biophysical characterization of a channel-forming protein from *Rhodococcus erythropolis*. J Bacteriol 182:764–770

Liu CW, Chang WN, Liu HS (2009) Bioremediation of n-alkanes and the formation of biofloccules by *Rhodococcus erythropolis* NTU-1 under various saline conditions and sea water. Biochem Eng J 45:69–75

Liu CW, Liang MS, Chen YC, Sayavedra-Soto LA, Liu HS (2012) Biodegradation of n-alkanes at high concentration and correlation to the accumulation of H+ ions in *Rhodococcus erythropolis* NTU-1. Biochem Eng J 63:124–128

Margesin R, Labbé D, Schinner F, Greer CW, Whyte LG (2003) Characterization of hydrocarbon-degrading microbial populations in contaminated and pristine alpine soils. Appl Environ Microbiol 69:3085–3092

Margesin R, Fonteyne PA, Redl B (2005) Low-temperature biodegradation of high amounts of phenol by *Rhodococcus* spp. and basidiomycetous yeasts. Res Microbiol 156:68–75

Marqués AM, Pinazo A, Farfan M, Aranda FJ, Teruel JA, Ortiz A, Manresa A, Espuny MJ (2009) The physicochemical properties and chemical composition of trehalose lipids produced by *Rhodococcus erythropolis* 51T7. Chem Phys Lipids 158:110–117

Martínková L, Uhnáková B, Pátek M, Nešvera J, Křen V (2009) Biodegradation potential of the genus *Rhodococcus*. Environ Int 35:162–177

McDonnell G, Russell AD (1999) Antiseptics and disinfectants: activity, action, and resistance. Clin Microbiol Rev 12:147–179

McLeod MP, Warren RL, Hsiao WWL, Araki N, Myhre M, Fernandes C et al (2006) The complete genome of *Rhodococcus* sp. RHA1 provides insights into a catabolic powerhouse. Proc Natl Acad Sci 103:15582–15587

Melchior DL (1982) Lipid phase transitions and regulation of membrane fluidity in prokaryotes. Curr Top Membr Transp 17:263–307

Metz JG, Roessler P, Facciotti D, Levering C, Dittrich F, Lassner M et al (2001) Production of polyunsaturated fatty acids by polyketide synthases in both prokaryotes and eukaryotes. Science 293:290–293

Monticello DJ (2000) Biodesulfurization and the upgrading of petroleum distillates. Curr Opin Biotechnol 11:540–546

Møretrø T, Sharifzadeh S, Langsrud S, Heir E, Rickard AH (2015) Coaggregation between *Rhodococcus* and *Acinetobacter* strains isolated from the food industry. Can J Microbiol 61:503–512

Mosqueda G, Ramos-Gonzalez M, Ramos J (1999) Toluene metabolism by solvent-tolerant *Pseudomonas putida* DOT-T1 strain and its role in solvent impermeabilization. Gene 232:69–76

Na KS, Kuroda A, Takiguchi N, Ikeda T, Ohtake H, Kato J (2005) Isolation and characterization of benzene-tolerant *Rhodococcus opacus* strains. J Biosci Bioeng 99:378–382

Nagasawa T, Shimizu H, Yamada H (1993) The superiority of the third-generation catalyst, *Rhodococcus rhodochrous* J1 nitrile hydratase, for industrial production of acrylamide. Appl Microbiol Biotechnol 40:189–195

Nielsen LE, Kadavy DR, Rajagopal S, Drijber R, Nickerson KW (2005) Survey of extreme solvent tolerance in gram-positive cocci: membrane fatty acid changes in *Staphylococcus haemolyticus* grown in toluene. Appl Environ Microbiol 71:5171–5176

Ohshiro T, Hirata T, Izumi Y (1995) Microbial desulfurization of dibenzothiophene in the presence of hydrocarbon. Appl Microbiol Biotechnol 44:249–252

Osborne SJ, Leaver J, Turner MK, Dunnill P (1990) Correlation of biocatalytic activity in an organic /aqueous two-liquid phase system with solvent concentration in the cell membrane. Enzyme Microb Technol 12:281–291

Pacífico C, Fernandes P, de Carvalho CCCR (2018) Mycobacterial response to organic solvents and possible implications on cross-resistance with antimicrobial agents. Front Microbiol 9:961

Parales RE, Haddock JD (2004) Biocatalytic degradation of pollutants. Curr Opin Biotechnol 15:374–379

Patel SB, Kilbane JJ, Webster DA (1997) Biodesulphurisation of dibenzothiophene in hydrophobic media by *Rhodococcus* sp. strain IGTS8. J Chem Technol Biotechnol 69:100–106

Pen Y, Zhang ZJ, Morales-García AL, Mears M, Tarmey DS, Edyvean RG et al (2015) Effect of extracellular polymeric substances on the mechanical properties of *Rhodococcus*. Biochim Biophys Acta 1848:518–526

Piddock LJ (2006) Multidrug-resistance efflux pumps—not just for resistance. Nat Rev Microbiol 4:629–636

Pini F, Grossi C, Nereo S, Michaud L, Giudice AL, Bruni V, Baldi F, Fani R (2007) Molecular and physiological characterization of psychrotrophic hydrocarbon-degrading bacteria isolated from Terra Nova Bay (Antarctica). Eur J Soil Biol 43:368–379

Poole K (2008) Bacterial multidrug efflux pumps serve other functions. Microbe 3:179–185

Portevin D, de Sousa-D'Auria C, Houssin C, Grimaldi C, Chami M, Daffé M et al (2004) A polyketide synthase catalyzes the last condensation step of mycolic acid biosynthesis in mycobacteria and related organisms. Proc Natl Acad Sci USA 101:314–319

Porto B, Maass D, Oliveira JV, de Oliveira D, Yamamoto CI, Ulson de Souza AA et al (2017) Heavy gas oil biodesulfurization by *Rhodococcus erythropolis* ATCC 4277: optimized culture medium composition and evaluation of low-cost alternative media. J Chem Technol Biotechnol 92:2376–2382

Pospíšilová D, Schreiberová O, Jirků V, Lederer T (2015) Effects of magnetic field on phenol biodegradation and cell physiochemical properties of *Rhodococcus erythropolis*. Biorem J 19:201–206

Prieto MB, Hidalgo A, Rodriguez-Fernandez C, Serra JL, Llama MJ (2002) Biodegradation of phenol in synthetic and industrial wastewater by *Rhodococcus erythropolis* UPV-1 immobilized in an air-stirred reactor with clarifier. Appl Microbiol Biotechnol 58:853–859

Pucci OH, Bak MA, Peressutti SR, Klein I, Hartig C, Alvarez HM, Wunsche L (2000) Influence of crude oil contamination on the bacterial community of semiarid soils of Patagonia (Argentina). Acta Biotechnol 20:129–146

Ramos JL, Duque E, Gallegos MT, Godoy P, Ramos-Gonzalez MI, Rojas A, Teran W, Segura A (2002) Mechanisms of solvent tolerance in gram-negative bacteria. Annu Rev Microbiol 56:743–768

Rapp P, Bock H, Wray V, Wagner F (1979) Formation, isolation and characterization of trehalose dimycolates from *Rhodococcus erythropolis* grown on *n*-alkanes. Microbiology 115:491–503

Rehfuss M, Urban J (2005) *Rhodococcus phenolicus* sp nov., a novel bioprocessor isolated actinomycete with the ability to degrade chlorobenzene, dichlorobenzene and phenol as sole carbon sources. Syst Appl Microbiol 28:695–701

Rieβ FG, Elflein M, Benk M, Schiffler B, Benz R, Garton N et al (2003) The cell wall of the pathogenic bacterium *Rhodococcus equi* contains two channel-forming proteins with different properties. J Bacteriol 185:2952–2960

Rodgers RP, Blumer EN, Emmett MR, Marshall AG (2000) Efficacy of bacterial bioremediation: demonstration of complete incorporation of hydrocarbons into membrane phospholipids from *Rhodococcus* hydrocarbon degrading bacteria by electrospray ionization fourier transform ion cyclotron resonance mass spectrometry. Environ Sci Technol 34:535–540

Rodrigues CJC, de Carvalho CCCR (2015) *Rhodococcus erythropolis* cells adapt their fatty acid composition during biofilm formation on metallic and non-metallic surfaces. FEMS Microbiol Ecol 91:fiv135

Rubashko GE, Kolomytseva MP, Golovleva LA (2006) Improvement of the process of ntarcti degradation by *Rhodococcus rhodochrous* strain 172. Appl Biochem Microbiol 42:396–398
Russell NJ (1988) Functions of lipids: structural roles and membrane functions. In: Ratledge C, Wilkinson SG (eds) Microbial lipids, vol 2. Academic Press, London, pp 279–365
Russell AD (1995) Mechanisms of bacterial resistance to biocides. Int Biodeter Biodegrad 36:247–265
Saier MH (2000) A functional-phylogenetic classification system for transmembrane solute transporters. Microbiol Mol Biol Rev 64:354–411
Sameshima Y, Honda K, Kato J, Omasa T, Ohtake H (2008) Expression of *Rhodococcus opacus* alkB genes in anhydrous organic solvents. J Biosci Bioeng 106:199–203
Sardessai Y, Bhosle S (2002) Tolerance of bacteria to organic solvents. Res Microbiol 153:263–268
Sardessai YN, Bhosle S (2004) Industrial potential of organic solvent tolerant bacteria. Biotechnol Prog 20:655–660
Segura A, Duque E, Mosqueda G, Ramos JL, Junker F (1999) Multiple responses of Gram-negative bacteria to organic solvents. Environ Microbiol 1:191–198
Serrano-González MY, Chandra R, Castillo-Zacarias C, Robledo-Padilla F, Rostro-Alanis MJ, Parra-Saldivar R (2018) Biotransformation and degradation of 2,4,6-trinitrotoluene by microbial metabolism and their interaction. Def Technol 14:151–164
Sikkema J, Weber FJ, Heipieper HJ, de Bont JAM (1994) Cellular toxicity of lipophilic compounds: mechanisms, implications, and adaptations. Biocatalysis 10:113–122
Sikkema J, de Bont JAM, Poolman B (1995) Mechanisms of membrane toxicity of hydrocarbons. Microbiol Rev 59:201–222
Sinensky M (1974) Homeoviscous adaptation—a homeostatic process that regulates the viscosity of membrane lipids in *Escherichia coli*. Proc Natl Acad Sci USA 71:522–525
Sokolovská I, Rozenberg R, Riez C, Rouxhet PG, Agathos SN, Wattiau P (2003) Carbon source-induced modifications in the mycolic acid content and cell wall permeability of *Rhodococcus erythropolis* E1. Appl Environ Microbiol 69:7019–7027
Soleimani M, Bassi A, Margaritis A (2007) Biodesulfurization of refractory organic sulfur compounds in fossil fuels. Biotechnol Adv 25:570–596
Solyanikova IP, Mulyukin AL, Suzina NE, El-Registan GI, Golovleva LA (2011) Improved xenobiotic-degrading activity of *Rhodococcus opacus* strain 1cp after dormancy. J Environ Sci Health B 46:638–647
Sonnleitner B (1998) Dynamic adaptation of microbes. J Biotechnol 65:47–60
Stachurski J, Michalek M (1996) The effect of the zeta potential on the stability of a non-polar oil-in-water emulsion. J Colloid Interface Sci 184:433–436
Stancu MM (2014) Physiological cellular responses and adaptations of *Rhodococcus erythropolis* IBB_{Po1} to toxic organic solvents. J Environ Sci (China) 26:2065–2075
Stewart PS, Costerton JW (2001) Antibiotic resistance of bacteria in biofilms. Lancet 358:135–138
Thomassin-Lacroix EJM, Yu ZT, Eriksson M, Reimer KJ, Mohn WW (2001) DNA-based and culture-based characterization of a hydrocarbon-degrading consortium enriched from Arctic soil. Can J Microbiol 47:1107–1115
Tischler D, Niescher S, Kaschabek SR, Schlömann M (2013) Trehalose phosphate synthases OtsA1 and OtsA2 of *Rhodococcus opacus* 1CP. FEMS Microbiol Lett 342:113–122
Torres S, Pandey A, Castro GR (2011) Organic solvent adaptation of Gram positive bacteria: applications and biotechnological potentials. Biotechnol Adv 29:442–452
Tsitko IV, Zaitsev GM, Lobanok AG, Salkinoja-Saloneni MS (1999) Effect of aromatic compounds on cellular fatty acid composition of *Rhodococcus opacus*. Appl Environ Microbiol 65:853–855
Urai M, Yoshizaki H, Anzai H, Ogihara J, Iwabuchi N, Harayama S, Sunairi M, Nakajima M (2007) Structural analysis of an acidic, fatty acid ester-bonded extracellular polysaccharide produced by a ntarcti-assimilating marine bacterium, *Rhodococcus erythropolis* PR4. Carbohydr Res 342:933–942
van der Geize R, Dijkhuizen L (2004) Harnessing the catabolic diversity of rhodococci for environmental and biotechnological applications. Curr Opin Microbiol 7:255–261
van Oss CJ (1995) Hydrophobicity of biosurfaces—origin, quantitative determination and interaction energies. Colloids Surf B Biointerfaces 5:91–110

Vermuë M, Sikkema J, Verheul A, Bakker R, Tramper J (1993) Toxicity of homologous series of organic solvents for the gram-positive bacteria *Arthrobacter* and *Nocardia* sp. and the gram-negative bacteria *Acinetobacter* and *Pseudomonas* sp. Biotechnol Bioeng 42:747–758
Vorbeck C, Lenke H, Fischer P, Spain JC, Knackmuss H-J (1998) Initial reductive reactions in aerobic microbial metabolism of 2,4,6-trinitrotoluene. Appl Environ Microbiol 64:246–252
Voss I, Steinbüchel A (2001) High cell density cultivation of *Rhodococcus opacus* for lipid production at a pilot-plant scale. Appl Microbiol Biotechnol 55:547–555
Wang L, Qiao N, Sun FQ, Shao ZZ (2008) Isolation, gene detection and solvent tolerance of benzene, toluene and xylene degrading bacteria from nearshore surface water and Pacific Ocean sediment. Extremophiles 12:335–342
Warhurst AM, Fewson CA (1994) Biotransformations catalyzed by the genus *Rhodococcus*. Crit Rev Biotechnol 14:29–73
Weathers TS, Higgins CP, Sharp JO (2015) Enhanced biofilm production by a toluene-degrading *Rhodococcus* observed after exposure to perfluoroalkyl acids. Environ Sci Technol 49:5458–5466
Weber FJ, de Bont JAM (1996) Adaptation mechanisms of microorganisms to the toxic effects of organic solvents on membranes. Biochim Biophys Acta 1286:225–245
Weber FJ, Isken S, de Bont JAM (1994) Cis/trans isomerization of fatty acids as a defence mechanism of *Pseudomonas putida* strains to toxic concentrations of toluene. Microbiology 140:2013–2017
Whyte LG, Slagman SJ, Pietrantonio F, Bourbonnière L, Koval SF, Lawrence JR, Inniss WE, Greer CW (1999) Physiological adaptations involved in alkane assimilation at low temperatures by *Rhodococcus* sp. strain Q15. Appl Environ Microbiol 65:2961–2968
Whyte LG, Schultz A, van Beilen JB, Luz AP, Pellizari D, Labbé D, Greer CW (2002) Prevalence of alkane monooxygenase genes in arctic and antarctic hydrocarbon-contaminated and pristine soils. FEMS Microbiol Ecol 41:141–150
Withell ER (1942) The significance of variation in the shape of the time—survivor curves. J Hyg 42:124–132
Yamashita S, Satoi M, Iwasa Y, Honda K, Sameshima Y, Omasa T, Kato J, Ohtake H (2007) Utilization of hydrophobic bacterium *Rhodococcus opacus* B-4 as whole-cell catalyst in anhydrous organic solvents. Appl Microbiol Biotechnol 74:761–767
Yoneda A, Henson WR, Goldner NK, Park KJ, Forsberg KJ, Kim SJ et al (2016) Comparative transcriptomics elucidates adaptive phenol tolerance and utilization in lipid-accumulating *Rhodococcus opacus* PD630. Nucleic Acids Res 44:2240–2254
Yoon JH, Cho YG, Kang SS, Kim SB, Lee ST, Park YH (2000) *Rhodococcus koreensis* sp. nov., a 2,4-dinitrophenol-degrading bacterium. Int J Syst Evol Microbiol 50:1193–1201
Zhang J, Sun Z, Li Y, Peng X, Li W, Yan Y (2009) Biodegradation of p-nitrophenol by *Rhodococcus* sp. CN6 with high cell surface hydrophobicity. J Hazard Mater 163:723–728

Degradation of Alkanes in *Rhodococcus*

Martina Cappelletti, Stefano Fedi, and Davide Zannoni

Contents

Abstract Alkanes are widely distributed in the environment as they not only constitute the large fraction of crude oil but are also produced by many living organisms. They are saturated hydrocarbons of different sizes and structures, which pose a variety of challenges to degradative microorganisms due to their physicochemical properties, i.e., the extremely limited solubility and the high energy required for activation. The hydrophobic cell surface of *Rhodococcus* spp., the ability to produce biosurfactants, and the possession of a wide range of oxygenases allow coping with such challenges. In particular, monooxygenase enzymes are involved in the activation of alkanes by converting them into alcohols, which undergo a series of oxidation steps before being converted to fatty acids. *Rhodococcus* alkane monooxygenases belong to different families (i.e., AlkB-like monooxygenase, soluble di-iron monooxygenase, cytochrome P450), have different genetic organization, and are subject to different regulatory mechanisms, which are poorly known. Because of their long-term survival capacity, broad catabolic abilities, and effective contact mechanisms with hydrocarbon molecules, alkanotrophic *Rhodococcus* strains have biotechnology applications and potential in bioremediation and biotransformation reactions.

M. Cappelletti (✉) · S. Fedi · D. Zannoni
Department of Pharmacy and Biotechnology, University of Bologna, Bologna, Italy
e-mail: martina.cappelletti2@unibo.it; stefano.fedi@unibo.it; davide.zannoni@unibo.it

H. M. Alvarez (ed.), *Biology of Rhodococcus*, Microbiology Monographs 16,
https://doi.org/10.1007/978-3-030-11461-9_6

1 Introduction

Alkanes are saturated hydrocarbons with the general formula C_nH_{2n+2} consisting of carbon and hydrogen atoms; they can be linear (*n*-alkanes), cyclic (cycloalkanes), or branched (*iso*-alkanes). The *n*-alkanes, methane (C_1), ethane (C_2), propane (C_3), and butane (C_4) are gaseous at room temperature, while *n*-alkanes ranging between C_5 (pentane) and C_{17} (heptadecane) are liquid, and those $>C_{18}$ (octadecane) are solid (the subindex indicates the number of carbon atoms of the *n*-alkane molecule).

Alkanes are the most abundant hydrocarbons (with an estimated abundance of 20–50%) in crude oil. They are also naturally produced by living organisms (plants, algae, and bacteria) as waste products and chemoattractants or for structural and defense purposes (van Beilen et al. 2003). In relation to this, alkanes have been detected in most soil and water systems; bacterial oxidation of alkanes is a very common phenomenon in natural environments and a major process in geochemical terms (van Beilen et al. 2003).

As alkanes are apolar and chemically inert compounds, their utilization by microorganisms faces significant challenges in relation to the energy required to activate them, their limited solubility, and their tendency to accumulate in cell membranes (Rojo 2009). Due to this, microorganisms with the capacity to degrade alkanes are of great interest in addition to the fact that microbial degradation of long-chain *n*-alkanes remains the main process for the remediation of oil-contaminated areas (Leahy and Colwell 1990; Binazadeh et al. 2009).

The solubility of alkanes decreases as molecular weight increases and aliphatic linear alkanes are normally biodegraded more rapidly as compared to multiple-branched alkanes such as pristane (2,6,10,14-tetramethylpentadecane) and phytane (2,6,10,14-tetramethylhexadecane) (Leahy and Colwell 1990).

Rhodococci are frequently isolated from petroleum hydrocarbon-contaminated environments where they play an important role in the degradation of aliphatic hydrocarbons including alkanes (Hamamura et al. 2006; Yuste et al. 2000). In this respect, many *Rhodococcus* species have been described to be able to use a wide range of alkanes as carbon and energy sources with high efficiency and growth rate (Table 1) (Binazadeh et al. 2009). Molecular methods, targeting 16S rRNA and key functional genes, further highlighted the wide distribution of *Rhodococcus* spp. in diverse hydrocarbon-contaminated environments (Hamamura et al. 2006, 2008), including oil-contaminated soils, rhizosphere, wastewater, coastal sediments, and seawater (Hassanshahian et al. 2013; Acosta-González et al. 2016; Viggor et al. 2015; Mikolasch et al. 2015). Despite the low temperatures generally limit hydrocarbon biodegradation, alkanotrophic *Rhodococcus* spp. were found in cold petroleum-polluted ecosystems (antarctic and alpine soils) (Margesin et al. 2003; Luz et al. 2004; Habib et al. 2018), and many *Rhodococcus* strains were able to metabolize hydrocarbons at $T < 5$ °C (Whyte et al. 1998, 1999). A large diversity can be observed among different *Rhodococcus* strains in terms of substrate specificity, enzymatic systems, and physiological response. The wide distribution and high presence of *Rhodococcus* spp. in oil-contaminated environments have been

Table 1 List of the main alkanotrophic *Rhodococcus* strains described in the literature along with their genetic and functional characteristics

Strain	Alkanes used as growth substrate[a]	Microbial features related to alkane oxidation		References
		Genetic data[a]	Functional data	
R. jostii RHA1	C_3	One *alkB* gene One *prm* gene cluster	RHA1 *prm* gene is transcriptionally induced during the growth on C_3 and cometabolizes NDMA	Sharp et al. (2007) and Ekprasert (2014)
R. ruber SP2B	C_5–C_{21}	Two *alkB* genes	The *n*-hexane degradation capacity of SP2B is inducible	Amouric et al. (2010)
R. qingshengii Q15	C_8–C_{32}, branched alkanes, substituted cyclohexane	Four *alkB* genes Two CYP153 genes	Q15 possesses multiple *alkB* genes with quite sequence divergence; it performs both terminal and subterminal oxidation of alkanes; it has low temperature alkane degradation range	Whyte et al. (1998, 2002a, b)
R. erythropolis NRRL B-16531	C_6–C_{36}	Four *alkB* genes Two CYP153 genes	NRRL B-16531 is able to grow on medium- to long-chain *n*-alkanes and possesses multiple *alkB* genes with quite sequence divergence	Whyte et al. (2002b) and van Beilen et al. (2002b)
R. erythropolis T7-2	C_{12}–C_{36}	ND	T7-2 is able to use diesel oil as its sole carbon and energy source at 15 °C	Huang et al. (2008)
Rhodococcus sp. strain SoD	C_{12}–C_{28}	One *alkB* gene	SoD is able to use diesel oil as sole carbon and energy source	Quatrini et al. (2008)
R. aetherivorans BCP1	C_2–C_{28}	Two *alkB* genes One *smo* gene cluster One *prm* gene cluster	BCP1 has at least three different alkane monooxygenases involved in the oxidation of *n*-alkanes C_3-C_{22}; it performs terminal and subterminal oxidation of propane and terminal oxidation of butane	Cappelletti et al. (2011, 2015)
R. opacus R7	C_{12}, C_{14}, C_{20}, C_{24}	One *alkB* gene One *prm*-like gene	R7 *alkB* gene is involved in the oxidation of medium- and long-chain *n*-alkanes	Di Gennaro et al. (2014) and Zampolli et al. (2014)
R. erythropolis PR4	C_8, C_{12}, C_{19}, C_{16}	Four *alkB* genes Two CYP153 genes and other mono- and di-oxygenase genes	The multiple monooxygenase genes of PR4 are differently induced after the growth on diesel oil, hexadecane, or hydrocarbon mixtures	Sekine et al. (2006) and van Beilen et al. (2006)

(continued)

Table 1 (continued)

Strain	Alkanes used as growth substrate[a]	Microbial features related to alkane oxidation		References
		Genetic data[a]	Functional data	
R. rhodochrous NCIMB 12566	C_6–C_{16}	One CYP450 gene	NCIMB 12566 produces a cytochrome P450 able to hydroxylate *n*-octane	Cardini and Jurtshuk (1970)
Rhodococcus sp. strain 1BN	C_6, C_8, C_{16}–C_{28}	One *alkB* gene	1BN degrades medium- (C_6) and long-chain alkanes (C_{16}–C_{28})	Andreoni et al. (2000)
Rhodococcus sp. strain T12	Pristane	ND	T12 degrades pristane at low temperatures	Kunihiro et al. (2005)
Rhodococcus sp. strain TMP2	C_9–C_{24}	Five *alkB* genes	Three *alkB* genes in TMP2 are transcriptionally induced by *n*-alkanes and pristane. TMP2 strain is able to degrade pristane at low temperatures	Takei et al. (2008)
Rhodococcus sp. strain RR12	C_{14}–C_{34}	ND	RR12 is able to grow on high molecular mass hydrocarbon mixture	Yuste et al. (2000)
Rhodococcus sp. strain RR14	C_{14}–C_{34}, pristane	ND	RR14 is able to grow on high molecular mass hydrocarbon mixture	Yuste et al. (2000)
Rhodococcus sp. strain NCIM 5126	C_{13}–C_{17}, pristane	ND	NCIM 5126 metabolizes *n*-alkanes through both bi-terminal and mono-terminal oxidation pathway while it metabolizes pristane only through mono-terminal oxidation pathway	Sharma and Pant (2000)
R. fascians 154-S	C_6–C_{24}	Four *alkB* genes	154-S is able to grow on medium- to long-chain *n*-alkanes and possesses multiple *alkB* genes	Van Beilen et al. (2003)
R. fascians 115-H	C_6–C_{32}	Three *alkB* genes	Growth on long-chain *n*-alkanes	Van Beilen et al. (2003)
R. erythropolis 42-O	C_6–C_{32}	Five *alkB* genes	42-O is able to grow on medium- to long-chain *n*-alkanes and possesses multiple *alkB* genes	Van Beilen et al. (2003)
R. erythropolis 50-V	C_6–C_{32}	Four *alkB* genes	50-V is able to grow on medium- to long-chain *n*-alkanes and possesses multiple *alkB* genes	Van Beilen et al. (2003)
R. erythropolis NRRL B-16531	C_6–C_{36}	Four *alkB* genes	NRRL B-16531 is able to grow on medium- to long-chain *n*-alkanes and possesses multiple *alkB* genes	Van Beilen et al. (2003)

(continued)

Table 1 (continued)

Strain	Alkanes used as growth substrate[a]	Microbial features related to alkane oxidation		References
		Genetic data[a]	Functional data	
R. erythropolis 23-D	C_6–C_{36}	Five *alkB* genes	23-D is able to grow on medium- to long-chain *n*-alkanes and possesses multiple *alkB* genes	Van Beilen et al. (2003)
R. erythropolis 35-O	C_6–C_{16}	Five *alkB* genes	35-O is able to grow on medium- to long-chain *n*-alkanes and possesses multiple *alkB* genes	Van Beilen et al. (2003)
R. erythropolis S+14He	C_9–C_{30}, pristane	ND	S+14He is able to degrade specific *n*-alkanes within a mixture; *n*-alkane molecules seem to be accumulated inside the S+14He cell	Kim et al. (2002)
R. opacus B4	C_8, C_{10}	Two *alkB* genes	The activity of the promoter of the two B4 *alkB* genes was investigated in anhydrous organic solvents	Sameshima et al. (2008)
Rhodococcus sp. strain p52	C_{14}, C_{24}, C_{32}, pristine	Two *alkB* genes; One CYP185 gene	p52 *CYP*185 gene is transcriptionally induced on long-chain *n*-alkanes and pristine	Yang et al. (2014)
R. rhodochrous PNKb1	C_3	ND	PNKb1 performs both terminal and subterminal propane oxidation	Woods and Murrell (1989)
R. erythropolis SK121	ND	Six *alkB* genes Five CYP153 genes	SK121 possesses multiple *alkB* genes	Nie et al. (2014)
R. erythropolis ATCC 4277	C_5–C_{16}	ND	ATCC 4277 performs only subterminal *n*-alkane oxidation	Ludwig et al. (1995)
Rhodococcus sp. strain EH831	C_6	ND	EH831 performs only subterminal *n*-alkane oxidation	Lee et al. (2010)
Rhodococcus sp. strain Moj-3449	C_{10}–C_{35}	ND	Moj-3449 performs only subterminal *n*-alkane oxidation	Binazadeh et al. (2009)
Rhodococcus sp. strain BPM 1613	Pristane ($C_{19}H_{40}$)	ND	BPM 1613 converts pristane to terminal oxidation metabolic intermediates which are then degraded through either β- or ω-oxidation.	Nakajima et al. (1985)

(continued)

Table 1 (continued)

Strain	Alkanes used as growth substrate[a]	Microbial features related to alkane oxidation		References
		Genetic data[a]	Functional data	
R. ruber SBUG 82	Pristane ($C_{19}H_{40}$)	ND	SBUG 82 performs both terminal and subterminal oxidation of pristane	Nhi-Cong et al. (2009)

[a]The subindex indicates the number of carbon atoms of the *n*-alkane molecule; *NDMA* N-nitrosodimethylamine; pristine = $C_{19}H_{40}$

associated not only to their wide catabolic abilities but also to their hydrophobic cell surfaces and their capacity to respond to various stress conditions for long-term survival in harsh environments, i.e., metal resistance, high salinity, desiccation, high concentration of organic solvents, and biosurfactant production (de Carvalho et al. 2005; Quatrini et al. 2008; Sameshima et al. 2008; Cappelletti et al. 2016). Further, in some cases (mainly with alkanes >C_5), it was shown that *Rhodococcus* alkane degradation pathways were not affected by catabolic repression, so that alkanes were efficiently degraded also in the presence of alternative carbon sources (Warhust and Fewson 1994). Several *Rhodococcus* strains were also shown to produce high value-added compounds during the growth on *n*-alkanes, i.e., triacylglycerols and wax esters. Taken together, these features identify *Rhodococcus* genus as an ideal candidate for bioremediation, microbial-enhanced oil recovery, and hydrocarbon bioconversion to valuable products. Considerable interest has therefore been devoted to the study of the physiology of alkanotrophic rhodococci and their enzymatic systems involved in alkane transformation.

2 Physical Interaction Between *Rhodococcus* Cells and Alkanes

To be metabolized, alkanes must cross first the outer hydrophilic cell wall of Gram-positive bacteria and then penetrate the periplasmic cell membrane to get access to cytosol. The alkane bioavailability is therefore strongly limited by the hydrophobic character of these compounds so that the cell surface needs to favor the contact between these water-insoluble molecules for an efficient biodegradation process (Stroud et al. 2007). Apparently, all microorganisms able to utilize hydrocarbons, like *Rhodococcus* spp., have overcome this bottleneck through several adaptive mechanisms, here briefly outlined and summarized in Fig. 1.

While the entry into cells of alkane molecules from the water phase is thought to be possible only for low-molecular-weight alkanes that are partly soluble in water, e.g., propane and butane, *Rhodococcus* cells gain access to medium- and long-chain *n*-alkanes by adhering to alkane droplets and/or by surfactant-facilitated access

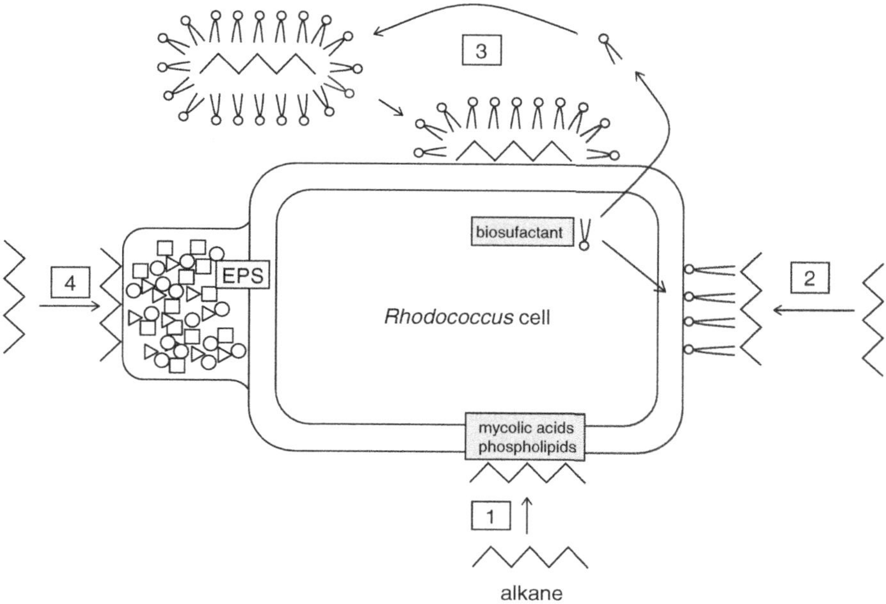

Fig. 1 Mechanisms of *Rhodococcus* cell contact with alkanes. (1) The contact is associated to an increase of cell membrane hydrophobicity due to the change of mycolic acids and phospholipids composition. (2) The contact is mediated by biosurfactants that after being produced remain associated to the cell wall. (3) The biosurfactants are excreted from the cell and allow the emulsification of the alkane molecules. (4) The production of extracellular polymeric substances (EPS) enhances the alkane-cell contact

through alkane emulsification. In these latter cases, a first mechanism would imply the direct contact of cells with large hydrocarbon droplets; alternatively, the biosurfactant activity would allow the entry of alkanes into cells by supporting the cell contact with fine oil droplets formed through emulsification process (Hua and Wang 2014).

Growth of *Rhodococcus* cells on alkanes has frequently been reported to induce the formation of surface-active compounds of diverse chemical nature, which improve the utilization of these hydrophobic compounds as growth substrates (Lang and Philp 1998; Yakimov et al. 1999; Rapp and Gabriel-Jürgens 2003; Peng et al. 2007). Rhodococci typically produce trehalose-containing glycolipid biosurfactants as extracellular or cell wall-associated compounds. In those *Rhodococcus* spp. where the biosurfactants are excreted from the cell, emulsification of the alkane with aqueous culture medium results in a very large surface area of contact between cell and hydrocarbon. More frequently, however, molecules with biosurfactant properties are retained by the *Rhodococcus* cell and localized on the outer cell surface, so that the cell surface increases its hydrophobic character facilitating the attachment and subsequent entry of alkanes to be used as growth substrate (Kretschmer et al. 1982; Bredholt et al. 1998; Lang and Philp 1998). On the other hand, the direct contact between bacterial cells and hydrocarbons depends on the interfacial area extent, which in turn is strictly related to cell hydrophobicity. For example, it has been reported that in *Rhodococcus* sp. strain TMP2, a strain unable to produce biosurfactants, the direct contact of its cell surface with pristane was fundamental for an effective assimilation and growth on this branched alkane (Takei et al. 2008). The high hydrophobicity of the cell surface has been related to (1) the aggregation of *Rhodococcus* cells in flocs, which controls the interfacial alkane entry mechanisms (Bouchez-Naïtali et al. 2001), and (2) the ability of some *Rhodococcus* strains to get into the oil phase when they are grown in aqueous/organic biphasic systems, the organic phase being composed of long-chain liquid alkanes, e.g., *n*-hexadecane (Yamashita et al. 2007; Iwabuchi et al. 2009). An increase of the cell surface hydrophobicity has been widely reported in *Rhodococcus* spp. cultivated on alkanes, and an important role in this change has been assigned to the cell wall lipids and mycolic acids (Stratton et al. 2003; Kolouchová et al. 2012). In particular, both mycolic acids and phospholipids were proposed to affect the membrane permeability as well as the passive entry and/or active transport of hydrophobic substrates (Sokolovská et al. 2003). Several studies have demonstrated that the membrane composition and cell surface properties differed after feeding the cells with different alkanes (de Carvalho et al. 2005; Iwabuchi et al. 2009; Cortes and de Carvalho 2015). The growth on alkanes and change of the cell envelope permeability were also observed in parallel with an increase of the antibiotic resistance in cells of *R. ruber* and *R. erythropolis* (Kuyukina et al. 2000; Sokolovská et al. 2003). This effect was mainly linked to a decreased penetration of hydrophilic antibiotic molecules into *Rhodococcus* cells after they were grown on alkanes (Sokolovská et al. 2003).

In addition to cell wall components, the production of extracellular polymeric substances (EPS) with surface-active properties was shown to play an important role in the ability of *Rhodococcus* cells to adhere to hydrocarbons (Neu 1996; Whyte et al. 1999). EPS is known to increase the bacterial tolerance toward organic solvents and, more in general, toward various stress conditions, i.e., metals and antimicrobials (Iwabuchi et al. 2009). Growth on diesel fuel of *Rhodococcus* sp. strain Q15 [recently affiliated to *R. qingshengii* (Táncsics et al. 2015)] was linked to production of EPS clusters surrounding and connecting cells, so to enhance their contact with hydrophobic substrates (Whyte et al. 1999). In *R. erythropolis* PR4, the expression of genes involved in EPS biosynthesis was upregulated during the growth on alkanes and hydrocarbon mixtures (Laczi et al. 2015).

Hydrocarbons that are adsorbed on the surface of microbial cells can go across the cytoplasmic membrane through the use of three main mechanisms, namely: (1) passive diffusion; (2) passive-facilitated diffusion; or (3) energy-dependent uptake (Hua and Wang 2014). To date, very few studies have addressed the problem linked to the entry of hydrocarbons into *Rhodococcus* cells. Most of the studies have reported that the increase of cell surface hydrophobicity (induced by biosurfactants, mycolic acid, phospholipid, EPS production) might facilitate the attachment and subsequent transport of alkanes into the cell without the need of energy and specialized membrane component (Hommel 1990; Lang and Philps 1998). Only one study suggested the presence of a respiratory energy-driven selective transport of *n*-hexadecane (C_{16}) into

R. erythropolis strain S+14He cells, this process being repressed by the cytochrome oxidase inhibitor azide (N_3^-) when used at 30 mM (Kim et al. 2002). Further details and studies are needed to clarify the mechanisms of possible respiratory-dependent hydrocarbon uptake mechanisms in *Rhodococcus* spp.

3 Enzymes Responsible for Alkane Oxidation in *Rhodococcus*

Hydroxylation of alkanes by alkane monooxygenases (MOs) represents the first and most critical step in aerobic alkane metabolism. Oxygen acts as a reactant to activate alkane molecules so that alkanes are converted to alkanols, which can be further metabolized providing both carbon and an energy source to cells. In particular, the oxidation of *n*-alkanes ($>C_2$) occurs at the terminal and/or subterminal carbon of the aliphatic chain resulting in the production of a primary alcohol, of a secondary alcohol, or of a mixture of the two isomers, depending on the enzymatic activity and selectivity of the alkane MOs involved in the first alkane oxidation step.

Based on the number and type of cofactors, protein subunits, redox partners, and cellular location, different MO superfamilies can be distinguished in bacteria (Coleman et al. 2006, 2011). Among these, three MO superfamilies are typically involved in alkane oxidation in *Rhodococcus* spp., namely: (1) alkane hydroxylases (AlkB), (2) soluble di-iron monooxygenases (SDIMO), and (3) cytochromes P450 of the CYP153 ($CY_{tochrome}$ P_{450}) group. While AlkB-type MOs and CYP153 enzymes are typically involved in oxidation of liquid alkanes (C_5–C_{16}), SDIMO act on short-chain hydrocarbons [gaseous and short-chain liquid alkanes and alkanes ($<C_8$)] (Coleman et al. 2011). For the only gaseous alkane metabolism, an additional enzyme was proposed to be involved in the initial oxidation step of propane and isobutane in *R. rhodochrous* ATCC 21198. This enzyme was suggested to be an intermolecular dioxygenase catalyzing the reaction of two molecules of alkane with one molecule of oxygen (see Sect. 3.4) (Babu and Brown 1984); however, no further studies were performed on the genetics and biochemistry of this putative dioxygenase.

In recent years, to assess the cell biodegradation potential and activity in situ, different molecular methods, based on the use of degenerate primers or hybridization probes targeting the different types of alkane monooxygenase, have been largely exploited (Smits et al. 1999; Vomberg and Klinner 2000; Kohno et al. 2002). Through these methods, the high abundance and activity of *Rhodococcus* spp. in diverse oil-contaminated environments was revealed. Additionally, advances in genome sequencing procedures along with the parallel increase in the amount of genomic data related to *Rhodococcus* strains have exponentially extended our knowledge on the alkane MO enzymes in this genus.

3.1 Membrane Non-heme Di-iron Alkane Hydroxylases (AlkB)

AlkB is a membrane alkane monooxygenase, which contains a conserved non-heme di-iron cluster in the active site catalyzing the oxidation of *n*-alkanes to the corresponding alkanols. While one atom of molecular dioxygen is introduced in the substrate by the AlkB MO, the second oxygen atom is reduced to water using two electrons from NADH+H$^+$, these reducing equivalents delivered to MO through the activity of the electron transfer components rubredoxin and rubredoxin reductase (van Beilen et al. 2002a; van Beilen and Funhoff 2007). Together, the three components, AlkB MO, rubredoxin, and rubredoxin reductase, form the integral membrane non-heme iron alkane hydroxylase. In this respect, the primary amino acid sequences of AlkB proteins present six hydrophobic membrane-spanning regions along with eight highly conserved histidine residues, the latter being essential for the iron coordination in the active site of MO (Shanklin et al. 1994, 1997).

AlkB hydroxylase is encoded by *alkB* gene, which has been described as the most common and widely distributed gene among all the alkane degraders (Smits et al. 1999; Kohno et al. 2002; Kloos et al. 2006; Amouric et al. 2010; Viggor et al. 2015). The *alkB* gene has been therefore used as favorite target for the description of alkane-degrading community composition, determination of the abundance of oil-degrading bacteria, and evaluation of the alkane biodegradation potential (Smits et al. 1999, 2002; van Beilen et al. 2002b; Kloos et al. 2006; Hamamura et al. 2008; Nie et al. 2014; Cappelletti et al. 2017a).

In the literature, *Rhodococcus* is among the most cited bacterial genera in relation to *alkB* gene identification, and *Rhodococcus alkB* sequences have been often described as the most abundant in studies describing the diversity of alkane hydroxylase genes in environments contaminated with crude oil and related pollutants (Whyte et al. 2002a; Hamamura et al. 2008; Luz et al. 2004; Quatrini et al. 2008; Jurelevicius et al. 2013; Cappelletti et al. 2017a) and also in pristine soils (Kloos et al. 2006).

Starting from *Rhodococcus* strains containing more *alkB* homologous genes, AlkB MO have been distinguished into different groups (AlkB1–AlkB7) based on the phylogenetic analysis of their sequences, often being related to similar role in hydrocarbon assimilation in respect of preferred sizes of alkanes. Panicker et al. (2010) reported that the distribution of different *alkB* gene groups in rhodococcal population was correlated with the types of alkanes present in the soil.

Also based on the above reported observations, several authors concluded that the *alkB* gene belongs to *Rhodococcus* core genome, which includes those genes present in all rhodococci (Nie et al. 2014; Orro et al. 2015). The study by Táncsics et al. (2015) suggested that *alkB* gene can be used as phylogenetic marker for *Rhodococcus* taxonomy resolution. The possible use of *alkB* as a tool for *Rhodococcus* systematics highlights the importance and conservation of this gene during the evolution of *Rhodococcus* spp. and the wide distribution of the *alkB* gene within this genus.

3.1.1 Organization and Expression of *alkB* Genes in *Rhodococcus*

In *Rhodococcus* spp., the *alkB* gene is typically arranged in a cluster with genes coding for two separate rubredoxins (*alkG1*, *alkG2* or *rubA1*, *rubA2*) and a rubredoxin reductase (*alkT* or *rubB*) (Fig. 2a). These genes are organized in a head-to-tail module, and the ORFs have overlapping stop and start codons. The organization of the genes within the *alk* cluster suggested their translational coupling aimed at the production of stoichiometric amounts of proteins and at the coupled regulation of their activity (Whyte et al. 2002b).

Many *Rhodococcus* spp. possessing one or more *alk* gene cluster(s) showed additional *alkB* genes not comprised within an *alk* gene cluster. For instance, four homologous alkane monooxygenase genes (*alkB1*, *alkB2*, *alkB3*, and *alkB4*) were initially identified in *R. qingshengii* Q15 and *R. erythropolis* NRRL B-16531. Among these, *alkB1* and *alkB2* were parts of clusters containing both rubredoxin and rubredoxin reductase genes, whereas *alkB3* and *alkB4* were found as isolated genes in the genome (Whyte et al. 2002b). In *R. aetherivorans* BCP1, while a single *alkB* gene was initially identified within an orthodox *alk* gene cluster, an additional *alkB* gene, which was outside the rubredoxin/rubredoxin reductase gene cluster, was detected by genome analysis (Cappelletti et al. 2011, 2013). The presence of isolated *alkB* genes in *Rhodococcus* genomes was suggested to be functionally related to electron transfer redox components encoded by other *alk* gene clusters (Whyte et al. 2002b; van Beilen et al. 2002a) or, alternatively, to result from horizontal gene transfer events (Laczi et al. 2015).

Most of the expression studies concerning *alkB* gene clustered with rubredoxin and the rubredoxin reductase genes demonstrated the *alkB* inducibility by alkanes (Sameshima et al. 2008; Cappelletti et al. 2011). Some studies reported the expression of all the genes included in the *alkB* gene cluster (Laczi et al. 2015), and, when tested, the co-transcription of these genes within a single polycistronic mRNA was reported (Cappelletti et al. 2011). In *R. aetherivorans* BCP1, both *n*-alkanes ranging from C_6 to C_{22} and 1-alkanols were shown to be inducers of the *alkB* promoter (P_{alkB}), and the P_{alkB} induction level was proportional to the alkane concentration added to the growth medium (Cappelletti et al. 2011).

On the other hand, not always the expression of *alkB* gene was induced by alkanes in *Rhodococcus* strains. Indeed, Takei et al. (2008) have shown that only three out the five *alkB* genes detected in *Rhodococcus* sp. strain TMP2 were transcriptionally induced by alkanes, such as hexadecane (C_{16}) and pristane. In *Rhodococcus ruber* SP2B, despite the hexane degradation ability was inducible, the *alkB* gene expression was detected after cell growth on glucose, and this was associated with the absence of an alkane-inducible promoter (Amouric et al. 2010). This might be related to the occurrence of subtle regulatory mechanisms controlling the activity of some rhodococcal AlkB enzymes.

Putative regulatory binding sites were described upstream of *alkB* gene in *R. opacus* B4, *R. ruber* SP2B, and *R. aetherivorans* BCP1 (Sameshima et al. 2008; Amouric et al. 2010; Cappelletti et al. 2011). These binding sites had low

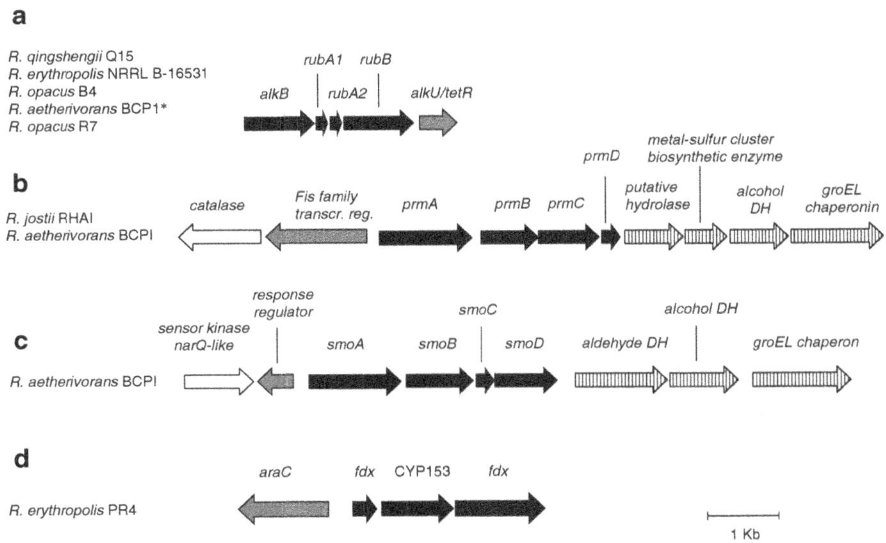

Fig. 2 Organization of the gene clusters encoding the alkane monooxygenases involved in alkane activation molecule in *Rhodococcus*. (**a**) *alkB* gene cluster encoding the membrane di-iron monooxygenase AlkB; (**b**) *prmABCD* gene cluster encoding the soluble di-iron monooxygenase Prm, standing for propane monooxygenase; (**c**) *smoABCD* gene cluster encoding the soluble di-iron monooxygenase Smo; and (**d**) the gene cluster encoding the heme-containing cytochrome P450 of CYP153 family. For each alkane monooxygenase gene cluster, the most representative *Rhodococcus* strains characterized in the literature are shown. Additionally, the flanking regions are displayed if they are conserved in *Rhodococcus* strains. Genes are represented by arrows, and the following colors/patterns were used: black, alkane monooxygenase gene clusters; gray, transcriptional regulator; stripes, genes involved in degradation reaction downstream of the initial alkane oxidation. The asterisk indicates that in BCP1 strain the *alkU/tetR* gene is transcribed in the opposite direction from that displayed

similarity with the target sequence recognized by AlkS which is the transcriptional regulator of the very well-known *alkB* gene cluster described in *P. putida* GPo1 (van Beilen et al. 2001).

A gene (*alkU* or *tetR*) encoding a putative TetR family transcriptional regulator (TetR-FTR) was found immediately downstream of the inducible *alk* gene cluster of a large number of *Rhodococcus* spp. and actinomycetes (Whyte et al. 2002b; Smits et al. 2002; Amouric et al. 2010; Liang et al. 2016a) (Fig. 2a). The conservation of the relative position of a TetR-FTR gene in proximity of the *alkB* gene cluster was taken as evidence of a role for TetR in the alkane-induced response (Cappelletti et al. 2011). The only functional study on the involvement of TetR-FTR on *alkB* gene expression in actinomycetes was reported in *Dietzia* sp. strain DQ12-45-1b (Liang et al. 2016a). In this strain, the TetR-FTR AlkX was found to repress the transcription of the *alkB*-like gene by binding to *alkB* promoter (Liang et al. 2016a). The repression of the transcription of the *alkB*-like gene was released when AlkX interacted with fatty acids, which are the intermediates in *n*-alkane degradation.

This was indicative of a product positive feedback mechanism regulating the *alkB* expression. Bioinformatic analyses revealed some conservation of the *Dietzia* AlkX binding site sequence in *Rhodococcus alkB* promoters (Liang et al. 2016a). Further, in *R. erythropolis* PR4, the *n*-hexadecane-induced transcriptome included both the transcripts belonging to an *alkB* gene cluster and the proximal TetR-FTR coding gene (Laczi et al. 2015). On the other hand, not in all *Rhodococcus* spp., TetR-FTR is transcribed in the same direction of *alkB* gene cluster, e.g., *R. aetherivorans* BCP1 (Cappelletti et al. 2011). These aspects, taken together, suggest that *Rhodococcus* probably shares with *Dietzia* the involvement of the TetR-FTR in *alkB* expression control, although possible divergences may exist in the regulation of TetR-FTR gene expression.

3.2 *Soluble Di-iron Monooxygenases (SDIMO)*

SDIMO are multicomponent enzymes able to hydroxylate different substrates ranging from methane, short-chain alkanes and alkenes, to phenol and toluene. SDIMO are found in phylogenetically and physiologically diverse bacteria and are classified into different groups (1–6) based on gene organization, substrate specificity, and sequence similarity (Leahy et al. 2003; Coleman et al. 2006, 2011; Holmes and Coleman 2008). In this respect, groups 3, 5, and 6 include SDIMO involved in alkane oxidation, and among these, the soluble methane monooxygenase from methanotrophic bacteria and the butane monooxygenase from *Thauera butanivorans* (formerly *Pseudomonas butanovora*) are the best characterized. These SDIMO, catalyzing the hydroxylation of methane and butane, show substrate unspecificity being able to co-oxidize a large range of short-chain hydrocarbons and structural analogous compounds (McDonald et al. 1997; Sluis et al. 2002; Coleman et al. 2006).

SDIMO are largely distributed throughout actinomycetes, mainly of *Mycobacterium* and *Rhodococcus* genera, and they were reported to be involved in the metabolism low-molecular-weight alkane and alkenes (Shennan 2006). The *Rhodococcus* SDIMO that have been functionally characterized are the alkene monooxygenase (AMO) from *Rhodococcus rhodochrous* (formerly *Nocardia corallina*) strain B-276 (Saeki and Furuhashi 1994), two propane monooxygenases encoded by *prmABCD* gene cluster from *Rhodococcus jostii* RHA1 (Sharp et al. 2007), and the SDIMO encoded by *smoABCD* cluster in *R. aetherivorans* BCP1 (Cappelletti et al. 2015). Except for AMO, the other SDIMO were found to be involved in the degradation of short-chain *n*-alkanes and in the co-oxidation of different contaminants.

3.2.1 The Organization and Expression of SDIMO Gene Clusters *prmABCD* and *smoABCD*

In *Rhodococcus*, the SDIMO gene clusters contain four genes coding for (1) the catalytic large and the structural small subunits of the oxygenase component, (2) the reductase containing both flavin and iron-sulfur clusters, and (3) a small "coupling protein" (Fig. 2b, c). According to the common biochemical features described for all SDIMO, the substrate oxidation is catalyzed by the oxygenase component, which contains the binuclear iron active site able to activate the molecular oxygen. The reductase component mediates the electron transfer from NADH to the active site of the oxygenase. The catalytic activity seems to be supported by a small "coupling protein," which is thought to facilitate the correct protein folding and di-iron center assembly (McCarl et al. 2018).

The different SDIMO gene clusters showed different substrate specificity and transcriptional inducer range. The *prmABCD* gene cluster (SDIMO group 5) in *Rhodococcus* spp. was induced by gaseous *n*-alkanes, mainly propane and, at a lower level, butane (Sharp et al. 2007; Cappelletti et al. 2015). Because of the initial correlation of this gene cluster with propane oxidation, the gene cluster was named *prmABCD*, and the encoded monooxygenase was the propane monooxygenase (Prm MO). In *R. jostii* RHA1, studies on the *prmA* gene mutant demonstrated the key role of Prm MO in the utilization of propane as sole source of carbon and energy and in the transformation of water contaminant *N*-nitrosodimethylamine (NDMA) (Sharp et al. 2007). Similar *prmABCD* gene clusters were described in other taxonomically related genera, i.e., *Gordonia*, *Mycobacterium*, and *Pseudonocardia* (Kotani et al. 2003, 2006; Furuya et al. 2011). Like in *Rhodococcus* spp., *prm* gene clusters were induced by gaseous *n*-alkanes, and they were found to play a key role in propane metabolism (Kotani et al. 2003; Furuya et al. 2011). Notably, in all the *prmABCD* expression studies, the genes within the *prm* cluster were induced as polycistronic operons (Kotani et al. 2003; Sharp et al. 2007; Cappelletti et al. 2015).

The flanking regions of the *prm* gene clusters were conserved among *Rhodococcus* spp. and other actinomycetes (Fig. 2b) (Cappelletti et al. 2015). Some of these genes were induced in *R. jostii* RHA1 cells grown on propane (Sharp et al. 2007). A gene encoding a FisR family transcriptional regulator (FisR-TR) was conserved upstream of *prm* gene cluster in different *Rhodococcus* spp. (Fig. 2b). The conservation of a FisR-like transcriptional regulator in the proximity of *prmABCD* gene suggested the involvement of this regulator in Prm MO expression. Accordingly, Cappelletti et al. (2015) described a palindromic sequence, overlapping the −35 region of the *prm* promoter, which shares similarities with FisR-type regulator binding site in *E. coli* (Cappelletti et al. 2015).

In *R. aetherivorans* BCP1, an additional SDIMO gene cluster, named *smoABCD*, was identified on the basis of sequence similarity in the database, and it was affiliated to SDIMO group 6. Unlike the *prmABCD* cluster that has a chromosomal localization in strain BCP1, the *smoABCD* gene cluster is harbored by the endogenous plasmid pBMC2 (Cappelletti et al. 2013). Further, *smoA* homologous genes (with amino acid sequence similarity >60%) are rare in the database (only 13 homologs are revealed by an updated pBLAST search running the BCP1 SmoA as a query sequence—July 2018), and they all belong to *Mycobacterium* spp. and few *Rhodococcus* spp. such as *R. aetherivorans*, *R. wratislaviensis*, and *R. ruber*. In particular, *R. aetherivorans* BCP1 *smoABCD* showed a high similarity (from 85% for *smoD* to 94% for *smoA*) with the *smo* gene cluster of *Mycobacterium chubuense*

(Coleman et al. 2011). This latter strain showed a peculiar level of SDIMO diversity because four distinct types of SDIMO (*smoXYB1C1Z*, *etnABCD*, *pmoABCD*, and *smoABCD*) were identified in its genome in addition to a membrane alkane MO of the AlkB family and a P450 cytochrome of the CYP153 family (Coleman et al. 2011). In BCP1 strain, the *smo* gene cluster expression was induced by short-chain ($<C_8$) *n*-alkanes and not by succinate (Cappelletti et al. 2015). The *smoA* homolog in *Mycobacterium chubuensis* NBB4 (Gene ID: *Mycch_*5395) was induced during the cell growth on acetate, ethene, and C_2–C_4 alkanes (Coleman et al. 2011). The apparent nonspecific inducer range for *smo* gene clusters in BCP1 (all the tested alkanes were inducers) and NBB4 (non-hydrocarbon substrates were inducers in addition to gaseous alkanes) might suggest sophisticated regulatory controls still unexplored. Recently, the disrupting mutation of *smoABCD* in *R. aetherivorans* BCP1 hampered the capacity of this strain to grow on short-chain *n*-alkanes and to degrade chlorinated *n*-alkanes indicating its key role in alkane metabolism and related cometabolic processes (data not published) (See Sect. 5).

A sensor kinase and a LuxR-like response regulator were coded by two genes immediately upstream of the *smoABCD* clusters described in both *R. aetherivorans* BCP1 and *M. chubuense* NBB4 (Fig. 2c). The tail-to-tail organization of the two genes was also conserved which suggest that these two genes might code for a two-component regulatory system involved in the control of *smoABCD* expression.

In strain BCP1, Cappelletti et al. (2015) have shown that analogously to the *prmABCD* gene cluster also *smoABCD* is transcribed as a single operon. Further, both the SDIMO gene clusters showed putative CRP (catabolite repression protein) binding sites, which were conserved among homologous regions. This was in line with the catabolite repression effect observed in the presence of glucose, succinate, and rich growth medium, i.e., LB, on the transcriptional induction of both the SDIMO gene clusters in BCP1 (Cappelletti et al. 2015).

Lastly, the heterologous expression of both Prm and Smo was supported by the co-expression of a chaperonin in the host strain. This consideration finds a connection with the conservation of a chaperonin-like gene downstream of each different SDIMO gene cluster in *Rhodococcus* (Fig. 2c). Chaperonins are supposed to have an essential role in the correct folding of the hydroxylase subunit of SDIMO system and are therefore necessary for the oxygenase activity (Furuya et al. 2013).

3.3 Cytochrome P450 of CYP153 Family

Cytochromes P450 form a superfamily of heme-thiolate proteins, which are ubiquitous among all life kingdoms and catalyze a wide range of oxidative reactions usually acting as terminal monooxygenases (Hannemann et al. 2007). It is noteworthy that in relation to the remarkable variety of chemical reactions catalyzed and the enormous number of substrates attacked, the sequence conservation among P450 proteins of different families can be less than 20% (Werck-Reichhart and Feyereisen 2000). Depending on the sequence similarity and redox partner, P450 cytochromes are categorized in families typically named as CYP ($CY_{tochome}$ P_{450}) followed by a number (Werck-Reichhart and Feyereisen 2000).

The genomes of environmental *Rhodococcus* strains generally possess many ORFs predicted to encode P450 cytochromes. *R. jostii* RHA1 possess 25 ORFs predicted to encode for P450 cytochromes (McLeod et al. 2006). *R. opacus* R7 genome, which consists of 10.1 Mb, is one of the largest bacterial genome described up to date and possesses 45 ORFs predicted to code for P450 monooxygenases (Di Gennaro et al. 2014).

However, the number of cytochrome P450 families involved in alkane degradation is limited. Cytochromes P450 of the CYP153 family function as soluble alkane hydroxylases involved in the degradation of short- and medium-chain-length *n*-alkanes (C_5–C_{10} alkanes) (Kubota et al. 2005; van Beilen et al. 2006). CYP153 enzymes require a ferredoxin and a ferredoxin reductase that transfer electrons from NAD(P)H to the cytochrome P450 in order to oxidize the alkane, enabling their host organisms to utilize it as carbon source (Koch et al. 2009). The gene cluster typically encoding the CYP153 enzyme in *Rhodococcus* and the accessory proteins are shown in Fig. 2d.

Members of CYP153 were reported to be commonly present in alkane-degrading bacteria which do not possess any AlkB MO or which have AlkB MO that do not act on medium-chain alkanes (Sekine et al. 2006; van Beilen et al. 2006). *R. rhodochrous* NCIMB 12566 was first described as having a cytochrome P450 enzyme able to hydroxylate *n*-octane (Cardini and Jurtshuk 1970). *R. erythropolis* NRRL B-16531 and *R. erythropolis* PR4 possess two related CYP153 and four AlkB (van Beilen et al. 2006). A *CYP153* gene located on a plasmid was overexpressed in *R. erythropolis* PR4 cells grown on C_{16} and on a mixture of hexadecane and branched alkanes (Laczi et al. 2015). In addition to CYP153, the study on PR4 cells grown on hexadecane hypothesized the involvement in hydrocarbon degradation of alternative families of cytochrome P450, i.e., CYP105 and CYP123 (Laczi et al. 2015). A gene encoding a P450 cytochrome of CYP185 family was transcriptionally induced in cells of *Rhodococcus* sp. strain p52 grown on the long-chain alkanes, C_{14}, C_{24}, C_{32}, and the branched alkane, pristine (Yang et al. 2014). These transcriptomic studies indicated the possible involvement of members of cytochrome P450 families different from CYP153 in *Rhodococcus* alkane metabolism, although functional studies are needed to define their role in alkane degradation pathway.

Sequence analysis revealed the presence of a putative AraC family transcriptional regulator (AFR) encoding gene located upstream of the *CYP153* gene cluster in *Proteobacteria* and *Actinobacteria*, including *Rhodococcus erythropolis* PR4 (Sekine et al. 2006) (Fig. 2d). The *AFR* gene was described to code for the CypR regulator in *Dietzia* strain DQ-12-45-1b. In this strain, the activator role of CypR in *CYP153* gene expression was identified (Liang et al. 2016b). The conservation of both gene arrangement of *AFR* and *CYP153* clusters and *CYP153* promoter region suggests that CypR may play similar roles in the *CYP153* gene regulation of *Dietzia* and *Rhodococcus* strains.

3.4 *Multiple Alkane Hydroxylase Systems in* Rhodococcus

In *Rhodococcus* genus, the presence of multiple alkane hydroxylase systems has been described as a widespread feature (Whyte et al. 2002b). Alkane hydroxylases sometimes exhibit overlapping substrate ranges, mainly when AlkB and cytochrome P450 enzymes are considered (van Beilen et al. 2006). A recent genomic and metagenomic survey targeting *alkB* and CYP153 genes in different environments reported that the simultaneous presence of *alkB* and CYP153 genes occurred only in selected bacterial genera including *Rhodococcus* and *Mycobacterium* and other specialized alkane degraders such as *Alcanivorax* (Nie et al. 2014). Further, in many cases, single *Rhodococcus* strains showed the presence of multiple genes encoding alkane MO of the same family, which in some cases showed quite divergent sequences. The coexistence of multiple homologous genes involved in catabolic pathways was proposed to contribute to the wide substrate range and environmental adaptation of *Rhodococcus*, and in some cases, this was associated to horizontal gene transfer events (van Beilen et al. 2003). Indeed, genes encoding transposases were often found in the vicinity of alkane hydroxylase-coding genes/operons (Cappelletti et al. 2015).

Typically, *Rhodococcus* strains are known to harbor multiple *alkB* genes. Indeed, most of the *Rhodococcus* isolates described by van Beilen et al. (2002b) were endowed with three to five *alkB* genes with quite divergent sequences. Each *alkB* homologous gene in *Rhodococcus* spp. was found to have different substrate ranges and induction styles, and the enzymes were suggested to belong to different AlkB groups. *R. qingshengii* Q15 and *R. erythropolis* NRRL B-16531 contained four AlkB (AlkB1-4) with quite sequence divergence (Whyte et al. 2002b; van Beilen et al. 2006). While AlkB3 and AlkB4 showed sequence similarity with *Acinetobacter* AlkB sequences, AlkB1 and AlkB2 were affiliated with AlkB sequence of other actinomycetes like *Mycobacterium*. Only genes *alkB*-1 and *alkB*-2 were located in clusters coding for the redox partner rubredoxins and rubredoxin reductase. Functional expression of AlkB2 showed the implication of this homolog in the initial oxidation of alkanes C_{12}–C_{16}, while AlkB3 and AlkB4 were associated to the degradation of very-long-chain alkanes ($>C_{20}$). Four of the five *alkB* genes found in *R. erythropolis* TMP2 showed the closest homology to the corresponding *alkB* genes of NRRL B-16531 and Q15. Among these genes, only *alkB*-1 and *alkB*-2 were functionally assigned to pristine metabolism, while *alkB*-3,4,5 were constitutively expressed (Takei et al. 2008). *R. aetherivorans* BCP1 possess two *alkB* genes, of which one is organized in cluster with *rubA1/2* and *rubB*, and two SDIMO gene clusters (*prmABCD* and *smoABCD*). While *alkB* gene cluster expression is induced during BCP1 growth on *n*-alkanes ranging from C_6 to C_{22}, *prmABCD* is transcriptionally induced during the growth on propane and at a lower extent on butane. SmoABCD results to be involved in the growth on short-chain alkanes (C_2–C_7) (Cappelletti et al. 2011, 2015).

R. erythropolis PR4 showed the ability to grow on a wide range of hydrocarbons, and the analysis of its genome revealed the presence of 4 *alkB* genes, 2 *CYP153* genes, 14 additional cytochrome P450 of different families, 3 multicopper oxidase (MCO), and many other mono- and dioxygenase genes. Transcriptomic analyses indicated that the expression of four *alkB* genes, two CYP153 genes, and other nine genes encoding P450 cytochromes of different families were differently induced after the growth on diesel oil, hexadecane, or hydrocarbon mixtures. In particular, two *alkB* genes were transcriptionally induced during the growth on C_{16} and diesel oil, while the other two *alkB* genes were not expressed under any tested conditions. One *CYP153* was induced during the growth on all the hydrocarbons under analysis, while the other *CYP153* was induced on diesel oil (Laczi et al. 2015). Similar to strain PR4, up to six *alkB* homologous genes and five *CYP153* homologous genes were found in *R. erythropolis* SK121 (Nie et al. 2014). Recently new isolates closely related to *R. erythropolis* species showed the presence of multiple copies of *alkB* genes that are probably involved in the terminal oxidation of long-chain alkanes (C_{12}–C_{29}) (Likhoshvay et al. 2014).

The coexistence of multiple alkane hydroxylase genes in *Rhodococcus* spp. is thought to be the result of an evolution process aimed at differentiating and widening the catabolic processes of members of this genus in response to environmental pressure, i.e., the presence of specific classes of hydrocarbons in the initial isolation source. Further functional studies are needed to get deep into the different selectivity, substrate specificity, and transcriptional regulation featuring the diverse alkane hydroxylase activities underlying the unique alkane degradation capacities of the members of *Rhodococcus* genus.

4 Alkane Metabolism After the Initial Oxidation

Depending on the enzymatic activity and selectivity of the alkane monooxygenase/s involved in the first oxidation step, alkanes are oxidized to a primary alcohol, a secondary alcohol, or a mixture of the two isomers. SDIMO enzymes have been reported to catalyze both the alkane oxidation reactions, whereas AlkB and CYP153 alkane MOs have been mainly described to catalyze terminal oxidations (van Beilen et al. 2006). However, the subterminal alkane oxidation ability was suggested for at least one of the AlkB MOs found in *Rhodococcus qingshengii* Q15, and the mutant cytochrome P450 from *Bacillus megaterium* was able to hydroxylate alkanes at the 2-position (Ji et al. 2013). Only a few numbers of studies have investigated the catalytic properties of the different AlkB and P450 enzymes in *Rhodococcus* spp.; in this regard, the high sequence divergence featuring the different alkane MOs might be associated to still unexplored substrate ranges and enzymatic activities.

After the initial oxidation step, the metabolism of medium- and long-chain alkanes by *Rhodococcus* spp. typically follows the terminal, the subterminal, or the bi-terminal oxidation pathway (Fig. 3). In the terminal oxidation pathway, the alkane is oxidized to the corresponding primary alcohol, which is further oxidized to the corresponding aldehyde by alcohol dehydrogenases and then to the corresponding fatty acid by aldehyde dehydrogenases (Rojo 2009). In the bi-terminal oxidation

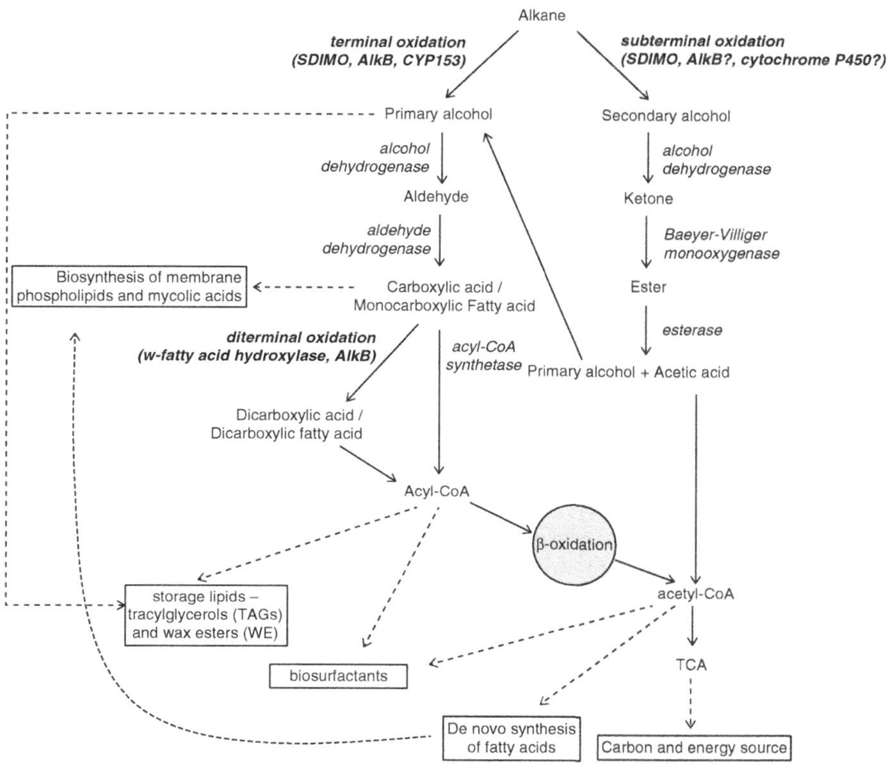

Fig. 3 The different alkane degradation pathways described in *Rhodococcus* spp. (terminal, subterminal, and bi-terminal oxidation pathways)

pathway, the fatty acid produced via terminal oxidation is further oxygenated by a ω-fatty acid monooxygenase to form dicarboxylic acid (Ji et al. 2013). In the subterminal pathway, the alkane is oxidized to the corresponding secondary alcohol, which is converted to the corresponding ketone and then oxidized to the ester by a Baeyer-Villiger monooxygenase. The ester is then hydrolyzed to an alcohol and a fatty acid molecule (Van Beilen et al. 2003). In all cases, the alkane oxidation pathways lead to the formation of CoA-activated fatty acids which enter the β-oxidation process that generates the tricarboxylic acid cycle (TCA) intermediates to be used as carbon and energy sources. Alternatively, the different metabolic intermediates produced during the alkane oxidation process can (1) be assimilated as membrane components, in this case the alkane chain can be incorporated in fatty acids composing phospholipids or can be used as precursors in the mycolic acid synthesis, and (2) be converted into storage lipids such as triacylglycerols (TAGs) and wax esters (WE) (Alvarez et al. 2013; Alvarez 2003) (Fig. 3).

Different alkanotrophic *Rhodococcus* strains typically possess a variety of alkane catabolic pathways. Several *Rhodococcus* strains showed the capacity to degrade alkanes following either the terminal or the subterminal oxidation pathways; further, the coexistence of the two oxidation pathways was also demonstrated in some *Rhodococcus* spp. metabolizing both short- and medium-/long-chain alkanes. Whyte et al. (1998) reported both terminal and subterminal oxidation pathways of n-C_{12} (dodecane) and n-C_{16} (hexadecane) in *R. qingshengii* Q15 cells producing 1-dodecanol and 2-dodecanone and 1-hexadecanol and 2-hexadecanol, respectively. *R. erythropolis* ATCC 4277 was described to degrade C_5–C_{16} alkanes using only the subterminal pathway (Ludwig et al. 1995) similarly to *Rhodococcus* sp. strain EH831 and *Rhodococcus* sp. strain Moj-3449 grown on C_6 and C_{12} (Binazadeh et al. 2009; Lee et al. 2010). *Rhodococcus* sp. strain BPM 1613 was described to convert the methyl-branched alkane pristane to terminal oxidation metabolic intermediates which were further degraded through either β- or ω-oxidation (via bi-terminal oxidation pathway) (Nakajima et al. 1985). *Rhodococcus ruber* sp. strain SBUG 82 could also metabolize pristane through the subterminal oxidation pathway (Nhi-Cong et al. 2009). Alkanotrophic *Rhodococcus* sp. strain NCIM 5126 produced primary alcohols that were converted into monocarboxylic acids when odd carbon number alkanes (C_{13} and C_{17}) were used as growth substrates. Conversely, when NCIM 5126 was grown on even carbon number n-alkanes (C_{14} and C_{16}), the primary alcohols were converted only to dicarboxylic acids (Sharma and Pant 2000). These data suggested the presence of both a bi-terminal and a mono-terminal oxidation pathway of n-alkanes in NCIM 5126. On the other hand, this strain catabolized pristane only through mono-terminal oxidation pathway (Sharma and Pant 2000).

A peculiar capability of members of *Rhodococcus* genus and of few other taxonomically related genera (e.g., *Mycobacterium*, *Nocardia*) concerns the capacity to metabolize gaseous alkanes (Shennan 2006). Like liquid alkanes, gaseous alkanes, i.e., ethane, propane, and butane, are oxidized to the primary and/or secondary alcohols that are further converted by the activities of alcohol and aldehyde DHs to short-chain carboxylic acids (e.g., propionic acid, butyric acid). These carboxylic acids can be directly converted to intermediates of TCA cycle (Fig. 4). While ethane metabolism is assumed to be oxidized to ethanol, then to acetaldehyde and acetate, which enter the central metabolic pathways in bacterial cells, butane oxidation has been mainly described to proceed via the terminal carbon atom oxidation pathway (1-butanol, butyraldehyde, butyrate) (Ashraf et al. 1994; Cappelletti et al. 2015). Conversely, propane can be oxidized via terminal, subterminal, or a mixture of the two pathways, depending on the *Rhodococcus* strain under analysis (Ashraf et al. 1994). In *R. rhodochrous* ATCC 21198, propane is oxidized to 1-propanol, propanal, and propanoate, which is carboxylated to methylmalonyl CoA and then converted to succinyl-CoA (MacMichael and Brown 1987). The subterminal oxidation pathway is more complicated with the possibility of three variations, two of which involve the metabolism of propane via 2-propanol to acetone, which can be converted to methyl acetate or to acetol. *R. rhodochrous* PNKb1 and *R. aetherivorans* BCP1 were shown to metabolize propane via the two oxidation pathways, terminal and subterminal (Woods and Murrell 1989; Ashraf et al. 1994; Cappelletti et al. 2015). In the latter case, the propane monooxygenase was suggested not to have preference in the insertion of oxygen into the propane

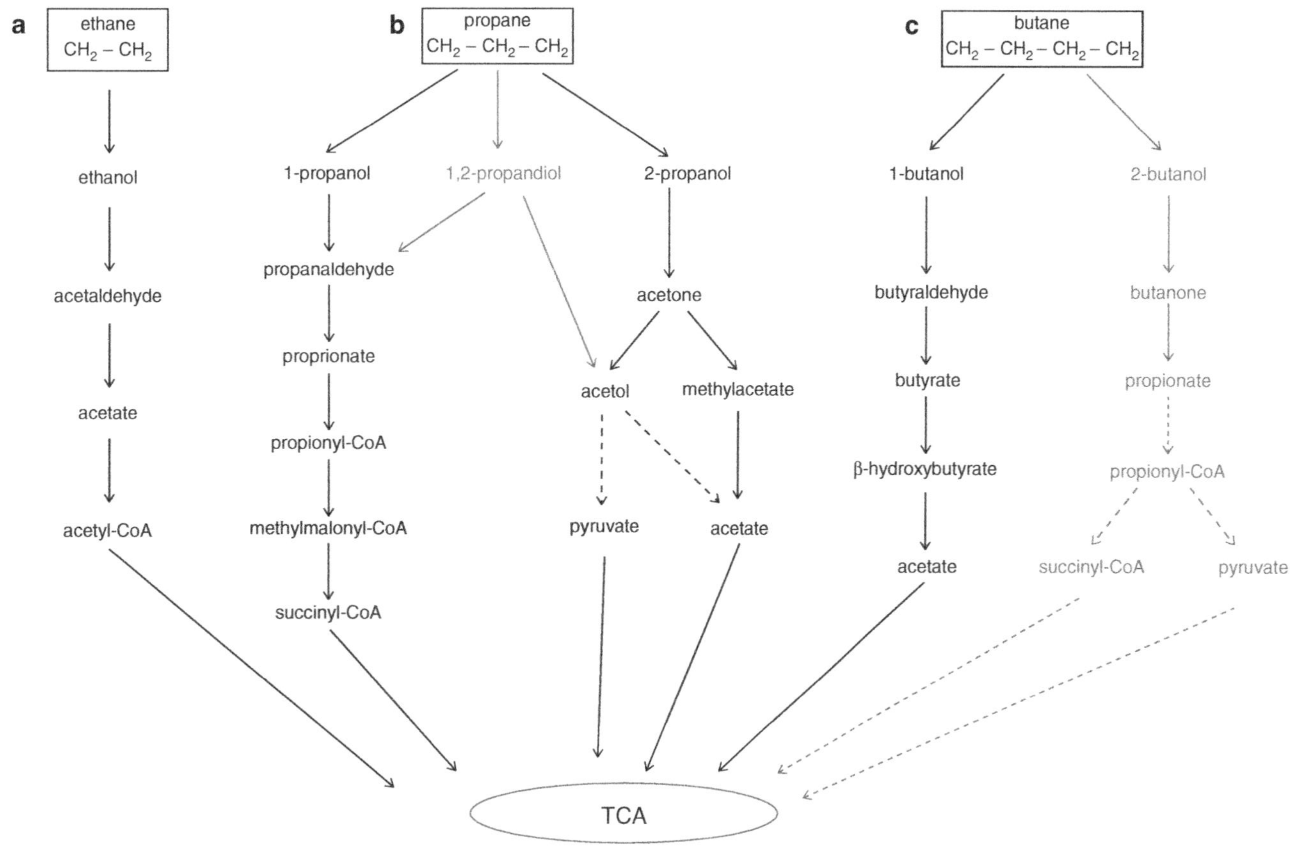

Fig. 4 Gaseous *n*-alkane [(**a**) ethane, (**b**) propane, (**c**) butane] oxidation pathways described in *Rhodococcus* spp. In gray, degradation pathways are displayed that have been only hypothesized or that have not been further characterized in *Rhodococcus*. Dashed lines correspond to more reactions. *TCA* tricarboxylic acid cycle

molecule (Ashraf and Murrell 1992), while specificity was observed at the stage of the alcohol dehydrogenase oxidizing either 1- or 2-propanol. An alternative pathway for propane oxidation was proposed to occur via 1,2-propandiol production through the activity of an intermolecular dioxygenase system (Babu and Brown 1984; Ashraf et al. 1994). The diol was proposed to be further converted to acetol or to propanal (Ashraf et al. 1994).

Several alcohol dehydrogenases (ADHs) have been described to be involved in the oxidation of the alcohol/s derived from alkane oxidation into the corresponding aldehyde or ketone in *Rhodococcus* spp. Among these, alcohol dehydrogenases involved in the alkanol oxidation were a NAD-dependent 2-propanol dehydrogenase from *R. rhodochrous* PNKb1 (Ashraf and Murrell 1990) and a NAD-dependent long-chain secondary alcohol dehydrogenase from *R. erythropolis* ATCC 4277 (Ludwig et al. 1995). NAD-dependent dehydrogenases showed secondary alcohols as preferential substrates, while primary alcohols are the preferred substrates for *N, N*-dimethyl-4-nitrosoaniline (NDMA)-dependent ADHs (Schenkels and Duine 2000). NDMA was used as an artificial cofactor for the latter alcohol dehydrogenase in *R. erythropolis* DSM 1069 incubated with alkanol derived from short-chain (both gaseous and liquid) alkane metabolism (Schenkels and Duine 2000), while both NAD^+- and NDsMA-dependent ADHs were found to be highly active during ethanol growth by *R. erythropolis* EK-1 (Pirog et al. 2008). Further, in *R. jostii* RHA1, several Baeyer-Villiger monooxygenases were shown to have a substrate range including intermediates of subterminal oxidation pathways, e.g., acetone, 2-butanone, and 2-octanone (Riebel et al. 2012). These enzymes are possibly involved in the oxidation steps downstream of the secondary alcohol production from alkane subterminal oxidation.

The overexpression of specific alcohol and aldehyde dehydrogenases (DHs) was seen in studies reporting transcriptomic and proteomic analyses of *Rhodococcus* spp. grown on alkanes as compared to orthodox carbon sources (e.g., glucose, succinate) (Sharp et al. 2007; Cappelletti et al. 2015). In these studies, the alkane-induced cells showed the overexpression of genes coding for alcohol and aldehyde DHs that were arranged in clusters with those genes encoding for alkane-induced monooxygenases (Sharp et al. 2007; Cappelletti et al. 2015). In particular, the gene encoding a secondary alcohol dehydrogenase in cluster with the *prmABCD* gene was transcriptionally induced during the growth *R. jostii* RHA1 cells on propane (Sharp et al. 2007). The proteomic profile of *R. aetherivorans* BCP1 cells grown on *n*-butane and *n*-hexane showed the upregulation of three alcohol dehydrogenases (DHs) and three aldehyde DHs which are likely to be involved in the oxidation steps occurring downstream of the monooxygenase hydroxylation (Cappelletti et al. 2015). Among these, two genes (one for alcohol DH and one for aldehyde DH) were consecutive to the *smoABCD* gene cluster encoding the alkane MO involved in the initial short-chain alkane oxidation in *R. aetherivorans* BCP1 (Cappelletti et al. 2015) (Fig. 2c).

5 Biotechnological Applications of Alkanotrophic Rhodococci

Alkanotrophic *Rhodococcus* are able to transform hydrocarbon contaminants into chemicals relevant for biotechnological and industrial applications (Larkin et al. 2005). In addition to regio- and stereoselective oxidations performed by some alkanotrophic *Rhodococcus* strains, these bacteria can also produce high-value compounds; storage lipids, biosurfactants, and carotenoids are some of the most significant biotechnological products that can be produced by *Rhodococcus* strains utilizing alkanes.

Several *Rhodococcus* strains produce biosurfactants in response to the presence of alkanes as growth substrate. They mainly consist of glycolipids containing trehalose as the carbohydrate. The physiological role of these glycolipids is involved in the uptake of hydrophobic and water-insoluble alkanes (see Sect. 2) (Lang and Philp 1998). Biosurfactants are surface-active, amphiphilic compounds characterized by a hydrophilic and a hydrophobic molecule part. In addition to the unique interfacial properties, trehalose-based biosurfactants produced by rhodococci also possess immunostimulating and cell-differentiating features along with low cytotoxicity (Kuyukina et al. 2015).

Recently, in *Rhodococcus erythropolis* SD-74, three genes were described to be involved in the biosynthesis of succinoyl trehalose lipids, including a putative acyl coenzyme A transferase, a fructose-bisphosphate aldolase, and an alkane monooxygenase (Inaba et al. 2013). Interestingly, the upregulation of the expression of the acyl coenzyme A transferase was shown to enhance the production of succinoyl trehalose lipids in this strain (Inaba et al. 2013). Further, *R. erythropolis* SD-74 cells produced succinoyl trehalolipids with acyl groups of the same carbon chain length as the *n*-alkanes used as growth substrate, suggesting the possibility to direct the synthesis of specific biosurfactants by supplying alkanes with specific chemical nature (Lang and Philp 1998; Inaba et al. 2013). This property is highly useful for the commercial production, as most biosurfactant producers do not produce specifically defined derivatives.

Triacylglycerol (TAG, triesters of glycerol and long-chain fatty acids) and wax esters (WE, esters of primary long-chain fatty acids and primary long-chain fatty alcohols) are in addition to polyhydroxyalkanoates (PHA), the most important lipids produced by bacteria, and they have relevant applications in the production of food additives, cosmetics, lubricants, oleo-chemicals, candles, and biofuels (Holder et al. 2011). PHAs are the main storage compounds produced by bacteria, while the production of TAGs and WEs has been documented in a few bacterial genera including strains of the genus *Rhodococcus* (Hernández et al. 2008; Alvarez et al. 2013). Indeed, several strains of rhodococci are able to store significant amount of TAGs, which can reach more than 70% of their biomass (Alvarez et al. 1996). TAG storage inside the cells is enhanced under nitrogen-limiting conditions when cells are grown with different carbon sources such as sugars and organic acids, but evidence that also alkane-utilizing species of *Rhodococcus* can accumulate TAGs has also been obtained in the last two decades (Alvarez 2003). The possibility to combine TAG production with hydrocarbon degradation would couple the environmental remediation with the production of high-value compounds.

R. opacus strain PD630 is taken as a model bacterial system regarding TAG production and storage (Alvarez et al. 1996). Several studies with strain PD630 showed that the use of alkanes as carbon source influenced the type of the predominant fatty acids present in the inclusions, since their chain lengths resembled that of the substrate, suggesting that alkanes were not completely metabolized to acetate when they were accumulated as cytosolic lipid inclusions. Furthermore, *R. opacus* strain B4 was described to be able to produce and accumulate high amounts of lipids using C_{16} as sole carbon source (Castro et al. 2016). Interestingly, while strains PD630 and B4 cultivated on C_{16} displayed similar fatty acid profiles in their cell membrane fractions, *R. opacus* B4 showed a higher variability in fatty acid composition as compared to *R. opacus* PD630 in the TAG fractions (Castro et al. 2016). In this respect, the biosynthesis of lipids with a higher variety of fatty acids can be an interesting advantage since they can be used in different types of industrial and commercial applications such as in personal care, food, and beverage industries (Castro et al. 2016). The effects of carbon sources on the composition of fatty acids, synthesized by two other *Rhodococcus* strains, *R. erythropolis* DCL14 and *R. opacus* PWD4, were recently reported by Cortes and Carvalho (2015). The chain length of the different *n*-alkanes used for growth by these two rhodococci influenced the percentage of each type of fatty acids stored inside the cells. Some *Rhodococcus* strains were able to produce wax esters, in addition to TAG, during growth on *n*-alkanes and phenylalkanes using fatty alcohols produced as intermediates during their degradation (Alvarez et al. 1996; Hernández et al. 2008). For instance, in addition to TAG, wax esters were detected in hexadecane- and hexadecane/hexadecanol-grown cells of *R. jostii* RHA1 (Hernández et al. 2008). Wax esters and fatty alcohols provide diverse industrial uses, such as additives for biolubricants and cosmetic and pharmaceutical products (Lanfranconi and Alvarez 2017).

Carotenoids form another important category of organic compounds produced by rhodococci. Carotenoids have antioxidant and anticancer properties; they can also activate the immune system and promote other beneficial effects on human health (Krinsky and Johnson 2005). For these reasons the use of carotenoids as natural food additives has greatly increased in the last decade (Krinsky and Johnson 2005). Many *Rhodococcus* strains can produce carotenoid pigments such as β-carotene, γ-carotene, and chlorobactene (Tao et al. 2006). Recently, it was reported that in a *R. erythropolis* strain able to use different alkanes such as cycloalkane, *n*-hexane, and *n*-decane, as sole carbon source, the exposure to these different hydrocarbons determined specific modifications of the cell carotenoid production. In particular, higher levels of lycopene, an important precursor involved in the biosynthetic pathway of γ-carotene and β-carotene, were found in extracts of cells grown on *n*-alkanes (Stancu 2015). The ability to form aggregates and to develop biofilms has also been documented to significantly influence carotenoid production in *Rhodococcus* spp. (Zheng et al. 2013).

Another extensively studied application of alkanotrophic rhodococci is their use in bioremediation of halogenated organic compounds through oxidative cometabolism. Halogenated volatile organic compounds, including chlorinated aliphatic hydrocarbons (CAHs), are widely applied in a variety of industrial processes and are among the most frequently occurring soil and groundwater contaminants released into the environment through accidents and/or non-correct industrial management (Frascari et al. 2008; Cappelletti et al. 2012). CAHs can be extremely harmful to human health, with some of them classified as either known or suspected carcinogens (McDermott and Heffron 2013).

Various *Rhodococcus* strains have been reported to aerobically degrade different CAHs via cometabolic degradation using several *n*-alkanes as primary carbon source mainly gaseous *n*-alkanes such as propane and butane (Table 2). TCE, VC, 1,1-dichloroethane, 1,1,1-trichloroethane, and 1,2,3-trichloropropane along with CF are the most frequently cometabolized CAHs by alkanotrophic rhodococci (Table 2). The aerobic mineralization of trichloroethylene (TCE) and vinyl chloride (VC) by two alkanotrophic *Rhodococcus* strains (Sm-1 and Wrink) was first described by Malachowsky et al. (1994).

R. aetherivorans BCP1 (formerly *Rhodococcus* sp. strain BCP1) was isolated from a microbial consortium that was subject to a prolonged chloroform exposure in bioreactors (Frascari et al. 2006). Strain BCP1 was able to grow on a wide range of *n*-alkanes (gaseous, liquid, and solid) (Cappelletti et al. 2011) and when BCP1 cells were grown on *n*-hexane (C_6) or *n*-butane (C_4) had the ability to degrade efficiently CF as well as VC and 1,1,2-TCA via cometabolism (Table 2) (Frascari et al. 2006; Cappelletti et al. 2012). The BCP1 cell growth on these *n*-alkanes induced the expression of alkane monooxygenase genes that are thought to be involved in the alkane/CAH cometabolism (Cappelletti et al. 2011, 2015). Further, CF was efficiently degraded by BCP1 cells grown on *n*-butane in a continuous-flow biofilm reactor, which has a series of operational advantages (Ciavarelli et al. 2012). Other *Rhodococcus* strains able to grow on gaseous alkanes and cometabolize CAHs were *Rhodococcus* sp. strain PB1 and *R. aetherivorans* TPA (Table 2). In particular, *R. aetherivorans* strain TPA, was able to cometabolically degrade 1,1,2,2-tetrachloroethane (1,1,2,2-TeCA) under propanotrophic growth conditions. This is the first isolate described to be able to degrade a high-chlorinated CAH like 1,1,2,2-TeCA under aerobic conditions, whereas most of the successful bioremediation methods of these compounds needed to be performed under anaerobic conditions and in the presence of microbial consortia (Cappelletti et al. 2017b).

6 Summary

During the last 20 years, many studies have shown that members of *Rhodococcus* genus are broadly distributed in hydrocarbon-contaminated sites and have shown to play a key role in oil biodegradation processes, even in cold temperature environments. Among hydrocarbons, alkanes are ubiquitous compounds released in the

Table 2 Main *Rhodococcus* strains characterized for their ability to grow on *n*-alkanes and to cometabolize chlorinated aliphatic hydrocarbons (CAHs) or other xenobiotics

Strain	Primary growth substrate	Cometabolized CAHs[a] or other xenobiotics	References
Rhodococcus sp. strain Sm-1 (ATCC 51239)	Propane	1,1-DCE, *cis*-1,2-DCE, CF, TCE, 1,1,2-TCA, 1,1,1-TCA and VC	Malachowsky et al. (1994)
Rhodococcus sp. strain Wrink (ATCC 51240)	Propane	TCE and VC	Malachowsky et al. (1994)
R. rhodochrous ATCC 21197	Propane	1,1-DCE, *cis*-1,2-DCE, CF, VC, CF, 1,1,2-TCA, 1,1,1-TCA	Malachowsky et al. (1994)
Rhodococcus sp. strain PB1	Propane	VC, *cis*-DCE, 1,1,2-TCA and CF	Frascari et al. (2008)
Rhodococcus sp. strain 179 BP *Rhodococcus* sp. strain 183 BP	*n*-Butane	1,1-DCA and 1,1,1-TCA	Semprini et al. (2009)
R. ruber ENV 425	Propane	NDMA and 1,2,3-TCP	Fournier et al. (2009) and Wang and Chu (2017)
R. jostii RHA1	Propane	NDMA and 1,2,3-TCP	Sharp et al. (2007) and Wang and Chu (2017)
R. aetherivorans BCP1	*n*-Butane, *n*-hexane	CF, VC, 1,1,2-TCA	Frascari et al. (2006)
R. aetherivorans TPA	Propane	1,1,2,2-TeCA	Cappelletti et al. (2017b)

[a]*CF* chloroform, *DCA* dichloroethane, *DCE* dichloroethylene, *TCA* trichloroethane, *TCE* trichloroethylene, *TeCA* tetrachloroethane, *TCP* trichloropropane, *NDMA* *N*-nitrosodimethylamine, *VC* vinyl chloride

environments from natural and anthropogenic sources. Members of *Rhodococcus* genus have been described for their unique degradation abilities toward a wide range of alkanes and have been indicated as ideal candidates for bioremediation and biotransformation strategies. This is associated to peculiar physiological and genetic aspects: (1) the ability to resist harsh environmental conditions and to tolerate the presence of high concentration of toxic compounds; (2) the hydrophobic *Rhodococcus* cell surface and the ability to produce biosurfactants, which facilitate the contact between the cell and the hydrophobic alkane and its subsequent internalization; and (3) several *Rhodococcus* spp. contain many genes encoding alkane hydroxylases which are involved in the alkane molecule activation through an initial oxidation reaction. The genes encoding these key enzymes showed in some cases high sequence divergence in *Rhodococcus*, and each expressed monooxygenase was commonly featured by distinct substrate specificity and selectivity. Further, although only few aspects concerning alkane metabolism regulatory mechanisms have been described in

Rhodococcus, expression studies have shown that the different alkane hydroxylase genes are induced under different conditions of growth and/or in the presence of different primary substrates, e.g., alkane with specific chain length.

In terms of potential industrial applications, alkanotrophic *Rhodococcus* spp. can be exploited as microbial cell factories able to convert alkanes into lipids and carotenoids, to produce biosurfactants, and to degrade chlorinated aliphatic hydrocarbons through alkane-induced cometabolic processes.

References

Acosta-González A, Martirani-von Abercron SM, Rosselló-Móra R, Wittich RM, Marqués S (2016) The effect of oil spills on the bacterial diversity and catabolic function in coastal sediments: a case study on the Prestige oil spill. Environ Sci Pollut Res Int 22:15200–15214. https://doi.org/10.1007/s11356-015-4458-y

Alvarez HM (2003) Relationship between β-oxidation pathway and the hydrocarbon-degrading profile in actinomycetes bacteria. Int Biodeterior Biodegr 52:35–42. https://doi.org/10.1016/S0964-8305(02)00120-8

Alvarez HM, Mayer F, Fabritius D, Steinbüchel A (1996) Formation of intracytoplasmic lipid inclusions by *Rhodococcus opacus* strain PD630. Arch Microbiol 165:377–386. https://doi.org/10.1007/s002030050341

Alvarez HM, Silva RA, Herrero M, Hernández MA, Villalba MS (2013) Metabolism of triacylglycerols in *Rhodococcus* species: insights from physiology and molecular genetics. J Mol Biochem 2:2119–2130. https://doi.org/10.1007/s00253-012-4360-1

Amouric A, Quéméneur M, Grossi V, Liebgott PP, Auria R, Casalot L (2010) Identification of different alkane hydroxylase systems in *Rhodococcus ruber* strain SP2B, an hexane-degrading actinomycete. J Appl Microbiol 108:1903–1916. https://doi.org/10.1111/j.1365-2672.2009.04592.x

Andreoni V, Bernasconi S, Colombo M, van Beilen JB, Cavalca L (2000) Detection of genes for alkane and naphthalene catabolism in *Rhodococcus* sp. strain 1BN. Environ Microbiol 2:572–577. http://doi.org/0.1046/j.1462-2920.2000.00134.x

Ashraf W, Murrell JC (1990) Purification and characterization of a NAD+-dependent secondary alcohol dehydrogenase from propane-grown *Rhodococcus rhodochrous* PNKb1. Arch Microbiol 153:163–168

Ashraf W, Murrell JC (1992) Genetic, biochemical and immunological evidence for the involvement of two alcohol dehydrogenases in the metabolism of propane by *Rhodococcus rhodochrous* PNKb1. Arch Microbiol 157:488–492

Ashraf W, Mihdhir A, Murrell JC (1994) Bacterial oxidation of propane. FEMS Microbiol Lett 122:1–6

Babu J, Brown L (1984) New type of oxygenase involved in the metabolism of propane and isobutane. Appl Environ Microbiol 48:260–264

Binazadeh M, Karimi IA, Li Z (2009) Fast biodegradation of long chain n-alkanes and crude oil at high concentrations with *Rhodococcus* sp. Moj-3449. Enzym Microb Technol 45:195–202. https://doi.org/10.1016/j.enzmictec.2009.06.001

Bouchez-Naïtali M, Blanchet D, Bardin V, Vandecasteele JP (2001) Evidence for interfacial uptake in hexadecane degradation by *Rhodococcus equi*: the importance of cell flocculation. Microbiology (Reading, England) 147:2537–2543. https://doi.org/10.1099/00221287-147-9-2537

Bredholt H, Josefsen K, Vatland A, Brunheim P, Eimhjellen K (1998) Emulsification of crude oil by an alkane-oxidizing *Rhodococcus* species isolated from seawater. Can J Microbiol 44:330–340. https://doi.org/10.1139/w98-005

Cappelletti M, Fedi S, Frascari D et al (2011) Analyses of both the *alkB* gene transcriptional start site and *alkB* promoter-inducing properties of *Rhodococcus* sp. strain BCP1 grown on n-alkanes. Appl Environ Microbiol 77:1619–1627. https://doi.org/10.1128/AEM.01987-10

Cappelletti M, Frascari D, Zannoni D, Fedi S (2012) Microbial degradation of chloroform. Appl Microbiol Biotechnol 96:1395–1409. https://doi.org/10.1007/s00253-012-4494-1

Cappelletti M, Di Gennaro P, D'Ursi P, Orro A, Mezzelani A, Landini M, Fedi S, Frascari D, Presentato A, Zannoni D, Milanesi L (2013) Genome sequence of *Rhodococcus* sp. strain BCP1, a biodegrader of alkanes and chlorinated compounds. Genome Announc 9:75. https://doi.org/10.1128/genomeA.00657-13

Cappelletti M, Presentato A, Milazzo G, Turner RJ, Fedi S, Frascari D, Zannoni D (2015) Growth of *Rhodococcus* sp. strain BCP1 on gaseous *n*-alkanes: new metabolic insights and transcriptional analysis of two soluble di-iron monooxygenase genes. Front Microbiol 6:393. https://doi.org/10.3389/fmicb.2015.00393

Cappelletti M, Fedi S, Zampolli J, Di Canito A, D'Ursi P, Orro A, Viti C, Milanesi L, Zannoni D, Di Gennaro P (2016) Phenotype microarray analysis may unravel genetic determinants of the stress response by *Rhodococcus aetherivorans* BCP1 and *Rhodococcus opacus* R7. Res Microbiol 1(9). https://doi.org/10.1016/j.resmic.2016.06.008

Cappelletti M, Frascari D, Pinelli D, Mezzetti F, Fedi S, Zannoni D (2017a) Aerobic cometabolism of 1,1,2,2-TeCA by a propane-growing microbial consortium (C2): diversity of alkane monooxygenase genes and design of an on-site bioremediation process. Int Biodeterior Biodegr 119:649–660. https://doi.org/10.1016/j.ibiod.2016.09.019

Cappelletti M, Pinelli D, Fedi S, Zannoni D, Frascari D (2017b) Aerobic co-metabolism of 1,1,2,2-tetrachloroethane by *Rhodococcus aetherivorans* TPA grown on propane: kinetic study and bioreactor configuration analysis. J Chem Technol Biotechnol 30:2420–2411. https://doi.org/10.1002/jctb.5335

Cardini G, Jurtshuk P (1970) The enzymatic hydroxylation of *n*-octane by *Corynebacterium* sp. strain 7E1C. J Biol Chem 245:2789–2796

Castro AR, Rocha I, Alves MM, Pereira MA (2016) *Rhodococcus opacus* B4: a promising bacterium for production of biofuels and biobased chemicals. AMB Express 6:35. https://doi.org/10.1186/s13568-016-0207-y

Ciavarelli R, Cappelletti M, Fedi S, Pinelli D, Frascari D (2012) Chloroform aerobic cometabolism by butane-growing *Rhodococcus aetherivorans* BCP1 in continuous-flow biofilm reactors. Bioprocess Biosyst Eng 35:667–681. https://doi.org/10.1007/s00449-011-0647-3

Coleman NV, Bui NB, Holmes AJ (2006) Soluble di-iron monooxygenase gene diversity in soils, sediments and ethene enrichments. Environ Microbiol 8:1228–1239. https://doi.org/10.1111/j.1462-2920.2006.01015.x

Coleman NV, Yau S, Wilson NL, Nolan LM, Migocki MD, Ly MA, Crossett B, Holmes AJ (2011) Untangling the multiple monooxygenases of *Mycobacterium chubuense* strain NBB4, a versatile hydrocarbon degrader. Environ Microbiol Rep 3:297–307. https://doi.org/10.1111/j.1758-2229.2010.00225.x

Cortes M, de Carvalho C (2015) Effect of carbon sources on lipid accumulation in *Rhodococcus* cells. Biochem Eng J 94:100–105. https://doi.org/10.1016/j.bej.2014.11.017

de Carvalho CC, Parreño-Marchante B, Neumann G, da Fonseca MM, Heipieper HJ (2005) Adaptation of *Rhodococcus erythropolis* DCL14 to growth on *n*-alkanes, alcohols and terpenes. Appl Microbiol Biotechnol 67:383–388. https://doi.org/10.1007/s00253-004-1750-z

Di Gennaro P, Zampolli J, Presti I, Cappelletti M, D'Ursi P, Orro A, Mezzelani A, Milanesi L (2014) Genome sequence of *Rhodococcus opacus* strain R7, a biodegrader of mono- and polycyclic aromatic hydrocarbons. Genome Announc 2:e00827–e00814. https://doi.org/10.1128/genomeA.00827-14

Ekprasert J (2014) Functional characterisation of alkane-degrading monooxygenases in *Rhodococcus jostii* strain 8. Doctoral thesis, University of East Anglia

Fournier D, Hawari J, Halasz A, Streger SH, McClay KR, Masuda H, Hatzinger PB (2009) Aerobic biodegradation of N-nitrosodimethylamine by the propanotroph *Rhodococcus ruber* ENV425. Appl Environ Microbiol 75:5088–5093. https://doi.org/10.1128/AEM.00418-09

Frascari D, Pinelli D, Nocentini M, Fedi S, Pii Y, Zannoni D (2006) Chloroform degradation by butane-grown cells of *Rhodococcus aetherovorans* BCP1. Appl Microbiol Biotechnol 73:421–428. https://doi.org/10.1007/s00253-006-0433-3

Frascari D, Pinelli D, Nocentini M, Baleani E, Cappelletti M, Fedi S (2008) A kinetic study of chlorinated solvent cometabolic biodegradation by propane-grown *Rhodococcus* sp PB1. Biochem Eng J 42:139–147. https://doi.org/10.1016/j.bej.2008.06.011

Furuya T, Hirose S, Semba H, Kino K (2011) Identification of the regulator gene responsible for the acetone-responsive expression of the binuclear iron monooxygenase gene cluster in Mycobacteria. J Bacteriol 193:5817–5823. https://doi.org/10.1128/JB.05525-11

Furuya T, Hayashi M, Semba H, Kino K (2013) The mycobacterial binuclear iron monooxygenases require a specific chaperonin-like protein for functional expression in a heterologous host. FEBS J 280:817–826. https://doi.org/10.1111/febs.12070

Habib S, Ahmad SA, Johari WLW, Shukor MYA, Alias SA, Khalil KA, Yasid NA (2018) Evaluation of conventional and response surface level optimisation of *n*-dodecane (n-C_{12}) mineralisation by psychrotolerant strains isolated from pristine soil at Southern Victoria Island, Antarctica. Microb Cell Factories 17:44. https://doi.org/10.1186/s12934-018-0889-8

Hamamura N, Olson SH, Ward DM, Inskeep WP (2006) Microbial population dynamics associated with crude-oil biodegradation in diverse soils. Appl Environ Microbiol 72:6316–6324. https://doi.org/10.1128/AEM.01015-06

Hamamura N, Fukui M, Ward DM, Inskeep WP (2008) Assessing soil microbial populations responding to crude-oil amendment at different temperatures using phylogenetic, functional gene (*alkB*) and physiological analyses. Environ Sci Technol 42:7580–7586. https://doi.org/10.1021/es800030f

Hannemann F, Bichet A, Ewen KM, Bernhardt R (2007) Cytochrome P450 systems-biological variations of electron transport chains. Biochim Biophys Acta 1770:330–344. https://doi.org/10.1016/j.bbagen.2006.07.017

Hassanshahian M, Ahmadinejad M, Tebyanian H, Kariminik A (2013) Isolation and characterization of alkane degrading bacteria from petroleum reservoir waste water in Iran (Kerman and Tehran provenances). Mar Pollut Bull 73:300–305. https://doi.org/10.1016/j.marpolbul.2013.05.002

Hernández MA, Mohn WW, Martínez E, Rost E, Alvarez AF, Alvarez HM (2008) Biosynthesis of storage compounds by *Rhodococcus jostii* RHA1 and global identification of genes involved in their metabolism. BMC Genomics 9:600. https://doi.org/10.1186/1471-2164-9-600

Holder JW, Ulrich JC, DeBono AC, Godfrey PA, Desjardins CA, Zucker J, Zeng Q, Leach AL, Ghiviriga I, Dancel C, Abeel T, Gevers D, Kodira CD, Desany B, Affourtit JP, Birren BW, Sinskey AJ (2011) Comparative and functional genomics of *Rhodococcus opacus* PD630 for biofuels development. PLoS Genet 7:e1002219. https://doi.org/10.1371/journal.pgen.1002219

Holmes AJ, Coleman NV (2008) Evolutionary ecology and multidisciplinary approaches to prospecting for monooxygenases as biocatalysts. Antonie Van Leeuwenhoek 94:75–84. https://doi.org/10.1007/s10482-008-9227-1

Hommel RK (1990) Formation and physiological role of biosurfactants produced by hydrocarbon-utilizing microorganisms. Biosurfactants in hydrocarbon utilization. Biodegradation 1:107–119

Hua F, Wang HQ (2014) Uptake and trans-membrane transport of petroleum hydrocarbons by microorganisms. Biotechnol Biotechnol Equip 28:165–175. https://doi.org/10.1080/13102818.2014.906136

Huang L, Ma T, Li D, Liang FL, Liu RL, Li GQ (2008) Optimization of nutrient component for diesel oil degradation by *Rhodococcus erythropolis*. Mar Pollut Bull 56:1714–1718. https://doi.org/10.1016/j.marpolbul.2008.07.007

Inaba T, Tokumoto Y, Miyazaki Y, Inoue N, Maseda H, Nakajima-Kambe T, Uchiyama H, Nomura N (2013) Analysis of genes for succinoyl trehalose lipid production and increasing production in *Rhodococcus* sp. strain SD-74. Appl Environ Microbiol 79:7082–7090. https://doi.org/10.1128/AEM.01664-13

Iwabuchi N, Sharma PK, Sunairi M et al (2009) Role of interfacial tensions in the translocation of *Rhodococcus erythropolis* during growth in a two phase culture. Environ Sci Technol 43:8290–8294. https://doi.org/10.1021/es901208s
Ji Y, Mao G, Wang Y, Bartlam M (2013) Structural insights into diversity and *n*-alkane biodegradation mechanisms of alkane hydroxylases. Front Microbiol 4:58. https://doi.org/10.3389/fmicb.2013.00058
Jurelevicius D, Alvarez VM, Peixoto R et al (2013) The Use of a combination of *alkB* primers to better characterize the distribution of alkane-degrading bacteria. PLoS One 8:e66565. https://doi.org/10.1371/journal.pone.0066565
Kim IS, Foght JM, Gray MR (2002) Selective transport and accumulation of alkanes by *Rhodococcus erythropolis* S+14He. Biotechnol Bioeng 80:650–659. https://doi.org/10.1002/bit.10421
Kloos K, Munch JC, Schloter M (2006) A new method for the detection of alkane-monooxygenase homologous genes (*alkB*) in soils based on PCR-hybridization. J Microbiol Methods 66:486–496. https://doi.org/10.1016/j.mimet.2006.01.014
Koch DJ, Chen MM, van Beilen JB, Arnold FH (2009) *In vivo* evolution of butane oxidation by terminal alkane hydroxylases AlkB and CYP153A6. Appl Environ Microbiol 75:337–344. https://doi.org/10.1128/AEM.01758-08
Kohno T, Sugimoto Y, Sei K, Mori K (2002) Design of PCR primers and gene probes for general detection of alkane-degrading bacteria. Microbes Environ 17:114–121. https://doi.org/10.1264/jsme2.17.114
Kolouchová I, Schreiberová O, Masák J, Sigler K, Řezanka T (2012) Structural analysis of mycolic acids from phenol-degrading strain of *Rhodococcus erythropolis* by liquid chromatography–tandem mass spectrometry. Folia Microbiol (Praha) 57:473–483. https://doi.org/10.1007/s12223-012-0156-z
Kotani T, Yamamoto T, Yurimoto H, Sakai Y, Kato N (2003) Propane monooxygenase and NAD+-dependent secondary alcohol dehydrogenase in propane metabolism by *Gordonia* sp. strain TY-5. J Bacteriol 185:7120–7128. https://doi.org/10.1128/JB.185.24.7120-7128.2003
Kotani T, Kawashima Y, Yurimoto H, Kato N, Sakai Y (2006) Gene structure and regulation of alkane monooxygenases in propane-utilizing *Mycobacterium* sp. TY-6 and *Pseudonocardia* sp. TY-7. J Biosci Bioeng 102:184–192. https://doi.org/10.1263/jbb.102.184
Kretschmer A, Bock H, Wagner F (1982) Chemical and physical characterization of interfacial-active lipids from *Rhodococcus erythropolis* grown on normal-alkanes. Appl Environ Microbiol 44:864–870
Krinsky NI, Johnson EJ (2005) Carotenoid actions and their relation to health and disease. Mol Asp Med 26:459–516. https://doi.org/10.1016/j.mam.2005.10.001
Kubota M, Nodate M, Yasumoto-Hirose M, Uchiyama T, Kagami O, Shizuri Y, Misawa N (2005) Isolation and functional analysis of cytochrome P450 CYP153A genes from various environments. Biosci Biotechnol Biochem 69:2421–2430. https://doi.org/10.1271/bbb.69.2421
Kunihiro N, Haruki M, Takano K, Morikawa M, Kanaya S (2005) Isolation and characterization of *Rhodococcus* sp. strains TMP2 and T12 that degrade 2,6,10,14-tetramethylpentadecane (pristane) at moderately low temperatures. J Biotechnol 115:129–136. https://doi.org/10.1016/j.jbiotec.2004.07.018
Kuyukina MS, Ivshina IB, Rychkova MI, Chumakov OB (2000) Effect of cell lipid composition on the formation of nonspecific antibiotic resistance in alkanotrophic rhodococci. Microbiology 69:62–69. https://doi.org/10.1007/BF02757257
Kuyukina MS, Ivshina IB, Baeva TA, Kochina OA, Gein SV, Chereshnev VA (2015) Trehalolipid biosurfactants from nonpathogenic *Rhodococcus* actinobacteria with diverse immunomodulatory activities. New Biotechnol 32:559–568. https://doi.org/10.1016/j.nbt.2015.03.006
Laczi K, Kis Á, Horváth B, Maróti G, Hegedüs B, Perei K, Rákhely G (2015) Metabolic responses of *Rhodococcus erythropolis* PR4 grown on diesel oil and various hydrocarbons. Appl Microbiol Biotechnol 99:9745–9759. https://doi.org/10.1007/s00253-015-6936-z
Lanfranconi MP, Alvarez HM (2017) Rewiring neutral lipids production for the de novo synthesis of wax esters in *Rhodococcus opacus* PD630. J Biotechnol 260:67–73. https://doi.org/10.1016/j.jbiotec.2017.09.009

Lang S, Philp JC (1998) Surface-active lipids in rhodococci. Antonie Van Leeuwenhoek 74:59–70. https://doi.org/10.1023/A:1001799711799

Larkin MJ, Kulakov LA, Allen CCR (2005) Biodegradation and *Rhodococcus*—masters of catabolic versatility. Curr Opin Biotechnol 16:282–290. https://doi.org/10.1016/j.copbio.2005.04.007

Leahy JG, Colwell RR (1990) Microbial degradation of hydrocarbons in the environment. Microbiol Rev 54:305–315

Leahy JG, Batchelor PJ, Morcomb SM (2003) Evolution of the soluble diiron monooxygenases. FEMS Microbiol Rev 27:449–479. https://doi.org/10.1016/S0168-6445(03)00023-8

Lee EH, Kim J, Cho KS, Ahn YG, Hwang GS (2010) Degradation of hexane and other recalcitrant hydrocarbons by a novel isolate, *Rhodococcus* sp. EH831. Environ Sci Pollut Res Int 17:64–77. https://doi.org/10.1007/s11356-009-0238-x

Liang J-L, Nie Y, Wang M, Xiong G, Wang Y-P, Maser E, Wu X-L (2016a) Regulation of alkane degradation pathway by a TetR family repressor via an autoregulation positive feedback mechanism in a Gram-positive *Dietzia* bacterium. Mol Microbiol 99:338–359. https://doi.org/10.1111/mmi.13232

Liang J-L, JiangYang J-H, Nie Y, Wu X-L (2016b) Regulation of the alkane hydroxylase CYP153 gene in a gram-positive alkane-degrading bacterium, *Dietzia* sp. strain DQ12-45-1b. Appl Environ Microbiol 82:608–619. https://doi.org/10.1128/AEM.02811-15

Likhoshvay A, Lomakina A, Grachev M (2014) The complete *alk* sequences of *Rhodococcus erythropolis* from Lake Baikal. SpringerPlus 3:621. https://doi.org/10.1186/2193-1801-3-621

Ludwig B, Akundi A, Kendall K (1995) A long-chain secondary alcohol dehydrogenase from *Rhodococcus erythropolis* ATCC 4277. Appl Environ Microbiol 61:3729–3733

Luz AP, Pellizari VH, Whyte LG, Greer CW (2004) A survey of indigenous microbial hydrocarbon degradation genes in soils from Antarctica and Brazil. Can J Microbiol 50:323–333. https://doi.org/10.1139/w04-008

MacMichael G, Brown L (1987) Role of carbon dioxide in catabolism of propane by *'Nocardia parafinicum' (Rhodococcus rhodochrous)*. Appl Environ Microbiol 53:65–69

Malachowsky KJ, Phelps TJ, Teboli AB, Minnikin DE, White DC (1994) Aerobic mineralization of trichloroethylene, vinyl chloride, and aromatic compounds by *Rhodococcus* species. Appl Environ Microbiol 60:542–548.

Margesin R, Labbé D, Schinner F, Greer CW, Whyte LG (2003) Characterization of hydrocarbon-degrading microbial populations in contaminated and pristine alpine soils. Appl Environ Microbiol 69:3085–3092. https://doi.org/10.1128/AEM.69.6.3085-3092.2003

McCarl V, Somerville MV, Ly MA, Henry R, Liew EF, Wilson NL, Holmes A, Coleman N (2018) Heterologous expression of alkene monooxygenases from *Mycobacterium* in Gram positive and Gram negative bacterial hosts. Appl Environ Microbiol. https://doi.org/10.1128/AEM.00397-18

McDermott C, Heffron JJ (2013) Toxicity of industrially relevant chlorinated organic solvents in vitro. Int J Toxicol 32:136–145. https://doi.org/10.1177/1091581813482006

McDonald IR, Uchiyama H, Kambe S, Yagi O, Murrell JC (1997) The soluble methane monooxygenase gene cluster of the trichloroethylene-degrading methanotroph Methylocystis sp. strain M. Appl Environ Microbiol 63:1898–1904

McLeod MP, Warren RL, Hsiao WWL, Araki N, Myhre M, Fernandes C, Miyazawa D, Wong W, Lillquist AL, Wang D, Dosanjh M, Hara H, Petrescu A, Morin RD, Yang G, Stott JM, Schein JE, Shin H, Smailus D, Siddiqui AS, Marra MA, Jones SJ, Holt R, Brinkman FS, Miyauchi K, Fukuda M, Davies JE, Mohn WW, Eltis LD (2006) The complete genome of *Rhodococcus* sp. RHA1 provides insights into a catabolic powerhouse. PNAS 103:15582–15587. https://doi.org/10.1073/pnas.0607048103

Mikolasch A, Omirbekova A, Schumann P, Reinhard A, Sheikhany H, Berzhanova R, Mukasheva T, Schauer F (2015) Enrichment of aliphatic, alicyclic and aromatic acids by oil-degrading bacteria isolated from the rhizosphere of plants growing in oil-contaminated soil from Kazakhstan. Appl Microbiol Biotechnol 99:4071–4084. https://doi.org/10.1007/s00253-014-6320-4

Nakajima K, Sato A, Takahara Y, Iida T (1985) Microbial oxidation of isoprenoid alkanes, phytane, norpristane and farnesane. Agric Biol Chem 49:1993–2002

Neu TR (1996) Significance of bacterial surface-active compounds in interaction of bacteria with interfaces. Microbiol Rev 60:151–166

Nhi-Cong LT, Mikolasch A, Klenk H-P, Schauer F (2009) Degradation of the multiple branched alkane 2,6,10,14-tetramethyl-pentadecane(pristane) in *Rhodococcus ruber* and *Mycobacterium neoaurum*. Int Biodeter Biodegr 63:201–207. https://doi.org/10.1016/j.ibiod.2008.09.002

Nie Y, Chi C-Q, Fang H, Liang JL, Lu SL, Lai GL, Tang YQ, Wu XL (2014) Diverse alkane hydroxylase genes in microorganisms and environments. Sci Rep 4:4968. https://doi.org/10.1038/srep04968

Orro A, Cappelletti M, D'Ursi P, Milanesi L, Di Canito A, Zampolli J, Collina E, Decorosi F, Viti C, Fedi S, Presentato A, Zannoni D, Di Gennaro P (2015) Genome and Phenotype Microarray analyses of *Rhodococcus* sp. BCP1 and *Rhodococcus opacus* R7: genetic determinants and metabolic abilities with environmental relevance. PLoS One 10:e0139467. https://doi.org/10.1371/journal.pone.0139467.s025

Panicker G, Mojib N, Aislabie J, Bej AK (2010) Detection, expression and quantitation of the biodegradative genes in Antarctic microorganisms using PCR. Antonie Van Leeuwenhoek 97:275–287. https://doi.org/10.1007/s10482-009-9408-6

Peng F, Liu Z, Wang L, Shao Z (2007) An oil-degrading bacterium: *Rhodococcus erythropolis* strain 3C-9 and its biosurfactants. J Appl Microbiol 102:1603–1611. https://doi.org/10.1111/j.1365-2672.2006.03267.x

Pirog TP, Korzh YV, Shevchuk TA, Tarasenko DA (2008) Peculiarities of C2 metabolism and intensification of the synthesis of surface-active substances in *Rhodococcus erythropolis* EK-1 grown in ethanol. Microbiology 77:665–673. https://doi.org/10.1134/S0026261708060039

Quatrini P, Scaglione G, De Pasquale C, Riela S, Puglia AM (2008) Isolation of Gram-positive *n*-alkane degraders from a hydrocarbon-contaminated Mediterranean shoreline. J Appl Microbiol 104:251–259. https://doi.org/10.1111/j.1365-2672.2007.03544.x

Rapp P, Gabriel-Jürgens LH (2003) Degradation of alkanes and highly chlorinated benzenes, and production of biosurfactants, by a psychrophilic *Rhodococcus* sp. and genetic characterization of its chlorobenzene dioxygenase. Microbiology 149:2879–28790. https://doi.org/10.1099/mic.0.26188-0

Riebel A, Dudek HM, de Gonzalo G, Stepniak P, Rychlewski L, Fraaije MW (2012) Expanding the set of rhodococcal Baeyer-Villiger monooxygenases by high-throughput cloning, expression and substrate screening. Appl Microbiol Biotechnol 95:1479–1489. https://doi.org/10.1007/s00253-011-3823-0

Rojo F (2009) Degradation of alkanes by bacteria. Environ Microbiol 11:2477–2490. https://doi.org/10.1111/j.1462-2920.2009.01948.x

Saeki H, Furuhashi K (1994) Cloning and characterization of a *Nocardia corallina* B-276 gene cluster encoding alkene monooxygenase. J Biosci Bioeng 78:399–406. https://doi.org/10.1016/0922-338X(94)90037-X

Sameshima Y, Honda K, Kato J, Omasa T, Ohtake H (2008) Expression of *Rhodococcus opacus alkB* genes in anhydrous organic solvents. J Biosci Bioeng 106:199–203. https://doi.org/10.1263/jbb.106.199

Schenkels P, Duine JA (2000) Nicotinoprotein (NADH-containing) alcohol dehydrogenase from *Rhodococcus erythropolis* DSM 1069: an efficient catalyst for coenzyme-independent oxidation of a broad spectrum of alcohols and the interconversion of alcohols and aldehydes. Microbiology 146:775–785. https://doi.org/10.1099/00221287-146-4-775

Sekine M, Tanikawa S, Omata S, Saito M, Fujisawa T, Tsukatani N, Tajima T, Sekigawa T, Kosugi H, Matsuo Y, Nishiko R, Imamura K, Ito M, Narita H, Tago S, Fujita N, Harayama S (2006) Sequence analysis of three plasmids harboured in *Rhodococcus erythropolis* strain PR4. Environ Microbiol 8(2):334–346. https://doi.org/10.1111/j.1462-2920.2005.00899.x

Semprini L, Dolan ME, Hopkins GD, McCarty PL (2009) Bioaugmentation with butane-utilizing microorganisms to promote in situ cometabolic treatment of 1,1,1-trichloroethane and 1,1-dichloroethene. J Contam Hydrol 103(3–4):157–167. https://doi.org/10.1016/j.jconhyd.2008.10.005

Shanklin J, Whittle E, Fox BG (1994) Eight histidine residues are catalytically essential in a membrane-associated iron enzyme, stearoyl-CoA desaturase, and are conserved in alkane hydroxylase and xylene monooxygenase. Biochemist 33:12787–12794
Shanklin J, Achim C, Schmidt H, Fox BG, Münck E (1997) Mössbauer studies of alkane omega-hydroxylase: evidence for a diiron cluster in an integral-membrane enzyme. Proc Natl Acad Sci USA 94:2981–2986. https://doi.org/10.1073/pnas.94.7.2981
Sharma SL, Pant A (2000) Biodegradation and conversion of alkanes and crude oil by a marine *Rhodococcus*. Biodegradation 11(5):289–294. https://doi.org/10.1023/A:1011185806974
Sharp JO, Sales CM, LeBlanc JC, Liu J, Wood TK, Eltis LD, Mohn WW, Alvarez-Cohen L (2007) An inducible propane monooxygenase is responsible for N-nitrosodimethylamine degradation by *Rhodococcus* sp. strain RHA1. Appl Environ Microbiol 73:6930–6938. https://doi.org/10.1128/AEM.01697-07
Shennan JL (2006) Utilisation of C2–C4 gaseous hydrocarbons and isoprene by microorganisms. J Chem Technol Biotechnol 81:237–256. https://doi.org/10.1002/jctb.1388
Sluis MK, Sayavedra-Soto LA, Arp DJ (2002) Molecular analysis of the soluble butane monooxygenase from "*Pseudomonas butanovora*". Microbiology 148:3617–3629. https://doi.org/10.1099/00221287-148-11-3617
Smits THM, Röthlisberger M, Witholt B, van Beilen JB (1999) Molecular screening for alkane hydroxylase genes in Gram-negative and Gram-positive strains. Environ Microbiol 1:307–317. https://doi.org/10.1046/j.1462-2920.1999.00037.x
Smits THM, Balada SB, Witholt B, van Beilen JB (2002) Functional analysis of alkane hydroxylases from gram-negative and gram-positive bacteria. J Bacteriol 184:1733–1742. https://doi.org/10.1128/JB.184.6.1733-1742.2002
Sokolovská I, Rozenberg R, Riez C, Rouxhet PG, Agathos SN, Wattiau P (2003) Carbon source-induced modifications in the mycolic acid content and cell wall permeability of *Rhodococcus erythropolis* E1. Appl Environ Microbiol 69:7019–7027. https://doi.org/10.1128/AEM.69.12.7019-7027.2003
Stancu MM (2015) Response of *Rhodococcus erythropolis* strain IBBPo1 to toxic organic solvents. Braz J Microbiol 46:1009–1018. https://doi.org/10.1590/S1517-838246420140462
Stratton HM, Brooks PR, Carr EL, Seviour RJ (2003) Effects of culture conditions on the mycolic acid composition of isolates of *Rhodococcus* spp. from activated sludge foams. Syst Appl Microbiol 26:165–171. https://doi.org/10.1078/072320203322345994
Stroud JL, Paton GI, Semple KT (2007) Microbe-aliphatic hydrocarbon interactions in soil: implications for biodegradation and bioremediation. J Appl Microbiol 102:1239–1253. https://doi.org/10.1111/j.1365-2672.2007.03401.x
Takei D, Washio K, Morikawa M (2008) Identification of alkane hydroxylase genes in *Rhodococcus* sp. strain TMP2 that degrades a branched alkane. Biotechnol Lett 30:1447–1452. https://doi.org/10.1007/s10529-008-9710-9
Táncsics A, Benedek T, Szoboszlay S, Veres PG, Farkas M, Máthé I, Márialigeti K, Kukolya J, Lányi S, Kriszt B (2015) The detection and phylogenetic analysis of the alkane 1-monooxygenase gene of members of the genus *Rhodococcus*. Syst Appl Microbiol 38:1–7. https://doi.org/10.1016/j.syapm.2014.10.010
Tao L, Wagner LW, Rouvière PE, Cheng Q (2006) Metabolic engineering for synthesis of aryl carotenoids in *Rhodococcus*. Appl Microbiol Biotechnol 70:222–228. https://doi.org/10.1007/s00253-005-0064-0
van Beilen JB, Funhoff EG (2007) Alkane hydroxylases involved in microbial alkane degradation. Appl Microbiol Biotechnol 74:13–21. https://doi.org/10.1007/s00253-006-0748-0
van Beilen JB, Panke S, Lucchini S, Franchini AG, Rothlisberger M, Witholt B (2001) Analysis of *Pseudomonas putida* alkane-degradation gene clusters and flanking insertion sequences: evolution and regulation of the *alk* genes. Microbiology 147:1621–1630. https://doi.org/10.1099/00221287-147-6-1621

van Beilen JB, Neuenschwander M, Smits TH, Roth C, Balada SB, Witholt B (2002a) Rubredoxins involved in alkane oxidation. J Bacteriol 184:1722–1732. https://doi.org/10.1128/JB.184.6.1722-1732.2002
van Beilen JB, Smits TH, Whyte LG, Schorcht S, Röthlisberger M, Plaggemeier T, Engesser KH, Witholt B (2002b) Alkane hydroxylase homologues in Gram-positive strains. Environ Microbiol 4:676–682. https://doi.org/10.1046/j.1462-2920.2002.00355.x
van Beilen JB, Li Z, Duetz WA, Smits THM, Witholt B (2003) Diversity of alkane hydroxylase systems in the environment. OGST 58:427–440. https://doi.org/10.2516/ogst:2003026
van Beilen JB, Funhoff EG, van Loon A, Just A, Kaysser L, Bouza M, Holtackers R, Röthlisberger M, Li Z, Witholt B (2006) Cytochrome P450 alkane hydroxylases of the CYP153 family are common in alkane-degrading eubacteria lacking integral membrane alkane hydroxylases. Appl Environ Microbiol 72:59–65. https://doi.org/10.1128/AEM.72.1.59-65.2006
Viggor S, Jõesaar M, Vedler E, Kiiker R, Pärnpuu L, Heinaru A (2015) Occurrence of diverse alkane hydroxylase *alkB* genes in indigenous oil-degrading bacteria of Baltic Sea surface water. Mar Pollut Bull 101:507–516. https://doi.org/10.1016/j.marpolbul.2015.10.064
Vomberg A, Klinner U (2000) Distribution of *alkB* genes within *n*-alkane-degrading bacteria. J Appl Microbiol 89:339–348. https://doi.org/10.1046/j.1365-2672.2000.01121.x
Wang B, Chu KH (2017) Cometabolic biodegradation of 1,2,3-trichloropropane by propane-oxidizing bacteria. Chemosphere 168:1494–1497. https://doi.org/10.1016/j.chemosphere.2016.12.007
Warhust AM, Fewson CA (1994) Biotransformations catalyzed by the genus *Rhodococcus*. Crit Rev Biotechnol 14:29–73
Werck-Reichhart D, Feyereisen R (2000) Cytochromes P450: a success story. Genome Biol 1: REVIEWS3003. https://doi.org/10.1186/gb-2000-1-6-reviews3003
Whyte LG, Hawari J, Zhou E, Bourbonnière L, Inniss WE, Greer CW (1998) Biodegradation of variable-chain-length alkanes at low temperatures by a psychrotrophic *Rhodococcus* sp. Appl Environ Microbiol 64:2578–2584
Whyte LG, Slagman SJ, Pietrantonio F et al (1999) Physiological adaptations involved in alkane assimilation at a low temperature by *Rhodococcus* sp. strain Q15. Appl Environ Microbiol 65:2961–2968
Whyte LG, Schultz A, van Beilen JB et al (2002a) Prevalence of alkane monooxygenase genes in Arctic and Antarctic hydrocarbon-contaminated and pristine soils. FEMS Microbiol Ecol 41:141–150. https://doi.org/10.1111/j.1574-6941.2002.tb00975.x
Whyte LG, Smits THM, Labbé D, Witholt B, Greer CW, van Beilen JB (2002b) Gene cloning and characterization of multiple alkane hydroxylase systems in *Rhodococcus* strains Q15 and NRRL B-16531. Appl Environ Microbiol 68:5933–5942. https://doi.org/10.1128/AEM.68.12.5933-5942.2002
Woods BN, Murrell JC (1989) The metabolism of propane in *Rhodococcus vhohchvous* PNKbl. J Gen Microbiol 135:2335–2344
Yakimov MM, Giuliano L, Bruni V, Scarfì S, Golyshin PN (1999) Characterization of antarctic hydrocarbon-degrading bacteria capable of producing bioemulsifiers. New Microbiol 22:249–256
Yamashita S, Satoi M, Iwasa Y, Honda K, Sameshima Y, Omasa T, Kato J, Ohtake H (2007) Utilization of hydrophobic bacterium *Rhodococcus opacus* B-4 as whole-cell catalyst in anhydrous organic solvents. Appl Microbiol Biotechnol 74:761–767. https://doi.org/10.1007/s00253-006-0729-3
Yang HY, Jia RB, Chen B, Li L (2014) Degradation of recalcitrant aliphatic and aromatic hydrocarbons by a dioxin-degrader *Rhodococcus* sp. strain p52. Environ Sci Pollut Res Int 21:11086–11093. https://doi.org/10.1007/s11356-014-3027-0
Yuste L, Corbella ME, Turiégano MJ, Karlson U, Puyet A, Rojo F (2000) Characterization of bacterial strains able to grow on high molecular mass residues from crude oil processing. FEMS Microbiol Ecol 32:69–75. https://doi.org/10.1111/j.1574-6941.2000.tb00700.x

Zampolli J, Collina E, Lasagni M, Di Gennaro P (2014) Biodegradation of variable-chain-length *n*-alkanes in *Rhodococcus opacus* R7 and the involvement of an alkane hydroxylase system in the metabolism. AMB Express 4:73. https://doi.org/10.1186/s13568-014-0073-4
Zheng YT, Toyofuku M, Nomura N, Shigeto S (2013) Correlation of carotenoid accumulation with aggregation and biofilm development in *Rhodococcus* sp. SD-74. Anal Chem 85:7295–7301. https://doi.org/10.1021/ac401188f

Biodegradation of Nitriles by *Rhodococcus*

Huimin Yu, Song Jiao, Miaomiao Wang, Youxiang Liang, and Lingjun Tang

Contents

Abstract Nitriles occur naturally in the environment, are produced by the metabolic pathways of organisms or are released by the chemical and pharmaceutical industries, from agricultural applications or from the processing of fossil fuels. Therefore, a variety of nitrile-converting bacterial species are used to alleviate this toxic effect. Among these bacteria, *Rhodococcus* species have proven to be a superior group for

H. Yu (✉)
Department of Chemical Engineering, Tsinghua University, Beijing, China

Key Laboratory of Industrial Biocatalysis, The Ministry of Education, Tsinghua University, Beijing, China

Center for Synthetic and Systems Biology, Tsinghua University, Beijing, China
e-mail: yuhm@tsinghua.edu.cn

S. Jiao · M. Wang · Y. Liang · L. Tang
Department of Chemical Engineering, Tsinghua University, Beijing, China

Key Laboratory of Industrial Biocatalysis, The Ministry of Education, Tsinghua University, Beijing, China

H. M. Alvarez (ed.), *Biology of Rhodococcus*, Microbiology Monographs 16,
https://doi.org/10.1007/978-3-030-11461-9_7

the clean-up of pollutants. Nitriles are converted into the less toxic carboxylic acid either by nitrilases or by nitrile hydratase (NHase)/amidase systems. Although NHases, nitrilases and amidases produced by different strains exhibit different catalytic characteristics towards aliphatic nitriles and aromatic nitriles, these enzymes exhibit considerable homology in amino acid sequence or structure. In contrast, the enzymes with different origins present different types of gene organization and regulatory patterns, although the amidase gene is always linked to the NHase gene. Due to the advantage of being resistant to toxic compounds, applications of *Rhodococcus* in pollutant biodegradation and biocatalytic processes are very promising. While studies on the biodegradation of nitrile pollutants focus on the screening and discovery of strains, the industrial application of these enzymes as biocatalysts focuses on engineering combined with immobilization of both *Rhodococcus* cells and enzymes to improve their performance under the adverse conditions in the catalytic process.

1 Introduction

Nitriles are widely distributed in nature and can be divided into three general categories: aliphatic nitriles, aromatic nitriles and heterocyclic nitriles. Nitriles may be produced by a variety of microorganisms and plants and exist as 3-indolylacetonitrile (an auxin precursor), 3-phenylpropionitrile (a growth hormone) and so on. In living organisms, these naturally occurring nitriles may be the product of the following pathway: amino acid → aldoxime → nitriles. Aldoxime dehydratase participates in the formation of nitrile from aldoxime. Nitriles are also products of the agricultural, pharmaceutical and chemical industries (Bhalla et al. 2012). These nitriles are important intermediates for the production of amides, acids and other compounds. Despite the importance of nitriles in organic synthesis, large amounts of nitrile compounds are released to the environment, including to soil, air and water, and are harmful and toxic to human beings; therefore, remediation of the polluted environment is an urgent issue.

Due to the widespread occurrence of nitriles in metabolic pathways and natural environments, nitrile-converting enzymes occur in a wide variety of bacterial species, such as *Pseudomonas* sp., *Bacillus*, *Corynebacterium*, *Nocardia*, *Micrococcus* and *Rhodococcus*. In microorganisms, nitriles are primarily degraded to carboxylic acids, which are then converted to other metabolites. The conversion of nitriles to carboxylic acids occurs via two types of pathways: (1) direct hydrolysis of nitriles to carboxylic acids by nitrilases and (2) cascade catalysis of nitriles by NHases and amidases via the pathway nitriles → amides → carboxylic acids. While some identified nitrile-converting strains have only one of the two pathways, others harbour both the pathways. Nitrilases and NHases produced by different strains also exhibit different catalytic preferences for aliphatic and aromatic nitriles. While some nitrilases are capable of catalysing aliphatic nitriles efficiently, the enzymes are not efficient at catalysing aromatic nitriles. A similar phenomenon has been observed for different types of NHases. In particular, *Rhodococcus rhodochrous* J1 can produce two types of NHases, namely, high-molecular-mass NHase and low-molecular-mass NHase, which exhibit different specificities for aliphatic nitriles (Komeda et al. 1996b, c).

The diverse catalytic properties of various NHases and nitrilases enable extensive application of these enzymes to transform a vast number of man-made nitriles for the production of many important bulk and fine chemicals or to remove pollutants from waste materials and environment. NHases, which catalyse the hydration of nitriles to the corresponding amides, have been widely used in the chemical industry for the production of acrylamide, nicotinamide and 5-cyanovaleramide. In addition, nitrilases are attractive biocatalysts for the production of various commodity chemicals and pharmaceutical intermediates, such as nicotinic acid, isonicotinic acid, mandelic acid, acrylic acid, glycolic acid and benzoic acid. Among nitrile-converting strains, *Rhodococcus* strains have proven to be the most powerful and successful in the industrial transformation of various nitriles. However, there are some bottlenecks in the application of these enzymes in the transformation of nitriles to the corresponding amides or carboxylic acids, such as activity inhibition at high concentrations of substrates and products, thermal susceptibility and low nitrilase activity. As a result, genetic engineering strategies combined with immobilization were applied to enhance the stability and activities of the *Rhodococcus* cells and enzymes to broaden the industrial application of these enzymes.

2 Nitrile-Degrading Enzymes

Nitrile compounds, which have a $-C{\equiv}N$ functional group, are ubiquitous in natural environments as intermediates synthesized by chemical processes or biological systems. Nitrile compounds have application in many areas, for example, 2,6-dichlorobenzonitrile is used as an herbicide (Tao et al. 2016), nitrile rubber is used in medical gloves, polyacrylonitrile fibres are used to manufacture clothing, acetonitrile is used in analytical chemistry, and citalopram is used as a pharmaceutical (Fleming et al. 2010). The widespread use of nitrile compounds inevitably causes water pollution, soil contamination and accumulation of nitrile-containing waste. Over the past few decades, nitrile degradation has become a cause for concern worldwide. Biodegradation is more promising than chemical and physical methods due to the complete conversion, low cost and low level of secondary pollution associated with this method (Alexander 2001). Usually, biodegradation is carried out by enzymatic catalysis. The key enzymes used for nitrile degradation are NHase, amidase and nitrilase (Gong et al. 2017). NHase catalyses the hydration of nitriles to the corresponding amides (Ma et al. 2010); amidase catalyses the hydrolysis of amides to the corresponding acids; and nitrilase catalyses the one-step hydrolysis of nitriles to the corresponding acids (Fig. 1).

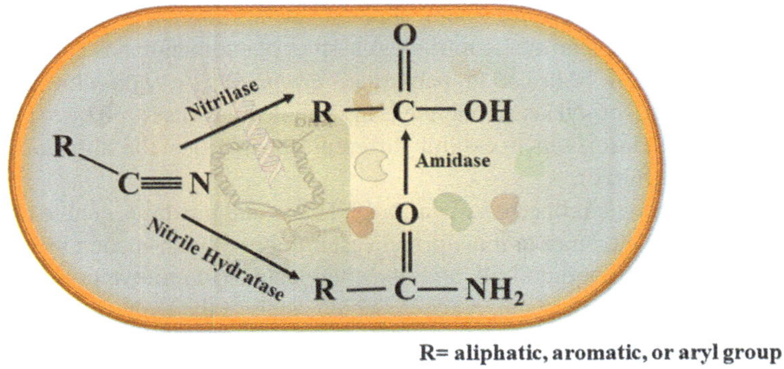

Fig. 1 Nitrile-biodegrading enzymes

2.1 NHase

NHase was first discovered in 1980 (Asano et al. 1980), and over the last few decades, many NHases have been identified and characterized. Various natural NHases are synthesized by plants and bacteria; these enzymes vary widely in length and exhibit structural diversity. However, microbial NHases usually exhibit considerable homology in amino acid sequence or structure (Cramp and Cowan 1999). Microbial NHase components are hetero-oligomers (αnβn), and typically, NHases exist as αβ dimers or tetramers (Cowan et al. 1998). Each αβ unit has a low-spin metal ion at the active centre (Huang et al. 1997), and NHases are classified into Co-type and Fe-type NHases according to the metal ions associated with the enzyme. Co-type NHase exhibits superior thermostability to Fe-type NHase in industrial biochemical synthetic processes (Cowan et al. 1998). Basically, structure determines function. As shown in Figs. 2 and 3, the sequence and structure characteristics of the α and β subunits of both classical Co-type NHases [PDB ID 1IRE from *Pseudonocardia thermophila* (Miyanaga et al. 2001) and PDB ID 1V29 from *Bacillus smithii* (Hourai et al. 2003)] and two Fe-type NHases [PDB ID 2AHJ from *Rhodococcus erythropolis* (Nagashima et al. 1998) and PDB 4FM4 from *Comamonas testosteroni* (Kuhn et al. 2012)] are illustrated.

The α subunits of Co-type and Fe-type NHases have highly homologous amino acid sequences and structures (Fig. 2). The size of the α subunit is 24–28 kDa. Structurally, three cysteine residues and one serine residue (-C-X-X-C-S-C-) in the cysteine cluster region constitute a fully conserved sequence in the α subunit of all Co- and Fe-type NHases (Fig. 2a). The three cysteine thiolates and the two main-chain amide nitrogens form a square-pyramidal geometry as ligands that co-ordinate the Fe or Co ion. The structures of the Co- and Fe-type NHase active centres are the same, except at one site. The sixth site is occupied by a water/hydroxide molecule as the active form in Co-type NHase, whereas a photolabile NO molecule is present at this site in Fe-type NHase (Endo et al. 2001).

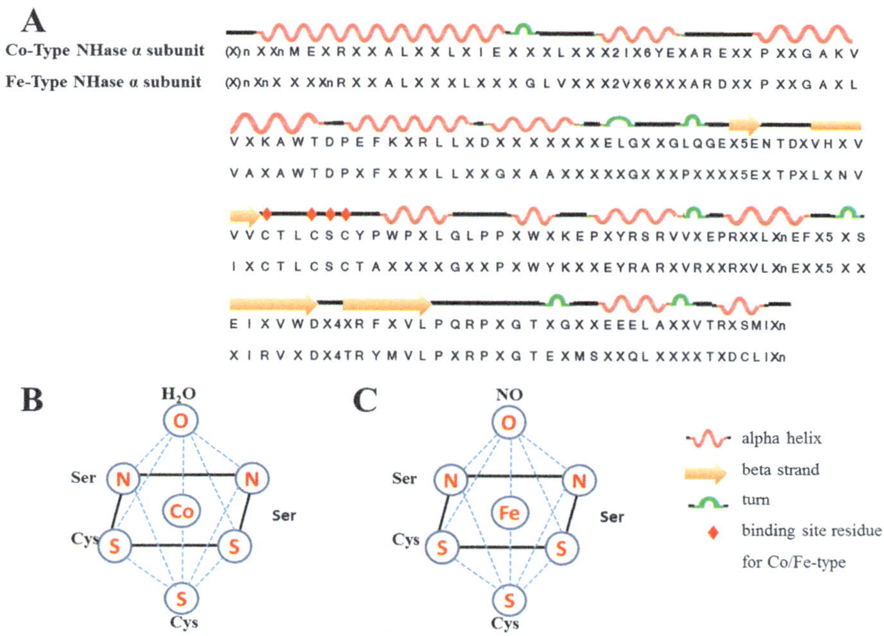

Fig. 2 Partial amino acid sequences and structure of Co- and Fe-type NHase α subunits. Co-type NHase: 1IRE (Miyanaga et al. 2001) and 1V29 (Hourai et al. 2003). Fe-type: 2AHJ (Kuhn et al. 2012) and 4FM4 (Endo et al. 2001). (**a**) Amino acid sequence alignment of Co- and Fe-type NHase α subunits. (**b**) Cobalt active centre of Co-type NHase. (**c**) Structure of the iron active centre of Fe-type NHase in the inactive state

The β subunits of NHases, ranging from 25 to 39 kDa in size, have lower amino acid sequence homology than the α subunits. Two Arg residues in the β subunit are highly conserved in both Co-type and Fe-type NHases (Fig. 3a, b). The guanidine groups of these two Arg residues react with the sulphur atoms of Cys in the α subunit and form a salt bridge network with electrostatic interactions (Fig. 3c). The formation of a salt bridge around the metal ion ensures the binding of the α and β subunits to stabilize the subunit interface. In addition, there exist several structural differences between Co-type and Fe-type NHases (Fig. 3). In particular, one α helix (blue in Fig. 3c, β111–β125) in Co-type NHase is replaced by a long loop (β95–β138 shown in Fig. 3c) in Fe-type NHase. This helix, which is composed in part of these residues (β111–β125), interacted with another helix belonging to the α subunit. This additional interaction may contribute to the better thermostability of Co-type NHase than Fe-type NHase (Miyanaga et al. 2001).

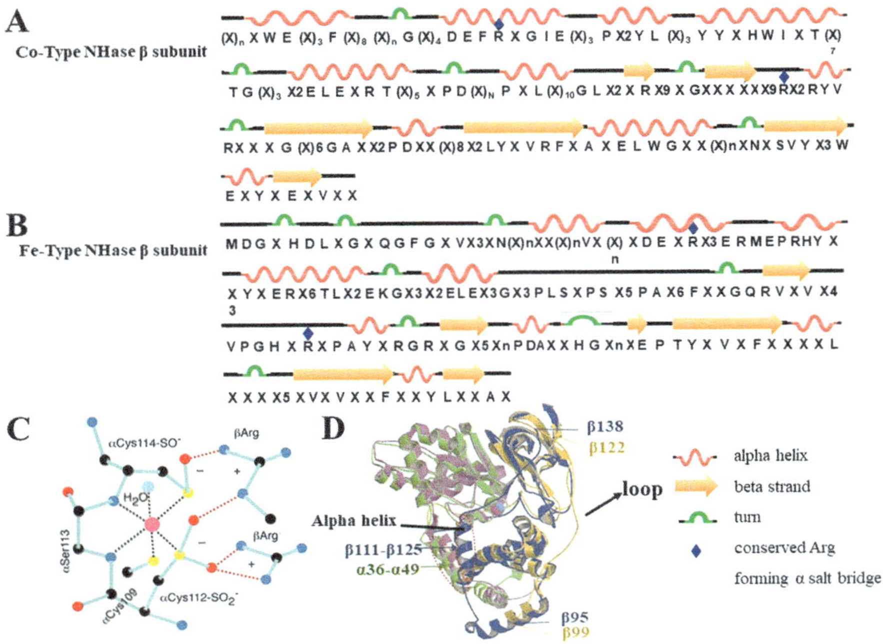

Fig. 3 Amino acid sequences and structures of Co- and Fe-type NHase β subunits. (**a**) Amino acid sequence alignment of the β subunit of Co-type NHase. Co-type NHase: 1IRE (Miyanaga et al. 2001) and 1V29 (Hourai et al. 2003). (**b**) Amino acid sequence alignment of the β subunit of Fe-type NHase. Fe-type NHase: 2AHJ (Kuhn et al. 2012) and 4FM4 (Endo et al. 2001). (**c**) Model of the noncorrin cobalt centre of NHase. Atoms are shown in different colours: pink for Co, black for C, red for O, yellow for S and blue for N. The salt bridge networks formed between the cysteines of the α subunit and the arginines of the β subunit are shown as red dotted lines (Zhou et al. 2008). (**d**) Ribbon diagram of Co-type and Fe-type NHases (Miyanaga et al. 2001)

2.2 *Amidase*

Amidase (EC 3.5.1.4) is widespread in nature and catalyses the hydrolysis of amides into the corresponding carboxylic acids and ammonia. Coupled with NHase, amidase has great potential in the degradation of toxic nitrile compounds. Based on amino acid sequence and structural homology, amidases have been classified into two groups: the nitrilase superfamily and the amidase signature (AS) family (Ohtaki et al. 2010).

The first group of amidases belongs to nitrilase superfamily. These enzymes have low amino acid sequence homology but high structural homology (Andrade et al. 2007; Kimani et al. 2007; Makhongela et al. 2007). Being members of the branch 2 nitrilase superfamily, amidases have the conserved catalytic Glu-Lys-Cys triad, and the amidase monomers have the typical nitrilase superfamily αββα sandwich fold, similar to nitrilase.

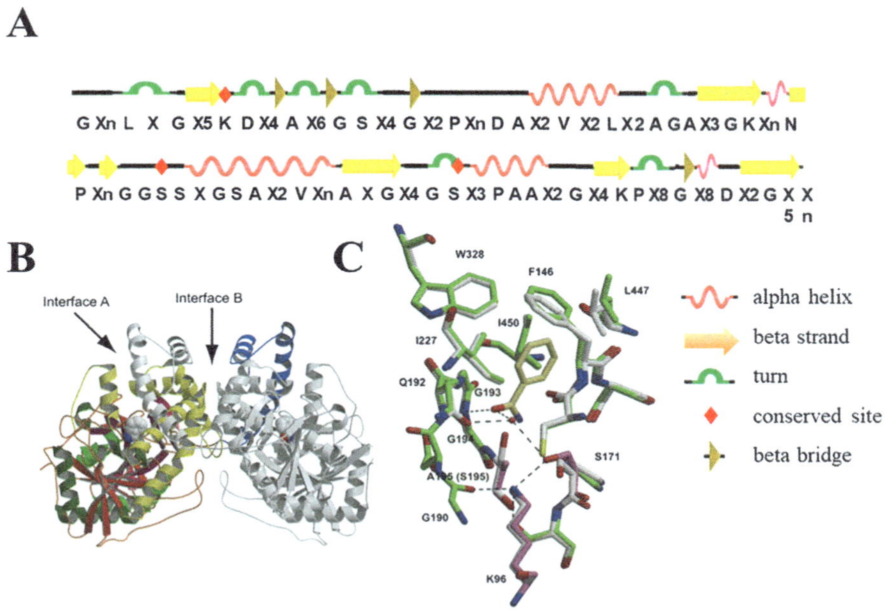

Fig. 4 Amino acid sequence and structure alignment of amidase. (**a**) Structural sequence alignment of the amidase signature (AS) region. The secondary structural elements identified in 3A1I (Ohtaki et al. 2010) are indicated in the top line. (**b**) Dimer structure of RhAmidase (Ohtaki et al. 2010). A monomer is shown in colour. (**c**) Structure of the active site of RhAmidase (Ohtaki et al. 2010)

The other group of amidases belongs to the AS family. AS family members have a highly conserved amino acid sequence, named the AS sequence, and a conserved Ser-Ser-Lys catalytic triad, and these proteins exist as homodimeric or homo-octameric complexes (Fig. 4). The amidase from *Rhodococcus* sp. N-771, namely, RhAmidase, can be taken as an example that has an active homodimeric structure and the active catalytic triad. A single monomer is shown in colour in Fig. 4b, and the two monomers binding crosswise. All helices of the small domain form interfaces A and B to participate in the formation of the homodimeric structure, and these interactions contribute to the formation of a closely packed dimer structure. The conserved catalytic Ser-Ser-Lys triad is located in the large domain. The first Lys96 can be deprotonated and acts as the catalytic base, and this residue also forms hydrogen bonds with Ser171 and Gly190; the second Ser171 is in an unusual *cis* conformation; and the third Ser195 forms a covalent bond with the substrate (Ohtaki et al. 2010; Lee et al. 2015; Valiña et al. 2004).

2.3 Nitrilase

Nitrilase was first described by Thimann and Mahadevan in 1964, and since then, over 200 nitrilases belonging to the nitrilase superfamily have been discovered and sequenced (Thuku et al. 2009). The nitrilase superfamily is characterized by a homodimeric building block with an αββα-αββα sandwich fold and is classified into 13 branches based on amino acid sequence similarity and the presence of additional domains (Thuku et al. 2009; Pace and Brenner 2001).

The nitrilases in branch 1 of the nitrilase superfamily hydrolyse nitrile (RCN) to ammonia and the corresponding carboxylic acid. Moreover, nitrilases are further categorized into aliphatic nitrilases, aromatic nitrilases and arylacetonitrilases according to substrate specificity, although some nitrilases exhibit broad substrate specificity, and the substrate specificity can also be changed by mutating key amino acid residues of nitrilases (Nigam et al. 2017). Natural nitrilases can be synthesized by plants, animals, fungi and bacteria. Among the nitrilases from different sources, microbial nitrilases are often exploited for biochemical syntheses and environmental remediation (Pace and Brenner 2001). The most well-known branch 1 nitrilases are from the most abundant nitrilase source, namely, bacteria of the *Rhodococcus* genus. To better depict their characteristics and structural homology, the sequences of the nitrilase from *R. rhodochrous* J1 (Thuku et al. 2007) and the nitrilase-related enzymes from *Agrobacterium* sp. KNK712 (PDB ID 1ERZ) (Nakai et al. 2000) and *Pyrococcus horikoshii* (PDB ID 1J31) (Sakai et al. 2004) are aligned in Fig. 5.

The majority of microbial nitrilases have a subunit size of 30–45 kDa and homomultimers with native compositions such as α_n (with n ranging from 4 to 22) (Thuku et al. 2009). The amino acid sequences of different nitrilases vary greatly, but all members exhibit secondary and tertiary homology upon alignment of their crystal structures (Thuku et al. 2007, 2009; Park et al. 2017) (Fig. 5a). Take the best characterized nitrilase from *R. rhodochrous* J1 as an example. The nitrilase in J1 is known to be inactive as a dimer in solution but to be active as an oligomeric complex (with 10–12 subunits) when subjected to heat treatment or in the presence of nitrile, ammonium sulphate or organic solvents (Thuku et al. 2009; Park et al. 2017). The dimer model of *R. rhodochrous* J1 nitrilase shown in Fig. 5b is built based on structural homology to nitrilase-related enzymes whose crystal structures have been solved (Thuku et al. 2007). The monomer association occurs via two interfaces, namely, the ‘A’ and ‘C’ surfaces, and leads to an eight-layered αββα-αββα dimer (Fig. 5b). As seen in Fig. 5a, the length of the amino acid sequence that forms ‘C’ surface varies much more than that of the sequence that forms the ‘A’ or ‘D’ surface (Thuku et al. 2007). The C surface is responsible for spiral elongation. All nitrilase superfamily enzymes have a conserved Glu-Lys-Cys catalytic triad, whereas the active site of the nitrilase in J1 has an extra glutamic acid and comprises the residues C165, K131, E48 and E138 (shown in Fig. 5c). C165 is speculated to initiate a nucleophilic attack on the substrate to form a tetrahedral intermediate, and K131 stabilizes the tetrahedral intermediate (Thuku et al. 2009). The two Glu residues are speculated to play an important role in positioning the substrate. E48 increases the

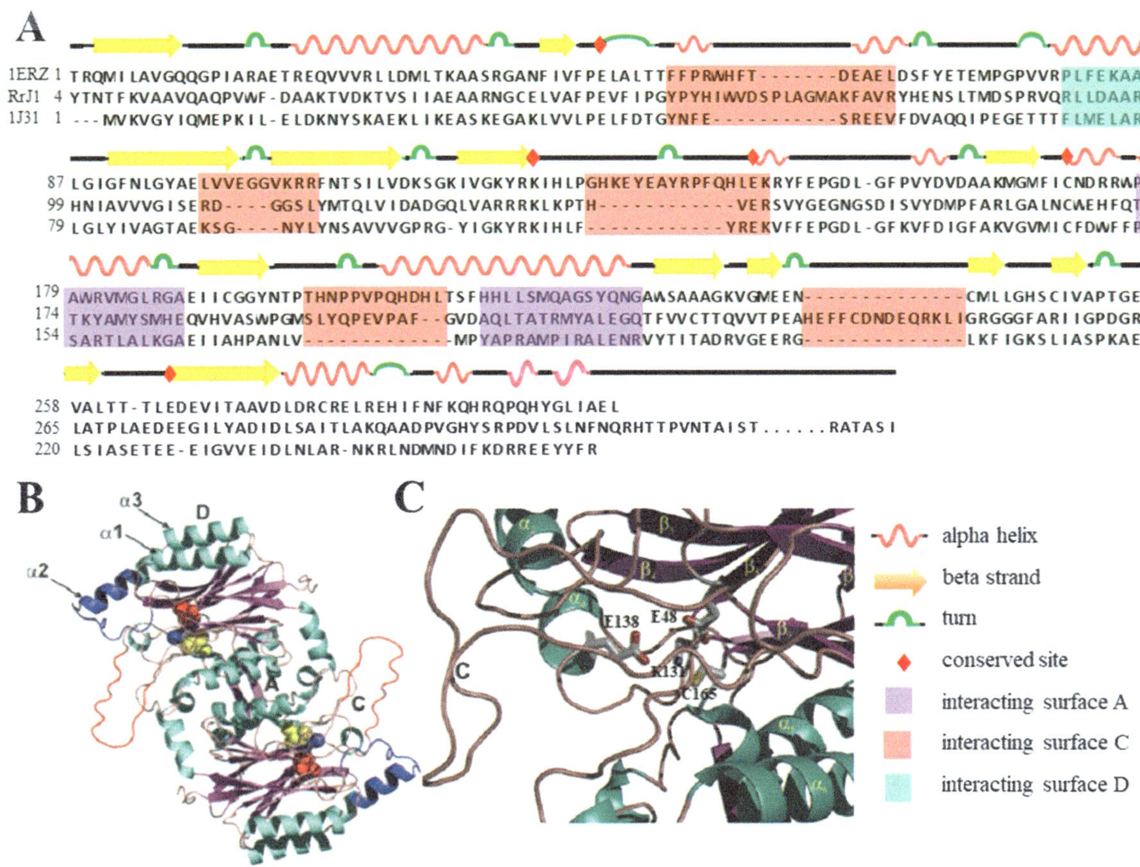

Fig. 5 Amino acid sequence and structural alignment of nitrilase. (**a**) Sequence alignment of the nitrilase from *Rhodococcus rhodochrous* J1 (RrJ1) with two nitrilase-related enzyme homologues [1ERZ (Nakai et al. 2000) and 1J31 (Sakai et al. 2004)]. The symbol in the top line indicates the position in the

nucleophilicity of the cysteine, and E138 is located in a 'C' surface loop, leading to the possibility that association of dimers moves this residue into the correct position for catalytic activity (Thuku et al. 2009).

3 Cluster Arrangement and Expression Regulation of Nitrile-Converting Genes

3.1 Nitrile-Converting Rhodococcus *Strains*

Nitrile-converting enzymes (NHases, nitrilases) exist in many *Rhodococcus* strains. Previous studies have reported that Fe-type and Co-type NHases were found in *R. erythropolis* and *R. rhodochrous* species and nitrilase-producing strains mainly belong to the *R. rhodochrous* species (Martínková et al. 2010). However, with the rapid development of next-generation sequencing technology, various *Rhodococcus* species have been revealed to be capable of producing NHase and nitrilase.

A genome-wide search of 283 *Rhodococcus* strains showed that all the *Rhodococcus* strains could produce amidase; 57% of the *Rhodococcus* strains harbour genes encoding NHases; and only 18% of the strains were identified as harbouring genes encoding nitrilases (Fig. 6a). Twenty-five of these strains exhibited NHase, amidase and nitrilase activities (Fig. 6b).

Except for a large number of strains not classified at the species level (designated *Rhodococcus* sp.), NHase could be produced by all the 36 strains of *R. equi*, 19 strains of *R. erythropolis* species, 8 strains of *R. opacus* and 7 strains of *R. qingshengii*. However, most of the NHases from these species have never been reported, except a number of NHases from *R. erythropolis* strains with high similarity and one from *R. equi* TG328-2, all of which bear the Fe^{3+} cofactor (Rzeznicka et al. 2010; Martínková et al. 2010). Co^{3+}-type NHases have been reported in *R. rhodochrous* J1 (Komeda et al. 1996b, c), *R. rhodochrous* M8 (Pogorelova et al. 1996; Pertsovich et al. 2005) and *R. ruber* TH (Ma et al. 2010). Specifically, in *R. rhodochrous* J1, two NHase subtypes were identified, i.e. high-molecular-mass NHase (H-NHase) and low-molecular-mass NHase (L-NHase), which exhibit different substrate specificities (Komeda et al. 1996b, c). In contrast, only H-NHase was identified in *R. ruber* TH, exhibiting 98% identity with the H-NHase from *R. rhodochrous* J1. A new type of NHase containing three types of metal ions (Co, Cu and Zn) was discovered and characterized in *Rhodococcus jostii* RHA1, which, however, shares no homology with the well-known Fe- and Co-type NHases (Okamoto and Eltis 2007).

Fig. 5 (continued) homologues. The approximate regions of the interacting surfaces A, C and D are indicated in purple, red and blue, respectively. The secondary structural elements identified in 1ERZ (Nakai et al. 2000) are indicated in the top line. (**b**) Stereo view of a dimer model of the nitrilase from *R. rhodochrous* J1 (Thuku et al. 2007). (**c**) Close-up view of the 'C' surface loop and the active site of the model of the nitrilase from *R. rhodochrous* J1 (Thuku et al. 2009)

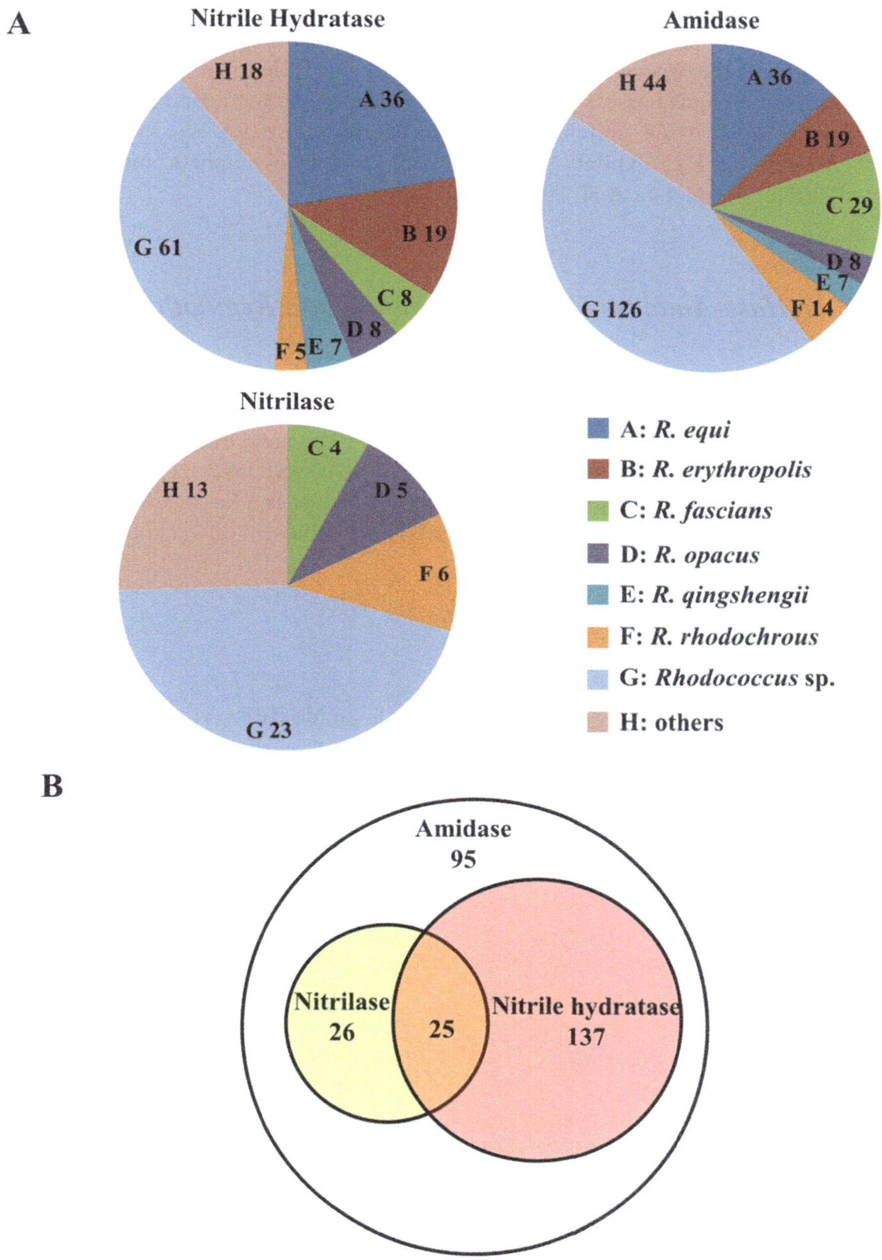

Fig. 6 *Rhodococcus* strains harbouring nitrile-converting enzymes. (**a**) NHase, amidase and nitrilase distribution in different *Rhodococcus* species. (**b**) Relationship of NHase-, amidase- and nitrilase-producing *Rhodococcus* strains

Unlike NHases, which can be produced by many *Rhodococcus* species, nitrilase-coding genes were detected in only a few species, mainly *R. fascians*, *R. opacus* and *R. rhodochrous*. In addition, many other nitrilase-containing strains have not been classified at the species level (*Rhodococcus* sp.). Nitrilases from different *Rhodococcus* strains possess individual preferences for aromatic or aliphatic nitriles. While most nitrilases prefer aromatic nitriles as substrates, the nitrilase from *R. rhodochrous* K22 exhibits significant activity towards aliphatic nitriles (Gong et al. 2012; Martínková et al. 2010).

3.2 NHase-Amidase Gene Organization and Regulation

Generally, genes that share a generalized function are often located in a gene cluster. In the metabolic pathway of nitrile degradation, nitrile is converted to amide by NHase and then transformed to carboxylic acid by amidase, so generally, the amidase gene is linked to the NHase gene, which has been detected in the majority of *Rhodococcus* strains regardless of whether the NHase is a Fe- or Co-type NHase. Despite this characteristic, Fe- and Co-type NHases represent different types of gene organization and regulation patterns.

3.2.1 Fe-Type NHase

The genes involved in the aldoxime-nitrile-amide-carboxylic acid pathway containing the Fe-type NHase are typically organized as shown in Fig. 7a, as seen in the *R. erythropolis* PR4 genome (GenBank: AP008957.1). NHase is expressed by *nha1* and *nha2*, coding for the α and β subunits, which are linked to the amidase gene (*ami*). The *nhr3* gene codes for an NHase activator, which activates the NHase via the oxidation of the iron centre and incorporation of an Fe ion into the protein (Nojiri et al. 2000). This activator gene is also essential for the function of NHase when expressing *nha1* and *nha2* genes in *E. coli*. The expression of NHase is probably regulated by *nhr4*, *nhr2* and *nhr1*, which encode regulators; however, to date, there has been no report regarding the exact functions of these genes. The gene coding for aldoxime dehydratase (*oxd*) was also found to be close to the *nha2*, *nha1* and *ami* genes in *R. erythropolis* PR4. Similar gene structures and organizations were also observed in *R. erythropolis* A4 (GenBank: AM946017.1), *R. globerulus* A-4 (GenBank: AB105912.1), *Rhodococcus* sp. N-771 (GenBank: AB016078.1) (Kato et al. 2004; Endo et al. 2001) and *Rhodococcus* sp. N-774 (Martínková et al. 2010). The NHase-amidase gene cluster was also identified in the *R. erythropolis* strains AJ270 and AJ300, whereas a copy of the insertion sequence IS1166 was present within the *nhr2* gene (O'mahony et al. 2005). In *R. jostii* RHA1, the genes coding for amidase and NHase share 79% and 90% homology, respectively, with those in *R. erythropolis* PR4; however, the regulatory genes *nhr1* and *nhr2* are absent, while the genes *oxd* and *nhr4* were detected (GenBank: CP000431.1).

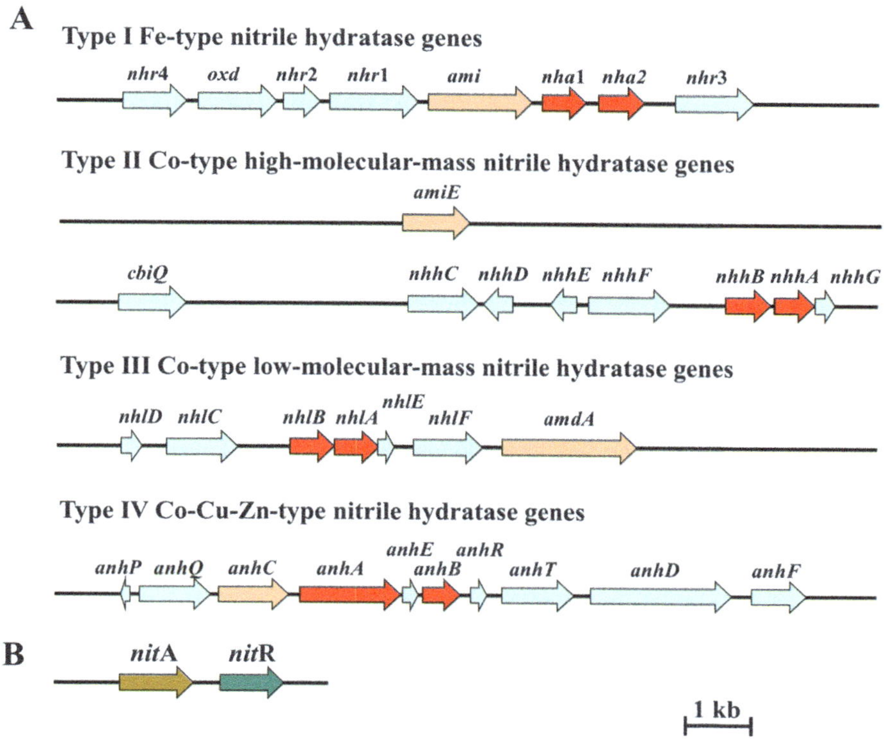

Fig. 7 Organization of nitrile-converting gene clusters from various *Rhodococcus* strains. (**a**) Four types of NHase-amidase gene cluster organizations. (**b**) Nitrilase gene cluster organization

In most cases, the NHase and amidase genes are inducible by amides (reaction products) instead of nitriles (reaction substrates) (Kobayashi and Shimizu 1998), which has been observed mostly with acetamide (Rucká et al. 2014; O'mahony et al. 2005). However, constitutive expression patterns occur in some other strains, such as *Rhodococcus* sp. N-771 and *Rhodococcus* sp. R312 (Prasad and Bhalla 2010).

3.2.2 Co-Type NHase

In contrast to the Fe-type NHase, two types of NHases, i.e. H-NHase and L-NHase, exist in *Rhodococcus* species and were found to coexist in *R. rhodochrous* J1.

The typical gene cluster of H-NHase contains *nhhC*, *nhhD*, *nhhE*, *nhhF*, *nhhB*, *nhhA* and *nhhG*, as determined in *R. rhodochrous* J1 (Fig. 7a) (GenBank: D67027.1) (Komeda et al. 1996b). *nhhB* and *nhhA* code for the subunits of NHase and are positively regulated by the essential *nhhC* and *nhhD* genes. The *nhhG* gene is homologous to *nhhB*, although the exact function of this gene has not been determined. *nhhF* encodes an insertion sequence IS1164, suggesting that horizontal gene transfer or genetic rearrangement of the H-NHase gene cluster occurred over the course of evolution. *nhhE* encodes a 14.6-kDa protein that does not share significant similarity with any reported protein and is not indispensable for the expression of the H-NHase gene. Genes homologous to *nhhC*, *nhhD*, *nhhB*, *nhhA* and *nhhG* are present in *Rhodococcus* sp. M8 (NCBI Reference Sequence: NZ_MLYX02000005.1), whereas *nhhE* and *nhhF* are absent. Despite the absence of *nhhF*, a gene coding for another IS481 family transposase has been identified immediately upstream of *nhhB*. In addition, the whole-genome sequence of *Rhodococcus* sp. M8 reveals a gene encoding amidase that is located 15.8 kb upstream of the H-NHase gene, which is consistent with the observation that amidase is always found together with NHase (Fig. 7a). As a result, whether amidase is present in the gene cluster of *R. rhodochrous* J1 H-NHase should be re-evaluated by cloning a sequence longer than the current 6555 bp. A gene encoding the cobalt ECF transporter T component (CbiQ) has also been identified upstream of *nhhBA* in *Rhodococcus* sp. M8 and may be involved in cobalt ion transport (Fig. 7a). Homologous H-NHase genes were also found in *Rhodococcus pyridinivorans* (Kohyama et al. 2006).

The subunits of L-NHase are encoded by *nhlB* and *nhlA*, which are clustered with *nhlD*, *nhlC*, *nhlE*, *nhlF* and *amdA* (amidase). *nhlC* and *nhlD* are located upstream of the *nhlBA* genes and code for negative and positive regulators, respectively. *nhlC* may function as an activator to inhibit the repression effect of *nhlD* in the presence of the inducer amide. *nhlE* codes for a protein that is homologous with the β subunit of L-NHase. *nhlF* encodes a markedly hydrophobic protein with eight hydrophobic putative membrane-spanning domains and is considered to be involved in cobalt uptake.

Expression of H-NHase and L-NHase also exhibits different preferences for various types of amide inducers. While H-NHase and L-NHase can both be induced by acetamide, propionamide, acrylamide and methacrylamide, H-NHase can be selectively induced by urea, and L-NHase can be selectively induced by cyclohexanecarboxamide (Komeda et al. 1996b).

3.2.3 Co-Cu-Zn-Type NHase

Other than the classical Fe- and Co-type NHases, a novel NHase that requires Co, Cu and Zn ions for its activity was discovered in the *R. jostii* RHA1 plasmid (Okamoto and Eltis 2007). Although the subunits of this NHase are encoded by *anhA* and *anhB*, similar to the Fe- and Co-type NHases, a gene, named *anhE*, with unknown function is present between *anhA* and *anhB*, and this gene has never been reported in any other *Rhodococcus* strains. The *anhC* gene encoding amidase is located upstream of *anhAB*. Three regulatory genes, namely, *anhP*, *anhQ* and *anhR*, and a possible cobalt transporter gene, namely, *anhT*, are clustered with the *anhAB* and *anhC* genes. A similar gene cluster, containing *anhP*, *anhQ*, *anhC*, *anhA*, *anhE*, *anhB*, *anhR* and *anhT*, has been identified in *R. opacus* 1CP (GenBank: CP009111.1). However, *anhD* is not present in this cluster, and *anhF* is missing in the genome. The expression of NHase and amidase is induced by acetamide.

Comparing the gene clusters of the three types of NHases, NHase genes are always associated with the amidase gene, although the gene arrangement differs; some amidase genes are adjacent to the NHase, while others may be 15.8 kb away from the NHase gene. Expression of NHase and amidase is always regulated by some genes present in the cluster.

3.3 Nitrilase Gene Organization and Regulation

Compared with the complex NHase-amidase gene cluster, the nitrilase gene cluster has been reported in only *R. rhodochrous* J1, which contains a *nitA* gene, encoding nitrilase, and a *nitR* gene, encoding a positive regulator of *nitA* expression (Fig. 7b) (Komeda et al. 1996a). These two genes are co-transcribed as a single mRNA and induced by isovaleronitrile. The nitrilase gene from *R. rhodochrous* K22 has also been found to be inducible by isovaleronitrile, but isobutyronitrile and fumaronitrile are the most favourable inducers for nitrilases from *R. rhodochrous* PA-34 and *R. rhodochrous* NCIMB11216 (Bhalla et al. 1992; Hoyle et al. 1998).

4 Biodegradation of Nitrile Pollutants

Nitriles are important compounds that are widely manufactured and used in the chemical industry and in agriculture. For example, acetonitrile is a commonly used solvent, and acrylonitrile is a precursor of acrylic fibres and plastics. Nitrile herbicides, such as dichlobenil, ioxynil and bromoxynil, are widely used for rice, corn, wheat and berry crops (Kobayashi and Shimizu 2000). These compounds, however, have been extensively released into our environments through industrial waste water or agricultural chemicals. Most nitriles are highly toxic and harmful to humans and the environment, resulting in an urgent need for remediation of nitrile-contaminated soil and water.

Chemical hydrolysis of nitriles requires harsh conditions such as extreme pH and elevated temperatures and leads to the creation of large amounts of by-products and secondary pollutants (Mukram et al. 2016). Bioremediation is a cost-effective technology to restore polluted environments by degrading these compounds into harmless intermediates by using microorganisms. A number of microorganisms, such as *Rhodococcus*, *Nocardia*, *Bacillus*, *Pseudomonas* and *Arthrobacter*, have been reported for the degradation of nitriles (Gong et al. 2012). Among these microorganisms, *Rhodococcus* species harbouring nitrilase or NHase are the most frequently used species. Nitrilase catalyses the hydrolysis of nitriles into their corresponding acids (ammonium salts), while NHase catalyses the hydration of nitriles to the corresponding amides. Mostly, NHase coexists in *Rhodococcus* with amidase, which converts amides into acids (ammonium salts) (Martínková et al. 2010). Nitrile pollutants include aliphatic nitriles and benzonitrile herbicides. The former, including acrylonitrile, acetonitrile, butyronitrile and adiponitrile, are primarily released by the chemical industry, while the latter are found in the residues of agricultural chemicals. The strains capable of degrading these compounds are summarized in Table 1.

Aliphatic nitrile pollutants, such as acrylonitrile and acetonitrile, can be degraded by *Rhodococcus* effectively via the NHase/amidase pathway or nitrilase pathway (Martínková et al. 2010). Strains harbouring NHases used for the industrial

Table 1 *Rhodococcus* species involved in the biodegradation of nitrile pollutants

Substrate	Organism	Enzyme involved	Product detected	References
Acrylonitrile vapour	*R. ruber* NCIMB 40757	Nitrilase	Ammonium acrylate	Roach et al. (2004)
Acrylonitrile	*R. erythropolis* AJ270	NHase	Amide	Baxter et al. (2006)
Acetonitrile	*R. erythropolis* BL1	NHase/ amidase	Acid; ammonium	Langdahl et al. (1996)
Acetonitrile	*R. pyridinivorans* S85-2; *B. diminuta* AM10-C-1	NHase/ amidase	Acid and amide	Kohyama et al. (2006)
Acetonitrile	*R. pyridinivorans* S85-2; *Rhodococcus* sp. S13-4	NHase/ amidase	Acid and amide	Kohyama et al. (2007)
Propionitrile, butyronitrile, valeronitrile	*Rhodococcus* sp. MTB5	NHase/ amidase	Acid and amide	Mukram et al. (2015)
Acetonitrile; acrylonitrile; crotononitrile	*R. rhodochrous* BX2	NHase/amidase (dominant); nitrilase	Ammonium	Fang et al. (2015) and An et al. (2018)
Butyronitrile	*Rhodococcus* sp. MTB5	NHase/ amidase	Acid and amide	Mukram et al. (2016)
Bromoxynil; ioxynil	*R. rhodochrous* NCIMB 11215	Nitrilase	Acid	Harper (1985)
Dichlobenil	*R. erythropolis* 9675; *R. erythropolis* 9685	NHase	Amide	Holtze et al. (2006)
Dichlobenil	*R. erythropolis* AJ270	NHase	Amide	Meth-Cohn and Wang (1997)
Bromoxynil; chloroxynil; ioxynil	*R. rhodochrous* PA-34; *Rhodococcus* sp. NDB 1165	Nitrilase	Acid	Veselá et al. (2010)
Chloroxynil; bromoxynil; ioxynil; dichlobenil	*R. erythropolis* A4; *R. rhodochrous* PA-34	NHase/ amidase	Amide; acid	Veselá et al. (2012)
Benzonitrile; indole-3-acetonitrile	*Rhodococcus* sp. MTB5	NHase/ amidase	Amide; acid	Mukram et al. (2015)

bioproduction of acrylamide can be used for the degradation of acrylonitrile in contaminated environments and waste waters (see Table 1). However, the corresponding amides can be highly toxic, for example, acrylamide is a neurotoxic compound with an LD_{50} (p.o.) of 107–203 mg/kg in rats (Martínková et al. 2010). As a result, amidases are required for further degradation of amides into the corresponding acids (ammonium salts). For the most part, the catalytic efficiencies of NHase and amidase in the same microorganism do not match, necessitating the combination of two different microorganisms for increased degradation efficiency. For example, *R. pyridinivorans* S85-2, with high NHase activity, and *Brevundimonas diminuta* AM10-C-1, with high amidase activity, were combined together for the biodegradation of acetonitrile. With the cascade reaction, 6 M acetonitrile could be converted to acetic acid with a conversion rate of over 90% in 10 h (Kohyama et al. 2006). In addition to the NHase/amidase pathway, nitrilases, which catalyse nitrile into acid (ammonium salt) in one step, also have great potential for the biodegradation of nitrile pollutants. Using *R. ruber* NCIMB 40757 immobilized in synthetic silicone polymer rings, Roach et al. achieved an acrylonitrile elimination capacity of over 7.2 kg/m^3/h with a removal efficiency of 90%. The nongrowing biocatalysts could work as long as 70 days, which showed the excellent tolerance of *Rhodococcus* against the toxic nitrile (Roach et al. 2004).

Benzonitrile herbicides include 3,5-diiodo-4-hydroxybenzonitrile (ioxynil), 3,5-dibromo-4-hydroxybenzonitrile (bromoxynil) and 2,6-dichlorobenzonitrile (dichlobenil) and their ammonium salts and ester analogues (Holtze et al. 2008; Martínková et al. 2010). The utilization of benzonitrile herbicides in agriculture has caused long-lasting pollution in groundwater. Chloroxynil, bromoxynil and ioxynil can be degraded into the corresponding amide or acid by several strains (Table 1). However, the biodegradation of dichlobenil to acid is much more difficult, and 2,6-dichlorobenzamide was considered to be the dead-end product. Though the use of dichlobenil was banned in Denmark in 1997, 2,6-dichlorobenzamide was still the most frequently detected contaminant in the groundwater in 2006 (Holtze et al. 2008). Recently, *R. erythropolis* A4 was reported to degrade 2,6-dichlorobenzamide to 2,6-dichlorobenzoic acid (Veselá et al. 2012). Using *R. erythropolis* A4 resting cells, 0.5 mM dichlobenil was degraded to amide with a conversion rate of 40% and to acid with a conversion rate of 41% in 3 days.

The biodegradation of nitrile pollutants focuses on the screening and discovery of strains, but a few studies have also focused on protein and strain engineering. The activity, substrate specificity and stability of functional enzymes involved in biodegradation can be enhanced by directed evolution and rational design. *Rhodococcus* strains with high tolerance against toxic nitriles and amides can be engineered as efficient and robust hosts for overexpression of NHase/amidase or nitrilase. It is expected that the combination of protein engineering and strain engineering will help improve the performance of *Rhodococcus* in the biodegradation of nitrile pollutants.

5 Bioconversion of Nitriles for Industrial Applications

Bioconversion of nitriles to the corresponding amides or carboxylic acids catalysed by NHase, amidase or nitrilase has been successfully applied in industrial scale. Among these, the most successful case is the application of NHase for production of amides, such as acrylamide and nicotinamide. Under the harsh industrial conditions, however, the natural state biocatalysts in terms of no matter whole cells or enzymes are not stable enough to maintain high productivity and multiple reuses. Hence, diverse cell engineering and enzyme engineering strategies have been applied to enhance the stability/activity/specificity performances of biocatalysts. We can expect that the industrial applications of *Rhodococcus* in bioconversion of nitriles will be further improved and accelerated by various new biotechnologies such as synthetic biology, enzyme rational design and genome editing tools.

5.1 Amide Production Using NHase

Amides are important chemicals due to their wide applications in enhanced oil recovery, nutritional supplements, medical treatment and so on. In particular, acrylamide and nicotinamide are manufactured at the kiloton scale. Owing to the advantages of this process, such as high productivity, high product purity and environmental friendliness, the biotransformation of nitriles to the corresponding amides with NHase is considered to be superior to the conventional chemical process. Currently, NHase is used to synthesize various amides, including acrylamide, nicotinamide, picolinamide, benzamide, indole-3-acetamide and 3-indolacetamide (Prasad and Bhalla 2010). Of these processes, NHase-mediated catalysis of acrylonitrile to acrylamide is the most successful and was the first case in which biotechnology was applied in the petrochemical industry (Kobayashi and Shimizu 1998). Using bioproduction of acrylamide as a typical example, the synthesis of amides with NHase will be discussed here.

Because of the considerable merits of the process, such as simplicity and ease of manipulation, biocatalysis using free resting cells harbouring NHase as the biocatalyst is the main method for the production of acrylamide to date. Many *Rhodococcus* strains have been industrialized to produce acrylamide, including *Rhodococcus* sp. N-774, *R. rhodochrous* J1 and *R. ruber* TH (Yamada and Kobayashi 1996). At the laboratory scale, the highest acrylamide content was obtained by fed-batch biotransformation of acrylonitrile at 10 °C for 10 h with wild-type *R. rhodochrous* J1, which could produce as high as 650 g/L acrylamide (Nagasawa et al. 1993).

However, many problems have emerged in the application of NHase. First, NHases are thermolabile, and their optimal operation temperature ranges from 20 to 35 °C. In contrast, the strong exothermal heat generated upon the conversion of acrylonitrile to acrylamide inhibits NHase activity. Second, NHase is susceptible to attack by polar organic solvents, i.e. the product acrylamide and substrate acrylonitrile, which hampers the accumulation of high concentrations of acrylamide. The NHase inactivation is enhanced with increasing reaction temperature and acrylamide concentration. Third, cell flocculation and sedimentation are common phenomena in most *Rhodococcus* species, resulting in very low cell density and in the failure of large-scale fermentation for preparation of *Rhodococcus* catalysts (Jiao et al. 2017). In addition, as mentioned above, in almost all *Rhodococcus* strains, inherent amidases are co-expressed with NHase, thus hydrolysing the amide product into carboxylic acid and ammonia, which not only leads to wastage of amides but also increases the cost of the subsequent purification process (Prasad and Bhalla 2010).

To meet the urgent demand for the production of acrylamide at high concentrations, various strategies have been proposed to evolve engineered strains at both the cell level and the enzyme level (Fig. 8). To solve the problem of by-product production, NHase may be heterologously expressed in a host lacking amidase activity; however, this solution is not ideal as amidases exist in all the known superior *Rhodococcus* strains (Prasad and Bhalla 2010). Therefore, amidase activity is generally eliminated via a gene knockout in the parent organism (Ma et al. 2010). In light of the hydrophobic properties of various *Rhodococcus* cells, cell flocculation and sedimentation are associated with the smooth-rough colony dimorphism phenomenon, and the smooth-type cells are preferred for industrial application to prevent cell flocculation in large-scale fermentation (Jiao et al. 2017).

Additional efforts have focused on how to enhance the thermal stability and acrylamide tolerance of in vivo NHases. Chaperones are a large family of proteins that have indispensable functions, including aiding the folding of newly synthesized polypeptide chains, oligomeric assembly, responding to the stress denaturation of proteins, assisting proteolytic degradation and transcription factor activation. Therefore, the *E.coli* chaperones GroEL-GroES were introduced to *R. ruber* TH3 to aid NHase folding and stabilize the NHase. Recently, it was discovered that the stress tolerance and cell integrity of *R. ruber* TH3 could be improved by overexpressing the small heat shock protein Hsp16 of *Rhodococcus*, thereby increasing the viability of *R. ruber* cells under high concentrations of acrylamide in the solution. Overexpression of the *Rhodococcus* chaperone GroEL2 was also reported to enhance the organic solvent tolerance of *Rhodococcus* (Takihara et al. 2014). In contrast to rational design with chaperones, a transcriptional regulation strategy by sigma factor random mutation and directed evolution selection was applied to increase the bioconversion productivity of acrylamide (Ma and Yu 2012).

The catalytic performance of the resting cells is ascribed to the activity and stability of the enzyme itself. Compared with the laborious random mutation method, rational design is suitable for enzymes whose crystal structures have been revealed, such as NHase. In addition, the identification of the thermosensitive regions and introduction of stable interactions have been highlighted. By RMSF calculations for thermophilic and mesophilic NHases, three deformation-prone thermosensitive regions were identified and stabilized by the introduction of salt bridges (Chen et al. 2013), and the stabilized C-terminus of the β subunit was the most powerful. In light of the instability of the C-terminus, a salt bridge-based design

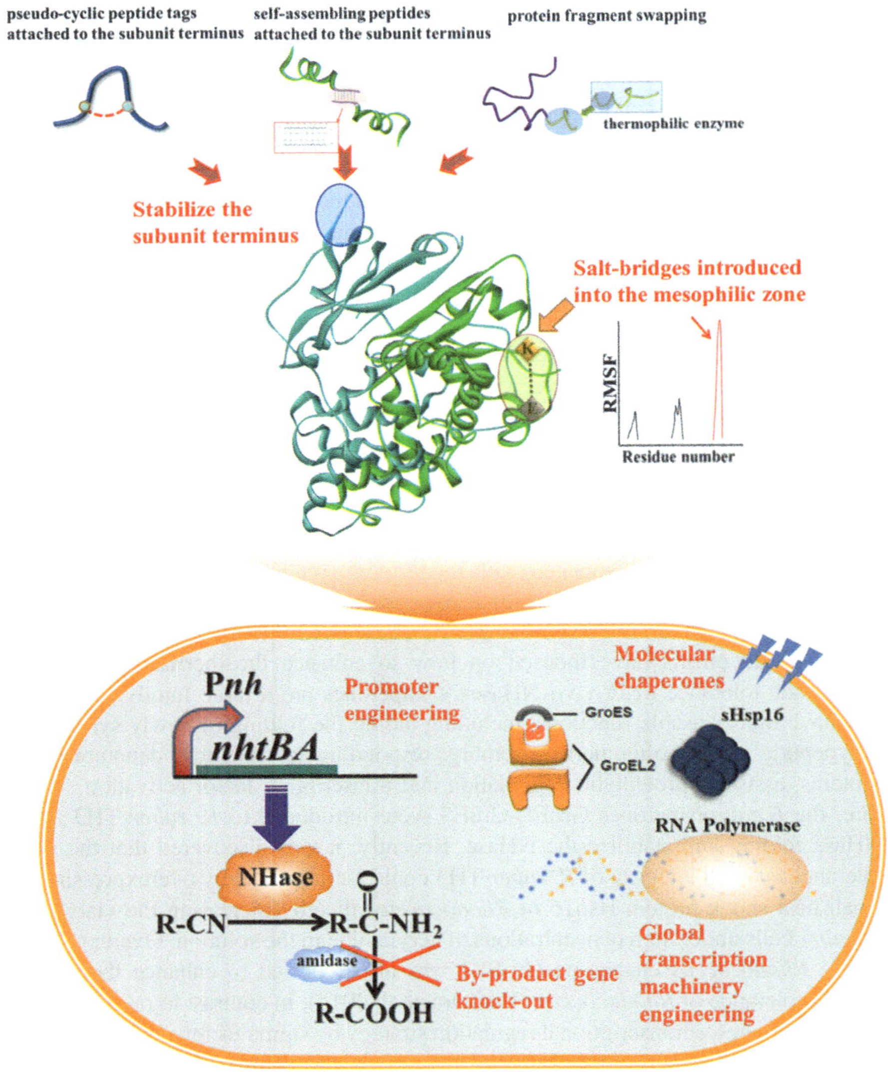

Fig. 8 Strategies used to evolve the engineered NHase-producing *Rhodococcus* strains at both the cell level and the enzyme level

of pseudocyclic peptide tags attached to the subunit terminus was recently applied to the NHase from *R. ruber* TH3, improving the NHase stability significantly without compromising activity. Using STAR (site-targeted amino recombination) software and molecular dynamics to determine the crossover sites for fragment recombination, a homologous protein fragment swapping strategy, involving the swapping of the mesophilic NHase fragments for the thermophilic NHase fragments, was proposed by Cui et al. (Cui et al. 2014). Similarly, Sun et al. constructed a chimeric NHase by swapping the corresponding C-domains of the β subunit from thermosensitive BpNHase and thermostable PtNHase (Sun et al. 2016b). Regardless of enzyme structure, Liu et al. attached the terminus of the NHase with two self-assembling peptides and obtained enzymes with enhanced stability (Liu et al. 2014). In total, all these efforts involve rigidifying flexible sites to enhance the stability of NHase (Yu and Huang 2014).

Finally, a synergistic evolution strategy for *Rhodococcus* cells and in-cell NHases promises to accomplish the recycling of recombinant *Rhodococcus* cells for production of high concentration acrylamide. Recently, some molecular toolkits for gene expression in *Rhodococcus* strains were developed (Jiao et al. 2018; DeLorenzo et al. 2018), and these may be potentially applied to NHase-producing strains as further modifications to improve catalytic performance.

The successful application of free cells as catalysts notwithstanding, immobilized cells and enzymes confer several benefits and are being used, as previously reviewed in detail (Velankar et al. 2010).

5.2 Carboxylic Acid Production Using Amidase

In addition to its widespread application in amide production, *Rhodococcus* has also been utilized for the synthesis of important carboxylic acids via the NHase/amidase pathway (Ismailsab et al. 2017; Ramteke et al. 2013; Maksimova et al. 2017). For example, whole cells of *R. equi* A4, with NHase and amidase activities, were utilized in the biotransformation of benzonitrile, 3-cyanopyridine, (*R,S*)-3-hydroxy-2-methylenebutanenitrile and (*R,S*)-3-hydroxy-2-methylene-3-phenylpropanenitrile to the corresponding acids (Kubáč et al. 2006). *R. erythropolis* ZJB-09149, harbouring NHase and amidase, was used to transform 2-chloro-3-cyanopyridine to 2-chloronicotinic acid (Jin et al. 2011). *Rhodococcus* sp. G20 was used for the transformation of β-aminopropionitrile to β-alanine (Liang et al. 2008). The enantioselectivity of NHase and amidase is also of great interest to organic chemists. For example, *R. erythropolis* NCIMB 11540 was found to have a highly active NHase/amidase enzyme system, which can be used to transform α-hydroxynitriles (cyanohydrins) to enantiopure α-hydroxy carboxylic acids. (*R*)-2-Chloromandelic acid and (*R*)-2-hydroxy-4-phenylbutyric acid were prepared at the gram scale with high optical (e.e. >99 and 98%, respectively) and chemical (98%) yields (Osprian et al. 2003). *R. erythropolis* AJ270, harbouring amidase, has been applied in the enantioselective desymmetrization of functionalized prochiral malonamides to afford carbamoylacetic acids with high yield and excellent enantioselectivity (Zhang et al. 2011). Biotransformation of nitrile with *R. erythropolis* AJ270 has been utilized for the enantioselective syntheses of diverse polyfunctionalized organic compounds that are not readily achievable by other methods (Wang 2005).

5.3 Carboxylic Acid Production Using Nitrilase

Recently, a variety of nitrilases from different microorganisms have been discovered to have applications in the fields of biosensing (Roach et al. 2003), bioremediation (Fang et al. 2015; Li et al. 2013) and biocatalysis. In particular, the high substrate specificity, enantioselectivity and regioselectivity of nitrilases make these enzymes attractive biocatalysts for the production of various commodity chemicals and pharmaceutical intermediates at the laboratory scale, such as nicotinic acid, isonicotinic acid, mandelic acid, acrylic acid, glycolic acid and benzoic acid (Bhalla et al. 2018). The biocatalytic reactions that use nitrilases and can be performed in aqueous solutions at moderate temperatures and pH are more efficient, economical and eco-friendly than traditional chemical synthesis methods. To date, *Rhodococcus* species harbouring nitrilase, which is considered to be a potential biocatalyst to hydrolyse nitriles for industrial applications, has been widely investigated (Table 2).

5.3.1 Important Aromatic and Aliphatic Carboxylic Acids

Aromatic acids, including nicotinic acid, isonicotinic acid, benzoic acid and hydroxybenzoic acid, have been synthesized using nitrilase as a biocatalyst. Among these syntheses, the biosynthesis of nicotinic acid (Lonza, China) has been successfully applied at the industrial scale (Gong et al. 2012). Free cells of *R. rhodochrous* J1 and *Rhodococcus* sp. NDB1165 were reported as biocatalysts for the conversion of 3-cyanopyridine to nicotinic acid with productivity values of 172 g/L and 196.8 g/L, respectively (Mathew et al. 1988; Prasad et al. 2007).

Among the aliphatic and aryl aliphatic acids, acrylic acid, glycolic acid, 3-hydroxyvaleric acid and mandelic acid have been widely investigated for their nitrilase-based biotransformation processes. Acrylic acid and its derivatives are commonly applied in surface coatings, textiles, adhesives, paper treatment, polymeric flocculants, dispersants and so on (Xiaobo et al. 2006). Currently, most commercial acrylic acid is produced by partial oxidation of propylene, which leads to energy wastage and the formation of by-products. To date, many subspecies of *Rhodococcus* have been isolated from the biosphere with high substrate specificity for acrylonitrile, including *R. rhodochrous* J1 (Nagasawa et al. 1990) and *R. ruber* AKSH-84 (Kamal et al. 2011). Moreover, *R. rhodochrous* tg1-A6, which was obtained after treatment with UV light combined with lithium chlorinate, was used to synthesize acrylic acid. The results exhibited that the highest concentration of acrylic acid obtained was 414.5 g/L during a 10-h continuous reaction (Luo et al. 2006).

However, most of the novel wild-type nitrilases have been found to not be suitable as potential biocatalysts for industrial applications due to their relatively low activity and stability. Immobilization has been considered to be an effective way to enhance the stability and reusability of biocatalysts. Yucai He et al. immobilized the cells of *Rhodococcus* sp. CCZU10-1 using calcium alginate, achieving efficient biocatalyst recycling with a product-to-biocatalyst ratio of 776 g terephthalic acid/g

Table 2 *Rhodococcus* harbouring nitrilase used for conversion of nitriles to the corresponding acids

Product	Substrate	Organism	Catalyst	Conversion (%); volumetric productivity (g/L/days); e.e. (%)	References
Nicotinic acid	3-Cyanopyridine	*R. rhodochrous* J1	Whole cells	100; 159; n.a.	Mathew et al. (1988)
Nicotinic acid	3-Cyanopyridine	*Rhodococcus* sp. NDB1165	Whole cells	94; 430; n.a.	Prasad et al. (2007)
Acrylic acid/ methacrylic acid	Acrylonitrile/ methacrylonitrile	*R. rhodochrous* J1	Whole cells	98.5; 392; n.a. 100; 260; n.a.	Nagasawa et al. (1990)
Acrylic acid	Acrylonitrile	*R. ruber* AKSH-84	Whole cells	63; 109; n.a.	Kamal et al. (2011)
Acrylic acid	Acrylonitrile	*R. rhodochrous* tg1-A6	Whole cells	n.a.; 994.8; n.a.	(Luo et al. 2006)
Acrylic acid	Acrylonitrile	Recombinant *R. ruber TH3dAdN(nit)*	Whole cells	n.a.; 2069.4; n.a.	Sun et al. (2016a)
Glycolic acid	Glycolonitrile	*R. rhodochrous* tg1-A6	Whole cells	n.a.; 14; n.a.	Luo et al. (2016)
Benzoylformic acid	Benzoyl cyanide	*Rhodococcus* sp. CCZU10-1	Immobilized cells	>91; 28.32; n.a.	He et al. (2012)
Terephthalic acid/ isophthalic acid	Tetrachloroterephthalonitrile/ isophthalonitrile	*Rhodococcus* sp. CCZU10-1	Immobilized cells	93.5; 64.8; n.a. 92.7; 62.4; n.a.	He et al. (2014)
(*R*)-Ethyl-3-hydroxyyglutarate	Ethyl 4-cyano-3-hydroxybutyate	*R. erythropolis* ZJB-0910	Whole cells	46.2; n.a.; 99	Dong et al. (2010)
(*S*)-2-Cyano-2-methylpentanoic acid	2-Methyl-2-propylmalononitrile	*R. rhodochrous* J1	Whole cells	97; 80; 96	Yoshida et al. (2013)

DCW and 630 g isophthalic acid/g DCW (He et al. 2014). In addition to immobilization, the use of efficient heterologous expression systems and desired variants generated by directed evolution or site-directed mutagenesis are also predominant strategies. Recently, the recombinant strain *R. ruber* TH3dAdN(Nit), overexpressing the nitrilase from *R. rhodochrous* tg1-A6, showed the highest activity to date (187.0 U/mL) and converted the acrylonitrile to acrylic acid with a volumetric productivity of 344.9 g/L/h (Sun et al. 2016a).

5.3.2 Synthesis of Enantiopure Carboxylic Acids

Nitrilases are attractive biocatalysts for the synthesis of chiral intermediates of various pharmaceuticals due to their inherent superiority in terms of enantioselectivity and regioselectivity. (*R*)-(−)-Mandelic acid (Mitsubishi Rayon, Japan; BASF, Germany) has been successfully produced at the industrial scale using nitrilase as a biocatalyst (Schmid et al. 2001). (*R*)-(−)-Mandelic acid is widely used as an optical resolving agent and as an intermediate in the preparation of various pharmaceuticals, such as semisynthetic penicillin, cephalosporin, antiobesity drugs and antitumour agents (Bhalla et al. 2018). Various bacteria have been reported to produce (*R*)-(−)-mandelic acid, such as *Pseudomonas putida* MTCC 5110, *Alcaligenes faecalis* ECU0401 and *Alcaligenes* sp. MTCC 10675.

Many research studies have focused on exploiting the biotransformation process of valuable compounds whose chemical synthesis is difficult. The nitrilase of *R. rhodochrous* J1, when expressed in *Escherichia coli*, enantioselectively hydrolysed 2-methyl-2-propylmalononitrile to form (*S*)-2-cyano-2-methylpentanoic acid (CMPA) with 96% e.e. in fed-batch mode (Yoshida et al. 2013). Optically pure (*R*)-ethyl-3-hydroxyyglutarate is a key precursor for the synthesis of the chiral side chain of rosuvastatin. Whole cells of *R. erythropolis* ZJB-0910 were used to produce (*R*)-ethyl-3-hydroxyyglutarate with a yield of 46.2% (e.e. >99%) (Dong et al. 2010).

In fact, the paucity of available enzymes and the relatively low enzyme activity and stability remain the main limitations of nitrilase-catalysed biotransformation for industrial applications. Virtually, all nitriles are water-insoluble and destructive to cells. In most studies, the reaction is often carried out in fed-batch mode. Nevertheless, this type of operating mode always has several challenges, which arise due to substrate inhibition, production inhibition and a severe scale-up effect in industrial applications. Therefore, the exploration of effective biocatalysts and design of novel bioreactors and downstream processes are two main areas to consider to exploit the full industrial potential of nitrilases.

6 Conclusions

Aerobic and nonsporulating *Rhodococcus* is a genus of Gram-positive bacteria with a relatively fast growth rate (http://en.wikipedia.org/wiki/Rhodococcus). *Rhodococcus* species are of great environmental, chemical and pharmaceutical importance owing to their powerful ability to degrade diverse pollutants in the environment and to produce valuable chemicals such as acrylamide. The level of nitriles in the environment is increasing due to their widespread use in diverse fields such as organic synthesis. Biodegradation of nitriles by *Rhodococcus* has thus become a focus of research over the years. The three major nitrile-degrading enzymes in *Rhodococcus*, i.e. NHase, amidase and nitrilase, with various reaction optima and substrate specificities, have been well described in terms of not only the organization and regulation of gene clusters but also the modification and evolution of enzyme structures and functions.

In comparison with other nitrile degraders, *Rhodococcus* species possess a number of advantages. For example, the species have the advantages of the varied fatty acid composition of their membrane lipids, the existence of several molecular chaperones, the presence of mycolic acids in their cell envelope and their sufficient extracellular polysaccharides (EPS), which lead to an improved resistance to toxic compounds, simple uptake of hydrophobic nitriles and large-scale high-cell-density fermentation without cell flocculation, respectively. Therefore, applications of *Rhodococcus* in pollutant biodegradation and biocatalytic processes are very promising. In particular, the use of nitrile-converting enzymes to produce some valuable chemicals and pharmaceuticals such as acrylamide has been commercialized, and further applications of nitrile biotransformation also have the potential to be scaled up.

The role of *Rhodococcus* in biocatalysis has been increasingly highlighted. These bacteria have been increasingly developed as advantageous biocatalysts, such as for the bioproduction of high-value chiral chemicals. The use of *Rhodococcus* in synthetic biology and metabolic engineering to obtain various interesting compounds and enzymes is also highly desirable. Therefore, new methods and tools for the genetic engineering of *Rhodococcus* are of great interest in the future.

Acknowledgements Support via the projects NSFC-21776157/21476126 (the National Natural Science Foundation of China) and 973-2013CB733600 (the National Key Basic Research Project 973) is gratefully acknowledged.

References

Alexander M (2001) Biodegradation and bioremediation. Q Rev Biol 2:1–2

An X, Cheng Y, Huang M, Sun Y, Wang H, Chen X, Wang J, Li D, Li C (2018) Treating organic cyanide-containing groundwater by immobilization of a nitrile-degrading bacterium with a biofilm-forming bacterium using fluidized bed reactors. Environ Pollut 237:908–916

Andrade J, Karmali A, Carrondo MA, Frazao C (2007) Structure of amidase from *pseudomonas aeruginosa* showing a trapped acyl transfer reaction intermediate state. J Biol Chem 282 (27):19598–19605

Asano Y, Tani Y, Yamada H (1980) A new enzyme "nitrile hydratase" which degrades acetonitrile in combination with amidase. Agric Biol Chem 44(9):2251–2252

Baxter J, Garton N, Cummings S (2006) The impact of acrylonitrile and bioaugmentation on the biodegradation activity and bacterial community structure of a topsoil. Folia Microbiol 51(6):591–597

Bhalla TC, Miura A, Wakamoto A, Ohba Y, Furuhashi K (1992) Asymmetric hydrolysis of α-aminonitriles to optically active amino acids by a nitrilase of *Rhodococcus rhodochrous* PA-34. Appl Microbiol Biotechnol 37(2):184–190

Bhalla TC, Sharma N, Bhatia RK (2012) Microbial degradation of cyanides and nitriles. In: Satyanarayana T, Narain JB, Prakash A (eds) Microorganisms in environmental management. Springer, Dordrecht, pp 569–587

Bhalla TC, Kumar V, Kumar V, Thakur N (2018) Nitrile metabolizing enzymes in biocatalysis and biotransformation. Appl Biochem Biotechnol 185:1–22

Chen J, Yu H, Liu C, Liu J, Shen Z (2013) Improving stability of nitrile hydratase by bridging the salt-bridges in specific thermal-sensitive regions. J Biotechnol 164(2):354–362

Cowan D, Cramp R, Pereira R, Graham D, Almatawah Q (1998) Biochemistry and biotechnology of mesophilic and thermophilic nitrile metabolizing enzymes. Extremophiles 2(3):207–216

Cramp RA, Cowan DA (1999) Molecular characterisation of a novel thermophilic nitrile hydratase. Biochim Biophys Acta 1431(1):249–260

Cui Y, Cui W, Liu Z, Zhou L, Kobayashi M, Zhou Z (2014) Improvement of stability of nitrile hydratase via protein fragment swapping. Biochem Biophys Res Commun 450(1):401–408

DeLorenzo DM, Rottinghaus AG, Henson WR, Moon TS (2018) Molecular toolkit for gene expression control and genome modification in *Rhodococcus opacus* PD630. ACS Synth Biol 7(2):727–738

Dong H-P, Liu Z-Q, Zheng Y-G, Shen Y-C (2010) Novel biosynthesis of (R)-ethyl-3-hydroxyglutarate with (R)-enantioselective hydrolysis of racemic ethyl 4-cyano-3-hydroxybutyate by *Rhodococcus erythropolis*. Appl Microbiol Biotechnol 87(4):1335–1345

Endo I, Nojiri M, Tsujimura M, Nakasako M, Nagashima S, Yohda M, Odaka M (2001) Fe-type nitrile hydratase. J Inorg Biochem 83(4):247–253

Fang S, An X, Liu H, Cheng Y, Hou N, Feng L, Huang X, Li C (2015) Enzymatic degradation of aliphatic nitriles by *Rhodococcus rhodochrous* BX2, a versatile nitrile-degrading bacterium. Bioresour Technol 185:28–34

Fleming FF, Yao L, Ravikumar P, Funk L, Shook BC (2010) Nitrile-containing pharmaceuticals: efficacious roles of the nitrile pharmacophore. J Med Chem 53(22):7902–7917

Gong J-S, Lu Z-M, Li H, Shi J-S, Zhou Z-M, Xu Z-H (2012) Nitrilases in nitrile biocatalysis: recent progress and forthcoming research. Microb Cell Factories 11(1):142

Gong J-S, Shi J-S, Lu Z-M, Li H, Zhou Z-M, Xu Z-H (2017) Nitrile-converting enzymes as a tool to improve biocatalysis in organic synthesis: recent insights and promises. Crit Rev Biotechnol 37 (1):69–81

Harper DB (1985) Characterization of a nitrilase from *Nocardia* sp. (*Rhodochrous* group) NCIB 11215, using p-hydroxybenzonitrile as sole carbon source. Int J Biochem 17(6):677–683

He Y-C, Zhou Q, Ma C-L, Cai Z-Q, Wang L-Q, Zhao X-Y, Chen Q, Gao D-Z, Zheng M, Wang X-D (2012) Biosynthesis of benzoylformic acid from benzoyl cyanide by a newly isolated *Rhodococcus* sp. CCZU10-1 in toluene–water biphasic system. Bioresour Technol 115:88–95

He Y-C, Wu Y-D, Pan X-H, Ma C-L (2014) Biosynthesis of terephthalic acid, isophthalic acid and their derivatives from the corresponding dinitriles by tetrachloroterephthalonitrile-induced *Rhodococcus* sp. Biotechnol Lett 36(2):341–347
Holtze MS, Sørensen J, Hansen HCB, Aamand J (2006) Transformation of the herbicide 2, 6-dichlorobenzonitrile to the persistent metabolite 2, 6-dichlorobenzamide (BAM) by soil bacteria known to harbour nitrile hydratase or nitrilase. Biodegradation 17(6):503–510
Holtze MS, Sørensen SR, Sørensen J, Aamand J (2008) Microbial degradation of the benzonitrile herbicides dichlobenil, bromoxynil and ioxynil in soil and subsurface environments–insights into degradation pathways, persistent metabolites and involved degrader organisms. Environ Pollut 154(2):155–168
Hourai S, Miki M, Takashima Y, Mitsuda S, Yanagi K (2003) Crystal structure of nitrile hydratase from a thermophilic *Bacillus smithii*. Biochem Biophys Res Commun 312(2):340–345
Hoyle AJ, Bunch AW, Knowles CJ (1998) The nitrilases of *Rhodococcus rhodochrous* NCIMB 11216. Enzym Microb Technol 23(7–8):475–482
Huang W, Jia J, Cummings J, Nelson M, Schneider G, Lindqvist Y (1997) Crystal structure of nitrile hydratase reveals a novel iron centre in a novel fold. Structure 5(5):691–699
Ismailsab M, Reddy PV, Nayak AS, Karegoudar TB (2017) Biotransformation of aromatic and heterocyclic amides by amidase of whole cells of *Rhodococcus* sp. MTB5: biocatalytic characterization and substrate specificity. Biocatalysis Biotransform 35(1):74–85
Jiao S, Chen J, Yu H, Shen Z (2017) Tuning and elucidation of the colony dimorphism in *Rhodococcus ruber* associated with cell flocculation in large scale fermentation. Appl Microbiol Biotechnol 101(16):6321–6332
Jiao S, Yu H, Shen Z (2018) Core element characterization of *Rhodococcus* promoters and development of a promoter-RBS mini-pool with different activity levels for efficient gene expression. New Biotechnol 44:41–49
Jin L-Q, Li Y-F, Liu Z-Q, Zheng Y-G, Shen Y-C (2011) Characterization of a newly isolated strain *Rhodococcus erythropolis* ZJB-09149 transforming 2-chloro-3-cyanopyridine to 2-chloronicotinic acid. New Biotechnol 28(6):610–615
Kamal A, Kumar MS, Kumar CG, Shaik TB (2011) Bioconversion of acrylonitrile to acrylic acid by *Rhodococcus ruber* strain AKSH-84. J Microbiol Biotechnol 21(1):37–42
Kato Y, Yoshida S, Xie S-X, Asano Y (2004) Aldoxime dehydratase co-existing with nitrile hydratase and amidase in the iron-type nitrile hydratase-producer *Rhodococcus* sp. N-771. J Biosci Bioeng 97(4):250–259
Kimani SW, Agarkar VB, Cowan DA, Sayed FR, Sewell BT (2007) Structure of an aliphatic amidase from geobacillus pallidus rapc8. Acta Crystallogr 63(10):1048–1058
Kobayashi M, Shimizu S (1998) Metalloenzyme nitrile hydratase: structure, regulation, and application to biotechnology. Nat Biotechnol 16(8):733
Kobayashi M, Shimizu S (2000) Nitrile hydrolases. Curr Opin Chem Biol 4(1):95–102
Kohyama E, Yoshimura A, Aoshima D, Yoshida T, Kawamoto H, Nagasawa T (2006) Convenient treatment of acetonitrile-containing wastes using the tandem combination of nitrile hydratase and amidase-producing microorganisms. Appl Microbiol Biotechnol 72(3):600–606
Kohyama E, Dohi M, Yoshimura A, Yoshida T, Nagasawa T (2007) Remaining acetamide in acetonitrile degradation using nitrile hydratase-and amidase-producing microorganisms. Appl Microbiol Biotechnol 74(4):829–835
Komeda H, Hori Y, Kobayashi M, Shimizu S (1996a) Transcriptional regulation of the *Rhodococcus rhodochrous* J1 *nit*A gene encoding a nitrilase. Proc Natl Acad Sci USA 93(20):10572–10577
Komeda H, Kobayashi M, Shimizu S (1996b) Characterization of the gene cluster of high-molecular-mass nitrile hydratase (H-NHase) induced by its reaction product in *Rhodococcus rhodochrous* J1. Proc Natl Acad Sci USA 93(9):4267–4272
Komeda H, Kobayashi M, Shimizu S (1996c) A novel gene cluster including the *Rhodococcus rhodochrous* J1 *nhlBA* genes encoding a low molecular mass nitrile hydratase (L-NHase) induced by its reaction product. J Biol Chem 271(26):15796–15802

Kubáč D, Čejková A, Masak J, Jirků V, Lemaire M, Gallienne E, Bolte J, Stloukal R, Martínková L (2006) Biotransformation of nitriles by *Rhodococcus equi* A4 immobilized in LentiKats®. J Mol Catal B Enzym 39(1–4):59–61

Kuhn ML, Martinez S, Gumataotao N, Bornscheuer U, Liu D, Holz RC (2012) The Fe-type nitrile hydratase from *Comamonas testosteroni* Ni1 does not require an activator accessory protein for expression in *Escherichia coli*. Biochem Biophys Res Commun 424(3):365–370

Langdahl BR, Bisp P, Ingvorsen K (1996) Nitrile hydrolysis by *Rhodococcus erythropolis* BL1, an acetonitrile-tolerant strain isolated from a marine sediment. Microbiology 142(1):145–154

Lee S, Park E-H, Ko H-J, Bang WG, Kim H-Y, Kim KH, Choi I-G (2015) Crystal structure analysis of a bacterial aryl acylamidase belonging to the amidase signature enzyme family. Biochem Biophys Res Commun 467(2):268–274

Li C, Li Y, Cheng X, Feng L, Xi C, Zhang Y (2013) Immobilization of *Rhodococcus rhodochrous* BX2 (an acetonitrile-degrading bacterium) with biofilm-forming bacteria for wastewater treatment. Bioresour Technol 131:390–396

Liang L-Y, Zheng Y-G, Shen Y-C (2008) Optimization of β-alanine production from β-aminopropionitrile by resting cells of *Rhodococcus* sp. G20 in a bubble column reactor using response surface methodology. Process Biochem 43(7):758–764

Liu Y, Cui W, Liu Z, Cui Y, Xia Y, Kobayashi M, Zhou Z (2014) Enhancement of thermo-stability and product tolerance of *Pseudomonas putida* nitrile hydratase by fusing with self-assembling peptide. J Biosci Bioeng 118(3):249–252

Luo H, Wang T, Yu H (2006) Expression and catalyzing process of the nirilase in *Rhodococcus rhodochrous* tg1-A6. Mod Chem Ind 26:109

Luo H, Ma J, Chang Y, Yu H, Shen Z (2016) Directed evolution and mutant characterization of Nitrilase from *Rhodococcus rhodochrous* tg1-A6. Appl Biochem Biotechnol 178(8):1510–1521

Ma Y, Yu H (2012) Engineering of *Rhodococcus* cell catalysts for tolerance improvement by sigma factor mutation and active plasmid partition. J Ind Microbiol Biotechnol 39(10):1421–1430

Ma Y, Yu H, Pan W, Liu C, Zhang S, Shen Z (2010) Identification of nitrile hydratase-producing *Rhodococcus ruber* TH and characterization of an *ami*E-negative mutant. Bioresour Technol 101(1):285–291

Makhongela HS, Glowacka AE, Agarkar VB, Sewell BT, Weber B, Cameron RA et al (2007) Novel thermostable nitrilase superfamily amidase from geobacillus pallidus showing acyl transfer activity. Applied Microbiology & Biotechnology 75(4):801–811

Maksimova YG, Gorbunova A, Demakov V (2017) Stereoselective biotransformation of phenylglycine nitrile by heterogeneous biocatalyst based on immobilized bacterial cells and enzyme preparation. Dokl Biochem Biophys 1:183–185

Martínková L, Pátek M, Veselá AB, Kaplan O, Uhnáková B, Nešvera J (2010) Catabolism of nitriles in *Rhodococcus*. In: Biology of *Rhodococcus* Springer Berlin, pp 171–206

Mathew CD, Nagasawa T, Kobayashi M, Yamada H (1988) Nitrilase-catalyzed production of nicotinic acid from 3-cyanopyridine in *Rhodococcus rhodochrous* J1. Appl Environ Microbiol 54(4):1030–1032

Meth-Cohn O, Wang M-X (1997) An in-depth study of the biotransformation of nitriles into amides and/or acids using *Rhodococcus rhodochrous* AJ270 1. J Chem Soc Perkin Trans 1(8):1099–1104

Miyanaga A, Fushinobu S, Ito K, Wakagi T (2001) Crystal structure of cobalt-containing nitrile hydratase. Biochem Biophys Res Commun 288(5):1169–1174

Mukram I, Nayak AS, Kirankumar B, Monisha T, Reddy PV, Karegoudar T (2015) Isolation and identification of a nitrile hydrolyzing bacterium and simultaneous utilization of aromatic and aliphatic nitriles. Int Biodeterior Biodegrad 100:165–171

Mukram I, Ramesh M, Monisha T, Nayak AS, Karegoudar T (2016) Biodegradation of butyronitrile and demonstration of its mineralization by *Rhodococcus* sp. MTB5. 3 Biotech 6(2):141

Nagasawa T, Nakamura T, Yamada H (1990) Production of acrylic acid and methacrylic acid using *Rhodococcus rhodochrous* J1 nitrilase. Appl Microbiol Biotechnol 34(3):322–324

Nagasawa T, Shimizu H, Yamada H (1993) The superiority of the third-generation catalyst, *Rhodococcus rhodochrous* J1 nitrile hydratase, for industrial production of acrylamide. Appl Microbiol Biotechnol 40(2-3):189–195

Nagashima S, Nakasako M, Dohmae N, Tsujimura M, Takio K, Odaka M, Yohda M, Kamiya N, Endo I (1998) Novel non-heme iron center of nitrile hydratase with a claw setting of oxygen atoms. Nat Struct Mol Biol 5(5):347

Nakai T, Hasegawa T, Yamashita E, Yamamoto M, Kumasaka T, Ueki T, Nanba H, Ikenaka Y, Takahashi S, Sato M (2000) Crystal structure of N-carbamyl-D-amino acid amidohydrolase with a novel catalytic framework common to amidohydrolases. Structure 8(7):729–738

Nigam VK, Arfi T, Kumar V, Shukla P (2017) Bioengineering of nitrilases towards its use as green catalyst: applications and perspectives. Indian J Microbiol 57(2):131–138

Nojiri M, Nakayama H, Odaka M, Yohda M, Takio K, Endo I (2000) Cobalt-substituted Fe-type nitrile hydratase of *Rhodococcus* sp. N-771. FEBS Lett 465(2–3):173–177

O'mahony R, Doran J, Coffey L, Cahill OJ, Black GW, O'reilly C (2005) Characterisation of the nitrile hydratase gene clusters of *Rhodococcus erythropolis* strains AJ270 and AJ300 and *Microbacterium* sp. AJ115 indicates horizontal gene transfer and reveals an insertion of IS1166. Antonie Van Leeuwenhoek 87(3):221–232

Ohtaki A, Murata K, Sato Y, Noguchi K, Miyatake H, Dohmae N, Yamada K, Yohda M, Odaka M (2010) Structure and characterization of amidase from *Rhodococcus* sp. N-771: insight into the molecular mechanism of substrate recognition. Biochim Biophys Acta 1804(1):184–192

Okamoto S, Eltis LD (2007) Purification and characterization of a novel nitrile hydratase from *Rhodococcus* sp. RHA1. Mol Microbiol 65(3):828–838

Osprian I, Fechter MH, Griengl H (2003) Biocatalytic hydrolysis of cyanohydrins: an efficient approach to enantiopure α-hydroxy carboxylic acids. J Mol Catal B Enzym 24:89–98

Pace HC, Brenner C (2001) The nitrilase superfamily: classification, structure and function. Genome Biol 2(1):reviews0001. 0001

Park JM, Sewell BT, Benedik MJ (2017) Cyanide bioremediation: the potential of engineered nitrilases. Appl Microbiol Biotechnol 101(8):3029–3042

Pertsovich S, Guranda D, Podchernyaev D, Yanenko A, Svedas V (2005) Aliphatic amidase from *Rhodococcus rhodochrous* M8 is related to the nitrilase/cyanide hydratase family. Biochem Mosc 70(11):1280–1287

Pogorelova TE, Ryabchenko LE, Sunzov NI, Yanenko AS (1996) Cobalt-dependent transcription of the nitrile hydratase gene in *Rhodococcus rhodochrous* M8. FEMS Microbiol Lett 144(2–3):191–195

Prasad S, Bhalla TC (2010) Nitrile hydratases (NHases): at the interface of academia and industry. Biotechnol Adv 28(6):725–741

Prasad S, Misra A, Jangir VP, Awasthi A, Raj J, Bhalla TC (2007) A propionitrile-induced nitrilase of *Rhodococcus* sp. NDB 1165 and its application in nicotinic acid synthesis. World J Microbiol Biotechnol 23(3):345–353

Ramteke PW, Maurice NG, Joseph B, Wadher BJ (2013) Nitrile-converting enzymes: an eco-friendly tool for industrial biocatalysis. Biotechnol Appl Biochem 60(5):459–481

Roach P, Ramsden D, Hughes J, Williams P (2003) Development of a conductimetric biosensor using immobilised *Rhodococcus ruber* whole cells for the detection and quantification of acrylonitrile. Biosens Bioelectron 19(1):73–78

Roach P, Ramsden D, Hughes J, Williams P (2004) Biocatalytic scrubbing of gaseous acrylonitrile using *Rhodococcus ruber* immobilized in synthetic silicone polymer (ImmobaSil™) rings. Biotechnol Bioeng 85(4):450–455

Rucká L, Volkova O, Pavlík A, Kaplan O, Kracík M, Nešvera J, Martínková L, Pátek M (2014) Expression control of nitrile hydratase and amidase genes in *Rhodococcus erythropolis* and substrate specificities of the enzymes. Antonie Van Leeuwenhoek 105(6):1179–1190

Rzeznicka K, Schätzle S, Böttcher D, Klein J, Bornscheuer UT (2010) Cloning and functional expression of a nitrile hydratase (NHase) from *Rhodococcus equi* TG328-2 in *Escherichia coli*, its purification and biochemical characterisation. Appl Microbiol Biotechnol 85(5):1417–1425

Sakai N, Tajika Y, Yao M, Watanabe N, Tanaka I (2004) Crystal structure of hypothetical protein PH0642 from *Pyrococcus horikoshii* at 1.6 Å resolution. Proteins 57(4):869–873

Schmid A, Dordick J, Hauer B, Kiener A, Wubbolts M, Witholt B (2001) Industrial biocatalysis today and tomorrow. Nature 409(6817):258

Sun J, Yu H, Chen J, Luo H, Shen Z (2016a) Ammonium acrylate biomanufacturing by an engineered *Rhodococcus ruber* with nitrilase overexpression and double-knockout of nitrile hydratase and amidase. J Ind Microbiol Biotechnol 43(12):1631–1639

Sun W, Zhu L, Chen X, Wu L, Zhou Z, Liu Y (2016b) The stability enhancement of nitrile hydratase from *Bordetella petrii* by swapping the C-terminal domain of β subunit. Appl Biochem Biotechnol 178(8):1481–1487

Takihara H, Matsuura C, Ogihara J, Iwabuchi N, Sunairi M (2014) *Rhodococcus rhodochrous* ATCC12674 becomes alkane-tolerant upon GroEL2 overexpression and survives in the n-octane phase in two phase culture. Microbes Environ 29(4):431–433

Tao Y, Han L, Li X, Han Y, Liu Z (2016) Molecular structure, spectroscopy (FT-IR, FT-Raman), thermodynamic parameters, molecular electrostatic potential and HOMO-LUMO analysis of 2, 6-dichlorobenzamide. J Mol Struct 1108:307–314

Thuku RN, Weber BW, Varsani A, Sewell BT (2007) Post-translational cleavage of recombinantly expressed nitrilase from *Rhodococcus rhodochrous* J1 yields a stable, active helical form. FEBS J 274(8):2099–2108

Thuku R, Brady D, Benedik M, Sewell B (2009) Microbial nitrilases: versatile, spiral forming, industrial enzymes. J Appl Microbiol 106(3):703–727

Valiña ALB, Mazumder-Shivakumar D, Bruice TC (2004) Probing the Ser-Ser-Lys catalytic triad mechanism of peptide amidase: computational studies of the ground state, transition state, and intermediate. Biochemistry 43(50):15657–15672

Velankar H, Clarke KG, du Preez R, Cowan DA, Burton SG (2010) Developments in nitrile and amide biotransformation processes. Trends Biotechnol 28(11):561–569

Veselá A, Franc M, Pelantová H, Kubáč D, Vejvoda V, Šulc M, Bhalla T, Macková M, Lovecká P, Janů P (2010) Hydrolysis of benzonitrile herbicides by soil actinobacteria and metabolite toxicity. Biodegradation 21(5):761–770

Veselá AB, Pelantová H, Šulc M, Macková M, Lovecká P, Thimová M, Pasquarelli F, Pičmanová M, Pátek M, Bhalla TC (2012) Biotransformation of benzonitrile herbicides via the nitrile hydratase–amidase pathway in rhodococci. J Ind Microbiol Biotechnol 39 (12):1811–1819

Wang M-X (2005) Enantioselective biotransformations of nitriles in organic synthesis. Top Catal 35(1–2):117–130

Xiaobo X, Jianping L, Peilin C (2006) Advances in the research and development of acrylic acid production from biomass1. Chin J Chem Eng 14(4):419–427

Yamada H, Kobayashi M (1996) Nitrile hydratase and its application to industrial production of acrylamide. Biosci Biotechnol Biochem 60(9):1391–1400

Yoshida T, Mitsukura K, Mizutani T, Nakashima R, Shimizu Y, Kawabata H, Nagasawa T (2013) Enantioselective synthesis of (S)-2-cyano-2-methylpentanoic acid by nitrilase. Biotechnol Lett 35(5):685–688

Yu H, Huang H (2014) Engineering proteins for thermostability through rigidifying flexible sites. Biotechnol Adv 32(2):308–315

Zhang L-B, Wang D-X, Wang M-X (2011) Microbial whole cell-catalyzed desymmetrization of prochiral malonamides: practical synthesis of enantioenriched functionalized carbamoylacetates and their application in the preparation of unusual α-amino acids. Tetrahedron 67(31):5604–5609

Zhou Z, Hashimoto Y, Shiraki K, Kobayashi M (2008) Discovery of posttranslational maturation by self-subunit swapping. Proc Natl Acad Sci USA 105(39):14849–14854

The Desulfurization Pathway in *Rhodococcus*

Shanshan Li and Ting Ma

Contents

Abstract The emission of sulfur oxides can have harmful effects on the environment. Biodesulfurization of fossil fuels is attracting more and more attention because such a bioprocess is environmentally friendly. Some bacteria, like *Rhodococcus*, have been used or studied to upgrading the fossil fuels on sulfur content limitation with their gentle desulfurization and high desulfurizing competence, without lowering the calorific value of the fuel. Recent advances have demonstrated the desulfurization pathway called "4S" pathway, including four enzymes, and the molecular mechanism for biodesulfurization has also been described. In addition, genetic

S. Li
School of Energy and Power Engineering, Xi'an Jiaotong University, Xi'an, China

T. Ma (✉)
College of Life Sciences, Nankai University, Tianjin, China
e-mail: tingma@nankai.edu.cn

H. M. Alvarez (ed.), *Biology of Rhodococcus*, Microbiology Monographs 16,
https://doi.org/10.1007/978-3-030-11461-9_8

manipulations, such as co-expression of flavin reductases, promoter modification, increasing the expression of key enzymes, expressing of desulfurization enzymes in heterologous hosts, and rearranging the *dsz* gene cluster were also used to improve sulfur removal efficiencies. In this chapter, we summarize the mechanism of biodesulfurization in *Rhodococcus*.

1 Introduction

Fossil fuels are humanity's most important source of energy. Many of the benefits that we used in our way of life are due to fossil fuel use. There are three major fuels—coal, oil, and natural gas. Oil leads with a proportion to near 40% of the total consumption in the world, followed by coal (25%) and natural gas (22%). Almost all oil is consumed by burning, which causes pollution because of the chemical gases released.

Sulfur is the third most abundant element in crude oil and can vary from 0.05 to 10% of the composition. In addition to elemental sulfur, sulfate, sulfite, thiosulfate, and sulfide, more than 200 sulfur-containing organic compounds have been identified in crude oils. These include sulfides, thiols, thiophenes, substituted benzo- and dibenzothiophenes, benzonaphthothiophene, and many considerably more complex molecules (Monticello et al. 1985). The condensed thiophenes are the most common form in which sulfur is present (Kropp and Gerber 1998). Dibenzothiophene (DBT), benzothiophene (BT), and their substitutes are the major sulfur-containing aromatic compounds in fuels, accounting for up to 70% of the sulfur content (Fig. 1) (Kertesz 2001) because they have higher boiling points (more than 200 °C), and it is difficult to remove them from atmospheric tower outlet streams (e.g., middle distillates) (Kawatra and Eisele 2001; Shennan 1996). Benzothiophene (BT), non-β, single β, and di-β-substituted benzothiophenes (B.P.N219 °C) are the typical thiophenic compounds that are found up to 30% in diesel oils (McFarland et al. 1998).

The strict new regulations to lower sulfur content in fossil fuels require new economic and more efficient methods for desulfurization, especially for removing organic sulfur. The concentrations of BT and DBT in fossil fuels are prominently decreased by hydrodesulfurization (HDS) process (Monticello 1998), which has been commercially used for a long time. HDS has several disadvantages: (1) For refractory sulfur compounds, it requires higher temperature, pressure, and longer residence time; (2) it removes relatively simple sulfur compounds such as thiols, sulfides, and disulfides effectively. However, some complex aromatic sulfur-containing compounds such as DBTs, BTs, and polyaromatic sulfur heterocycles are resistant to HDS and form the most abundant organ sulfur compounds after HDS (Monticello 1998; Ma et al. 1994); (3) the cost of sulfur removal in industrial factories in HDS process is expensive, although HDS is considered to be a cost-effective method for fossil fuel desulfurization. Atlas et al. (2001) estimated the cost of lowering the sulfur

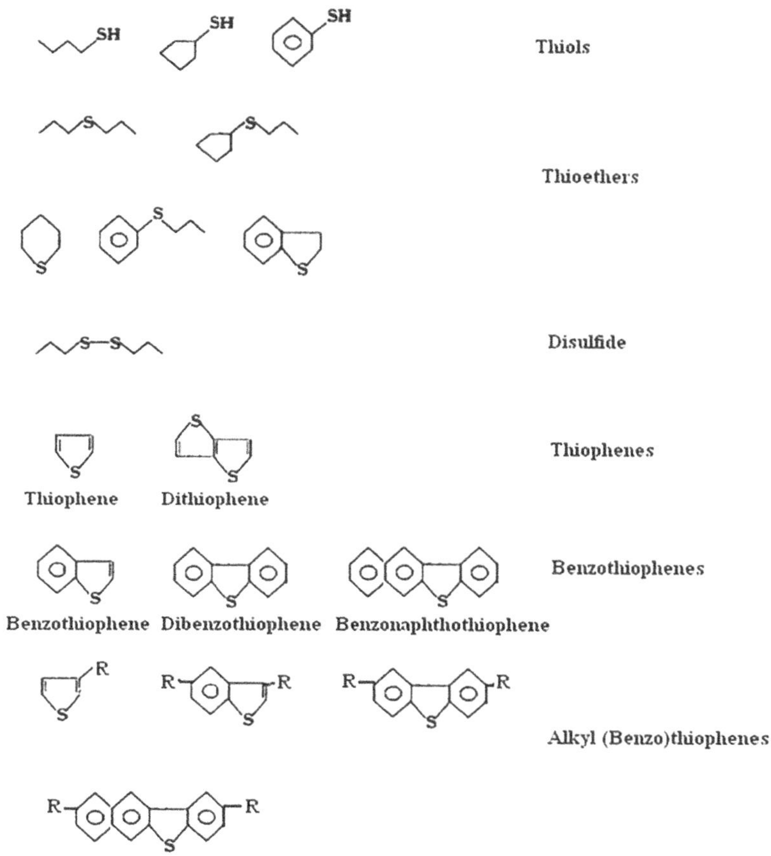

Fig. 1 Chemical structure of typical organic sulfur compounds in fossil fuel. The alkylated DBTs are different in type of substituents, number of substituents, and their bond position on benzene ring

content from 500 to 200 mg/kg to be approximately one cent per gallon. To reduce the sulfur content from 200 to 50 mg/kg, the desulfurization cost would be 4~5 times or higher.

Biodesulfurization (BDS) is a process that is being developed based on naturally occurring bacteria that can remove organically bound sulfur from diesel oil through their metabolism (Borgne and Quintero 2003). Intensive research has been conducted in BDS and has isolated many desulfurizing bacteria from many genera, such as *Rhodococcus* (Kilbane and Jackowski 1992; Izumi et al. 1994; Yu et al. 2006a, b; Ma et al. 2006a, b), *Microbacterium* (Li et al. 2005a), *Gordonia* (Rhee et al. 1998; Li et al. 2006a, b), *Mycobacterium* (Li et al. 2003, 2005b, 2007a, b; Chen et al. 2008), *Pseudomonas* (Gupta et al. 2005; Shan et al. 2003), and so on, of which *Rhodococcus* sp. is an important desulfurizing bacterium with a wide substrate range and deep desulfurizing activity. In this chapter, we introduce desulfurization by *Rhodococcus*.

2 Biodesulfurization Pathways in *Rhodococcus*

The genus *Rhodococcus* belongs to the phylum and class *Actinobacteria*, the order Actinomycetales, and the family Nocardiaceae. Rhodococci possess a variety of plasmids, which give them greater capability to remediating environment pollutions (Dosomer et al. 1988; Kayser 2002).

Most sulfur in fossil fuel can be removed easily by HDS. However, there is one type, known as refractory organic sulfur, which is very difficult to remove. The current methods that can remove the refractory part operate under extremely invasive conditions. They are very costly and produce considerable amounts of carbon dioxide. Microbial desulfurization is an environmentally friendly method that can remove sulfur from refractory organic compounds, such as DBTs and BTs, under ambient temperature and pressure without lowering the calorific value of the fuel. These features have been the reason to conduct extensive studies to develop methods by which desulfurization of refractory organic sulfur compounds under mild condition can be viable (Gupta et al. 2005; Soleimani and Bassi 2007).

2.1 DBT Biodesulfurization Pathway in Rhodococcus

Much effort has been put into the investigation of biological desulfurization systems using DBT or alkylated DBTs as model compounds. The pathway specifically cleaving the C–S bond during metabolic desulfurization has been termed the "4S" pathway (Fig. 2) (Gallagher et al. 1993; Kilbane 2006; Gray et al. 2003; Yan et al. 2000), because four different molecules are formed during DBT desulfurization.

The "4S" pathway for sulfur removal was first reported for *Rhodococcus erythropolis* IGTS8 in 1993 by Gallagher et al. (1993). Besides *R. erythropolis* IGTS8 (Kilbane and Jackowski 1992), other *Rhodococcus* that are also reported to follow this 4S pathway are *R. erythropolis* D1 (Izumi et al. 1994; Ohshiro et al. 1994), *R. erythropolis* DS-3 (Li et al. 2006a, b), *Rhodococcus* ECRD1 (Grossman et al. 1999), *Rhodococcus* B1 (Denis-Larose et al. 1997), *Rhodococcus* SY1 (Omori et al. 1992), *Rhodococcus* UM3 (Purdy et al. 1993), *Rhodococcus* sp. KT462, and *R. erythropolis* KA2-5-1 (Kobayashi et al. 2000). Among these, IGTS8 has been studied most extensively. *R. erythropolis* IGTS8 was isolated by Kilbane and Bielaga (1990) and was used by Energy Biosystems Corp. (EBC) for the development of their commercial microbial desulfurization plan. The strain IGTS8 is a Gram-positive rod-shaped bacterium approximately 0.5 μm long.

In the "4S" pathway, DBT is first converted to DBT sulfoxide (DBTO), then DBT sulfone ($DBTO_2$), then 2′-hydroxybiphenyl 2-sulfinic acid (HBPS), and finally 2-hydroxybiphenyl (2-HBP), releasing sulfate into the medium. Isotopic labeling experiments have shown that the oxygen atom of the hydroxyl group of 2-HBP originates from molecular oxygen, implicating a role for an oxygenase or oxygenases in the pathway (Oldfield et al. 1997). BDS processing using this strain

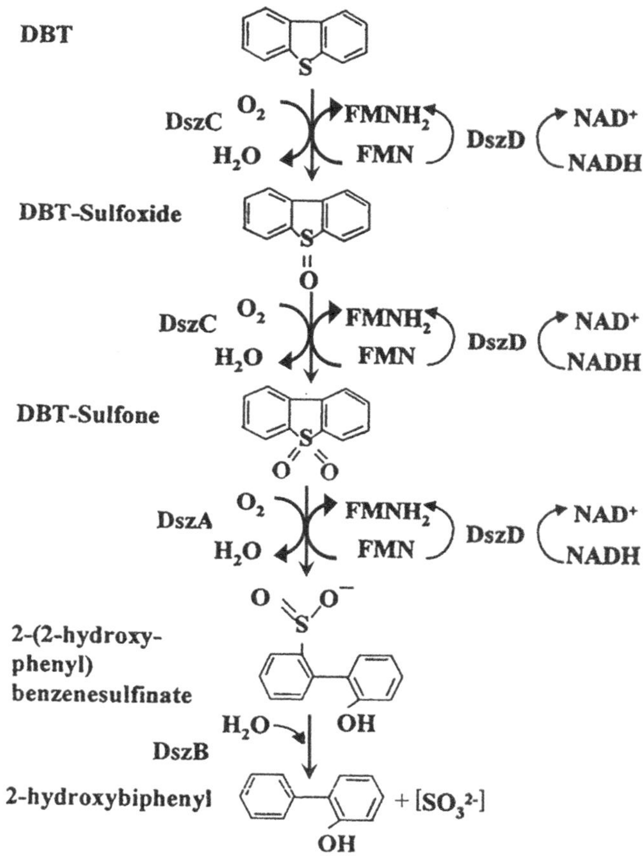

Fig. 2 "4S" pathway of microbial DBT desulfurization and involvement of relative enzymes and flavin reductase

will be ideal. In a BDS process the end product, 2-HBP and its derivatives, would partition back into the oil, thus preserving the fuel value.

2.2 *BT Biodesulfurization Pathway in* Rhodococcus

In contrast to DBT-desulfurizing bacteria, little is known about bacteria that can desulfurize BT. BT predominates in gasoline. Oil contamination may impact the organisms that live in contaminated ecosystems because some of these compounds, such as benzothiophene derivatives, have been reported to be mutagenic and carcinogenic (Kropp and Fedorak 1998). Therefore the degradation pathway of BTH was also studied.

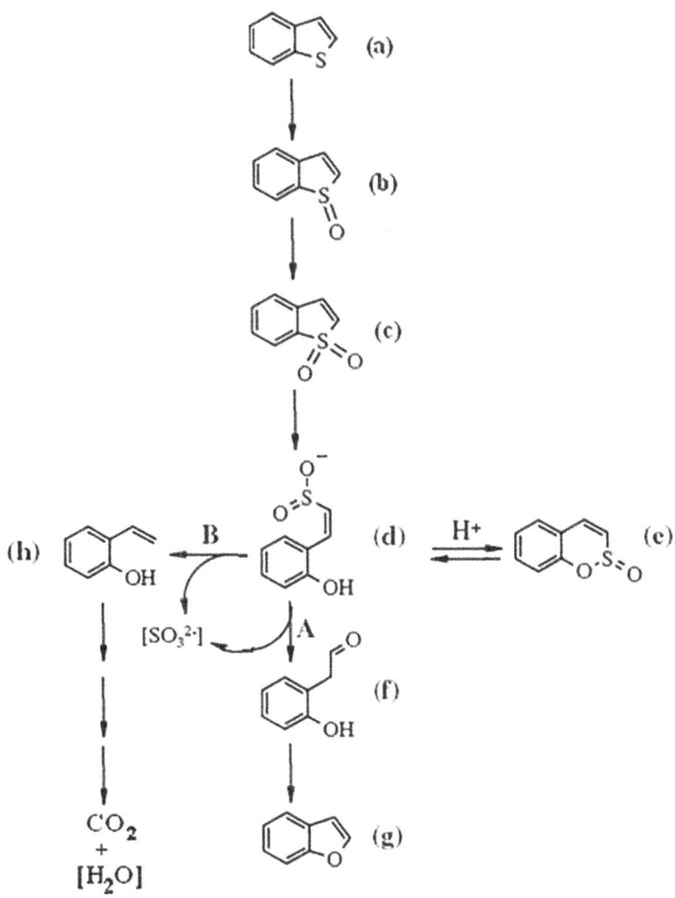

Fig. 3 Two possible degradation pathways of Benzothiophene (BT). (A) *G. desulfuricans* strain 213E (B) *Paenibacillus* sp. strain A11-2: (a) BT; (b) BT sulfoxide; (c) BT sulfone; (d) 2-(2′-hydroxyphenyl) ethen 1-sulfinate; (e) benzo[*e*][1,2]oxathiin *S*-oxide (sultine); (f) 2-(2′-hydroxyphenyl)ethan-1-al; (g) Benzofuran; (h) *o*-hydroxystyrene

Based on the mass spectral data, two different pathways of desulfurization of sulfur from benzothiophene were proposed, and each end product has been identified (Fig. 3). In both of the pathways, BT is firstly converted to BT sulfoxide, then BT sulfone, and following by 2-(2′-hydroxyphenyl) ethan-1-al. The next desulfurization steps can proceed along two separated pathways. In the first pathway, the sulfinate group is removed with oxygenation of the molecule 2-(2′-hydroxyphenyl) ethan-1-al (Gilbert et al. 1998). This product is recovered as benzofuran due to dehydration under acidic extraction conditions. In the second pathway, the final product is *o*-hydroxystyrene, produced through desulfination of the molecule, which finally may oxygenate the carbon atom to dioxide carbon (Konishi et al. 2000). Most microorganisms can only desulfurize BT with one pathway, but *Rhodococcus* sp. JVH1 and *Rhodococcus* sp. WU-K2R have been reported to produce both end products from the desulfurization of benzothiophene (Kirimura et al. 2002).

3 Enzymes Involved in Specific Desulfurization

The enzymes involved in specific DBT-desulfurizing pathways have been purified and characterized; their optimum reacting conditions and activities have also been studied. Compared to DBT-desulfurizing enzymes, the enzymes in BT-desulfurizing pathway were rarely known.

3.1 *Enzymes Involved in DBT Desulfurization of the 4S Pathway*

The complete removal of sulfur from DBT through the 4S pathway requires four enzymes. DBT monooxygenase (DszC or DBT-MO) catalyzes the stepwise *S*-oxidation of DBT, first to DBTO and then to $DBTO_2$. DBT-sulfone monooxygenase (DszA or $DBTO_2$-MO) catalyzes the conversion of $DBTO_2$ to HBPS. Both DBT-MO and $DBTO_2$-MO are flavin-dependent and require a third enzyme (the flavin reductase, DszD) for activity. The fourth enzyme, HPBS desulfinase (DszB), catalyzes the desulfurization of HBPS to give HBP and sulfate, completing the reaction sequence (Gray et al. 1996; Ohshiro and Izumi 2000).

3.1.1 DBT-MO

The first enzyme catalyzes the conversion of DBT to DBT sulfone in a two-step process with DBT sulfoxide being the intermediate compound (DBT→DBTO→DBTO2). The presence of DBTO is difficult to detect because it is readily consumed. The first oxidation step (rate constant 0.06 min^{-1}) is one-tenth of the rate of the second step (rate constant 0.5 min^{-1}). Purified DBT-MO is shown to have a peak absorption at 281 nm. This enzyme shows homology to the acyl coenzyme A enzyme and is a homotetramer with a subunit molecular weight of 50 kDa, as reported by Gray et al. (1996). Ohshiro et al. (1994) isolated DBT-MO from *R. erythropolis* D-1 and reported it to be a homohexamer with a subunit molecular weight of 45 kDa. Its activity is maximum at a temperature of 40 °C and a pH of 8.0.

DBT-MO can act on the derivatives of DBT such as 4,6-dimethyl DBT, 2,8-dimethyl DBT, and 3,4-benzo-DBT, but it does not show any activity on carbazole, dibenzofuran, and fluorine; i.e., DBT atoms are substituted for sulfur atoms. Isotopic labeling studies indicated that the two oxygen atoms were derived from molecular oxygen. The DBT-MO from *Rhodococcus*, compared with other genera, has been shown to have a higher specific reaction rate for sparsely alkylated DBTs (Arensdorf et al. 2002).

3.1.2 $DBTO_2$-MO

The $DBTO_2$-MO enzyme is widely studied. It is a monooxygenase that oxidizes $DBTO_2$ to HPBS. The enzyme isolated from *R. erythropolis* D-1, a thermophile, (Ohshiro et al. 1999) was found to have a molecular mass of 97 kDa and to consist of two subunits with identical masses of 50 kDa. The N-terminal amino acid sequence of the purified $DBTO_2$-MO completely coincided with the deduced amino acid sequence for $DBTO_2$-MO from *R. erythropolis* IGTS8 except for a methionine residue at the latter N-terminal. The optimal temperature and pH for $DBTO_2$-MO activity are 35 °C and about 7.5.

Oldfield et al. (1997) found that $DBTO_2$-MO from *R. erythropolis* IGTS8 catalyzed the conversion of dibenz[*c,e*][1,2]oxathiin 6,6-dioxide (sultone) to 2,2′-dihydroxybiphenyl (DHBP). Ohshiro et al. demonstrated that, by using $DBTO_2$-MO from *R. erythropolis* D-1, sultone showed 54% activity as a substrate compared with DBT sulfone, and DHBP was formed as a product. In addition, dibenz[*c,e*][1,2] oxathiin 6-oxide (sultine) showed 23% activity and yielded DHBP as a product. However, $DBTO_2$-MO did not act on DBT and HBPS. Although sultine was nonenzymatically hydrolyzed to form HBPS, it was also oxidized to sulfonic acid during shaking. It was thought that once sultone was nonenzymatically formed from sultine, it was immediately converted to DHBP by $DBTO_2$-MO. $DBTO_2$-MO may recognize the sulfone moiety within the structure of DBT sulfone and sultone.

Gray et al. (1996) demonstrated that 10 mM EDTA did not inhibit the activity of $DBTO_2$-MO from *R. erythropolis* IGTS8. On the contrary, the activity of $DBTO_2$-MO from *R. erythropolis* D-l was inhibited 50% by 1 mM EDTA. Moreover, 2,2′-bipyridine, 8-quinolinol, and the other metal-chelating reagents, such as Mn^{2+} and Ni^{2+}, also inhibited the activity of the enzyme, suggesting that a metal might be involved in its activity. $DBTO_2$-MO acted not only on DBT sulfone but also on dibenz[*c,e*][l,2]oxathiin 6,6-oxide and dibenz[*c,e*] (Ohshiro and Izumi 1999; Ohshiro et al. 1995), and oxathiin 6,6-dioxide. Dihydroxybiphenyl was formed from the latter two substrates.

3.1.3 HPBS Desulfinase

HPBS desulfinase is a novel enzyme, in that it can specifically cleave the carbon–sulfur bond of HBPSi to give 2-HBP and the sulfite ion without the aid of any other protein components or coenzymes. It has been demonstrated that the activity of HPBS desulfinase is the lowest among enzymes of desulfurization metabolism. It is the rate-limiting enzyme of 4S pathway. It is also the least studied enzyme since only a very small amount is produced.

It is a monomer with a subunit molecular weight of 40 kDa and shows enzyme activity over a wide temperature range (25–50 °C) with the optimum at 35 °C (Watkins et al. 2003). The working pH range for this enzyme is 6.0–7.5. Lee et al. (2004, 2006) elucidated the 3D structure of DszB, which was the first X-ray crystallographic study of enzymes involved in DBT desulfurization (Fig. 4). HPBS desulfinase does not require a metal cofactor for catalysis, and the inhibition by Zn^{2+} and Cu^{2+} is likely caused by interference of substrate binding or catalysis.

A Cys residue must be the catalytic center of DszB because SH reagents inhibited the enzyme activity. DszB has only one Cys residue, at position 27, and it was found that the C27S mutant enzyme lost its activity completely. Therefore, there is no doubt that this residue is the catalytic center.

Based on information about the 3D structure of DszB and a comparison of amino acid sequences between DszB and reported thermophilic and thermostable homologs (TdsB and BdsB), two amino acid residues, Tyr63 and Gln65, were selected as targets for mutagenesis to improve DszB. The promising mutant enzymes, which were replaced with these two residues by other amino acids, were purified and their properties examined. Among the wild-type and mutant enzymes, Y63F had higher catalytic activity but similar thermostability, and Q65H showed higher thermostability but less catalytic activity and affinity for the substrate. Furthermore, the double mutant enzyme Y63F-Q65H was purified and overcomes these drawbacks. This mutant enzyme had higher thermostability without loss of catalytic activity or affinity for the substrate.

Ohshiro et al. (2007) found that each mutation at positions 63 and 65 of DszB enhanced the maximum activity and thermal stability, respectively, and that the double mutation increased thermostability without losses in maximal activity or affinity for the substrate. For the purpose of developing microbial desulfurization as a practical process, it is necessary to improve DszB further by structural analysis of the mutant enzymes in the near future.

3.1.4 Flavin Reductase

The flavin reductases are associated with monooxygenases since monooxygenases cannot work in the absence of these reductases (DBT-MO and $DBTO_2$-MO). The purified flavin reductase from the thermophilic strain *R. erythropolis* D-l contains no chromogenic cofactors and was found to have a molecular mass of 86 kDa with four identical 22 kDa subunits (Matsubara et al. 2001). The enzyme catalyzed NADH-dependent reduction of flavin mononucleotide (FMN).

For the flavin reductases from *R. erythropolis* D-l, flavin adenine dinucleotide was a poor substrate, and NADPH was inert. The enzyme did not catalyze the reduction of any nitroaromatic compound. The optimal temperature and optimal pH for enzyme activity were 35 °C and 6.0, respectively, and the enzyme retained 30% of its activity after heat treatment at 80 °C for 30 min. The N-terminal amino acid sequence of the purified flavin reductase was identical to that of the flavin reductase from *R. erythropolis* IGTS8.

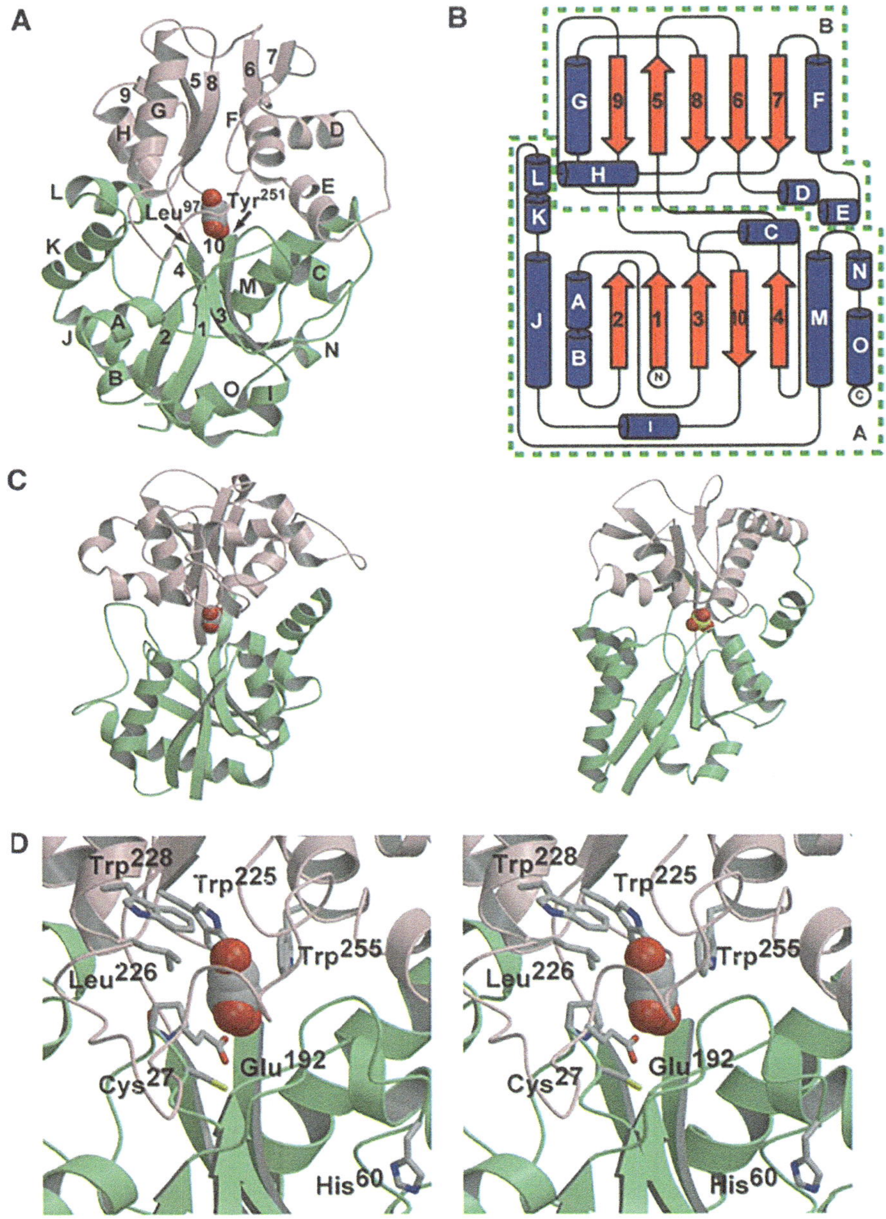

Fig. 4 The overall structure of DszB. (**a**) ribbon model of DszB. Domain A and B are colored in light green and pink, respectively. Two crossover residues that define the domains are labeled. (**b**) Topology diagram of DszB. Helices and strands are colored in blue and red, respectively. Dotted green lines designate domain A and domain B. (**c**) Ribbon models of proteins structurally related to DszB. Ovotransferrin (left, PDB code 1NNT) and a sulfate-binding protein (right, PDB code 1SBP) are depicted in ribbon models. Two domains of each protein are colored in a similar fashion to (**a**). Substrate ferric carbonate and sulfate ions are depicted in space-filling models. (**d**) Stereo view of

Xi et al. (1997) studied the enhanced desulfurization activity of DszC and DszA under in vitro conditions by increasing the concentrations of flavin reductase, suggesting that the two are terminal oxygenases. The reaction rate with 1 unit/ml of flavin reductase was linear for 10–15 min, whereas it was linear for more than 20 min with a lower concentration.

The inhibition experiments revealed that the flavin reductase activity of *R. erythropolis* D-1 was inhibited by 7-hydroxycoumarin but not by other coumarin derivatives, including dicoumarol, which inhibited FRase I activity and was used for analysis of its crystal structure (Koike et al. 1998). FRase I was a flavoprotein possessing FMN as a prosthetic group. The flavin reductase of *R. erythropolis* D-1 has no flavin cofactor.

3.2 *Enzymes Involved in BT-Desulfurizing Pathway*

In contrast to DBT-desulfurizing enzymes, little is known about enzymes involved in the BT-desulfurizing pathway. At present, there are no related reports on BT-desulfurizing enzymes at home or abroad. The purified enzymes involved in BTH degradation would provide a detailed explanation for the degradation of BT.

4 Specific Desulfurizing Genes in *Rhodococcus*

In order to obtain better control over the machinery of specific sulfur removal, related research has been conducted on the molecular biology of this and similar strains since the metabolic identification of *R. erythropolis* IGTS8.

The primary genes involved in DBT metabolism, which are called both *dsz* and *sox*, have been cloned and are fairly well characterized. Although the *sox* (sulfur oxidation) designation was used first, the *dsz* (desulfurization) designation has generally been adopted.

The *dsz* genes are arranged in an operon-regulated system in a 4-kb conserved region of a 150 kb mega-plasmid, pSOX, in *R. erythropolis* IGTS8 (Oldfield et al. 1998) and a 100 kb plasmid in other strains. An insertion sequence (IS1166) was found to be associated with the *dsz* gene. It is a cluster of three genes (*dsz*A, *dsz*B, *dsz*C) transcribed in the same direction, coding for three proteins Dsz A, Dsz B, and Dsz C, respectively (Piddington et al. 1995). The fourth gene, *dsz*D, was on the chromosome of *R. erythropolis* IGTS8 rather than on pSOX with *dsz*A, *dsz*B, and *dsz*C.

Fig. 4 (continued) the active site. A glycerol molecule is depicted in space-filling model. Residues mentioned in the text are depicted as sticks

```
-192  CGAAGGCGGC CAAGTCATCG GCACCGTCAC CGTCACCTTG ACCCGACGTG
         Protein Binding Site                     Promoter Region
-142  CCCCGTGGTT CAAGGCCTGA ATTTGGCTGG TGGAGCATTG AAATCAGGTG
              -35 Region     Promoter Region      -10 Region          +1
 -92  AAGTTTAACG GTGGGCACAC CCCGGGGGTG GGGGTGAGAC TGCTTAGCGA
                                  Ribosome Binding Site    Initial Codon
 -42  CAGGAATCTA GCCATGATTG ACATTTAAAG GACGCATACG CGATG-----
                                                            MET
```

Fig. 5 Sequence of the 385-bp *dsz* promoter-containing fragment that starts immediately after the *Hin*dIII site. Nucleotides below the line indicate the positions of mutations, including the deletion of a C at 210. Boxed sequences are the protein-binding domain and the promoter region deduced from deletion analysis. The arrow indicates the G residue at position 11 in the 5′ end of the mRNA, and the nucleotides in bold are the putative −10 and −35 regions of the promoter. The *Hin*dIII site at the 5′ end and the *Spe*I site near the 3′ end were added for cloning purposes. Sequence numbering is for the native fragment, where 21 is the base preceding the A of the ATG initiation codon of *dsz*A

The termination codon of *dsz*A and the initiation codon of *dsz*B overlap (ATGA, Fig. 5), indicating that there may be translational coupling of these two genes. Between *dszB* and *dszC*, there was a 13-bp gap. Potential ribosome binding sites were also present upstream of each putative ATG initiation codon. The spacing and orientation of the three genes suggest that they are expressed as an operon, a suggestion that was also supported by the results of subclone analyses and promoter replacement analyses. Although expressed in the operon, Dsz B is present at concentrations severalfold less in the cytoplasm, as compared with Dsz A and Dsz C (Li et al. 1996).

These genes, when cloned into a non-desulfurized strain (called dsz^-), confer the ability to desulfurize DBT to 2-HBP. The *dsz* operon was found on a large, 150 kb plasmid in *R. erythropolis* IGTS8 and on a 100 kb plasmid in other strains. An insertion sequence (IS1166) was found to be associated with the *dsz* gene.

To develop the biodesulfurization process, it was important to know under what conditions the desulfurization genes were expressed or repressed. Li et al. (1996) investigated the effect of various sulfur-containing compounds such as dimethyl sulfoxide (DMSO), cysteine, methionine, and sulfate on *dsz* gene expression. The results showed that desulfurization activity decreased when the concentration of cysteine, methionine, or sulfate in the media increased. In comparison, methionine caused the strongest repression in these substrates. When the concentration of these inhibitors reached more than 375 μM, desulfurization activity was strongly repressed. This repression was found to be due to the binding of a repressor protein next to the *dsz* promoter, which was located within the 385 bp region immediately upstream of *dsz*A. Deletion analysis showed that the promoter was located to the region between −121 and −44. The S1 nuclease protection assay confirmed that the 5′ end of the *dsz* mRNA was the G at −46. A possible −35 promoter region with the sequence AAGTTTAA and a −10 region of GGGTGA are similar to those of

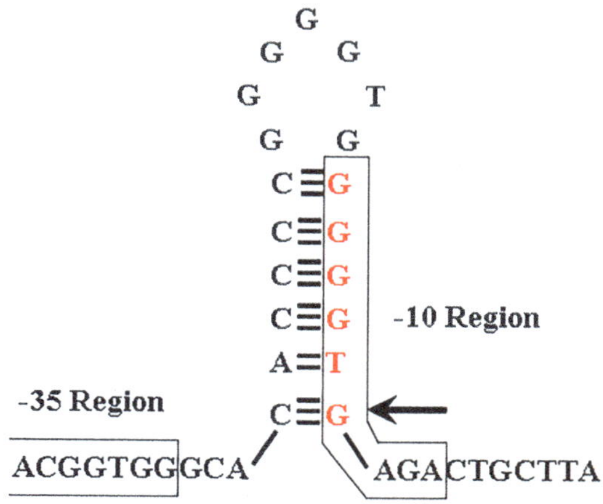

Fig. 6 Potential hairpin structure located in the *dsz* promoter. The hairpin is between −75 and −57 and has a free energy value of −15.4 kcal (1 cal = 4.184 J). Deletion of the G at −57 (indicated by the arrow), as in mutant R4, would reduce this to −13.2 kcal

Bacillus subtilis promoters that use the sigma factor σ^B (Li et al. 1996). The sequence at the transcription initiation site, TAG, is also the same as that of the *dsz* promoter, with two starting at the middle A and one starting at the G, as does the *dsz* promoter. The main difference between the two promoters is that the *Bacillus* promoters have a 14-bp spacer region between the −10 and −35 regions, where the *dsz* promoter is 23 bp.

The promoter region from −75 to −57 could be a potential case of dyad symmetry (Fig. 6). It is a strict inverted repeat sequence and could be part of an operator. An almost identical inverted repeat occurs within *dszB* from 1562 to 1578 and could be part of another operator.

Apart from promoter, the 385 bp fragment has at least three elements that affect Dsz activity (some overlapping the promoter region). The region from −263 to −244 proved to reduce *dsz* repression. However, deletion of the region did not affect repression or gene expression. The region from −144 to −121 could bind a protein such as an activator, and deletion of this region reduced gene expression, but not repression. The region between −98 and −57 may be a repressor-binding site (Li et al. 1996). It is possible that combinations of these mutations could further decrease repression.

5 Enhanced Biodesulfurization by Recombinant Bacteria

These specific bacteria can remove sulfur pollutants from petroleum and will reduce the amount of sulfur oxides released. However, genetic manipulations for the removal of harmful sulfur compounds from fossil fuels can be developed. In most times, engineered bacteria are required to remove more sulfur compounds with higher activities. Cultures with improved substrate ranges are also needed to better address the complicated mixture of chemicals present in petroleum.

5.1 Co-expression of Flavin Reductases

Since $FMNH_2$ is essential for the activities of DszC and DszA, the overexpression of flavin reductase in *Rhodococcus* or in recombinant bacteria will enhance the activities of DszC and DszA.

Lei et al. (1997) found enhanced desulfurizing activities of purified DszC and DszA protein from *R. erythropolis* IGTS8 in vitro when activated with flavin reductase from *Vibrio harveyi*. Hirasawa et al. (2001) purified the flavin reductase DszD from *R. erythropolis* IGTS8, and the enzyme was overexpressed in *Escherichia coli*. The specific activity in crude extracts of the overexpressed strain was about 275-fold that of the wild-type strain.

Reichmuth et al. (1999) studied the desulfurization ability of an engineered *E. coli* DH10B strain that contained the plasmids pDSR2 and pDSR3. The plasmid pDSR2 contained a *Vibrio harveyi* NADH:FMN oxidoreductase gene, and pDSR3 encoded all of the three enzymes that converted DBT to HBP. In plasmid pDSR3 the native desulfurization control element, which located in the promoter, had been removed. Therefore, *E. coli* DH10B/pDSR3 could express its desulfurization trait even in the presence of sulfate ion or rich media such as LB. However, the oxidoreductase level proved to be insufficient for the overexpressed *dsz*ABC. Designing an operon that expresses the proper amount of FMN:NADH reductase to existed *dsz*ABC enzymes is crucial to reach an optimum desulfurization activity. Insufficient FMN:NADH reductase would make NADH the limiting step in DBT oxidation. On the other hand, a high concentration of $FMNH_2$ will give rise to H_2O_2 formation, which would be lethal to cells (Gaudu et al. 1994; Galán et al. 2000).

In the search for the development of a method to provide the required amount of reduced flavin to DBT oxygenation, Galán et al. (2000) used *hpa*C, a flavin reductase from *E. coli* W, and connected it, in vitro, with a system of *dsz*ABC purified enzymes and an NADH source. They also used catalase in the desulfurization medium to minimize the probability of H_2O_2 formation, which might be produced by nonenzymatic reoxidation of $FMNH_2$ under high oxygen concentrations. The addition of *hpa*C flavin reductase increased DBT desulfurization 7–10 times over 30 min. The enzyme HpaC flavin reductase and the oxidoreductase originated from IGTS8 were from the same subfamily of flavin:NAD(P)H reductases.

All the experiments confirm that the expression of an oxidoreductase with the *dsz* genes caused an increase in the rate of DBT removal.

5.2 Promoter Modification

The expression of *dsz* genes in most desulfurizing bacteria is repressed by sulfate, which is the product of biodesulfurization, through a repressor-binding site that may be in the promoter. So, looking for a new promoter that cannot be repressed by sulfate will be a new pathway to increase the desulfurizing rate.

Gallardo et al. (1997) subcloned the *dsz* cassette into the broad-host-range plasmid pVLT31 under the control of a hybrid promoter, *Ptac*, that has been shown to be functional in a wide range of bacteria. The resulting plasmid was transferred into *Pseudomonas putida*. The recombinant bacteria with *dsz* were shown to keep its desulfurization phenotype even in sulfate-containing media.

Several 16S ribosomal RNA promoters of mycobacteria have also been studied and found to be functionally constitutive (Ji et al. 1994). Matsui et al. (2002) reported a 16S ribosomal RNA promoter applied to the expression of *dsz* enzymes in *Rhodococcus* sp. strain T09. The putative *Rhodococcus rrn* promoter region was cloned from the *Rhodococcus* sp. strain T09, and the dibenzothiophene-desulfurizing gene, *dsz*, was expressed under the control of the putative *rrn* promoter in strain T09 using a *Rhodococcus–E.coli* shuttle vector. Strain T09 harboring the expression vector could desulfurize dibenzothiophene in the presence of inorganic sulfate, methionine, or cysteine, while the Dsz phenotype was completely repressed in recombinant cells carrying the gene under the control of the native *dsz* promoter under the same conditions.

At the same time, Noda et al. (2002) constructed a promoter probe transposon using a promoterless red-shifted green fluorescence protein gene (rs*gfp*). A 340 bp putative promoter element, *kap1*, was isolated from a recombinant strain, KA2-5-1, that had been shown to have high fluorescence intensity. The promoter element of *kap1* was not repressed by 1 mM of sulfate, and it had about twofold greater activity than the *rrn* promoter from *R. erythropolis*. *Kap1* stimulated cell growth with biodesulfurization activity without the repression of sulfate. In conclusion, *kap1* is a convenient tool for improving biodesulfurization in *Rhodococcus*.

Otherwise, screening for recombinant bacteria that cannot be inhibited by sulfate is another substitutable method. Tanaka et al. (2002) isolated two mutants of the dibenzothiophene-desulfurizing *R. erythropolis* KA2-5-1 that express a high level of desulfurizing activity in the presence of sulfate using the transposome technique. The level of dibenzothiophene desulfurization by cell-free extracts prepared from mutants grown on sulfate was about fivefold higher than that by cell-free extracts from the wild type. Gene analysis of the mutants revealed that the same gene was disrupted and that the transposon-inserted gene in these strains was the gene for cystathionine β-synthase, *cbs*. The *cbs* mutants also expressed high levels of Dsz enzymes when methionine was used as the sole source of sulfur.

5.3 Increasing the Expression of Key Enzymes

The reaction catalyzed by DszC and DszB have been widely recognized as rate-limiting steps in the microbial desulfurization pathway. Several approaches have been performed by genetic engineering to improve desulfurizing enzyme activities, including those of DszC and DszB.

Coco et al. (2001) used random chimeragenesis on a transient template (RACHITT) to improve the activity of DszC by 20 times, and it must have increased the rate of the whole pathway.

As described previously, the rate of desulfurization is limited by the last enzyme in the pathway, DszB. In the native *dsz* operon, the ratio of mRNA of *dszA*, *dsz*B, and *dsz*C was 11:3.3:1, indicating that the translation levels of the desulfurization enzymes decreased according to their positions in the operon due to polar effects on *dsz* gene transcription; however, western blot analysis indicated that the expression level of *dsz*B was far lower than that of *dsz*C. These results suggest that the translation of *dsz*B mRNA was not as efficient as *dszA* or *dsz*B mRNA. Gene analysis revealed that the termination codon of *dsz*A and the initiation codon of *dsz*B overlapped, whereas there was a 13-bp gap between *dsz*B and *dsz*C. Potential ribosome binding sites were present upstream of each putative ATG initiation codon. In order to get a better, steady expression of DszB, Li et al. (2007a, b) removed the overlap structure by overlap polymerase chain reaction (PCR) and expressed the redesigned *dsz* operon in *R. erythropolis* without desulfurization activity, named *R. erythropolis* DR-2. Real-time PCR analysis confirmed that the transcription characteristics did not change in *R. erythropolis* DR-2 compared with *R. erythropolis* DR-1, which contains the original *dsz* operon. However, western blot analysis revealed that *R. erythropolis* DR-2 produced more DszB than *R. erythropolis* DR-1 did. The desulfurization activity of resting cells prepared from *R. erythropolis* DR-2 was about fivefold higher than that of *R. erythropolis* DR-1. That indicated that the enhanced expression level increased the metabolic rate of HBPS in the cells and contributed to the improved desulfurization rate of *R. erythropolis* DR-2.

To increase DszB production, Reichmuth et al. (2004) mutated the untranslated 5′ region of *dsz*B using degenerate oligonucleotides. Because neither DszB activity nor the amount of DszB protein produced could be directly measured, it was difficult to determine the exact cause for the lack of HBP production. To clarify the results of our genetic manipulations, they chose to tag the production of the desulfurization transcripts and proteins by creating transcriptional and translational fusions with a fluorescent protein. This permitted a quick, straightforward, and direct determination of the amount of the desulfurization protein produced. This technique does not measure the activity of the proteins; however, activity screens could be used after protein production was optimized using fluorescent fusion tags. The protein used for those fusions was GFP. GFP has been widely used for the quantitative measurement of protein production and is known to be stable for a period of several days, allowing an integrative and quantitative measure of protein production (Albano et al. 1998; Cha et al. 2000). After screening only 96 mutants, several showed increased green fluorescence, and two showed increased DszB activity. When cotransformed with the full *dsz*ABC operon, the mutant *dsz*B increased the rate of desulfurization ninefold relative to the native *dsz*B.

R. erythropolis KA2-5-1 can desulfurize DBT into 2-HBP through the 4S pathway. Hirasawa et al. (2001) constructed an *Escherichia coli-Rhodococcus* shuttle vector, and the desulfurization gene cluster, *dsz*ABC, and the related reductase gene, *dsz*D, were cloned from KA2-5-1, reintroduced into KA2-5-1, and efficiently expressed. The DBT desulfurization ability of the transformant carrying two *dsz*ABC and one *dsz*D on the vector was about fourfold higher than that of the parent strain, and the transformant also showed improved desulfurization activity for light gas oil. Matsui et al. (2001) also enhanced the desulfurization rate by 3.3 times, by increasing the copy number of *dsz* genes.

5.4 The Expression of Desulfurization Enzymes in Heterologous Hosts

In fossil fuels, there are many kinds of compounds inhibiting on desulfurization process of *Rhodococcus*. *Pseudomonas* was found to be an ideal candidate for biodesulfurization because they are organic solvent tolerant and have a high growth rate. *Pseudomonas* sp. are among the best studied and most abundant microorganisms found in crude oil (Leahy and Colwell 1990), and a wide variety of genetic tools are now available for their molecular manipulation (Lorenzo and Timmis 1994). Furthermore, several biotechnological properties for the design of biocatalysts targeted to industrial biodesulfurization processes are present in *Pseudomonas* species. For example, while the solvent tolerance of *Rhodococcus* is the lowest reported (log *P* values from 6.0 to 7.0), that of the genus *Pseudomonas* (log *P* values from 3.1 to 3.4) is the highest known (Inoue and Horikoshi 1991), and several *Pseudomonas* strains that are highly resistant to heavy metals present in fossil fuels have been reported (Atlas 1994).

With the properties noted, *dsz*ABC genes from *R. erythropolis* XP were cloned into *P. putida* Idaho to construct a solvent-tolerant, desulfurizing *P. putida* A4. This strain, when contacted with sulfur refractory compounds dissolved in hydrocarbon solvent, maintained the same substrate desulfurization traits as observed in *R. erythropolis* XP. Resting cells of *P. putida* A4 could desulfurize 86% of DBT at 10% (v/v) *p*-xylene in 6 h. In the first 2 h, the desulfurization occurred with a rate of 1.29 mM DBT (gdw cell)$^{-1}$ h^{-1}. No DBT reduction was noticed when the experiment was repeated with *R. erythropolis* or *P. putida* Idaho under identical conditions (Tao et al. 2006).

In the development of engineered strains with potential industrial or environmental applications, a high degree of predictability in their performance and behavior is desirable. To achieve this goal, stable chromosomal insertion of the genes conferring the new trait is required. Therefore, Gallardo et al. (1997) constructed *P. putida* EGSOX, which carried *dsz* genes stably inserted into the chromosome of the host cell.

To improve the biodesulfurization process, it would be interesting to design a recombinant biocatalyst that combines the Dsz phenotype with another trait of potential interest, such as the production of biosurfactants. To accomplish this goal, Gallardo et al. (1997) also transferred *dsz* genes into *P. aeruginosa* PG201 (Ochsner et al. 1995), which cannot use DBT as the sole carbon and/or sulfur source and produces rhamnolipid biosurfactants. These are of increasing industrial relevance because of their applications in emulsification, wetting, phase separation, and viscosity reduction. The final recombinant bacteria were named *P. aeruginosa* EGSOX and carried *dsz* genes stably inserted into the chromosome of the host cell.

At 48 h of incubation, cultures of strain IGTS8 still contained DBT; however, this compound was exhausted by the two engineered *Pseudomonas* strains. *P. aeruginosa* EGSOX had the fastest metabolism of DBT, transforming 95% of the DBT at 24 h of incubation. Only 18% of the DBT was transformed by *R. erythropolis* IGTS8, and 40% was transformed by *P. putida* EGSOX. Remarkably, DBT depletion was concomitant with 2-HBP accumulation in all three strains, indicating that 2-HBP is a dead-end metabolite that cannot be further catabolized or used as a carbon source. These data demonstrated that the IGTS8-derived *dsz* cassette was efficiently expressed, allowing the elimination of sulfur with no loss of DBT carbon atoms, both in *P. putida* EGSOX and *P. aeruginosa* EGSOX. Moreover, in comparison with wild-type *R. erythropolis* IGTS8, the two recombinant biocatalysts showed enhanced biodesulfurization ability.

However, many *Pseudomonas* strains were unable to desulfurize DBT in the oil phase, and this will restrict their application in industry. Darzins et al. (1999) found that *P. fluorescens* with *dsz*ABC genes cannot desulfurize DBT in the oil phase; but the whole cell lysate with the cell wall removed can. The results showed a lack of DBT uptake ability from the oil phase to the inside of the recombinant *Pseudomonas* strains. Noda et al. (2003) transferred the *dsz* desulfurization gene cluster from *R. erythropolis* IGTS8 into the chromosome of *P. aeruginosa* NCIMB9571 using a transposon vector. All of the recombinant strains completely desulfurized 1 mM DBT in *n*-tetradecane (*n*-TD) except one, named PARM1. PARM1 was unable to desulfurize DBT in *n*-TD but was able to desulfurize it in water. The transposon tagging analysis indicated that the transposon is inserted into *hcu*A of the open reading frames *hcu*ABC. The full-length *hcu*ABC genes, when transformed into PARM1, achieved 87% recovery of the desulfurization activity of DBT in *n*-TD, but partial *hcu*ABC genes achieved only 0–12%. These results indicated that DBT desulfurization in the oil phase by recombinant *P. aeruginosa* NCIMB9571 required the full-length *hcu*ABC gene cluster. The *hcu*ABC gene cluster is related to DBT uptake from the oil phase into the cell.

5.5 Rearranging the dsz Gene Cluster

As described before, the levels of transcription and translation of *dsz*A, *dsz*B, and *dsz*C decreased according to the positions of the genes in the *dsz* operon. Furthermore, the translation of *dsz*B was repressed by an overlapping structure in the *dsz* operon. In order to get better and steady expression of the Dsz enzymes and optimize the metabolic flux of DBT, the overlapped structure was removed, and the expression level of *dsz*B was increased. The DBT desulfurization rate was 5 times faster than that of the native *dsz* operon (Li et al. 2007a, b), but this is still low in comparison to the requirements of a commercial process.

The rate of an enzyme catalytic reaction is determined by the catalytic activity, the quantity of the enzyme, and the substrate concentration. Higher levels of mRNA are the precondition for higher levels of the encoded protein. Therefore, rearranging these genes according to the catalytic capabilities of the enzymes and their reaction orders could not only balance the catalytic capabilities but also increase the substrate concentrations for the enzymes. Li et al. introduced a genetic rearrangement strategy for optimizing the metabolic pathway of DBT. By using recombinant PCR, the *dsz* operon of *R. erythropolis* DS-3 was rearranged according to the catalytic capabilities of the Dsz enzymes and their reaction orders in the 4S pathway (Fig. 7).

The catalytic capabilities of the Dsz enzymes were approximately 25:1:5 (DszA: DszB:DszC). Hence, the *dsz* operon was rearranged according to the catalytic capabilities of the enzymes. The expression levels of *dsz*B and *dsz*C were improved by rearranging the order of the *dsz* genes to generate the operon *dsz*BCA, which contained *dsz*B, *dsz*C, and *dsz*A in tandem. After rearrangement, the ratio of *dszA*, *dszB*, and *dszC* mRNAs in the cells was changed, from 11:3.3:1 to 1:16:5. The desulfurization rate of the recombinant strain containing the rearranged *dsz* operon was 12 times faster than that of the native *dsz* operon. The maximum desulfurization rate was only about 26 μmol DBT/g DCW/h for the strain containing the native *dsz* operon. After removing the overlapped structure before the initiation codon of *dsz*B, the rate was 120 μmol DBT/g DCW/h. The recombinant strain containing the rearranged *dsz* operon had the highest desulfurization rate, about 320 μmol DBT/g DCW/h. Therefore, the enhanced expression levels of DszC and DszB increased the desulfurization rate of the recombinant strain.

Feng et al. (2006) found that the function of the surfactant Tween 80 in the desulfurization was to decrease the product concentration associated with the cells, reducing product inhibition. The *dsz* genes of *R. erythropolis* DS-3 were also integrated into the chromosome of *Bacillus subtilis* ATCC 21332, which can secrete biosurfactant, yielding the recombinant strain *B. subtilis* M29, which has higher desulfurization efficiency than *R. erythropolis* DS-3 and showed no product inhibition (Ma et al. 2006a, b). It should be noted that the biosurfactant secreted from *B. subtilis* M29 significantly varied the interfacial tension of the supernatant. The biosurfactant therefore has an important function in the degradation of DBT.

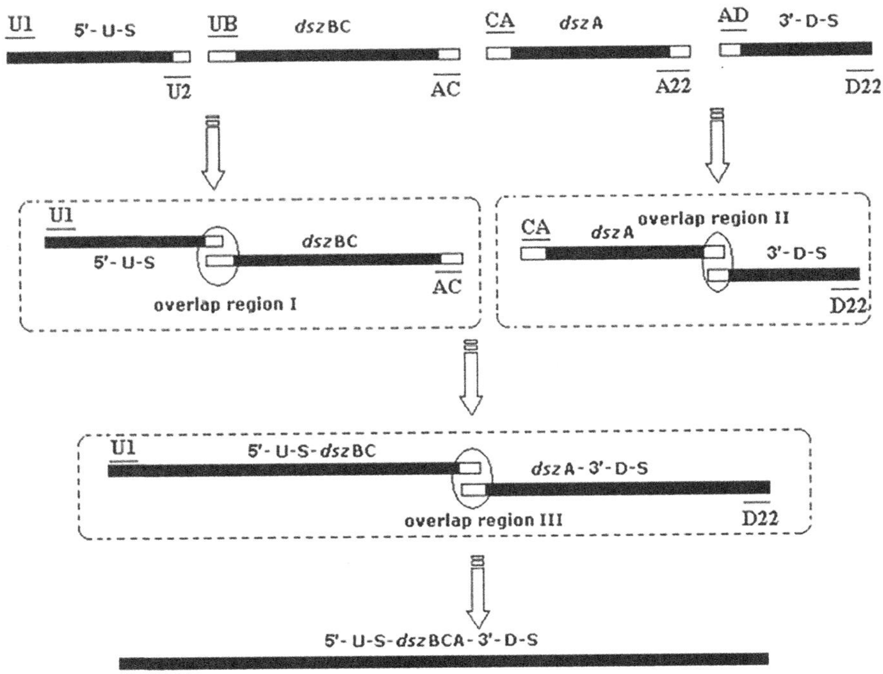

Fig. 7 Rearrangement of the *dsz* operon by overlap PCR. Fragments of the 400-bp 5′ upstream segment (5′-U-S) and the 400-bp 3′ downstream segment (3′-D-S) of *dsz*ABC and the *dsz*A and *dsz*BC segments, including the overlap regions, were amplified by PCR, then the ligated 5′ upstream-*dsz*BC segment (5′-U-S-*dsz*BC) and the ligated *dsz*A-3′ downstream segment (*dsz*A-3′-D-S) were produced by overlap PCR via their overlap regions, and finally, the 5′ upstream-*dsz*BC segment and the *dsz*A-3′ downstream segment were linked together by overlap PCR via their overlap region to yield the reconstructed *dsz* operon. Black bars represent genes, and white bars represent overlap regions

6 Process Engineering for BDS with *Rhodococcus*

To apply the BDS process from laboratory to industrial level, it is necessary to improve the BDS process engineering. In this section, we summarized the main advances in process design for BDS with *Rhodococcus* strains from two parts: the application of immobilized cells during BDS process and the design of the novelty bioreactor for BDS.

6.1 The Application of Immobilized Cells

Though biodesulfurization with growing cells is a simple and widely accepted methods with different *Rhodococcus* strains, the application of immobilized cells during BDS process could help improve the desulfurization of DBT with a higher concentration, increase desulfurization efficiency, and enhance the separation of cells from the final products.

Rhodococcus spp. strains were immobilized for BDS process on silica (Si), alumina (Al), and sepiolite (Sep) or encapsulated in polymeric materials such as alginates with the advantages of biocompatibility and low-cost. However, the immobilization leads to the reduction mass transference of oil to the bacterial cells. One considerable solution to address this problem is the use of surfactants. The synthetic surfactants such as Tween 80, Span 80, and Triton X-100 have been confirmed to enhance biodesulfurization effectively. Derikvand and Etemadifar (2014) demonstrated that the addition of (Tween 80 and Span 80) increased the dissolved DBT concentration in the aqueous phase and facilitate its close to *R. erythropolis* R1 in the alginate beads and resulted in the increase of biodesulfurization. Taking into account the toxicity of synthetic surfactants, biological surfactants (biosurfactants) with the advantages of non-toxicity, biodegradation, and adaption to extreme pH and temperature were also studied. The biosurfactant from a marine bacteria strain was purified and added in the BDS system of *R. erythropolis*. The results showed that adding biosurfactants increased the higher BDS activity when compared with synthetic surfactant Tween 80. It may be due to the formation of micelles increases the solubilization of DBT (Dinamarca et al. 2014).

Generally, the bacterial surface was negatively charged which will lead to the adsorption with positively charged nano-ɣ Al2O3. The production of 2-HBP increased twofolds after 24 h in the *Rhodococcus* spp. strains alginate beads containing nano-ɣ Al2O3 when compared with the controls. The combination of nano-ɣ Al2O3 and alginate encapsulated cells could be an effective approach to enhance BDS process (Derikvand et al. 2014).

Hassan et al. (2013) synthesized the Fe3O4 magnetite nanoparticles, which could be magnetically separated from oil/water biphasic system conveniently. The magnetite nanoparticle–coated *R. erythropolis* FMF and *R. erythropolis* IGTS8 cells exhibited similar desulfurization activity with the free cells (67 ± 3 and 69 ± 4, respectively). Furthermore, the coated Fe3O4 nanoparticles facilitate the recovery of cells from the biodesulfurization systems and then increase the recycled times of the immobilized cells.

6.2 Bioreactor Design

A vertical rotating immobilized cell reactor (VRICR) was designed and investigated for its BDS activity with the *R. erythropolis* (Amin 2011). The maximum desulfurization rate was up to 167 mM 2-HBP/Kg/h, and 100% of sulfur could be removed from the model oil (dibenzothiophene in hexadecane) within a 120 h period. Another research studied the influence of bed lengths and support particle size on the desulfurization efficiency of immobilized *R. rhodochrous* cells in a catalytic bed reactor packed with silica. The results suggested that longer bed, lower substrate flow, and large particle size would be benefited for the desulfurization (Alejandro et al. 2014).

To increase the utilizing efficiency of bacteria cells for desulfurization, a new aqueous–organic two-layer partitioning and continuous process was designed. Different from the batch and fed-batch processes, the biphasic bioreactor showed the efficient biodesulfurization activity for a long period (Yang et al. 2007). In this bioreactor, the oil and the cells were kept in organic and aqueous phase, respectively. DBT transferred from the oil phase to the aqueous phase and desulfurized by the bacterial cells. And then the produced 2-HBP was washed out of the bioreactor to reduce the inhibition.

7 Future Perspectives

Our understanding of how microorganisms metabolize sulfur heterocyclic compounds in petroleum has increased rapidly. All the studies outlined above are significant steps to explore the biotechnological potential for developing an efficient biodesulfurization process. However, these technologies have not yet been available for large-scale applications. We still need a much better understanding of more aspects of this pathway to turn it into a commercial process earlier. Any progress that provides the possibility to remove sulfur in crude oil at higher temperature, with higher rate, or longer stability of desulfurization activity is considered to be a significant step toward industry level biodesulfurization. Microorganisms with a wider substrate range and higher substrate affinity in biphasic reaction containing toxic solvents or higher biodesulfurization activities could be engineered if the biocatalysts were to be used for petroleum treatment.

Acknowledgements We gratefully acknowledge the support for our research from the Natural Science Foundation of Tianjin, China (grant numbers 05YFJMJC00700, 09JCZDJC03200). We also thank the publisher of the American Society of Microbiology and the American Society for Biochemistry and Molecular Biology to offer kindly help with useful pictures.

References

Albano C, Randers-Eichhorn L, Bentley W, Rao G (1998) Green fluorescent protein as a real time quantitative reporter of heterologous protein production. Biotechnol Prog 14:351–354

Alejandro DM, Orellana L, Aguirre J, Baeza P, Espinoza G, Canales C, Ojeda J (2014) Biodesulfurization of dibenzothiophene and gas oil using a bioreactor containing a catalytic bed with Rhodococcus rhodochrous immobilized on silica. Biotechnol Lett 36: 1649–1652

Amin GA (2011) Integrated two-stage process for biodesulfurization of model oil by vertical rotating immobilized cell reactor with the bacterium Rhodococcus erythropolis. J Pet Environ Biotechnol 2:107–111

Arensdorf JJ, Loomis AK, DiGrazia PM, Monticello DJ, Pienkos PT (2002) Chemostat approach for the directed evolution of biodesulfurization gain-of-function mutants. Appl Environ Microbiol 68:691–698

Atlas RM (1994) Petroleum microbiology. Macmillan Publishing, New York

Atlas RM, Boron DJ, Deever WR, Johnson AR, McFarland BL, Meyer JA (2001) Method for removing organic sulfur from heterocyclic sulfur containing organic compounds. US patent number H1, p 986

Borgne S, Quintero R (2003) Review: biotechnological processes for the refining of petroleum. Fuel Process Technol 81:155–169

Cha H, Wu C, Valdes J, Rao G, Bentley W (2000) Observations of green fluorescent protein as a fusion partner in genetically engineered *Escherichia coli*: monitoring protein protein expression and solubility. Biotechnol Bioeng 67:565–574

Chen H, Zhang WJ, Chen JM, Cai YB, Li W (2008) Desulfurization of various organic sulfur compounds and the mixture of DBT + 4,6-DMDBT by *Mycobacterium* sp. ZD-19. Bioresour Technol 99:3630–3634

Coco WM, Levinson WE, Crist MJ, Hektor HJ, Darzins A, Pienkos PT, Squires CH, Monticello DJ (2001) DNA shuffling method for generating highly recombined genes and evolved enzymes. Nat Biotechnol 19:354–359

Darzins A, Xi L, Childs JD, Monticello DJ, Squires CH (1999) DSZ gene expression in pseudomonas hosts. US Patent No 5952208

Denis-Larose C, Labbe D, Bergeron H, Jones AM, Greer CW, al-Hawari J, Grossman MJ, Sankey BM, Lau PC (1997) Conservation of plasmid-encoded dibenzothiophene desulfurization genes in several *Rhodococci*. Appl Environ Microbiol 63:2915–2919

Derikvand P, Etemadifar Z (2014) Improvement of biodesulfurization rate of alginate immobilized Rhodococcus erythropolis R1. Jundishapur J Microbiol 7:e9123

Derikvand P, Etemadifar Z, Biria D (2014) Taguchi optimization of dibenzothiophene biodesulfurization by Rhodococcus erythropolis, R1 immobilized cells in a biphasic system. Int Biodeter Biodegr 86:343–348

Dinamarca MA, Rojas A, Baeza P, Espinoza G, Ibacache-Quiroga C, Ojeda J (2014) Optimizing the biodesulfurization of gas oil by adding surfactants to immobilized cell systems. Fuel 116:237–241

Dosomer JP, Dhaese P, Montagu MV (1988) Conjugative transfer of cadmium resistance plasmids in *Rhodococcus fascians* strains. J Bacteriol 170:2401–2405

Feng J, Zeng Y, Ma C, Cai X, Zhang Q, Tong M, Yu B, Xu P (2006) The surfactant tween 80 enhances biodesulfurization. Appl Environ Microbiol 72:7390–7393

Galán B, Díaz E, García JL (2000) Enhancing desulfurization by engineering a flavin reductase-encoding gene cassette in recombinant biocatalyst. Environ Microbiol 2:687–694

Gallagher JR, Olson ES, Stanley DC (1993) Microbial desulphurization of dibenzothiophene: a sulfur-specific pathway. FEMS Microbiol Lett 107:31–36

Gallardo ME, Ferrandez A, De Lorenzo V, Garcia JL, Diaz E (1997) Designing recombinant *Pseudomonas* strains to enhance biodesulfurization. J Bacteriol 179:7156–7160

Gaudu P, Touati D, Niviere V, Fontecave M (1994) The NAD(P)H: flavin oxidoreductase from *Escherichia coli* as a source of superoxide radicals. J Biol Chem 269:8182–8185

Gilbert SC, Morton J, Buchanan S, Oldfield C, McRoberts A (1998) Isolation of a unique benzothiophene-desulphurizing bacterium, *Gordona* sp. strain 213E (NCIMB 40816), and characterization of the desulphurization pathway. Microbiology 144:2545–2553

Gray KA, Pogrebinsky OS, Mrachko GT, Xi L, Monticello DJ, Squires CH (1996) Molecular mechanisms of biocatalytic desulfurization of fossil fuels. Nat Biotechnol 14:1705–1709

Gray KA, Mrachkoyz GT, Squiresy CH (2003) Biodesulfurization of fossil fuels. Curr Opin Microbiol 6:229–235

Grossman MJ, Lee MK, Prince RC, Garrett KK, George GN, Pickering IJ (1999) Microbial desulfurization of a crude oil middle-distillate fraction: analysis of the extent of sulfur removal and the effect of removal on remaining sulfur. Appl Environ Microbiol 65: 181–188

Gupta N, Roychoudhury PK, Deb JK (2005) Biotechnology of desulfurization of diesel: prospects and challenges. Appl Microbiol Biotechnol 66:356–366

Hassan B, Jamshid R, Hossein M, Behnam R, Ayyoob A (2013) Desulfurization activity and reusability of magnetite nanoparticle-coated *Rhodococcus erythropolis* FMF and *R. erythropolis* IGTS8 bacterial cells. Biotechnol Appl Biochem 60:323–329

Hirasawa K, Ishii Y, Kobayashi M, Koizumi K, Maruhashi K (2001) Improvement of desulfurization activity in *Rhodococcus erythropolis* KA2-5-1 by genetic engineering. Biosci Biotechnol Biochem 65:239–246

Inoue A, Horikoshi K (1991) Estimation of solvent-tolerance of bacteria by the solvent parameter log P. J Ferment Bioeng 71:194–196

Izumi Y, Ohshiro T, Ogino H, Hine Y, Shinao M (1994) Selective desulphurisation of dibenzothiophene by *R. erythropolis* D-1. Appl Environ Microbiol 60:223–226

Ji YE, Colston MJ, Cox RA (1994) The ribosomal RNA (*rrn*) operons of fast-growing mycobacteria: primary and secondary structures and their relation to *rrn* operons of pathogenic slow-growers. Microbiology 140:2829–2840

Kawatra SK, Eisele TC (2001) Coal desulfurization, high-efficiency preparation methods. Taylor & Francis, New York

Kayser KJ (2002) Molecular biological characterization and enhancement of the biodesulfurization (DSZ) pathway. PhD thesis. Illinois Institute of Technology, 106 pp

Kertesz L (2001) Building a scientific foundation for prevention. Healthplan 42:44–47

Kilbane JJ (2006) Microbial biocatalyst developments to upgrade fossil fuels. Curr Opin Biotechnol 17:305–314

Kilbane JJ, Bielaga BA (1990) Toward sulfur-free fuels. Chem Tech 20:747–751

Kilbane JJ, Jackowski K (1992) Biodesulphurisation of watersoluble coal-derived material by *Rhodococcus rhodochrous* IGTS8. Biotechnol Bioeng 40:1107–1114

Kirimura K, Furuya T, Sato R, Ishii Y, Kino K, Usami S (2002) Biodesulfurization of naphthothiophene and benzothiophene through selective cleavage of carbon-sulfur bonds by *Rhodococcus* sp. strain WU-K2R. Appl Environ Microbiol 68:3867–3872

Kobayashi M, Onaka T, Ishii Y, Konishi J, Takaki M, Okada H, Ohta Y, Koizumi K, Suzuki M (2000) Desulfurization of alkylated forms of both dibenzothiophene and benzothiophene by single bacterial strain. FEMS Microbiol Lett 187:123–126

Koike H, Sasaki H, Kobori T, Zenno S, Saigo K, Murphy MEP, Adman ET, Tanokura M (1998) 1.8Å crystal structure of the major NAD-(P)H:FMN oxidoreductase of a bioluminescent bacterium, *Vibrio fischeri*: overall structure, cofactor and substrate- analog binding, and comparison with related flavoproteins. J Mol Biol 280:259–273

Konishi J, Onaka T, Ishii Y, Suzuki M (2000) Demonstration of the carbon-sulfur bond-targeted desulfurization of benzothiophene by thermophile *Paenibacillus* sp. strain A11-2 capable of desulfurizing dibenzothiophene. FEMS Microbiol Lett 187:151–154

Kropp KG, Fedorak PM (1998) A review of the occurrence, toxicity, and biodegradation of condensed thiophenes found in petroleum. Can J Microbiol 44:605–622

Kropp P, Gerber WD (1998) Prediction of migraine attacks using a slow cortical potential, the contingent negative variation. Neurosci Lett 257:73–76

Leahy JG, Colwell RR (1990) Microbial degradation of hydrocarbons in the environment. Microbiol Rev 54:305–315

Lee WC, Ohshiro T, Matsubara T, Izumi Y, Tanokura M (2004) Crystallization and preliminary X-ray analyses of desulfurization enzyme DszB and its C27S mutant complexed with biphenyl-2-sulfinic acid. Acta Crystallogr D Biol Crystallogr 60:1636–1638

Lee WC, Ohshiro T, Matsubara T, Izumi Y, Tanokura M (2006) Crystal structure and desulfurization mechanism of 2′-hydroxybiphenyl-2- sulfinic acid desulfinase. J Boil Chem 281:32534–32539

Lei X, Squires CH, Monticello DJ, Child D (1997) A flavin reductase stimulates DszA and DszC proteins of *Rhodococcus erythropolis* IGTS8 *in vitro*. Biochem Biophys Res Commun 230: 73–75

Li ZM, Squires CH, Monticello DJ, Childs JD (1996) Genetic analysis of the *dsz* promoter and associated regulatory region of *Rhodococcus erythropolis* IGTS8. J Bacteriol 178:6409–6418

Li FL, Xu P, Ma CQ, Luo LL, Wang XS (2003) Deep desulfurization of hydrodesulfurization treated diesel oil by a facultative thermophilic bacterium *Mycobacterium* sp. X7B. FEMS Microbiol Lett 223:301–307

Li W, Zhang Y, Wang MD, Shi Y (2005a) Biodesulfurization of dibenzothiophene and other organic sulfur compounds by a newly isolated *Microbacterium* strain ZD-M2. FEMS Microbiol Lett 247:45–50

Li FL, Xu P, Feng JH, Meng L, Zheng Y, Luo LL, Ma CQ (2005b) Microbial desulfurization of gasoline in a *Mycobacterium goodii* X7B immobilized-cell system. Appl Environ Microbiol 71:276–281

Li GQ, Ma T, Li JH, Li H, Liu RL (2006a) Co-expression of *Rhodococcus* sp. DS-3 *dsz*ABC and *dsz*D gene with incompatible plasmids in *Escherichia coli*. Acta Microbiol Sin 46:275–279

Li W, Wang MD, Chen H, Chen JM, Shi Y (2006b) Biodesulfurization of dibenzothiophene by growing cells of Gordonia sp. in batch cultures. Biotechnol Lett 28:1175–1179

Li GQ, Ma T, Li SS, Li H, Liang FL, Liu RL (2007a) Improvement of dibenzothiophene desulfurization activity by removing the gene overlap in the *dsz* operon. Biosci Biotechnol Biochem 71:849–854

Li FL, Zhang ZZ, Feng JH, Cai XF, Xu P (2007b) Biodesulfurization of DBT in tetradecane and crude oil by a facultative thermophilic bacterium *Mycobacterium goodii* X7B. J Biotechnol 127:222–228

Lorenzo V, Timmis KN (1994) Analysis and construction of stable phenotypes in Gram-negative bacteria with Tn5 and Tn10-derived minitransposons. Methods Enzymol 235:386–405

Ma X, Sakanishi K, Mochida I (1994) Hydrodesulfurization reactivities of various sulfur compounds in diesel fuel. Ind Eng Chem Res 33:218–222

Ma CQ, Feng JH, Zeng YY, Cai XF, Sun BP, Zhang ZB, Blankespoor HD, Xu P (2006a) Methods for the preparation of a biodesulfurization biocatalyst using *Rhodococcus* sp. Chemosphere 65:165–169

Ma T, Li GQ, Li J, Liang FL, Liu RL (2006b) Desulfurization of dibenzothiophene by *Bacillus subtilis* recombinants carrying *dsz*ABC and *dsz*D genes. Biotechnol Lett 28:1095–1100

Matsubara T, Ohshiro T, Nishina Y, Izumi Y (2001) Purification, characterization, and overexpression of flavin reduced involved in dibenzothiophene desulfurization by *Rhodococcus erythropolis* D-1. Appl Environ Microbiol 67:1179–1184

Matsui T, Hirasawa K, Koizumi KI, Maruhashi K, Kurane R (2001) Optimization of the copy number of dibenzothiophene desulfurizing genes to increase the desulfurization activity of recombinant *Rhodococcus* sp. Biotechnol Lett 23:1715–1718

Matsui T, Noda K, Tanaka Y, Maruhashi K, Kurane R (2002) Recombinant *Rhodococcus* sp. strain T09 can desulfurize DBT in the presence of inorganic sulfate. Curr Microbiol 45:240–244

McFarland BL, Boron DJ, Deever W, Meyer JA, Johnson AR, Atlas RM (1998) Biocatalytic sulfur removal from fuels: applicability for producing low sulfur gasoline. Crit Rev Microbiol 24: 99–147

Monticello DJ (1998) Riding the fossil fuel biodesulfurization wave. Chem Tech 28:38–45

Monticello DJ, Bakker D, Finnerty WR (1985) Plasmid-mediated degradation of dibenzothiophene by *Pseudomonas* species. Appl Environ Microbiol 49:756–760

Noda K, Kimiko W, Kenji M (2002) Cloning of a rhodococcal promoter using a transposon for dibenthiophene biodesulfurization. Biotechnol Letters 24:1875–1882

Noda K, Watanabe K, Maruhashi K (2003) Isolation of the *Pseudomonas aeruginosa* gene affecting uptake of dibenzothiophene in n-tetradecane. J Biosci Bioeng 95:504–511

Ochsner UA, Reiser J, Fiechter A, Witholt B (1995) Production of *Pseudomonas aeruginosa* rhamnolipid biosurfactants in heterologous hosts. Appl Environ Microbiol 61:3503–3506

Ohshiro T, Izumi Y (1999) Microbial desulfurization of organic sulfur compounds in petroleum. Biosci Biotechnol Biochem 63:l–9

Ohshiro T, Izumi Y (2000) Purification, characterization and crystallization of enzymes for dibenzothiophene desulfurization. Bioseparation 9:185–188

Ohshiro T, Hine Y, Izumi Y (1994) Enzymatic desulfurization of dibenzothiophene by a cell-free system of *Rhodococcus erythropolis* D-1. FEMS Microbiol Lett 118:341–344

Ohshiro T, Hirata T, Izumi Y (1995) Microbial desulfurization of dibenzothiophene in the presence of hydrocarbon. Appl Microbiol Biotechnol 44:249–252

Ohshiro T, Kojima T, Torii K, Kawasoe H, Izumi Y (1999) Purification and characterization of dibenzothiophene (DBT) sulfone monooxygenase, an enzyme involved in DBT desulfurization, from *Rhodococcus erythropolis* D-l. J Biosci Bioeng 88:610–616

Ohshiro T, Ohkita R, Takikawa T, Manabe M, Lee WC, Tanokura M, Izumi Y (2007) Improvement of 2′-Hydroxybiphenyl-2-sulfinate Desulfinase, an Enzyme Involved in the Dibenzothiophene Desulfurization Pathway, from *Rhodococcus erythropolis* KA2-5-1 by Site-Directed Mutagenesis. Biosci Biotechnol Biochem 71:2815–2821

Oldfield C, Pogrebinsky O, Simmonds J, Olson ES, Kulpa CF (1997) Elucidation of the metabolic pathway for dibenzothiophene desulfurization by *Rhodococcus* sp. IGTS8 (ATCC 53968). Microbiology 143:2961–2973

Oldfield C, Wood NT, Gilbert SC, Murray FD, Faure FR (1998) Desulfurization of benzothiophene by actinomycete organisms belonging to the genus *Rhodococcus*, and related taxa. Antonie Van Leeuwenhoek 74:119–132

Omori T, Monna L, Saiki Y, Kodama T (1992) Desulfurization of dibenzothiophene by *Corynebacterium* sp. Strain SY1. Appl Environ Microbiol 58:911–915

Piddington CS, Kovacevich BR, Rambosek J (1995) Sequence and molecular characterization of a DNA region encoding the dibenzothiophene desulfurization operon of *Rhodococcus* sp. strain IGTS8. Appl Environ Microbiol 61:468–475

Purdy RF, Lepo JE, Ward B (1993) Biodesulfurization of organicsulfur compounds. Curr Microbiol 27:219–222

Reichmuth DS, Hittle JL, Blanch HW, Keasling JD (1999) Biodesulfurization of dibenzothiophene in *Escherichia coli* is enhanced by expression of *Vibrio harveyi* oxidoreductase gene. Biotechnol Bioeng 67:72–79

Reichmuth DS, Blanch HW, Keasling JD (2004) Dibenzothiophene biodesulfurization pathway improvement using diagnostic GFP fusions. Biotechnol Bioeng 88:94–99

Rhee SK, Chang JH, Chan YK, Chang HN (1998) Desulfurization of dibenzothiophene and diesel oils by a newly isolated *Gordona* strain, CYKS1. Appl Environ Microbiol 64:2327–2331

Shan GB, Xing JM, Luo MF, Liu HZ, Chen JY (2003) Immobilization of *Pseudomonas delafieldii* with magnetic polyvinyl alcohol beads and its application in biodesulfurization. Biotechnol Lett 25:1977–1981

Shennan JL (1996) Microbial attack on sulfur-containing hydrocarbons, implications for the biodesulphurization of oils and coals. J Chem Technol Biotechnol 67:109–123

Soleimani M, Bassi A (2007) Biodesulfurization of refractory organic sulfur compounds in fossil fuels. Biotechnol Adv 25:570–596

Tanaka Y, Matsui T, Konishi J, Maruhashi K, Kurane R (2002) Biodesulfurization of benzothiophene and dibenzothiophene by a newly isolated *Rhodococcus* strain. Appl Microbiol Biotechnol 59:325–328

Tao F, Yu B, Xu P, Ma CQ (2006) Biodesulfurization inbiophasic systems containing organic solvents. Appl Environ Microbiol 72:4604–4609

Watkins LM, Rodriguez R, Schneider D, Broderick R, Cruz M, Chambers R, Ruckman E, Cody M, Mrachko GT (2003) Purification and characterization of the aromatic desulfinase, 2-(2′-hydroxyphenyl) benzenesulfinate desulfinase. Arch Biochem Biophys 415:14–23

Xi L, Squires CH, Monticello DJ, Childs JD (1997) A flavin reductase stimulates DszA and DszC proteins of *Rhodococcus erythropolis* IGTS8 in vitro. Biochem Biophys Res Commun 230: 73–75

Yan H, Kishimoto M, Omasa T, Katakura Y, Suga K, Okumura K, Yoshikawa O (2000) Increase in desulfurization activity of *Rhodococcus erythropolis* KA2-5-1 using ethanol feeding. J Biosci Bioeng 89:361–366

Yang J, Hu Y, Zhao D, Wang S, Lau PCK, Marison IW (2007) Two-layer continuous-process design for the biodesulfurization of diesel oils under bacterial growth conditions. Biochem Eng J 37:212–218

Yu B, Xu P, Shi Q, Ma C (2006a) Deep desulfurization of diesel oil and crude oils by a newly isolated Rhodococcus erythropolis strain. Appl Environ Microbiol 72:54–58

Yu B, Ma CQ, Zhou WJ, Wang Y, Cai XF, Tao F, Zhang Q, Tong MY, Qu JY, Xu P (2006b) Microbial desulfurization of gasoline by free whole-cells of *Rhodococcus erythropolis* XP. FEMS Microbiol Lett 258:284–289

Bioremediation of Contaminated Environments Using *Rhodococcus*

Maria S. Kuyukina and Irena B. Ivshina

Contents

Abstract Environmental pollution with anthropogenic organic compounds is the global problem of our planet. Bioremediation has a great potential to effectively restore polluted environments by using biodegradative activities of microorganisms. The genus *Rhodococcus* is a promising group of bacteria suitable for biodegradation of recalcitrant contaminants, such as petroleum hydrocarbons, chlorinated, nitrogenated, and other complex organics. *Rhodococcus* species are ubiquitous in pristine and contaminated environments, survive under harsh environmental conditions, compete successfully in complex bacterial populations, and therefore could be efficiently used in bioremediation applications. Some success in bioremediation of contaminated soils, waters, and air has been achieved using rhodococci either as bioaugmentation agents or members of indigenous microbial communities stimulated by nutrient and oxygen amendments. Laboratory and field-scale studies on *Rhodococcus* application in

M. S. Kuyukina (✉) · I. B. Ivshina
Institute of Ecology and Genetics of Microorganisms, Perm Scientific Centre of the Ural Branch of the Russian Academy of Sciences, Perm State University, Perm, Russia
e-mail: kuyukina@iegm.ru

H. M. Alvarez (ed.), *Biology of Rhodococcus*, Microbiology Monographs 16,
https://doi.org/10.1007/978-3-030-11461-9_9

cleanup technologies are reviewed relating to in situ subsurface and groundwater remediation, on-site treatments of contaminated soils, sludges, wastewaters, and gaseous emissions.

1 Introduction

Environmental pollution with anthropogenic organic compounds has become one of the most urgent problems worldwide. This negatively impacts economical and social developments and poses significant threats to human health and natural biodiversity. Potential sources of organic contaminants include industrial leaks and spills, improper application of pesticides and fire retardants, negligent disposal of industrial and domestic wastes, and landfills and garbage dumps. Oil spillage and oily waste deposits represent the major part of hydrocarbon contamination in our planet. Petroleum hydrocarbons are widespread soil and groundwater pollutants resulting from leaking of underground storage sites, spillage from the storage tanks, and damaged pipelines. There are thousands of sites that have been seriously contaminated by petroleum products in oil-producing regions around the world (Etkin 2001; Ivshina et al. 2015a). Moreover, marine oil spills from crashed tankers are responsible for the massive contamination of seawater and shorelines. Apart from oil industry, major sources of hydrocarbon contaminants, such as alkanes and polycyclic aromatic hydrocarbons (PAHs), are coal-mining sites and coking plants, gas processing plants, solid fuels for domestic heating, aircraft and car exhausts, and forest fires. Also, chlorinated hydrocarbons such as chlorobenzenes, chlorophenols, and polychlorinated biphenyls (PCBs) that are used commercially for a variety of purposes, including production of solvents, paint additives, pesticides, fire retardants, and insulating fluids, represent a large proportion in long-term persistent contamination of soils and groundwater. It should be noted that soil contamination by organic compounds is a complex process and difficult to treat due to many reasons, for example, a tendency to sorption of contaminants into the soil matrix, low water solubility, and limited rate of mass transfer. Remediation activities are often hampered by remoteness and low accessibility of contaminated sites, harsh environmental conditions, as well as high pollution levels and large amounts of contaminated material to be treated. Bioremediation is considered to be a nondestructive, cost- and treatment-effective, and sometimes logistically favorable cleanup technology capable of accelerating naturally occurring biodegradation of contaminants through the optimization of limiting conditions (Alexander 1999). Currently, methods of biological remediation of contaminated sites gain ever increasing popularity due to their sustainability, relatively low cost, and environmental safety (Fig. 1).

Bioremediation of polluted environments is based on contaminant biodegradation, that is, metabolic abilities of microorganisms to transform or mineralize organic

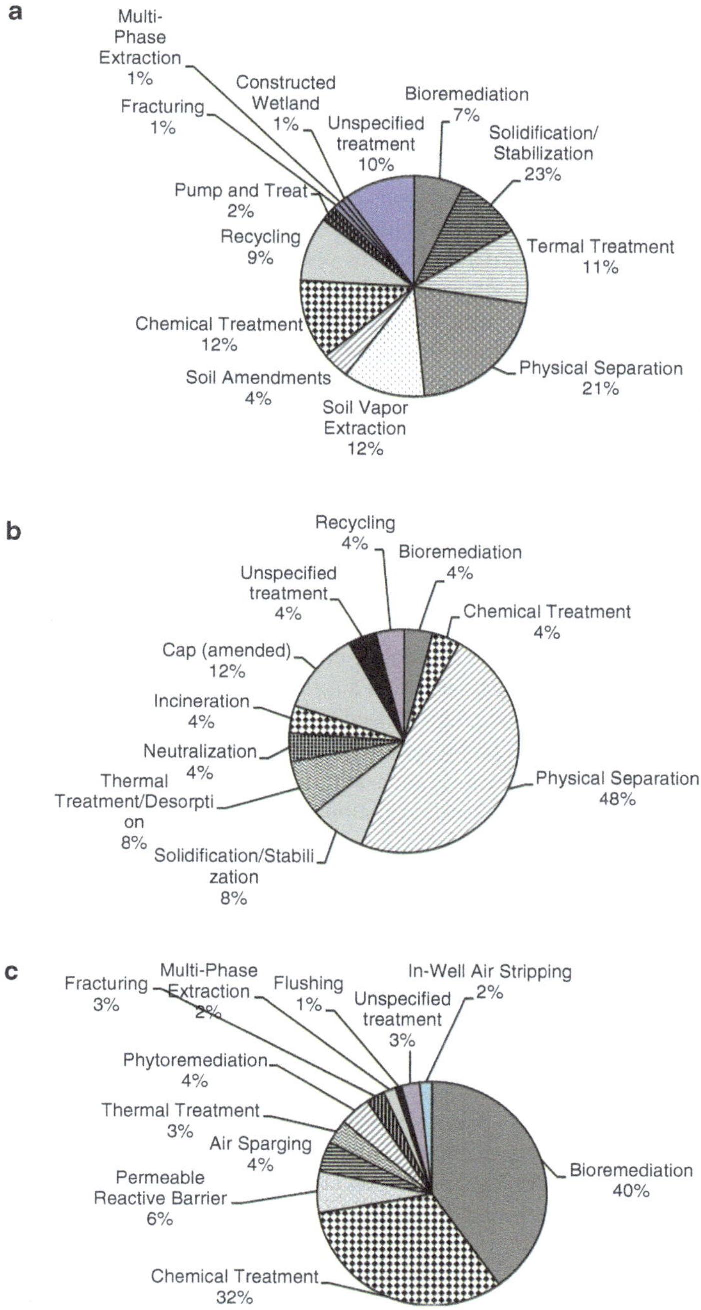

Fig. 1 Proportion of bioremediation among remediation technologies reported in the US EPA Superfund Remedy Report (15th Edition) for 2012–2014. (**a**) Source [soil, sludge, sediment, solid waste, debris, drummed waste, leachate, and any nonaqueous phase liquid both light (LNAPL) and

contaminants into less harmful, nonhazardous substances, which are further integrated into natural biogeochemical cycles (Alexander 1999; Whyte et al. 2001). In most cases, treatment of contaminated environments has involved biostimulation, addition of nutrients and other factors to stimulate spontaneous enrichment of the indigenous contaminant-degrading microbial population. However, there has been considerable debate over the efficacy of bioaugmentation (Aislabie et al. 2006), the addition of either indigenous or exogenous microorganism's cultures to enhance the remediation process. An application area for bioaugmentation could be contaminated environments deficient in microorganisms due to extreme climate conditions (e.g., polar and desert soils, low-oxygen groundwaters) or high levels of toxic contaminants (e.g., industrial waste dumps and wastewater streams). For example, low numbers of hydrocarbon-oxidizing bacteria in cold climate pristine soils coupled with short summer seasons may limit the spontaneous enrichment of oil-contaminated soils with autochthonous hydrocarbon oxidizers when biostimulation alone is applied (Ruberto et al. 2005). Bioaugmentation, in the context of bioremediation, should be considered as inoculation of contaminated soil or water with specific strains or consortia of microorganisms to improve the biodegradation capacity of the system for specific organic pollutant(s) (Alexander 1999). There are two fundamental approaches to bioaugmentation of a contaminated site. The first involves increasing the genetic diversity of the residential microbial community by inoculation with allochthonous microorganisms, which may enhance the catabolic potential, and thereby the biodegradation rate (Dejonghe et al. 2001). The second approach applied in most bioremediation projects involves a serial enrichment of indigenous microbial populations isolated from the contaminated samples using laboratory cultivation with the contaminant(s) as the sole carbon source. This enriched inoculum containing mostly fast-growing organisms with exceptional degradative capacities is then returned back to the site to increase the rate of biodegradation. Various factors are known to influence the success or failure of bioaugmentation; the predictability, however, is beyond our scope yet. In many cases, contaminated environments are hostile to the introduction of allochthonous microorganisms, and large numbers will never survive unless they have a selective advantage. The controlling environmental factors are both biotic and abiotic (for reviews, see van Veen et al. 1997; Watanabe and Hamamura 2003). So, assuming bioaugmentation as a feasible bioremediation approach, it seems to be important to find the "proper" microorganisms for the bioaugmentation consortium. These microorganisms should possess high biodegrading abilities, be highly competitive in natural bacteriocenoses, as well as be nonpathogenic and environmentally friendly.

The genus *Rhodococcus* was regarded as one of the most promising groups of microorganisms suitable for the biodegradation of compounds not readily transformed by other organisms. The biochemical potential of rhodococci has been increasingly explored because of their broad catabolic versatility and unique enzymatic capabilities

Fig. 1 (continued) dense (DNAPL)] treatment. (**b**) Sediment treatment. (**c**) In situ groundwater treatment. Figures generated from US EPA (2017)

(Van der Geize and Dijkhuizen 2004; Martínková et al. 2009; Ivshina et al. 2017). Xenobiotic compounds metabolized by rhodococci cover a wide range of structural groups, including aliphatic and aromatic hydrocarbons, oxygenated and halogenated compounds, nitroaromatics, heterocyclic compounds, nitriles, various pharmaceuticals, and pesticides. Many of these substrates are complex synthetic molecules with high chemical stabilities and toxicity. *Rhodococcus* species are ubiquitous in pristine and contaminated environments, possess remarkable metabolic activities, can persist under harsh environmental conditions, compete successfully in complex bacterial populations, and therefore could be considered as having great potential in bioremediation applications. The significance of rhodococci in environmental biotechnology was discussed in early reviews characterizing the genus *Rhodococcus* (Warhurst and Fewson 1994; Bell et al. 1998). This chapter describes the roles of rhodococci as members of natural hydrocarbon-oxidizing consortia and promising bioremediation agents and also reviews laboratory and field-scale studies on *Rhodococcus* application in cleanup technologies for contaminated environments.

2 Why Are Rhodococci Considered as Most Suitable for Environment Bioremediation?

Rhodococcus species are naturally present in diverse temperate and extreme environments. They can persist and grow in highly contaminated soils and waters and even under oxygen- and nutrient-limited conditions. *Rhodococcus* strains can be readily isolated from environmental samples and enriched in mixed or pure cultures by incubation with a particular contaminant of interest, which is important for preparing inocula for bioremediation. Their natural occurrence in contaminated environments, broad catabolic versatility, and physiological and ecological adaptations to extreme environmental conditions imply that they may play a significant role in both natural degradation of persistent pollutants and bioremediation of contaminated ecosystems.

2.1 *Pristine and Contaminated Environments Are Common Habitats for* Rhodococcus *Species*

Rhodococci have been isolated from a large variety of natural sources, including clean and contaminated soils and rocks, surface and groundwaters, and marine sediments, from animals and plants. *Rhodococcus* strains are often isolated from environments where hydrocarbons are present. Since petroleum hydrocarbons are most widespread contaminants of terrestrial and marine ecosystems, a large variety of studies were undertaken attempting isolation, characterization, and cleanup application of petroleum-degrading bacteria (for a review, see Van Hamme et al. 2003).

Some *Rhodococcus* species, particularly *R. rhodochrous* and *R. ruber*, are able to grow using gaseous hydrocarbons (such as propane, butane, and acetylene) as a sole carbon source (Ivshina et al. 1994). Ivshina et al. (1981) found large numbers of propane- and butane-degrading *R. rhodochrous* and *R. ruber* isolates in soil, rock, and groundwater associated with oil-bearing sites, suggesting that these gas-oxidizing rhodococci are indicative of the presence of subterranean hydrocarbon deposits and thus could be used in oil prospecting. Crude oil-contaminated soil and marine samples collected from different localities in Kuwait were screened for microorganisms capable of oil degradation (Sorkhoh et al. 1990). It was found that, among many bacterial isolates, *Rhodococcus* strains were the most abundant mesophilic hydrocarbon-oxidizing bacteria and most efficient oil degraders. Cold-tolerant *Rhodococcus* strains were isolated from oil-contaminated soils in Antarctica; they grew on a range of alkanes from hexane (C_6) through at least eicosane (C_{20}) and the isoprenoid compound pristane (2,6,10,14-tetramethyl-pentadecane) and retained metabolic activities at sub-zero temperatures of −2 °C (Bej et al. 2000). Several cold-tolerant *Rhodococcus* strains isolated from a deep Greenland glacier ice core grew rapidly at 2 °C (Miteva et al. 2004). A psychrotrophic *Rhodococcus* sp. from Arctic soil (Whyte et al. 1998) utilized a broad range of aliphatics (C_{10}–C_{21} alkanes, branched alkanes, and a substituted cyclohexane) present in diesel oil at 5 °C. The strain mineralized short-chain alkanes (C_{10} and C_{16}) to a significantly greater extent (by a factor of about 2–3) than long-chain alkanes (C_{28} and C_{32}) at 0 and 5 °C. The psychrotrophic halotolerant oil-degrading strain *Rhodococcus* sp. YHLT-2 isolated from gasoline-contaminated groundwater was able to grow at 7% NaCl and utilized short-chain alkenes (C_9–C_{12}) as well as a broad range of long-chain alkenes (C_{19}–C_{32}) present in crude oil at 4 °C (Ryu et al. 2006). The study of Mergaert et al. (2001) showed that a large proportion (34%) of facultative oligotrophic and psychrotrophic strains isolated from Arctic and Antarctic seawaters were grouped according to their fatty acid profiles into the *Rhodococcus fascians* cluster. A recent study of Sinha et al. (2017) revealed a seasonal dominance of *R. fascians* in Arctic fjord heterotrophic bacterial communities. Phylogenetic analysis of 16S rRNA genes from alkane-degrading bacterial isolates indicated that *Rhodococcus* spp. from cold regions mostly group with *R. erythropolis* or *R. fascians* (Aislabie et al. 2006). We have isolated a large number of alkanotrophic *Rhodococcus* strains identified as *R. erythropolis*, *R. fascians*, *R. opacus*, *R. rhodochrous*, and *R. ruber* from soil, surface and spring water, snow, air, and core samples taken from ecologically and geographically diverse regions of the former Soviet Union (Ivshina et al. 1994, 1995).

However, most of the recent environmental microbiology studies using molecular genetic techniques focused on bacterial community structure and dynamics rather than on culture isolation. The occurrence of four alkane monooxygenase genotypes (*Pseudomonas putida* GPo1, Pp *alkB*; *Rhodococcus* sp. strain Q15, Rh *alkB1* and Rh *alkB2*; and *Acinetobacter* sp. strain ADP-1, Ac *alkM*) in hydrocarbon-contaminated and pristine soils from the Arctic and Antarctica was determined by both culture-independent (PCR hybridization analyses) and culture-dependent (colony hybridization analyses) molecular methods (Whyte et al. 2002a). PCR hybridization of the total soil community DNA revealed that Rh *alkB1* and Rh *alkB2* genes are common in both contaminated and clean soils, whereas Pp *alkB* is common in contaminated soil, while Ac *alkM* is rare. Furthermore, Rh *alkB1* was prevalent in culturable psychrotolerant bacteria, suggesting that *Rhodococcus* is the predominant alkane degrader in both pristine and contaminated polar soils. Similar results for Rh *alkB1* and Rh *alkB2* prevalence in polar (Antarctica), Alpine (Austria), and tropical (Brazil) soils (Margesin et al. 2003; Luz et al. 2004) suggested that rhodococci are typical alkanotrophic soil bacteria through various ecological and climatic regions. However, in a recent study of Nie et al. (2014), only a few *alkB* sequences retrieved from freshwater, marine, and terrestrial metagenomic databases were closely related to the sequences found in previously identified microbial genomes, suggesting the presence of numerous novel *alkB* genes, including those from *Rhodococcus*, in different environments. Since rhodococci are known to harbor multiple *alkB* genes for different alkane 1-monooxygenases, similarity levels of *alkB* genes with the unique nucleotide sequence encoding a conserved amino acid motif:WLG(I/V/L)D(G/D) GL can be useful for revealing the evolution and improving systematic of this genus (Táncsics et al. 2015).

Phylogenetic comparison of pristine and hydrocarbon-contaminated Alpine soils using DGGE fingerprinting of PCR-amplified 16S rRNA gene sequences indicated the abundance of the *Actinobacteria* phylum members, *Rhodococcus* and *Mycobacterium* (Labbé et al. 2007). Microbial communities of heavy fuel-impacted shoreline in north Spain analyzed by DGGE of PCR-amplified 16S rDNA also contained high proportions of *Rhodococcus* members associated with weathered and biotreated contaminations, suggesting that this genus may be important for biodegradation of high-molecular-weight hydrocarbons (Jiménez et al. 2007).

2.2 Rhodococci Can Be Successfully Enriched in Laboratory Hydrocarbon-Oxidizing Consortia

There are considerable numbers of studies showing that rhodococci play a leading role in aerobic biodegradation of mono- and polyaromatic compounds, highly toxic to many bacterial species (for a review, see Martínková et al. 2009). Taki et al. (2007) found that in pristine and trichloroethylene-contaminated soils incubated with *o*-xylene (the most recalcitrant isomer of xylenes) and mineral nutrients, the *Rhodococcus opacus* was abundant, increasing by almost two orders of magnitude during an active *o*-xylene biodegradation as it was estimated by competitive PCR using a primer set specific for *R. opacus* and *R. koreensis*. These authors also isolated *o*-xylene-degrading *Rhodococcus* strains that may be effective in the bioaugmentation of soil polluted with BTEX (benzene, toluene, ethylbenzene, and xylene). A *Rhodococcus* sp. isolated from PAH-contaminated river sediment utilized anthracene, phenanthrene, pyrene, and fluoranthene as a sole source of carbon and energy (Dean-Ross et al. 2001). Fahy et al. (2008a) have investigated two groundwater samples from a BTEX-contaminated aquifer located below a petrochemical plant using 16S rDNA fingerprinting and found that aerobic benzene-degrading communities contain *Actinobacteria*, including *Rhodococcus* and *Arthrobacter*, which were enriched at high benzene concentrations. Alkali-tolerant benzene-degrading *R. erythropolis* strains were isolated; they have potential applications in bioremediation or natural attenuation of aromatic-contaminated alkaline waters (Fahy et al. 2008b). The prevalence of *Rhodococcus* in benzene- and toluene-degrading bacterial communities of compost-based biofilters treating air polluted with aromatic compounds was confirmed by cultivation-dependent (plate counts and isolated strain identification) and cultivation-independent [automated ribosomal intergenic spacer analysis (ARISA) and PCR-DGGE of 16s rRNA gene] methods (Juteau et al. 1999; Borin et al. 2006). It was hypothesized that *Rhodococcus* cells originally present in compost at very low concentrations were enriched during biofilter operation at high benzene/toluene load and they have outcompeted other aromatic-degrading bacteria, such as *Pseudomonas*.

Soils, sediments, and waters contaminated with chlorinated hydrocarbons, nitroaromatic and complex heterocyclic compounds could also be a source for isolation of metabolically active *Rhodococcus* strains (Coleman et al. 1998; Wagner-Döbler et al. 1998; Poelarends et al. 2000; Seth-Smith et al. 2002; Petrić et al. 2007; Ito et al. 2016). For example, in the biphenyl-enriched microbial communities isolated from soils and sediments contaminated with PCBs, the majority of strains was identified as *R. opacus*, which outcompeted other biphenyl-mineralizing bacteria in the microcosms during long-term enrichment, thereby demonstrating a great potential for use in bioremediation requiring long-term survival of inocula (e.g., for recalcitrant xenobiotic compounds, such as PCBs) (Wagner-Döbler et al. 1998). In a similar study of Petrić et al. (2007), the isolated *R. erythropolis* strain was characterized by the highest PCB-transformation potential comparable with that of the parental mixed culture obtained from PCB-contaminated soil. Several *Rhodococcus* strains able to aerobically degrade 2,4,6-trinitrophenol and hexahydro-1,3,5-trinitro-1,3,5-triazine were isolated from soils heavily contaminated with nitrophenols and explosives (Coleman et al. 1998; Seth-Smith et al. 2002; Shen et al. 2009b). An endosulfan sulfate-degrading *R. koreensis* strain S1-1 with a new metabolic pathway was isolated from soil long-term contaminated with this pesticide (Ito et al. 2016).

A large body of research showed that rhodococci are abundant and often predominant components of natural and industrial biofilms developing upon the contact with hydrocarbon contamination (Sorkhoh et al. 1995; Tresse et al. 2002; Di Lorenzo et al. 2005). For example, *Rhodococcus* members were most abundant among indigenous oil-degrading bacteria immobilized in cyanobacterial mats on crude oil-contaminated coasts of the Arabian Gulf (Sorkhoh et al. 1995). Microbial consortia of artificially developed oil-degrading biofilms on gravel particles and glass plates included hydrocarbon-oxidizing bacteria, namely, nocardioforms (a group to which rhodococci belong) and *Acinetobacter calcoaceticus* partly attached to filaments of cyanobacteria (Al-Awadhi et al. 2003). The authors discussed a potential use of these biofilms for preparing trickling filters (gravel particles) and in bioreactors (glass plates) for biotreatment of oily wastes and oil-contaminated waters. Complex toluene-degrading biofilms developed on pumice granules in the laboratory bioreactor, following the inoculation with a microbial consortium obtained by enrichment of toluene-contaminated water (Di Lorenzo et al. 2005). Interestingly, the identification of the species present in the biofilm based on 16S rDNA comparative analysis revealed that the majority (85%) of the attached cells was represented by *R. erythropolis*, whereas *Pseudomonas marginalis* represented only 10% of the entire consortium. On the contrary, in the inoculum used for biofilm development, *P. marginalis* was predominant (86%), and *R. erythropolis* was only 10% of the consortium. Apparently, adhesion to the pumice support promoted the growth of *R. erythropolis*, modifying the initial ratio between the two species. Masy et al. (2016a) demonstrated *R. erythropolis* biofilm formation on gravel and such biofilm was stable during 15-month experiment in real car parking conditions and it was able to limit hydrocarbon leaching from artificial rainfall.

From the results of numerous studies referenced above on the occurrence and frequent dominance of *Rhodococcus* members in diverse pristine and human-impacted ecosystems, it is clear that this genus is ubiquitous in the environment, associating with the presence of hydrocarbons and their substituted derivatives. There are increasing numbers of *Rhodococcus* strains isolated from clean and contaminated environments able to degrade/transform hydrocarbon contaminants and xenobiotics; these strains are available from the culture collections worldwide (Home Pages of Culture Collections in the World 2018).

2.3 *Outstanding Physiological, Biochemical, and Ecological Properties of* Rhodococcus

Physiologo-biochemical and ecological properties of *Rhodococcus* suitable for environment bioremediation are listed in Table 1. First, rhodococci are characterized by high catabolic diversity and unique enzymatic capabilities. Chemical pollutants degraded by rhodococci range from aliphatic and aromatic hydrocarbons through chlorinated hydrocarbons and nitroaromatics to complex poly- and heterocyclic compounds. The reader is referred to the fundamental review of Warhurst and Fewson (1994) and more recent reviews (van der Geize and Dijkhuizen 2004; Larkin et al. 2005; Martínková et al. 2009). A number of studies on the degradation of most abundant environmental contaminants, crude oil and its refinery products, by rhodococci have been carried out (Whyte et al. 1998, 2001; Sharma and Pant 2000), and although many other bacteria can also degrade petroleum hydrocarbons, some novel catalytic pathways have been described in *Rhodococcus* (Whyte et al. 2002a, b; Kim et al. 2004; van der Geize and Dijkhuizen 2004). Metabolic studies were focused on the rhodococcal pathways for alkane, cycloalkane, mono- and polyaromatic hydrocarbon biodegradation (Dean-Ross et al. 2001; Kim et al. 2004; Whyte et al. 2002a, b; Larkin et al. 2005; Lee and Cho 2008), transformations of nitrogen and sulfur compounds (Xu et al. 2006; Shen et al. 2009a, b), and the

Table 1 Physiological, biochemical, and ecological properties of *Rhodococcus* suitable for environmental bioremediation

Characteristics	Description	Advantage in bioremediation	References
High catabolic diversity	Degradation of wide range of chemicals, including aliphatic and aromatic hydrocarbons and their nitro- and halogenated derivatives, oxygenates, and heterocyclic compounds	Biodegradation of complex contaminations, e.g., crude oil and industrial wastes. Biodegradation of recalcitrant xenobiotics	Warhurst and Fewson (1994), van der Geize and Dijkhuizen (2004), and Martínková et al. (2009)
Aerobic and microaerophilic metabolism	Persistence in low-oxygen environments and activation upon oxygen supply	Bioremediation under well-aerated and oxygen-limited conditions (e.g., in high-density soils, sediments and hypoxic aquifers)	Travkin et al. (2002), Fahy et al. (2006, 2008a), Vogt et al. (2004), and Joshi et al. (2008)
Oligotrophy and nitrogen fixation	Growth under nutrient limitation conditions (C/N limitation) and resistance to long-term starvation	Bioremediation of resource-limited environments, e.g., low-nutrient soils and groundwaters	Ivshina et al. (1981), Elo et al. (2000), Mergaert et al. (2001), Priestley et al. (2006), and Ohhata et al. (2007)
Lack of catabolic repression	Biodegradation of contaminants is not repressed by the presence of easily assimilable nutrients	Bioremediation of organic-rich environments, e.g., wastewaters	Warhurst and Fewson (1994)
Adaptation to hydrophobic substrates	Cell surface hydrophobicity, adhesion to hydrocarbons, and biosurfactant production	Enhanced biodegradation of hydrophobic pollutants	Lang and Philp (1998) and Whyte et al. (1999)
Adhesion and biofilm formation	Adhesion to solid surfaces, cell aggregation, and biofilm formation	Application as immobilized or self-immobilized biocatalysts	Sorkhoh et al. (1995), Di Lorenzo et al. (2005), and Borin et al. (2006)
Resistance to environmental stresses	Psychrotrophy, thermo-, xero-, and galotolerance, pH-resistance, resistance to heavy metals and xenobiotics	Bioremediation in extreme environments and harsh industrial conditions	Whyte et al. (1999), Bej et al. (2000), Aislabie et al. (2006), Ryu et al. (2006), and Fahy et al. (2008b)
Ecological behavior	*K*-strategy and high competitiveness in communities under crowded, substrate-limited conditions	Survival in bioaugmentation inoculaand indigenous communities	Juteau et al. (1999), Margesin et al. (2003), and Borin et al. (2006)

(continued)

Table 1 (continued)

Characteristics	Description	Advantage in bioremediation	References
Nonpathogenicity and lack of antagonistic properties	Low number of pathogenic species (only *R. equi* and *R. fascians*), lack of antimicrobial activity, plant biosafing	Biosafety and environmental safety	Bell et al. (1998), Nagy et al. (1995), Uroz et al. (2003), and Aoshima et al. (2007)

interactions between rhodococci and other oil degraders (Van Hamme et al. 2003; Hamamura et al. 2006). Chlorophenols, chlorobenzenes, and polychlorinated biphenyls can be degraded by rhodococci (for review see Martínková et al. 2009), as well as other recalcitrant and toxic pollutants including nitriles (Baxter et al. 2006), sulfonated azo dyes (Joshi et al. 2008), pesticides (Nagy et al. 1995; Ito et al. 2016), pharmaceuticals (Ivshina et al. 2015b; Homklin et al. 2012), cyanides (Baxter and Cummings 2006; An et al. 2018), benzothiazoles (Besse et al. 2001), hydrofurans (Daye et al. 2003; Tajima et al. 2012), and many others.

2.3.1 Adaptation to Hydrocarbon Assimilation

Apart from remarkable biodegradative abilities, rhodococcal adaptation to hydrophobic contaminants is the important matter for bioremediation (Pieper and Reineke 2000; Ivshina et al. 2017). It is assumed that prolonged persistence of hydrophobic organic compounds in the environment is mainly determined by their solubilization-limited bioavailability for microorganisms. Many bacteria can assimilate hydrophobic substances, such as hydrocarbons, only in solubilized or emulsified forms. On the contrary, the hydrocarbon uptake by *Rhodococcus* occurs via the direct cell contact with large oil drops (Lang and Philp 1998). Thus, Whyte et al. (1999) observed physiological adaptations involved in alkane assimilation by *Rhodococcus* cells at a low temperature; these included production of cell-bound biosurfactants, increase in cell surface hydrophobicity, production of intracellular inclusions and extracellular polymers, and alteration of membrane fluidity. The authors suggested that high cell hydrophobicity and cell-associated biosurfactants promote the adhesion of rhodococcal cells not only to liquid alkanes but also to hydrophobic solid surfaces, allowing direct uptake from sorbed/crystalline hydrocarbons, another important mechanism of microbial assimilation of hydrophobic pollutants in soils and sediments (Wattiau 2002). In our experiments, *Rhodococcus* biosurfactants desorbed crude oil and PAHs from soil, thus facilitating its biodegradation by soil microorganisms (Kuyukina et al. 2005; Ivshina et al. 2016). It should be noted that many hydrocarbons, for example, low-molecular-weight alkanes, monoaromatics, and chlorinated aliphatic compounds, are toxic to microorganisms primarily due to the solvent effect on cell membranes. They destroy microbial cells and therefore abolish the desired biodegradative activity. Several *Rhodococcus* strains resistant to organic solvents have been isolated, and possible mechanisms of solvent tolerance, such as alterations in the composition of cell envelope, have been reported (see chapter "Adaptation of *Rhodococcus* to Organic Solvents" by de Carvalho).

2.3.2 Ecological Plasticity

Another important *Rhodococcus* feature is the ability to persist and metabolize in microaerophilic and oligotrophic conditions. Rhodococci could be isolated from microaerophilic environments, for example, high-density soils, deep-sea sediments, and hypoxic aquifers (Colquhoun et al. 1998; Hendrickx et al. 2005; Fahy et al. 2006). The study of Travkin et al. (2002) reported the isolation of an enrichment culture and a *Rhodococcus* strain derived from it, transforming 3,4-halogenated anilines under nitrate-reducing conditions. Anaerobic bioconversion of these haloanilines by *Rhodococcus* sp. started with reductive deamination, resulting in production of dihalobenzene intermediates, which were further dehalogenated in the biodegradation pathway. A natural bacterial consortium consisting of *Aeromonas caviae*, *Proteus mirabilis*, and *R. globerulus* was reported to decolorize azo dyes under microaerophilic condition in the presence of the organic carbon source (Joshi et al. 2008). The consortium decolorized 14 azo dyes individually as well as in simulated mixed wastewater, suggesting its possible application in industrial wastewater treatment. Although the abovementioned studies reveal the possibility of *Rhodococcus* application in microaerophilic/anaerobic biodegradation of contaminants, the vast majority of extensive researches resume that rhodococci can persist well in low-oxygen environments and rapidly increase in number upon the oxygen supply, thus contributing to aerobic (oxidative) environmental decontamination (Vogt et al. 2004; Fahy et al. 2008a). Ohhata et al. (2007) isolated a *R. erythropolis* strain N9T-4 from a crude oil sample and found that this strain and some other collection *R. erythropolis* strains grew in extremely oligotrophic conditions, suggesting that the oligotrophy could be a common feature of *Rhodococcus* (Mergaert et al. 2001). Additionally, there are several reports on the abilities of rhodococci to grow and degrade organic contaminants under carbon- and nitrogen-limiting conditions (Priestley et al. 2006), to fix atmospheric nitrogen (Ivshina et al. 1981; Elo et al. 2000), and to oxidize complex pollutants even in the presence of more easily assimilable carbon sources (Warhurst and Fewson 1994). These features make rhodococci the promising candidates for bioremediation of both resource-limited and organic-rich environments.

Many contaminated sites are characterized by harsh environmental conditions, for example, low or elevated temperatures, acidic or alkaline pH, high salt concentrations, or high pressure. Apparently, extremotolerant *Rhodococcus* members adapted to grow and thrive in these environments play an important role in bioremediation of polluted extreme habitats (Sorkhoh et al. 1990; Whyte et al. 1999; Aislabie et al. 2006; Ryu et al. 2006; Fahy et al. 2008b). A hydrocarbon-degrading potential of cold-adapted rhodococci was mentioned earlier in this chapter. It is assumed that low temperature greatly influences the process of hydrocarbon

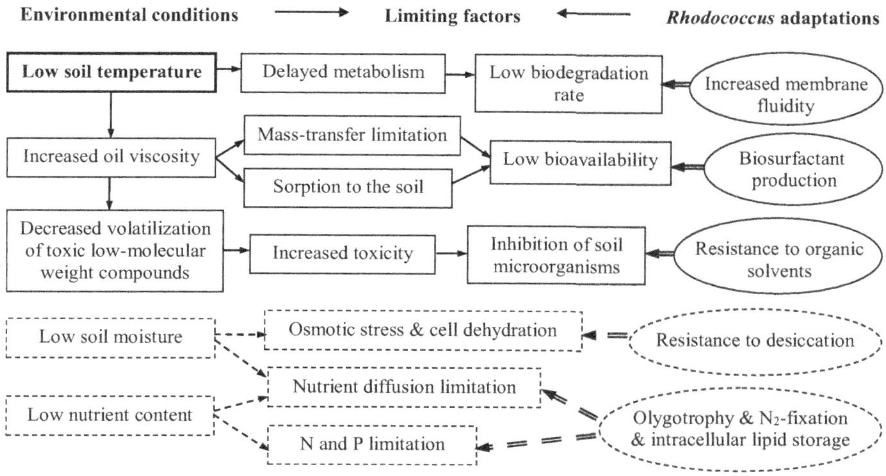

Fig. 2 Adaptations of *Rhodococcus* to cold soil conditions. Optional environmental conditions, factors, and corresponding adaptations are shown by dotted lines

biodegradation in soil by affecting both the physical nature of spilled oil and microbial metabolism (Fig. 2). Particularly, at low temperatures, the oil viscosity is increased, and the volatilization of toxic low-molecular-weight compounds is reduced, thus decreasing the bioavailability and biodegradation of hydrocarbon pollutants. Additionally, soils of cold regions are often subject to drying because of low precipitation and long freezing periods. The low soil moisture results in nutrient diffusion limitations and, in addition to typically low available N and P contents of these soils, leads to development of oligotrophic conditions for soil microorganisms (Aislabie et al. 2006). On the one hand, it is important that complex physiological adaptations of rhodococci (see Fig. 2) allow them to survive in such extreme cold environments as well as to contribute to microbiological degradation of hydrophobic pollutants (Bej et al. 2000). On the other hand, indigenous mesophilic hydrocarbon-oxidizing rhodococci were suggested as promising bioremediation agents for hot climate regions, for example, for the Kuwaiti desert soil heavily contaminated with crude oil (Sorkhoh et al. 1990). Moreover, there are increasing numbers of reports on the isolation of heat-resistant enzymes, such as specific dioxygenases, catalyzing initial steps of degradation of (poly)aromatic compounds by *Rhodococcus* cells (Gakhar et al. 2005; Yang et al. 2008). These enzymes, due to their high chemical and thermal stability, offer powerful tools for biological treatment of polluted environments and industrial wastes at elevated temperatures. One more important aspect of bioremediation is rhodococcal resistance to heavy metal ions and their bioaccumulation, including those that are radioactive (Bell et al. 1998; Ivshina et al. 2002, 2013). Since heavy metals are often present in hydrocarbon-contaminated soils associated with oil spills, petrochemical and other chemical waste discharges, and in industrial effluents, rhodococci can be used for bioremediation of such complex metallo-organic contaminations (Kuyukina et al. 2017).

The ability of rhodococci to adhere to different surfaces and to form biofilms is widely used to develop immobilized biocatalysts suitable for various eco-biotechnological applications, namely, bioreactor treatment of contaminated waters and sludges, and soil bioremediation (Prieto et al. 2002a; Di Lorenzo et al. 2005; Podorozhko et al. 2008; Kuyukina et al. 2009, 2017; Hatzinger et al. 2017). It is assumed that microorganisms in biofilms are more resistant to environmental stressors than in a free-living state. Therefore, rhodococci with high adhesive activities, especially toward hydrophobic surfaces, tend to be more successful in colonizing of hydrocarbon-contaminated sites compared to less adhering microorganisms (Masák et al. 2004). Relative prevalence of *Rhodococcus* species in many natural and laboratory contaminant-degrading microbial consortia may be also explained by their ecological behavior, particularly the *r-K* scheme, which suggests that evolution favors either adaptation to high rates of reproduction (*r* strategists) or optimal utilization of environmental resources (*K* strategists) (Margesin et al. 2003). Bacteria, such as pseudomonads, which rapidly grow in nutrient-rich media, are *r* strategists. Others, such as rhodococci, tend to be more successful in resource-limited, crowded environments, are *K* strategists. Apparently, populations of *K* strategists would be more stable and permanent members of the communities of chronically contaminated biotopes or bioreactor microcosms, when easy-degradable substrates are depleted and the competition for nutrients is intensive.

2.3.3 Biosafety Aspects

One potential problem with bioaugmentation is that the inocula may contain microorganisms harmful for the human health or environment, for example, human, animal, and plant pathogens or strains producing toxins or antibiotics, which were present in indigenous populations or came from laboratory media contamination and were enriched during the inoculum growth. To prevent the pathogen occurrence in bioaugmentation inocula, such consortia should contain only taxonomically defined microorganisms belonging to species known as nonpathogenic. Moreover, in some countries, national environmental and health authorities require an assessment of biosafety and environment safety of microbial inocula intended for bioremediation applications (Aoshima et al. 2007). The study of Aoshima et al. (2007) evaluated the safety of the hydrocarbon-oxidizing soil isolate *R. erythropolis* C2 for the application in open oil-contaminated ecosystems and found that this strain demonstrates the lack of pathogenicity, mutagenicity, or ecotoxicity. It therefore requires no special occupational health precautions during application processes and has low environmental impact. These results are in agreement with other literature data indicating that only two *Rhodococcus* species, *R. equi* and *R. fascians*, are associated with pathogenicity for animals and plants, respectively (chapters "Genetics of the Virulence of *Rhodococcus equi*" by Vázquez-Boland and "Phytopathogenic Strategies of *Rhodococcus fascians*" by Vereecke). Such relatively small proportion of pathogenic species is rare within the micolata group of actinobacteria harboring genera *Corynebacterium*, *Gordonia*, *Mycobacterium*, *Nocardia*, and *Tsukamurella*, characterized by abundant presence of human pathogens. Additionally, literature and our research data suggest that rhodococci are unlikely to produce any toxins or antimicrobial compounds (Kitamoto et al. 2002; Kuyukina et al. 2007); this is another strong advantage of *Rhodococcus* applications in environment bioremediation technologies. There are only two reports by Kitagawa and Tamura (2008a, b) on *R. erythropolis* producing antibiotics active against Gram-positive bacteria, including *Rhodococcus* and related genera. Furthermore, the rhizospheric *R. erythropolis* strain W2 degrading *N*-acylhomoserine lactones was shown to be effective in quenching of quorum-sensing-regulated functions of plant pathogenic bacteria, thereby reducing their pathogenicity (Uroz et al. 2003). This bacterium and herbicide-degrading *Rhodococcus* strains applicable for plant biosafing (Nagy et al. 1995; Hongming et al. 2015) could be used in phytoremediation projects for contaminated agricultural soils. Additionally, Shagol et al. (2014) described an arsenic-tolerant plant-growth-promoting *R. aetherivorans* strain isolated from smelter-polluted soil, which consistently increased root length of maize in the presence of 100 and 200 μM As(V) and can therefore enhance the efficiency of phytoremediation in As-polluted soils.

3 *Rhodococcus* Applications in Bioremediation Technologies

Biological remediation of terrestrial and aquatic habitats contaminated with hazardous compounds received increasing attention in early 1990s, with enhanced awareness of the potential harmful effects on human health and the environment (Alexander 1999). Various bioremediation techniques have been used at a large number of sites contaminated with organic pollutants since the most well-known cleanup of oil spilt from the Exxon Valdez in Prince William Sound, Alaska, in 1989 (Van Hamme et al. 2003). A historical aspect of *Rhodococcus* application in environment bioremediation could be addressed to late 1990s, when first attempts were made to remediate oil-contaminated lands using naturally accruing rhodococci or laboratory *Rhodococcus* cultures (Sorkhoh et al. 1995; Koronelli et al. 1997; Christofi et al. 1998). Upon revealing new catabolic abilities of *Rhodococcus* species and isolation of environmental strains degrading a wide range of contaminants, these bacteria have been increasingly explored for bioremediation of soils, waters, and air polluted with different recalcitrant and toxic organic chemicals (Fig. 3).

3.1 *In Situ Treatment*

In situ bioremediation comprises various techniques, which treat contaminated material in place (without excavation and transfer) and keep the material treated

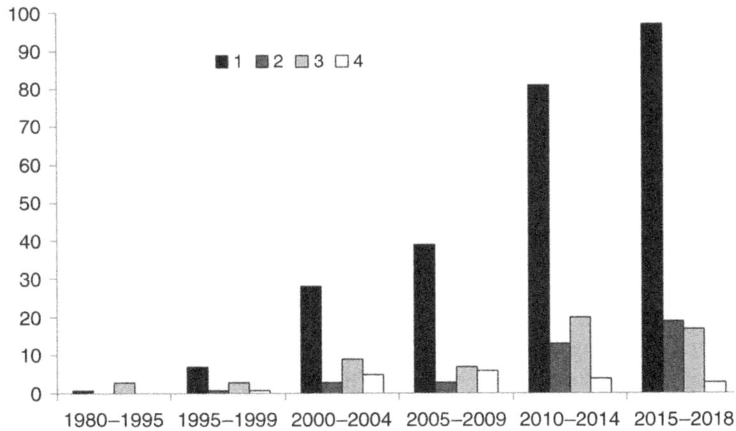

Fig. 3 Number of research articles concerned to *Rhodococcus* applications in environment bioremediation (according to http://www.scopus.com). Queries: Title/Abstract/Keywords: *Rhodococcus* and (1) soil bioremediation; (2) groundwater bioremediation; (3) bioreactor degradation; (4) air biofilter. Nonrelevant papers were removed from the query results

essentially undisturbed. These techniques are usually applied to the remote or difficult of access polluted environments, such as forest soils and wetlands, aquatic sediments, and subsurface zones, as well as in cases when intrusive methods are inappropriate, for example, for remediation of unique landscapes and valued soils. Most in situ processes involve stimulation of indigenous microbial populations through the addition of nutrients and other factors enhancing biodegradation (e.g., surfactants and oxygen source) and sometimes the augmentation with specifically adapted microbial cultures possessing high biodegradative abilities (Van Hamme et al. 2003). The limited number of laboratory and field studies referenced in Table 2 suggested that rhodococci, either as parts of bioaugmentation inocula or members of indigenous communities, can be successfully used for reducing in situ organic contaminant levels. Thus, laboratory soil microcosm study using an indigenous psychrotolerant alkane-degrading *Rhodococcus* strain ADH reported a positive effect of bioaugmentation on the biodegradation of diesel fuel (Ruberto et al. 2005). Although the natural microflora responded significantly to the pollutants, *Rhodococcus*-inoculated microcosms showed enhanced biodegradation compared to non-inoculated soil and sterilized controls. The survival and activity of *Rhodococcus* sp. strain 1BN introduced into naphthalene-contaminated sandy-loam soil were studied by Cavalca et al. (2002). The naphthalene consumption and CO_2 production rates were the highest in the *Rhodococcus*-amended sterilized soil, although inoculation of non-sterile soil did not enhance significantly the biodegradation process, indicating a considerable bioremediation potential of the indigenous naphthalene-degrading bacteriocenosis. Nevertheless, the introduced *Rhodococcus* strain was well-established in the contaminated soil even in the presence of native naphthalene-degrading bacteria. In our experiments, poly(vinyl alcohol) cryogel-immobilized *R. ruber* and *R. erythropolis* survived successively in oil-contaminated

Table 2 Selected examples of *Rhodococcus* application for bioremediation of contaminated environments

Contaminated substrate	Bioremediation method	*Rhodococcus* application mode	Effect of treatment	References
In situ treatment—treatability studies				
Antarctic soil contaminated with diesel fuel	Laboratory bioaugmented soil system exposed to natural climate conditions of Antarctica	Liquid culture of *Rhodococcus* sp.	+	Ruberto et al. (2005)
Naphthalene-contaminated soil from industrial area	Laboratory bioaugmented soil systems	Liquid culture of *Rhodococcus* sp.	+	Cavalca et al. (2002)
PAH-contaminated soil from petroleum refinery site	Laboratory bioaugmented soil systems	Liquid co-culture of *Rhodococcus* sp. and *Aspergillus terreus*/*Penicillium* sp.	+	Kim and Lee (2007)
Diesel-contaminated soil	Soil inoculated with diesel-degrading rhizobacterium and seeded with diesel-resistance *Zea mays*	Liquid culture of *Rhodococcus* sp.	+	Hong et al. (2007)
Soil contaminated with crude oil	Laboratory systems with preheated and non-heated soils	Poly(vinyl alcohol) cryogel-immobilized culture of *R. ruber* and *R. erythropolis*	+	Kuyukina et al. (2013)
Aroclor 1242-contaminated soil	Laboratory soil column with mineral nutrient addition, bioaugmented with bacteria and earthworms	Liquid culture of *Ralstonia eutrophus* and *Rhodococcus* sp.	+	Luepromchai et al. (2002)
Aroclor 1242-contaminated sediment	Laboratory two-stage anaerobic/aerobic biotreatment (aerobic stage)	Liquid culture of recombinant *Burkholderia xenovorans* and *Rhodococcus* sp.	+	Rodrigues et al. (2006)
Groundwater artificially contaminated with hexahydro-1,3,5-trinitro-1,3,5-triazine (RDX)	Repacked aquifer sediment columns operated at 15 °C	Liquid culture of *Gordonia* sp., *Pseudomonas fluorescens*, and *R. jostii*	+	Fuller et al. (2015)
Tap water artificially contaminated with gasoline	Laboratory aerobic biobarriers filled with volcanic pumice	Liquid culture of *Rhodococcus* sp. and *Methylibium petroleiphilum*	±[a]	Daghio et al. (2015)

(continued)

Table 2 (continued)

Contaminated substrate	Bioremediation method	*Rhodococcus* application mode	Effect of treatment	References
Benzene-contaminated groundwater	Laboratory groundwater microcosms	Indigenous bacterial community containing *Rhodococcus*	+	Fahy et al. (2006, 2008a)
In situ treatment—field studies				
Shoreline contaminated with heavy fuel oil	Plots on the cobblestone beach spread with oleophilic fertilizer S200 (IEP Europe)	Indigenous bacterial community containing *Rhodococcus*	+	Jiménez et al. (2007)
Crude oil-polluted river bay, lakes, wetland, and marshy peat soil	Preliminary mechanical collection of spilled oil using skimmers followed by spraying with biopreparation and mineral nutrients	Liquid biopreparation *Rhoder* consisting of *R. erythropolis* and *R. ruber*	+	Murygina et al. (2000)
Groundwater polluted with methyl *t*-butyl ether and *t*-butyl alcohol	Biobarrier plots with O_2 or air sparging	Mixed culture or pure *Rhodococcus* sp. culture isolated from polluted groundwater	+	Salanitro et al. (2001)
Chlorobenzene-polluted groundwater	Reactive barrier supplied with hydrogen peroxide and nitrate solution	Indigenous bacterial community containing *Rhodococcus*	+	Vogt et al. (2004)
Petroleum hydrocarbon polluted groundwater	Closed bipolar system (one extraction and two injection wells)	Pre-grown consortium of zymogenous microorganisms containing *Rhodococcus*	+	Beškoski et al. (2017)
On-site treatment—treatability studies				
Crude oil-contaminated Arctic soil spiked with *n*-hexadecane	Laboratory bioaugmented soil systems	Liquid culture of *Rhodococcus* sp.	+	Whyte et al. (1998)
Arctic soil contaminated with *n*-alkanes or diesel fuel	Laboratory soil systems with additions of inorganic/organic fertilizers and bulking agents	Indigenous bacterial community containing *Rhodococcus*	+	Whyte et al. (2001)
Arctic soil contaminated with weathered diesel fuel	Laboratory bioaugmented soil system	Liquid enriched indigenous culture containing *Rhodococcus*	±[a]	Thomassin-Lacroix et al. (2002)

Soil contaminated with crude oil	Laboratory soil systems with mineral nutrient addition	Indigenous bacterial community containing *Rhodococcus*	+	Peressutti et al. (2003)
Soils from geographically distinct areas contaminated with weathered crude oil	Laboratory soil systems with mineral nutrient addition	Indigenous bacterial community containing *Rhodococcus*	+	Hamamura et al. (2006)
Soil contaminated with fuel oil	Laboratory bioaugmented soil plots with bulking agent and mineral nutrient addition	Liquid biopreparation *Devoroil* containing *R. erythropolis* and *Rhodococcus* sp.	+	Sidorov et al. (1998)
Soil contaminated with BTEX	Laboratory bioaugmented soil systems with mineral nutrient addition	Liquid culture of *R. opacus*	+	Taki et al. (2004)
Soil contaminated with BTEX	Laboratory soil systems with mineral nutrient addition	Indigenous bacterial community with dominating *Rhodococcus*	+	Taki et al. (2007)
Soil contaminated with disulfide oil	Laboratory bioaugmented soil systems with mineral nutrient and glucose addition	Vermiculite-immobilized culture of *Rhodococcus* sp.	+	Taheri et al. (2008)
On-site treatment—field studies				
Crude oil-contaminated desert soil	Bioaugmented landfarming cells with addition of mineral nutrients and lime	Liquid mixed culture containing *Rhodococcus*, removed from cyanobacterial mats floating in oil-polluted waters	+	Sorkhoh et al. (1995)
Tundra soil artificially contaminated with crude oil	Bioaugmented soil plots with mineral nutrient addition	Lyophilized culture of *R. erythropolis*	+	Koronelli et al. (1997)
Soil artificially contaminated with crude oil and oil-field brine	Bioaugmented landfarming cells with mineral nutrient addition	Liquid biopreparation *Devoroil* containing *R. erythropolis* and *Rhodococcus* sp.	+	Sidorov et al. (1997)
Crude oil-contaminated soil	Composted inoculated biopiles with addition of mineral nutrients and straw as bulking agent	Liquid culture of biosurfactant-producing *R. ruber*	+	Christofi et al. (1998)
Arctic soil contaminated with weathered diesel fuel	Inoculated biopiles with addition of fertilizer (granular urea and diammonium phosphate), surfactant, and cocoa-fiber bulking agent	Liquid enriched indigenous culture containing *Rhodococcus*	±[a]	Thomassin-Lacroix et al. (2002)

(continued)

Table 2 (continued)

Contaminated substrate	Bioremediation method	*Rhodococcus* application mode	Effect of treatment	References
Crude oil-contaminated soil	Landfarming cells with addition of woodchips and oleophilic biofertilizer	Oleophilic biofertilizer containing *R. erythropolis* and *R. ruber*	+	Kuyukina et al. (2003)
Polar marshy wetland polluted with crude oil	Landfarming + phytoremediation with addition of biopreparation, inorganic fertilizer, and lime	Liquid biopreparation *Rhoder* consisting of *R. erythropolis* and *R. ruber*	+	Murygina et al. (2005)
Dehydrated oil sludge	Inoculated biopiles with addition of sand, woodchips, and inorganic fertilizer	Liquid biopreparation *Rhoder* consisting of *R. erythropolis* and *R. ruber*	+	De-qing et al. (2007)
Soil contaminated with fuel hydrocarbons	Inoculated biopile with addition of $(NH_4)_2HPO_4$	Liquid cultures of *Pseudomonas putida*, *Acinetobacter johnsonii*, and *R. erythropolis*	+	Genovese et al. (2008)
Soil artificially contaminated with diesel oil and fuel oil	Inoculated biopiles with addition of rhamnolipid biosurfactant	Liquid cultures of *Gordonia alkanivorans*, *R. erythropolis*, *Acinetobacter junii*, and *Exiguobacterium aurantiacum*	+	Lin et al. (2010)
Olive mill waste (humid husk)	Pilot biopile amended with waste wool, olive leaves and twigs, wheat straw, and pigeon manure	Indigenous tannin-degrading bacterial community containing *R. rhodochrous*	+	Federici et al. (2011)
PAH-contaminated soil	On-site landfarming plots added with nutrients	Liquid culture of *R. ruber*	+	Sun et al. (2012)
Aged fuel-contaminated soil	Pilot tank mimicking groundwater flow conditions and two bioreactors	Liquid culture of *R. erythropolis*	+	Masy et al. (2016b)
Bioreactor/biofilter treatment—laboratory studies				
Soil from former coke works site contaminated with $K_2Ni(CN)_4$	Laboratory shaking-flask experiments (1 g soil/100 ml minimal salt medium) with glucose addition	Liquid culture of *Rhodococcus* sp.	±[a]	Baxter and Cummings (2006)

Soil contaminated with bis (2-ethylhexyl)phthalate	Slurry-phase batch and sequencing batch bioreactors with mineral nutrient addition	Mixed culture isolated from contaminated soil and containing *R. fascians*	+	Juneson et al. (2001)
Aged polychlorinated biphenyl-contaminated soils	Aerobic solid-phase and packed-bed loop reactors with addition of mineral nutrients, biphenyl, and methyl-β-cyclodextrins	Indigenous bacterial community containing *Rhodococcus*	+	Fava et al. (2003)
Groundnut cake naturally contaminated with aflatoxin B1	Shake-flask bioreactors	Liquid culture of *R. erythropolis*	+	Dogan et al. (2017)
Sediment contaminated with polycyclic aromatic hydrocarbons	Aerobic slurry-phase bioreactors with mineral nutrient addition	Indigenous community containing *Rhodococcus*	+	Ringelberg et al. (2001)
Phenol- and formaldehyde-contaminated synthetic and industrial wastewaters	Laboratory shaking-flask experiments with addition of phosphate, ammonium, and micronutrients	Liquid culture of *R. erythropolis*	+	Hidalgo et al. (2002a, b)
Phenol- and formaldehyde-contaminated synthetic and industrial wastewaters	Air-stirred and packed-bed column bioreactors with addition of nitrogen source and micronutrients	Biolite- and diatomaceous earth-immobilized cultures of *R. erythropolis*	+	Prieto et al. (2002a, b)
Phenol-contaminated water	Stirred-tank bioreactor	Liquid culture of *Rhodococcus* sp.	+	Yaacob et al. (2016)
Ammonium- and 2-fluorophenol-contaminated synthetic wastewater	Sequencing batch reactor	Aerobic sludge granules and liquid culture of *Rhodococcus* sp.	+	Ramos et al. (2017)
Aroclor 1242- and biphenyl-contaminated water	Granular biofilm reactor with limited aeration and addition of mineral nutrients, yeast extract, and microelements	*Rhodococcus* sp. culture immobilized on anaerobic sludge granules	±[b]	Tartakovsky et al. (2001)
Di-*n*-octyl phthalate-contaminated water	Sequencing batch reactor	Liquid cultures of *Arthrobacter* sp. and *Rhodococcus* sp.	+	Zhang et al. (2018)
1,1,2,2-Tetrachloroethane-contaminated water	Continuous stirred-tank reactor	Liquid cultures of *R. aetherivorans*	+	Cappelletti et al. (2018)
Simulated azo dye- contaminated wastewater	Laboratory flask experiments under microaerophilic condition, with addition of yeast extract and glucose	Mixed culture from dye-contaminated soil and sludge and containing *R. globerulus*	+	Joshi et al. (2008)

(continued)

Table 2 (continued)

Contaminated substrate	Bioremediation method	*Rhodococcus* application mode	Effect of treatment	References
Model petroleum- contaminated water	Fluidized-bed column bioreactor	Sawdust-immobilized culture of *R. ruber* and *R. opacus*	+	Kuyukina et al. (2009)
Oil-field wastewater	Fluidized-bed column bioreactor	Sawdust- and polyvinyl alcohol cryogel- immobilized cultures of *R. ruber* and *R. opacus*	+	Kuyukina et al. (2017) and Serebrennikova et al. (2017)
2,4-Dinitrophenol- contaminated water	Column plug-flow bioreactor	Agar granule-immobilized culture of *R. erythropolis*	+	Kitova et al. (2004)
2,4,6-Trinitrophenol or picric acid-polluted wastewater or groundwater	Sequencing batch bioreactor	Liquid culture of *R. opacus*	+	Weidhaas et al. (2007)
Artificial groundwater contaminated with *N*-nitrosodimethylamine	Propane-fed membrane bioreactor	Liquid cultures of *R. ruber*	+	Hatzinger et al. (2011)
Organic cyanide-contaminated groundwater	Fluidized-bed reactors	Coconut granular activated carbon immobilized *R. rhodochrous* and *Bacillus mojavensis*	+	An et al. (2018)
Synthetic pharmaceutical wastewater contaminated with ibuprofen and ketoprofen	Aerobic suspension-sequencing batch reactor	Indigenous agricultural community containing *R. ruber*	+	Hasan et al. (2016)
Methylhydrazine/hydrazine-contaminated wastewater	Fixed-film trickle-bed bioreactor	Coarse sand-immobilized culture of *Rhodococcus* sp.	+	Nwankwoala et al. (2001)
Artificial toluene-contaminated waste gas	Rotating biological contactor with mineral nutrient addition	Liquid culture of *Pseudomonas putida* and *R. erythropolis*	+	Vinage and von Rohr (2003)
Artificial waste air containing 2-chlorotoluene	Biotrickling filter	Polyurethane foam-immobilized culture of *Rhodococcus* sp.	+	Dobslaw and Engesser (2018)
Air polluted with benzene	Compost-packed biofilter	Indigenous compost community containing *Rhodococcus*	+	Borin et al. (2006)
Air polluted with isopropylbenzene	Two-phase partitioning bioreactor with addition of ethanol	Liquid culture of *R. erythropolis*	+	Aldric and Thonart (2008)

Volatile compounds of tobacco leaves	Polyamide fiber-packed microbioreactors	Mixed consortium of *Pseudomonas*, *Rhodococcus*, *Nocardia*, and *Micrococcus*	+	Zagustina et al. (2012)
Gas-phase mixture of methanol, α-pinene and H_2S	Biotrickling filter	Autotrophic H_2S-degrading culture, *Candida boidinii*, *R. erythropolis*, and *Ophiostoma stenoceras*	+	López et al. (2013)
Air polluted with toluene, benzo(*a*) pyrene, and formaldehyde	Vermiculite-packed column biofilter	Consortium of *R. erythropolis* and *Fusarium solani*	+	Vergara-Fernández et al. (2018)
Bioreactor/biofilter treatment—pilot/field studies				
Crude oil-contaminated soil	Field slurry-phase bioreactor with addition of oleophilic biofertilizer	Oleophilic biofertilizer containing *R. erythropolis* and *R. ruber*	+	Kuyukina et al. (2003)
Waste oil contaminated with polychlorinated biphenyls	Pilot plant consisting of UV-irradiation equipment and two successive bioreactors	Liquid culture of *Comamonas testosteroni* and *R. opacus*	+	Kimbara et al. (1998)
Industrial wastewater polluted with 1,3-dichloropropene and other chlorinated aliphatics	Pilot extractive membrane biofilm reactor	Mixed biofilm culture isolated from bioreactor and containing *R. erythropolis*	+	Katsivela et al. (1999)
Synthetic tetrahydrofuran-contaminated wastewater	Pilot membrane bioreactor	Mixed culture isolated from the industrial wastewater treatment plant and containing *R. ruber*	+	Daye et al. (2003)
Model wastewater containing 2,4,6-trinitrophenol	Pilot biological aerated filter with mineral nutrient addition	Liquid culture of *Rhodococcus* sp.	+	Shen et al. (2009a)
Industrial wastewater contaminated with hydrocarbons, surfactants, and heavy metals	Pilot biotreatment installation	Immobilized algal-bacterial co-culture containing *Rhodococcus* sp.	+	Safonova et al. (2004)
Groundwater polluted with ethyl *t*-butyl ether and gasoline	On-site pilot bioreactor	Liquid culture of *R. wratislaviensis*, *R. aetherivorans*, and *Aquincola tertiaricarbonis*	+	Fayolle-Guichard et al. (2012)

(continued)

Table 2 (continued)

Contaminated substrate	Bioremediation method	*Rhodococcus* application mode	Effect of treatment	References
Groundwater polluted with *N*-nitrosodimethylamine and *N*-nitrodimethylamine	Field-scale propane-fed fluidized-bed bioreactor	Coconut granular activated carbon immobilized *R. ruber*	+	Hatzinger et al. (2017)

[a]±, effect of bioaugmentation did not exceed that of biostimulation; [b]±, effect of *Rhodococcus* immobilized on anaerobic sewage did not exceed that of anaerobic sewage alone. *ND* not determined

soil under drought conditions, contributing to efficient alkane biodegradation (Kuyukina et al. 2013). Several treatability studies showed positive effects of *Rhodococcus* co-cultures with other bacteria and fungi, as well as rhodococcal associations with higher plants and earthworms on biodegradation of petroleum hydrocarbons and polychlorinated biphenyls in soil and sediments (Luepromchai et al. 2002; Rodrigues et al. 2006; Hong et al. 2007; Kim and Lee 2007). Additionally, in situ bioremediation can be an economically advisable and environmentally harmless approach to cleanup shorelines contaminated as a result of marine oil spills. During the field-scale trial carried out by Jiménez et al. (2007), some success in the removal of heavy fuel oil from the cobble beach on the Cantabrian coast (north Spain) polluted after the oil tanker *Prestige* crush in 2002 was achieved using the oleophilic fertilizer, and the biodegradative potential of the indigenous microbial community, including *Rhodococcus* representatives, was established.

It is assumed that engineered in situ bioremediation is a feasible and effective method for treating contaminants within the saturated zone of soil and contaminated groundwater. Engineering in situ technologies involve drilling a series of wells for direct injection of appropriate solutions into the subsurface or the construction of reactive permeable barriers allowing the passage of groundwater while promoting the biodegradation of contaminants. Although polluted subsurface and groundwater systems are often very low in oxygen and nutrients and therefore characterized by slow biological oxidation rates, several laboratory and field-scale studies described successful applications of reactive biobarriers inoculated with *Rhodococcus* pure or mixed cultures (Salanitro et al. 2001) and supplied with oxygen and mineral nutrients (Vogt et al. 2004) for degradation of recalcitrant groundwater contaminants such as chlorinated benzene, methyl *t*-butyl ether, and *t*-butyl alcohol. In a recent field study of Beškoski et al. (2017), groundwater contaminated with petroleum hydrocarbons from an underground storage tank was treated using a closed bipolar system (one extraction and two injection wells) by bioaugmentation with pre-grown indigenous hydrocarbon-degrading consortia containing *Rhodococcus* and biostimulation with nutrients. After 60 days of bioremediation, more than 95% of *n*-alkanes, terpanes, and steranes were biodegraded; phenanthrene and its methyl-, dimethyl-, and trimethyl-isomers were removed completely, suggesting the applicability of this technology for the in situ treatment of PAH-contaminated groundwater.

3.2 On-Site Treatment

As evident from Table 2, the on-site bioremediation using *Rhodococcus* mostly involves the landfarming and biopile treatment of complex petroleum hydrocarbon mixtures, including crude oil, diesel fuel, and fuel oil. Historically, landfarming was one of the first forms of on-site contamination treatment and was widely used in oil industry for the disposal and neutralization of oily wastes. The waste material is applied to clean soil and landfarming area managed by tilling, fertilization, watering, and addition of bulking agents to maintain optimum soil conditions of nutrients, moisture, and pH. Microorganisms used in the biodegradation process are mostly indigenous soil populations. However, high concentrations of toxic contaminants present in oily wastes often hinder the development of resident oil-oxidizing microbial consortia. Therefore, the bioaugmentation with selected cultures of oil-degrading microorganisms adapted to high toxicant content and harsh environmental conditions may be used to enhance bioremediation process. It should be noted that while landfarming of oily sludges is no longer considered environmentally acceptable due to large volumes of clean soil contaminated during the process and the potential risk of contaminant leaching and emission from the treatment area (Van Hamme et al. 2003), it is still being used in many countries. More recently, ventilated and composting biopiles, which involve a greater degree of engineering and containment, have been developed for the oil-contaminated soil and sludge treatment. The contaminated material is removed to a specifically prepared area, which is usually lined with low permeability material such as high-density polyethylene or clay to minimize contaminant movement off-site. Soil biopile systems often include leachate-collecting and sometimes emission-control facilities. Construction of composting biopiles involving a succession of mesophilic and thermophilic microorganisms consists of piling the contaminated soil and mixing with an organic bulking agent such as composted agricultural waste, straw, or woodchips. The piles are aerated by either passive or forced ventilation or pile turning, and the temperature, pH, moisture, and nutrient contents are controlled.

Several laboratory studies have reported favorable effects of *Rhodococcus* augmentation on petroleum hydrocarbon biodegradation in soils at low (Whyte et al. 1998) and mesophilic (Sidorov et al. 1998; Taki et al. 2004; Taheri et al. 2008; Kuyukina et al. 2013) temperatures. Thus, mineralization of ^{14}C-labelled hexadecane at 5 °C was significantly greater in both crude oil-contaminated and pristine soil microcosms seeded with *Rhodococcus* sp. Q15 cells compared to non-inoculated control soil microcosms (Whyte et al. 1998). Moreover, efficient removal of less degradable contaminants such as aromatic hydrocarbons (including most recalcitrant *o*-xylene) and fuel oil from soil was obtained upon soil inoculation with either pure culture of *R. opacus* (Taki et al. 2004) or mixed culture of hydrocarbon-oxidizing bacteria containing *R. erythropolis* and *Rhodococcus* sp. strains (Sidorov et al. 1998). Taheri et al. (2008) performed a feasibility study for the soil polluted with disulfide oil, a waste product of liquefied petroleum gas desulfurization, and found that a vermiculite-immobilized *Rhodococcus* sp. strain previously isolated from disulfide oil-contaminated soil has a great potential for its bioremediation, although no comparison of immobilized and liquid forms of inoculum was made. There are also few reports on considerable bioremediation potential of indigenous bacterial communities inhabiting hydrocarbon-contaminated soils and containing large proportions of *Rhodococcus* representatives (Whyte et al. 2001; Hamamura et al. 2006), which could be enriched in laboratory microcosms during bioremediation process (Peressutti et al. 2003; Taki et al. 2007). However, the laboratory microcosm study of Thomassin-Lacroix et al. (2002) demonstrated low effect of bioaugmentation of fuel-contaminated Arctic tundra soil with enriched bacterial culture originated from the same soil and containing *Rhodococcus* members, indicating that biostimulation with a mineral nitrogen source was sufficient for the soil naturally rich in hydrocarbon-degrading microorganisms (including *Rhodococcus*).

It is now generally agreed that results of field bioremediation can differ significantly from the laboratory studies due to much stronger and more complex influence of environmental factors, both abiotic and biotic, on contaminant biodegradation process. Since bench-scale feasibility results often lack representativeness to field situations, field bioremediation trials must be conducted to corroborate findings of laboratory experiments. Small-scale field experiments conducted by Koronelli et al. (1997) reported that introduction of a hydrocarbon-degrading strain of *R. erythropolis* into tundra soil artificially contaminated with crude oil resulted in increased counts of hydrocarbon-degrading bacteria and an increased rate of hydrocarbon degradation. Christofi et al. (1998) found that inoculation with biosurfactant-producing *R. ruber* increased counts of hydrocarbon-oxidizing bacteria persisted in composted crude oil-contaminated soil and enhanced oil biodegradation. More recently Sun et al. (2012) applied a liquid *R. ruber* culture for on-site landfarming of heavily PAH-contaminated soil at the abandoned cooking plant site. They found out that a combination of biostimulation and bioaugmentation significantly enhanced the removal of PAHs from the contaminated soil. Sorkhoh et al. (1995) have used naturally occurring bacterial consortia removed from cyanobacterial mats floating in oil-polluted waters in the Arabian Gulf to inoculate oil-contaminated sand. This increased removal of oil from the sand, and rhodococci appeared to predominate in microbial populations. Bacterial preparations consisting of two-component *R. erythropolis* and *R. ruber* cultures (Kuyukina et al. 2003; Murygina et al. 2005; De-qing et al. 2007); a three-component bacterial culture of *R. erythropolis*, *Pseudomonas putida*, and *Acinetobacter johnsonii* (Genovese et al. 2008); and a complex bacterial-yeast consortium of *Dietzia* (former *Rhodococcus*) *maris*, *R. erythropolis*, *Rhodococcus* sp., *Pseudomonas stutzeri*, and *Candida* sp. (Sidorov et al. 1997) were successfully used in field trials on bioremediation of soils contaminated with crude oil, fuel, and oily wastes. Masy et al. (2016b) applied electrical resistivity tomography to detect soil heterogeneities and to monitor *R. erythropolis* biodegradation activity during a pilot-scale bioremediation of aged fuel-contaminated clay loam soil. It should be noted that in most field studies referenced above, contaminated soils were seeded with bacterial inocula and amended by addition of mineral fertilizers, bulking agents, and other factors stimulating biodegradation process, thus suggesting that combination of two bioremediation approaches, bioaugmentation and biostimulation, is essential for the cleanup of hydrocarbon-contaminated soils.

3.3 Bioreactor Treatment

Bioreactors are widely used to degrade toxic compounds in industrial effluents to prevent environmental pollution. Furthermore, bioreactor treatment of contaminated soils and sludge is not yet mainstream, but growing technology that overcomes some rate-limiting and variability factors observed in landfarming and biopile processes.

Particularly, bioreactor-based technologies allow more precise control and management of biodegradation parameters such as temperature, pH, oxygen, nutrient and water contents, and homogenous distribution of contaminated material and biomass in the reactor volume, which leads to increased mass transfer and reaction rates (Van Hamme et al. 2003). Several laboratory studies reported the application of indigenous bacterial communities containing *Rhodococcus* representatives in bioreactors of different types for treatment of contaminated soils and sediments; these included solid-phase, slurry-phase, and packed-bed loop reactors and resulted in reducing the levels of phthalate esters and polychlorinated and polycyclic aromatic compounds (Juneson et al. 2001; Ringelberg et al. 2001; Fava et al. 2003). We have used an oleophilic biofertilizer in the form of concentrated emulsion of hydrocarbon-grown *R. erythropolis* and *R. ruber* cultures and mineral salt solution stabilized by a *Rhodococcus* biosurfactant for successive treatment of crude oil-contaminated soil in a field slurry bioreactor and landfarming plots and found out that high biodegradation rate for petroleum hydrocarbons can be achieved following stimulation of the degradation process in a slurry bioreactor (Kuyukina et al. 2003).

Rhodococci are also candidate organisms for use as inocula in contaminated water treatments, demonstrating promising results in laboratory simulations (see Table 2 for references). For example, inoculations with suspended and biolite/diatomaceous earth-immobilized *R. erythropolis* cells were shown to be efficient in the biotreatment of phenol- and formaldehyde-contaminated synthetic and industrial wastewaters (Hidalgo et al. 2002a, b; Prieto et al. 2002a, b). Sawdust-immobilized *R. ruber* IEGM 615 and *R. opacus* IEGM 249 cells degraded petroleum hydrocarbons (including aliphatic from C_{10} to C_{19} and polyaromatic) to a great extent when applied to a fluidized-bed bioreactor and retained high catalytic activity during repeated bioreactor cycles (Kuyukina et al. 2009, 2017). A coarse sand-immobilized *Rhodococcus* sp. culture was successfully used in a fixed-film trickle-bed bioreactor treating wastewaters contaminated with methylhydrazine/hydrazine (Nwankwoala et al. 2001). It has been reported that a mixed culture isolated from azo dye-contaminated soil and containing *R. globerulus* cells was able to decolorize azo dyes under microaerophilic conditions (Joshi et al. 2008). However, another microaerophilic biodegradation study conducted by Tartakovsky et al. (2001) showed no significant difference in reduction of Aroclor 1242 levels in *Rhodococcus*-bioaugmented and non-bioaugmented reactors, although identification of indigenous bacterial populations of the non-bioaugmented reactor by 16S rDNA sequencing revealed *Rhodococcus* members among other biphenyl-degrading bacteria. Liquid and immobilized *Rhodococcus* spp. cultures were efficiently applied to different type bioreactors treating water contaminated with phenol, di- and trinitrophenol, fluorophenol, phthalic acid ester, nitrosodimethylamine, and organic cyanide (Kitova et al. 2004; Weidhaas et al. 2007; Shen et al. 2009a; Hatzinger et al. 2011; Yaacob et al. 2016; Ramos et al. 2017; An et al. 2018; Zhang et al. 2018). Moreover, complex emergent pollutants, such as aflatoxin and pharmaceuticals, can be efficiently degraded by *Rhodococcus* in bioreactors (Hasan et al. 2016; Dogan et al. 2017). As a promising approach to mitigate biofouling in membrane bioreactors, the abovementioned quorum quenching activity of *Rhodococcus* was used alone or in combination with chlorination, resulting in reduced levels of *N*-acyl homoserine lactones and extracellular biopolymers in the biofilm (Maqbool et al. 2015; Weerasekara et al. 2016).

Several pilot-scale bioreactor studies referenced in Table 2 involved *Rhodococcus* applications to industrial waste and wastewater treatments (Kimbara et al. 1998; Katsivela et al. 1999; Daye et al. 2003; Safonova et al. 2004; Fayolle-Guichard et al. 2012; Hatzinger et al. 2017). Particularly, pilot membrane bioreactors inoculated with *Rhodococcus*-containing mixed cultures were used to treat tetrahydrofuran and dichloropropene waste streams (Katsivela et al. 1999; Daye et al. 2003). Industrial wastewater inoculation with an algal-bacterial co-culture containing a *Rhodococcus* sp. strain Ac-1267 and immobilized on capron fibers resulted in the formation of a stable microbial consortium and significant decrease of petroleum hydrocarbon, phenol, anionic surfactant, and heavy metal concentrations (Safonova et al. 2004). A pilot plant consisting of UV-irradiation equipment and two successive bioreactors was constructed to treat PCB-contaminated waste oil from high-voltage transformers and condensers, and liquid cultures of *Comamonas testosteroni* TK102 and *R. opacus* TSP 203 were used as inocula for bioreactors providing complete biodegradation of PCBs partially dechlorinated by the UV pretreatment (Kimbara et al. 1998). Groundwater polluted with ethyl *t*-butyl ether (ETBE) and gasoline was treated in the on-site bioreactor inoculated with *R. wratislaviensis*, *R. aetherivorans*, and *Aquincola tertiaricarbonis* cultures, and a 100-fold increase in the abundance of the *ethB* gene encoding a cytochrome P450 involved in ETBE biodegradation was detected by *q*-PCR, thus reflecting the groundwater colonization by the relevant microorganism (*R. aetherivorans*) (Fayolle-Guichard et al. 2012). However, more recently Hatzinger et al. (2017) carried out a pilot propane-fed bioreactor treatment of groundwater contaminated with *N*-nitrosodimethylamine and *N*-nitrodimethylamine left from the use of liquid rocket propellant, and they demonstrated the replacement of initially inoculated propane-oxidizing *R. ruber* ENV425 with native propanotrophs along with the significant increase in microbial diversity and propane monooxygenase abundance over the year of experiment resulted in the efficient contaminant biodegradation.

Biological oxidation of volatile organic carbon vapors by microbial biofilms formed on a solid support in biofilters/bioreactors provides an effective and inexpensive alternative to physicochemical methods (Vinage and von Rohr 2003). A modified rotating biological contactor inoculated with a suspension of *Pseudomonas putida* F1 and *R. erythropolis* PWD1 was proposed by Vinage and von Rohr (2003) for the biological treatment of artificial waste gas polluted with toluene vapors. The proposed system allowed proper control of the biofilm growth and long-term bioremediation performance for a year indicating its feasibility for industrial applications. Borin et al. (2006) investigated microbial succession in a compost-packed biofilter treating benzene-contaminated air and found out that the maximum benzene removal rate strongly correlated with the prevalence of *Rhodococcus* representatives in the bacterial community, thus suggesting their major role in benzene degradation. Aldric and Thonart (2008) evaluated the performance of a water/silicone oil two-phase partitioning bioreactor inoculated with *R. erythropolis* T902.1 cells for removing volatile organic compounds from gaseous effluents. They reported simultaneous degradation of isopropylbenzene and ethanol by rhodococci, suggesting that ethanol improves contaminant biodegradation process in the bioreactor. Bacterial-fungal consortia containing *Rhodococcus* were successfully tested in column biofilters to treat most common industrial and residential air pollutants, such as volatile organic compounds, PAHs, and hydrogen sulfide (López et al. 2013; Vergara-Fernández et al. 2018). These studies also revealed the ability of rhodococci to survive for a long time in two-component and complex microbial populations of biofilters and to contribute considerably to the contaminant degradation. However, in earlier study of Borràs et al. (2010), a co-culture of *Pseudomonas aeruginosa* CCM 1960 and *R. erythropolis* CCM 2595 inhibited both laccase production and PAH degradation by white-rot fungi in the case of removal of acenaphthylene, benzo[*a*]pyrene, dibenzo[*a*,*h*]anthracene, and benzo[*g*,*h*,*i*]perylene from contaminated soil, thus indicating the need for further research of rhodococcal-fungal interactions during bioremediation.

4 Concluding Remarks

It could be resumed that some success in bioremediation of contaminated environments has been achieved using actinobacteria of the genus *Rhodococcus* either as bioaugmentation agents or members of indigenous microbial communities stimulated by nutrient amendments. Although possible bioremediation applications of genetically modified *Rhodococcus* were not discussed in this review, it should be noted that advanced methods and powerful tools for genetic engineering of rhodococci were developed (see chapter "Diversity and Plasticity of Rhodococcus Genomes" by Cappelletti and Di Gennaro), which could be used in constructing recombinant strains for improved bioremediation inocula. Indeed, the complete nucleotide sequences of numerous *Rhodococcus* genomes provided new insights that could facilitate biotechnological exploitation of this genus. For example, amidase genes from *R. erythropolis* MP50 and *Agrobacterium tumefaciens* D3 co-expressed along with nitrile hydratase from *Bradyrhizobium japonicum* USDA 110 were used for biodegradation of dihalogenated benzonitrile herbicides by recombinant *Escherichia coli* (Pei et al. 2017). Furthermore, *Arabidopsis thaliana* (*Arabidopsis*) was transformed with the hexahydro-1,3,5-trinitro-1,3,5-triazine (RDX)-degrading *xplA* and associated reductase *xplB* from *R. rhodochrous* 11Y in combination with the 2,4,6-trinitrotoluene (TNT)-detoxifying nitroreductase *nfsI* from *Enterobacter cloacae*, resulting in transgenic plants beneficial for remediating RDX- and TNT-contaminated soil and groundwater. However, up today, applications of genetically modified rhodococci in bioremediation fields are limited to laboratory biodegradation studies attempting to reveal their multitudinous catabolic pathways and regulatory mechanisms for different organic contaminants. In the context of future applications of genetically engineered *Rhodococcus* strains in environmental biotechnology, it could be feasible to use biosynthesis products (e.g., enzymes and biosurfactants) rather than whole rhodococcal cells, which may help to overcome biosafety limitations associated with release of genetically modified microorganisms into open environments. Another possible perspective for molecular genetic approach to be applied to bioremediation is the use of oligonucleotide primers and DNA probes constructed for rhodococcal biodegradation genes to estimate their in situ functional activities (Whyte et al. 2002a; Coffey et al. 2010; Táncsics et al. 2015). These techniques, as well as novel genomic and proteomic methods, could be used to predict bacterial metabolism in contaminated environments and to enhance bioremediation. Moreover, correct prognosis of rhodococcal survival and biodegradation activity in contaminated environments would require further fundamental studies of interactions between *Rhodococcus* cultures introduced and indigenous micro- and macroorganisms using physiological and molecular approaches (Watanabe and Hamamura 2003). Clearly, *Rhodococcus* application in bioremediation of contaminated sites is a promising and evolving field of environmental biotechnology, and its success depends on the increase in our fundamental knowledge of these remarkable bacteria.

Acknowledgments This research was funded by the Ministry of Science and Higher Education of the Russian Federation (State Task Registration No. 01201353247 for IEGM) and the Russian Science Foundation grant 18-14-00140 for PSU.

References

Aislabie J, Saul DJ, Foght JM (2006) Bioremediation of hydrocarbon-contaminated polar soils. Extremophiles 10:171–179

Al-Awadhi H, Al-Hasan RH, Sorkhoh NA, Salamah S, Radwan SS (2003) Establishing oil-degrading biofilms on gravel particles and glass plates. Int Biodeterior Biodegrad 51:181–185

Aldric J-M, Thonart P (2008) Performance evaluation of a water/silicone oil two-phase partitioning bioreactor using *Rhodococcus erythropolis* T902.1 to remove volatile organic compounds from gaseous effluents. J Chem Technol Biotechnol 83:1401–1408

Alexander M (1999) Biodegradation and bioremediation, vol 2. Academic Press, London

An X, Cheng Y, Huang M, Sun Y, Wang H, Chen X, Wang J, Li D, Li C (2018) Treating organic cyanide-containing groundwater by immobilization of a nitrile-degrading bacterium with a biofilm-forming bacterium using fluidized bed reactors. Environ Pollut 237:908–916

Aoshima H, Hirase T, Tada T, Ichimura N, Kato H, Nagata Y, Myoenzono T, Tagauchi M, Takuzumi T, Aoki T, Makino S, Hagita K, Ishiwata H (2007) Safety evaluation of a heavy oil-degrading bacterium *Rhodococcus erythropolis* C2. J Toxicol Sci 32:69–78

Baxter J, Cummings SP (2006) The impact of bioaugmentation on metal cyanide degradation and soil bacteria community structure. Biodegradation 17:207–217

Baxter J, Garton NJ, Cummings SP (2006) The impact of acrylonitrile and bioaugmentation on the biodegradation activity and bacterial community structure of a topsoil. Folia Microbiol 51:591–597

Bej AK, Saul D, Aislabie J (2000) Cold-tolerant alkane-degrading *Rhodococcus* species from Antarctica. Polar Biol 23:100–105

Bell KS, Philp JC, Aw DWJ, Christofi N (1998) The genus *Rhodococcus*. A review. J Appl Microbiol 85:195–210

Beškoski VP, Miletić S, Ilić M, Gojgić-Cvijović G, Papić P, Marić N, Šolević-Knudsen T, Jovančićević BS, Nakano T, Vrvić MM (2017) Biodegradation of isoprenoids, steranes, terpanes and phenanthrenes during *in situ* bioremediation of petroleum contaminated groundwater. Clean Soil Air Water 45:1600023. https://doi.org/10.1002/clen.201600023

Besse P, Combourieu B, Boyse G, Sancelme M, de Wever H, Delort A-M (2001) Long-range 1H-15N heteronuclear shift correlation at natural abundance: a tool to study benzothiazole biodegradation by two *Rhodococcus* strains. Appl Environ Microbiol 67:1412–1417

Borin S, Marzorati M, Brusetti L, Zilli M, Cherif H, Hassen A, Converti A, Sorlini C, Daffonchio D (2006) Microbial succession in a compost-packed biofilter treating benzene-contaminated air. Biodegradation 17:79–89

Borràs E, Caminal G, Sarrà M, Novotný C (2010) Effect of soil bacteria on the ability of polycyclic aromatic hydrocarbons (PAHs) removal by Trametes versicolor and Irpex lacteus from contaminated soil. Soil Biol Biochem 42:2087–2093

Cappelletti M, Pinelli D, Fedi S, Zannoni D, Frascari D (2018) Aerobic co-metabolism of 1,1,2,2-tetrachloroethane by *Rhodococcus aetherivorans* TPA grown on propane: kinetic study and bioreactor configuration analysis. J Chem Technol Biotechnol 93:155–165

Cavalca L, Colombo M, Larcher S, Gigliotti C, Collina E, Andreoni V (2002) Survival and naphthalene-degrading activity of *Rhodococcus* sp. strain 1BN in soil microcosms. J Appl Microbiol 92:1058–1065

Christofi N, Ivshina IB, Kuyukina MS, Philp JC (1998) Biological treatment of crude oil contaminated soil in Russia. In: Lerner DN, London NRG (eds) Contaminated land and groundwater: future directions, vol 14. Geological Society Engineering Geology Publications, London, pp 45–51

Coffey L, Owens E, Tambling K, O'Neill D, O'Connor L, O'Reilly C (2010) Real-time PCR detection of Fe-type nitrile hydratase genes from environmental isolates suggests horizontal gene transfer between multiple genera. Antonie Van Leeuwenhoek 98:455–463

Coleman NV, Nelson DR, Duxbury T (1998) Aerobic biodegradation of hexahydro-1,3,5-trinitro-1,3,5-triazine (RDX) as a nitrogen source by a *Rhodococcus* sp., strain DN22. Soil Biol Biochem 30:1159–1167

Colquhoun JA, Heald SC, Tamaoka LLJ, Kato C, Horikoshi K, Bull AT (1998) Taxonomy and biotransformation activities of some deep-sea actinomycetes. Extremophiles 2:269–277

Daghio M, Tatangelo V, Franzetti A, Gandolfi I, Papacchini M, Careghini A, Sezenna E, Saponaro S, Bestetti G (2015) Hydrocarbon degrading microbial communities in bench scale aerobic biobarriers for gasoline contaminated groundwater treatment. Chemosphere 130:34–39. https://doi.org/10.1016/j.chemosphere.2015.02.022

Daye KJ, Groff JC, Kirpekar AC, Mazumder R (2003) High efficiency degradation of tetrahydrofuran (THF) using a membrane bioreactor: identification of THF-degrading cultures of *Pseudonocardia sp.* strain M1 and *Rhodococcus ruber* isolate M2. J Ind Microbiol Biotechnol 30:705–714

Dean-Ross D, Moody JD, Freeman JP, Doerge DR, Cerniglia CE (2001) Metabolism of anthracene by a *Rhodococcus* species. FEMS Microbiol Lett 204:205–211

Dejonghe W, Boon N, Seghers D, Top EM, Verstraete W (2001) Bioaugmentation of soils by increasing microbial richness: missing links. Environ Microbiol 3:649–657

De-qing S, Jian Z, Zhao-long G, Jian D, Tian-li W, Murygina V, Kalyuzhnyi S (2007) Bioremediation of oil sludge in Shengli oilfield. Water Air Soil Pollut 185:177–184

Di Lorenzo A, Varcamonti M, Parascandola P, Vignola R, Bernardi A, Sacceddu P, Sisto R, de Alteriis E (2005) Characterization and performance of a toluene-degrading biofilm developed on pumice stones. Microb Cell Factories 4:4

Dobslaw D, Engesser K-H (2018) Biodegradation of gaseous emissions of 2-chlorotoluene by strains of *Rhodococcus* sp. in polyurethane foam packed biotrickling filters. Sci Total Environ 639:1491–1500

Dogan OB, Onal-Ulusoy B, Bozoglu F, Sagdicoglu-Celep AG, Cekmecelioglu D (2017) Detoxification of groundnut cake naturally contaminated with aflatoxin B1 using *Rhodococcus erythropolis* in shake flask bioreactors. Waste Biomass Valoriz 8:721–731

Elo S, Maunuksela L, Salkinoja-Salonen M, Smolander A, Haahtela K (2000) Humus bacteria of Norway spruce stands: plant growth promoting properties and birch, red fescue and alder colonizing capacity. FEMS Microbiol Ecol 31:143–152

Etkin DS (2001) Analysis of oil spill trends in the United States and worldwide. Proceeding of the International Oil Spill Conference. American Petroleum Institute, Washington, pp 1291–1300

Fahy A, McGenity TJ, Timmis KN, Ball AS (2006) Heterogeneous aerobic benzene-degrading communities in oxygen-depleted groundwaters. FEMS Microbiol Ecol 58:260–270

Fahy A, Ball AS, Lethbridge G, McGenity TJ, Timmis KN (2008a) High benzene concentrations can favour Gram-positive bacteria in groundwaters from a contaminated aquifer. FEMS Microbiol Ecol 65:526–533

Fahy A, Ball AS, Lethbridge G, Timmis KN, McGenity TJ (2008b) Isolation of alkali-tolerant benzene-degrading bacteria from a contaminated aquifer. Lett Appl Microbiol 47:60–66

Fava F, Bertin L, Fedi S, Zannoni D (2003) Methyl-β-cyclodextrin-enhanced solubilization and aerobic biodegradation of polychlorinated biphenyls in two aged-contaminated soils. Biotechnol Bioeng 81:381–390

Fayolle-Guichard F, Durand J, Cheucle M, Rosell M, Michelland RJ, Tracol J-P, Le Roux F, Grundman G, Atteia O, Richnow HH, Dumestre A, Benoit Y (2012) Study of an aquifer contaminated by ethyl tert-butyl ether (ETBE): site characterization and on-site bioremediation. J Hazard Mater 201–202:236–243

Federici E, Pepi M, Esposito A, Scargetta S, Fidati L, Gasperini S, Cenci G, Altieri R (2011) Two-phase olive mill waste composting: community dynamics and functional role of the resident microbiota. Bioresour Technol 102:10965–10972

Fuller ME, Hatzinger PB, Condee CW, Andaya C, Vainberg S, Michalsen MM, Crocker FH, Indest KJ, Jung CM, Eaton H, Istok JD (2015) Laboratory evaluation of bioaugmentation for aerobic treatment of RDX in groundwater. Biodegradation 26:77–89. https://doi.org/10.1007/s10532-014-9717-y

Gakhar L, Malik ZA, Allen CCR, Lipscomb DA, Larkin MJ, Ramaswamy S (2005) Structure and increased thermostability of *Rhodococcus* sp. naphthalene 1,2-dioxygenase. J Bacteriol 187:7222–7231

Genovese M, Denaro R, Cappello S, Di Marco G, La Spada G, Giuliano L, Genovese L, Yakimov MM (2008) Bioremediation of benzene, toluene, ethylbenzene, xylenes-contaminated soil: a biopile pilot experiment. J Appl Microbiol 105:1694–1702

Hamamura N, Olson SH, Ward DM, Inskeep WP (2006) Microbial population dynamics associated with crude-oil biodegradation in diverse soils. Appl Environ Microbiol 72:6316–6324

Hasan HA, Abdullah SRS, Al-Attabi AWN, Nash DAH, Anuar N, Rahman NA, Titah HS (2016) Removal of ibuprofen, ketoprofen, COD and nitrogen compounds from pharmaceutical wastewater using aerobic suspension-sequencing batch reactor (ASSBR). Sep Purif Technol 157:215–221

Hatzinger PB, Condee C, McClay KR, Togna AP (2011) Aerobic treatment of N-nitrosodimethylamine in a propane-fed membrane bioreactor. Water Res 45:254–262

Hatzinger PB, Lewis C, Webster TS (2017) Biological treatment of N-nitrosodimethylamine (NDMA) and N-nitrodimethylamine (NTDMA) in a field-scale fluidized bed bioreactor. Water Res 126:361–371

Hendrickx B, Dejonghe W, Boenne W, Brennerova M, Cernik M, Lederer T, Bucheli-Witschel M, Bastiaens L, Verstraete W, Top EM, Diels L, Springa D (2005) Dynamics of an oligotrophic bacterial aquifer community during contact with a groundwater plume contaminated with benzene, toluene, ethylbenzene, and xylenes: an in situ mesocosm study. Appl Environ Microbiol 71:3815–3825

Hidalgo A, Jaureguibeitia A, Prieto MB, Rodríguez-Fernández C, Serra JL, Llama MJ (2002a) Biological treatment of phenolic industrial wastewaters by *Rhodococcus erythropolis* UPV-1. Enzyme Microb Technol 31:221–226

Hidalgo A, Lopategi A, Prieto M, Serra JL, Llama MJ (2002b) Formaldehyde removal in synthetic and industrial wastewater by *Rhodococcus erythropolis* UPV-1. Appl Microbiol Biotechnol 58:260–263

Home Pages of Culture Collections in the World (2018) World federation for culture collections. http://www.wfcc.info/index.php/collections/. Cited 20 Jul 2018

Homklin S, Kee Ong S, Limpiyakorn T (2012) Degradation of 17α-methyltestosterone by *Rhodococcus* sp. and *Nocardioides* sp. isolated from a masculinizing pond of Nile tilapia fry. J Hazard Mater 221–222:35–44

Hong SH, Hae LP, Ko U-R, Jae JY, Cho K-S (2007) Bioremediation of oil-contaminated soil using an oil-degrading rhizobacterium *Rhodococcus* sp. 412 and Zea mays. Korean J Microbiol Biotechnol 35:150–157

Hongming L, Xu L, Zhaojian G, Fan Y, Dingbin C, Jianchun Z, Jianhong X, Shunpeng L, Qing H (2015) Isolation of an aryloxyphenoxy propanoate (AOPP) herbicide-degrading strain *Rhodococcus ruber* JPL-2 and the cloning of a novel carboxylesterase gene (*feh*). Braz J Microbiol 46:425–432

Ito K, Kawashima F, Takagi K, Kataoka R, Kotake M, Kiyota H, Yamazaki K, Sakakibara F, Okada S (2016) Isolation of endosulfan sulfate-degrading *Rhodococcus koreensis* strain S1-1 from endosulfan contaminated soil and identification of a novel metabolite, endosulfan diol monosulfate. Biochem Biophys Res Commun 473:1094–1099

Ivshina IB, Oborin AA, Nesterenko OA, Kasumova SA (1981) Bacteria of the *Rhodococcus* genus from the ground water of oil-bearing deposits in the Perm region near the Urals. Microbiology (Moscow) 50:709–717

Ivshina IB, Kamenskikh TN, Liapunov YE (1994) IEGM catalogue of strains of regional specialized collection of alkanotrophic microorganisms. Nauka, Moscow

Ivshina IB, Berdichevskaya MV, Zvereva LV, Rybalka LV, Elovikova EA (1995) Phenotypic characterization of alkanotrophic rhodococci from various ecosystems. Microbiology (Moscow) 64:507–513

Ivshina IB, Peshkur TA, Korobov VP (2002) Effective uptake of cesium ions by *Rhodococcus* cells. Microbiology (Moscow) 71:357–361

Ivshina IB, Kuyukina MS, Kostina LV (2013) Adaptive mechanisms of nonspecific resistance to heavy metal ions in alkanotrophic actinobacteria. Russ J Ecol 44:123–130

Ivshina IB, Kuyukina MS, Krivoruchko AV, Elkin AA, Makarov SO, Cunningham CJ, Peshkur TA, Atlas RM, Philp JC (2015a) Oil spill problems and sustainable response strategies through new technologies. Environ Sci Process Impacts 17:1201–1219

Ivshina IB, Mukhutdinova AN, Tyumina HA, Suzina NE, El'-Registan GI, Mulyukin AL (2015b) Drotaverine hydrochloride degradation using cyst-like dormant cells of *Rhodococcus ruber*. Curr Microbiol 70:307–314

Ivshina I, Kostina L, Krivoruchko A, Kuyukina M, Peshkur T, Anderson P, Cunningham C (2016) Removal of polycyclic aromatic hydrocarbons in soil spiked with model mixtures of petroleum hydrocarbons and heterocycles using biosurfactants from *Rhodococcus ruber* IEGM 231. J Hazard Mater 312:8–17

Ivshina IB, Kuyukina MS, Krivoruchko AV (2017) Hydrocarbon-oxidizing bacteria and their potential in eco-biotechnology and bioremediation. In: Kurtböke I (ed) Microbial resources: from functional existence in nature to industrial applications. Elsevier, Amsterdam, pp 121–148

Jiménez N, Viñas M, Bayona JM, Albaiges J, Solanas AM (2007) The Prestige oil spill: bacterial community dynamics during a field biostimulation assay. Appl Microbiol Biotechnol 77:935–945

Joshi T, Iyengar L, Singh K, Garg S (2008) Isolation, identification and application of novel bacterial consortium TJ-1 for the decolourization of structurally different azo dyes. Bioresour Technol 99:7115–7121

Juneson C, Ward OP, Singh A (2001) Biodegradation of bis(2-ethylhexyl)phthalate in a soil slurry-sequencing batch reactor. Proc Biochem 37:305–313

Juteau P, Larocque R, Rho D, LeDuy A (1999) Analysis of the relative abundance of different types of bacteria capable of toluene degradation in a compost biofilter. Appl Microbiol Biotechnol 52:863–868

Katsivela E, Bonse D, Krüger A, Strömpl C, Livingston C, Wittich R-M (1999) An extractive membrane biofilm reactor for degradation of 1,3-dichloropropene in industrial waste water. Appl Microbiol Biotechnol 52:853–862

Kim J-D, Lee C-G (2007) Microbial degradation of polycyclic aromatic hydrocarbons in soil by bacterium-fungus co-cultures. Biotech Bioproc Eng 12:410–416

Kim D, Chae JC, Zylstra GJ, Kim YS, Kim SK, Nam MH, Kim YM, Kim E (2004) Identification of a novel dioxygenase involved in metabolism of o-xylene, toluene, and ethylbenzene by *Rhodococcus sp.* strain DK17. Appl Environ Microbiol 70:7086–7092

Kimbara K, Hayakawa T, Shimura M (1998) Remediating PCB wastes using microorganisms. Jpn Railway Transp Rev 17:17–20

Kitagawa W, Tamura T (2008a) A quinoline antibiotic from *Rhodococcus erythropolis* JCM 6824. J Antibiot 61:680–682

Kitagawa W, Tamura T (2008b) Three types of antibiotics produced from *Rhodococcus erythropolis* strains. Microbes Environ 23:167–171

Kitamoto D, Isoda H, Nakahara T (2002) Functions and potential applications of glycolipid biosurfactants—from energy-saving materials to gene delivery carriers. J Biosci Bioeng 94:187–201

Kitova AE, Kuvichkina TN, Arinbasarova AY, Reshetilov AN (2004) Degradation of 2,4-dinitrophenol by free and immobilized cells of *Rhodococcus erythropolis* HL PM-1. Appl Biochem Microbiol 40:258–261

Koronelli TV, Komarova TI, Il'inskii VV, Kuz'min YI, Kirsanov NB, Yanenko AS (1997) Introduction of bacteria of the genus *Rhodococcus* into oil-contaminated tundra soils. Appl Biochem Microbiol 33:172–175

Kuyukina MS, Ivshina IB, Ritchkova MI, Philp JC, Cunningham JC, Christofi N (2003) Bioremediation of crude oil-contaminated soil using slurry-phase biological treatment and land farming techniques. Soil Sediment Contam 12:85–99

Kuyukina MS, Ivshina IB, Makarov SO, Litvinenko LV, Cunningham CJ, Philp JC (2005) Effect of biosurfactants on crude oil desorption and mobilization in a soil system. Environ Int 31:155–161

Kuyukina MS, Ivshina IB, Gein SV, Baeva TA, Chereshnev VA (2007) *In vitro* immunomodulating activity of biosurfactant glycolipid complex from *Rhodococcus ruber*. Bull Exp Biol Med 144:326–330

Kuyukina MS, Ivshina IB, Serebrennikova MK, Krivorutchko AV, Podorozhko EA, Ivanov RV, Lozinsky VI (2009) Petroleum-contaminated water treatment in a fluidized-bed bioreactor with immobilized *Rhodococcus* cells. Int Biodeter Biodegr 63:427–432

Kuyukina MS, Ivshina IB, Kamenskikh TN, Bulicheva MV, Stukova GI (2013) Survival of cryogel-immobilized *Rhodococcus* cells in crude oil-contaminated soil and their impact on biodegradation efficiency. Int Biodeter Biodegr 84:118–125

Kuyukina MS, Ivshina IB, Serebrennikova MK, Krivoruchko AV, Korshunova IO, Peshkur TA, Cunningham CJ (2017) Oilfield wastewater biotreatment in a fluidized-bed bioreactor using co-immobilized *Rhodococcus* cultures. J Environ Chem Eng 5:1252–1260

Labbé D, Margesin R, Schinner F, Whyte LG, Greer CW (2007) Comparative phylogenetic analysis of microbial communities in pristine and hydrocarbon-contaminated Alpine soils. FEMS Microbiol Ecol 59:466–475

Lang S, Philp JC (1998) Surface-active lipids in rhodococci. Antonie Van Leeuwenhoek 74:59–70

Larkin MJ, Kulakov LA, Allen CCR (2005) Biodegradation and *Rhodococcus*—masters of catabolic versatility. Curr Opin Biotechnol 16:282–290
Lee E-H, Cho K-S (2008) Characterization of cyclohexane and hexane degradation by *Rhodococcus* sp. EC1. Chemosphere 71:1738–1744
Lin T-C, Pan P-T, Cheng S-S (2010) Ex situ bioremediation of oil-contaminated soil. J Hazard Mater 176:27–34
López ME, Rene ER, Malhautier L, Rocher J, Bayle S, Veiga MC, Kennes C (2013) One-stage biotrickling filter for the removal of a mixture of volatile pollutants from air: performance and microbial community analysis. Bioresour Technol 138:245–252
Luepromchai E, Singer AC, Yang C-H, Crowley DE (2002) Interactions of earthworms with indigenous and bioaugmented PCB-degrading bacteria. FEMS Microbiol Ecol 41:191–197
Luz AP, Pellizari VH, Whyte LG, Greer CW (2004) A survey of indigenous microbial hydrocarbon degradation genes in soils from Antarctica and Brazil. Can J Microbiol 50:323–333
Maqbool T, Khan SJ, Waheed H, Lee C-H, Hashmi I, Iqbal H (2015) Membrane biofouling retardation and improved sludge characteristics using quorum quenching bacteria in submerged membrane bioreactor. J Membr Sci 483:75–83
Margesin R, Labbé D, Schinner F, Greer CW, Whyte LG (2003) Characterization of hydrocarbon-degrading microbial populations in contaminated and pristine alpine soils. Appl Environ Microbiol 69:3085–3092
Martínková L, Uhnáková B, Pátek M, Nešvera J, Křen V (2009) Biodegradation potential of the genus *Rhodococcus*. Environ Int 35:162–177
Masák J, Čejková A, Jirků V, Kotrba D, Hron P, Siglová M (2004) Colonization of surfaces by phenolic compounds utilizing microorganisms. Environ Int 31:197–200
Masy T, Bertrand C, Xavier P-M, Vreuls C, Wilmot A, Cludts M, Renard P, Mawet P, Smets S, Dethy B, Thonart P, Jacques P, Hiligsmann S (2016a) Stable biofilms of *Rhodococcus erythropolis* T902.1 in draining pavement structures for runoff water decontamination. Int Biodeter Biodegr 112:108–118
Masy T, Caterina D, Tromme O, Lavigne B, Thonart P, Hiligsmann S, Nguyen F (2016b) Electrical resistivity tomography to monitor enhanced biodegradation of hydrocarbons with *Rhodococcus erythropolis* T902.1 at a pilot scale. J Contam Hydrol 184:1–13
Mergaert J, Verhelst A, Cnockaert MC, Tan T-L, Swings J (2001) Characterization of facultative oligotrophic bacteria from polar seas by analysis of their fatty acids and 16S rDNA sequences. Syst Appl Microbiol 24:98–107
Miteva VI, Sheridan PP, Brenchley JE (2004) Phylogenetic and physiological diversity of microorganisms isolated from a deep Greenland glacier ice core. Appl Environ Microbiol 70:202–213
Murygina V, Arinbasarov M, Kalyuzhnyi S (2000) Bioremediation of oil polluted aquatic systems and soils with novel preparation 'Rhoder'. Biodegradation 11(6):385–389
Murygina VP, Markarova MY, Kalyuzhnyi SV (2005) Application of biopreparation Rhoder for remediation of oil polluted polar marshy wetlands in Komi Republic. Environ Int 31:163–166
Nagy I, Schoofs G, Compernolle F, Proost P, Vanderleyden J, de Mot R (1995) Degradation of the thiocarbamate herbicide EPTC (*s*-ethyldipropylcarbamothioate) and biosafening by *Rhodococcus* sp. strain NI86/21 involve an inducible cytochrome P-450 system and aldehyde dehydrogenase. J Biotechnol 177:676–687
Nie Y, Chi CQ, Fang H, Liang JL, Lu SL, Lai GL, Tang YQ, Wu XL (2014) Diverse alkane hydroxylase genes in microorganisms and environments. Sci Rep 4:4968. https://doi.org/10.1038/srep04968
Nwankwoala AU, Egiebor NO, Nyavor K (2001) Enhanced biodegradation of methylhydrazine and hydrazine contaminated NASA wastewater in fixed-film bioreactor. Biodegradation 12:1–10
Ohhata N, Yoshida N, Egami H, Katsuragi T, Tani Y, Takagi H (2007) An extremely oligotrophic bacterium, *Rhodococcus erythropolis* N9T-4, isolated from crude oil. J Biotechnol 189:6824–6831

Pei X, Wang J, Guo W, Miao J, Wang A (2017) Efficient biodegradation of dihalogenated benzonitrile herbicides by recombinant Escherichia coli harboring nitrile hydratase-amidase pathway. Biochem Eng J 125:88–96

Peressutti SR, Alvarez HM, Pucci OH (2003) Dynamics of hydrocarbon-degrading bacteriocenosis of an experimental oil pollution in Patagonian soil. Int Biodeter Biodegr 52:21–30

Petrić I, Hršak D, Fingler S, Vončina E, Ćetković H, Kolar AB, Kolić NU (2007) Enrichment and characterization of PCB-degrading bacteria as potential seed cultures for bioremediation of contaminated soil. Food Technol Biotechnol 45:11–20

Pieper DH, Reineke W (2000) Engineering bacteria for bioremediation. Curr Opin Biotechnol 11:262–270

Podorozhko EA, Lozinsky VI, Ivshina IB, Kuyukina MS, Krivorutchko AV, Philp JC, Cunningham CJ (2008) Hydrophobised sawdust as a carrier for immobilisation of the hydrocarbon-oxidizing bacterium *Rhodococcus ruber*. Bioresour Technol 99:2001–2008

Poelarends GJ, Zandstra M, Bosma T, Kulakov LA, Larkin MJ, Marchesi JR, Weightman AJ, Janssen DB (2000) Haloalkane-utilizing *Rhodococcus* strains isolated from geographically distinct locations possess a highly conserved gene cluster encoding haloalkane catabolism. J Bacteriol 5:2725–2731

Priestley JT, Coleman NV, Duxbury T (2006) Growth rate and nutrient limitation affect the transport of *Rhodococcus sp.* strain DN22 through sand. Biodegradation 17:571–576

Prieto MB, Hidalgo A, Rodríguez-Fernández C, Serra JL, Llama MJ (2002a) Biodegradation of phenol in synthetic and industrial wastewater by *Rhodococcus erythropolis* UPV-1 immobilized in an air-stirred reactor with clarifier. Appl Microbiol Biotechnol 58:853–859

Prieto MB, Hidalgo A, Serra JL, Llama MJ (2002b) Degradation of phenol by *Rhodococcus erythropolis* UPV-1 immobilized on Biolite® in a packed-bed reactor. J Biotechnol 97:1–11

Ramos C, Amorim CL, Mesquita DP, Ferreira EC, Carrera J, Castro PML (2017) Simultaneous partial nitrification and 2-fluorophenol biodegradation with aerobic granular biomass: reactor performance and microbial communities. Bioresour Technol 238:232–240

Ringelberg DB, Talley JW, Perkins EJ, Tucker SG, Luthy RG, Bouwer EJ, Fredrickson HL (2001) Succession of phenotypic, genotypic, and metabolic community characteristics during in vitro bioslurry treatment of polycyclic aromatic hydrocarbon-contaminated sediments. Appl Environ Microbiol 67:1542–1550

Rodrigues JLM, Kachel CA, Aiello MR, Quensen JF, Maltseva OV, Tsoi TV, Tiedje JM (2006) Degradation of Aroclor 1242 dechlorination products in sediments by *Burkholderia xenovorans* LB400(*ohb*) and *Rhodococcus* sp. strain RHA1(*fcb*). Appl Environ Microbiol 72:2476–2482

Ruberto LAM, Vazquez S, Lobalbo A, Mac Cormack WP (2005) Psychrotolerant hydrocarbon-degrading *Rhodococcus* strains isolated from polluted Antarctic soils. Antarct Sci 17:47–56

Ryu HW, Yang HJ, Youn-J A, Kyung-Suk C (2006) Isolation and characterization of psychrotrophic and halotolerant *Rhodococcus sp.* YHLT-2. J Microbiol Biotechnol 16:605–612

Safonova E, Kvitko KV, Iankevitch MI, Surgko LF, Afti IA, Reisser W (2004) Biotreatment of industrial wastewater by selected algal-bacterial consortia. Eng Life Sci 4:347–353

Salanitro JP, Spinnler GE, Maner PM, Tharpe DL, Pickle DW, Wisniewski HL, Johnson PC, Bruce C (2001) In situ bioremediation of MTBE using biobarriers of single or mixed cultures. In: Leeson A, Alleman BC, Alvarez PJ, Magar VS (eds) Bioaugmentation, biobarriers and biogeochemistry. Battelle Press, Columbus, OH, pp 1–7

Serebrennikova MK, Golovina EE, Kuyukina MS, Ivshina IB (2017) A consortium of immobilized rhodococci for oilfield wastewater treatment in a column bioreactor. Appl Biochem Microbiol 53:435–440

Seth-Smith HMB, Rosser SJ, Basran A, Travis ER, Dabbs ER, Nicklin S, Bruce NC (2002) Cloning, sequencing, and characterization of the hexahydro-1,3,5-trinitro-1,3,5-triazine degradation gene cluster from *Rhodococcus rhodochrous*. Appl Environ Microbiol 68(10):4764–4771

Shagol CC, Krishnamoorthy R, Kim K, Sundaram S, Sa T (2014) Arsenic-tolerant plant-growth-promoting bacteria isolated from arsenic-polluted soils in South Korea. Environ Sci Pollut Res 21:9356–9365

Sharma SL, Pant A (2000) Biodegradation and conversion of alkanes and crude oil by a marine *Rhodococcus* sp. Biodegradation 11:289–294

Shen J, He R, Yu H, Wang L, Zhang J, Sun X, Li J, Han W, Xu L (2009a) Biodegradation of 2,4,6-trinitrophenol (picric acid) in a biological aerated filter (BAF). Bioresour Technol 100:1922–1930

Shen J, Zhang J, Zuo Y, Wang L, Sun X, Li J, Han W, He R (2009b) Biodegradation of 2,4,6-trinitrophenol by *Rhodococcus* sp. isolated from a picric acid-contaminated soil. J Hazard Mater 163:1199–1206

Sidorov DG, Borzenkov IA, Ibatullin RR, Milekhina EI, Khramov IT, Belyaev SS, Ivanov MV (1997) A field experiment on decontamination of oil-polluted soil employing hydrocarbon-oxidizing microorganisms. Appl Biochem Microbiol 33:441–445

Sidorov DG, Borzenkov IA, Milekhina EI, Belyaev SS, Ivanov MV (1998) Microbial destruction of fuel oil in soil induced by the biological preparation Devoroil. Appl Biochem Microbiol 34:255–260

Sinha RK, Krishnan KP, Hatha AA, Rahiman M, Thresyamma DD, Kerkar S (2017) Diversity of retrievable heterotrophic bacteria in Kongsfjorden, an Arctic fjord. Braz J Microbiol 48:51–61

Sorkhoh NA, Ghannoum MA, Ibrahim AS, Stretton RJ, Radwan SS (1990) Crude oil and hydrocarbon-degrading strains of *Rhodococcus rhodochrous* isolated from soil and marine environments in Kuwait. Environ Pollut 65:1–17

Sorkhoh NA, Al-Hasan RH, Khanafer M, Radwan SS (1995) Establishment of oil-degrading bacteria associated with cyanobacteria in oil-polluted soil. J Appl Bacteriol 78:194–199

Sun G-D, Xu Y, Jin J-H, Zhong Z-P, Liu Y, Luo M, Liu Z-P (2012) Pilot scale ex-situ bioremediation of heavily PAHs-contaminated soil by indigenous microorganisms and bioaugmentation by a PAHs-degrading and bioemulsifier-producing strain. J Hazard Mater 233–234:72–78

Taheri HE, Hatamipour MS, Emtiazi G, Beheshti M (2008) Bioremediation of DSO contaminated soil. Proc Saf Environ Protect 86:208–212

Tajima T, Hayashida N, Matsumura R, Omura A, Nakashimada Y, Kato J (2012) Isolation and characterization of tetrahydrofuran-degrading *Rhodococcus aetherivorans* strain M8. Process Biochem 47:1665–1669

Taki H, Syutsubo K, Mattison RG, Harayama S (2004) Biodegradation of *o*-xylene in soil using bioaugmentation technology. Proceedings of the Second International Conference on Remediation of Contaminated Sediments, pp 279–283

Taki H, Syutsubo K, Mattison RG, Harayama S (2007) Identification and characterization of o-xylene-degrading *Rhodococcus spp.* which were dominant species in the remediation of o-xylene-contaminated soils. Biodegradation 18:17–26

Táncsics A, Benedek T, Szoboszlay S, Veres PG, Farkas M, Máthé I, Márialigeti K, Kukolya J, Lányi S, Kriszt B (2015) The detection and phylogenetic analysis of the alkane1-monooxygenase gene of members of the genus *Rhodococcus*. Syst Appl Microbiol 38:1–7

Tartakovsky B, Michotte A, Cadieux J-CA, Lau PCK, Hawari J, Guiot SR (2001) Degradation of Aroclor 1242 in a single-stage coupled anaerobic/aerobic bioreactor. Water Res 35:4323–4330

Thomassin-Lacroix EJM, Eriksson M, Reimer KJ, Mohn WW (2002) Biostimulation and bioaugmentation for on-site treatment of weathered diesel fuel in Arctic soil. Appl Microbiol Biotechnol 59:551–556

Travkin V, Baskunov BP, Golovlev EL, Boersma MG, Boeren S, Vervoort J, van Berkel WJH, Rietjens MCM, Golovleva LA (2002) Reductive deamination as a new step in the anaerobic microbial degradation of halogenated anilines. FEMS Microbiol Lett 209:307–312

Tresse O, Lorrain M-J, Rho D (2002) Population dynamics of free-floating and attached bacteria in a styrene-degrading biotrickling filter analyzed by denaturing gradient gel electrophoresis. Appl Microbiol Biotechnol 59:585–590

Uroz S, D'Angelo-Picard C, Carlier A, Elasri M, Sicot C, Petit A, Oger P, Faure D, Dessaux Y (2003) Novel bacteria degrading N-acylhomoserine lactones and their use as quenchers of quorum-sensing-regulated functions of plant-pathogenic bacteria. Microbiology 149:1981–1989

US EPA 2017 Superfund remedy report, 15th edn. EPA-542-R-17-001. Office of Superfund Remediation and Technology Innovation (OSRTI)

Van der Geize R, Dijkhuizen L (2004) Harnessing the catabolic diversity of rhodococci for environmental and biotechnological applications. Curr Opin Microbiol 7:255–261

Van Hamme JD, Singh A, Ward OP (2003) Recent advances in petroleum microbiology. Microbiol Mol Biol Rev 67:503–549

Van Veen JA, van Overbeek LS, van Elsas JD (1997) Fate and activity of microorganisms introduced into soil. Microbiol Mol Biol Rev 61:121–135

Vergara-Fernández A, Yánez D, Morales P, Scott F, Aroca G, Diaz-Robles L, Moreno-Casas P (2018) Biofiltration of benzo[α]pyrene, toluene and formaldehyde in air by a consortium of *Rhodococcus erythropolis* and *Fusarium solani*: effect of inlet loads, gas flow and temperature. Chem Eng J 332:702–710

Vinage P, von Rohr R (2003) Biological waste gas treatment with a modified rotating biological contactor. I. Control of biofilm growth and long-term performance. Bioprocess Biosyst Eng 26:69–74

Vogt C, Alfreidera A, Lorbeerb H, Hoffmanna D, Wuenschea L, Babel W (2004) Bioremediation of chlorobenzene-contaminated ground water in an in situ reactor mediated by hydrogen peroxide. J Contam Hydrol 68:121–141

Wagner-Döbler I, Bennasar A, Vancanneyt M, Strömpl C, Brümmer I, Eichner C, Grammel I, Moore ERB (1998) Microcosm enrichment of biphenyl-degrading microbial communities from soils and sediments. Appl Environ Microbiol 64:3014–3022

Warhurst AM, Fewson CA (1994) Biotransformations catalyzed by the genus *Rhodococcus*. Crit Rev Biotechnol 14:29–73

Watanabe K, Hamamura N (2003) Molecular and physiological approaches to understanding the ecology of pollutant degradation. Curr Opin Biotechnol 14:289–295

Wattiau P (2002) Microbial aspects in bioremediation of soils polluted by polyaromatic hydrocarbons. In: Agathos SN, Reineke W (eds) Biotechnol for the environment: strategy and fundamentals. Kluwer Academic Publishers, Netherlands, pp 69–89

Weerasekara NA, Choo K-H, Lee C-H (2016) Biofouling control: bacterial quorum quenching versus chlorination in membrane bioreactors. Water Res 103:293–301

Weidhaas JL, Schroeder ED, Chang DPY (2007) An aerobic sequencing batch reactor for 2,4,6-trinitrophenol (picric acid) biodegradation. Biotechnol Bioeng 97:1408–1414

Whyte LG, Hawari J, Zhou E, Bourbonniere L, Inniss WE, Greer CW (1998) Biodegradation of variable-chain-length alkanes at low temperatures by a psychrotrophic *Rhodococcus* sp. Appl Environ Microbiol 64:2578–2584

Whyte LG, Slagman SJ, Pietrantonio F, Bourbonniere L, Koval SF, Lawrence JR, Innis WE, Greer SW (1999) Physiological adaptations involved in alkane assimilation at a low temperature by *Rhodococcus* sp. strain Q15. Appl Environ Microbiol 65:2961–2968

Whyte LG, Goalen B, Hawari J, Labbé D, Greer CW, Nahir M (2001) Bioremediation treatability assessment of hydrocarbon-contaminated soils from Eureka, Nunavut. Cold Reg Sci Technol 32:121–132

Whyte LG, Schultz A, van Beilen JB, Luz AP, Pellizari V, Labbé D, Greer CW (2002a) Prevalence of alkane monooxygenase genes in Arctic and Antarctic hydrocarbon-contaminated and pristine. FEMS Microb Ecol 41:141–150

Whyte LG, Smits TH, Labbé D, Witholt B, Greer CW, van Beilen JB (2002b) Gene cloning and characterization of multiple alkane hydroxylase systems in *Rhodococcus* strains Q15 and NRRL B-16531. Appl Environ Microbiol 68:5933–5942

Xu P, Yu B, Li FL, Cai XF, Ma CQ (2006) Microbial degradation of sulfur, nitrogen and oxygen heterocycles. Trends Microbiol 14:397–404

Yaacob NS, Mohamad R, Ahmad SA, Abdullah H, Ibrahim AL, Ariff AB (2016) The influence of different modes of bioreactor operation on the efficiency of phenol degradation by *Rhodococcus* UKMP-5M. Rend Lincei 27:749–760. https://doi.org/10.1007/s12210-016-0567-x

Yang X, Xie F, Zhang G, Shi Y, Qian S (2008) Purification, characterization, and substrate specificity of two 2,3-dihydroxybiphenyl 1,2-dioxygenase from *Rhodococcus* sp. R04, showing their distinct stability at various temperature. Biochimie 90:1530–1538

Zagustina NA, Misharina TA, Veprizky AA, Zhukov VG, Ruzhitsky AO, Terenina MB, Krikunova NI, Kulikova AK, Popov VO (2012) Elimination of volatile compounds of leaf tobacco from air emissions using biofiltration. Appl Biochem Microbiol 48:385–395

Zhang K, Liu Y, Chen Q, Luo H, Zhu Z, Chen W, Chen J, Mo Y (2018) Biochemical pathways and enhanced degradation of di-n-octyl phthalate (DOP) in sequencing batch reactor (SBR) by *Arthrobacter* sp. SLG-4 and *Rhodococcus* sp. SLG-6 isolated from activated sludge. Biodegradation 29(2):171–185. https://doi.org/10.1007/s10532-018-9822-4

Production of Trehalolipid Biosurfactants by *Rhodococcus*

Maria S. Kuyukina and Irena B. Ivshina

Contents

Abstract Members of the genus *Rhodococcus* produce biosurfactants in response to the presence of liquid hydrocarbons in the growth medium. These biosurfactants are predominantly cell-bound glycolipids containing trehalose as the carbohydrate. Physiological roles of these glycolipids are diverse and involve participation in the uptake of water-insoluble substrates, promotion of the cell adhesion to hydrophobic surfaces, and increased rhodococcal resistance to physicochemical influences. In terms of surfactant characteristics (e.g., surface and interfacial tension, critical micelle concentration, emulsifying activity), *Rhodococcus* biosurfactants compete favorably with other microbial and synthetic surfactants. Additionally, biological activities of trehalolipids from rhodococci were revealed, including immunomodulating, antitumor, and anti-adhesive properties. Recently developed optimization procedures for their biosynthesis and recovery would broaden potential applications of *Rhodococcus* biosurfactants in new advanced technologies, such as environmental bioremediation, improved material construction, and biomedicine. The present chapter summarizes recent research on *Rhodococcus* biosurfactants and focuses on biosynthesis features, physicochemical and bioactive properties, and their potential applications.

M. S. Kuyukina (✉) · I. B. Ivshina
Institute of Ecology and Genetics of Microorganisms, Perm Scientific Centre of the Ural Branch of the Russian Academy of Sciences, Perm State University, Perm, Russia
e-mail: kuyukina@iegm.ru

H. M. Alvarez (ed.), *Biology of Rhodococcus*, Microbiology Monographs 16,
https://doi.org/10.1007/978-3-030-11461-9_10

1 Introduction

Biosurfactants, surface-active compounds of biological origin, are of increasing interest for many industries because of their chemical diversity, multifunctional characteristics, and low toxicity in comparison to synthetic, petrochemical-derived surfactants (Maier 2003). All biosurfactants are amphiphilic in nature, containing at least one hydrophilic (usually sugar or amino acid) moiety and one hydrophobic (usually fatty acid) moiety in the molecule. Owing to their amphiphilic nature, surfactants tend to accumulate at surfaces and interfaces and, as a result, reduce surface (liquid–air) and interfacial (liquid–liquid and liquid–solid) tension. When surfactant monomers are added into a solution, the surface or interfacial tension will decrease until the surfactant concentration reaches what is known as the critical micelle concentration (CMC). At the CMC, surfactant monomers begin to spontaneously associate into structured aggregates, called micelles. The CMC of a biosurfactant could be regarded as an efficiency measure and is dependent on the surfactant structure as well as on pH, ionic strength, and temperature of the solution.

Microbial biosurfactants are a structurally diverse group of surface-active molecules synthesized by various microorganisms, and they appear to play a role in many physiological processes occurring at interfaces, ranging from hydrophobic substrate uptake, cell adhesion, and surface colonization to microbial antagonism and pathogenesis (Neu 1996; Ron and Rosenberg 2001; Van Hamme et al. 2006). According to Maier (2003), biosurfactants of phylogenetically distant microorganisms are functionally convergent, thus indicating their significant role in the vital functions of biosurfactant producers. At the same time, biosurfactants synthesized by the bacteria of different species belonging to the same genus are often different in terms of their structure and function. Unlike chemically synthesized surfactants, which are classified according to the nature of their polar grouping, biosurfactants are categorized mainly by their chemical composition and their microbial origin. In general, the major classes of biosurfactants include glycolipids, lipopeptides and lipoproteins, phospholipids and fatty acids, polymeric surfactants, and particulate surfactants.

Fields of application or potential use of biosurfactants are diverse and include not only oil, cleaning, food, and cosmetic industries and environmental protection but also agriculture and medicine (Marchant and Banat 2012; Sachdev and Cameotra 2013). The global biosurfactant market is estimated to grow at a compound annual growth rate (CAGR) of 5.6% from 2017 to 2022. Based on application, the agricultural chemical segment is projected to grow at highest CAGR due to increasing demand for green surfactants to improve soil quality and promote plant growth and plant-microbe interactions (Biosurfactants market 2018).

The most promising applications of biosurfactants are the environmental remediation technologies since product purity is of less concern (Lang and Philp 1998; Christofi and Ivshina 2002; Banat et al. 2010). In recent years, biosurfactants traditionally used as emulsifying and solubilizing agents for hydrophobic substances are attracting an increasing attention as possible biomedical agents with expressed biological activities (Ryll et al. 2001; Kitamoto et al. 2002; Cameotra and Makkar 2004; Gudina et al. 2013; Kuyukina et al. 2015). It should be noted that, at present, the industrial use of biosurfactants is not generally competitive with synthetic surfactants because of their higher production cost. However, among biogenic surfactants, the products of microbial synthesis are the most promising candidates for biotechnological applications because of simple mineral media and available carbon sources used in fermentation processes as well as a shorter generation time compared to animal or plant growth. Thereby, possible optimization of a biosurfactant synthesis process would allow increasing the product yield at relatively low material and energy expenditures. The possibility of in situ production of biosurfactants is essential for the bioremediation of water and soil contaminated with organic pollutants and heavy metals (Christofi and Ivshina 2002).

In terms of applied aspects, glycolipid biosurfactants structurally represented by mono- and disaccharide-based complexes having ester bonds with fatty acids are being thoroughly studied due to high productivity from renewable resources (e.g., carbohydrate biomass or agro-industrial wastes) and versatile biochemical properties compared to other types of biosurfactants (Kitamoto et al. 2002; Lang 2002; Franzetti et al. 2010). Most well-known glycolipid biosurfactants include rhamnolipids from *Pseudomonas aeruginosa*, sophorolipids from *Candida bombicola*, cellobiose lipids from *Ustilago maydis*, mannosylerythritol lipids from *Pseudozyma antarctica*, and trehalolipids from *Rhodococcus erythropolis* (Kitamoto et al. 2002; Paulino et al. 2016). Members of *Rhodococcus* species possessing unique biological properties and wide catabolic abilities are perspective candidates for biosurfactant producers (Van der Geize and Dijkhuizen 2004; Ivshina et al. 2017).

2 Surfactant Production by *Rhodococcus* Species and Related Actinobacteria

The production of biosurfactants by rhodococci and related coryneform and nocardioform actinobacteria was studied intensively, but most of the research has been done on trehalolipid surfactants formed by *R. erythropolis* (Lang and Philp 1998; Franzetti et al. 2010), while other *Rhodococcus* species are also reported as active biosurfactant producers (Table 1). Comparing the data on surface-active properties of the presented *Rhodococcus* species (*R. erythropolis*, *R. fascians*, "*R. longus*," *R. opacus*, *R. ruber*, and *R. wratislaviensis*), it can be suggested that biosurfactant production is generalized throughout this genus. When grown on liquid alkanes, rhodococci produce surfactants capable of reducing the surface and interfacial tension of water to low values (Ivshina et al. 1998). Alkane growth of the most potent biosurfactant producers, *R. erythropolis* DSMZ 43215 and *R. ruber* IEGM 231, resulted in the biosurfactant yield of 32 and 10 g l^{-1}, respectively (Kim et al. 1990; Kuyukina et al. 2001). Bouchez-Naïtali et al. (1999) described several soil isolates belonging to *Rhodococcus equi* and

Table 1 Biosurfactant-producing *Rhodococcus* species and their product characteristics

Microorganism	Surfactant product	Surface tension (mNm^{-1})	Interfacial tension (mNm^{-1})	CMC (mgl^{-1}) or CMD*	References
R. erythropolis (14 strains)	Whole culture broth	27.1 (0.55)[a]	6.3 (3.57)	60*	Ivshina et al. (1998) and Philp et al. (2002)
R. erythropolis DSM 43215	Trehalose dicorynomycolates	36	17	4	Kim et al. (1990)
R. erythropolis DSM 43215	Trehalose monocorynomycolates	32	14	4	Kim et al. (1990)
R. erythropolis DSM 43215	Trehalose-2,2′,3,4-tetraester	26	<1	15	Kim et al. (1990)
R. erythropolis DSM 51T7	Trehalose tetraester	27.9	5	37	Marqués et al. (2009)
R. erythropolis MTCC 2794	Crude organic extract	33.8	n.d.	100	Pal et al. (2009)
R. erythropolis EK-1	Cell-free culture broth	30–39	n.d.	6*	Pirog et al. (2004)
R. erythropolis P6-4P	Cell-free culture broth	28	n.d.	10*	Cai et al. (2014)
R. erythropolis Z25	Crude organic extract	29.5	4.5	70	Xia et al. (2011)
R. fascians A3	Cell-free culture broth	27	n.d.	n.d.	Gesheva et al. (2010)
"*R. longus*" (3 strains)	Whole culture broth	27.2 (0.40)	1.8 (0.1)	90*	Ivshina et al. (1998) and Philp et al. (2002)
R. opacus (3 strains)	Whole culture broth	26.5 (0.61)	3.0 (1.75)	95*	Ivshina et al. (1998) and Philp et al. (2002)
R. ruber (15 strains)	Whole culture broth	27.4 (0.57)	2.7 (1.61)	72*	Ivshina et al. (1998) and Philp et al. (2002)
R. ruber IEGM 231	Crude organic extract	28.5–30.1	0.3–1.6	86–173	Kuyukina et al. (2001)
R. ruber IEGM 235	Glycolipid complex	26.8	0.9	54	Ivshina et al. (1998)

R. ruber Z25	Crude organic extract	29.5	1.0	57–133	Zheng et al. (2012)
R. wratislaviensis BN38	Whole culture broth	28.6	5.3	n.d.	Tuleva et al. (2008)
R. wratislaviensis BN38	Trehalose tetraester	24.4	1.3	5	Tuleva et al. (2008)
R. wratislaviensis P1-5P	Cell-free culture broth	28	n.d.	11*	Cai et al. (2014)
Rhodococcus sp. (6 strains)	Cell-free culture broth	27.3–28.6	n.d.	n.d.	Malavenda et al. (2015)
Rhodococcus sp. 51T7	Cell-free culture broth	30	n.d.	n.d.	Espuny et al. (1995)
Rhodococcus sp. BS-15	Tri-glucose lipid	29.5	n.d.	2.4 (2.3×10^{-6} M)	Konishi et al. (2014)
Rhodococcus sp. H13-A	Octaacyl-trehalose	n.d.	0.02	1.5	Singer and Finnerty (1990)
Rhodococcus sp. PML026	Trehalolipid	29	n.d.	250	White et al. (2013)
Rhodococcus sp. SD-74	Succinoyl trehalolipid	19	n.d.	5.6×10^{-6} M	Tokumoto et al. (2009)
Rhodococcus sp. TA6	Whole culture broth	29.8	n.d.	30*	Shavandi et al. (2011)

CMC critical micelle concentration, *CMD* critical micelle dilution, *n.d.* none determined

[a]Standard deviations

Corynebacterium spp., which were able to produce biosurfactants when grown on hexadecane. The *Rhodococcus rhodochrous* strain CF222 was reported to produce a surface-active complex of acidic polysaccharides and lipids, which increased bacterial tolerance to hexadecane (Iwabuchi et al. 2000). A cold-adapted Antarctic soil isolate *R. fascians* A3 was found to produce rhamnose-containing biosurfactant when grown on glucose and kerosene (Gesheva et al. 2010). Recently, biosurfactant production of alkanothrophic strains *Rhodococcus wratislaviensis* BN38 and P1-5P has been characterized (Tuleva et al. 2008; Cai et al. 2014). It should be noted that among 53 valid *Rhodococcus* species, only 6 species (and 1 invalid species—"*R. longus*") are reported to produce biosurfactants so far. However, members of these species (*R. equi*, *R. erythropolis*, *R. opacus*, *R. rhodochrous*, *R. ruber*, and *R. wratislaviensis*) are the most common environmental isolates, and they represent the major portion of *Rhodococcus* holdings in culture collections. Nevertheless, many biosurfactant-producing *Rhodococcus* sp. strains are still waiting for species identification. Biosurfactant production is also described in *Rhodococcus*-related genera, namely, *Nocardia* (Kim et al. 2000), *Gordonia* (Arenskötter et al. 2004), *Dietzia* (Kavyanifard et al. 2016), *Corynebacterium* (Cooper et al. 1979), *Mycobacterium* (Abdelhay et al. 2009), *Tsukamurella* (Choi et al. 1999), and *Arthrobacter* (Passeri et al. 1991).

3 Structures and Physicochemical Properties

It is now assumed that the type of a biosurfactant produced is dictated by the producing microorganism. Trehalolipids are produced by members of closely related actinobacterial genera, including *Rhodococcus*, *Nocardia*, *Gordonia*, *Dietzia*, *Corynebacterium*, *Mycobacterium*, *Tsukamurella*, and *Arthrobacter*. These compounds include α,ά-D-trehalose, a nonreducing disaccharide, which is linked by an ester bond to long-chain fatty acids. Among the latter *α*-branched β-hydroxy acids, so-called mycolic acids play a preferential role (Lang and Philp 1998). Trehalolipids from rhodococci are characterized by high structural diversity, and they often occur as a complex mixture, the composition of which varies depending on strain physiology and growth conditions. Trehalose monomycolates (Batrakov et al. 1981; Kretschmer and Wagner 1983), trehalose dimycolates (Rapp et al. 1979; Batrakov et al. 1981; Philp et al. 2002; Niescher et al. 2006), trehalose trimycolates (Tomiyasu et al. 1986), as well as mono-, di-, tetra-, hexa-, and octa-acylated derivatives of trehalose (Kretschmer and Wagner 1983; Singer and Finnerty 1990; Philp et al. 2002) represent the nonionic trehalolipid type (Fig. 1a). Trehalose tetraesters (Ristau and Wagner 1983; Espuny et al. 1995; Rapp and Gabriel-Jürgens 2003; Tuleva et al. 2008; Marqués et al. 2009) and succinoyl trehalolipids (Tokumoto et al. 2009), which additionally harbor one or two succinoyl residues, represent the anionic trehalolipid type (Fig. 1b). Other biosurfactant types were described in rhodococci, such as glucolipids (Kurane et al. 1995; Peng et al. 2007; Konishi et al. 2014) and polysaccharides (Iwabuchi et al. 2000), and the existence of additional chemical structures could be expected.

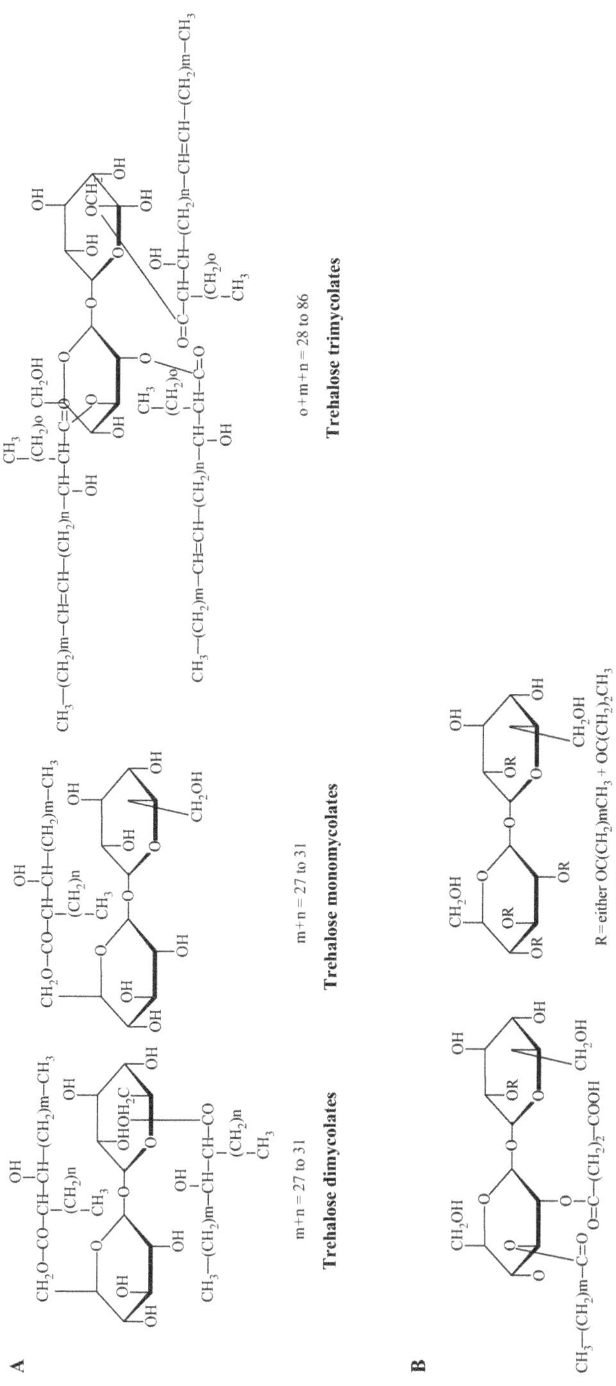

Fig. 1 Chemical structures of mostly known trehalolipids from *Rhodococcus*. **(a)** Nonionic and **(b)** anionic trehalolipids

Unlike synthetic surfactants having relatively simple chemical structure, biosurfactants are usually quite complex molecules. So, physicochemical properties of biosurfactants are more complex and "flexible" compared to synthetic surfactants. Evaluating the properties of *Rhodococcus* biosurfactants, it was observed that they lowered the surface tension of water from 72 mN m^{-1} to values between 19 and 43 mN m^{-1} and the interfacial tension of water/*n*-hexadecane system from 43 mN m^{-1} to values between 0.02 and 15 mN m^{-1} (Ivshina et al. 1998; Philp et al. 2002; Marqués et al. 2009; Tokumoto et al. 2009; Zheng et al. 2012; Konishi et al. 2014). Considering surface tension, interfacial tension, and CMC values, *Rhodococcus* biosurfactants compare well with other microbial and some synthetic surfactants. Another essential feature of biosurfactants is their emulsifying activity, i.e., the ability to emulsify various hydrocarbons and oils. High emulsion indices with *n*-hexadecane were recorded for biosurfactants produced by *R. erythropolis*, "*R. longus*," *R. opacus*, and *R. ruber* with a maximum E_{24} of 62.5% (Ivshina et al. 1998). Similar E_{24} values ranging from 20% to 60% were registered by Bicca et al. (1999) for the *R. ruber* biosurfactant with *n*-alkanes (pentane, hexane, heptane), aromatic hydrocarbons (benzene, toluene, xylene), and petroleum fractions (diesel, petrol, kerosene). Tuleva et al. (2008) showed high emulsification activities (E_{24} = 23–69%) of the purified trehalose tetraester from *R. wratislaviensis* BN38 toward various water-immiscible substrates including *n*-alkanes, toluene, benzene, xylene, kerosene, crude oil, mineral oils, sunflower oil, and almond oil. Cold-adapted *Rhodococcus* sp. strains isolated from Arctic and Antarctic shoreline sediments produced stable emulsions (E_{24} = 50–67%) with kerosene (Malavenda et al. 2015). A *Rhodococcus* sp. strain TA6 emulsified efficiently (E_{24} = 50–70%) kerosene, gas, and motor oils but not *n*-pentane, toluene, or light crude oil, suggesting its specificity toward long-chain hydrocarbons (Shavandi et al. 2011). Hydrophilic–lipophilic balance (HLB) of biosurfactants, a relative hydrophobicity measure, is the important matter for their industrial applications. We calculated the HLB value of 8 from the chemical structure of purified trehalolipids produced by *R. ruber* IEGM 231, while the corresponding HLB of 6.4–7.5 was determined for the crude biosurfactant from this strain (Kuyukina et al. 2006). These HLB values, which are relatively low when compared to other bacterial and synthetic surfactants (Noordman et al. 2002), indicate prevailing hydrophobic properties of the *R. ruber* biosurfactant and confirm its potential as an oil-in-water emulsifier. Trehalolipids from *R. erythropolis* 51T7 with the HLB value of 11 were shown to produce stable emulsions with water and paraffin or isopropyl myristate (Marqués et al. 2009). Thermal, osmotic, and pH stabilities of *Rhodococcus* biosurfactants were also reported (Kuyukina et al. 2006; Marqués et al. 2009; Shavandi et al. 2011; Kundu et al. 2013; White et al. 2013). However, there is still very little information available on physicochemical properties of *Rhodococcus* biosurfactants, thereby hindering their introduction to industrial fields.

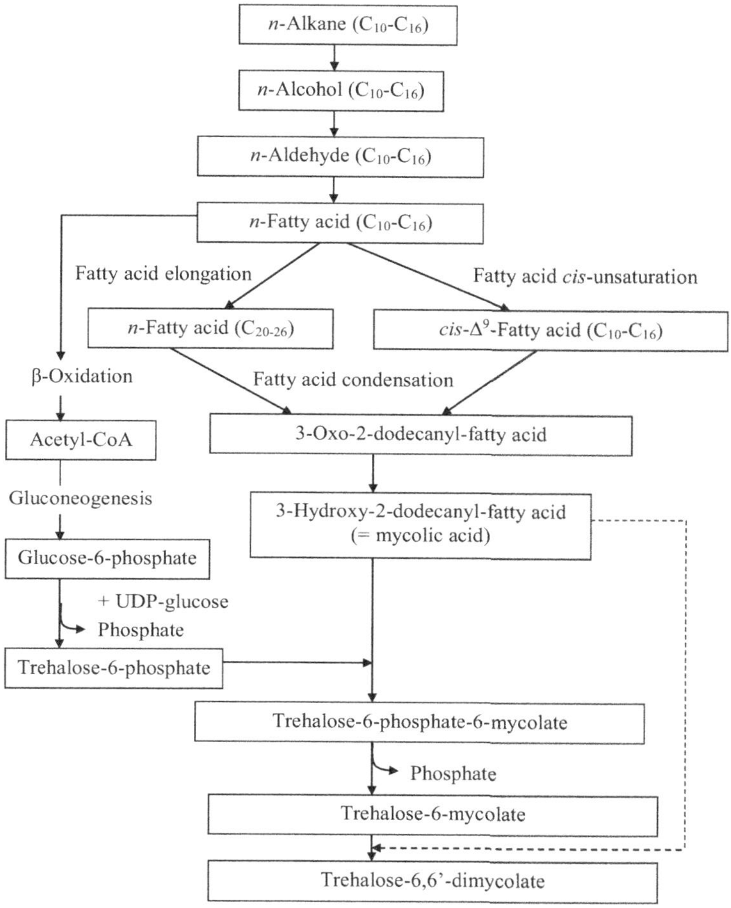

Fig. 2 Scheme of trehalose mono- and dimycolate synthesis from *n*-alkanes. (Modified from Lang and Philp 1998)

4 Biosynthesis and Recovery

Metabolic pathways involved in trehalolipid synthesis in *Rhodococcus* are reviewed by Lang and Philp (1998). It seems that the trehalose moiety and the fatty (mycolic) acid moiety of trehalolipid molecules are synthesized independently and are subsequently etherified (Fig. 2). Trehalose monomycolate was postulated to be an intermediate of trehalose dimycolate biosynthesis (Kretschmer and Wagner 1983). The formation of the mycolate is considered to be a Claisen-type condensation of two fatty acids, a carboxylated acyl-coenzyme A and an activated acyl chain, to yield a 3-oxo intermediate, which would then be reduced to form mycolic acid (Lang and Philp 1998). For the formation of the various mycolates, first the long-chain fatty acid constituting the major part of the molecule is synthesized before condensation with the second fatty acid. Fatty acids are synthesized in many metabolic pathways, for example, via the successive oxidation of alkane substrates to alcohols, aldehydes, and carbonic acids followed by carbon chain elongation (for review, see Alvarez and Steinbüchel, this volume). Intermediates of the fatty acid ß-oxidation could also be involved in the trehalolipid synthesis. The synthesis of the final resulting sugar residue, trehalose-6-phosphate, from UDP-glucose and glucose-6-phosphate proceeds apparently via the OtsA–OtsB pathway described for coryneform actinobacteria (Tzvetkov et al. 2003), although alternative trehalose biosynthesis pathways (TreY-TreZ pathway and TreS pathway) could also be involved (De Smet et al. 2000). *n*-Alkane substrates are first degraded to acetyl-CoA before glucose synthesis (see Fig. 2). Although the key pathways and enzymes of fatty (mycolic) acid and trehalose syntheses are described in *Rhodococcus* and related bacteria, only little is known about the final step of trehalolipid synthesis, trehalose acylation. This reaction is probably catalyzed by acyltransferases specific to particular fatty acid types. For example, Belisle et al. (1997) described mycolyltransferases responsible for the transfer of mycolic acids to α,ά-trehalose to form α,ά-trehalose monomycolate and α,ά-trehalose dimycolate in *Mycobacterium tuberculosis*. Similar enzymes and corresponding genes were later described in other mycobacteria, corynebacteria (Kacem et al. 2004), and rhodococci (Sydor et al. 2008; Ivshina et al. 2014). Three novel types of genes involved in the biosynthesis of succinoyl trehalolipids, including a putative acyl-coenzyme A (acyl-CoA) transferase (*tls*A), fructose-bisphosphate aldolase (*fda*), and alkane monooxygenase (*alk*B), were identified in *Rhodococcus* sp. SD-74. An overexpression of *tls*A, but not *fda*, resulted in twofold increase in the glycolipid production, thus suggesting that this final conversion of acyl-CoA, succinate, and trehalose is the rate-limiting step of succinoyl trehalolipid biosynthesis (Inaba et al. 2013). However, precise biochemical mechanisms of acylation leading to the formation of anionic trehalose tetraesters remain unknown.

More research was performed on physiological aspects of biosurfactant production by *Rhodococcus* aimed to optimize growth conditions for surfactant overproduction. Rhodococci were reported to synthesize trehalolipids in growth-associated, growth-limited, and resting cell conditions (Kim et al. 1990; Philp et al. 2002; Tuleva et al. 2008; Haddadin et al. 2009). It is assumed that biosurfactants cannot be produced in considerable amounts when *Rhodococcus* cells are grown with water-soluble substrates. We tested various hydrocarbons as the substrates for *R. ruber* biosurfactant production (Philp et al. 2002) and found that short-chain *n*-alkanes ranging from hexane to nonane as sole carbon sources failed to support the growth of *R. ruber* IEGM 231 (Table 2). Decane appeared to allow limited growth, but no biosurfactant was produced. The limited growth observed with dec-1-ene was visually identical to the growth on decane. Only from undecane upward did the organism produce significant amounts of biomass and biosurfactant. The surfactant production of *R. ruber* increased with the increase in *n*-alkane chain length. Thus, during growth on undecane and dodecane, the surfactant concentrations were minimal, but increased from 1.8–2.2 g l^{-1} on tridecane and tetradecane to 4.0 g l^{-1} on pentadecane, reaching

Table 2 Growth, surface-active properties, and surfactant production of *R. ruber* IEGM 231 on different hydrocarbons

Growth substrate	Biomass, gl^{-1}	Surface tension, mNm^{-1}	Surfactant concentration, gl^{-1}
Hexane	–	63.1	–
Cyclohexane	–	65.9	–
Heptane	0.68	67.9	–
Octane	–	56.2	–
Nonane	0.34	56.7	–
Decane	1.24	56.5	–
Dec-1-ene	1.00	55.9	–
Undecane	2.15	26.9	0.18
Dodecane	4.53	29.7	0.59
Tridecane	4.93	32.1	2.23
Tetradecane	5.67	31.8	1.79
Pentadecane	4.53	31.2	4.01
Hexadecane	4.80	29.6	9.90

From Philp et al. (2002)

the maximal level (9.9 g l^{-1}) on hexadecane. These results are consistent with the data on biosurfactant production by *R. wratislaviensis* BN38, which also started from undecane and reached the maximal level of 3.1 g l^{-1} on hexadecane (Tuleva et al. 2008). Among four carbon sources tested (diesel, naphthalene, crude oil, benzene), the optimal biosurfactant production of *R. erythropolis* and *R. ruber* strains was obtained using naphthalene and diesel (Haddadin et al. 2009). Since the acyl group composition of *Rhodococcus* trehalolipids is affected by the *n*-alkane used as the substrate, this opens possibilities to obtain specifically defined derivatives for commercial applications (Inaba et al. 2013).

Kinetics of biosurfactant production by rhodococci grown with hydrocarbons was studied thoroughly in numerous batch and continuous culture experiments. Nonionic glycolipid production by *R. erythropolis* and *R. ruber* under non-limiting conditions correlated with biomass formation and *n*-hexadecane consumption (Figs. 3 and 4), thus suggesting its growth-associated mode (Rapp et al. 1979; Philp et al. 2002; Haddadin et al. 2009). Nitrogen limitation and temperature shift (from 30 to 22 °C) were shown to favor the formation of anionic trehalose tetraesters (up to 32 g l^{-1}) during growth of *R. erythropolis* DSM 43215 on hydrocarbons (Ristau and Wagner 1983; Kim et al. 1990), while *R. ruber* and *R. opacus* synthesized only nonionic trehalolipids under these conditions (Philp et al. 2002; Niescher et al. 2006). Since the overproduction began at a low nitrogen concentration, the complete separation of growth and production phase was attempted. Resting cell experiments with *R. erythropolis* DSM 43215 resulted in the efficient conversion of the technical-grade C10 *n*-alkane to more than 20 g l^{-1} of trehalolipids (Kim et al. 1990). However, in the study of Peng et al. (2007), resting *R. erythropolis* 3C-9 cells incubated in a medium containing hexadecane, but lacking a nitrogen source, were not able to produce biosurfactants.

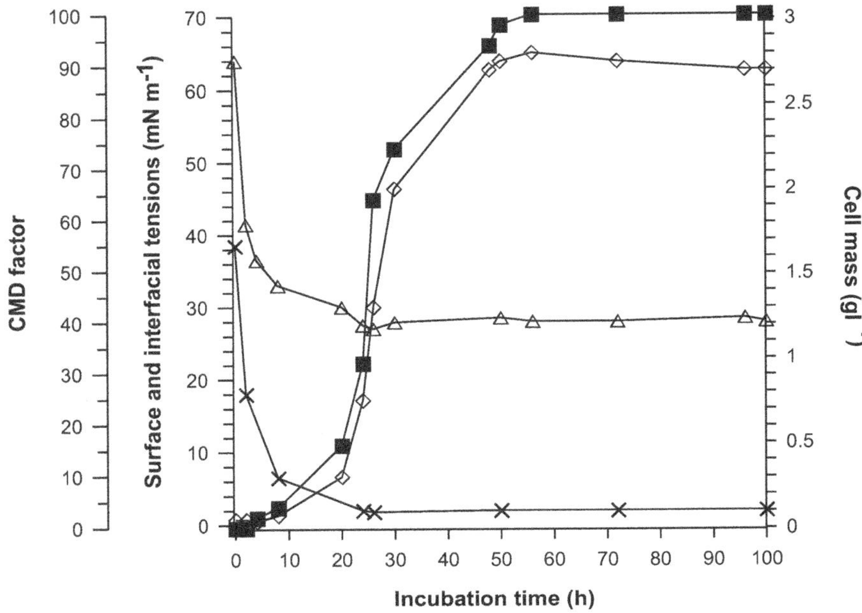

Fig. 3 Surfactant properties of *Rhodococcus ruber* IEGM 231 during batch culture growth on *n*-hexadecane. ■, CMD factor; Δ, surface tension; x, interfacial tension; ◇, cell mass. (From Philp et al. 2002)

Importantly, cheap renewable substrates, such as technical glycerol (waste product of biodiesel production) and sunflower and rapeseed oils, were tested successfully for the biosurfactant synthesis by *Rhodococcus*, thus suggesting sustainable alternatives to petroleum-derived hydrocarbons (Ruggeri et al. 2009; Pirog et al. 2015; White et al. 2013; Malavenda et al. 2015).

Mathematical modeling was applied to enhance the biosurfactant production of *R. erythropolis* MTCC 2794 (Pal et al. 2009). Two optimization techniques, such as the artificial neural network (ANN) coupled with genetic algorithm (GA) and the response surface methodology (RSM), were used for the optimization of medium components (sucrose, yeast extract, meat peptone, and toluene), resulting in the increased biosurfactant yield.

The recovery and concentration of biosurfactants from the fermentation broth largely determine their production cost. Often, low concentration and the amphiphilic nature of microbial surfactants limit their recovery (Desai and Banat 1997). Various techniques used for biosurfactant isolation include high-speed centrifugation, dia- and ultrafiltration, acid and salt precipitation, solvent extraction, and adsorption chromatography (Bryant 1990; Desai and Banat 1997). However, surfactants produced by rhodococci under unrestricted growth conditions are predominantly cell-associated trehalolipids (Lang and Philp 1998), which can be effectively isolated only by the organic solvent extraction. A wide variety of organic solvents, for example, methanol,

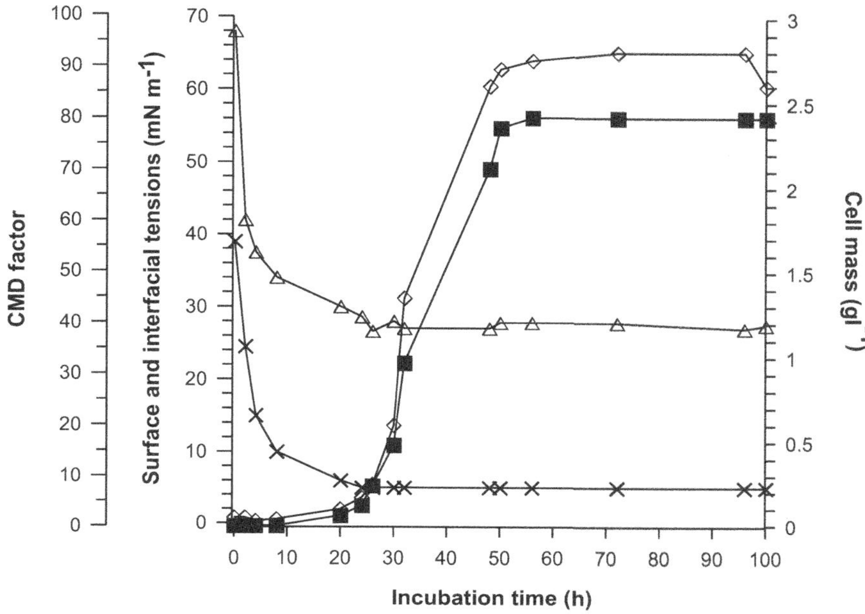

Fig. 4 Surfactant properties of *Rhodococcus erythropolis* IEGM 20 during batch culture growth on *n*-hexadecane. ■, CMD factor; Δ, surface tension; x, interfacial tension; ◇, cell mass. (From Philp et al. 2002)

ethanol, diethyl ether, pentane, acetone, chloroform, and dichloromethane, have been used, either singly or in combination, for biosurfactant extraction (Desai and Banat 1997). We proposed methyl tertiary-butyl ether (MTBE), a less toxic and inexpensive solvent with relatively low polarity for the extraction of biosurfactants from *Rhodococcus* cultures (Kuyukina et al. 2001). The combination of extraction procedure with an ultrasonic treatment of the extraction mixture resulted in a good trehalolipid recovery (Table 3).

5 Physiological Roles and Biological Activity

Rhodococci are able to utilize aliphatic and aromatic hydrocarbons of extremely low water solubility as carbon and energy sources. The role of biosurfactants is related to the low water solubility of *n*-alkanes as growth substrates and is determined by their ability to reduce interfacial tension between hydrocarbons and an aqueous phase (Ron and Rosenberg 2001). Although many bacteria can assimilate hydrophobic substrates in solubilized or emulsified forms, the hydrocarbon uptake by *Rhodococcus* occurs via the direct cell contact with large oil drops (Lang and Philp 1998). Production of cell-bound biosurfactants, together with an increase in cell surface hydrophobicity, and

Table 3 Yield and surface-active properties of biosurfactants from *R. ruber* IEGM 231 extracted using different solvent systems

Solvent system	Surfactant concentration, gl^{-1}	Surface tension, mNm^{-1}	Interfacial tension, mNm^{-1}	CMC, mgl^{-1}
MTBE	10.1	29.2	0.9	173
MTBE[a]	8.6	30.1	1.5	135
MTBE[b]	9.1	29.5	1.5	444
CH_2Cl_2	9.4	35.0	0.5	180
CH_2Cl_2[a]	8.1	29.9	1.4	86
CH_2Cl_2[b]	9.3	32.4	1.3	463
$CHCl_3$:CH_3OH (1:2)	9.8	28.9	1.0	171
$CHCl_3$:CH_3OH (1:2)[a]	9.6	30.0	1.6	90
$CHCl_3$:CH_3OH (1:1)[b]	12.2	30.9	2.2	119
$CHCl_3$:CH_3OH (2:1)	10.7	28.5	0.3	97
MTBE:$CHCl_3$ (1:1)	10.1	29.2	1.2	140
MTBE:$CHCl_3$ (1:1)[b]	10.7	30.3	7.6	320

MTBE tertiary-butyl ether, *CH_2Cl_2* methylene chloride, *$CHCl_3$* chloroform, *CH_3OH* methanol
[a]Extraction was performed by ultrasonic treatment (23 kHz, 10 min)
[b]Extraction was performed from the whole culture broth. (From Kuyukina et al. 2001)

formation of intracellular inclusions and extracellular polymers were considered as specific adaptive mechanisms of alkane assimilation by *Rhodococcus* cells, especially at a low temperature when the substrate bioavailability is extremely low (Whyte et al. 1999; Ivshina et al. 2017). Cell-associated biosurfactants promote the adhesion of rhodococcal cells not only to liquid hydrocarbons but also to hydrophobic solid surfaces (Neu 1996; Ivshina et al. 2013b), allowing effective cell colonization and direct uptake from sorbed/crystalline hydrocarbons (Whyte et al. 1999). A possible mechanism of biosurfactant-mediated cell detachment from used oil droplets proposed by Rosenberg and Ron (1999) for the *Acinetobacter calcoaceticus* strain RAG-1 could also act in rhodococcal cells.

Trehalose mycolates along with other cell wall lipids are involved in cellular tolerance to antibiotics and organic solvents (Kuyukina et al. 2000; Sokolovska et al. 2003; Nguyen et al. 2005; de Carvalho, this volume), as well as to physical factors, for example, high temperature and desiccation (Sung et al. 2004). We studied the protective effect of a trehalolipid biosurfactant on propane-oxidizing *R. ruber* cells during a long-term storage of freeze-dried cultures (Kamenskikh et al. 2004) and observed the two to threefold viability increase, when cell suspensions were added with 10% biosurfactant, 1.5% gelatin, and 0.1% agar prior to freeze-drying compared to the commonly used cryoprotectant (10% trehalose, 1.5% gelatin, and 0.1% agar). Interestingly, no viability increase of *R. ruber* cells treated with trehalose was registered, suggesting that although trehalose was reported among endogenous osmolytes protecting rhodococcal cells from the desiccation stress (Alvarez et al. 2004), the acylated trehalose would be a preferable form of the exogenous cell protectant. LeBlanc et al. (2008) performed transcriptomic analyses of air-dried *Rhodococcus jostii* RHA1 to identify genes involved in the regulatory response to desiccation. Among desiccation-specific up-regulated genes, 23 genes of lipid metabolism and cell envelope modification were revealed as well as the α,α-trehalose-phosphate synthase gene responsible for the synthesis of trehalose moiety of trehalolipids. In the study of Sung et al. (2004), genes of trehalose mycolyltransferases (Ag85B and alpha antigen) involved in the trehalolipid formation were strongly expressed in the heat stress-exposed cells and thus were associated with the heat *resistance of Mycobacterium paratuberculosis*.

There are considerable numbers of studies showing that biosurfactants play a significant role in fundamental biological processes, such as bacterial cell signaling and quorum sensing, biofilm formation and cellular differentiation, and bacterial pathogenesis and antagonism (for reviews, see Peypoux et al. 1999; Kitamoto et al. 2002; Lang 2002; Cameotra and Makkar 2004; Van Hamme et al. 2006). Although glycolipids from *Rhodococcus* and related bacteria showed no antagonistic activity against Gram-negative bacteria and yeasts (for review, see Kitamoto et al. 2002), trehalolipids from *R. erythropolis* DSMZ 43215 inhibited conidia germination of the fungus *Glomerella cingulata* (Kitamoto et al. 2002). Trehalose dimycolate from *M. tuberculosis* potentiated non-specific resistance in mice to influenza virus infection (Hoq et al. 1997).

Many biological activities of trehalolipids from *Rhodococcus* and related bacteria are determined by their amphiphilic nature and resulted from the interaction with cellular membranes (Ortiz et al. 2009). Thus, Retzinger et al. (1981) hypothesized that biological responses evoked by trehalose dimycolate emulsions do not involve any recognition of specific chemical structures of the glycolipid, but are determined solely by the geometry and behavior of trehalose dimycolate at water-hydrophobic interfaces. The studies of Aranda et al. (2007) and Ortiz et al. (2008, 2009) showed that *Rhodococcus* trehalolipid interactions with cellular membranes result in the altered structural and rheological membrane properties, which might contribute to the molecular mechanisms of trehalolipid biological activity. More recent studies of the same group (Zaragoza et al. 2013; Teruel et al. 2014) revealed particular mechanisms of trehalolipid interactions with membrane phospholipids and enzymes, demonstrating their destabilizing and inhibitory effects on enzyme activities.

It should be noted that cell wall trehalose di- and monomycolates of pathogenic (*M. tuberculosis*, *Corynebacterium diphtheriae*) and opportunistic (*Mycobacterium avium/Mycobacterium intracellulare* group, *Nocardia asteroides*, *Corynebacterium matruchotii*, *Corynebacterium xerosis*) mycolata play a key role in the pathogenesis of infections caused by these actinobacteria, and they are characterized by high immunomodulating activity. Particularly, they stimulate innate, early adaptive, and both humoral and cellular adaptive immunity by inducing the production of cytokines (IL-12, IFN-γ, TNF-α, IL-4, IL-6, IL-10) and chemokines (MCP-1, MIP-1α, IL-8) (for review, see Ryll et al. 2001; Kuyukina et al. 2015). Biological activity of

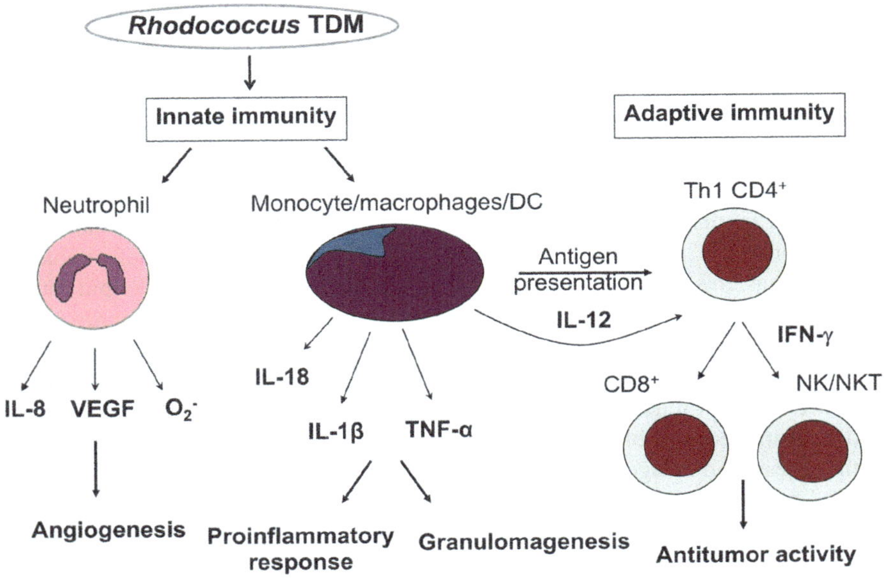

Fig. 5 Immunomoregulating properties of *Rhodococcus* trehalolipids. *DC* dendritic cell, *IL-1b* interleukin-1 beta, *IL-18* interleukin-18, *TNF-a* tumor necrosis factor alpha, *IL-12* interleukin-12, *IL-8* interleukin-8, *NK* natural killer cell, *NKT* natural killer T cell, *IFN-g* interferon gamma, *VEGF* vascular endothelial growth factor, *Th1* type 1 T-helper, O_2^- superoxide anion. (From Kuyukina et al. 2015)

these trehalolipids is significant; however, the apparent or potential pathogenicity of the producer strains and high toxicity of produced glycolipids limit their biomedical applications (Sakaguchi et al. 2000). Therefore, the search for trehalolipid producers among nonpathogenic actinobacteria is of importance. In our experiments, a trehalolipid biosurfactant from *R. ruber* IEGM 231 consisting of trehalose dimycolate, diacyltrehalose, and monoacyltrehalose stimulated interleukin-1β (IL-1β), IL-6, and tumor necrosis factor-α (TNF-α) production of human monocytes (Kuyukina et al. 2007; Gein et al. 2011). At the same time, it stimulated the production of anti-inflammatory cytokines IL-12 and IL-18 by monocytes and mononuclear cells, as well as the secretion of IL-8 by neutrophils (Chereshnev et al. 2010; Baeva et al. 2014). Interestingly, TLB exhibited cytokine-stimulating activity only when applied as an ultrasonic o/w emulsion, but not as an oleophilic film, presumably due to its inhibitory effect on the monocyte adhesion (Gein et al. 2011). Additionally, a trehalolipid from *R. ruber* displayed no cytotoxicity against human lymphocytes and therefore could be proposed as a potential immunomodulating and antitumor agent (Fig. 5) (Kuyukina et al. 2015).

6 Industrial Potential

Current and potential application fields for microbial surfactants, including oil, pharmaceutical, food, and cosmetic industries as well as agriculture and environmental technologies, are reviewed by Christofi and Ivshina (2002), Kitamoto et al. (2002), Lang (2002), Mulligan (2005), Singh et al. (2007), Banat et al. (2010), Marchant and Banat (2012), Sachdev and Cameotra (2013), and Paulino et al. (2016). Biosurfactants can be used as emulsifiers, de-emulsifiers, wetting and foaming agents, functional food ingredients, agriculture chemicals, detergents, and flocculating agents. In particular, biosurfactants from cold-adapted microorganisms, active at low and freezing temperatures, were reviewed recently as promising candidates for sustainable and low-energy impact products and processes, such as gas hydrate technologies, cold-active detergents, and biodiesel (Perfumo et al. (2018). However, despite numerous advantages of biosurfactants, namely, functional stability, low toxicity, and environmental safety, their proportion in the industrial surfactant market, currently occupied by synthetic, mainly petrochemical surfactants, does not exceed a few percent; the major factor limiting the wide use of biosurfactants is their production cost. The largest application field for biosurfactants is the oil industry, including microbial enhanced oil recovery (MEOR), improvement of technological characteristics of refinery products, and cleanup of oil contaminations (Singh et al. 2007; Banat et al. 2010; Marchant and Banat 2012). Enhanced oil recovery methods utilizing microorganisms and/or their metabolites for the residual oil extraction were developed in the late 1970s, when Rapp et al. (1977) reported 30% increase in oil recovery from sandy rocks using trehalolipids from *Rhodococcus* (former *Nocardia*) *rhodochrous*. However, since then, a constantly increasing global oil production has been associated with the growing environmental contamination by crude oil and refinery products.

6.1 Environmental Applications

In natural conditions, oil penetration through soil is an extremely complex process involving physical, chemical, and biological factors. Crude oil is a highly hydrophobic material with most of its components having low water solubility. These components bind to soil particles and become non-bioavailable to microorganisms. To increase the bioavailability of hydrocarbon pollutants, surfactants may be used, allowing desorption and solubilization of petroleum hydrocarbons and thus facilitating their assimilation by microbial cells. There are two mechanisms of surfactant-enhanced soil washing. One occurs below the CMC value, when surfactant monomers increase the contact angle between the soil and hydrophobic contaminant, thereby promoting the separation of contaminant from soil particles and finally displacing oil from the soil (roll-up mechanism). The other mechanism, solubilization, occurs above the CMC, when contaminants are partitioned from the soil into the hydrophobic core of surfactant micelles (Deshpande et al. 1999). Micellar phase bioavailability of hydrophobic organics means that contaminants partitioned into the micellar phase are biodegradable without being transferred to the dissolved phase first (Deshpande et al. 1999). Solubilization using surfactants at concentrations above their CMC values is currently explored in soil remediation (Mulligan 2005; Paria 2008). However, the inhibition of contaminant biodegradation in soil systems at surfactant concentrations above CMC has been reported (Billingsley et al. 2002). Moreover, many commonly used synthetic surfactants derived from petrochemicals are toxic and poorly biodegradable; their application may lead to the accumulation of ecologically harmful compounds in soil (Mulligan 2005). In recent decades, many studies have shown a great potential of biosurfactants for desorption and solubilization of hydrophobic organic contaminants in soil and their subsequent biodegradation (for reviews, see Christofi and Ivshina 2002; Ron and Rosenberg 2002; Makkar and Rockne 2003; Mulligan 2005; Singh et al. 2007; Banat et al. 2010). In the use of biosurfactants, there is less need for product purity, and benefits are likely from in situ production by indigenous or introduced microbial populations (Christofi and Ivshina 2002).

Kanga et al. (1997) compared the effect of (bio)surfactants on solubilization of naphthalene and methyl-substituted naphthalenes and found that the synthetic surfactant (Tween 80) has a lower solubilization potential and higher toxicity than the glycolipid biosurfactant from *Rhodococcus* sp. H13-A. This result was later confirmed by Page et al. (1999) showing increased biosurfactant-mediated solubilization of three- and four-ring polyaromatic constituents of crude oil. Greater hydrocarbon solubilization was suggested to attribute to the larger micellar volume of the glycolipid biosurfactant (Kanga et al. 1997). Biosurfactant produced by various *Rhodococcus* species was found to partition high amount (up to 99%) of crude oil from the sand slurries into the aqueous phase under shaking conditions (Ivshina et al. 1998). Moreover, oil-removal activity of the *R. ruber* IEGM 231 biosurfactant tested in model soil-packed columns heavily contaminated with crude oil was 1.9–2.3 times greater than that of a synthetic surfactant Tween 60 (Kuyukina et al. 2005). The biosurfactant produced by *R. ruber* grown on dodecane was most effective for the oil removal from contaminated soil in colder conditions (at 15 °C), but at higher temperatures (22 °C and 28 °C), the biosurfactant from hexadecane-grown cells was most effective (Fig. 6). Importantly, the glycolipid complex of *R. ruber* was less toxic than the purified biosurfactants of other bacterial strains, synthetic surfactants, and bioremediation formulations, such as Inipol EAP 22 (oleophilic fertilizer that is a microemulsion of urea and synthetic surfactants used to treat oil-contaminated soil) (Table 4). Considering their low toxicity and biodegradability, crude glycolipids from *R. ruber* can be used in soil washing technologies for simultaneous removal of PAHs and sulfur heterocycles (PASHs) (Ivshina et al. 2016).

Pacwa-Płociniczak et al. (2016) revealed plant growth-promoting traits (1-aminocyclopropane-1-carboxylic acid deaminase activity, phytohormone and siderophore production, and phosphate solubilization) coupled with biosurfactant production in two *R. erythropolis* strains (CD 106 and CD 111) isolated from soil heavily contaminated with petroleum hydrocarbons, suggesting their applications in microbe-assisted phytoremediation. However, despite of the promising laboratory results, only limited field studies have been performed so far on application of

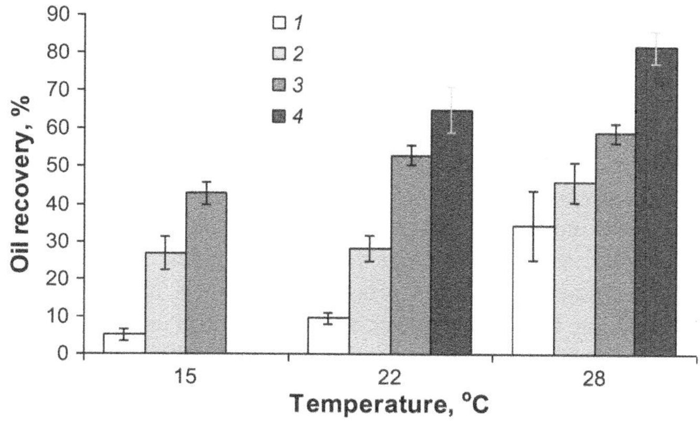

Fig. 6 (Bio)surfactant-enhanced oil recovery from the model soil at different temperatures. Surfactants used: 1, water (control); 2, Tween 60 (synthetic surfactant); 3, *Rhodococcus* biosurfactant produced on *n*-dodecane; 4, *Rhodococcus* biosurfactant produced on *n*-hexadecane. (From Kuyukina et al. 2005)

Table 4 The toxicity of biosurfactants compared to synthetic surfactants

Surfactant	IC_{50} of *Vibrio fischeri* (mgl^{-1})
Glycolipid complex from *R. ruber* IEGM 235	650
Trehalose dicorynomycolate from *R. erythropolis*	49
Trehalose tetraester from *R. erythropolis*	286
Rhamnolipids from *P. aeruginosa*	50
Nonylphenol-(ethylenoxide)$_9$-acetate (EQ 9)	78
Sucrose stearate (DK 50)	67
Finasol OSR-5	7
Corexit 9597	5
Inipol EAP 22	0.4

The toxicity results were calculated as IC_{50}—inhibition concentration leading to a 50% light decrease measured in Microtox toxicity test. (From Ivshina et al. 1998)

microbially produced biosurfactants in soil remediation processes (Christofi and Ivshina 2002; Ławniczak et al. 2013).

Microbial surfactants were also used to remove heavy metals from soil (Mulligan 2005; Ławniczak et al. 2013), although this potential for *Rhodococcus* biosurfactants is still to be revealed. In our experiments, non-specific resistance of *Rhodococcus* cultures to heavy metals correlated positively with their growth on hydrocarbons and biosurfactant production (estimated as emulsifying activity) (Ivshina et al. 2013a). Furthermore, the biosurfactant produced by *R. ruber* IEGM 231 was found to form complexes with Ni^{2+} ions, demonstrating conditional stability constants comparable to those determined for rhamnolipids, most intensively used in biosurfactant-assisted bioremediation studies (Ochoa-Loza et al. 2001; Mulligan 2005; Banat et al. 2010).

6.2 Other Potential Applications

The emulsifying activity of *Rhodococcus* biosurfactants was discussed earlier in this chapter. Considering a high physicochemical stability and a hydrophobic nature of the glycolipids produced by rhodococci, they can be effectively used as stabilizers of oleophilic emulsions and bases for creams, oily films and pastes widely used in food and pharmaceutical industries, and cosmetics (Lang and Philp 1998; Paulino et al. 2016). Additionally, several *Rhodococcus* (*R. globerulus*, *Rhodococcus* sp.) were shown to possess de-emulsification properties (Singh et al. 2007), which can be effective in breaking both oil-in-water and water-in-oil emulsions, for example, formed during oil recovery and processing (Kosaric 1992). Kurane et al. (1995) isolated from the *R. erythropolis* S-1 culture and purified trehalo- and glucolipids with high flocculating activity; the authors suggested their possible application as bioflocculants widely used in many industrial processes, including wastewater treatment, downstream processing, and food and fermentation technologies (Singh et al. 2007). Our study revealed anti-adhesive effects of the *R. ruber* IEGM 231 biosurfactant against Gram-positive and Gram-negative bacteria with different percentages of inhibition (30–76%) (Kuyukina et al. 2016). Interestingly, the biosurfactant was more active against growing bacteria rather than resting cells, thus showing biofilm-preventing properties potentially useful for food processing applications (Nitschke and Silva 2016). But herein, rhodococci along with other mycolata group actinobacteria are reported as foaming-causative agents owing to their filamentous growth and biosurfactant production (Lang and Philp 1998). It should be noted that microbial foaming is among the most frequent and widespread problems associated with activated sludge wastewater treatment plants (Stainsby et al. 2005).

Utilization of biosurfactants in the production of new materials with improved physicochemical characteristics is another possible application area, at present covered by chemical surfactants. In our experiments, a glycolipid biosurfactant from *R. ruber* changed significantly thermal and rheological properties of the poly (vinyl alcohol) (PVA) cryogel, when added to the composite mixture prior to cryotropic gelation (Kuyukina et al. 2006). Particularly, biosurfactant addition distinctly increased the gel strength, elasticity, and plasticity (Table 5). Thus, instantaneous shear modulus (G_0) values of PVA cryogel increased from 3.1–4.7 to 5.6–6.0 kPa, and dynamic shear modulus (G_{30}) measured after 30-min exposure increased from 2.3–3.2 to 4.1–4.7 kPa, depending on biosurfactant concentration. In addition, gel fusion temperatures (T_f,) did not change even in the presence of the highest (15%, v/v) biosurfactant concentration. Similar tests with synthetic surfactants, for example, sodium dodecyl sulfate (SDS) and cetyl trimethylammonium bromide (CTAB), revealed weakening effects of surfactant additives on PVA cryogels (data not shown). It was also found that incorporation of oleophilic polypropylene material Drizit into PVA cryogel weakened the mechanical strength of the resulting gel (Cunningham et al., 2004). Presumably, strengthening effect of biosurfactant additives could be used in polymeric gel formulations designed for

Table 5 Effect of the biosurfactant from *R. ruber* IEGM 231 on rheological properties of the PVA cryogel

PVA cryogel sample	T_f, °C	G_0, kPa	G_{30}, kPa
Control PVA cryogel (no additions)	69.5 ± 0.3	4.65 ± 0.28	3.21 ± 0.31
PVA cryogel + 5% (v/v) of water	68.6 ± 0.2	3.93 ± 0.72	3.19 ± 0.50
PVA cryogel + 5% (v/v) of biosurfactant	69.5 ± 0.1	6.04 ± 0.70	4.65 ± 0.27
PVA cryogel + 10% (v/v) of water	68.3 ± 0.2	3.72 ± 0.28	2.69 ± 0.10
PVA cryogel + 10% (v/v) of biosurfactant	69.6 ± 0.1	5.63 ± 0.60	4.14 ± 0.30
PVA cryogel + 15% (v/v) of water	67.0 ± 0.1	3.07 ± 0.21	2.30 ± 0.21
PVA cryogel + 15% (v/v) of biosurfactant	69.7 ± 0.1	5.63 ± 0.69	4.39 ± 0.33

From Kuyukina et al. (2006)

various industrial and biotechnological applications. Moreover, biosurfactants were recently suggested as useful chemical aids in the synthesis of nanostructured materials with tunable pore size and surface hydrophilicity (Boffa et al. 2014).

Singh et al. (2007) reviewed biotechnological applications of microbial surfactants, including bioprocessing and biocatalysis, which are determined by the surfactant effect on cellular membranes. Particularly, by interaction with cell membranes, surfactants can promote increased production of extracellular products during the fermentation and recovery of intracellular products due to cell lysis. In addition, they can facilitate the oxygen and mass transfer through the membrane and modulate the membrane-associated enzyme activity, specificity, and stability. Although all these applications have so far been confined to chemical surfactants, a growing volume of research on microbial surfactants, for example, the abovementioned studies of *Rhodococcus* trehalolipid interactions with phospholipid membranes and enzymes (Zaragoza et al. 2013; Teruel et al. 2014), would lead to the biosurfactant penetration into this biotechnology market.

As mentioned earlier in this chapter, glycolipid surfactants produced by *Rhodococcus* and related bacteria exhibit biological activities toward various organisms ranging from prokaryotes to mammals and thus have been investigated for their possible biomedical applications (Kitamoto et al. 2002). A challenge for the future in this field would be the identification of glycolipids with high biological (e.g., immunomodulating, antitumor) activity, but less toxic than the trehalose dimycolate from *M. tuberculosis* (Ryll et al. 2001). This could be achieved by the introduction of nontoxic trehalolipids produced by nonpathogenic *Rhodococcus* species (Sakaguchi et al. 2000; Kuyukina et al. 2015). Although many fundamental issues of the trehalolipid influence on a human immune system and their interactions with immune cells and signal molecules still need more detailed investigations, it is reasonable to assume that these rhodococcal products will be used in clinical applications in the near future.

7 Conclusion

Actinobacteria of the genus *Rhodococcus* and closely related genera synthesize glycolipid (mostly trehalolipid) biosurfactants with diverse physicochemical properties and biological activities comparable with other microbial surfactants, which determine their potential applications in environmental and industrial biotechnologies. The only limitation for successful biosurfactant penetration in the biotechnology market is their high production cost, including bacterial fermentation and product recovery expenditures. Biosurfactant production from renewable resources is an attractive option being currently developed for *Rhodococcus* surfactants. Several studies showed that rhodococci are able to grow on vegetable oils and glycerol waste, forming stable emulsions due to intensive surfactant production (Haba et al. 2000; Sadouk et al. 2008; Ruggeri et al. 2009; Pirog et al. 2015; White et al. 2013; Malavenda et al. 2015). It should be noted that development of a cost-effective biotechnology for *Rhodococcus* biosurfactant production requires further fundamental research on the metabolic pathways involved in the glycolipid synthesis from these alternative substrates and on the regulation mechanisms allowing a surfactant overproduction. This research would involve molecular biology tools currently available for *Rhodococcus* and developed on the base of complete genome sequences determined for several rhodococcal species (see Cappelletti et al., this volume). Another research area includes the modeling of cell growth conditions (e.g., mineral and organic medium components, temperature, pH and other physical factors, and chemical inducers) to obtain biosurfactants with target properties. For example, in our experiments, a higher biosurfactant yield was recorded for *R. ruber* cells grown on hexadecane, although greater emulsification activity was registered for the biosurfactant produced using dodecane as a growth substrate (Kuyukina et al. 2005). In summary, development of cheaper fermentation processes for *Rhodococcus* biosurfactants using low-cost renewable raw materials and increased product yields through genetically engineered strains would make them economically attractive and competitive with synthetic surfactants.

Acknowledgments This research was funded by the Ministry of Science and Higher Education of the Russian Federation (State Task Registration No. 01201353246 for IEGM and State Task 6.3330.2017/4.6 for PSU).

References

Abdelhay A, Magnin J-P, Gondrexon N, Baup S, Willison J (2009) Adaptation of a *Mycobacterium* strain to phenanthrene degradation in a biphasic culture system: influence on interfacial area and droplet size. Biotechnol Lett 31:57–63

Alvarez HM, Silva RA, Cesari AC, Zamit AL, Peressutti SR, Reichelt R, Keller U, Malkus U, Rasch C, Maskow T, Mayer F, Steinbüchel A (2004) Physiological and morphological responses of the soil bacterium *Rhodococcus opacus* strain PD630 to water stress. FEMS Microbiol Ecol 50:75–86

Aranda FJ, Teruel JA, Espuny MJ, Marqués A, Manresa Á, Palacios-Lidon E, Ortiz A (2007) Domain formation by a *Rhodococcus* sp. biosurfactant trehalose lipid incorporated into phosphatidylcholine membranes. Biochim Biophys Acta 1768:2596–2604

Arenskötter M, Bröker D, Steinbüchel A (2004) Biology of the metabolically diverse genus *Gordonia*. Appl Environ Microbiol 7:3195–3204

Baeva TA, Gein SV, Kuyukina MS, Ivshina IB, Kochina OA, Chereshnev VA (2014) Effect of glycolipid *Rhodococcus* biosurfactant on secretory activity of neutrophils *in vitro*. Bull Exp Biol Med 157:238–242

Banat IM, Franzetti A, Gandolfi I, Bestetti G, Martinotti MG, Fracchia L, Smyth TJ, Marchant R (2010) Microbial biosurfactants production, applications and future potential. Appl Microbiol Biotechnol 87:427–444

Batrakov SG, Rozynov BV, Koronelli TV, Bergelson LD (1981) Two novel types of trehalose lipids. Chem Phys Lipids 29:241–266

Belisle JT, Vissa VD, Sievert T, Takayama K, Brennan PJ, Besra GS (1997) Role of the major antigen of *Mycobacterium tuberculosis* in cell wall biogenesis. Science 276:1420–1422

Bicca FC, Fleck LC, Ayub ZMA (1999) Production of biosurfactant bt hydrocarbon degrading *Rhodococcus ruber* and *Rhodococcus erythropolis*. Rev Microbiol 30:231–236

Billingsley KA, Backus SM, Wilson S, Singh A, Ward OP (2002) Remediation of PCBs in soil by surfactant washing and biodegradation in the wash by *Pseudomonas* sp. LB400. Biotechnol Lett 24:1827–1832

Biosurfactants market estimated to be worth $5.52 bn by 2022 (2018). Focus Surfactants 2018(2):4. https://doi.org/10.1016/j.fos.2018.02.014

Boffa V, Perrone DG, Magnacca G, Montoneri E (2014) Role of a waste-derived polymeric biosurfactant in the sol–gel synthesis of nanocrystalline titanium dioxide. Ceram Int 40: 12161–12169

Bouchez-Naïtali M, Rakatozafy H, Marchal R, Leveau JY, Vandecasteele J-P (1999) Diversity of bacterial strains degrading hexadecane in relation to the mode of substrate uptake. J Appl Microbiol 86:421–428

Bryant F (1990) Improved method for the isolation of biosurfactant glycolipids from *Rhodococcus* sp. strain H13A. Appl Environ Microbiol 56:494–149

Cai Q, Zhang B, Chen B, Zhu Z, Lin W, Cao T (2014) Screening of biosurfactant producers from petroleum hydrocarbon contaminated sources in cold marine environments. Mar Pollut Bull 86:402–410

Cameotra SS, Makkar RS (2004) Recent applications of biosurfactants as biological and immunological molecules. Curr Opin Microbiol 7:262–266

Chereshnev VA, Gein SV, Baeva TA, Galkina TV, Kuyukina MS, Ivshina IB (2010) Modulation of cytokine secretion and oxidative metabolism of innate immune effectors by *Rhodococcus* biosurfactant. Bull Exp Biol Med 149:734–738

Choi K-S, Kim S-H, Lee T-H (1999) Purification and characterization of biosurfactant from *Tsukamurella* sp. 26A. J Microbiol Biotechnol 9:32–38

Christofi N, Ivshina IB (2002) Microbial surfactants and their use in field studies of soil remediation. J Appl Microbiol 93:915–929

Cooper DG, Zajic JE, Gerson DF (1979) Production of surface-active lipids by *Corynebacterium lepus*. Appl Environ Microbiol 37:4–10

Cunningham CJ, Ivshina IB, Lozinsky VI, Kuyukina MS, Philp JC (2004) Bioremediation of diesel contaminated soil by microorganisms immobilised in a polyvinyl alcohol cryogel. Int Biodeterior Biodegrad 54:167–174

Desai JD, Banat IM (1997) Microbial production of surfactants and their commercial potential. Microbiol Mol Biol Rev 61:47–64

De Smet KA, Weston A, Brown IN, Young DB, Robertson BD (2000) Three pathways for trehalose biosynthesis in mycobacteria. Microbiology 146:199–208

Deshpande S, Shiau BJ, Wade D, Sabatini DA, Harwell JH (1999) Surfactant selection for enhancing ex situ soil washing. Wat Res 33:351–360

Espuny MJ, Egido S, Mercade ME, Manresa A (1995) Characterization of trehalose tetraester produced by a waste lube oil degrader *Rhodococcus* sp. Toxicol Environ Chem 48:83–88

Franzetti A, Gandolfi I, Bestetti G, Smyth TJP, Banat IM (2010) Production and applications of trehalose lipid biosurfactants. Eur J Lipid Sci Technol 112:617–627

Gein SV, Kuyukina MS, Ivshina IB, Baeva TA, Chereshnev VA (2011) *In vitro* cytokine stimulation assay for glycolipid biosurfactant from *Rhodococcus ruber*: role of monocyte adhesion. Cytotechnology 63:559–566

Gesheva V, Stackebrandt E, Vasileva-Tonkova E (2010) Biosurfactant production by halotolerant *Rhodococcus fascians* from Casey Station, Wilkes Land, Antarctica. Curr Microbiol 61: 112–117

Gudina EJ, Rangarajan V, Sen R, Rodrigues LR (2013) Potential therapeutic applications of biosurfactants. Trends Pharm Sci 34:667–675

Haba E, Bresco O, Ferrer C, Marqués A, Busquets M, Manresa A (2000) Isolation of lipase-secreting bacteria by deploying used frying oil as selective substrate. Enzym Microb Technol 26:40–44

Haddadin MSY, Arqoub AAA, Reesh IA, Haddadin J (2009) Kinetics of hydrocarbon extraction from oil shale using biosurfactant producing bacteria. Energy Convers Manag 50:983–990

Hoq MM, Suzutani T, Toyoda T, Horiike G, Yoshida I, Azuma M (1997) Role of γδ TCRM lymphocytes in the augmented resistance of trehalose 6,6-dimycolate-treated mice to influenza virus infection. J Gen Virol 78:1597–1603

Inaba T, Tokumoto Y, Miyazaki Y, Inoue N, Maseda H, Nakajima-Kambe T, Uchiyama H, Nomura N (2013) Analysis of genes for succinoyl trehalose lipid production and increasing production in *Rhodococcus* sp. strain SD-74. Appl Environ Microbiol 79:7082–7090

Ivshina I, Kostina L, Krivoruchko A, Kuyukina M, Peshkur T, Anderson P, Cunningham C (2016) Removal of polycyclic aromatic hydrocarbons in soil spiked with model mixtures of petroleum hydrocarbons and heterocycles using biosurfactants from *Rhodococcus ruber* IEGM 231. J Hazard Mater 312:8–17

Ivshina IB, Kuyukina MS, Kostina LV (2013a) Adaptive mechanisms of nonspecific resistance to heavy metal ions in alkanotrophic actinobacteria. Russ J Ecol 44:123–130

Ivshina IB, Kuyukina MS, Krivoruchko AV (2017) Hydrocarbon-oxidizing bacteria and their potential in eco-biotechnology and bioremediation. In: Kurtböke I (ed) Microbial resources: from functional existence in nature to industrial applications. Elsevier, London, pp 121–148

Ivshina IB, Kuyukina MS, Krivoruchko AV, Barbe V, Fischer C (2014) Draft genome sequence of propane and butane oxidizing actinobacterium *Rhodococcus ruber* IEGM 231. Genome Announc 2:6

Ivshina IB, Kuyukina MS, Krivoruchko AV, Plekhov OA, Naimark OB, Podorozhko EA, Lozinsky VI (2013b) Biosurfactant-enhanced immobilization of hydrocarbon-oxidizing *Rhodococcus ruber* on sawdust. Appl Microbiol Biotechnol 97:5315–5327

Ivshina IB, Kuyukina MS, Philp JC, Christofi N (1998) Oil desorption from mineral and organic materials using biosurfactant complexes produced by *Rhodococcus* species. World J Microbiol Biotechnol 14:711–717

Iwabuchi N, Sunairi M, Anzai H, Nakajima M, Harayama S (2000) Relationships between colony morphotypes and oil tolerance in *Rhodococcus rhodochrous*. Appl Environ Microbiol 66:5073–5077

Kacem R, De Sousa-D'Auria C, Tropis M, Chami M, Gounon P, Leblon G, Houssin C, Daffé M (2004) Importance of mycoloyltransferases on the physiology of *Corynebacterium glutamicum*. Microbiology 150:73–84

Kamenskikh TN, Kuyukina MS, Ivshina IB (2004) Some features in preserving actinobacteria of the genus *Rhodococcus*. Perm Univ Her Biol (2):110–113

Kanga SA, Bonner JS, Page CA, Mills MA, Autenrieth RL (1997) Solubilization of naphthalene and methyl-substituted naphthalenes from crude oil using biosurfactant. Environ Sci Technol 31: 556–561

Kavyanifard A, Ebrahimipour G, Ghasempour A (2016) Structure characterization of a methylated ester biosurfactant produced by a newly isolated Dietzia cinnamea KA1. Microbiology 85(4):430–435

Kim J-S, Powalla M, Lang S, Wagner F, Lunsdorf H, Wray V (1990) Microbial glycolipid production under nitrogen limitation and resting cell conditions. J Biotechnol 13:257–266

Kim SH, Lim EJ, Lee SO, Lee JD, Lee TH (2000) Purification and characterization of biosurfactants from *Nocardia* sp. L-417. Biotechnol Appl Biochem 31:249–253

Kitamoto D, Isoda H, Nakahara T (2002) Functions and potential applications of glycolipid biosurfactants – from energy-saving materials to gene delivery carriers. J Biosci Bioeng 94: 187–201

Konishi M, Nishi S, Fukuoka T, Kitamoto D, Watsuji T-O, Nagano Y, Yabuki A, Nakagawa S, Hatada Y, Horiuchi J (2014) Deep-sea *Rhodococcus* sp. BS-15, lacking the phytopathogenic fas genes, produces a novel glucotriose lipid biosurfactant. Mar Biotechnol 16:484–493

Kosaric N (1992) Biosurfactants in industry. Pure Appl Chem 64:1731–1737

Kretschmer A, Wagner F (1983) Characterization of biosynthetic intermediates of trehalose dicorynomycolates from *Rhodococcus erythropolis* grown on *n*-alkanes. Appl Environ Microbiol 44:864–870

Kundu D, Hazra C, Dandi N, Chaudhari A (2013) Biodegradation of 4-nitrotoluene with biosurfactant production by *Rhodococcus pyridinivorans* NT2: metabolic pathway, cell surface properties and toxicological characterization. Biodegradation 24:775–793

Kurane R, Hatamochi K, Kakuno T, Kiyohara M, Tajima T, Hirano M, Taniguchi Y (1995) Chemical structure of lipid bioflocculant produced by *Rhodococcus erythropolis*. Biosci Biotechnol Biochem 59:1652–1656

Kuyukina MS, Ivshina IB, Baeva TA, Gein SV, Chereshnev VA (2015) Trehalolipid biosurfactants from nonpathogenic *Rhodococcus* actinobacteria with diverse immunomodulatory activities. New Biotechnol 32:559–568

Kuyukina MS, Ivshina IB, Gavrin YA, Podorozhko EA, Lozinsky VI, Jeffree CE, Philp JC (2006) Immobilization of hydrocarbon-oxidizing bacteria in poly(vinyl alcohol) cryogels hydrophobized using a biosurfactant. J Microbiol Methods 65:596–603

Kuyukina MS, Ivshina IB, Gein SV, Baeva TA, Chereshnev VA (2007) *In vitro* immunomodulating activity of biosurfactant glycolipid complex from *Rhodococcus ruber*. Bull Exp Biol Med 144:326–330

Kuyukina MS, Ivshina IB, Korshunova IO, Stukova GI, Krivoruchko AV (2016) Diverse effects of a biosurfactant from *Rhodococcus ruber* IEGM 231 on the adhesion of resting and growing bacteria to polystyrene. AMB Express 6:14. https://doi.org/10.1186/s13568-016-0186-z

Kuyukina MS, Ivshina IB, Makarov SO, Litvinenko LV, Cunningham CJ, Philp JC (2005) Effect of biosurfactants on crude oil desorption and mobilization in a soil system. Environ Int 31:155–161

Kuyukina MS, Ivshina IB, Philp JC, Christofi N, Dunbar SA, Ritchkova MI (2001) Recovery of *Rhodococcus* biosurfactants using methyl tertiary-butyl ether extraction. J Microbiol Methods 46:149–156

Kuyukina MS, Ivshina IB, Ritchkova MI, Chumakov OB (2000) Effect of cell lipid composition on the formation of non-specific antibiotic resistance in alkanotrophic rhodococci. Microbiology 69:51–57

Lang S (2002) Biological amphiphiles microbial biosurfactants. Curr Opin Colloid Interface Sci 7: 12–20

Lang S, Philp JC (1998) Surface-active lipids in rhodococci. Antonie van Leeuwenhoek Int J Gen Mol Microbiol 74:59–70

Ławniczak Ł, Marecik R, Chrzanowski Ł (2013) Contributions of biosurfactants to natural or induced bioremediation. Appl Microbiol Biotechnol 97:2327–2339

LeBlanc JC, Gonçalves ER, Mohn WW (2008) Global response to desiccation stress in the soil actinomycete *Rhodococcus jostii* RHA1. Appl Environ Microbiol 74:2627–2636

Maier RM (2003) Biosurfactants: evolution and diversity in bacteria. Adv Appl Microbiol 52: 101–121

Makkar RS, Rockne KJ (2003) Comparison of synthetic surfactants and biosurfactants in enhancing biodegradation of polycyclic aromatic hydrocarbons. Environ Toxicol Chem 22:2280–2292

Malavenda R, Rizzo C, Michaud L, Gerçe B, Bruni V, Syldatk C, Hausmann R, Giudice AL (2015) Biosurfactant production by Arctic and Antarctic bacteria growing on hydrocarbons. Polar Biol 38:1565–1574

Marchant R, Banat IM (2012) Microbial biosurfactants: challenges and opportunities for future exploitation. Trends Biotechnol 30:558–565

Marqués AM, Pinazo Farfan AM, Aranda FJ, Teruel JA, Ortiz A, Manresa A, Espuny MJ (2009) The physicochemical properties and chemical composition of trehalose lipids produced by *Rhodococcus erythropolis* 51T7. Chem Phys Lipids 158:110–117

Mulligan CN (2005) Environmental applications for biosurfactants. Environ Pollut 133:183–198

Neu TR (1996) Significance of bacterial surface-active compounds in interaction of bacteria with interfaces. Microbiol Rev 60:151–166

Nguyen L, Chinnapapagari S, Thompson CJ (2005) FbpA-dependent biosynthesis of trehalose dimycolate is required for the intrinsic multidrug resistance, cell wall structure, and colonial morphology of *Mycobacterium smegmatis*. J Bacteriol 187:6603–6611

Nitschke M, Silva SSE (2016) Recent food applications of microbial surfactants. Crit Rev Food Sci Nutr 58:631–638

Niescher VW, Lang S, Kaschabek SR, Schlömann M (2006) Identification and structural characterisation of novel trehalose dinocardiomycolates from n-alkane-grown *Rhodococcus opacus* 1CP. Appl Microbiol Biotechnol 70:605–611

Noordman WH, Wachter JHJ, de Boer GJ, Janssen DB (2002) The enhancement by surfactants of hexadecane degradation by *Pseudomonas aeruginosa* varies with substrate availability. J Biotechnol 94:195–212

Ochoa-Loza FJ, Artiola JF, Maier RM (2001) Stability constants for the complexation of various metals with a rhamnolipid biosurfactant. J Environ Qual 30:479–485

Ortiz A, Teruel JA, Espuny MJ, Marqués A, Manresa A, Aranda FJ (2008) Interactions of a *Rhodococcus sp.* biosurfactant trehalose lipid with phosphatidylethanolamine membranes. Biochim Biophys Acta 1778:2806–2813

Ortiz A, Teruela JA, Espuny MJ, Marqués A, Manresa A, Aranda FJ (2009) Interactions of a bacterial biosurfactant trehalose lipid with phosphatidylserine membranes. Chem Phys Lipids 158:46–53

Pacwa-Płociniczak M, Płociniczak T, Iwan J, Zarska M, Chorazewski M, Dzida M, Piotrowska-Seget Z (2016) Isolation of hydrocarbon-degrading and biosurfactant-producing bacteria and assessment their plant growth-promoting traits. J Environ Manag 168:175–184

Page CA, Bonner JS, Kanga SA, Mills MA, Autenrieth RL (1999) Biosurfactant solubilization of PAHs. Environ Eng Sci 16:465–474

Pal MP, Vaidya BK, Desai KM, Joshi RM, Nene SN, Kulkarni BD (2009) Media optimization for biosurfactant production by *Rhodococcus erythropolis* MTCC 2794: artificial intelligence versus a statistical approach. J Ind Microbiol Biotechnol 36:747–756

Paria S (2008) Surfactant-enhanced remediation of organic contaminated soil and water. Adv Colloid Interface Sci 138:24–58

Passeri A, Lang S, Wagner F, Wray V (1991) Marine biosurfactants, II. Production and characterization of an anionic trehalose tetraester from the marine bacterium *Arthrobacter* sp. EK 1. J Biosci 46:204–209

Paulino BN, Pessôa MG, Mano MC, Molina G, Neri-Numa IA, Pastore GM (2016) Current status in biotechnological production and applications of glycolipid biosurfactants. Appl Microbiol Biotechnol 100:10265–10293

Perfumo A, Banat IM, Marchant R (2018) Going green and cold: biosurfactants from low-temperature environments to biotechnology applications. Trends Biotechnol 36:277–289

Peng F, Liu Z, Wang L, Shao Z (2007) An oil-degrading bacterium: *Rhodococcus erythropolis* strain 3C-9 and its biosurfactants. J Appl Microbiol 102:1603–1611

Peypoux F, Bonmatin JM, Wallach J (1999) Recent trends in the biochemistry of surfactin. Appl Microbiol Biotechnol 51:553–563

Philp JC, Kuyukina MS, Ivshina IB, Dunbar SA, Christofi N, Lang S, Wray V (2002) Alkanotrophic *Rhodococcus ruber* as a biosurfactant producer. Appl Microbiol Biotechnol 59:318–324

Pirog TP, Shevchuk TA, Voloshina IN, Karpenko EV (2004) Production of surfactants by *Rhodococcus erythropolis* strain EK-1, grown on hydrophilic and hydrophobic substrates. Appl Biochem Microbiol 40:470–475

Pirog TP, Shulyakova M, Sofilkanych A, Shevchuk T, Mashchenko O (2015) Biosurfactant synthesis by *Rhodococcus erythropolis* IMV Ac-5017, *Acinetobacter calcoaceticus* IMVB-7241 and *Nocardia vaccinii* IMV B-7405 on byproduct of biodiesel production. Food Bioprod Process 93:11–18

Rapp P, Bock H, Urban E, Wagner F, Gebetsbergwer W, Schulzw W (1977) Mikrobielle Bildung eines Trehaloselipids und seine Anwendung in Modellversuchen zum Tensidfluten von Erdollagerstatten. Dechema-Monographien 81:177–186

Rapp P, Bock H, Wray V, Wagner F (1979) Formation, isolation and characterization of trehalose dimycolates from *Rhodococcus erythropolis* grown on *n*-alkanes. J Gen Microbiol 115:491–503

Rapp P, Gabriel-Jürgens LHE (2003) Degradation of alkanes and highly chlorinated benzenes, and production of biosurfactants, by a psychrophilic *Rhodococcus* sp. and genetic characterization of its chlorobenzene dioxygenase. Microbiology 149:2879–2890

Retzinger GS, Meredith SC, Takayama K, Hunter RL, Kezdy FJ (1981) The role of surface in the biological activities of trehalose 6,6-dimicolate. J Biol Chem 256:8208–8216

Ristau E, Wagner F (1983) Formation of novel anionic trehalose tetraesters from *Rhodococcus erythropolis* under growth-limiting conditions. Biotechnol Lett 5:95–100

Ron EZ, Rosenberg E (2001) Natural roles of biosurfactants. Environ Microbiol 3:229–236

Ron EZ, Rosenberg E (2002) Biosurfactants and oil bioremediation. Curr Opin Biotechnol 13: 249–252

Rosenberg E, Ron EZ (1999) High- and low-molecular-mass microbial surfactants. Appl Microbiol Biotechnol 52:154–162

Ruggeri C, Franzetti A, Bestetti G, Caredda P, La Colla P, Pintus M, Sergi S, Tamburini E (2009) Isolation and characterisation of surface active compound producing bacteria from hydrocarbon-contaminated environments. Int Biodeterior Biodegrad 63:936–942

Ryll R, Kumazawa Y, Yano I (2001) Immunological properties of trehalose dimycolate cord factor and other mycolic acid-containing glycolipids – a review. Microbiol Immunol 45:801–811

Sachdev DP, Cameotra SS (2013) Biosurfactants in agriculture. Appl Microbiol Biotechnol 97:1005–1016

Sadouk Z, Hacene H, Tazerouti A (2008) Biosurfactants production from low cost substrate and degradation of diesel oil by a *Rhodococcus* strain. Oil Gas Sci Technol 63:747–753

Sakaguchi I, Ikeda N, Nakayama N, Kato Y, Yano I, Kaneda K (2000) Trehalose 6,6-dimycolate cord factor neovascularization trough vascular endothelial growth factor production by neutrophiles and macrophages. Infect Immun 68:2043–2052

Shavandi M, Mohebali G, Haddadi A, Shakarami H, Nuhi A (2011) Emulsification potential of a newly isolated biosurfactant-producing bacterium, *Rhodococcus* sp. strain TA6. Colloids Surf B Biointerfaces 82:477–482

Singer MEV, Finnerty WR (1990) Physiology of biosurfactant synthesis by *Rhodococcus* species H13-A. Can J Microbiol 36:741–745

Singh A, Van Hamme JD, Ward OP (2007) Surfactants in microbiology and biotechnology: Part 2. Application aspects. Biotechnol Adv 25:99–121

Sokolovska I, Rozenberg R, Riez C, Rouxhet PG, Agathos SN, Wattiau P (2003) Carbon source-induced modifications in the mycolic acid content and cell wall permeability of *Rhodococcus erythropolis* E1. Appl Environ Microbiol 69:7019–7027

Stainsby FM, Philp JC, Dunbar S, Ivshina IB, Kuyukina MS (2005) Microbial foaming and bulking in activated sludge plants. In: Lehr JH, Keeley J, Lehr J, Kingery TB III (eds) Water encyclopedia: domestic, municipal, and industrial water supply and waste disposal. Wiley, Hoboken, NJ, pp 844–848

Sung N, Takayama K, Collins MT (2004) Possible association of GroES and Antigen 85 proteins with heat resistance of *Mycobacterium paratuberculosis*. Appl Environ Microbiol 70: 1688–1697

Sydor T, von Bargen K, Becken U, Spuerck S, Nicholson VM, Prescott JF, Haas A (2008) A mycolyl transferase mutant of *Rhodococcus equi* lacking capsule integrity is fully virulent. Vet Microbiol 128:327–341

Teruel JA, Ortiz A, Aranda FJ (2014) Interactions of a bacterial trehalose lipid with phosphatidylglycerol membranes at low ionic strength. Chem Phys Lipids 181:34–39

Tokumoto Y, Nomura N, Uchiyama H, Imura T, Morita T, Fukuoka T, Kitamoto D (2009) Structural characterization and surface-active properties of succionyl trehalose lipid produced by *Rhodococcus* sp SD-74. J Oleo Sci 58:97–102

Tomiyasu I, Yoshinaga J, Kurano F, Kato Y, Kaneda K, Imaizumi S, Yano I (1986) Occurrence of a novel glycolipid, 'trehalose 2,3,6′-trimycolate' in a psychrophilic, acid-fast bacterium, *Rhodococcus aurantiacus* (*Gordona aurantiaca*). FEBS Lett 203:239–242

Tuleva B, Christova N, Cohen R, Stoev G, Stoineva I (2008) Production and structural elucidation of trehalose tetraesters (biosurfactants) from a novel alkanothrophic *Rhodococcus wratislaviensis* strain. J Appl Microbiol 104:1703–1710

Tzvetkov M, Klopprogge C, Zelder O, Liebl W (2003) Genetic dissection of trehalose biosynthesis in *Corynebacterium glutamicum*: inactivation of trehalose production leads to impaired growth and an altered cell wall lipid composition. Microbiology (SGM) 149:1659–1673

Van der Geize R, Dijkhuizen L (2004) Harnessing the catabolic diversity of rhodococci for environmental and biotechnological applications. Curr Opin Microbiol 7:255–261

Van Hamme JD, Singh A, Ward OP (2006) Physiological aspects. Part 1 in a series of papers devoted to surfactants in microbiology and biotechnology. Biotechnol Adv 24:604–620

White DA, Hird LC, Ali ST (2013) Production and characterization of a trehalolipid biosurfactant produced by the novel marine bacterium *Rhodococcus* sp., strain PML026. J Appl Microbiol 115:744–755

Whyte LG, Slagman SJ, Pietrantonio F, Bourbonniere L, Koval SF, Lawrence JR, Innis WE, Greer SW (1999) Physiological adaptations involved in alkane assimilation at a low temperature by *Rhodococcus* sp. strain Q15. Appl Environ Microbiol 65:2961–2968

Xia W-J, Dong H-P, Yu L, Yu D-F (2011) Comparative study of biosurfactant produced by microorganisms isolated from formation water of petroleum reservoir. Colloids Surf A Physicochem Eng Asp 392:124–130

Zaragoza A, Teruel JA, Aranda FJ, Ortiz A (2013) Interaction of a trehalose lipid biosurfactant produced by *Rhodococcus erythropolis* 51T7 with a secretory phospholipase A2. J Colloid Interface Sci 408:132–137

Zheng C, Yu L, Huang L, Xiu J, Huang Z (2012) Investigation of a hydrocarbon-degrading strain, *Rhodococcus ruber* Z25, for the potential of microbial enhanced oil recovery. J Petrol Sci Eng 81:49–56

Biology of Triacylglycerol Accumulation by *Rhodococcus*

Héctor M. Alvarez and Alexander Steinbüchel

Contents

H. M. Alvarez (✉)
Institute of Bioscience of Patagonia-National Scientific and Technical Research Council (INBIOP-CONICET) and Faculty of Natural Sciences and Health Sciences, University of Patagonia San Juan Bosco, Comodoro Rivadavia, Chubut, Argentina
e-mail: halvarez@unpata.edu.ar

A. Steinbüchel
Institut für Molekulare Mikrobiologie und Biotechnologie der Westfälischen Wilhelms-Universität, Münster, Germany

Environmental Sciences Department, King Abdulaziz University, Jeddah, Saudi Arabia
e-mail: steinbu@uni-muenster.de

H. M. Alvarez (ed.), *Biology of Rhodococcus*, Microbiology Monographs 16,
https://doi.org/10.1007/978-3-030-11461-9_11

Abstract Members of the genus *Rhodococcus* are specialist in the accumulation of triacylglycerols (TAG). Some of them can be considered oleaginous microorganisms since they are able to produce significant amounts of those lipids under certain conditions. In this context, *R. opacus* strain PD630 and *R. jostii* RHA1 became models among prokaryotes in this research area. The basic knowledge generated for rhodococci could be also extrapolated to related microorganisms with clinical importance, such as mycobacteria. The biosynthesis and accumulation of TAG by species of the genus *Rhodococcus* and other actinomycetes seems to be a process linked to the stationary growth phase or as a response to stress. The chemical structure of rhodococcal TAG can be controlled by the composition of the carbon source used. The biosynthesis and accumulation of novel TAG containing unusual components, such as aromatic and isoprenoid fatty acids, by members of *Rhodococcus* and related genera has been reported. The low specificity of wax ester synthase/diacylglycerol acyltransferase enzymes (WS/DGAT), which catalyze TAG biosynthesis in prokaryotes, may contribute to the high variability of TAG composition. The occurrence of genes coding for WS/DGAT enzymes is highly redundant in rhodococcal genomes. The enrichment of genes and enzymes involved in TAG metabolism in rhodococci suggests an important role of these lipids in the physiology of these microorganisms. Genomic, transcriptomic, and proteomic data from TAG-accumulating rhodococci are now available, and some genes coding for enzymes of the central metabolism, the Kennedy pathway, lipid transporter proteins, structural lipid inclusion body-associated proteins, and transcriptional regulatory proteins have been identified and characterized. This article aims to summarize the most relevant achievements of basic research in this field, including the most recent knowledge emerged from studies on TAG accumulation by rhodococci.

1 Introduction

Triacylglycerols (TAG) are nonpolar, water-insoluble fatty acid triesters of glycerol, which are accumulated in most eukaryotic organisms, including animals, plants, yeast, and fungi. These compounds are the main reserve material in eukaryotes for energy and fatty acids required for membrane biosynthesis (Sorger and Daum 2002). Similarly, poly(3-hydroxybutyric acid) (PHB) or other polyhydroxyalkanoic acids (PHA) mainly function as a carbon and energy-reserve material in most bacteria (Anderson and Dawes 1990; Steinbüchel 1991). PHA are polyesters of alkanoic acids containing a hydroxyl group as a functional group in addition to the carboxyl group, which are accumulated by diverse bacteria as intracellular inclusions (Steinbüchel 1991). More than 150 different hydroxyalkanoic acids have been reported as constituents of bacterial PHA (Steinbüchel 1991; Steinbüchel and Valentin 1995). Despite the wide occurrence of PHA among prokaryotes, TAG also occurs as storage lipids in several groups of prokaryotes (Alvarez and Steinbüchel 2002; Alvarez 2006).

Within the last decades, the reports on new TAG-accumulating bacteria have been considerable increased. Gram-negative bacteria are able to accumulate neutral lipids composed of wax esters (WS) as main lipids and TAG only as minor components. WS and TAG have been reported for Gram-negative members of the genera *Acinetobacter*, *Alcanivorax*, and *Marinobacter* (Makula et al. 1975; Alvarez et al. 1997a; Bredemeier et al. 2003; Rontani et al. 2003). Gram-positive bacteria belonging to the actinomycetes group seem to be the TAG-accumulating specialists among prokaryotes. TAG accumulation has been reported for sporulating-actinomycete genera, such as *Streptomyces*, as well as for non-sporulating members, such as *Rhodococcus*, *Nocardia*, *Dietzia*, and *Mycobacterium* (Olukoshi and Packter 1994; Alvarez and Steinbüchel 2002; Alvarez 2006; Kaddor et al. 2009). Some members of these genera are able to accumulate significant amounts of TAG as intracellular inclusions. The majority of the published research on basic aspects of bacterial TAG has been derived from studies on species of the genera *Rhodococcus* and *Acinetobacter*, with *R. opacus* PD630 and *A. baylii* ADP1 being the preferred bacterial models for these studies. Recently, new knowledge has emerged from research mainly focused on the molecular characterization of genes and enzymes involved in TAG formation, for bacteria belonging to *Mycobacterium*, *Streptomyces*, *Alcanivorax*, and *Marinobacter* genera (Arabolaza et al. 2008; Daniel et al. 2004; Kalscheuer et al. 2007; Holtzapple and Schmidt-Dannert 2007). The knowledge of the physiology of such microorganisms may be useful for clinical and for environmental biotechnology purposes, respectively.

In this review, we will summarize the current knowledge on the TAG metabolism, physiology, and molecular biology in members of the *Rhodococcus* genus.

2 Triacylglycerol Accumulation by Species of the Genus *Rhodococcus*

The ability to accumulate TAG is a widespread feature among species of the genus *Rhodococcus*. They are able to accumulate variable amounts of TAG during cultivation on diverse substrates (Table 1). Some strains can be considered as oleaginous bacteria since they accumulate more than 20% of their biomass as lipids (Table 1). *R. opacus* PD630, which is the best known TAG-accumulating member of the *Rhodococcus* genus, is able to accumulate very high levels of TAG in the cells after cultivation on gluconate and other substrates (Alvarez et al. 1996). Figure 1 shows a cell of strain PD630 containing several TAG granules in the cytoplasm. Voss and Steinbüchel (2001) used *R. opacus* strain PD630 for high cell density cultivation to obtain high concentrations of TAG in bioreactors, which contained sugar beet molasses and sucrose as sole carbon sources. This work demonstrated that inexpensive feedstock, such as organic wastes or residual materials from industry, can be also used for lipid production. In this context, Gouda et al. (2008) reported the TAG accumulation by *R. opacus* and *Gordonia* sp. from agro-industrial wastes, such as carob and orange wastes and sugar cane molasses. *R. opacus* was also able to produce high yields of cell biomass and lipids from whey, which is a waste of the

Table 1 Biosynthesis and accumulation of TAG by members of the *Rhodococcus* genus

Bacterial strains	Carbon source	TAG content[a]	References
R. opacus PD630 (DSMZ 44193)	Gluconate	76.0	Alvarez et al. (1996)
	Fructose	40.0	
	Acetate	31.0	
	Propionate	18.0	
	Pentadecane	39.0	
	Hexadecane	38.0	
	Heptadecane	28.0	
	Octadecane	39.0	
	Olive oil	87.0	
	Sugar beet molasses	68.1	Voss and Steinbüchel (2001)
	Carob wastes	88.9*	Gouda et al. (2008)
	Sesame oil	11.3*	
	Whey	45.1	Herrero and Alvarez (2016)
	Lactose	38.0	
	Galactose	36.7	
R. opacus MR22 (DSMZ 3346)	Gluconate	48.0	Alvarez et al. (1997b)
	Hexadecane	43.0	
	Valerate	42.5	
	Whey	46.1	Herrero and Alvarez (2016)
	Lactose	36.5	
	Galactose	35.2	
R. jostii RHA1	Gluconate	56.9	Hernández et al. (2008)
	Glucose	48.4	
	Acetate	21.2	
	3-Hydroxybutyric acid	32.5	
	Hexadecane	30.4	
R. ruber NCIMB 40126	Glucose	19.0	Alvarez et al. (1997b)
	Hexadecane	26.0	
	Valerate	12.2	
R. fascians D188-5	Glucose	3.8	Alvarez et al. (1997b)
	Hexadecane	18.1	
	Valerate	1.8	
R. fascians 123	Gluconate	3.8	Alvarez (2003)
	Pentadecane	4.8	
	Hexadecane	12.9	
R. fascians F7	Glycerol	44.6	Herrero et al. (2016)
R. erythropolis DSMZ 43060	Gluconate	21.0	Alvarez et al. (1997b)
	Hexadecane	17.6	
	Valerate	15.1	

(continued)

Table 1 (continued)

Bacterial strains	Carbon source	TAG content[a]	References
R. erythropolis 17	Gluconate	7.7	Alvarez (2003)
	Pentadecane	56.8	
	Hexadecane	43.4	
R. aetherivorans IAR1	Toluene	24.0	
Rhodococcus sp. 20	Gluconate	7.6	Alvarez (2003)
	Hexadecane	8.1	
Rhodococcus sp. 602	Gluconate	71.2	Silva et al. (2010)
	Benzoate	64.9	
	Hexadecane	22.3	
Rhodococcus sp. A27	Fructose	30.0	Röttig et al. (2016)
R. rhodochrous	Glucose	50.0	Shields-Menard et al. (2015)
R. corynebacterioides DSM 20151	Glucose	9.2	Bequer Urbano et al. (2013)
	Hexadecane	17.9	
Rhodococcus sp. A5	Glucose	11.5	Bequer Urbano et al. (2013)
	Hexadecane	35.3	
R. opacus B4	Hexadecane	140.0*	Castro et al. (2016)

[a]Expressed as % of total fatty acids by cellular dry weight, except in*, which is expressed as mg/L

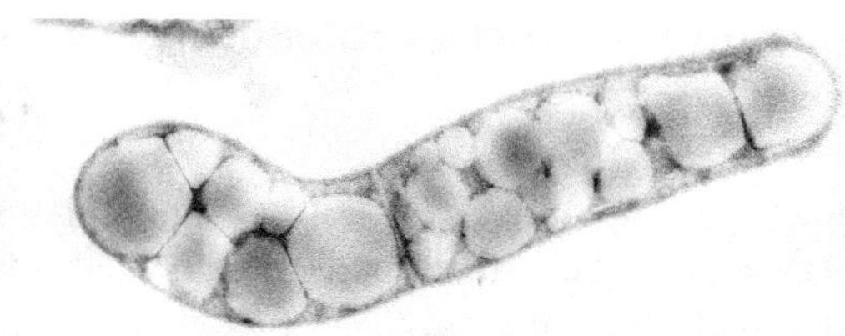

Fig. 1 Micrograph showing a cell of *Rhodococcus opacus* PD630 containing several TAG granules during growth on gluconate as sole carbon and energy source. Picture: F. Mayer from the Georg-August University of Göttingen, Germany

dairy industry generated worldwide in enormous quantities (Herrero and Alvarez 2016). On the other hand, lubricant-based wastewater was successfully used as sole carbon source by *R. opacus* PD630 for its conversion into TAG (Da Silva et al. 2016). All these studies demonstrated that cultivation of rhodococci on a cheap residual carbon source from agricultural or industrial products could be applied to the biotechnological production of interesting single cell oils and probably other lipid-derived products as well.

In addition to TAG, rhodococci are able to produce other storage compounds, such as wax esters, PHA, and glycogen, generally as minor compounds. The accumulation of small amounts of wax esters was reported for *R. opacus* PD630 and *R. jostii* RHA1 after cultivation on phenyldecane and a mixture of hexadecane and hexadecanol, respectively (Alvarez et al. 2002; Hernández et al. 2008). Rhodococci are able to accumulate PHA containing short-chain length monomer units, such as 3-hydroxybutyric acid (C_4) (3HB) and/or 3-hydroxyvaleric acid (C_5) (3HV) (Anderson et al. 1995; Pieper and Steinbüchel 1992; Alvarez et al. 1997b; Alvarez 2003). In general, PHAs represent minor components of the storage lipids accumulated by most rhodococci, with the exception of *R. ruber* and the related *Nocardia corallina*, which produced large amounts of both storage lipids, TAG and the copolyester poly (3HB-*co*-3HV) during growth on glucose. Although the PHA content and composition vary among strains, most of rhodococci produce poly(3HB-*co*-3HV) with 3HV as major monomer unit of the copolyester. Some strains belonging to *R. erythropolis* and *R. fascians* accumulate a polyester containing only 3HB monomer units (Alvarez et al. 1997b).

Hernández et al. (2008) reported the occurrence of glycogen in *R. jostii* cells during growth on gluconate, in addition to TAG and PHA. The accumulation of glycogen seems to be a usual feature among rhodococci, since this material has been also identified in cells of *R. erythropolis*, *R. fascians*, *R. opacus*, and *R. equi* (Hernández and Alvarez 2010). These rhodococcal species were able to produce glycogen up to 0.2–5.6% of cellular dry weight principally during exponential growth phase.

3 Composition and Structure of Rhodococcal Triacylglycerols

Rhodococci are able to produce a diversity of TAG with a high variability of fatty acid composition depending of the carbon source used for cell cultivation. Chemical analyses of TAG accumulated by diverse *Rhodococcus* species revealed the occurrence of saturated and unsaturated straight long-chain fatty acids, principally with a chain length between C_{14} and C_{18} (Alvarez and Steinbüchel 2002; Alvarez 2006). In general, palmitic acid ($C_{16:0}$) and octadecenoic acid ($C_{18:1}$) are the major fatty acids synthesized from non-related substrates such as glucose, gluconate, or acetate. Some strains belonging to *R. opacus*, *R. jostii*, and *R. erythropolis* produce significant amounts of odd-numbered fatty acids during growth on those substrates (from 25 to 40% of the total fatty acids). Substrates as citrate and succinate, which are also intermediates of the tricarboxylic acid cycle (TCA), or acetate, which is fed to the TCA cycle, and odd-numbered organic acids such as propionate or valerate, promote an increase of the fraction of odd-numbered fatty acids in TAG compared to lipids occurring in cells cultivated on glucose or gluconate (Alvarez et al. 1997b; Alvarez 2003). The mentioned strains possess an efficient mechanism for production of the intermediate propionyl-CoA, which is presumably utilized as precursor for the biosynthesis of fatty acids containing an odd number of carbon atoms. Cells are able to produce substantial amounts of propionyl-CoA during growth on diverse substrates from succinyl-CoA via the methylmalonyl-CoA pathway (Anderson et al. 1995; Alvarez et al. 1997b). On the other hand, during cultivation of rhodococcal cells on *n*-alkanes, the main fatty acids produced are related to the chain length of the substrate, as well as to other fatty acids derived from the β-oxidation pathway. Thus, the degradation pathways of hydrocarbons are well coupled to the lipid metabolism in these hydrocarbon-degrading microorganisms.

Previous studies demonstrated that the biosynthetic pathway of TAG is very flexible in rhodococci and related bacteria, being able to accept acyl residues with various chemical structures. During cultivation of *R. opacus* PD630 cells on phenyldecane as sole carbon source, a mixture of TAG containing phenyldecanoic acid residues was detected (Alvarez et al. 2002). In addition, cells produced the wax ester phenyldecylphenyldecanoate by condensation of phenyldecanoic acid and phenyldecanol formed as intermediate during the catabolism of phenyldecane. Other related microorganisms were also able to incorporate unusual fatty acids into TAG or wax esters. *Nocardia globerula* strain 432 accumulated TAG containing the branched fatty acid 4,8,12-trimethyl tridecanoic acid after cultivation of the cells on the recalcitrant branched alkane, pristane (Alvarez et al. 2001), whereas *Mycobacterium ratisbonense* strain SD4 was able to produce a mixture of wax esters containing isoprenoid fatty acids and fatty alcohols, such as 2,6,10,14-tetramethylhexadecanoic acid and 2,6,10,14-tetramethylhexadecan-1-ol among others, after cultivation of cells on phytane (Silva et al. 2007). In another study, cells of *Rhodococcus* sp. 602, an indigenous strain isolated from a soil sample in Patagonia (Argentina), were cultivated under nitrogen-limiting conditions in the presence of naphthyl-1-dodecanoate as sole carbon source. After 6 days of incubation, a mixture of novel TAG containing only medium-chain length fatty acids (C_8, C_{10}, and C_{12}) was identified in the cells (Silva et al. 2010). The results suggested the formation of 1-naphthol and dodecanoic acid residues by an esterase and subsequent β-oxidation of the fatty acid during catabolism of naphthyl-1-dodecanoate. Thus, the TAG biosynthesis pathway of strain 602 was able to incorporate the catabolic intermediates into the storage lipids structure.

The composition and the properties of storage lipids can be also changed by alteration of genes/enzymes involved in lipid metabolism. One example of this is the mutant UFA4 of *R. opacus* PD630, which exhibited a defect in the fatty acid desaturation system. This mutant accumulated increased amounts of stearic acid (C18:0) and lacked odd-numbered fatty acids in TAG during cultivation on gluconate, thus producing a cocoa butter-like oil containing about 74% saturated fatty acids with a relative high content of stearic acid (>18%) (Wältermann and Steinbüchel 2000). All these results demonstrated that the content and composition of rhodococcal TAG can be influenced by the carbon source used for growth of cells or manipulated by engineering procedures.

Wältermann et al. (2000) determined by stereospecific analysis the distribution of fatty acids in TAG for *R. opacus* PD630. The final acyl composition of TAG and the distribution of diverse acyl groups on the hydroxyl groups of the glycerol backbone depend on the differing specificities of the acyltransferases involved in the sequential acylation of the *sn*-1,2 and 3 positions of glycerol-3-phosphate during TAG biosynthesis. This study demonstrated that the enzymes involved in TAG biosynthesis in strain PD630 exhibit specificity for the acyl-CoAs different from the corresponding enzymes in eukaryotes. In eukaryotic TAG (from mammals, plants, and yeasts), unsaturated fatty acids are found in position *sn*-2, and saturated fatty acids are almost totally excluded from this central position. In contrast, *R. opacus* PD630 preferentially incorporated the shorter and saturated fatty acids in the *sn*-2 carbon atom, and the unsaturated fatty acids were predominantly found at position 3. Brennan (1988) reported that fatty acids with more than 20 carbon atoms were predominantly located in the *sn*-3-position of the glycerol molecule with C16-fatty acids occupying the 2-position and either octadecanoate, octadecenoate, or 10-methyloctadecanoate at the 1-position by TAG-accumulating mycobacteria.

4 Conditions for Triacylglycerol Accumulation and Mobilization

The biosynthesis and accumulation of TAG by members of the genus *Rhodococcus* and by other actinomycetes seems to be a process linked to the stationary growth phase or as a response to stress (Olukoshi and Packter 1994; Alvarez et al. 2000). In general, the total content of TAG accumulated by rhodococci depends on both the strain and the carbon source used for growth. However, the nutritional stress seems to be the main condition that influences TAG accumulation by rhodococci. Nitrogen-limiting conditions in the presence of an excess of a carbon source promote significantly TAG biosynthesis and accumulation by *Rhodococcus* members (Alvarez and Steinbüchel 2002). Almost a fourfold increase in the cellular TAG content occurred during cultivation of *R. opacus* PD630 on gluconate with only 0.05 g/L ammonium in the medium, as compared to cells cultivated in a medium containing 1 g/L ammonium (Alvarez et al. 2000). When the N source is lacking in the medium, the biosynthesis of N-containing compounds, such as proteins and nucleotides, is impaired; thus, the biosynthesis of compounds containing only C, O, and H, such as lipids or carbohydrates, is favored. In general, cells accumulate TAG principally during the stationary growth phase. This is logic considering that the fatty acids necessary for TAG biosynthesis are indispensable intermediates for biosynthesis of phospholipids and membranes, which are essential for cell growth and proliferation. Thus, TAG biosynthesis competes with cell growth. In contrast to many bacteria, which block lipid metabolism under growth-restricting conditions (Huisman et al. 1993), rhodococci are able to maintain an active de novo fatty acid biosynthesis pathway under such conditions generating acyl-residues from the available carbon source, which are used for TAG formation.

Another nutritional stress that affects TAG metabolism in rhodococci is C-starvation. When cells of *R. opacus* PD630 and *R. ruber* were incubated in the presence of a nitrogen source and in the absence of any carbon source, they were able to mobilize the stored TAG (Alvarez et al. 2000). This indicated that TAG serve as endogenous carbon and energy sources during incubation of cells under starvation conditions.

Some studies suggest that conditions of limited aeration also promote TAG biosynthesis and accumulation by *Rhodococcus* members (Hernández and Alvarez 2018). In this context, Daniel et al. (2004) reported that several genes involved in TAG biosynthesis in *Mycobacterium tuberculosis* are induced under oxygen-limiting conditions, when cells go into the non-replicative drug-resistance state. Some of these genes show the highest induction and activity by hypoxia (Daniel et al. 2004). The authors concluded that TAG may be the form of energy storage for use during long-term dormancy in this microorganism. However, TAG may act also as a sink for reducing equivalents under these conditions, since the fatty acid-biosynthetic pathway includes pyridine nucleotide-dependent reduction reactions. Thus, TAG biosynthesis may avoid accumulation of reduced pyridine nucleotides in the cells under oxygen-limiting conditions, which may inhibit some key enzymes of the central metabolism (Alvarez and Steinbüchel 2002).

Whether TAG accumulation by rhodococci is also promoted by other stress conditions remains to be investigated.

5 Triacylglycerol Biosynthesis by *Rhodococcus*

Despite the knowledge obtained on the biochemistry of TAG biosynthesis in rhodococci is still fragmentary, some generalizations can be made in this section based on experimental and genomic data. In this section, we subdivide the biosynthesis of TAG into three steps: (1) production of key metabolic precursors for fatty acids and TAG biosynthesis, (2) biosynthesis of fatty acids, and (3) sequential esterification of the glycerol moiety with fatty acyl-residues.

5.1 *Production of Key Metabolic Precursors for Fatty Acid Biosynthesis*

Biosynthesis of TAG requires an efficient metabolic network able to produce the necessary precursors and energy for the specific reactions. In general, the central metabolism of rhodococci possesses a great flexibility and diversity of metabolic reactions, which supports the energy-demanding TAG biosynthesis process under certain conditions from a diversity of carbon sources, as is shown in Table 1. The pathways of rhodococcal central metabolism are able to efficiently convert diverse carbon sources to the key metabolic intermediates, such as pyruvate, acetyl-CoA, and glycerol-3-phosphate, to create reducing equivalents that are required by lipid biosynthesis pathways and to produce the necessary energy as ATP. For more detailed information on the central metabolism of rhodococci see Chapter "Central Metabolism of Species of *Rhodococcus* Genus" by Hernández et al. in this volume. However, many bacteria, which are not able to accumulate TAG, are also able to produce these metabolic intermediates, reducing equivalents and ATP. Thus, an oleaginous microorganism must also be able to maintain a high carbon flux toward the lipid production pathways. Since TAG accumulation is a carbon- and energy-expensive process, rhodococcal cells are able to arrest cell growth and replication and shift their metabolism and carbon flux to lipid biosynthesis pathway. Such changes in cell metabolism depend on the stimuli from the environment, as mentioned above.

Diverse pathways may contribute to produce the acetyl-CoA pool in rhodococci. The conversion of acetyl-CoA from glycolysis-derived pyruvate might be the major route of carbon flux to fatty acid biosynthesis. In general, sugars support significant TAG accumulation by oleaginous *Rhodococcus* members (Table 1). The intermediate acetyl-CoA might be produced alternatively by the reaction catalyzed by citrate lyase enzyme. Citrate lyase, which converts citric acid into acetyl-CoA and oxalacetate, is one of the key enzymes of the reductive TCA cycle. The presence of citrate lyase and 2-oxoglutarate synthase in genome databases of *R. jostii* RHA1 and *R. opacus* B4 suggests that these microorganisms are able to drive the TCA cycle in the reductive direction. This permits the metabolism to incorporate CO_2 for synthesis of intermediates, which may feed the lipid biosynthesis pathways under growth-restricting conditions. On the other hand, free acetate could be activated to acetyl-CoA by acetyl-CoA synthetase in an ATP-dependent reaction. This enzyme together with acetate kinase and phosphotransacetylase enzymes, which were detected in genome databases of *R. jostii* RHA1, *R. opacus* PD630, and *R. opacus* B4, may be involved in the maintenance of the intracellular pools of acetyl-CoA and acetyl-P in these microorganisms. Other metabolic intermediate required for fatty acid biosynthesis in cells of rhodococci is propionyl-CoA, which is generally used for the synthesis of odd-numbered fatty acids (Anderson et al. 1995; Alvarez et al. 1997b). Feisthauer et al. (2008) reported that *R. opacus* 1CP possesses an essential dependence on heterotrophic CO_2 fixation by anaplerotic reactions. Using $^{13}CO_2$ for cultivation experiments, the authors demonstrated that during growth on glucose, the fixed CO_2 was directed principally to the biosynthesis of odd-numbered fatty acids probably via the methyl malonyl-CoA pathway using TCA cycle intermediates as precursors. Fatty acid containing odd number of carbon atoms may account up to 20–30% of the total fatty acids in many *Rhodococcus* strains (Alvarez et al. 1997b; Alvarez 2003).

The synthesis of fatty acids requires stoichiometric amounts of ATP and acetyl-CoA, NADPH, and NADH for each C_2 addition to a growing acyl chain in the reactions catalyzed by acetyl-CoA carboxylase and fatty acid synthetase (Rawsthorne 2002). The necessary ATP might be generated by substrate level phosphorylation in rhodococci through glycolytic routes, among other possible ATP-generating reactions. One of the sources of reducing equivalents for fatty acid biosynthesis in rhodococci (at least in *R. jostii* RHA1) is the Entner-Doudoroff (ED) pathway, which was significantly up-regulated during TAG-accumulation conditions (Dávila Costa et al. 2015; Juarez et al. 2017). Enzymatic analyses also demonstrated the induction of ED enzymes in strain RHA1 during lipogenesis (Juarez et al. 2017), suggesting that ED pathway might be one potential source of NADPH. On the other hand, MacEachran and Sinskey (2013) demonstrated that the NADPH-depending reaction catalyzed by the non-phosphorylative glyceraldehyde dehydrogenase enzyme (GAPN) provides reducing equivalents for fatty acid biosynthesis in *R. opacus* PD630. GAPN catalyzes the reversible oxidation of glyceraldehyde-3-phosphate to 3-phosphoglycerate with the production of NADPH, which can be used for lipid synthesis. Malic enzyme might also be involved in the generation of NADPH in oleaginous rhodococci (Table 2).

Proteomic studies performed in *R. jostii* RHA1 revealed an intensive metabolic reorganization to generate an oleaginous physiological state, which includes the activation of the ED pathway, glycogen mobilization, induction of glyceroneogenesis to generate glycerol-3-phosphate precursor, degradation of amino acids to produce acetyl-CoA, propionyl-CoA, and NADPH, and inhibition of L-ectoine biosynthesis, which consumes acetyl-CoA and reducing equivalents, among other changes (Dávila Costa et al. 2015).

Little is known about the interaction of pathways that occur in cells of oleaginous rhodococci. Previous studies using inhibitors of lipid metabolism such as cerulenin and acrylic acid revealed that the biosynthesis pathways of PHA and TAG in cells of *R. ruber* and *N. corallina* compete for the common intermediates acetyl-CoA and propionyl-CoA during cultivation of cells under nitrogen-limiting conditions (Alvarez et al. 1997b). The inhibition of fatty acid synthesis by the addition of cerulenin in medium caused an increase in the PHA content and altered the composition of the copolyester with an increase of the 3HB monomer units. In contrast, some mutants of *R. ruber* impaired in PHA accumulation produced increasing amounts of TAG in comparison with the wild type (Alvarez et al. 1997b).

5.2 *Biosynthesis of Fatty Acids*

The first step for fatty acid biosynthesis in animals, plants, and prokaryotes is the synthesis of the intermediate malonyl-CoA by the acetyl-CoA carboxylase enzymatic complex (ACC). Malonyl-CoA is the central carbon donor for fatty acid biosynthesis (Wakil et al. 1983). The ACC are highly conserved enzymes which catalyze the carboxylation of acetyl-CoA to produce malonyl-CoA in eukaryotic and prokaryotic organisms (Wakil et al. 1983). The ACC complex is formed by three functional components, such as biotin carboxylase, biotin carboxyl carrier protein, and carboxyltransferase (Cronan and Waldrop 2002). In general, the ACC complex found in chloroplasts and most prokaryotes is an enzyme formed by multiple subunits, whereas a unique enzyme with multiple domains is found in eukaryotic organisms. Interestingly, *R. jostii* RHA1 possesses a eukaryotic-like ACC with multiple domains (RHA1_RS20530), which increased its abundance 1.9-fold during cultivation under TAG-accumulating conditions as revealed by proteomic analyses

Table 2 Enzymes/proteins related to TAG accumulation identified and characterized in TAG-accumulating rhodococci

Strain and protein ID	Protein name	Function	Source
Central metabolism			
Rhodococcus opacus PD630			
OPAG_03892 (TadD)	Non-phosphorylative glyceraldehyde dehydrogenase	Oxidation of glyceraldehyde-3-phosphate to 3-phosphoglycerate	MacEachran and Sinskey (2013)
Rhodococcus jostii RHA1			
RHA1_RS44255	NADPH-dependent malic enzyme	Decarboxylation of malate to pyruvate	Hernández and Alvarez (2018)
Rhodococcus fascians F7			
ACG96_05175 (GlpK1)	Glycerol kinase	Transfer of a phosphate from ATP to glycerol	Herrero et al. (2016)
ACG96_05170 (GlpD1)	Glycerol-3-phosphate dehydrogenase	Reversible conversion of dihydroxyacetone-P to glycerol-3-P	Herrero et al. (2016)
Fatty acid synthesis			
Rhodococcus opacus PD630			
OPAG_00508	Thioesterase	Release of fatty acid from acyl-ACP	Huang et al. (2016)
Kennedy pathway (TAG synthesis)			
Rhodococcus opacus PD630			
OPAG_07257 (Atf1)	WS/DGAT	Acylation of DAG	Alvarez et al. (2008)
OPAG_00138 (Atf2)	WS/DGAT	Acylation of DAG	Hernández et al. (2013)
Rhodococcus jostii RHA1			
RHA1_RS00400 (PAP2)	Phosphatidic acid phosphatase	Synthesis of DAG from phosphatidic acid	Hernández et al. (2015)
RHA1_RS26160 (Atf8)	WS/DGAT	Acylation of DAG	Amara et al. (2016)
Lipid transporters			
Rhodococcus jostii RHA1			
RHA1_RS27545 (Ltp1)	ABC transporter protein	Importer of fatty acids	Villalba and Alvarez (2014)
Lipid inclusion body-associated proteins			
Rhodococcus opacus PD630			
OPAG_00658 (TadA)	Structural Protein	TAG body-associated protein	MacEachran et al. (2010)
Rhodococcus jostii RHA1			
RHA1_RS10270 (TadA/MLDS)	Structural protein	TAG body-associated protein	Ding et al. (2012)

(continued)

Table 2 (continued)

Strain and protein ID	Protein name	Function	Source
Transcriptional regulator proteins			
Rhodococcus jostii RHA1			
RHA1_RS31140 (NlpR)	Regulatory protein	Modulate gene expression	Hernández et al. (2017a)
RHA1_RS10275 (MLDSR)	Regulatory protein	Modulate gene expression	Zhang et al. (2017)
Rhodococcus opacus PD630			
OPAG_03371 (NlpR)	Regulatory protein	Modulate gene expression	Hernández et al. (2017a)

WE wax esters, *TAG* triacylglycerols, *DAG* diacylglycerols, *WS* wax ester synthase, *DGAT* diacylglycerol acyltransferase, *P* phosphate

(Dávila Costa et al. 2015). The activity of this ACC, which may be involved in fatty acid biosynthesis in strain RHA1 during TAG accumulation, seemed to be controlled at the transcriptional level. However, redox proteomic analyses suggested that the activity of this eukaryotic-like ACC may also be finely modulated by redox status of the cell (Dávila Costa et al. 2015). The role of this ACC in fatty acid and TAG synthesis and accumulation in *R. jostii* RHA1 should be investigated in the future.

The biosynthesis of fatty acids is performed by a multienzymatic complex known as fatty acid synthase (FAS). This complex catalyzes the successive reaction of condensation, dehydration, and reduction. Two alternative FAS complexes exist in organisms. The FAS type II is present in most prokaryotes, and some eukaryote organelles, such as mitochondria and chloroplasts, consist of independent proteins encoded by different genes (Bloch 1977). In contrast, the FAS type I consists in a unique large protein with the different catalytic activities. FASI enzymes are found in the cytoplasm of eukaryotic cells and in a subgroup of actinobacteria. FASI is responsible for fatty acid biosynthesis in mycobacteria, which are used for phospholipids and TAG synthesis or for mycolic acid production after an elongation process mediated by FASII (Bloch 1977; Zimhony et al. 2004). FASII uses medium-chain length fatty acids (C_{16}–C_{24}) as primers for synthesizing long-chain length mycolic acids (Shweizer and Hofmann 2004). The FASI multienzyme gene of mycobacteria and rhodococci seems to be structurally very similar. All rhodococcal enzymes are similar in size and amino acid sequences, comprising 3128 amino acids in *R. jostii* RHA1 (RHA1_RS06915), 3107 in *R. opacus* B4 (*ROP_11350*), 3100 in *R. erythropolis* PR4 (*RER_38730*), and 3103 in *R. erythropolis* SK121 (*RHOER0001_5412*), among others. The main products of rhodococcal FASI may be C_{16}–C_{18} fatty acids, which may be utilized for phospholipids and TAG biosynthesis. Proteomic analyses indicated that the FASI system increased 3.9-fold its abundance in *R. jostii* RHA1 cells under TAG-accumulation conditions (Dávila Costa et al. 2015). In addition, three components of the FASII system, such as malonyl-CoA-[acyl-carrier protein] (ACP) transferase, β-ketoacyl-[ACP] reductase, and β-ketoacyl-[ACP] synthase, also increased their abundances during lipogenesis. The contribution of FASII to TAG accumulation by strain RHA1 and other rhodococci must be investigated in the future.

Thioesterases (TE) are enzymes that play an essential role in fatty acid biosynthesis, hydrolyzing the thioester bond between a carbonyl group and a sulfur atom of acyl-ACP to release the fatty acids. They are then converted to fatty acyl-CoAs that are further transformed to TAG and other lipids in cells. Thus, production and composition of fatty acids are determined by acyl-ACP TEs. Huang et al. (2016) reported the occurrence of four genes coding for putative TE enzymes in *R. opacus* PD630. After a molecular and physiological characterization of these genes/enzymes, the authors concluded that the putative acyl-ACP TE2 (OPAG_00508) and TE4 (WP_012687673.1) contribute to the production of total fatty acids and also have a specific influence on the fatty acid composition in *R. opacus* PD630 (Huang et al. 2016). However, the role of TE1 (EHI47208.1) and TE3 (WP_005241865.1) in the lipid content and fatty acid composition could not be determined in strain PD630.

5.3 Biosynthesis of Triacylglycerols

The TAG biosynthesis in rhodococci has been proposed to occur via sequential acyl-CoA-dependent reactions referred to as the "Kennedy pathway," which has been described for yeast and plants (Fig. 2). The pathway involves the sequential acylation of the *sn*-1, 2 positions of glycerol-3-phosphate, resulting in the formation of phosphatidic acid. The removal of the phosphate group catalyzed by the phosphatidic acid phosphatase enzyme occurs before the final acylation step. In the third acylation reaction, an acyl-residue is transferred to the vacant position of diacylglycerol, which is the final step of TAG biosynthesis (Fig. 2). The three acylation reactions are catalyzed by different acyltransferases. The differing specificities of the acyltransferase determine the distribution of diverse acyl groups on the hydroxyl groups of the glycerol backbone and, thus, the final acyl composition of TAG. Phosphatidic acid and diacylglycerol generated in the Kennedy pathway are also used for the synthesis of phospholipids occurring in the membranes. Thus, the third acylation step of the glycerol backbone is the unique enzymatic reaction to TAG biosynthesis. This reaction is catalyzed by a diacylglycerol acyltransferase enzyme (DGAT). Kalscheuer and Steinbüchel (2003) identified as the first prokaryotic DGAT in *Acinetobacter baylyi* ADP1, which exhibited simultaneously both DGAT and acyl-CoA:fatty alcohol acyltransferase (wax ester synthase, WS) activities. Strain ADP1 accumulates mainly wax esters and TAG as minor compounds, amounting up to 6.9 and 1.4% of cellular dry weight, respectively. Interestingly, WS/DGAT from *A. baylyi* ADP1 represents a new class of TAG synthesizing enzyme, which exhibits no extended sequence similarity to any known eukaryote acyltransferase (Kalscheuer and Steinbüchel 2003). A highly conserved motif HHxxxDG, which may be the catalytic site responsible for ester bond formation, is found in WS/DGAT from strain ADP1 and related proteins from other

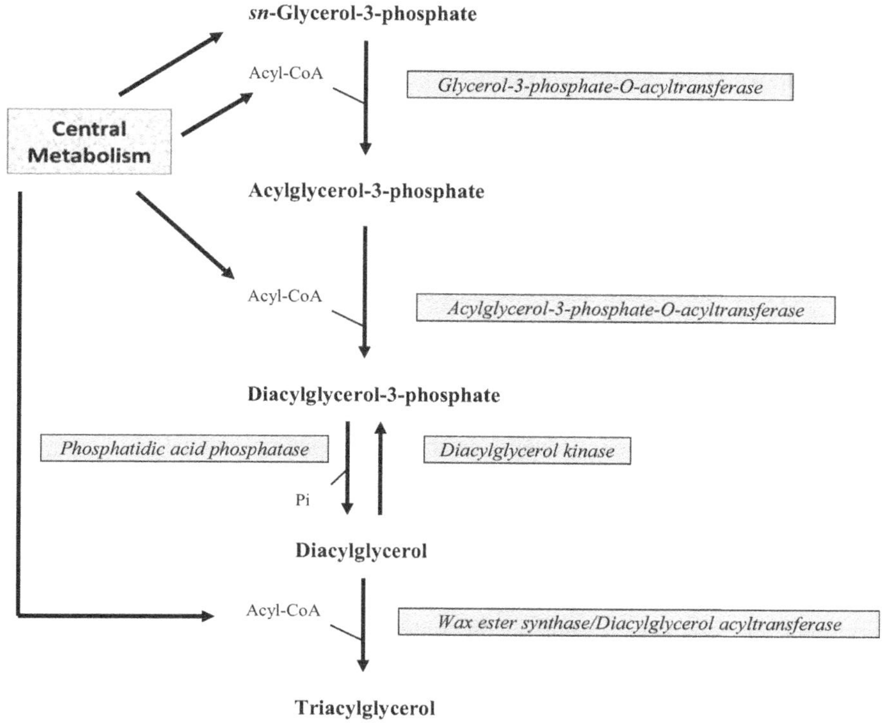

Fig. 2 Pathway for TAG biosynthesis (Kennedy pathway) in rhodococci

microorganisms (Kalscheuer and Steinbüchel 2003). Later, several WS/DGATs were described in various TAG- or WS-accumulating bacteria. Whereas only one or few WS/DGATs occur in Gram-negative bacteria able to produce WS and TAG, a high redundancy of these enzymes occurs in most TAG-accumulating actinomycetes bacteria, such as the genera *Mycobacterium*, *Nocardia*, and *Rhodococcus*. Daniel et al. (2004) identified 15 genes as putative WS/DGAT in *M. tuberculosis* strain H37Rv, which exhibited acyltransferase activity when expressed in *E. coli*. Eleven of these genes have the conserved active-site motif HHxxxDG, whereas three of them have modified versions of this motif, and one has no recognizable motif (Daniel et al. 2004). Alvarez et al. (2008) identified and cloned the first WS/DGAT gene (called *atf1*) in a *Rhodococcus* member, *R. opacus* PD630, when any rhodococcal genomic database was available. They obtained an 800 bp PCR product from chromosomal DNA of strain PD630 by using degenerate primers designed from conserved stretches of WS/DGAT proteins of *A. baylyi* ADP1 and *M. smegmatis* mc2155. The *atf* gene fragment was used as a probe for a strain PD630 gene library, resulting in the identification of a 3948 bp chromosomal DNA fragment containing the complete *atf1* gene (Alvarez et al. 2008). ATF1 exhibited high WS activity and only scant DGAT activity when expressed in *E. coli*. When *atf1* gene was disrupted in strain PD630, cells of the mutant showed a significant reduction of DGAT activity and accumulated up to 50% less fatty acids in comparison to the wild type during cultivation on gluconate under nitrogen-limiting conditions (Alvarez et al. 2008). Although the results of this study demonstrated that ATF1 is mainly responsible for TAG biosynthesis in *R. opacus* PD630, it was clear that additional WS/DGAT contributes to the total DGAT activity and TAG content in this strain. Interestingly, TAG accumulated by the *atf1*-disrupted mutant showed a significant reduction of oleic acid content in comparison to TAG produced by the wild type, after cultivation on gluconate and oleic acid. These results suggested that WS/DGAT isoenzymes in actinomycetes are specialized for the selective incorporation of different fatty acyl residues into TAG (Alvarez et al. 2008). When the genome database of *R. jostii* RHA1 was publicly available, nine additional *atf*-homologous genes were identified in strain PD630 (*atf2–atf10*) using nondegenerate primers deduced from strain RHA1 sequence data. WS/DGATs of strain PD630 exhibited 88–99% sequence identity to the corresponding strain RHA1 enzymes (Alvarez et al. 2008). All deduced proteins showed the complete putative active-site motif HHxxxDG described for bacterial WS/DGAT enzymes. Those of Atf5 and Atf10 exhibited a modified active-site motif, in which the second histidine was replaced by serine or lysine, respectively (Alvarez et al. 2008). All WS/DGAT of strain PD630 were heterologously expressed in *E. coli* for analyzing their acyltransferase activities. In general, all crude extracts of recombinant *E. coli* strains exhibited only low enzymatic activities compared to those obtained from the *R. opacus* (Alvarez et al. 2008). In addition to *atf1* as mentioned above, recombinant *E. coli* harboring plasmid pBluescriptSK::*atf2* exhibited WS as well as significant DGAT activities. However, crude protein extracts of *E. coli* strains expressing *atf3–atf10* exhibited no or only slightly increased WS/DGAT activities in comparison to the vector control cultivated under conditions used in that study (Alvarez et al. 2008). The gene coding for Atf2 from *R. opacus* PD630 was also cloned and characterized. The disruption of *atf2* gene resulted in a decrease of approximately 30% of TAG content and in any evident modification in the fatty acid composition of lipids in comparison to the wild-type PD630 (Hernández et al. 2013). The results of that study demonstrated an active role of Atf2 in the TAG accumulation process in *R. opacus* PD630 (Table 2).

R. jostii RHA1 is also able to accumulate significant amounts of TAG, in addition to other storage compounds, such as PHA, glycogen, and polyphosphate (Hernández et al. 2008). This strain possesses all necessary genes/enzymes for TAG biosynthesis via the Kennedy pathway. Amara et al. (2016) reported the occurrence of 16 *atf* genes potentially encoding DGAT enzymes in *R. jostii* RHA1. Transcriptomic analyses revealed that the *atf8* transcripts were the most abundant during cultivation of cells under lipid-accumulating conditions. Interestingly, the disruption of *atf8* promoted a 70% decrease in TAG accumulation when compared to the wild-type strain (Amara et al. 2016). These results suggested that Atf8 was the main DGAT enzyme involved in TAG accumulation by *R. jostii* RHA1 under the conditions used in the study. The WS/DGAT genes of strain RHA1 are not located in operons with other genes involved in TAG metabolism, and they are widely distributed throughout the genome, which seems to be common in TAG-accumulating actinomycetes (Daniel et al. 2004; Wältermann et al. 2007; Amara et al. 2016). However, some of the RHA1 WS/DGAT genes are adjacent or proximal to other genes likely involved in TAG or lipid metabolism (Hernández et al. 2008). Bioinformatic analysis of the available genomic databases showed the occurrence of varying numbers of putative WS/DGAT genes in the genome of different species, such as *R. opacus*, *R. equi*, *R. jostii*, *R. fascians*, and *R. erythropolis* (Villalba et al. 2013). The WS/DGAT gene number found in the rhodococcal genomes seems to be a strain-dependent feature.

Phosphatidic acid phosphatase (PAP) is other key enzyme involved in the Kennedy pathway for phospholipids and TAG synthesis. Bioinformatic analysis of the *R. jostii* RHA1 genome showed the occurrence of several genes coding for putative PAP proteins (Hernández et al. 2015). A similar situation was observed in the genomes of *R. opacus* and *R. erythropolis*. The high diversity of PAP enzymes in rhodococci suggests differing or specialized functions of these enzymes within lipid metabolism for adapting cells to diverse environmental conditions. The functional role of RHA1_RS00400 as a PAP enzyme was analyzed by cloning and expressing its gene in *E.coli* and *R. jostii* RHA1. Results of this study demonstrated that RHA1_RS00400 play an active role in TAG biosynthesis and accumulation in strain RHA1 catalyzing the desphosphorylation of phosphatidic acid to yield diacylglycerol (DAG), which is the direct precursor for TAG synthesis in rhodococci (Fig. 2). Interestingly, Amara et al. (2016) suggested that three genes encoding phosphatases of the haloacid dehalogenase superfamily, which were significantly expressed during lipid accumulation, can catalyze this enzymatic step in TAG biosynthesis pathway in *R. jostii* RHA1.

The Kennedy pathway seems to be the main TAG biosynthesis pathway in rhodococci; however, alternative acyl-CoA-independent routes for TAG synthesis could occur in these microorganisms. Dahlqvist et al. (2000) reported a pathway that uses phospholipids as acyl donors and diacylglycerols as acceptor for TAG biosynthesis in plants and yeast. This reaction is catalyzed by a phospholipids: diacylglycerol acyltranferase enzyme (PDAT). Interestingly, Arabolaza et al. (2008) demonstrated that phospholipids could act as acyl donors for TAG biosynthesis in *Streptomyces coelicolor* and that this reaction could be catalyzed by a PDAT enzyme. The absence of sequence similarities of eukaryotic PDATs to any of the genomic sequences makes it difficult to study such enzymes in TAG-accumulating actinomycetes and to establish their physiological role in cells. It has been proposed that the PDAT enzyme might function to modulate membrane lipid composition (Dahlqvist et al. 2000; Arabolaza et al. 2008). The occurrence of acyl-CoA-independent routes for TAG synthesis and PDAT like enzymes in rhodococci remains to be investigated.

6 Transporter Proteins Related with TAG Accumulation

Lipid transporters may play a key role in the maintenance of the lipid homeostasis of oleaginous rhodococci. Villalba and Alvarez (2014) identified and characterized a novel ATP-binding cassette transporter involved in long-chain fatty acid import in *R. jostii* RHA1 named as Ltp1. Interestingly, *ltp1* gene was clustered with other genes coding for the three putative acyltransferase enzymes of the Kennedy pathway for TAG synthesis. Results of this study suggested that Ltp1 transporter plays a role in lipid homeostasis in rhodococcal cells and in the distribution of fatty acids between different metabolic pathways and lipid species, since overexpression of *ltp1* in the RHA1 promoted a significant increase of TAG and cellular biomass formation (Villalba and Alvarez 2014). The study demonstrated the relevance of lipid transporters as molecular tools for improving TAG accumulation through genetic engineering in rhodococci (Table 2).

7 TAG Inclusion Bodies and Their Associated Proteins

Detailed studies on the formation of lipid inclusions in bacteria have been made in *Acinetobacter baylyi* strain ADP1 and in *R. opacus* strain PD630. Lipid biosynthesis starts at the inner leaflet of the cytoplasmic membrane to which the acyltransferase is bound as revealed by cytoimmunological studies using polyclonal antibodies raised against this enzyme and ultrathin sections of cells just starting lipid biosynthesis (Wältermann et al. 2005). Also in transmission electron micrographs, a thin film of material emerged on the surface of the inner leaflet. From this film, small lipid droplets arose which conglomerated to lipid prebodies. When these prebodies had reached a certain size, they were released from the cytoplasm membrane and became separate, discrete structures which further matured to the lipid inclusions in their final stage. These steps were in principle also indirectly observed when in vitro studies using an artificial membrane and the purified acyltransferase protein were beside the other necessary compounds (substrates, etc.) used in combination with a quartz crystal microbalance in combination with scanning force microscopy. The changes of the frequency of the quartz crystal and the changes at the surface, respectively, could be interpreted as similar steps occurring in vitro (Wältermann et al. 2005).

There are evidences that lipid bodies are surrounded by a half-unit membrane of phospholipids with several proteins associated. In this context, Chen et al. (2014) performed an integrated omic study including genome and transcriptomic analyses and a proteome of isolated lipid inclusions from *R. opacus* PD630. They identified 177 proteins involved in lipid metabolism and lipid body dynamics. Among several enzymes and proteins, the authors identified a dominant structure-like protein (LPD06283) and several dynamin and SNARE-like proteins probably involved in lipid body dynamics. LPD06283 is identical to TadA of strain PD630 and its orthologue RHA1_RS10270 from *R. jostii* RHA1. TadA had been previously identified and characterized by MacEachran et al. (2010) as a lipid body-associated protein in *R. opacus* PD630. Gene disruption or overexpression promoted alterations in the size of the inclusion bodies and in TAG content in cells. Its orthologue in *R. jostii* RHA1 (RHA1_RS10270) was identified as one of the major proteins co-purified with the lipid inclusion bodies (Ding et al. 2012) (Table 2). Deletion of this gene induced the formation of super-sized lipid bodies in strain RHA1. Proteomic analyses performed with this strain revealed that RHA1_RS10270 was one of the three most abundant proteins during TAG accumulation (Dávila Costa et al. 2015). Interestingly, Zhang et al. (2017) proposed that this inclusion body-associated protein (TadA) increases survival rate of cells under nutritional and genotoxic stress. The authors renamed TadA protein as MLDS (microorganism lipid droplet small protein) (Table 2). The study demonstrated that TadA/MLDS binds DNA and recruits DNA to lipid inclusion bodies (Zhang et al. 2017). Protein binding to DNA occurs without sequence specificity through its positively charged C-terminus domain. The authors suggested that the association of DNA to lipid inclusion bodies mediated by the TadA/MLDS protein contributes to protection against environmental stresses, such as UV radiation.

More studies are necessary to understand the biogenesis and the structural and functional dynamics of TAG inclusion bodies in rhodococci.

8 Regulation of Lipid Accumulation in *Rhodococcus*

Several genes and proteins are required for the transition from vegetative growth to oleaginous phenotype. The control of metabolic transition between these physiological states might require a complex regulatory network involving pleiotropic global regulators and other regulatory proteins working at different hierarchical levels. The first regulatory protein leading to activation of lipogenesis and TAG accumulation was identified and characterized in *R. jostii* RHA1 (Hernández et al. 2017a). The transcriptional regulator NlpR (Nitrogen Lipid Regulator) participates in the modulation of nitrogen and lipid metabolisms in response to nitrogen limitation in the environment. More specifically, NlpR contributes in *R. jostii* RHA1 and also in *R. opacus* PD630 to the allocation of carbon into the different lipid fractions, such as TAG, DAG, fatty acids, and phospholipids, in response to nitrogen levels, increasing the rate of carbon flux into lipid metabolism (Hernández et al. 2017b). NlpR, which seems to be part of GlnR regulon, is significantly up-regulated during cultivation of cells under nitrogen-limiting conditions (Hernández et al. 2017a). This regulatory protein is a global regulator that controls the expression of several genes involved in nitrogen and lipid metabolisms, such as those genes implicated in the NO_3^-/NO_2^- assimilation, fatty acid synthesis (FASI and FASII), the Kennedy pathway for TAG, and phospholipid synthesis, among others. Thus, NlpR modulates large modules of lipid metabolism controlling the carbon flux within the module in response to nitrogen limitation. However, *nlpR* gene is not essential for lipid synthesis and accumulation since *nlpR*-disrupted mutant is still able to accumulate significant amounts of TAG (Hernández et al. 2017a, b).

These results suggested that additional regulatory components are simultaneously involved in TAG accumulation during nitrogen-limiting conditions. In this context, Zhang et al. (2017) reported the characterization of a transcriptional regulatory protein (MLDSR) in *R. jostii* RHA1, which regulates the expression of gene coding for the MLDS protein described above (Sect. 7) as a lipid inclusion body-associated protein. MLDS seems to participate in the TAG accumulating dynamics and lipid inclusion body biogenesis in rhodococcal cells. The MLDSR protein is induced at low nitrogen conditions, stimulating the expression of its own gene and *mlds* gene. Nevertheless, MLDSR is able to regulate transcription both positively and negatively (Zhang et al. 2017). At low concentration of MLDSR, it positively regulates expression of *mlds* gene and itself, whereas when MLDSR becomes high, this protein represses the expression of these genes. MLDSR concentration in cellular cytosol seems to be controlled by the lipid inclusion body recruitment, thus modulating MLDS expression.

Juarez et al. (2017) suggested the involvement of a regulatory mechanism mediated by the cAMP-dependent CRP regulator for TAG accumulation in *R. jostii* RHA1. They identified putative CRP-binding sites in some genes significantly up-regulated during TAG accumulation, such as those of the Entner-Doudoroff pathway and the WS/DGAT Atf8, which is the main WS/DGAT enzyme involved in TAG accumulation in *R. jostii* RHA1 according to a previous study (Amara et al. 2016). These results suggested that TAG accumulation in strain RHA1 is probably driven by an increase of cAMP concentration in cells that activates the expression of genes involved in lipogenesis. However, this hypothesis should be experimentally confirmed.

Proteome analyses revealed differential expression of proteins involved in TAG accumulation in *R. jostii* RHA1 after addition of methyl viologen (MV), a potent prooxidant (Dávila Costa et al. 2017). The presence of MV in the culture medium promoted a significant decrease in the abundance of key proteins of fatty acid synthesis and the Kennedy pathway for TAG biosynthesis under nitrogen-limiting conditions in comparison to those cells cultivated at the same conditions in the absence of MV. These results suggested the occurrence of a NADPH-mediated process regulating TAG accumulation in *R. jostii* RHA1, since the antioxidant response against MV competes with lipogenesis for NADPH pools. Finally, redox proteomic analyses suggested that the activities of some key enzymes probably involved in lipogenesis, such as the eukaryotic-like acetyl-CoA carboxylase, FabF, and fructose 1,6-biphosphatase, are regulated posttranscriptionally by reversible thiol modifications in response to changes in the redox status of rhodococcal cells (Dávila Costa et al. 2015). These results suggested that TAG accumulation by oleaginous rhodococci could be regulated not only at transcriptional level but also by posttranscriptional mechanisms (Fig. 3). Indeed, we still have a very limited understanding on how TAG accumulation is controlled at molecular and metabolic level in rhodococci.

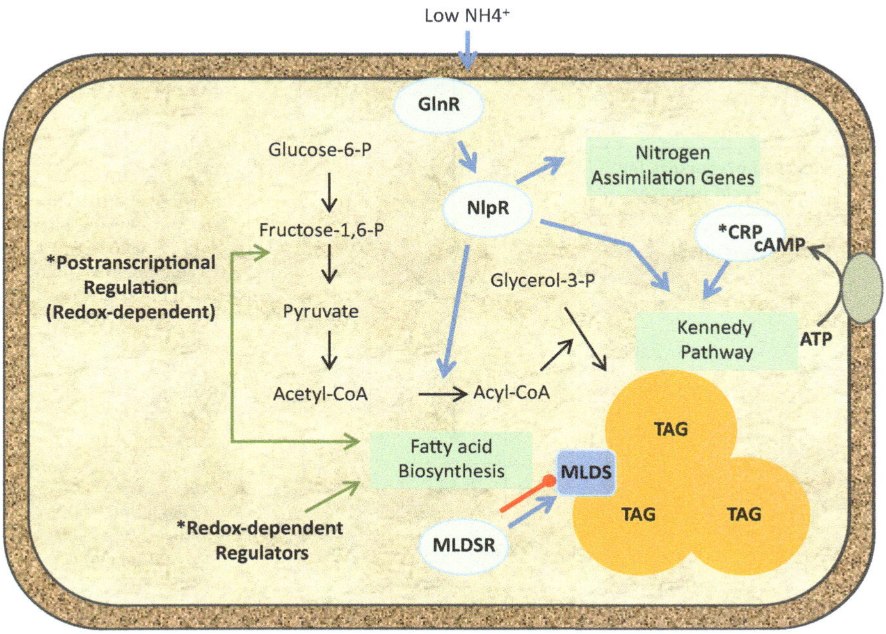

Fig. 3 Some components of the regulatory network controlling TAG biosynthesis and accumulation in oleaginous rhodococci. Blue arrows indicate activation, red bars indicate repression, and green arrows indicate control of gene expression or enzyme activities by unknown mechanisms (activation or repression). During growth under nitrogen-limiting conditions, the global regulator NlpR is probably activated via GlnR. NlpR positively modulates the expression of genes involved in nitrogen assimilation and lipid metabolism. The transcriptional regulator MLDSR activates the expression of *mlds* gene, which codes for a lipid inclusion body-associated protein and itself. At high concentrations of MLDSR in the cytosol, this protein represses expression of the respective genes. Adenylate cyclases generate cAMP which binds to CRP protein. Activated CRP probably modulates the expression of genes involved in TAG biosynthesis. Additional redox-dependent regulatory mechanisms at transcriptional and posttranscriptional level are probably involved in the control of TAG accumulation in rhodococci. (*) Only indirect evidences are available and experimental confirmation is necessary

9 Physiological Functions of TAG in *Rhodococcus*

Rhodococcus species, which are enriched in a particular class of lipids, such as TAG, may be highly dependent on these compounds and their functions, for successful survival in the environment. In this context, TAG seem to play a key role for the cells under growth-restricting conditions that frequently predominate in the environment (Fig. 4).

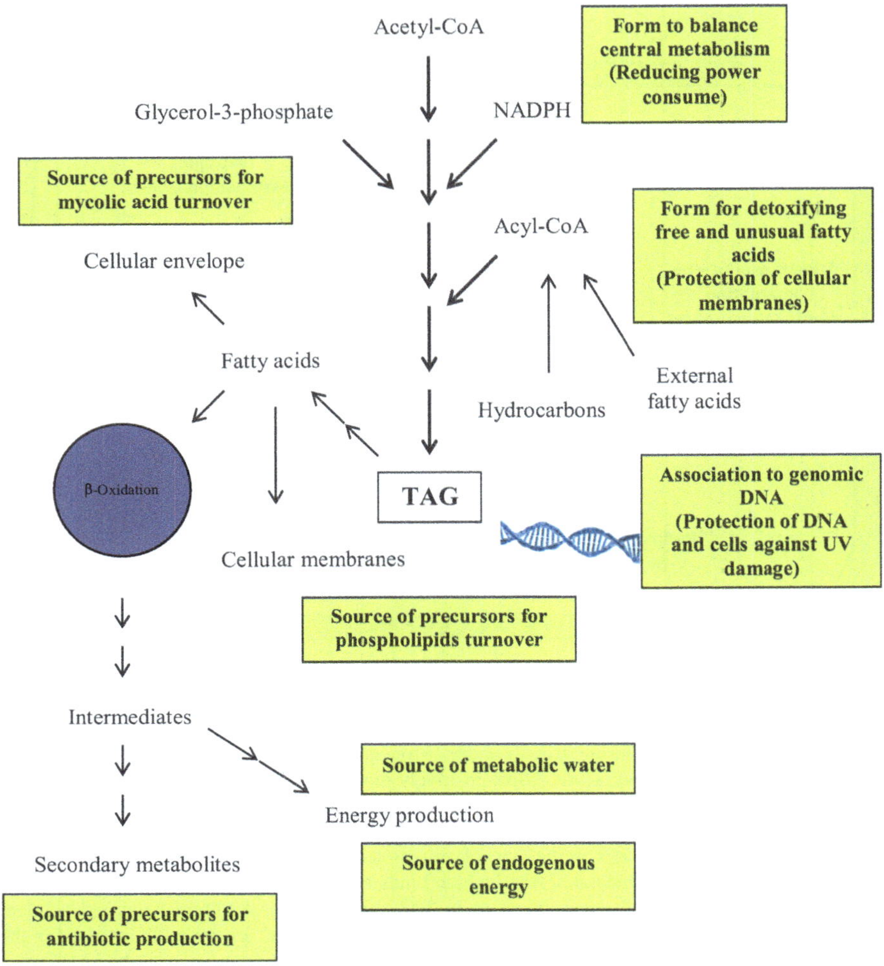

Fig. 4 Physiological functions of TAG proposed for TAG-accumulating rhodococci

9.1 TAG as Endogenous Carbon and Energy Sources

Rhodococci have been detected in different natural environments, such as tropical, arctic, and arid soils, as well as in marine and very deep-sea sediments (Whyte et al. 1998; Heald et al. 2001; Peressutti et al. 2003; Alvarez et al. 2004; Luz et al. 2004; Peng et al. 2008). Interestingly, these microorganisms and other related actinomycetes are frequently dominant components of microbial communities of arid environments (Skujins 1984).

Previous studies revealed that species of the genera *Rhodococcus* and *Gordonia* belong to the autochthonous population in pristine and crude oil-contaminated soils in semiarid Patagonia (Argentina), exhibiting high persistence in these environments (Pucci et al. 2000; Peressutti et al. 2003). In other study, Warton et al. (2001) identified 11 isolates as *Rhodococcus* spp. among a total of 18 Gram-positive bacteria, which were responsible for the biodegradation of the fumigant metham sodium in soil on a farm located in Western Australia. These strains were able to resist dry heat treatments and to recover their degrading ability following dehydration (Warton et al. 2001). The frequent occurrence of rhodococci in arid sites around the world may reflect adaptation to environments with poor nutritional conditions and tolerance to other extreme stresses. The accumulation of significant amounts of TAG by rhodococci is a carbon-intensive and energy-demanding process, which compete with cell growth. Thus, the occurrence of such storage compounds in microorganisms which habits energy-poor environments must be an important feature for their physiology. It is known that TAG are excellent reserve materials due to their extremely hydrophobic properties, which allow their accumulation in large amounts in cells without changing the osmolarity of cytoplasm. In addition, oxidation of TAG produce the maximum yields of energy in comparison with other storage compounds such as carbohydrates and PHA, since the carbon atoms of the acyl moieties of TAG are in their most reductive form (Alvarez and Steinbüchel 2002). Previous studies revealed that TAG serve as carbon and energy sources during incubation of *R. opacus* PD630 cells under starvation and water stress conditions (Alvarez et al. 2000, 2004). In addition, the metabolic activity of cells dropped after incubation under those conditions, whereas the cell counts remained constant. Profound metabolic suppression during unfavorable growth conditions, allows a slow utilization of stored lipids, which are likely mobilized in a programmed manner. The energy obtained by the slow mobilization of stored TAG may support the necessary biochemical and physiological adaptation mechanisms. This process may provide cells of energetic autonomy and a temporal independence from the environment and contribute for cell survival when they do not have access to energy resources in soil.

A similar function has been postulated for the virulent bacterium *M. tuberculosis* which may use TAG as a storage form of energy for its long-term survival under dormancy (Daniel et al. 2004). This microorganism survives for decades within the host in a state of non-replicative, drug resistance dormancy. This state results probably in a diminution in basal metabolic rate, which facilitates survival of cells at expenses of the accumulated TAG.

In addition, TAG may play other important role in TAG-accumulating bacteria such as rhodococci, which occur frequently in arid environments. These lipids may serve also as a reservoir of metabolic water under dry conditions, since fatty acid oxidation releases large amounts of water. Thus, the stored lipids in actinomycetes may be important not only for their energy potential, but also for their metabolic water content (Hauschild et al. 2017). Different rhodococcal strains with the ability to accumulate TAG were isolated from desert soil samples (Bequer Urbano et al. 2013; Röttig et al. 2016). In this context, TAG mobilization enhanced cell survival of the indigenous *Rhodococcus* sp. A5 during cultivation under carbon starvation and desiccation conditions (Bequer Urbano et al. 2013).

9.2 TAG as Source of Precursors for Membranes and Cell Envelope

TAG may serve as precursors for mycolic acid biosynthesis during adaptation of mycolic acid-producing actinomycetes to environmental stresses. Mycolic acids are long-chain length fatty acids produced by elongation of normal fatty acids, which are key components for the integrity and function of the cellular envelope in these bacteria. We investigated the physiological and morphological responses of *R. opacus* PD630 to water stress conditions. During incubation of strain PD630 cells under desiccation conditions, no significant changes in the ultrastructure of the cellular envelope could be detected; thus, the adaptation of its fluidity and permeability may be the result of the variation of the lipid content in response to water stress by a controlled turnover of mycolic acids. Since mycolic acids are produced by elongation of fatty acids by the type II fatty acid synthase complex (FAS II), we studied the effect of isoniazid, which is an inhibitor of the FAS II-system, on the survival of water-stressed cells. Cells pre-treated with isoniazid (40 μg/mL) exhibited lower survival percentages, which were approximately 18% less than those of non-treated cells after 22 days under dehydration conditions. These results suggested that mycolic acid turnover using the pre-formed fatty acids contained into TAG, contributed to cell envelope adaptation under water stress conditions in *R. opacus* PD630 (Alvarez et al. 2004).

TAG may also play a role in regulating the fatty acid composition of membrane lipids, in order to adapt their fluidity to the environment. TAG may serve as a donor of fatty acid for phospholipid biosynthesis under nutritional fluctuating conditions.

So far there is only indirect evidence on the role of TAG as source of precursors, such as pre-formed fatty acids, for biosynthesis or turnover of membranes and cell envelope lipids. Specific studies on this topic are necessary to confirm this function in TAG-accumulating bacteria.

9.3 TAG as a Form to Detoxify Free Fatty Acids

TAG formation may act to protect cells from sudden increases in fluxes of fatty acids in cells. In this context, Garton et al. (2002) proposed that the biosynthesis of TAG may be a form to detoxify free fatty acids, since they observed a rapid accumulation of lipid inclusion bodies by *Mycobacterium* species after transfer of the cells to oleic acid-containing media. This may be relevant for pathogenic actinomycetes, since *M. tuberculosis* and *R. equi* normally sequester fatty acids from the host cells during the infection.

Another interesting aspect for considering in hydrocarbon-degrading rhodococci is the role of TAG as acceptor of unusual fatty acids, which may be generated by the catabolism of cells, protecting the integrity and functionality of cellular membranes (Alvarez and Steinbüchel 2002; Alvarez 2006). We reported that *N. globerula*

432 and *R. opacus* PD630 were able to degrade pristane and phenyldecane, respectively, and synthesize from them TAG containing unusual fatty acids, under unbalanced growth conditions (Alvarez et al. 2001, 2002). In addition, cells of *Mycobacterium ratisbonense* SD4 were able to produce wax esters containing isoprenoid acyl and alcohol residues during incubation of cells on phytane under nitrogen-starved conditions (Silva et al. 2007), whereas *Rhodococcus* sp. 602 accumulated a mixture of TAG containing medium-chain length fatty acids (C_8–C_{12}) after cultivation on naphthyl-1-dodecanoate as sole carbon source (Silva et al. 2010). All these results suggest that TAG serve as acceptor of unusual fatty acids, which would otherwise disturb membrane fluidity during degradation of hydrocarbons under conditions that normally occur in the environment (Alvarez and Steinbüchel 2002; Alvarez 2006). Thus, *Rhodococcus* spp. and related actinomycetes seem to possess metabolic mechanisms that permit cells to maintain the physiological conditions of cytoplasmic membranes during degradation of hydrocarbons under growth-restricting conditions.

9.4 TAG as a Form to Balance Central Metabolism

The biosynthesis of TAG by rhodococci may be also a form to balance the central metabolism dealing with an eventual excess of intermediates, such as acetyl-CoA, or reductive power, under fluctuating conditions as frequently found in natural environments. Previous studies revealed that oxygen-limiting conditions promote TAG accumulation by members of *Mycobacterium* and *Rhodococcus* genera (Daniel et al. 2004; Hernández and Alvarez 2018). When the terminal electron acceptor is not sufficiently supplied during cultivation of cells under conditions of limited aeration, TAG may serve as a sink for reducing equivalents in cells. Under oxygen-limiting conditions, the excess of reducing power may inhibit some key enzymes of central metabolism in cells. The biosynthesis of fatty acids for TAG production, which consumes reduced pyridine nucleotides, may avoid their accumulation in cells. Thus, the biosynthesis of TAG allows cells to balance their metabolism according to the changes of environmental conditions (Alvarez and Steinbüchel 2002; Alvarez 2006).

9.5 TAG as Source of Intermediates for Secondary Metabolism

There is some evidence that TAG may serve as source of intermediates for the synthesis of compounds, which are not essential for growth but for survival of cells in the environment. Some authors demonstrated that TAG act as carbon source for the biosynthesis of antibiotics from acetyl-CoA or malonyl-CoA precursors as has been described by *Streptomyces* strains (Olukoshi and Packter 1994). Storage lipids accumulated by *S. coelicolor* provided carbon for the subsequent synthesis of the acetate-derived antibiotic, actinorhodin, during nutrient deprivation (Banchio and Gramajo 2002). Whether this process also occurs in antibiotic-producing rhodococci must be investigated in the future.

On the other hand, TAG may serve as a source of intermediates for the biosynthesis of the extracellular polymeric substance (EPS) produced as response of diverse stress conditions, such as desiccation, in *Rhodococcus* members. Previous studies revealed that *R. opacus* PD630 was able to progressively accumulate an EPS at the surface of cells during incubation under desiccation conditions (Alvarez et al. 2004). Since the biosynthesis of polysaccharides is a carbon- and energy-intensive process and any external carbon source was available, cells must produce the protective EPS using an endogenous carbon and energy source, such as TAG, among other possible.

9.6 TAG Seem to Protect Cells Against UV Radiation

There are evidences that TAG contribute to withstand the stress exerted by UV radiation in rhodococci. The inhibition of TAG mobilization by the addition of Orlistat promoted a dramatic decrease in cell survival after UV radiation of rhodococcal cells (Bequer Urbano et al. 2013). It has been proposed that the inhibition of TAG degradation by the addition of the metabolic inhibitor might generate an imbalance of the NADPH/NADH ratio affecting the functionality of defense mechanisms against oxidative stress. Interestingly, Zhang et al. (2017) demonstrated that the lipid inclusion body-associated protein (MLDS) provides a survival advantage to bacterial cells under UV-mediated stress by interaction of inclusion bodies with genomic DNA. The authors suggested that the association of lipid droplets with DNA promoted a protection of cells in *R. jostii* RHA1 from UV damage (Zhang et al. 2017). Altogether, these studies suggested that TAG metabolism influences the responses of rhodococci to UV-mediated stress.

10 Biotechnological Significance of Rhodococcal TAG

The world is currently facing a severe energy crisis. On the one side, the known and accessible sources of crude oil and other fossil resources are being slowly but continuously depleted, and on the other side, the demand for fossil resources is rising due to a continuing global industrialization in particular also in countries with large populations like China and India. Therefore, the possibilities to exploit alternative energies are currently intensively investigated. This includes regenerative energies and energy generation from renewable resources. One prominent example is ethanol that is currently mainly produced from liquefied corn starch or sugar cane in particular in North and South America, respectively. TAG are currently produced at large scale by agriculture for synthesis of fatty acid methyl esters. They are currently in Europe the preferred products from renewable resources and are referred to as "Biodiesel." Biodiesel is produced from synthetic methanol of the chemical industry and from TAG by chemical transesterification yielding beside FAME about 10% (wt/wt) glycerol as a by-product (Röttig et al. 2010). Very little amounts of biodiesel are also enzymatically produced (Adamczak et al. 2009). Biodiesel and bioethanol are currently capturing about 90% of the biofuel market (Antoni et al. 2007; Uthoff et al. 2009).

TAG for biodiesel production are currently exclusively produced by agriculture; comparably very little amounts are obtained from the use of frying oil of fast-food restaurants. The main crops for TAG production are rapeseed in Europe, palm oil trees in Southeast Asia, and Soja in North America. TAG could, however, in principle also be produced in bacteria. One of the probably most suitable candidates is *R. opacus* due to its extraordinary high lipid content and the good growth of the cells. As already outlined in other parts of this chapter, *R. opacus* strain PD630 has been investigated in much detail. Lipid contents of as high as 87% have been described for cells cultivated on a small scale (Alvarez et al. 1996). In first attempts cells of *R. opacus* were also grown on a scale of 30 liter and even of 500 liter (Voss and Steinbüchel 2001). Using a mineral salts medium supplemented with beet molasses and sucrose, a cell density of 37.5 g cell dry matter per liter with a lipid content as high as 52% (wt/wt) was obtained at the 30-L scale. At the 500-L scale, which was only done once, a cell density of 18.4 g cell dry matter per liter and a lipid content of 38.4% (wt/wt) were obtained (Voss and Steinbüchel 2001).

This oleaginous bacterium is therefore a promising candidate for the biotechnological production of TAG from renewable resources. Production in bacteria gives a greater flexibility in comparison to plants because various renewable resources and in particular also residual carbon, which is not directly used for production of food and feed, may be used. If the residual carbon cannot be utilized by a strain, the metabolism of this strain may be engineered toward the utilization of such carbon and energy sources for growth and lipid production. This is important in order to avoid a competition between feed and food industry on one side and chemical and energy industry on the other side and also for a sustainable production of lipids and also to avoid a further increase of emission of greenhouse gases (Searchinger et al. 2008; Fargione et al. 2008). In this context, one of the biggest challenges at present is the degradation and conversion of lignocellulosic wastes into lipids by rhodococci. Microbial degradation of lignocellulosic material is almost exclusive to fungi. Kosa and Ragauskas (2012) reported the ability of *R. opacus* strains to convert 4-hydroxybenzoic and vanillic acid as lignin model compounds, into TAG using the β-ketoadipate pathway. In order to efficiently combine lignocellulose catabolism with TAG synthesis in rhodococci, the oxidative machinery of lignocellulose material should be improved by genetic and metabolic engineering. Different studies demonstrated that the production of TAG from the main components of lignocellulosic material is feasible. Xiong et al. (2012) reported the accumulation of lipids by engineered *R. jostii* RHA1 and *R. opacus* PD630 from xylose, an important component of lignocellulose, under nitrogen-limited conditions. The heterologous expression of two genes from *Streptomyces lividans* TK23, *xylA* encoding xylose isomerase and *xylB* encoding xylulokinase, promoted not only growth but also lipid accumulation (52.5% in strain RHA1 and 68.3% in strain PD630 of CDW, respectively). In a different study, Hetzler and Steinbüchel (2013) conferred the ability to utilize cellobiose for growth and production of TAG to *R. opacus* PD630. In this study, recombinant PD630 accumulated fatty acids up to 39.5 ± 5.7% of CDW from cellobiose. *R. opacus* PD630 is not able to utilize L-arabinose present in lignocellulosic hydrolysates as sole carbon source for growth and TAG synthesis. For this reason, Kurosawa et al. (2015) introduced *araB*, *araD*, and *araA* genes derived from a *Streptomyces* species, into strain PD630. After 3 days of cultivation, recombinant cells produced 39.7% of CDW of lipids from L-arabinose. The oleaginous bacterium *R. jostii* RHA1 has also been engineered to utilize L-arabinose derived from lignocellulosic biomass for TAG accumulation. The heterologous expression of the operon, *araBAD*, and *araFGH* genes encoding the arabinose transporter from *Escherichia coli*, as well as the additional expression of *atf1* gene encoding a WS/DGAT from *R. opacus* PD630, promoted the production of 56.8% (CDW) of lipids in the recombinant *R. jostii* RHA1 (Xiong et al. 2016).

The ability of these oleaginous bacteria to convert organic compounds into lipids of interest in biotechnology can be extended to other industrial wastes, such as whey or glycerol. Herrero and Alvarez (2016) reported the efficient production of cellular biomass (6.1–6.3 g/L) and lipids (45–48% of CDW) from whey by different strains of *R. opacus*. On the other hand, *R. fascians*, *R. erythropolis*, and engineered cells of *R. opacus* were able to produce significant amounts of TAG from glycerol, which is the main by-product from biodiesel industry (Herrero et al. 2016).

The key enzyme of triacylglycerol or wax ester biosynthesis in bacteria is a novel type of an acyltransferase which was so far not known in other groups of organisms. They occur frequently in multiple copies in bacteria, and also *R. opacus* strain PD630 possesses several of these acyltransferases (Alvarez et al. 2008). One common feature of all of these acyltransferases is the low substrate specificity. The enzyme from *A. baylyi* seems to transfer acyl moieties of varying carbon chain length from the corresponding acyl coenzyme A thioesters to almost any hydroxyl group and even to some thiol groups (see above). Although no detailed biochemical studies on the substrate ranges of the acyltransferases from *R. opacus* have been made (Alvarez et al. 2008), yet, data from preliminary enzymatic studies and from physiological experiments clearly indicate also for the acyltransferases of this bacterium a low substrate specificity. This makes the enzymes from *R. opacus* and *R. jostii* also to putative candidates for the synthesis of fine chemicals or oleochemicals comprising organic alcohols and thiols to which acyl moieties were covalently attached (Stöveken and Steinbüchel 2008).

11 Concluding Remarks

The accumulation of TAG is a common feature among rhodococci. Some of them can be considered as oleaginous microorganisms because they produce significant amounts of TAG as intracellular inclusion bodies. Although the knowledge acquired during the last decade about the production of TAG in rhodococci has been considerable, many fundamental aspects remain to be clarified. The understanding of this topic in rhodococci is important because two members of this genus, *R. opacus* PD630 and *R. jostii* RHA1, have been the preferred research models in this field, which can be extrapolated also to other actinobacteria with clinical importance, such as *M. tuberculosis*. The occurrence of storage lipids seems to be relevant for the survival of this microorganism in the host cells and, thus, for the development of the disease. Therefore, TAG biosynthesis may be a new target for developing drugs to prevent this important disease. Basic knowledge on this field is also relevant for predicting biotechnological applications of oleaginous bacteria in the industry, for example, for the production of cosmetic products, biofuels, oleochemicals, lubricants, and other manufactured products. In addition, the advances in rhodococcal TAG research will permit the better understanding of their physiology and relation with the environment. TAG may permit cells to survive under fluctuating and unfavorable conditions as occur normally in natural environments. In this context, the occurrence of TAG could be one of the factors that determine the high water stress resistance of rhodococci and their wide distribution in arid environments. The current availability of appropriate molecular tools and methods of analysis, as well as the availability of omic data bases of TAG-accumulating strains, will permit interesting advances in our understanding of the biology of *Rhodococcus* genus. Moreover, several studies have demonstrated the feasibility of producing interesting lipids from diverse industrial wastes by genetic modification of the rhodococcal species and strains.

References

Adamczak M, Bornscheuer UT, Bednarski W (2009) The application of biotechnological methods for the synthesis of biodiesel. Eur J Lipid Sci Technol 111:808–813

Alvarez HM (2003) Relationship between β-oxidation pathway and the hydrocarbon-degrading profile in actinomycetes bacteria. Int Biodeter Biodegr 52:35–42

Alvarez HM (2006) Bacterial triacylglycerols. In: Welson LT (ed) Triglycerides and cholesterol research, vol 6. Nova Science Publishers, New York, pp 159–176

Alvarez HM, Steinbüchel A (2002) Triacylglycerols in prokaryotic microorganisms. Appl Microbiol Biotechnol 60:367–376

Alvarez HM, Mayer F, Fabritius D, Steinbüchel A (1996) Formation of intracytoplasmic lipid inclusions by *Rhodococcus opacus* strain PD630. Arch Microbiol 165:377–386

Alvarez HM, Pucci OH, Steinbüchel A (1997a) Lipid storage compounds in marine bacteria. Appl Microbiol Biotechnol 47:132–139

Alvarez HM, Kalscheuer R, Steinbüchel A (1997b) Accumulation of storage lipids in species of *Rhodococcus* and *Nocardia* and effect of inhibitors and polyethylene glycol. Fett-Lipid 99:239–246

Alvarez HM, Kalscheuer R, Steinbüchel A (2000) Accumulation and mobilization of storage lipids by *Rhodococcus opacus* PD630 and *Rhodococcus ruber* NCIMB 40126. Appl Microbiol Biotechnol 54:218–223

Alvarez HM, Souto MF, Viale A, Pucci OH (2001) Biosynthesis of fatty acids and triacylglycerols by 2,6,10,14-tetramethyl pentadecane-grown cells of *Nocardia globerula* 432. FEMS Microbiol Lett 200:195–200

Alvarez HM, Luftmann H, Silva RA, Cesari AC, Viale A, Wältermann M, Steinbüchel A (2002) Identification of phenyldecanoic acid as constituent of triacylglycerols and wax ester produced by *Rhodococcus opacus* PD630. Microbiology (SGM) 148:1407–1412

Alvarez HM, Silva RA, Cesari AC, Zamit AL, Peressutti SR, Reichelt R, Keller U, Malkus U, Rasch C, Maskow T, Mayer F, Steinbüchel A (2004) Physiological and morphological responses of the soil bacterium *Rhodococcus opacus* strain PD630 to water stress. FEMS Microbiol Ecol 50:75–86

Alvarez AF, Alvarez HM, Kalscheuer R, Wältermann M, Steinbüchel A (2008) Cloning and characterization of a gene involved in triacylglycerol biosynthesis and identification of additional homologous genes in the oleogenous bacterium *Rhodococcus opacus* PD630. Microbiology (SGM) 154:2327–2335

Amara S, Seghezzi N, Otani H, Diaz-Salazar C, Liu J, Eltis LD (2016) Characterization of key triacylglycerol biosynthesis processes in rhodococci. Sci Rep 6:24985

Anderson AJ, Dawes EA (1990) Occurrence, metabolism, metabolic role, and industrial uses of bacterial polyhydroxyalkanoates. Microbiol Rev 54:450–472

Anderson AJ, Williams D, Dawes EA, Ewing D (1995) Biosynthesis of poly(3-hydroxybutyrate-*co*-3-hydroxyvalerate) in *Rhodococcus ruber*. Can J Microbiol 41:4–13

Antoni D, Zverlov VV, Schwarz WH (2007) Biofuels from microbes. Appl Microbiol Biotechnol 77:23–35

Arabolaza A, Rodriguez E, Altabe S, Alvarez H, Gramajo H (2008) Multiple pathways for triacylglycerol biosynthesis in *Streptomyces coelicolor*. Appl Environ Microbiol 74:2573–2582

Banchio C, Gramajo H (2002) A stationary-phase acyl-coenzyme A synthetase of *Streptomyces coelicolor* A3 (2) is necessary for the normal onset of antibiotic production. Appl Environ Microbiol 68:4240–4246

Bequer Urbano S, Albarracín VH, Ordoñez OF, Farías ME, Alvarez HM (2013) Lipid storage in high-altitude Andean Lakes extremophiles and its mobilization under stress conditions in *Rhodococcus* sp. A5, a UV-resistant actinobacterium. Extremophiles 17:217–227

Bloch K (1977) Control mechanisms for fatty acid synthesis in *Mycobacterium smegmatis*. Adv Enzymol Relat Areas Mol Biol 45:1–84

Bredemeier R, Hulsch R, Metzger JO, Berthe-Corti L (2003) Submersed culture production of extracellular wax esters by the marine bacterium *Fundibacter jadensis*. Mar Biotechnol 52:579–583

Brennan PJ (1988) *Mycobacterium* and other actinomycetes. In: Ratledge C, Wilkinson SG (eds) Microbial lipids, vol 1. Academic, London, pp 203–298

Castro AR, Rocha I, Alves MM, Pereira MA (2016) *Rhodococcus opacus* B4: a promising bacterium for production of biofuels and biobased chemicals. AMB Express 6:35

Chen Y, Ding Y, Yang L, Yu J, Liu G, Wang X, Zhang S, Yu D, Song L, Zhang H, Zhang C, Huo L, Huo C, Wang Y, Du Y, Zhang H, Zhang P, Na H, Xu S, Zhu Y, Xie Z, He T, Zhang Y, Wang G, Fan Z, Yang F, Liu H, Wang X, Zhang X, Zhang MQ, Li Y, Steinbüchel A, Fujimoto T, Cichello S, Yu J, Liu P (2014) Integrated omics study delineates the dynamics of lipid droplets in *Rhodococcus opacus* PD630. Nucleic Acids Res 42:1052–1064

Cronan JE Jr, Waldrop GL (2002) Multi-subunit acetyl-CoA carboxylases. Prog Lipid Res 41:407–435

Dahlqvist A, Stahl U, Lanman M, Banas A, Lee M, Sandager L, Ronne H, Stymne S (2000) Phospholipid:diacylglycerol acyltransferase: an enzyme that catalyzes the acyl-CoA-independent formation of triacylglycerol in yeast and plants. Proc Natl Acad Sci U S A 12:6487–6492

Daniel J, Deb C, Dubey VS, Sirakova TD, Abomoelak B, Morbidoni HR, Kolattukudy PE (2004) Induction of a novel class of diacylglycerol acyltransferases and triacylglycerol accumulation in *Mycobacterium tuberculosis* as it goes into a dormancy-like state in culture. J Bacteriol 186:5017–5030

Da Silva PDMP, Lima F, Alves MM, Bijmans MFM, Pereira MA (2016) Valorization of lubricant-based wastewater for bacterial neutral lipids production: growth-linked biosynthesis. Water Res 101:17–24

Dávila Costa JS, Herrero OM, Alvarez HM, Leichert L (2015) Label-free and redox proteomic analyses of the triacylglycerol-accumulating *Rhodococcus jostii* RHA1. Microbiology 161:593–610

Dávila Costa JS, Silva RA, Leichert L, Alvarez HM (2017) Proteome analysis reveals differential expression of proteins involved in triacylglycerol accumulation by *Rhodococcus jostii* RHA1 after addition of methyl viologen. Microbiology (SGM) 163:343–354

Ding Y, Yang L, Zhang S, Wang Y, Du Y, Pu J, Peng G, Chen Y, Zhang H, Yu J, Hang H, Wu P, Yang F, Yang H, Steinbüchel A, Liu P (2012) Identification of the major functional proteins of prokaryotic lipid droplets. J Lipid Res 53:399–3411

Fargione J, Hill J, Tilman D, Polasky S, Hawthorne P (2008) Land clearing and the biofuel carbon debt. Science 319:1325–1238

Feisthauer S, Wick LY, Kästner M, Kaschabek SR, Schlömann M, Richnow HH (2008) Differences of heterotrophic $^{13}CO_2$ assimilation by *Pseudomonas knackmussii* strain B13 and *Rhodococcus opacus* 1CP and potential impact on biomarker stable isotope probing. Environ Microbiol 10:1641–1651

Garton NJ, Christensen H, Minnikin DE, Adegbola RA, Barer MR (2002) Intracellular lipophilic inclusions of mycobacteria in vitro and in sputum. Microbiology 148:2951–2958

Gouda MK, Omar SH, Aouad LM (2008) Single cell oil production by *Gordonia* sp. DG using agroindustrial wastes. World J Microbiol Biotechnol 24:1703–1711

Hauschild P, Röttig A, Madkour MH, Al-Ansari AM, Almakishah NH, Steinbüchel A (2017) Lipid accumulation in prokaryotic microorganisms from arid habitats. Appl Microbiol Biotechnol 101:2203–2216

Heald SC, Brandão PF, Hardicre R, Bull AT (2001) Physiology, biochemistry and taxonomy of deep-sea nitrile metabolising *Rhodococcus* strains. Antonie Van Leeuwenhoek 80:169–183

Hernández MA, Mohn WW, Martínez E, Rost E, Alvarez AF, Alvarez HM (2008) Biosynthesis of storage compounds by *Rhodococcus jostii* RHA1 and global identification of genes involved in their metabolism. BMC Genomics 12(9):600

Hernández MA, Alvarez HM (2010) Glycogen formation by *Rhodococcus* species and effect of inhibition of lipid biosynthesis on glycogen accumulation in *R. opacus* PD630. FEMS Microbiol Lett 312:93–99

Hernández MA, Arabolaza A, Rodríguez E, Gramajo H, Alvarez HM (2013) The *atf2* gene is involved in triacylglycerol biosynthesis and accumulation in the oleaginous *Rhodococcus opacus* PD630. Appl Microbiol Biotechnol 97:2119–2130

Hernández MA, Comba S, Arabolaza A, Gramajo H, Alvarez HM (2015) Overexpression of a phosphatidic acid phosphatase type 2 leads to an increase in triacylglycerol production in oleaginous *Rhodococcus* strains. Appl Microbiol Biotechnol 99:2191–2207

Hernández MA, Lara J, Gago G, Gramajo H, Alvarez HM (2017a) The pleiotropic transcriptional regulator NlpR contributes to the modulation of nitrogen metabolism, lipogenesis and triacylglycerol accumulation in oleaginous rhodococci. Mol Microbiol 103:366–385

Hernández MA, Gleixner G, Sachse D, Alvarez HM (2017b) Carbon allocation in *Rhodococcus jostii* RHA1 in response to disruption and overexpression of *nlpR* regulatory gene, based on $_{13}$C-labeling analysis. Front Microbiol 8:1992

Hernández MA, Alvarez HM (2018) Increasing lipid production using an $NADP_+$-dependent malic enzyme from *Rhodococcus jostii*. Microbiology (SGM). https://doi.org/10.1099/mic.0.000736
Herrero OM, Alvarez HM (2016) Whey as a renewable source for lipid production by *Rhodococcus* strains: physiology and genomics of lactose and galactose utilization. Eur J Lipid Sci Technol 118:262–272
Herrero OM, Moncalián G, Alvarez HM (2016) Physiological and genetic differences amongst *Rhodococcus* species for using glycerol as a source for growth and triacylglycerol production. Microbiology (SGM) 162:384–397
Hetzler S, Steinbüchel A (2013) Establishment of cellobiose utilization for lipid production in *Rhodococcus opacus* PD630. Appl Environ Microbiol 79:3122–3125
Holtzapple E, Schmidt-Dannert C (2007) Biosynthesis of isoprenoid wax ester in *Marinobacter hydrocarbonoclasticus* DSM 8798: identification and characterization of isoprenoid coenzyme A synthetase and wax ester synthases. J Bacteriol 189:3804–3812
Huang L, Zhao L, Zan X, Song Y, Ratledge C (2016) Boosting fatty acid synthesis in *Rhodococcus opacus* PD630 by overexpression of autologous thioesterases. Biotechnol Lett 38:999–1008
Huisman GW, Siegele DA, Zambrano MM, Kolter R (1993) Morphological and physiological changes during stationary phase. In: Neidhardt FC, Ingraham JL, Low KB, Magsanik B, Schaechter M, Umbarger HE (eds) *Escherichia coli* and *Salmonella typhimurium*: cellular and molecular biology, vol 2. American Society for Microbiology, Washington DC, pp 1672–1682
Juarez A, Villa JA, Lanza VF, Lázaro B, de la Cruz F, Alvarez HM, Moncalián G (2017) Nutrient starvation leading to triglyceride accumulation activates the Entner Doudoroff pathway in *Rhodococcus jostii* RHA1. Microb Cell Factories 16:35
Kaddor C, Biermann K, Kalscheuer R, Steinbüchel A (2009) Analysis of neutral lipid biosynthesis in *Streptomyces avermitilis* MA-4680 and characterization of an acyltransferase involved herein. Appl Microbiol Biotechnol 84:143–155
Kalscheuer R, Steinbüchel A (2003) A novel bifunctional wax ester synthase/acyl-CoA: diacylglycerol acyltransferase mediates wax ester and triacylglycerol biosynthesis in *Acinetobacter calcoaceticus* ADP1. J Biol Chem 278:8075–8082
Kalscheuer R, Stöveken T, Malkus U, Reichelt R, Golyshin PN, Sabirova JS, Ferrer M, Timmis KN, Steinbüchel A (2007) Analysis of storage lipid accumulation in *Alcanivorax borkumensis*: evidence for alternative triacylglycerol biosynthesis routes in bacteria. J Bacteriol 189:918–928
Kosa M, Ragauskas AJ (2012) Bioconversion of lignin model compounds with oleaginous rhodococci. Appl Microbiol Biotechnol 93:891–900
Kurosawa K, Plassmeier J, Kalinowski J, Rückert C, Sinskey AJ (2015) Engineering L-arabinose metabolism in triacylglycerol-producing *Rhodococcus opacus* for lignocellulosic fuel production. Metab Eng 30:89–95
Luz AP, Pellizari VH, Whyte LG, Greer CW (2004) A survey of indigenous microbial hydrocarbon degradation genes in soils from Antarctica and Brazil. Can J Microbiol 50:323–333
Makula RA, Lockwood PJ, Finnerty WR (1975) Comparative analysis of lipids of *Acinetobacter* species grown on hexadecane. J Bacteriol 121:250–258
MacEachran DP, Prophete ME, Sinskey AJ (2010) The *Rhodococcus opacus* PD630 heparin-binding hemagglutinin homolog TadA mediates lipid body formation. Appl Environ Microbiol 76:7217–7225
MacEachran DP, Sinskey AJ (2013) The *Rhodococcus opacus* TadD protein mediates triacylglycerol metabolism by regulating intracellular NAD(P)H pools. Microb Cell Factories 12:104
Olukoshi ER, Packter NM (1994) Importance of stored triacylglycerols in *Streptomyces*: possible carbon source for antibiotics. Microbiology (SGM) 140:931–943
Peressutti SR, Alvarez HM, Pucci OH (2003) Dynamic of hydrocarbon-degrading bacteriocenosis of an experimental oil pollution on patagonic soil. Int Biodeterior Biodegrad 52:21–30
Peng F, Wang Y, Sun F, Liu Z, Lai Q, Shao Z (2008) A novel lipopeptide produced by a Pacific Ocean deep-sea bacterium, *Rhodococcus* sp. TW53. J Appl Microbiol 105:698–705

Pieper U, Steinbüchel A (1992) Identification, cloning and sequence analysis of the poly (3-hydroxyalkanoic acid) synthase gene of the Gram-positive bacterium *Rhodococcus ruber*. FEMS Microbiol Lett 96:73–80

Pucci OH, Bak MA, Peressutti SR, Klein SR, Härtig C, Alvarez HM, Wünsche L (2000) Influence of crude oil contamination on the bacterial community of semi-arid soils of Patagonia (Argentina). Acta Biotechnol 10:129–146

Rawsthorne S (2002) Carbon flux and fatty acid synthesis in plants. Prog Lipid Res 41:182–196

Rontani JF, Mouzdahir A, Michotey V, Caumette P, Bonin P (2003) Production of a polyunsaturated isoprenoid wax ester during aerobic metabolism of squalene by *Marinobacter squalenivorans* sp. nov. Appl Environ Microbiol 69:4167–4176

Röttig A, Wenning L, Bröker D, Steinbüchel A (2010) Fatty acid alkyl esters: perspectives for production of alternative biofuels. Appl Microbiol Biotechnol 85:1713–1733

Röttig A, Hauschild P, Madkour MH, Al-Ansari AM, Almakishah NH, Steinbüchel A (2016) Analysis and optimization of triacylglycerol synthesis in novel oleaginous *Rhodococcus* and *Streptomyces* strains isolated from desert soil. J Biotechnol 225:48–56

Searchinger T, Heimlich R, Houghton RA, Dong F, Elobeid A, Fabiosa J, Tokgoz S, Hayes D, Yu T-H (2008) Use of U.S. Croplands for biofuels increases greenhouse gases through emissions from land-use change. Science 319:1238–1240

Shields-Menard SA, Amirsadeghi M, Sukhbaatar B, Revellame E, Hernandez R, Donaldson JR, French WT (2015) Lipid accumulation by *Rhodococcus rhodochrous* grown on glucose. J Ind Microbiol Biotechnol 42:693–699

Shweizer E, Hofmann J (2004) Microbial type I fatty acid synthases (FAS): major players in a network of cellular FAS systems. Microbiol Mol Biol Rev 68:501–517

Silva RA, Grossi V, Alvarez HM (2007) Biodegradation of phytane (2,6,10,14-tetramethylhexadecane) and accumulation of related isoprenoid wax esters by *Mycobacterium ratisbonense* strain SD4 under nitrogen-starved conditions. FEMS Microbiol Lett 272:220–228

Silva RA, Grossi V, Olivera N, Alvarez HM (2010) Characterization of the indigenous *Rhodococcus* sp. 602, a strain able to accumulate triacylglycerides from naphthyl-compounds under nitrogen-starved conditions. Res Microbiol 161(3):198–207

Skujins J (1984) Microbial ecology of desert soils. Adv Microb Ecol 7:49–91

Sorger D, Daum G (2002) Synthesis of triacylglycerols by the acyl-Coenzyme A:diacyl-glycerol acyltransferase Dga1p in lipid particles of the yeast *Saccharomyces cerevisiae*. J Bacteriol 184:519–524

Steinbüchel A (1991) Polyhydroxyalkanoic acids. In: Byrom D (ed) Biomaterials. MacMillan, London, pp 123–213

Steinbüchel A, Valentin HE (1995) Diversity of bacterial polyhydroxyalkanoic acids. FEMS Microbiol Lett 128:219–228

Stöveken T, Steinbüchel A (2008) Bacterial acyltransferases as an alternative for lipase-catalyzed acylation for the production of oleochemicals and fuels. Angew Chem Int Ed 47:3688–3694

Uthoff S, Bröker D, Steinbüchel A (2009) Current state and perspectives of producing biodiesel-like compounds by biotechnology. Microb Biotechnol 2:551–565

Villalba MS, Hernández MA, Silva RA, Alvarez HM (2013) Genome sequences of triacylglycerol in *Rhodococcus* as a platform for comparative genomics. J Mol Biochem 2:94–105

Villalba MS, Alvarez HM (2014) Identification of a novel ATP-binding cassette transporter involved in long-chain fatty acid import and its role in triacylglycerol accumulation in *Rhodococcus jostii* RHA1. Microbiology (SGM) 160:1523–1532

Voss I, Steinbüchel A (2001) High cell density cultivation of *Rhodococcus opacus* for lipid production at a pilot scale. Appl Microbiol Biotechnol 55:547–555

Wakil SJ, Stoops JK, Joshi VC (1983) Fatty acid synthesis and its regulation. Annu Rev Biochem 52:537–579

Wältermann M, Steinbüchel A (2000) In vitro effects of sterculic acid on lipid biosynthesis in *Rhodococcus opacus* strain PD630 and isolation of mutants defective in fatty acid desaturation. FEMS Microbiol Lett 190:45–50

Wältermann M, Luftmann H, Baumeister D, Kalscheuer R, Steinbüchel A (2000) *Rhodococcus opacus* PD630 as a source of high-value single cell oil? Isolation and characterisation of triacylglycerols and other storage lipids. Microbiology (SGM) 146:1143–1149

Wältermann M, Hinz A, Robenek H, Troyer D, Reichelt R, Malkus U, Galla H-J, Kalscheuer R, Stöveken T, von Landenberg P, Steinbüchel A (2005) Mechanism of lipid-body formation in prokaryotes: how bacteria fatten up. Mol Microbiol 55:750–763

Wältermann M, Stöveken T, Steinbüchel A (2007) Key enzymes for biosynthesis of neutral lipid storage compounds in prokaryotes: properties, function and occurrence of wax ester synthases/ acyl-CoA:diacylglycerol acyltransferases. Biochimie 89:230–242

Warton B, Matthiessen JN, Roper MM (2001) The soil organisms responsible for the enhanced biodegradation of metam sodium. Biol Fertil Soils 34:264–269

Whyte LG, Hawari J, Zhou E, Bourbonniere L, Inniss WE, Greer CHW (1998) Biodegradation of variable-chain length alkanes at low temperatures by a psychrotrophic *Rhodococcus* sp. Appl Environ Microbiol 64:2578–2584

Xiong X, Wang X, Chen S (2012) Engineering of a xylose metabolic pathway in *Rhodococcus* strains. Appl Environ Microbiol 78:5483–5491

Xiong X, Wang X, Chen S (2016) Engineering of an L-arabinose metabolic pathway in *Rhodococcus jostii* RHA1 for biofuel production. J Ind Microbiol Biotechnol 43:1017–1025

Zhang C, Yang L, Ding Y, Wang Y, Lan L, Ma Q, Chi X, Wei P, Zhao Y, Steinbüchel A, Zhang H, Liu P (2017) Bacterial lipid droplets bind to DNA via an intermediary protein that enhances survival under stress. Nat Commun 8:15979

Zimhony O, Vilchèze C, Jacobs WR Jr (2004) Characterization of *Mycobacterium smegmatis* expressing the *Mycobacterium tuberculosis* fatty acid synthase I (fas1) gene. J Bacteriol 186:4051–4055

Interaction of *Rhodococcus* with Metals and Biotechnological Applications

Alessandro Presentato, Elena Piacenza, Martina Cappelletti, and Raymond J. Turner

Contents

Abstract In studies of environmental stresses caused by metals, *Rhodococcus* species are routinely identified as part of a beneficial microbial rhizosphere community. These bacterial strains, inhabiting diverse ecological niches, possess a variety of enzymatic activities to carry out relevant biodegradation reactions, such as degradation of organic pollutants in some cases using them for both carbon and energy. In this context, most *Rhodococcus* strains have been found to have very high

A. Presentato (✉)
Environmental Microbiology Laboratory, Department of Biotechnology, University of Verona, Verona, Italy
e-mail: alessandro.presentato@univr.it

E. Piacenza · R. J. Turner (✉)
Department of Biological Sciences, University of Calgary, Calgary, AB, Canada
e-mail: elena.piacenza@ucalgary.ca; turnerr@ucalgary.ca

M. Cappelletti
Unit of General and Applied Microbiology, Department of Pharmacy and Biotechnology, University of Bologna, Bologna, Italy
e-mail: martina.cappelletti2@unibo.it

H. M. Alvarez (ed.), *Biology of Rhodococcus*, Microbiology Monographs 16,
https://doi.org/10.1007/978-3-030-11461-9_12

levels of metal resistance. Thus, these microorganisms are not only capable of metabolizing various organic pollutants in the presence of co-contaminating heavy metals, but they can also bioadsorb and/or bioconvert various metals and metalloids [metal(loid)s]. Indeed, some *Rhodococcus* exploit these metal(loid) compounds to generate biogenic nanoscale materials of intriguing physical-chemical properties, which can find applications in biotechnology.

This book chapter has the focus in overviewing the biotechnological relevance of the *Rhodococcus* genus relationship with metal(loid)s, the bioprocesses elicited by these microorganisms in handling metal(loid)s' toxicity, and the importance of these actinomycetes in the context of the bioremediation and bionanotechnology fields.

1 Introduction

Metal elements such as aluminum (Al), cadmium (Cd), cesium (Cs), chromium (Cr), cobalt (Co), copper (Cu), iron (Fe), lead (Pb), manganese (Mn), mercury (Hg), molybdenum (Mo), nickel (Ni), silver (Ag), strontium (Sr), zinc (Zn), and uranium (U) and the metalloids arsenic (As), selenium (Se), and tellurium (Te) are naturally occurring throughout the Earth's crust, being defined "heavy" as both their atomic weight and density are higher than water (Fergusson 1990). Although some of these elements (e.g., Fe, Se, Zn, Cu) are essential in traces for living organisms, the overdose of these and exposure to some others (e.g., As, Pb, Hg, whose toxicity is undisputed) can be severely toxic (Chang et al. 1996; Tchounwou et al. 2008). Indeed, the global human concern toward these contaminants relies mostly on their mutagenicity and carcinogenicity, negatively impacting the quality of life of both flora and fauna in different habitats (i.e., aquatic, terrestrial, and atmospheric) (Chakraborty et al. 2017). This aspect is exacerbated by the intense industrial progress in several fields (i.e., engineering, biomedicine, agriculture, metallurgy, microelectronics, mining operations, etc.), as well as unjustified waste disposal processes, which inexorably add to environmental pollution. The result is that there are sites in almost every community on the planet with a buildup of substantial amounts of recalcitrance toxic metal(loid)s in diverse ecological niches (Chakraborty et al. 2017). In this context, a prolonged exposure to metal(loid) compounds is often associated to human poisoning, as well as to their biomagnification at successively high levels in the food chain (Phillips and Rainbow 2013).

Toxicity of a metal to an organism depends on a variety of physical-chemical and biological parameters influencing the metal ion speciation and bioavailability. Chemical parameters include temperature, pH, redox potential, and presence/absence of compounds capable of chelating or binding to the metal. Some biological parameters include the metal-lipid solubility, physiological state of the organism, presence/absence of key metal interacting proteins, and any physiological adaptation of the organisms. The combination of such parameters and their interplay contribute to the environmental resiliency of metal(loid)s (Hamelink et al. 1994). Focusing the attention at the molecular level, it is generally believed that proteins are the primary targets of metal(loid)s. Indeed, malfunctioning of proteins in their native conformation can be attributed to (1) the binding of metals to protein ligands such as free thiols (RSHs), imidazoles, amines, or carboxylates, (2) the displacement of essential metal ions in metalloenzymes, (3) the oxidation of the amino acid side chains, and (4) their interaction with non-folded proteins causing their aggregation and, therefore, impairing protein homeostasis in living organisms (Beyersmann and Hartwig 2008; Sharma et al. 2011; Lemire et al. 2013).

As dramatic as the above picture may appear, human beings have evolved the concept of *environmental remediation*, which combines physical-chemical and biological knowledge for strategies aimed to decrease the concentration of metal (loid) compounds into the environment, restoring its ecological fitness. Overall, biological methods, exploiting the natural catabolic abilities of bacteria, fungi, and plants to clean up diverse contaminated matrices in a *green* and cost-effective manner (Strong and Burgess 2008), are generally preferred over physical-chemical approaches, as the latter often have cost-prohibitive and non-efficient results at low metal(loid) environmental load (Ali et al. 2013).

Aerobic non-sporulating microorganisms belonging to the *Rhodococcus* genus are known for their environmental robustness and persistence, as they can physiologically adapt (i.e., modification of the cell membrane composition, formation of intracellular inclusions) to several stressful conditions (e.g., pH, desiccation, low nutrient availability, presence of xenobiotic compounds, etc.) (Martínková et al. 2009; Alvarez and Steinbüchel 2010; Presentato et al. 2018a). This allows members of this genus to populate diverse ecological niches. Although much less is known about the *Rhodococcus* genus potential to resist toxic metal(loid) compounds, recently the unique capacity of some rhodococci to overcome environmental stresses like the presence of antimicrobials and very high concentrations of metal(loid)s such as tellurite [Te (IV)], arsenate [As (V)], and selenite [Se (IV)] has been described (Orro et al. 2015; Cappelletti et al. 2016; Presentato et al. 2016, 2018b). Such findings highlight how these microorganisms are worth exploring as candidates for bioremediation of metal(loid)-polluted sites. Further, it has been reported that bacteria are able to convert toxic metal(loid) ion/oxyanions into less/nontoxic metal forms, generating discrete structures at the nanometer scale (1×10^{-9} m) with different shapes, i.e., nanoparticles (NPs) and nanorods (NRs) (Turner et al. 2012). To this aim, different *Rhodococcus* strains have been studied and exploited as cell factories for nanotechnological purposes, proving the specific proficiency of rhodococci to biosynthesize high-quality biogenic metal(loid) nanostructures with relevant biotechnological applications (Ahmad et al. 2003; Otari et al. 2012; Kundu et al. 2014; Presentato et al. 2016, 2018b, c). Thus, this chapter will overview the interaction between *Rhodococcus* genus and metal(loid)s in terms of bacterial resistance and strategies adopted to counteract metal(loid) toxicity, as well as their remarkable ability in generating biogenic nanomaterials.

2 Microbial Interaction with Metal(loid)s

A large number of organisms are capable of thriving in the presence of high metal (loid) load, being this feature ascribed to intrinsic and/or induced biochemical mechanisms combined with environmental factors (Gadd 1992a). In this respect, bacteria are defined to be tolerant as they are not killed by the toxicity exerted by metal(loid) compounds through their intrinsic properties yet do not grow or proliferate, while those that are resistant show the ability to live because of the induction of detoxification mechanisms implemented in response to the presence of metal(loids) (Gadd 1992b). Bacteria have evolved uptake mechanisms in order to exploit diverse nutrients, which enable them to survive even under extreme and adverse environmental conditions (Brooks et al. 2011). Because microorganisms have coevolved with the geochemistry of the planet and inhabit all possible niches, many can utilize even toxic metal(loid) compounds (i.e., As, Cd, Te, Se, etc.) (Gadd 2010). For instance, bacterial cells can encounter metal(loid) ions from either aqueous or solid matrices, using them as (1) energy source, (2) terminal electron acceptor during microbial respiration, and (3) cofactors in metalloproteins and enzymes (Ehrlich 1997; Valls and de Lorenzo, 2002). Thus, microbial scavenging and biotransformation of metal(loid) elements can be related to either detoxification processes or resistance phenomena, which are carried out as parallel events related or not with bacterial growth (Park et al. 2018).

The different mechanisms exploited by microbes to cope with metal(loid)s' toxicity can rely on their transformation by means of either redox processes or alkylation reactions (Ledin 2000). Bacterial cells can sequester metal(loid) ions onto their cell wall either in a passive and metabolism-independent manner (biosorption) or through active mechanisms (bioaccumulation), which is a metabolism-dependent process based on the actual metal(loid) uptake (Ledin 2000). Further, metal(loid) ions can be trapped outside bacterial cells through interaction with (1) surface proteins, (2) carbohydrates, (3) cell wall polymers, and (4) extracellular polymeric substance (EPS) often produced when bacteria grow as a biofilm. Finally, there may be chelation (precipitation and/or complexation) through the production of either siderophores or specific metalloproteins (Gupta and Diwan 2017). All these strategies occur in the microbial world; hence, the exploitation of the microbial catabolic abilities represents a powerful tool to abate the metal(loid) load polluting the environment.

2.1 Rhodococcus *Tolerance/Resistance Toward Metal(loid)s*

Bacterial strains belonging to the *Actinomycetales* order, which include *Rhodococcus* genus, exploit different mechanisms enabling them to highly tolerate metal(loid) compounds (Pavel et al. 2013). For instance, the reduction of cellular sensitivity, the intracellular sequestration of metal ions and oxyanions, their complexation with siderophores, the alteration of the membrane permeability, mutations, and repairing mechanisms of the DNA responsible for both plasmid and chromosomal DNA stability are some of the mechanisms implemented by bacteria to tolerate and/or resist to metalloids' toxicity (Stillman 1995; Garbisu and Alkorta 2003; Figueira et al. 2005). A primary mechanism through which metal(loid) compounds can exert their toxicity toward bacteria relies on the alteration of the cellular thiol chemistry. Indeed, the reaction of metal(loid)s with cell thiols leads to the depletion of glutaredoxin and thioredoxin from their activity of reducing metal (loid) compounds and becoming oxidized (Turner 2001). Thus, such reactions can cause an increase in the intracellular content of reactive oxygen species (ROS), which can ultimately cause cellular death (Turner et al. 1999). This process is even more emphasized if we consider the presence of oxygen as terminal electron acceptor in the case of strictly aerobic microorganisms, as in the case of *Rhodococcus* strains. This process is even more emphasized if we consider the presence of oxygen as terminal electron acceptor in the case of strictly aerobic microorganisms, such as in the case of *Rhodococcus* strains. However, instead of having the redox buffer molecule glutathione (GSH), actinomycetes are mostly characterized by the presence of genes involved in the synthesis of mycothiols (MSHs), which are more stable and less prone to oxidation (Newton et al. 1996). Thus, the potential of actinomycetes in resisting the toxicity exerted by metal(loid)s overall derives from the *redox stability* of these microorganisms. Rhodococci investigated for metal tolerance to date are listed in Table 1 along with their resistance levels. It is noteworthy to highlight that among the metal(loid) elements listed in Table 1, tellurium in the form of oxyanion tellurite (TeO_3^{2-}) is typically the most toxic, with minimal inhibitory concentrations (MIC) as low as 1 $\mu g\ mL^{-1}$ (Taylor 1999). This concentration is several orders of magnitude lower than other metal(loid) compounds (Nies 1999; Harrison et al. 2004), and yet *R. aetherivorans* BCP1 strain showed to be highly resistant to this oxyanion (Presentato et al. 2016, 2018c).

In nature microorganisms are found in a close relationship to each other and adhering to a surface forming a complex community known as biofilm. Within a biofilm, the bacterial cells are overprotected by a heterogeneous surrounding matrix defined as extracellular polymeric substance (EPS), which is mostly constituted by water, proteins, polysaccharides, lipids, and extracellular DNA (e-DNA) (Harrison et al. 2005). Thus, a bacterium living within a biofilm has the advantage of resisting and adapting to harsh environmental conditions (this topic has been reviewed by Harrison et al. 2007). In line with this, Adhami et al. (2017) described two different *Rhodococcus* strains with an enhanced metal resistance against Cd, Cr, Cu, Pb, and Zn supplied as salts when these bacterial cells were grown as biofilms in comparison with those grown as planktonic cultures (Adhami et al 2017).

Table 1 Minimal inhibitory concentration (MIC) of different metal(loid)s of *Rhodococcus* species

MICs of metal(loid)s of *Rhodococcus* spp. (mM)										
Strain		Cd	Cr	Cu	Hg	Pb	Se	Te	Zn	References
R. RS67		ND[a]	ND	ND	ND	0.5	ND	ND	0.2	Perelomov et al. (2018)
R. rhodochrous		8	1	4	ND	8	ND	ND	8	Adhami et al. (2017)
	MBEC[b]	16	4	8	ND	16	ND	ND	16	
	MBIC[c]	4	0.5	1	ND	4	ND	ND	2	
R. rhodnii		8	1	4	ND	8	ND	ND	8	
	MBEC	16	4	8	ND	16	ND	ND	16	
	MBIC	4	0.5	2	ND	2	ND	ND	1	
R. aetherivorans BCP1		ND	ND	ND	ND	ND	500	11.2	ND	Presentato et al. (2016, 2018b, c)
R. strain 3		8–10	ND	8–16	ND	>16	ND	ND	8–16	Vela-Cano et al. (2014)
R. ruber	AC239	ND	>1.8	3.7	0.28	ND	ND	ND	ND	Fleck et al. (2000)
	AC74	ND	>1.8	1.5	0.17	ND	ND	ND	ND	
	AC87	ND	>1.8	3.5	0.28	ND	ND	ND	ND	
R. erythropolis	AC272	ND	>1.8	3	0.21	ND	ND	ND	ND	
	AC265	ND	>1.8	0.1	0.03	ND	ND	ND	ND	

[a]Not determined
[b]Minimum biofilm eradicating concentration
[c]Minimum biofilm inhibitory concentration

3 Metal(loid) Biosorption in *Rhodococcus*

Biosorption is recognized as an emerging and *eco-friendly* approach for metal(loid)s' or radionuclides' removal from polluted environments or in biomining, as it is cost-effective, is highly efficient, and can be specific. This technique relies on the application of different microbial biomasses (i.e., microalgae or bacterial cells of the genera including *Bacillus*, *Pseudomonas*, *Streptomyces*, etc.) as biosorbents, which possess a natural affinity toward metal(loid) compounds, contributing to a feasible and economic way toward the purification of metals from contaminated matrices (Volesky 1990; Wilde and Benemann 1993). From an engineering perspective, bacteria are considered good biosorbents as their small size offer a high surface-to-volume ratio, conferring a large surface area of interaction with metal(loid)s. Several molecular mechanisms are possible for biosorption performed by bacteria (Gadd and White 1993), dependent on factors, such as capacity, affinity, and specificity of the biomass and/or biosorbent considered, as well as the physical-chemical conditions of a given environmental niche and the metal of interest.

3.1 Parameters Influencing Metal(loid) Biosorption

The biosorption of metal(loid)s by bacteria depends on the characteristics of the surface structural chemistry of the cell wall and membrane (i.e., charge and orientation of the metal(loid) binding functional groups), as well as the chemistry and speciation of the considered metal(loid) (Ledin 2000). A given bacterial cell surface can be composed of lipids, peptidoglycan, lipopolysaccharides, lipoproteins, and enzymes, being featured by carboxyl (COO^-), sulfate (SO_4^{2-}), phosphate (PO_4^{3-}), amino (NH_2^+/NH_3^+), carbonyl (CO^-), and hydroxyl (OH^-) functional groups as well as the acyl chains of membranes. Thus, the amphoteric behavior of a cell wall depends on the presence of these anionic and cationic groups, being those anionic predominant over the cationic ones (Plette et al. 1995; van der Wal et al. 1997).

Restricting the focus onto rhodococci, it has been reported that these microorganisms have a cell wall characterized by polysaccharides, carboxylic acids, lipid groups, and mycolic acids, which are responsible for the amphoteric behavior of the cellular surface (Stratton et al. 2002; Botero et al. 2007). Moreover, Plette et al. (1995) showed that the isolated cell wall material deriving from *R. erythropolis* A177 was featured by a chemical heterogeneity due to the existence of at least three different proton group binding sites (i.e., carboxylic, phosphate, and amino types), therefore leading to a surface charge dependent on the pH and salt level (Plette et al. 1995). A reasonable explanation for the effect of the pH on the sorption capability of a given biosorbent relies on the pKa of the biosorbent molecules and the combined isoelectric point of all biosorbent molecules and the number of active sites present as well as the speciation state of the metal ion. Therefore, at pH values lower than the biosorbent isoelectric point, its surface will be positively charged because of the association with hydronium (H_3O^+) ions, determining a repulsion of metal cations. By contrast, the biosorbent surface will be negatively charged for pH values higher than its isoelectric point; hence, it favors the adsorption of cations. Furthermore, at pH values above the pKa of a given biosorbent, its functional groups will be dissociated, enabling proton exchange with the metal(loid) in solution (Cayllahua and Torem 2010). In line with this, Cayllahua and Torem (2010) reported that the biosorption capability of a *R. opacus* strain toward Al (III) ions was optimal at pH of 5, as its isoelectric point was c.a. 3.26. Similar observations were also reported for ions of Pb (II), Cr (III), Cu (II), Cd (II), and Zn (II) (Sheng et al. 2004; Vasquez et al. 2007; Bueno et al. 2008). Thus, depending on the type of metal(loid)s to be removed (i.e., anions or cations), it must be considered the characteristics of the biosorbent and the pH of the system at which the metal(loid)s' sorption occurs.

In natural settings, a variety of different metal(loid) ions other than those targets of bioremediation can be found, leading to a competition for the available binding sites present on the surface of bacterial cells, therefore interfering with the remediation of polluted sites. Calcium [Ca (II)], for example, is known as the major bivalent ion competitor for metal(loid) sorption in soil (Doyle et al. 1980; Xue et al. 1988; Flemming et al. 1990). However, it is noteworthy to mention that the cell wall material isolated from *R. erythropolis* A177 showed an enhanced Cd (II) sorption in the presence of Ca (II) ions. Indeed, the binding of Ca (II) ions to the cell wall can cause (1) the breaking of cross-linkages that normally occurs between different groups (e.g., COO^- and NH_3^+) of the cell wall and (2) the disruption of bonds occurring between phosphate groups of the phospholipids and others present in the peptidoglycan layer. These effects cause a drastic change of the cell surface and wall morphology leading to an increased availability of the number of accessible and active binding sites for other bivalent cations, such as Cd (II). Thus, since the binding of Ca (II) to the cell wall enhances the binding of Cd (II), in this case Ca (II) can be considered as a cooperative acting ion toward Cd (II), influencing in a salt-dependent manner the sorption potential of *R. erythropolis* A177-derived cell wall (Plette et al. 1996).

The concentration of the biosorbent in a given environment is another parameter of crucial importance, as the sorption capability of a biosorbent tends to decrease when its concentration rises. This aspect might be related to the cross-linkages occurring among the functional groups as described above (Koch 1990). Indeed, the cell wall material recovered from *R. opacus* strain was capable of efficiently adsorbing Al (III), Cd (II), and Zn (II) ions, while their sorption dropped down when the biosorbent concentration was increased (Vasquez et al. 2007; Cayllahua and Torem 2010).

3.2 EPS-Mediated Extracellular Biosorption

Another modality through which microorganisms can interface with metal(loid) compounds relies on the production of biological exudates featured by a high adsorption potential. Overall, these biologically active compounds are known as EPS (see Sect. 2), which includes polysaccharides and their derivatives, proteins, peptides, metabolites, nucleic acids, and lipids. These microbial-derived compounds showed to have an efficient adsorptive property (Jixian et al. 2015; Zheng-bo et al. 2015), due to the presence of different functional reactive groups [i.e., OH^-, COO^-, acetamido ($AcNH^+$), NH_3^+] characterizing the polymer chains, which in turn confer to EPS: (1) high hydrophilicity, (2) polarity, and (3) structural flexibility (Crini 2005). Moreover, the amount of functional binding groups featuring EPS reflects its actual biosorption potential, working also as enhancer for the metal(loid) binding to microbial cell surfaces (Liu and Fang 2002; Wei et al. 2011). Since the treatment of metal(loid)-polluted sites via EPS is based on the binding of ions, the pH values at which the sorption process occurs are of fundamental importance as discussed above with the cell surface. Indeed, the competition between protons and metal(loid) ions to bioadsorb onto the active and accessible sites present in the EPS surface should be controlled, in order to maintain the appropriate electronic state allowing for chelation or complexation (Lamelas et al. 2006; Guibaud et al. 2008).

The EPS composition will be highly dependent on the bacterial species/strain, and thus its adsorption properties might be influenced by different protocols applied, reagents utilized, and physical-chemical (e.g., sonication, ionic exchange resin) method adopted for the extraction (Liu and Fang 2002; Comte et al. 2005). This leads toward the challenge that every system will be different and optimization strategy will need to be empirically determined. This reflects the importance of fundamental microbiology research to expand the knowledge on the possible microbial factories implied for this purpose.

3.3 Parameters Influencing EPS Biosorption

As biosorption EPS-mediated appears an attractive and *green* choice to develop new eco-sustainable strategies of environmental remediation, researchers have focused their attention in exploiting a variety of microorganisms belonging to diverse families (i.e., *Cyanobacteriaceae*, *Bacillaceae*, *Pseudomonadaceae*, *Myxococcaceae*) as cell factories for the production and extraction of EPS to test against different metals (e.g., Cd, Cu, Pb, Mn, Ni, Zn, Hg) (Merroun et al. 2003; Salehizadeh and Shojaosadati, 2003; Freire-Nordi et al. 2005; Lau et al. 2005; Paperi et al. 2006). Among rhodococci, Dobrowolski et al. (2017) studied the sorption capability of EPS material derived from *R. opacus* and *R. rhodochrous*, revealing a fast adsorption rate (between 1 and 30 min) of Ni (II), Pb (II), Co (II), Cd (II), and Cr (VI) for EPS material from both strains, which supports their application in flow adsorption systems (Dobrowolski et al. 2017).

The application of microbial EPS is highly influenced by the pH values at which the metal(loid)s' sorption occurs. Indeed, the best biosorption of Ni (II) mediated by the EPS recovered from either *R. opacus* or *R. rhodochrous* strains occurred at pH values ranging from 3 to 4, as, decreasing the content of H_3O^+ ions, the EPS surfaces resulted more negatively charged and thus more prone to bind this divalent cation. By contrast, an increase of the pH value (>4) resulted in a decreased EPS sorption potential, as the two prominent species of Ni (II) ions in solution (i.e., $Ni(OH)_2$ and $Ni(OH)_3^-$) could not be adsorbed onto the negatively charged surfaces characterizing EPS. Similar observations can be drawn for Pb (II) and Co (II), while sorption up to a pH value of 7.5 was observed in the case of Cd (II), as cadmium hydroxyl [$Cd_x(OH)_y$] was the major species of Cd present in solution at this pH (Dobrowolski et al. 2017).

Temperature is another parameter to experimentally control for the optimization of a biosorption process, as it can influence the ionization of the EPS functional groups and influence the formed biosorbent-metal(loid) complex, as well as the diffusion and equilibrium kinetics. Generally, an increase in the temperature can cause (1) a decrease of the biosorption extent in the case of an exothermal process, (2) a destruction of the biosorbent's active sites, and (3) a faster ion motion, which enhances off-rates of the metal(loid)s to bioadsorb (Dobrowolski et al. 2017).

Other than EPS, humic substances (HSs) present in the soil and aquatic environments can mitigate the toxicity of metal(loid)s and organic compounds, which is due to the complex chemistry of HSs thus providing sorption properties (Boguta and Sokołowska 2016). Further, the network generated among metal(loid) compounds, bacteria, and environment can influence (1) the speciation and bioavailability of metal(loid)s, (2) the sorption phenomenon, (3) the metabolic capacity of bacteria toward xenobiotics, and (4) the biochemical/physiological functions of bacteria (Perminova and Hatfield 2005). Although the comprehension of this process is difficult and poorly investigated, Perelomov et al. (2018) described different bacteria capable of tolerating higher amounts of Zn (II) and Pb (II) consequently to the copresence of HSs and derivative fractions (i.e., humic acids, hymatomelanic acids). Indeed, a study employing infrared spectroscopy revealed that humic acids and hymatomelanic acids possess different functional active sites represented by alcohols, phenols, amines, amides, and carboxylic acids, which can be either free or intermolecularly bonded (Rodrigues et al. 2009), enabling HSs to bind cations (Kar et al. 2011). Thus, the addition of 200 mg L^{-1} of HSs (mixture of humic and hymatomelanic acids) resulted in an enhanced tolerance to Zn (II) in the case of *Pseudomonas fluorescence* 142NF and *R.* sp. RS67, which were five- and fourfold more tolerant, respectively, as compared to the non-HSs amended cultures (Perelomov et al. 2018). This effect originates from the stable HS-Zn complex formed (Perminova et al. 2006), which is not bioavailable to the bacteria. Similar results were achieved in the case of Pb (II), therefore highlighting the importance of HSs in attenuating metal(loid)s' toxicity and bioavailability. However, this also limits the ability to remediate a site of the metal pollutant.

As discussed in this section, biosorption is a powerful biotechnological tool that can be exploited for the remediation of metal(loid)-polluted sites. The process relies on the binding, complexation, ion exchange, crystallization, and/or precipitation of the metal(loid) in the biochemical context of a given cell wall. The parameters (i.e., pH, salt level, temperature, and biosorbent concentration) described are crucial and must be optimized according to the physical-chemical characteristics of both metal (loid) and biosorbent, particularly focusing on its amphoteric behavior, in order to take advantage of biosorbents for bioremediation purposes.

4 Metal(loid) Bioaccumulation in *Rhodococcus*

As opposed to biosorption, metal(loid) accumulation is an energy-demanding and metabolic-dependent process, as it requires an active uptake transporter by a living organism and, therefore, it occurs slower as compared to the above described biosorption phenomenon (Mowll and Gadd 1984). In this respect, Mirimanoff and Wilkinson (2000) showed that the initial Zn adsorption performed by *R. opacus* C125 occurred in less than 1 min; on the other hand, microorganisms have evolved specialized uptake systems for essential metal(loid)s that are needed for the cell biochemistry. Specific ion channels or carriers facilitate uptake for the ion down its concentration gradient and active transporters using either ATP or the electrochemical gradient as energy for transport. Such transporters are key to charge metalloenzymes that exploit transition metals as cofactors. For example, molybdenum (Mo) is actively taken up, in the form of oxyanion (MoO_4^{2-}), either by an ATP-binding cassette (ABC) transporter with high affinity or through low affinity carrier systems, which have been both evolved by those microorganisms that use Mo for nitrogenases and/or oxidoreductases featured by molybdopterin as cofactor, respectively (Pau et al. 1997).

A common route for toxic heavy metal accumulation is mediated by transporters of essential metals, which facilitate the transport of closely related elements. For example, the molybdate ABC transporter will also transport tungsten (W) as WO_4^{2-}, in this way nonessential elements can be bioaccumulated. Bacteria can take up divalent cations such as Co (II), Zn (II), and Ni (II) exploiting the magnesium [Mg (II)] transport system, although this strategy only weakly contributes to the uptake of transition metals in the presence of magnesium (Smith and Maguire 1998). Thus, bacterial strains featured by genes coding for enzymes dependent on Ni (II) as cofactor exploit specific transporters for Ni (II), which constitute a family of permeases featured by eight transmembrane segments with high affinity for this cation but a very low transport capacity, while *Escherichia coli* cells exploiting an ABC transporter represents an exception (Hausinger 1997; Eitinger and Friedrich 1997). These permeases for Ni (II) have been identified in different bacterial genera (i.e., *Bradyrhizobium*, *Alcaligenes*, *Helicobacter*) (Fu et al. 1994; Eitinger et al. 1997; Fulkerson et al. 1998). Among actinomycetes, *Mycobacterium tuberculosis* relies on this permease to accumulate Ni (II), supporting the activity of a Ni-dependent urease, which is also an important virulence factor for this microorganism (Cole et al. 1998).

Concerning rhodococci, Komeda et al. (1997) studied in *R. rhodochrous* strain J1 the NhlF transporter for Co (II), which is responsible to provide Co (II) for nitrile hydratases containing non-corrin Co as cofactor (Komeda et al. 1997). Nitrile utilization by *R. rhodochrous* strain M8 was dependent on the nitrile hydratase activity as well, whose expression was enhanced by the presence of Co (II) in the growth medium as compared to Co-limiting conditions, suggesting that Co (II) deficiency could lead to either a weak expression of the nitrile hydratase coding gene or to a fast nitrile hydratase transcript (mRNA) degradation (Pogorelova et al. 1996). However, the NhlF transporter is highly similar at amino acid sequence level to the Ni (II) transporter HoxN of *Alcaligenes eutrophus* (Wolfram et al. 1995). Although both NhlF and HoxN transporters share similar membrane topology, Co (II) uptake NhlF-mediated was impaired by the copresence of Ni (II), thus indicating that NhlF could also transport Ni (II) (Komeda et al. 1997). Later, Degen et al. (1999) demonstrated that HoxN was able to mediate the transport of Ni (II) and incapable of taking up Co (II), while NhlF had a promiscuous binding ability toward these divalent cations (Degen et al. 1999). Indeed, NhlF transporter could mediate Ni (II) uptake, as demonstrated by the enhanced Ni-dependent urease activity of recombinant *Escherichia coli* clones expressing *nhlF* gene grown under Ni-limiting conditions, as well as cross-complementing Ni deficiency due to the deletion of the *hoxN* gene in *Alcaligenes eutrophus*. Nevertheless, the Ni (II) uptake by NhlF was markedly reduced when *R. rhodochrous* strain J1 was cultured in the presence of both nickel and cobalt salts, therefore suggesting that Co (II) was the preferred cation-binding substrate, as well as the NhlF transporter was featured by a certain level of ion selectivity toward two similar transition metals (Degen et al. 1999).

4.1 *Metal(loid) Homeostasis in Microbial Bioaccumulation*

In addition to metal(loid) uptake, the efflux system is a biologically relevant phenomenon responsible for the control of toxin accumulation in the cell, and examples exist to control metal(loid) accumulation, as well as its homeostasis between intra- and extracellular environment, highlighting a sort of "microbial metal toxicity care". This is the case for *R. opacus* C125 cells that accumulated at most c.a. 6 μmol g^{-1} (dry weight) of Zn (II) after 20 min exposure to 5 μM Zn (II), which with time drastically decreased. When Zn-loaded *R. opacus* C125 cells were exposed to either additional 5 μM Zn (II) or Cd (II), cellular Zn concentrations decreased, therefore suggesting both an inducible and cross-inducible efflux mechanism by Zn (II) and Cd (II), respectively, as well as the existence of Zn-binding ligands capable of mediating the accumulation of this transition metal in the first place (Mirimanoff and Wilkinson, 2000).

Overall, nonessential metal(loid)s can be taken up by microorganisms, and the amount of these compounds accumulated varies as function of the species and the type of metal(loid) considered. An example is given by cesium (Cs), which became a metal of pollutant interest after the Chernobyl accident in 1986, attracting the research interest in both the removal and the fate of radioactive Cs. In this respect, Tomioka et al. (1994) investigated two different *Rhodococcus* strains (i.e., *R. erythropolis* CS98 and *R.* sp. CS402) capable of tolerating and growing in the presence of Cs and rubidium (Rb) supplied in a mineral salt medium up to concentrations of 1 mM. It was found that Cs uptake by both *Rhodococcus* strains was negatively influenced by the presence of monovalent cations such as K (I) and Rb (I). Reasonably, Cs (I) and Rb (I) may substitute K (I) in bacteria growing under K-limiting conditions due to the poor specificity of the K (I) transport systems (Avery et al. 1991, 1992). This idea was reported earlier in the case of other bacterial strains (i.e., *Rhodopseudomonas capsulata*, *Escherichia coli*, and *Rhodococcus* spp.; Rhoads and Epstein 1977; Jasper 1978; Tomioka et al. 1994).

If the biosorption phenomenon is the first event occurring when bacterial cells interface metal(loid)s, the active bioaccumulation of these ions within the intracellular environment would be a complementary event. To date we are still at early stages to appreciate the diversity and mechanisms for most metal uptake systems. However, the microbial potential in handling metal(loid)s' toxicity goes beyond the above depicted strategies, as metal(loid) bioconversion and/or biotransformation reactions may take place, leading to either biomineralization or production of valuable metal(loid) products. Indeed, the next section will overview this aspect of the microbial world that crosses and merges itself with the nanotechnology field.

5 Nanotechnology: A Physical-Chemical Perspective

Nanotechnology is the science related to the world of *very small material*, which implies the manipulation of matter at molecular or atomic level (Horikoshi and Serpone 2013). The prefix *nano* is referred to structures defined as intermediate states between molecules and bulk materials (Yuwen and Wang 2013). The uniqueness of nanostructures relies on the possibility to manipulate physical-chemical properties of the materials scaled down to the nanorange (1–100 nm) (Yuwen and Wang 2013), conferring them fundamental and peculiar features as high surface-to-volume ratio and large surface energy with high spatial confinement (Cao 2004a). These peculiar characteristics allow in turn innovative chemical, catalytic, mechanical, electrical, and opto-magnetic properties in the nanomaterials (Appenzeller 1991; Yuwen and Wang 2013), making them suitably manipulated for applications in multidisciplinary fields, such as biomedicine/biotechnologies, energy production, environmental engineering, material science, and optoelectronics (Cao 2004a; Horikoshi and Serpone 2013).

In the past few decades, the nanoscience field has exponentially grown in terms of study and implementation, leading to the development of new methods for nanomaterial production (Rao et al. 2004). In this context, we have witnessed the generation of nanostructures having various compositions and morphologies, of which the most exploited to date are *zero-dimensional* (0D) nanoparticles (NPs), nanocrystals and cluster quantum dots (QDs), *one-dimensional* (1D) nanorods (NRs), nanowires (NWs), nanotubes (NTs) and nanobelts (NBs) (or ribbons), nanoporous solids, *two-dimensional* (2D) arrays of NPs, thin films, and *three-dimensional* (3D) structures (superlattices) (Rao et al. 2004).

Although the physical-chemical methods to synthesize nanomaterials are efficient and available for a scaled-up production, they require the use of toxic substances that are converted during the fabrication process into hazardous waste requiring specialized disposal or are released to the environment (Zhang et al. 2006, ACS Nano 2011; CDC 2014). Moreover, several of these production procedures require expensive equipment and chemicals, affecting the production costs. Therefore, these drawbacks led to a demand for the generation of novel cost-effective and *eco-friendly* methods of nanomaterial production (Ankamwar et al. 2005).

5.1 Microbial Nanotechnology

To date, the new and *green* frontier in nanostructure synthesis is constituted by the exploitation of the natural ability of biological systems (i.e., plants, bacteria, fungi, yeasts, and algae) to bioconvert toxic metal(loid) ions into their less toxic elemental forms producing nanostructures (Suresh et al. 2004; Bhainsa and D'Souza 2006; Song and Kim 2009). Particularly, bacteria are among the organisms that have been the most explored for nanomaterial generation, due to their ability to populate different environmental niches that has led to unique biochemistry and physiology features (Li et al. 2011; Pantidos and Horsfall 2014). Indeed, in nature, the production of metal(loid), metal sulfide, and oxide nanostructures by bacterial cells is achieved through either aerobic detoxification processes or the use, under anaerobic conditions, of metal(loid) ions as terminal electron acceptors to produce energy (Li et al. 2011).

Among the diverse bacterial species that can biosynthesize nanostructures, those belonging to the *Actinomycetales* order recently gained interest in the nanoscience field, as they previously showed their proficiency to produce bioactive compounds, such as antibiotics and biocides (Duraipandiyana et al. 2010). Thus, several *Streptomyces*, *Thermoactinomyces*, and *Nocardia* strains have been investigated for the production of intra- or extracellular gold (Au), silver (Ag), zinc oxide (ZnO), copper oxide (CuO), or selenium (Se) NPs upon their growth and/or exposure to the metal(loid) precursor salts (Forootanfar et al. 2014; Manimaran and Kannabiran 2017). For example, Fig. 1 shows the different cellular localizations of metalloid nanomaterials generated by *R. aetherivorans* BCP1 as a result of metal(loid) oxyanion (i.e., SeO_3^{2-} and TeO_3^{2-}) bioconversion into their elemental forms (i.e., Se^0 and Te^0). The intracellular biosynthesis of metal(loid) NPs by actinomycetes is mainly ascribed to either (1) electrostatic interactions occurring between metal cations and R-COO^- groups of enzymes associated with the bacterial cell membrane (Manimaran and Kannabiran 2017) or (2) a Painter-type reaction between RSH groups of cellular enzymes and metalloid oxyanions (Tan et al. 2016). As a result, these reactions can reduce the metal(loid) ions, leading to the formation of elemental metal(loid) nuclei, which in turn aggregate forming nanomaterials of defined shape (Ahmad et al. 2003; Sunitha et al. 2013). On the other hand, it is proposed that soluble secreted enzymes are responsible for the reduction of metal cations into their elemental forms in the case of the extracellular generation of metal-based nanomaterials (Manivasagan et al. 2016).

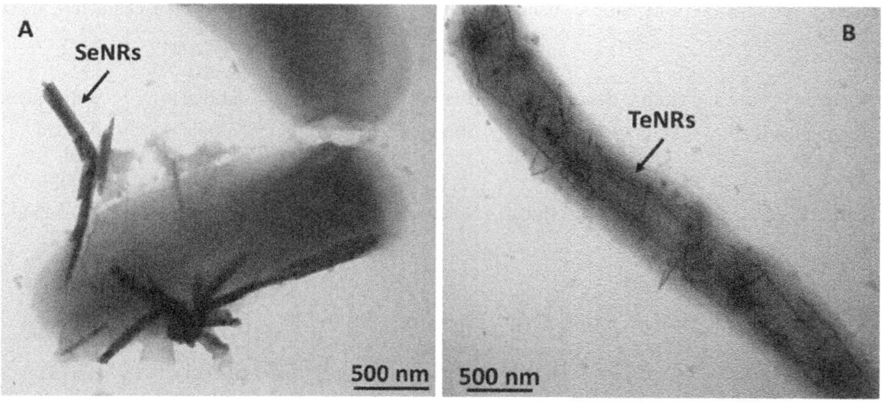

Fig. 1 Electron micrographs of *R. aetherivorans* BCP1 cells featured by the presence of (**a**) selenium nanorods (SeNRs) and (**b**) tellurium nanorods (TeNRs). Data not published, Dr. Alessandro Presentato, University of Calgary (2017)

5.2 Rhodococcus *Strains as Cell Factories for Nanotechnological Purposes*

Considering the great potential of the *Rhodococcus* genus for biotechnological purposes, recently a few research groups started to evaluate the ability of these microorganisms to produce metal(loid) nanomaterials through the bioconversion of metal(loid) salts. Although this research field is only in its infancy, various *Rhodococcus* strains revealed their capability of generating both intra- and extracellular Au-, Ag-, ZnO-, Se-, and Te-based nanostructures (Ahmad et al. 2003; Otari et al. 2012; Kundu et al. 2014; Presentato et al. 2016, 2018b, c), whose key features are summarized in Table 2, while an example of different biogenic nanomaterial shapes is shown in Fig. 2. A common property of these biogenic metal(loid) nanomaterials is their natural thermodynamic stability, which is in contrast with the behavior of those chemically synthesized (Piacenza et al. 2018). Indeed, materials at the nanoscale are generally unstable from a thermodynamic point of view, as they possess a high surface energy (Goldstein et al. 1992). To overcome their instability, nanostructures tend to either combine forming bigger aggregates (Ostwald ripening process) or agglomerate each other without altering individual particles (Cao 2004b). Since the unique properties of nanosized materials are strictly dependent on their small size, it is imperative to avoid their aggregation, making it necessary to use expensive and hazardous substances to stabilize chemically produced nanostructures prior their use (Cao 2004a; Piacenza et al. 2018). By contrast, the natural thermodynamic stability of metal(loid) nanomaterials generated by *Rhodococcus* species is a fundamental feature for the application of these nanostructures in different fields. Moreover, these biogenic nanomaterials resulted in being stabilized by biomolecules (e.g., proteins, enzymes, or surfactant-like molecules) produced by the bacterial strains themselves (Kundu et al. 2014,

Table 2 Production of metal(loid) nanomaterials by *Rhodococcus* spp. and their characterization

Strain	Core	Localization	Shape	Size (nm)	Reducing agents/ stabilizers	References
Rhodococcus sp.	Au	Intracellular	Spherical	5–15	Cellular proteins	Ahmad et al. (2003)
R. NCIM 2891	Ag	Intracellular	Spherical	ca. 100	NADH-dependent nitrate reductases	Otari et al. (2012)
R. pyridinivorans NT2	ZnO	Extracellular	Spherical porous	100–120	Extracellular proteins, heterocyclic compounds	Kundu et al. (2014)
R. aetherivorans BCP1	Te	Intracellular	Rod-shape	200–450	MSHs, biosurfactants	Presentato et al. (2016)
R. aetherivorans BCP1	Se	Intracellular	Spherical	50–90	MSHs, biosurfactants	Presentato et al. (2018a)
			Rod-shape	400–600		
R. aetherivorans BCP1	Te	Intracellular	Spherical	100–900	MSHs, surfactant-like molecules	Presentato et al. (2018b)

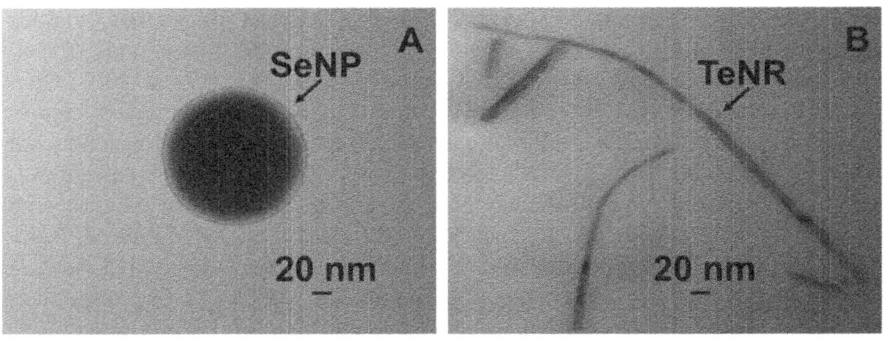

Fig. 2 Electron micrographs of (**a**) selenium nanoparticles (SeNPs) and (**b**) tellurium nanorods (TeNRs) recovered from *R. aetherivorans* BCP1 selenite- and tellurite-grown cells, respectively. Data not published, PhD Student Elena Piacenza, University of Calgary (2017)

Presentato et al. 2016, 2018b, c), therefore constituting a *greener* choice as compared to the compounds used for the stabilization of chemical nanostructures.

To date, metal(loid) nanomaterials generated by *Rhodococcus* spp. have been mainly investigated for their potential use in biomedicine. Indeed, AgNPs produced by *R.* sp. NCIM 2891 were found to be efficient as both antimicrobials toward

Escherichia coli and *Bacillus subtilis* pathogenic indicator strains and anticancer agents limiting the proliferation of A549 lung cancer cell line (Subbaiya et al. 2014). Similarly, ZnO NPs biosynthesized using *R. pyridinivorans* NT2 have been used as coating for cotton textile surfaces intended for the production of footwear (Kundu et al. 2014). As a result of this conjugation process, the cotton fibers acquired long-term UV-protective, photocatalytic, and antimicrobial properties, whose efficacy was evaluated against *Staphylococcus epidermidis* ATCC 12228, an opportunistic pathogen of the sole of the foot (Kundu et al. 2014). Moreover, ZnO NPs showed their proficiency as an in vitro drug delivery system of an anticancer compound (i.e., anthraquinone) toward HT-29 colon carcinoma cell line, without exerting cytotoxic effects on normal peripheral blood mononuclear cells (PBMCs) (Kundu et al. 2014). Finally, the potential of using metal(loid) nanomaterials generated by *Rhodococcus* strains in other fields rather than biomedicine has been recently explored (Presentato et al. 2018c). Particularly, since Te is a well-known narrow band gap *p*-type semiconductor (Araki and Tanaka 1972) exhibiting high photoconductivity, piezo, and thermoelectricity (Araki and Tanaka 1972; Tangney and Fahy 2002; Suchand Sandeep et al. 2010), TeNRs obtained from resting (not growing) cells of *R. aetherivorans* BCP1 were studied for the ability to conduct electrical current. As a result, these TeNRs revealed electrical conductivity approaching the values established for chemically produced TeNRs, highlighting the potential of these biogenic nanomaterials in the development of electrical components or devices (Presentato et al. 2018c).

6 Summary

Rhodococcus is a bacterial genus with strong resilience in handling metal(loid)s' toxicity, applying different strategies to tolerate and/or resist metal(loid) compounds (i.e., biosorption, bioaccumulation, efflux, extracellularly chelation) regardless their form (i.e., anions, cations, oxyanions). This is in part related with the rhodococci's innate and adaptive mechanisms, enabling them to survive and thrive, even under typically unfavorable conditions. The intense study and efforts devoted to comprehending how bacteria interface metal(loid)s led to understanding that *Mother Nature* can be respectfully exploited for different purposes, such as the bioremediation of metal(loid)-polluted environmental matrices and the generation of *eco-friendly* strategies to expand the nanotechnology field without negatively affecting the environment. Thus, this kind of research implies that the knowledge of both the microbiology and the physical-chemical characteristics of metal(loid)s must be used in a conscious way, as the system represented by the *real life* is complicated through a variety of different factors constantly interacting each other. Only by awareness of this complex picture it is possible to predict the biological effects occurring in natural systems.

Acknowledgments Natural Science and Engineering Research Council of Canada (NSERC) is gratefully acknowledged for the support of this study (Grant/Award Number: 216887-2010).

References

Adhami E, Aghaei SS, Zolfaghari MR (2017) Evaluation of heavy metals resistance in biofilm cells of native *Rhodococcus* spp. isolated from soil. Arch Hyg Sci 6:235–243

Ahmad A, Senapati S, Khan MI, Kumar R, Ramani R, Srinivas V, Sastry M (2003) Intracellular synthesis of gold nanoparticles by a novel alkalotolerant actinomycete, *Rhodococcus* species. Nanotechnology 14:824–828. https://doi.org/10.1088/0957-4484/14/7/323

Ali H, Khanb E, Sajad AM (2013) Phytoremediation of heavy metals-concepts and applications. Chemosphere 91:869–881. https://doi.org/10.1016/j.chemosphere.2013.01.075

Alvarez HM, Steinbüchel A (2010) Physiology, biochemistry and molecular biology of triacylglycerol accumulation by *Rhodococcus*. In: Alvarez HM (ed) Biology of *Rhodococcus*, Microbiology monographs, vol 16. Springer, Heidelberg, pp 263–290. https://doi.org/10.1007/978-3-642-12937-7_10

American Chemistry Society (ACS) NANO(2011) Green nanotechnology challenges and opportunities. http://greennano.org/sites/greennano2.uoregon.edu/files/GCI_WP_GN10.pdf

Ankamwar B, Chaudhary M, Sastry M (2005) Gold nanoparticles biologically synthesized using tamarind leaf extract and potential application in vapor sensing. Synth React Inorg Met Org Nano Met Chem 35:19–26. https://doi.org/10.1081/SIM-200047527

Appenzeller T (1991) The man who dared to think small. Science 254:1300–1300. https://doi.org/10.1126/science.254.5036.1300

Araki K, Tanaka T (1972) Piezoelectric and elastic properties of single crystalline Se-Te alloys. Appl Phys Expr 11:472–479. https://doi.org/10.1143/JJAP.11.472

Avery SV, Codd GA, Gadd GM (1991) Cesium accumulation and interactions with other monovalent cations in the cyanobacterium *Synechocystis* PCC 6803. J Gen Microbiol 137:405–413. https://doi.org/10.1099/00221287-137-2-405

Avery SV, Codd GA, Gadd GM (1992) Replacement of cellular potassium by cesium in *Chlorella emersonii*: differential sensitivity of photoautotrophic and chemoheterotrophic growth. J Gen Microbiol 138:69–76. https://doi.org/10.1099/00221287-138-169

Beyersmann D, Hartwig A (2008) Carcinogenic metal compounds: recent insight into molecular and cellular mechanisms. Arch Toxicol 82:493–512. https://doi.org/10.1007/s00204-008-0313-y

Bhainsa KC, D'Souza SF (2006) Extracellular biosynthesis of silver nanoparticles using the fungus *Aspergillus fumigatus*. Colloids Surf B Biointerfaces 47:160–164. https://doi.org/10.1016/j.colsurfb.2005.11.026

Boguta P, Sokołowska Z (2016) Interactions of Zn(II) ions with humic acids isolated from various type of soils. Effect of pH, Zn concentrations and humic acids chemical properties. PLoS One 11:1–20. https://doi.org/10.1371/journal.pone.0153626

Botero AEC, Torem ML, de Mesquita LMS (2007) Fundamental studies of *Rhodococcus opacus* as a biocollector of calcite and magnesite. Miner Eng 20:1026–1032. https://doi.org/10.1016/j.mineng.2007.03.017

Brooks AN, Turkarslan S, Beer KD, Lo FY, Baliga NS (2011) Adaptation of cells to new environments. Wiley Interdiscip Rev Syst Biol Med 3:544–561. https://doi.org/10.1002/wsbm.136

Bueno BYM, Torem ML, Molina F, de Mesquita LMS (2008) Biosorption of lead (II), chromium (III) and copper (II) by *R. opacus*: equilibrium and kinetic studies. Miner Eng 21:65–75. https://doi.org/10.1016/j.mineng.2007.08.013

Cao G (2004a) Chapter 1: Introduction. In: Cao G (ed) Nanostructures & nanomaterials, synthesis, properties and applications. Imperial College Press, London, pp 1–14

Cao G (2004b) Chapter 2, Physical chemistry of solid surfaces. In: Cao G (ed) Nanostructures & nanomaterials, synthesis, properties and applications. Imperial College, London, pp 15–48

Cappelletti M, Fedi S, Zampolli J, Di Canito A, D'ursi P, Orro A, Viti C, Milanesi L, Zannoni D, Di Gennaro P (2016) Phenotype microarray analysis may unravel genetic determinants of the stress response by *Rhodococcus aetherivorans* BCP1 and *Rhodococcus opacus* R7. Res Microbiol 167:766–773. https://doi.org/10.1016/j.resmic.2016.06.008

Cayllahua JEB, Torem ML (2010) Biosorption of aluminum ions onto *Rhodococcus opacus* from wastewaters. Chem Eng J 161:1–8. https://doi.org/10.1016/j.cej.2010.03.025

Centers for Disease Control and Prevention National Institute for Occupational Safety and Health (CDC) (2014) Current strategies for engineering controls in nanomaterial production and downstream handling processes. https://www.cdc.gov/niosh/docs/2014-102/pdfs/2014-102.pdf

Chakraborty J, Dash AR, Das S (2017) Metals and their toxic effects. An introduction to noxious elements. In: Das S, Dash HR (eds) Handbook of metal-microbe interactions and bioremediation. CRC, Boca Raton, FL, pp 3–17. ISBN: 9781498762434

Chang LW, Magos L, Suzuki T (1996) Toxicology of metals. CRC, Boca Raton, FL. ISBN: 9780873718035

Cole ST, Broch R, Parkhill J, Garnier T, Churcher C, Harris D, Gordon SV, Eiglmeir K, Gas S, Barry CE 3rd, Tekaia F, Badcock K, Basham D, Brown D, Chillingworth T, Connor R, Davies R, Devlin K, Feltwell T, Gentles S, Hamlin N, Holroyd S, Hornsby T, Jagels K, Krogh A, McLean J, Moule S, Murphy L, Oliver K, Osborne J, Quail MA, Rajandream MA, Rogers J, Rutter S, Seeger K, Skelton J, Squares R, Squares S, Sulston JE, Taylor K, Whitehead S, Barrell BG (1998) Deciphering the biology of *Mycobacterium tuberculosis* from the complete genome sequence. Nature 393:537–544. https://doi.org/10.1038/31159

Comte S, Guibaud G, Baudu M (2005) Relation between extraction protocols of the activated sludge extracellular polymeric substances (EPS) and EPS complexation properties. Part I. Comparison of the efficiency of eight EPS extraction properties. Enzyme Microb Technol 38:237–245. https://doi.org/10.1016/j.enzymictec.2005.06.016

Crini G (2005) Recent developments in polysaccharide-based materials used as adsorbents in wastewater treatment. Prog Polym Sci 30:38–70. https://doi.org/10.1016/j.progpolymsci.2004.11.002

Degen O, Kobayashi M, Shimizu S, Eitinger T (1999) Selective transport of divalent cations by transition metal permeases: the *Alcaligenes eutrophus* HoxN and the *Rhodococcus rhodochrous* NhlF. Arch Microbiol 171:139–145. https://doi.org/10.1007/s002030050

Dobrowolski R, Szcześ A, Czemierska M, Jarosz-Wikołazka A (2017) Studies of cadmium(II), lead (II), nickel(II), cobalt(II) and chromium(VI) sorption on extracellular polymeric substances produced by *Rhodococcus opacus* and *Rhodococcus rhodochrous*. Bioresour Technol 225:113–120. https://doi.org/10.1016/j.biortech.2016.11.040

Doyle RJ, Matthews TH, Streips UN (1980) Chemical basis for the selectivity of metal ions by the *Bacillus subtilis* wall. J Bacteriol 143:471–480

Duraipandiyana V, Sasi AH, Islam VHI, Valanarasu M, Ignacimuthu S (2010) Antimicrobial properties of actinomycetes from the soil of Himalaya. J Med Mycol 20:15–20. https://doi.org/10.1016/j.mycmed.2009.11.002

Ehrlich HL (1997) Microbes and metals. Appl Microbiol Biotechnol 48:687–692. https://doi.org/10.1007/s002530051

Eitinger T, Friedrich B (1997) Microbial nickel transport and incorporation into hydrogenases. In: Winkelmann G, Carrano CJ (eds) Transition metals in microbial metabolism. Harwood, Amsterdam, pp 235–256

Eitinger T, Wolfram L, Degen O, Anthon C (1997) A Ni^{2+} binding motif is the basis of high affinity transport of the *Alcaligenes eutrophus* nickel permease. J Biol Chem 272:17139–17144. https://doi.org/10.1074/jbc.272.27.17139

Fergusson JE (1990) The heavy elements: chemistry, environmental impact and health effects. Pergamon, Oxford. https://doi.org/10.1016/0269-7491(91)90124-F

Figueira EMAP, Lima AIG, Pereira SIA (2005) Cadmium tolerance plasticity in *Rhizobium leguminosarum* bv. *viciae*: glutathione as a detoxifying agent. Can J Microbiol 51:7–14. https://doi.org/10.1139/w04-101

Fleck LC, Bicca FC, Ayub MAZ (2000) Physiological aspects of hydrocarbon emulsification, metal resistance and DNA profile of biodegrading bacteria isolated from oil polluted sites. Biotechnol Lett 22:285–289. https://doi.org/10.1023/A:1005607112566

Flemming CA, Ferris FG, Beveridge TJ, Bailey GW (1990) Remobilization of toxic heavy metals absorbed to wall-clay composites. Appl Environ Microbiol 56:3191–3209

Forootanfar H, Zare B, Fasihi-Bam H, Amirpour-Rostami S, America A, Shakibaie M, Nami MT (2014) Biosynthesis and characterization of selenium nanoparticles produced by terrestrial actinomycete *Streptomyces microflavus* strain FSHJ31. RRJMB 3:47–53. e-ISSN: 2320-3528

Freire-Nordi CS, Vieira AAH, Nakaie CR, Nascimento OR (2005) Effect of polysaccharide capsule of the microalgae *Staurastrum iversenii* var. *americanum* on diffusion of charged and uncharged molecules, using EPR technique. Braz J Phys 36:75–82. https://doi.org/10.1590/S0103-97332006000100013

Fu C, Javedan S, Moshiri F, Maier RJ (1994) Bacterial genes involved in incorporation of nickel into a hydrogenase enzyme. Proc Natl Acad Sci USA 91:5099–5103. https://doi.org/10.1073/pnas.91.11.5099

Fulkerson JF Jr, Garner RM, Mobley HLT (1998) Conserved motifs and residues in the NixA protein of *Helicobacter pylori* are critical for the high affinity transport of nickel ions. J Biol Chem 273:235–241. https://doi.org/10.1074/jbc.273.1.235

Gadd GM (1992a) Microbial control of heavy metal pollution. In: Fry J, Gadd GM, Herbert RA, Jones CW, Watson-Craik IA (eds) Forty-eighth symposium of the society for general microbiology. Cambridge Univ. Press, The University of Cardiff, Cardiff, pp 59–88

Gadd GM (1992b) Metals and microorganisms: a problem of definition. FEMS Microbiol Lett 100:197–204. https://doi.org/10.1111/j.1574-6968.1992.tb14040.x

Gadd GM (2010) Metals, minerals and microbes: geomicrobiology and bioremediation. Microbiology 156:609–643. https://doi.org/10.1099/mic.0.037143-0

Gadd GM, White C (1993) Microbial treatment of metal pollution—a working biotechnology? Trends Biotechnol 11:353–359. https://doi.org/10.1016/0167-7799(93)90158-6

Garbisu C, Alkorta I (2003) Basic concepts on heavy metal soil bioremediation. Eur J Miner Process Environ Prot 3:58–66

Goldstein AN, Echer CM, Alivisatos AP (1992) Melting in semiconductor nanocrystals. Science 256:1425–1427. https://doi.org/10.1126/science.256.5062.1425

Guibaud G, Bordas F, Saaid A, D'abzac P, Hullebusch EV (2008) Effect of pH on cadmium and lead binding by extracellular polymeric substances (EPS) extracted from environmental bacterial strains. Colloid Surf B Biointerfaces 63:48–54. https://doi.org/10.1016/j.colsurfb.2007.11.002

Gupta P, Diwan B (2017) Bacterial Exopolysaccharide mediated heavy metal removal: a review on biosynthesis, mechanism and remediation strategies. Biotechnol Rep 13:58–71. https://doi.org/10.1016/j.btre.2016.12.006

Hamelink J, Landrum PF, Bergman H, Benson WH (1994) Bioavailability: physical, chemical and biological interactions. CRC, Boca Raton, FL. ISBN: 9781566700863

Harrison JJ, Ceri H, Stremick CA, Turner RJ (2004) Biofilm susceptibility to metal toxicity. Environ Microbiol 6:1220–1227. https://doi.org/10.1111/j.1462-2920.2004.00656.x

Harrison JJ, Turner RJ, Marques LLR, Ceri H (2005) Biofilms: a new understanding of these microbial communities is driving a revolution that may transform the science of microbiology. Am Sci (6):508–515

Harrison JJ, Ceri H, Turner RJ (2007) Multimetal resistance and tolerance in microbial biofilms. Nat Rev Microbiol 5:928–938. https://doi.org/10.1038/nrmicro1774

Hausinger RP (1997) Metallocenter assembly in nickel-containing enzymes. J Biol Inorg Chem 2:279–286. https://doi.org/10.1007/s007750050

Horikoshi S, Serpone N (2013) Chapter 1, General introduction to nanoparticles. In: Horikoshi S, Serpone N (eds) Microwaves in nanoparticle synthesis: fundamentals and applications. Wiley-VCH, Weinheim, pp 1–24

Jasper P (1978) Potassium transport system of *Rhodopseudomonas capsulate*. J Bacteriol 133:1314–1322

Jixian Y, Wei W, Shanshan P, Fang M, Ang L, Dan W, Jie X (2015) Competitive adsorption of heavy metals by extracellular polymeric substances extracted from *Klebsiella* sp. J1. Bioresour Technol 196:533–539. https://doi.org/10.1016/j.biortech.2015.08.011

Kar S, Maity JP, Jean JS, Liu CC, Nath B, Lee YC, Bundschuh J, Chen CY, Li Z (2011) Role of organic matter and humic substances in the binding and mobility of arsenic in a Gangetic aquifer. J Environ Sci Health A 46:1231–1238. https://doi.org/10.1080/10934529.2011.598796

Koch AL (1990) Growth and form of the bacterial cell wall. Am Sci 78:327–341

Komeda H, Kobayashi M, Shimizu S (1997) A novel transporter involved in cobalt uptake. Proc Natl Acad Sci USA 94:36–41. https://doi.org/10.1073/pnas.94.1.36

Kundu D, Hazra C, Chatterjee A, Chaudhari A, Mishra S (2014) Extracellular biosynthesis of zinc oxide nanoparticles using *Rhodococcus pyridinivorans* NT2: multifunctional textile finishing, biosafety evaluation and in vitro drug delivery in colon carcinoma. J Photochem Photobiol B 140:194–204. https://doi.org/10.1016/j.jphotobiol.2014.08.001

Lamelas C, Benedetti M, Wilkinson KJ, Slaveykova VI (2006) Characterization of H^+ and Cd^{2+} binding properties of the bacterial exopolysaccharides. Chemosphere 65:1362–1370. https://doi.org/10.1016/j.chemosphere.2006.04.021

Lau TC, Wu XA, Chua H, Qian PY, Wong PK (2005) Effect of exopolysaccharides on the adsorption of metal ions by *Pseudomonas* sp. CU-1. Water Sci Technol 52:63–68

Ledin M (2000) Accumulation of metals by microorganisms-processes and importance for soil systems. Earth Sci Rev 51:1–31. https://doi.org/10.1016/S0012-8252(00)00008-8

Lemire JA, Harrison JJ, Turner RJ (2013) Antimicrobial activity of metals: mechanisms, molecular targets and applications. Nat Rev Microbiol 11:371–384. https://doi.org/10.1038/nrmicro3028

Li X, Xu H, Chen ZS, Chen ZS, Chen G (2011) Biosynthesis of nanoparticles by microorganisms and their applications. J Nanomater. https://doi.org/10.1155/2011/270974. Article ID 270974

Liu H, Fang HH (2002) Characterization of electrostatic binding sites of extracellular polymers by linear programming analysis of titration data. Biotechnol Bioeng 30:806–811. https://doi.org/10.1002/bit.10432

Manimaran M, Kannabiran K (2017) Actinomycetes-mediated biogenic synthesis of metal and metal oxide nanoparticles: progress and challenges. Lett Appl Microbiol 64:401–408. https://doi.org/10.1111/lam.12730

Manivasagan P, Venkatesan J, Sivakumar K, Kim SK (2016) Actinobacteria mediated synthesis of nanoparticles and their biological properties: a review. Crit Rev Microbiol 42:209–221. https://doi.org/10.3109/104084X.2014.917069

Martínková L, Uhnáková B, Pátek M, Nesvera J, Kren V (2009) Biodegradation potential of the genus *Rhodococcus*. Environ Int 35:162–177. https://doi.org/10.1016/j.envint.2008.07.018

Merroun ML, Ben Chekroun K, Arias JM, González-Muñoz MT (2003) Lanthanum fixation by *Myxococcus xanthus*: cellular location and extracellular polysaccharide observation. Chemosphere 52:113–120. https://doi.org/10.1016/S0045-6535(03)00220-0

Mirimanoff N, Wilkinson KJ (2000) Regulation of Zn accumulation by a freshwater Gram-positive bacterium (*Rhodococcus opacus*). Environ Sci Technol 34:616–622. https://doi.org/10.1021/es990744g

Mowll JL, Gadd GM (1984) Cadmium uptake by *Aureobasidium pullulans*. J Gen Microbiol 130:279–284. https://doi.org/10.1099/00221287-130-2-279

Newton GL, Arnold K, Price MS, Sherrill C, delCardayre SB, Aharonowitz Y, Cohen G, Davies J, Fahey RC, Davis C (1996) Distribution of thiols in microorganisms: mycothiol is a major thiol

in most *actinomycetes*. J Bacteriol 178:1990–1995. https://doi.org/10.1128/jb.178.7.1990-1995.1996
Nies D (1999) Microbial heavy-metal resistance. Appl Microbiol Biotechnol 51:730–750. https://doi.org/10.1007/s002530051
Orro A, Cappelletti M, D'Ursi P, Milanesi L, Di Canito A, Zampolli J, Collina E, Decorosi F, Viti C, Fedi S, Presentato A, Zannoni D, Di Gennaro P (2015) Genome and phenotype microarray analyses of *Rhodococcus* sp. BCP1 and *Rhodococcus opacus* R7: genetic determinants and metabolic abilities with environmental relevance. PLoS One 10(10):e0139467. https://doi.org/10.1371/journal.pone.0139467
Otari SV, Patil RM, Nadaf NH, Ghosh SJ, Pawar SH (2012) Green biosynthesis of silver nanoparticles from an actinobacteria *Rhodococcus* sp. Mater Lett 72:92–94. https://doi.org/10.1016/j.matlet.2011.12.109
Pantidos N, Horsfall LE (2014) Biological synthesis of metallic nanoparticles by bacteria, fungi and plants. J Nanomed Nanotechnol 5:5. https://doi.org/10.4172/2157-7439.1000233
Paperi R, Micheletti E, De Philips R (2006) Optimization of copper sorbing-desorbing cycles with confined cultures of the exopolysaccharide-producing cyanobacterium *Cyanospira capsulata*. J Appl Microbiol 101:1351–1356. https://doi.org/10.1111/j.1365-2672.2006.03021.x
Park JH, Kim BS, Chon CM (2018) Characterization of iron and manganese minerals and their associated microbiota in different mine sites to reveal the potential interactions of microbiota with mineral formation. Chemosphere 191:245–252. https://doi.org/10.1016/j.chemosphere.2017.10.050
Pau RN, Klipp W, Leimkühler S (1997) Molybdenum transport, processing and gene regulation. In: Winkelmann G, Carrano CJ (eds) Transition metals in microbial metabolism. Harwood, Amsterdam, pp 217–234
Pavel VL, Sobariu DL, Tudorache-Fertu ID, Statescu F, Gaverilescu M (2013) Symbiosis in the environment biomanagement of soils contaminated with heavy metals. Eur J Sci Theol 9:211–224
Perelomov LV, Sarkarb B, Sizovad OI, Chilachava KB, Shvikina AY, Perelomova IV, Atroshchenkoa YM (2018) Zinc and lead detoxifying abilities of humic substances relevant to environmental bacterial species. Ecotoxicol Environ Saf 151:178–183. https://doi.org/10.1016/j.ecoenv.2018.01.018
Perminova IV, Hatfield K (2005) Remediation chemistry of humic substances: theory and implications for technology. In: Perminova IV, Hatfield K, Hertcorn N (eds) Use of humic substances to remediate polluted environments: from theory to practice. Springer, Dordrecht, pp 3–36. https://doi.org/10.1007/1-4020-3252-8_1
Perminova IV, Kulikova NA, Zhilin D, Grechischeva M, Kovalevskii DV, Lebedeva GF, Matorin DN, Venediktov PS, Konstantinov AI, Kholodov VA, Petrosyan VS, (2006) Mediating effects of humic substances in the contaminated environments. Concepts, results, and prospects. In: Twardowska I, Allen HE, Haggblom MH, Stefaniak S (Eds) Viable methods of soils and water pollution monitoring, protection and remediation, Krakow, Poland, pp 249–273. https://doi.org/10.1007/978-1-4020-4728-2_17
Phillips DJ, Rainbow PS (2013) Biomonitoring of trace aquatic contaminants, vol 37. Springer, New York. ISBN: 978-94-011-2122-4
Piacenza E, Presentato A, Turner RJ (2018) Stability of biogenic metal(loid) nanomaterials related to the colloidal stabilization theory of chemical nanostructures. Crit Rev Biotechnol 25:1–20. https://doi.org/10.1080/07388551.2018.1440525
Plette ACC, van Riemsdijk WH, Benedetti MF, van der Wal A (1995) pH dependent charging behavior of isolated cell walls of a Gram-positive soil bacterium. J Colloid Interface Sci. 173:354–363. https://doi.org/10.1006/jcis.1995.1335
Plette ACC, Benedetti MF, Vanriemsdijk WH (1996) Competitive binding of protons, calcium, cadmium, and zinc to isolated cell walls of a Gram-positive soil bacterium. Environ Sci Technol 30:1902–1910. https://doi.org/10.1021/es9505681

Pogorelova TE, Ryabchenko LE, Sunzow NI, Yanenko AS (1996) Cobalt-dependent transcription of nitrile hydratase gene in *Rhodococcus rhodochrous* M8. FEMS Microbiol Lett 144:191–195. https://doi.org/10.1016/0378-1097(96)00361-8

Presentato A, Piacenza E, Anikovskiy M, Cappelletti M, Zannoni D, Turner RJ (2016) *Rhodococcus aetherivorans* BCP1 as cell factory for the production of intracellular tellurium nanorods under aerobic conditions. Microb Cell Fact 15:204. https://doi.org/10.1186/s12934-016-0602-8

Presentato A, Cappelletti M, Sansone A, Ferreri C, Piacenza E, Demeter MA, Crognale S, Petruccioli M, Milazzo G, Fedi S, Steinbüchel A, Turner RJ, Zannoni D (2018a) Aerobic growth of *Rhodococcus aetherivorans* BCP1 using selected naphthenic acids as the only carbon and energy sources. Front Microbiol 9:672. https://doi.org/10.3389/fmicb.2018.00672

Presentato A, Piacenza E, Anikovskiy M, Cappelletti M, Zannoni D, Turner RJ (2018b) Biosynthesis of selenium-nanoparticles and -nanorods as a product of selenite bioconversion by the aerobic bacterium *Rhodococcus aetherivorans* BCP1. New Biotechnol 41:1–8. https://doi.org/10.1016/j.nbt.2017.11.02

Presentato A, Piacenza E, Darbandi A, Anikovskiy M, Cappelletti M, Zannoni D, Turner RJ (2018c) Assembly growth and conductive properties of tellurium nanorods produced by *Rhodococcus aetherivorans* BCP1. Sci Rep. https://doi.org/10.1038/s41598-018-22320-x

Rao CNR, Muller A, Cheetham AK (2004) Chapter 1, Nanomaterials. In: Rao CNR, Muller A, Cheetham AK (eds) The chemistry of nanomaterials: synthesis, properties and applications. WILEY-VCH, Weinheim, pp 1–11

Rhoads DB, Epstein W (1977) Energy coupling to net K+ transport in *Escherichia coli* K-12. J Biol Chem 252:1394–1401

Rodrigues A, Brito A, Janknecht P, Proenca MF, Nogueira R (2009) Quantification of humic acids in surface water: effects of divalent cations, pH, and filtration. J Environ Monit 11:377–382. https://doi.org/10.1039/B811942B

Salehizadeh H, Shojaosadati SA (2003) Removal of metal ions from aqueous solution by polysaccharide produced from *Bacillus firmus*. Water Res 37:4231–4235. https://doi.org/10.1016/S0043-1354(03)00418-4

Sharma SK, Goloubinoff P, Christen P (2011) Non-native proteins as newly identified targets of heavy metals and metalloids. In: Bánfalvi G (ed) Cellular effects of heavy metals. Springer, Heidelberg, pp 263–274. https://doi.org/10.1007/978-94-007-0428-2_12

Sheng PX, Tan LH, Chen JP, Ting YP (2004) Biosorption performance of two brown marine algae for removal of chromium and cadmium. J Disper Sci Technol 25:679–686. https://doi.org/10.1081/DIS-200027327

Smith RL, Maguire ME (1998) Microbial magnesium transport: unusual transporters searching for identity. Mol Microbiol 28:217–226. https://doi.org/10.1046/j.1365-2958.1998.00810.x

Song JY, Kim BS (2009) Rapid biological synthesis of silver nanoparticles using plant leaf extracts. Bioprocess Biosyst Eng 32:79–84. https://doi.org/10.1007/s00449-008-0224-6

Stillman MJ (1995) Metallothioneins. Coord Chem Rev 144:461–571. https://doi.org/10.1016/0010-8545(95)01173-M

Stratton H, Brooks P, Griffiths P, Seviour R (2002) Cell surface hydrophobicity and mycolic acid composition of *Rhodococcus* strains isolated from activated sludge foam. J Ind Microbiol Biotechnol 28:264–267. https://doi.org/10.1038/sj/jim/7000241

Strong PJ, Burgess JE (2008) Treatment methods for wine-related and distillery wastewaters: a review. Bioremediat J 12:70–87. https://doi.org/10.1080/10889860802060063

Subbaiya R, Preetha L, Gayathril S, Swarnalatha WA, Selvam MM (2014) Synthesis and characterization of silver nanoparticles from *Rhodococcus*-2891 and its antitumor activity against lung cancer cell line (A549). In: International conference on science, engineering and management research (ICSEMR 2014), ISBN: 978-1-4799-7613-3

Suchand Sandeep CS, Samal AK, Pradeep T, Philip R (2010) Optical limiting properties of Te and Ag_2Te nanowires. Chem Phys Lett 485:326–330. https://doi.org/10.1016/j.cplett.2009.12.065

Sunitha A, Isaac RSR, Geo S, Sornalekshmi S, Rose A, Praseetha PK (2013) Evaluation of antimicrobial activity of biosynthesized iron and silver nanoparticles using the fungi *Fusarium oxysporum* and *Actinomycetes* sp. on human pathogens. Nano Biomed Eng 5:39–45. https://doi.org/10.5101/nbe.v5i1.p39-45

Suresh K, Prabagaran SR, Sengupta S, Shivaji S (2004) *Bacillus indicus* sp. nov., an arsenic-resistant bacterium isolated from an aquifer in West Bengal, India. J Syst Evol Microbiol 54:1369–1375. https://doi.org/10.1099/ijs.0.03047-0

Tan Y, Yao R, Wang R, Wang D, Wang G, Zheng S (2016) Reduction of selenite to Se (0) nanoparticles by filamentous bacterium *Streptomyces* sp. ES2-5 isolated from a selenium mining soil. Microb Cell Fact 15:157. https://doi.org/10.1186/s12934-016-0554-z

Tangney P, Fahy S (2002) Density-functional theory approach to ultrafast laser excitation of semiconductors: application to the A1 phonon in tellurium. Phys Rev B 14:279. https://doi.org/10.1103/PhysRevB.65.054302

Taylor DE (1999) Bacterial tellurite resistance. Trends Microbiol 7:111–115. https://doi.org/10.1016/S0966-842X(99)01454-7

Tchounwou P, Newsome C, Williams J, Glass K (2008) Copper-induced cytotoxicity and transcriptional activation of stress genes in human liver carcinoma cells. Metal Ions Biol Med 10:285–290

Tomioka N, Uchiyama H, Yagi O (1994) Cesium accumulation and growth characteristics of *Rhodococcus erythropolis* CS98 and *Rhodococcus* sp. strain CS402. Appl Environ Microbiol 60:2227–2231

Turner RJ (2001) Tellurite toxicity and resistance in Gram-negative bacteria. Recent Res Dev Microbiol 5:69–77

Turner RJ, Weiner JH, Taylor DE (1999) Tellurite-mediated thiol oxidation in *Escherichia coli*. Microbiology 145:2549–2557. https://doi.org/10.1099/00221287-145-9-2549

Turner RJ, Borghese R, Zannoni D (2012) Microbial processing of tellurium as a tool in biotechnology. Biotechnol Adv 30:954–963. https://doi.org/10.1016/j.biotechadv.2011.08.018

Valls M, de Lorenzo V (2002) Exploiting the genetic and biochemical capacities of bacteria for the remediation of heavy metal pollution. FEMS Microbiol Rev 4:327–338. https://doi.org/10.1111/j.1574-6976.2002.tb0018.x

van der Wal A, Norde W, Zehnder AJB, Lyklema J (1997) Determination of total charge in the cell walls of Gram-positive bacteria. J Colloids Surf B Biointerfaces 9:81–100. https://doi.org/10.1016/S0927-7765(96)01340-9

Vasquez TGP, Botero AEC, de Mesquita LMS, Torem ML (2007) Biosorptive removal of Cd and Zn from liquid streams with a *Rhodococcus opacus* strain. Miner Eng 20:939–944. https://doi.org/10.1016/j.mineng.2007.03.014

Vela-Cano M, Castellano-Hinojosa A, Vivas AF, Toledo MVM (2014) Effect of heavy metals on the growth of bacteria isolated from sewage sludge compost tea. Adv Microbial 4:644–655. https://doi.org/10.4236/aim.2014.410070

Volesky B (1990) Biosorption and biosorbents. In: Volesky B (ed) Biosorption of heavy metals. CRC, Boca Raton, FL, pp 3–6. ISBN: 9780849349171

Wei X, Fang L, Cai P, Huang Q, Chen H, Liang W, Rong X (2011) Influence of extracellular polymeric substances (EPS) on Cd adsorption by bacteria. Environ Pollut 159:1369–1374. https://doi.org/10.1016/j.envpol.2011.01.006

Wilde EW, Benemann JR (1993) Bioremoval of heavy metals by the use of microalgae. Biotechnol Adv 4:781–812. https://doi.org/10.1016/0734-9750(93)9000-6

Wolfram L, Friedrich B, Eitinger T (1995) The *Alcaligenes eutrophus* protein HoxN mediates nickel transport in *Escherichia coli*. J Bacteriol 177:1840–1843. https://doi.org/10.1128/jb.177.7.1840-1843.1995

Xue HB, Stumm W, Sigg L (1988) The binding of heavy metals to algal surfaces. Water Res 22:917–626. https://doi.org/10.1016/0043-1354(88)90029-2

Yuwen L, Wang L (2013) Chapter 11.5, Nanoparticles and quantum dots. In: Devillanova F, Du Mont WW (eds) Handbook of chalcogen chemistry: new perspectives in sulfur, selenium and tellurium, 2nd edn. The Royal Society of Chemistry, Cambridge, pp 232–260
Zhang B, Ye X, Dai W, Hou W, Zuo F, Xie Y (2006) Biomolecule-assisted synthesis of single-crystalline selenium nanowires and nanoribbons via a novel flake-cracking mechanism. Nanotechnology 17:385–390. https://doi.org/10.1088/0957-4484/17/2/007
Zheng-Bo Y, Qing L, Chuan-chuan L, Tian-hu C, Jin W (2015) Component analysis and heavy metal adsorption ability of extracellular polymeric substances (EPS) from sulfate reducing bacteria. Bioresour Technol 194:399–402. https://doi.org/10.1016/j.biortech.2015.07.042

Plant-Associated *Rhodococcus* Species, for Better and for Worse

Isolde M. Francis and Danny Vereecke

Contents

Abstract *Rhodococcus fascians*, causative agent of the leafy gall syndrome, produces a mixture of cytokinins to modify the hormone landscape of its broad range of plant hosts leading to tissue deformations and developmental alterations. Recent developments indicate that the pathogenic nature of these bacteria is superimposed on its plant growth-promoting effect. In the last two decades, its unique position as the only species within the genus able to interact with plants has been overthrown. Indeed, Pistachio Bushy Top Syndrome is an emerging disease linked to the presence of two *Rhodococcus* species, *R. fascians* and *R. corynebacterioides*. Both bacteria would act synergistically to cause the symptoms, giving the prospect of virulence strategies that differ from those of the leafy gall inducers. Additionally, as a result of microbiome research, it is clear that many *Rhodococcus* species live in close association with plants, and several of them exhibit plant growth-promoting

I. M. Francis
Department of Biology, California State University Bakersfield, Bakersfield, CA, USA
e-mail: ifrancis@csub.edu

D. Vereecke (✉)
Department of Plant and Crop, Faculty of Bioscience Engineering, Ghent University, Ghent, Belgium

H. M. Alvarez (ed.), *Biology of Rhodococcus*, Microbiology Monographs 16,
https://doi.org/10.1007/978-3-030-11461-9_13

activities. Finally, genome analyses of a collection of *R. fascians* isolates imply that the taxonomic position of this group of bacteria within the genus will have to be reevaluated, and likely a new genus consisting of several species will be proposed soon.

1 Introduction

As illustrated by most chapters in this book, members of the genus *Rhodococcus* are renowned for their specialized metabolic capacities, and most are harmless, environmental, free-living organisms (Ceniceros et al. 2017). The genus, however, also comprises two pathogens: *R. equi*, the causative agent of pneumonia in foals leading to major losses worldwide in the equine industry (von Bargen and Haas 2009; Anastasi et al. 2016; MacArthur et al. 2017; Vázquez-Boland et al. this issue), and *R. fascians*, a plant pathogen causing the leafy gall syndrome mainly affecting ornamental production facilities all over the world (Depuydt et al. 2008b). Nevertheless, based on recent developments, the monopoly of *R. fascians* within the genus to interact with plants appears to be broken.

As a result of a growing environmental awareness, the use of agrochemicals to assure optimal crop yield is now considered less favorable, and considerable efforts are made toward more integrated agricultural approaches in which crop productivity and resilience against abiotic and biotic stress can be achieved with a reduced input. In that context, the use of biostimulants has gained a lot of attention, and the search for microorganisms that can stimulate plant productivity and protect crops from adverse conditions has become a major endeavor (Calvo et al. 2014). As an interesting side effect of these studies, a much broader view is obtained on the composition of rhizospheric and phyllospheric microbiomes, and it has become clear that diverse actinobacteria, including *Rhodococcus* species, readily associate with plants (Hamedi and Mohammadipanah 2015). Additionally, a novel disease termed Pistachio Bushy Top Syndrome has been reported which appears to be caused by two *Rhodococcus* species (Stamler et al. 2015a, 2015b).

Here, we summarize the latest progress on the *R. fascians* leafy gall pathology and provide an overview of the novel insights on the capacity of other *Rhodococcus* species to beneficially or detrimentally interact with plants.

2 *Rhodococcus fascians*, Causative Agent of the Leafy Gall Syndrome

2.1 Symptoms and Host Range

R. fascians, unlike most pathogens, is a genuine biotroph that does not truly damage its host but uses subtler strategies to infect plants. It was first recognized as a plant pathogen in 1936 as the causative agent of fasciation on pea (Brown 1927; Tilford 1936). However, on most hosts, fasciations are not induced, but rather the disease is apparent as a combination of several symptoms, including deformed leaves and flowers, swollen tissues due to secondary growth of vascular tissues, loss of apical dominance, excessive shoots, witches' brooms, and/or differentiated galls covered with small shoots, collectively called the leafy gall syndrome (Stes et al. 2011b, 2013; Dolzblasz et al. 2018). The host range of *R. fascians* has reached over 150 species across 50 plant families, including primarily herbaceous dicotyledonous plants and some monocots (Putnam and Miller 2007). The extensive host range of leafy gall-inducing *R. fascians* isolates illustrates that the virulence factors of these bacteria target widely conserved pathways in plants. Since typical symptoms associated with the leafy gall syndrome lead to disfigured plants, *R. fascians* has mainly been problematic in the ornamentals industry where plants are grown for their aesthetic value (Depuydt et al. 2008b).

2.2 The Trick-with-the-Cytokinin-Mix

The genome of the leafy gall-inducing model strain *R. fascians* D188 consists of three replicons: the circular chromosome, a circular plasmid pD188 encoding cadmium resistance and not involved in pathogenicity, and a conjugative linear virulence plasmid pFiD188 (Fig. 1). Over 25 years of research on the interaction between strain D188 and the model plants *Arabidopsis thaliana* and *Nicotiana tabacum* resulted in a molecular and developmental framework bringing together the pathways of microbe and plant that contribute to the disease (reviewed by Stes et al. 2011b, 2013).

In essence, key to symptom development is the bacterial production of an array of cytokinins via genes of the *fas* operon encoded on pFiD188 (Crespi et al. 1992; Francis et al. 2012; Pertry et al. 2009, 2010) and the concomitant modification of the hormone landscape of the infected plant (Depuydt et al. 2008a, 2009a; Stes et al. 2011a, 2012, 2015; Dhandapani et al. 2018). The interaction starts with the epiphytic colonization of the plant. During this phase, no symptoms are induced, but the bacteria secrete low levels of a mixture of cytokinins and of auxin produced via chromosomal pathways that stimulate the metabolism of the plant cells resulting in the release of nutrients to the benefit of bacterial colonization (Cornelis et al. 2001; Depuydt et al. 2009b; Dhandapani et al. 2017, 2018; Francis et al. 2016). As a

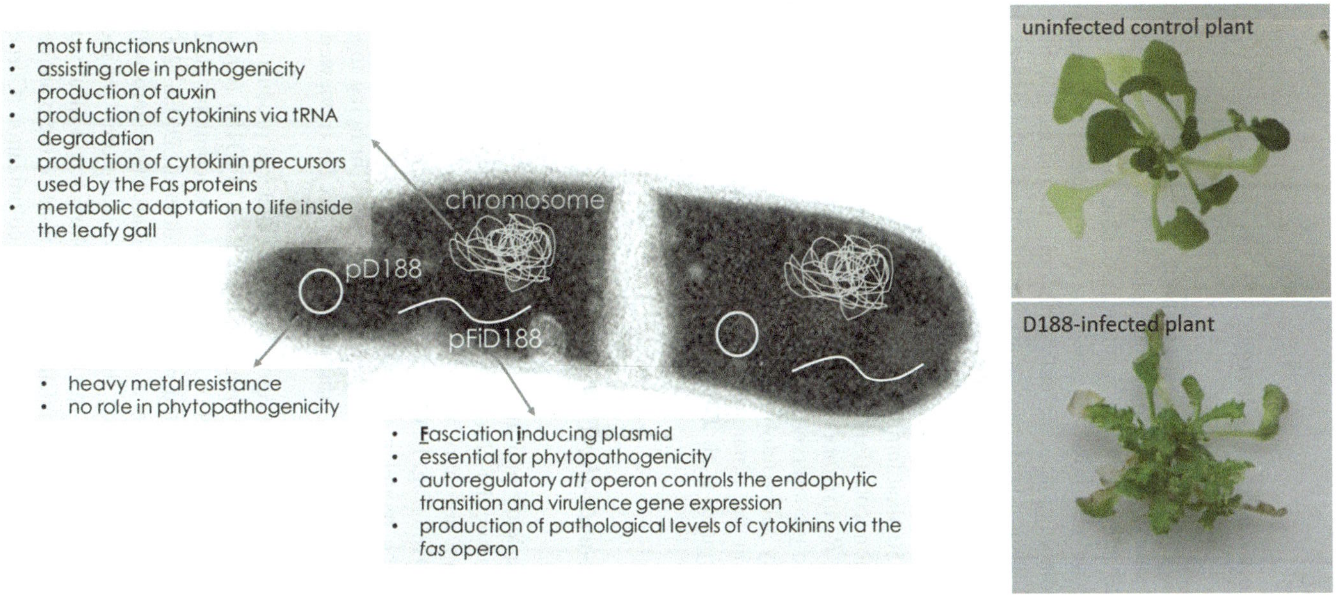

Fig. 1 Genome composition of the leafy gall-inducing model *R. fascians* strain D188 and symptoms on *Arabidopsis thaliana*

consequence of the shift in the primary metabolism of the plant upon perception of the bacteria (Depuydt et al. 2009b), a physiological change is triggered in the bacteria that will ultimately lead to the second phase of the interaction: the pathogenic endophytic phase. This phase is initiated by the production of an autoregulatory molecule encoded by the *att* locus located on pFiD188, establishing a positive regulatory loop resembling quorum sensing. The *att* system activates the transition to the endophytic phase possibly by stimulating the production of cell wall-degrading enzymes allowing active penetration of plant tissues and results in the production of higher pathological levels of a mixture of cytokinins encoded by the *fas* locus inducing the formation of an abundance of meristematic tissues in the plant that develop into shoots (Temmerman et al. 2000; Maes et al. 2001; Pertry et al. 2009; Fig. 1). The physiology of the symptomatic tissues is very different from uninfected plants, and the high concentration of meristematic cells leads to the provision of specific nutrients (Depuydt et al. 2009a, 2009b). These nutrients are believed to be metabolized via the glyoxylate shunt, since the bacteria require a functional malate synthase encoded by the chromosomal *vicA* gene for efficient survival inside the plant tissues (Vereecke et al. 2002). Hence, the success of the interaction with the host depends on the co-option of chromosomal and linear-plasmid-encoded functions, and eventually, through hormone secretion, the bacteria trigger the plant to provide it with a unique niche in terms of food and protection (Depuydt et al. 2009b; Dhandapani et al. 2017, 2018).

The virulence mechanism described above for *R. fascians* D188 is most likely similar for most other pathogenic leafy gall-inducing isolates because the *fas* and *att* genes are highly conserved (Creason et al. 2014a; Radhika et al. 2015). Nevertheless, the central position of cytokinins in the mode of action of the leafy gall inducers has been debated since their discovery because of the very low levels produced by in vitro-grown *R. fascians* cultures (Thimann and Sachs 1966; Eason et al. 1995; Pertry et al. 2009; Radhika et al. 2015). Even today, newcomers in the field question if cytokinins are essential for the leafy gall syndrome (Creason et al. 2014a; Savory et al. 2017), but until now, no valid alternative function for the *fas* operon has been proposed, and no other mode of action has been experimentally demonstrated (Vereecke, 2018). In contrast, there is a wealth of information that supports the indispensable role of *fas*-derived cytokinins in leafy gall formation:

1. The implication of cytokinins in symptom development has been recognized since 1966 (Klämbt et al. 1966; Helgeson and Leonard, 1966; Rathbone and Hall 1972; Scarbrough et al. 1973; Balázs and Sziráki 1974; Armstrong et al. 1976; Murai et al. 1980; Crespi et al. 1992; Eason et al. 1996; Dhandapani et al. 2017).
2. Mutations in the *fasA* and *fasD* genes render *R. fascians* nonpathogenic (Crespi et al. 1992, 1994; Pertry et al. 2009, 2010).
3. There is a 100% correlation between the presence of *fasD* and virulence (Eason et al. 1995, 1996; Stange et al. 1996; Nikolaeva et al. 2009; Creason et al. 2014a; Savory et al. 2017).

4. The biochemical activity of several Fas proteins in the production of cytokinins has been experimentally demonstrated (Crespi et al. 1992; Pertry et al. 2009, 2010; Radhika et al. 2015).
5. The cytokinins produced by *R. fascians* in culture are recovered in a higher amount in infected plants versus non-infected ones (Pertry et al. 2009, 2010; Radhika et al. 2015; Dhandapani et al. 2017).
6. *R. fascians*-infected plants activate cytokinin homeostasis mechanisms in the same way as if they were subjected to an excess of exogenous cytokinins (Depuydt et al. 2008a, 2009b; Motte et al. 2013).
7. The transcriptome of infected *Arabidopsis* and sweet pea plants clearly shows an enhanced cytokinin response (Depuydt et al. 2008a, 2009b; Dhandapani et al. 2017, 2018).

One challenging aspect of the trick-with-the-cytokinin-mix model was that the same mix of *fas*-dependent cytokinins was also detected in the supernatants of the plasmid-free strain D188-5. The cytokinin levels in this nonpathogenic derivative are much lower than those produced by the wild-type strain D188, and they are believed to result from tRNA degradation (Pertry et al. 2009, 2010). Because the bacteria cause symptoms only when the linear plasmid pFiD188 is present (Crespi et al. 1992), it was postulated that virulence does not depend on the exposure of the plant to one specific kind of cytokinin molecule but rather to particular concentrations and ratios of several of these plant growth regulators (Pertry et al. 2009, 2010). The higher pathological levels of cytokinins could only be produced through a co-option of chromosomal and linear plasmid genes (Francis et al. 2016). Although the cytokinin-mix production strategy used by *R. fascians* has been extended to plant-insect interactions (Giron and Glevarec 2014), Radhika et al. (2015) provided the missing link in the model. Indeed, the *mtr* genes encoding SAM-dependent methyltransferases and located upstream of the *fas* operon are involved in methylating the precursor of 2-iP used by FasD. These methylated cytokinins, which are not produced by plants, are resistant to plant cytokinin dehydrogenase-oxidases, are recognized by the plant cytokinin receptors, and have high morphogenic activities, most likely represent unique virulence-associated cytokinins (Radhika et al. 2015).

2.3 Epidemiology of Leafy Gall Disease in Ornamental Nurseries

As early as 1950, modes of spread of *R. fascians* within commercial nurseries of ornamental plants have been documented, and infield survival of up to 5 years in soil under favorable conditions has been reported (Baker 1950; Faivre-Amiot 1967; Oduro 1975; Miller et al. 1980), but overall information on the epidemiology of leafy gall-inducing *R. fascians* isolates remains scarce. Until recently, no in-depth information on the composition of the *R. fascians* populations on infected plants was available, and the relatedness between the bacteria on different plants within a nursery or between nurseries had not been studied. The phylogenetic relationship of 60 isolates collected from symptomatic tissues from single plants or from different plants in the same production facility was analyzed using a genomic epidemiological approach. The data showed that single nurseries could be infected by up to seven genotypes, that single plants could be colonized by at least three genotypes, and that different nurseries harbored highly related genotypes. Altogether, it could be concluded that the pathogens had been introduced in nurseries at multiple and independent occasions. Additionally, evidence was provided for point source outbreaks likely as a result of the acquisition of diseased plants from a common provider, but the presence of reservoir populations in some nurseries was apparent as well. When the linear plasmids of these isolates were compared, two major plasmid types could be distinguished, but the distribution of these plasmids indicated that plasmid exchange between genotypes occurred frequently. Additionally, the pathogenic isolates coexisted with nonpathogenic ones (Savory et al. 2017). Thus, as in other plant-associated bacteria, including *Pantoea agglomerans* (Weinthal et al. 2007), *Streptomyces scabies* (Zhang et al. 2016; Zhang and Loria 2017), *Pseudomonas syringae* (Monteil et al. 2016), and *Pseudomonas savastanoi* pv. *savastanoi* (Buonaurio et al. 2015), also in leafy gall-associated *Rhodococcus* species, the acquisition of a virulence plasmid drives pathogen evolution.

2.4 **Rhodococcus fascians,** ***a Troubled Species***

The taxonomic position of *R. fascians* has been troublesome since it was first isolated from diseased peas. At that time, the causative agent of the observed fasciations was named *Phytomonas fascians*, which was thought to be a specialized form of *Phytomonas (Agrobacterium) tumefaciens* (Tilford 1936). Few years later, the bacterium was renamed *Bacterium fascians*, because the generic name "*Phytomonas*" for plant pathogenic bacteria was no longer valid (Lacey 1939). Nonetheless, also this generic name led to persistent controversy, and based on post-fission movements of the bacteria and the presence of intracellular bodies, it was eventually recognized as *Corynebacterium fascians* (Dowson 1942). In the following years, it became apparent however that the leafy gall-inducing bacteria were more similar to *Rhodococcus* species than to *Corynebacterium*, and thus, more than 40 years later, in 1984, Goodfellow reclassified the pathogen as *Rhodococcus fascians* based on genetic, phenetic, and chemical characteristics (Goodfellow 1984). Now, as a consequence of the data obtained by (next-generation) sequencing techniques, once again doubt has been casted on the taxonomy of this microbe, and after 34 years, phylogenomic evidence suggests that the systematics of the genus *Rhodococcus* should be revised, and the *R. fascians* species likely belong to a distinct genus encompassing several species (Creason et al. 2014b; Sangal et al. 2016; Savory et al. 2017).

A strong indication that the genus *Rhodococcus* probably needs to be divided into more genera was based on 16S rRNA phylogenetic analysis of 217 actinomycete isolates, which revealed the occurrence of at least four *Rhodococcus* groups that were as phylogenetically distinct from each other as from *Nocardia* and *Gordonia* (Gürtler et al. 2004). However, an example that 16S rRNA-based classification is probably not sufficient to resolve the *Rhodococcus* genus is the inclusion into the species *R. fascians* of new isolates from diverse extreme environments unrelated to plants, such as Antarctic soils, marine waters, tidal flats, sediments contaminated with nuclear waste, polychlorinated dioxins, crude oil and hydrocarbons, deep glacier ice cores, and even carp intestines (Gürtler and Seviour 2010). Genome analysis of 19 isolates obtained from diseased ornamental plants, 2 isolates from a glacial ice core and 1 from permafrost, and 2 endophytes isolated from *Arabidopsis* indicated that these bacteria grouped into two clades representing up to seven different species (Creason et al. 2014b). Both clades contained pathogenic as well as nonpathogenic isolates. Even more, *Next Generation Systematics* based on the genome analysis of 100 rhodococcal strains and 15 representatives of related genera has provided a genomic framework that revealed seven distinct species-groups and three singletons (Sangal et al. 2016). Species-group E consisting of 28 genomes represents the *R. fascians* cluster which can be further subdivided in two subgroups corresponding to the two clades identified by Creason et al. (2014b). Subgroup E1 corresponds to clade II and would comprise two species, whereas subgroup E2, including the type strain and the model strain D188, corresponds to clade I and encompasses six predicted species (Sangal et al. 2016). Based on this information and as discussed by Sutcliffe and colleagues in this issue, a formal proposal of a new and hopefully final genus name for *R. fascians* is to be expected soon.

3 Pistachio Bushy Top Syndrome, a Novel *Rhodococcus*-Associated Disease

In 2011, a novel disease emerged on pistachio "UCB-1" rootstock trees, termed Pistachio Bushy Top Syndrome (PBTS) which significantly affected the pistachio industry in California, Arizona, and New Mexico, causing massive economic losses (Stamler et al. 2015a, 2015b). PBTS is characterized by stunted growth of the rootstock trees, the formation of additional shoots, the strongly reduced capacity of the rootstock to accept a *Pistacia vera* graft, the development of cracked tissue at graft junctions weakening the graft union, the emergence of ectopic shoots from gall-like tissue formed at lateral buds, and, most typically, a complete disturbance of regular root development (Stamler et al. 2015a, 2015b) (Fig. 2). The causative agents of PBTS were found to be two *Rhodococcus* species: *Rhodococcus* sp. strain 1 (PBTS1) genetically related to *R. corynebacterioides* and *Rhodococcus* sp. strain 2 (PBTS2) with high genetic similarity to *R. fascians* (Stamler et al. 2015b, 2016). Little information is available on the former bacterium, but it has been found as an endophyte in xylem of eggplant and leaves of *Arabidopsis* (Traw et al. 2007; Achari and Ramesh 2014). Concerning the *R. fascians* isolate, few woody plants were

Fig. 2 Typical symptoms associated with Pistachio Bushy Top Syndrome on "UCB-1" pistachio rootstock. (**a**) Loss of apical dominance leads to bushy top appearance; (**b**) deformed leaves; (**c**) deformed root system; (**d**) unsuccessful grafting attempts; (**e**) bark cracking at the graft union

shown to be sensitive to leafy gall-inducing *R. fascians* strains (Putnam and Miller 2007), but the Anacardiaceae or cashew family to which pistachio belongs is a new host of this bacterium. While the strategy utilized by the model strain *R. fascians*

D188 to cause leafy galls on the model plant *Arabidopsis* is largely understood and the major players involved in the initiation and maintenance of the symptoms have been identified, as summarized above, to date, virtually nothing is known on the pathogenic strategies deployed by the PBTS strains. Nevertheless, both isolates generated an amplicon with *fasD*-specific primers, and PBTS2 also generated an amplicon with *fasA*-specific primers, suggesting that cytokinins might play a role. Interestingly, while performing Koch's postulates, it appeared that both PBTS *Rhodococcus* strains work in synergism to alter plant development on clonal "UCB-1" pistachio (Stamler et al. 2015b).

Whereas the current knowledge on the mode of action of the PBTS *Rhodococcus* strains is still in its infancy (Stamler et al. 2015a, 2015b, 2016), Koch's postulates were clearly fulfilled. Still the diagnosis of PBTS1 and PBTS2 as outbreak strains of PBTS on pistachio has been challenged because the presence of the virulence genes could not be confirmed, and consequently, the PBTS isolates analyzed were not pathogenic. Hence, the hypothesis has been put forward that the PBTS *Rhodococcus* strains would actually be beneficial bacteria naturally associated with pistachio (Savory et al. 2017). In our opinion, it is premature to assume that leafy gall and PBTS inducers necessarily use the same virulence mechanism and that all aspects of the biology of both syndromes are shared. Additionally, concerning the epidemiology of PBTS, very strong conclusions were drawn based on the analysis of two single PBTS isolates, whereas the epidemiology of leafy gall inducers in ornamental orchards could only be revealed by studying 60 different isolates (Savory et al. 2017). One can only hope that good and integer science will resolve this issue in the future (Randall et al. 2018; Vereecke 2018).

4 Beneficial and Undefined Associations of *Rhodococcus* Species with Plants

The concept of plants serving as a habitat for complex microbial communities interacting with each other and with the plant host is now widely accepted and gaining more research interest as to how we can use these organismal interactions in plant growth and health management. The composition of the microbial communities interacting with different plant parts can vary greatly and reflects niche adaptation. Within these microbiomes, several *Rhodococcus* species, apart from the well-known plant pathogen *R. fascians* (discussed above), have recently been found living in close association with various plant parts. For example, *Rhodococcus* isolates were found as part of the rhizosphere of *Origanum vulgare* in a sub-Himalayan region (Bafana 2013), the phyllosphere of *Arabidopsis thaliana* plants (Bai et al. 2015; Ritpitakphong et al. 2016), and as an endophyte of leaves of *Cercidiphyllum japonicum* (Li et al. 2008) and stem tissues of *Artemisia annua* and *Sorghum bicolor* (Zhao et al. 2012; Maropola et al. 2015). Although the ecological role of most of these plant-associated *Rhodococcus* species remains to be discovered, a number of isolates have been shown to display plant growth-promoting properties and can confer resistance to plant pathogens.

4.1 Rhodococcus fascians *Beyond Leafy Gall Formation*

Recent work has shown that upon loss of its linear plasmid, the resulting nonpathogenic derivative D188-5 of *R. fascians* strain D188 can stimulate plant growth (Francis et al. 2016). This bacterium is very well equipped for an epiphytic lifestyle and likely acts as a phytostimulator mainly through the production of the plant hormones cytokinin and auxin (IAA) and the degradation of ethylene via 1-aminocyclopropane-1-carboxylate (ACC) deaminase (Francis et al. 2016). The delicate interplay between cytokinin and auxin regulates meristem formation and development in both shoots and roots from early embryogenesis (Su et al. 2011), while ethylene inhibits shoot and root proliferation at high concentrations which generally occurs when the plant is under environmental stress (Sing et al. 2015). The fact that this plasmid-free, nonpathogenic derivative of D188 promotes plant growth suggests that pathogenic isolates may actually be plant growth-promoting bacteria gone astray. This reasoning is supported by the conservation of the plant growth-promoting determinants, especially the chromosomal genes involved in cytokinin and auxin metabolism, in strains covering both clades of the *R. fascians* species group discussed above (Francis et al. 2016). Additionally, plant growth-promoting activity has been attributed to other leafy gall-associated *R. fascians* strains (Savory et al. 2017), but the experimental data to support this conclusion were thought to be insufficient (Randall et al. 2018; Vereecke 2018).

The plant pathogen *R. fascians* was also identified as one of the dominant bacterial species on different cultivars of spring barley in combination with various other pathogenic and nonpathogenic bacteria, yeasts, and fungi (Newton et al. 2010) and as one of the *Rhodococcus* isolates in the rhizosphere of the nickel hyperaccumulator *Thlaspi goesingense* (Idris et al. 2004). The latter isolates displayed moderate to high siderophore production, and some also produced ACC deaminase (Idris et al. 2004). Although the later subjective synonym *R. luteus* (Klatte et al. 1994) was used, *R. fascians* was also isolated as a xylem-inhabiting bacterium from grape cultivars in Nova Scotia (Bell et al. 1995). Importantly, none of these hosts displayed disease symptoms suggesting that the pathogenic lifestyle of *R. fascians* might be strongly condition dependent.

4.2 *Plant-Associated* Rhodococcus *Species Have Plant Growth- Promoting Activities and Confer Plant Resistance to Pathogens*

Apart from the plant growth-promoting ability of the plasmid-free derivative of *R. fascians* D188 (Francis et al. 2016), several other *Rhodococcus* species were shown to stimulate plant growth and development. An isolate from postglacial clay and positive for ACC deaminase significantly increased the root biomass of pea plants in a pot experiment by 25% (Belimov et al. 2001). An endophytic species of tomato roots also tested positive for IAA production, synthesis of ACC deaminase, and had high siderophore activity (Abbamondi et al. 2016). Although not tested on plants, *Rhodococcus* aff. *qingshengii*, isolated from a polluted river in India, produced high levels of IAA (Hasuty et al. 2018). As part of its plant growth-promoting properties, *R. globerulus*, an endophyte of the medicinal plant *Plectranthus amboinicus*, was capable of phosphate solubilization making insoluble organic phosphorus available to plants. This trait opens up the possibility of using the isolate as a biofertilizer (Murugappan et al. 2017).

In a diversity study of the nodule microbiomes of *Lotus japonicus* and *Anthyllis vulneraria* in Sweden, two *Rhodococcus* isolates have been found with the ability to nodulate these two plants. These bacteria probably obtained this exceptional ability through horizontal gene transfer from other nodule-inhabiting bacteria, as their *nodA* gene was closely related to that of *Mesorhizobium loti* (Ampomah and Huss-Danell 2011). Moreover, it is speculated that the combined activity of the nitric oxide synthase of *Rhodococcus* isolate APG1, an endophyte of the aquatic fern *Azolla pinnata*, together with symbiotic *Arthrobacter* species could be beneficial to plants by providing nitrogen oxides for regulatory functions (Cohen and Yamasaki 2003; Cohen et al. 2004). *Rhodococcus* was also found as one of the main genera colonizing the shoot endosphere of a willow cultivar that is particularly well-suited for phytoremediation of heavy metal-contaminated soils. This property was attributed to the very diverse endophytic bacterial population among which a high number of metal-resistant plant growth-promoting bacteria, although for the identified *Rhodococcus* species no specific plant-stimulating properties were reported (Weyens et al. 2013).

Rhodococcus species themselves are commonly used for bioremediation due to their unique biochemical properties, such as biodegradation of hydrophobic natural compounds and xenobiotics, including polychlorinated biphenyls (PCB) (van der Geize and Dijkhuizen 2004; Larkin et al. 2005; Ceniceros et al. 2017; and further discussed in this volume). These bacteria are often found to live in close association with plants grown at such contaminated sites, and it is this unique plant—*Rhodococcus* combination that improves soil bioremediation (Leigh et al. 2006). *R. erythropolis* U23A, isolated from the rhizosphere of plants grown in PCB-contaminated soils, showed increased PCB degradation due to chemotaxis toward plant secondary metabolites (Toussaint et al. 2012; Pham et al. 2015). *R. erythropolis* strain CD 106 not only significantly increased the effectiveness of the phytoremediation of petroleum hydrocarbon-contaminated soil by 31.2% in combination with ryegrass but also increased the biomass of roots and shoots by 30–49%, respectively (Płociniczak et al. 2017). Also, plant growth inhibition by chromium, more specifically Cr^{6+}, can be alleviated by the psychrotroph *R. erythropolis* MTCC 7905 due to its ability to reduce Cr^{6+} to the nontoxic Cr^{3+} and several plant growth-promoting activities, such as ACC deaminase production, indole and siderophore production, although these latter activities decreased with increasing Cr^{6+} concentrations (Trivedi et al. 2007).

Finally, microbiome studies show that the natural inhabitants of plants such as leaf epi- and endophytic bacteria can confer resistance to plant pathogens. In that respect, *R. kyotonensis* and *R. corynebacterioides*, endophytes isolated from *Arabidopsis*, were shown to confer mild disease suppression against *Pseudomonas syringae* pv. tomato DC3000 (Hong et al. 2015), and another endophyte, *Rhodococcus* sp. KB6, significantly repressed black rot disease in sweet potato caused by *Ceratocystis fimbriata* (Hong et al. 2016). *R. erythropolis* can intercept and degrade the quorum-sensing signals produced by *Pectobacterium atrosepticum* and *P. carotovorum*, the causative agents of blackleg and soft rot disease in potato plants and tubers. These signals are important in the onset of virulence gene expression, and hence, degradation of the quorum-sensing N-acyl homoserine lactone signals limits disease development (Jafra et al. 2006; Cirou et al. 2007; Latour et al. 2013; Kwasiborski et al. 2015).

5 Concluding Remarks

Although the leafy gall-inducing model *R. fascians* strain D188 has been studied for over a quarter of a century, detailed information on the genes implicated in the pathogenicity is only available for the *fas* and the *att* operon. However, the linear plasmid, around 200 kb in size, is highly conserved in all other leafy gall inducers analyzed to date, suggesting that many other determinants encoded by the linear plasmid must be essential for a successful interaction with the plant host. Future experimentation will hopefully give more insight into the role of these other genes in the pathogenicity of *R. fascians*, and time will tell if the *fas* operon and the trick-with-the-cytokinin-mix are indeed determinative for leafy gall formation.

With the outbreak of PBTS on pistachio, the genus *Rhodococcus* acquired an additional pathogenic member, and with the expanding research on plant-associated microbiomes, it is clear that the capacity of members of this genus to interact with plants has been largely underestimated. Given that the genus *Rhodococcus* is renowned for its highly specialized metabolic capacities, the comparison of the genomes of all plant-interacting *Rhodococcus* species, beneficial and adverse, might reveal novel functions specifically utilized by this group of bacteria to interact with a plant host.

References

Abbamondi GR, Tommonaro G, Weyens N, Thijs S, Sillen W, Gkorezis P, Iodice C, de Melo Rangel W, Nicolaus B, Vangronsveld J (2016) Plant growth-promoting effects of rhizospheric and endophytic bacteria associated with different tomato cultivars and new tomato hybrids. Chem Biol Technol Agric 3:1

Achari GA, Ramesh R (2014) Diversity, biocontrol, and plant growth promoting abilities of xylem residing bacteria from solanaceous crops. Int J Microbiol 2014:296521

Ampomah OY, Huss-Danell K (2011) Genetic diversity of root nodule bacteria nodulating *Lotus corniculatus* and *Anthyllis vulneraria* in Sweden. Syst Appl Microbiol 34:267–275

Anastasi E, MacArthur I, Scortti M, Alvarez S, Giguère S, Vázquez-Boland JA (2016) Pangenome and phylogenomic analysis of the pathogenic actinobacterium *Rhodococcus* equi. Genome Biol Evol 8(10):3140–3148

Armstrong DJ, Scarbrough E, Skoog F, Cole DL, Leonard NJ (1976) Cytokinins in *Corynebacterium fascians* cultures. Isolation and identification of 6-(4-hydroxy-3-methyl-cis-2-butenylamino)-2-methylthiopurine. Plant Physiol 58:749–752

Balázs E, Sziráki I (1974) Altered levels of indoleacetic acid and cytokinin in geranium stems infected with *Corynebacterium fascians*. Acta Phytopathol Acad Sci Hungaricae 9:287–292

Bafana A (2013) Diversity and metabolic potential of culturable root-associated bacteria from *Origanum vulgare* in sub-Himalayan region. World J Microbiol Biotechnol 29:63–74

Bai Y, Müller DB, Srinivas G, Garrido-Oter R, Potthoff E, Rott M, Dombrowski N, Münch PC, Spaepen S, Remus-Emsermann M, Hüttel B, McHardy AC, Vorholt JA, Schulze-Lefert P (2015) Functional overlap of the Arabidopsis leaf and root microbiota. Nature 528:364–369

Baker KF (1950) Bacterial fasciation disease of ornamental plants in California. Plant Dis Rep 34:121–126

Belimov AA, Safronova VI, Sergeyeva TA, Egorova TN, Matveyeva VA, Tsyganov VE, Borisov AY, Tikhonovich IA, Kluge C, Preisfeld A, Dietz KJ, Stepanok VV (2001) Characterization of plant growth promoting rhizobacteria isolated from polluted soils and containing 1-aminocyclopropane-1-carboxylate deaminase. Can J Microbiol 47:642–652

Bell CR, Dickie GA, Harvey WLG, Chan JWYF (1995) Endophytic bacteria in grapevine. Can J Microbiol 41:46–53

Brown NA (1927) Sweet pea fasciation, a form of crowngall. Phytopathology 17:29–30

Buonaurio R, Moretti C, da Silva DP, Cortese C, Ramos C, Venturi V (2015) The olive knot disease as a model to study the role of interspecies bacterial communities in plant disease. Front Plant Sci 6:434

Calvo P, Nelson L, Kloepper JW (2014) Agricultural uses of plant biostimulants. Plant Soil 383:3–41

Ceniceros A, Dijkhuizen L, Petrusma M, Medema MH (2017) Genome-based exploration of the specialized metabolic capacities of the genus *Rhodococcus*. BMC Genomics 18:593

Cirou A, Diallo S, Kurt C, Latour X, Faure D (2007) Growth promotion of quorum-quenching bacteria in the rhizosphere of *Solanum tuberosum*. Environ Microbiol 9(6):1511–1522

Cohen MF, Meziane T, Yamasaki H (2004) A photocarotenogenic *Rhodococcus* sp. isolated from the symbiotic fern *Azolla*. Endocytobiosis Cell Res 15:350–355

Cohen MF, Yamasaki H (2003) Involvement of nitric oxide synthase in sucrose-enhanced hydrogen peroxide tolerance of *Rhodococcus* sp. strain APG1, a plant-colonizing bacterium. Nitric Oxide 9:1–9

Cornelis K, Ritsema T, Nijsse J, Holsters M, Goethals K, Jaziri M (2001) The plant pathogen *Rhodococcus fascians* colonizes the exterior and interior of the aerial parts of plants. Mol Plant Microbe Interact 14:599–608

Creason AL, Vandeputte OM, Savory EA, Davis EW II, Putnam ML, Hu E, Swader-Hines D, Mol A, Baucher M, Prinsen E, Zdanowska M, Givan SA, El Jaziri M, Loper JE, Mahmud T, Chang JH (2014a) Analysis of genome sequences from plant pathogenic *Rhodococcus* reveals genetic novelties in virulence loci. PLoS One 9:e101996

Creason AL, Davis EW II, Putnam ML, Vandeputte OM, Chang JH (2014b) Use of whole genome sequences to develop a molecular phylogenetic framework for *Rhodococcus fascians* and the *Rhodococcus* genus. Front Plant Sci 5:406
Crespi M, Messens E, Caplan AB, Van Montagu M, Desomer J (1992) Fasciation induction by the phytopathogen *Rhodococcus fascians* depends upon a linear plasmid encoding a cytokinin synthase gene. EMBO J 11:795–804
Crespi M, Vereecke D, Temmerman W, van Montagu M, Desomer J (1994) The *fas* operon of *Rhodococcus fascians* encodes new genes required for efficient fasciation of host plants. J Bacteriol 176(9):2492–2501
Depuydt S, Doležal K, Van Lijsebettens M, Moritz T, Holsters M, Vereecke D (2008a) Modulation of the hormone setting by *Rhodococcus fascians* results in ectopic *KNOX* activation in Arabidopsis. Plant Physiol 146:1267–1281
Depuydt S, Putnam M, Holsters M, Vereecke D (2008b) *Rhodococcus fascians*, an emerging threat for ornamental crops. In: Teixeira da Silva JA (ed) Floriculture, ornamental, and plant biotechnology: advances and topical issues, vol 5. Global Science Books, Isleworth, pp 480–489
Depuydt S, De Veylder L, Holsters M, Vereecke D (2009a) Eternal youth, the fate of developing Arabidopsis leaves upon *Rhodococcus fascians* infection. Plant Physiol 149:1387–1398
Depuydt S, Trenkamp S, Fernie AR, Elftieh S, Renou J-P, Vuylsteke M, Holsters M, Vereecke D (2009b) An integrated genomics approach to define niche establishment by *Rhodococcus fascians*. Plant Physiol 149:1366–1386
Dhandapani P, Song J, Novak O, Jameson P (2017) Infection by *Rhodococcus fascians* maintains cotyledons as a sink tissue for the pathogen. Ann Bot 119(5):841–852
Dhandapani P, Song J, Novak O, Jameson P (2018) Both epiphytic and endophytic strains of *Rhodococcus fascians* influence transporter gene expression and cytokinins in infected *Pisum sativum* L. seedlings. Plant Growth Regul 85:231–242
Dolzblasz A, Banasiak A, Vereecke D (2018) Neovascularization during leafy gall formation on *Arabidopsis thaliana* upon *Rhodococcus fascians* infection. Planta 247:215–228
Dowson WJ (1942) On the generic name of the Gram-positive bacterial plant pathogens. Trans Br Mycol Soc 25:311–314
Eason JR, Jameson PE, Bannister P (1995) Virulence assessment of *Rhodococcus fascians* strains on pea cultivars. Plant Pathol 44:141–147
Eason JR, Morris RO, Jameson PE (1996) The relationship between virulence and cytokinin production by *Rhodococcus fascians*. Plant Pathol 45:323–331
Faivre-Amiot A (1967) Quelques observations sur la presence de *Corynebacterium fascians* (Tilford) Dowson dans les cultures maraichères et florals en France. Phytiatr Phytopharm 16:165–176
Francis I, De Keyser A, De Backer P, Simón-Mateo C, Kalkus J, Pertry I, Ardiles-Diaz W, De Rycke R, Vandeputte OM, El Jaziri M, Holsters M, Vereecke D (2012) pFiD188, the linear virulence plasmid of *Rhodococcus fascians* D188. Mol Plant Microbe Interact 25(5):637–647
Francis IM, Stes E, Zhang Y, Rangel D, Audenaert K, Vereecke D (2016) Mining the genome of *Rhodococcus fascians*, a plant growth-promoting bacterium gone astray. New Biotechnol 33 (5 Pt B):706–717
Giron D, Glevarec G (2014) Cytokinin-induced phenotypes in plant-insect interactions: learning from the bacterial world. J Chem Ecol 40(7):826–835
Goodfellow M (1984) Reclassification of *Corynebacterium fascians* (Tilford) Dowson in the genus *Rhodococcus*, as *Rhodococcus fascians* comb. nov. Syst Appl Microbiol 5(2):225–229
Gürtler V, Mayall BC, Seviour R (2004) Can whole genome analysis refine the taxonomy of the genus *Rhodococcus*? FEMS Microbiol Rev 28(3):377–403
Gürtler V, Seviour R (2010) Systematics of members of the genus *Rhodococcus* (Zopf 1891) Emend Goodfellow et al. 1998. The past, present and future. In: Alvarez HM (ed) Biology of *Rhodococcus*, Microbiology Monographs, vol 16. Springer, Berlin, pp 1–28
Hamedi J, Mohammadipanah F (2015) Biotechnological application and taxonomical distribution of plant growth promoting actinobacteria. J Ind Microbiol Biotechnol 42(2):157–171

Hasuty A, Choliq A, Hidayat I (2018) Production of indole acetic acid (IAA) by *Serratia marcescens* subsp. *marcescens* and *Rhodococcus* aff. *Qingshengii*. Int J Agric Technol 14 (3):299–312

Helgeson JP, Leonard NJ (1966) Cytokinins: indentification of compounds isolated from *Corynebacterium fascians*. Proc Natl Acad Sci U S A 56:60–63

Hong CE, Jeong H, Jo SH, Jeong JC, Kwon SY, An D, Park JM (2016) A leaf-inhabiting endophytic bacterium, *Rhodococcus* sp. KB6, enhances sweet potato resistance to black rot disease caused by *Ceratocystis fimbriata*. J Microbiol Biotechnol 26(3):488–492

Hong CE, Jo SH, Moon JY, Lee JS, Kwon SY, Park JM (2015) Isolation of novel leaf-inhabiting endophytic bacteria in *Arabidopsis thaliana* and their antagonistic effects on phytopathogens. Plant Biotechnol Rep 9(6):451–458

Idris R, Trifonova R, Puschenreiter M, Wenzel WW, Sessitsch A (2004) Bacterial communities associated with flowering plants of the Ni hyperaccumulator *Thlaspi goesingense*. Appl Environ Microbiol 70(5):2667–2677

Jafra S, Przysowa J, Czajkowski R, Michta A, Garbeva P, Van de Wolf JM (2006) Detection and characterization of bacteria from the potato rhizosphere degrading *N*-acyl-homoserine lactone. Can J Microbiol 52:1006–1015

Klämbt D, Thies G, Skoog F (1966) Isolation of cytokinins from *Corynebacterium fascians*. Proc Natl Acad Sci U S A 56:52–59

Klatte S, Jahnke JD, Kroppenstedt RM, Rainey F, Stackebrandt E (1994) *Rhodococcus luteus* is a later subjective synonym of *Rhodococcus fascians*. Int J Syst Bacteriol 44(4):627–630

Kwasiborski A, Mondy S, Chong TM, Barbey C, Chan KG, Beury-Cirou A, Latour X, Faure D (2015) Transcriptome of the quorum-sensing signal-degrading *Rhodococcus erythropolis* responds differentially to virulent and avirulent *Pectobacterium atrosepticum*. Heredity 114:476–484

Lacey MS (1939) Studies on a bacterium associated with leafy galls, fasciation and 'cauliflower' disease of various plants. Part III. Further isolations, inoculation experiments and cultural studies. Ann Appl Biol 26:262–278

Larkin MJ, Kulakov LA, Allen CCR (2005) Biodegradation and *Rhodococcus* – masters of catabolic versatility. Curr Opin Biotechnol 16:282–290

Latour X, Barbey C, Chane A, Groboillot A, Burini JF (2013) *Rhodococcus erythropolis* and its γ-lactone catabolic pathway: an unusual biocontrol system that disrupts pathogen quorum sensing communication. Agronomy 3:816–838

Leigh MB, Prouzová P, Macková M, Macek T, Nagle DP, Fletcher JS (2006) Polychlorinated biphenyl (PCB)-degrading bacteria associated with trees in a PCB-contaminated site. Appl Environ Microbiol 72(4):2331–2341

Li J, Zhao GZ, Chen HH, Qin S, Xu LH, Jiang CL, Li WJ (2008) *Rhodococcus cercidiphylli* sp. nov., a new endophytic actinobacterium isolated from a *Cercidiphyllum japonicum* leaf. Syst Appl Microbiol 31:108–113

MacArthur I, Anastasi E, Alvarez S, Scortti M, Vázquez-Boland JA (2017) Comparative genomics of *Rhodococcus equi* virulence plasmids indicates host-driven evolution of the *vap* pathogenicity island. Genome Biol Evol 9(5):1241–1247

Maes T, Vereecke D, Ritsema T, Cornelis K, Ngo Thi Thu H, Van Montagu M, Holsters M, Goethals K (2001) The *att* locus of *Rhodococcus fascians* strain D188 is essential for full virulence on tobacco through the production of an autoregulatory compound. Mol Microbiol 42:13–28

Maropola MKA, Ramond JB, Trindade M (2015) Impact of metagenomics DNA extraction procedures on the identifiable endophytic bacterial diversity in *Sorghum bicolor* (L. Moench). J Microbiol Methods 112:104–117

Miller HJ, Janse JD, Kamerman W, Muller PJ (1980) Recent observations of leafy gall in Liliaceae and some other families. Neth J Plant Pathol 86:55–68

Monteil CL, Yahara K, Studholme DJ, Mageiros L, Méric G, Swingle B, Morris CE, Vinatzer BA, Sheppard SK (2016) Population-genomic insights into emergence, crop adaptation and dissemination of *Pseudomonas syringae* pathogens. Microb Genom 2:e000089

Motte H, Galuszka P, Spíchal L, Tarkowski P, Plíhal O, Šmehilová M, Jaworek P, Vereecke D, Werbrouck S, Geelen D (2013) Phenyl-adenine, identified in a *LIGHT-DEPENDENT SHORT HYPOCOTYLS4*-assisted chemical screen, is a potent compound for shoot regeneration through the inhibition of CYTOKININ OXIDASE/DEHYDROGENASE activity. Plant Physiol 161:1229–1241

Murai N, Skoog F, Doyle ME, Hanson RS (1980) Relationships between cytokinin production, presence of plasmids, and fasciation caused by strains of *Corynebacterium fascians*. Proc Natl Acad Sci U S A 77:619–623

Murugappan RM, Benazir Begun S, Usha C, Lok Kirubahar S, Karthikeyan M (2017) Growth promoting and probiotic potential of the endophytic bacterium *Rhodococcus globerulus* colonizing the medicinal plant *Plectranthus amboinicus* (Lour.) Spreng. Int J Curr Res Rev 9 (14):7–13

Newton AC, Gravouil C, Foutaine JM (2010) Managing the ecology of foliar pathogens: ecological tolerance in crops. Ann Appl Biol 157:343–359

Nikolaeva EV, Park S-Y, Kang S, Olson TN, Kim SH (2009) Ratios of cells with and without virulence genes in *Rhodococcus fascians* populations correlate with degrees of symptom development. Plant Dis 93:499–506

Oduro KA (1975) Factors affecting epidemiology of bacterial fasciation of *Chrysanthemum maximum*. Phytopathology 65:719–721

Pertry I, Václavíková K, Depuydt S, Galuszka P, Spíchal L, Temmerman W, Stes E, Schmülling T, Kakimoto T, Van Montagu M, Strnad M, Holsters M, Tarkowski P, Vereecke D (2009) Identification of *Rhodococcus fascians* cytokinins and their modus operandi to reshape the plant. Proc Natl Acad Sci U S A 106:929–934

Pertry I, Václavíková K, Gemrotová M, Spíchal L, Galuszka P, Depuydt S, Temmerman W, Stes E, De Keyser A, Riefler M, Biondi S, Novák O, Schmülling T, Strnad M, Tarkowski P, Holsters M, Vereecke D (2010) *Rhodococcus fascians* impacts plant development through the dynamic Fas-mediated production of a cytokinin mix. Mol Plant Microbe Interact 23:1164–1174

Pham TTM, Pimo Rodriguez NJ, Hijri M, Sylvestre M (2015) Optimizing polychlorinated biphenyl degradation by flavonoid-induced cells of the rhizobacterium *Rhodococcus erythropolis* U23A. PLoS One 10(5):e0126033

Putnam ML, Miller ML (2007) *Rhodococcus fascians* in herbaceous perennials. Plant Dis 91 (9):1064–1076

Płociniczak T, Fic E, Pacwa-Płociniczak M, Pawlik M, Piotrowska-Seget Z (2017) Improvement of phytoremediation of an aged petroleum hydrocarbon-contaminated soil by *Rhodococcus erythropolis* CD 106 strain. Int J Phytoremediation 19(7):614–620

Radhika V, Ueda N, Tsuboi Y, Kojima M, Kikuchi J, Kudo T, Sakakibara H (2015) Methylated cytokinins from the phytopathogen *Rhodococcus fascians* mimic plant hormone activity. Plant Physiol 169:1118–1126

Randall JJ, Stamler RA, Kallsen CE, Fichtner EJ, Heerema RJ, Cooke P, Francis I (2018) Comment on "Evolutionary transitions between beneficial and phytopathogenic *Rhodococcus* challenge disease management". eLife 7:e35272

Rathbone MP, Hall RH (1972) Concerning the presence of the cytokinin, N6–(12–isopentenyl) adenine in cultures of *Corynebacterium fascians*. Planta 108:93–102

Ritpitakphong U, Falquet L, Vimoltust A, Berger A, Métraux JP, L'Haridon F (2016) The microbiome of the leaf surface of *Arabidopsis* protects against a fungal pathogen. New Phytol 210:1033–1043

Sangal V, Goodfellow M, Jones AL, Schwalbe EC, Blom J, Hoskisson PA, Sutcliffe I (2016) Next-generation systematics: an innovative approach to resolve the structure of complex taxa. Sci Rep 6:38392

Savory EA, Fuller SL, Weisberg AJ, Thomas WJ, Gordon MI, Stevens DM, Creason AL, Belcher MS, Serdani M, Wiseman MS, Grünwald NJ, Putnam ML, Chang JH (2017) Evolutionary transitions between beneficial and phytopathogenic *Rhodococcus* challenge disease management. eLife 6:e30925
Scarbrough E, Armstrong DJ, Skoog F, Frihart CR, Leonard NJ (1973) Isolation of *cis*-zeatin from *Corynebacterium fascians* cultures. Proc Natl Acad Sci U S A 70:3825–3829
Sing RP, Shelke GM, Kumar A, Jha PN (2015) Biochemistry and genetics of ACC deaminase: a weapon to "stress ethylene" produced in plants. Front Microbiol 6:937
Stamler RA, Heerema R, Randall JJ (2015a) First report of phytopathogenic *Rhodococcus* isolates on Pistachio Bushy Top Syndrome 'UCB-1' rootstock in New Mexico. Plant Dis 99:1854
Stamler RA, Kilcrease J, Kallsen C, Fichtner EJ, Cooke P, Heerema RJ, Randall JJ (2015b) First report of *Rhodococcus* isolates causing Pistachio Bushy Top Syndrome on 'UCB-1' rootstock in California and Arizona. Plant Dis 99:1468–1476
Stamler RA, Vereecke D, Zhang Y, Schilkey F, Devitt N, Randall JJ (2016) Complete genome and plasmid sequences for *Rhodococcus fascians* D188 and draft sequences for *Rhodococcus* isolates PBTS 1 and PBTS 2. Genome Announc 4:e00495
Stange RR, Jeffares D, Young C, Scott DB, Eason JR, Jameson PE (1996) PCR amplification of the fas-1 gene for detection of virulent strains of *Rhodococcus fascians*. Plant Pathol 45:407–417
Stes E, Biondi S, Holsters M, Vereecke D (2011a) Bacterial and plant signal integration via D3-type cyclins enhances symptom development in the Arabidopsis-*Rhodococcus fascians* interaction. Plant Physiol 156:712–725
Stes E, Depuydt S, De Keyser A, Matthys C, Audenaert K, Yoneyama K, Werbrouck S, Goormachtig S, Vereecke D (2015) Strigolactones as an auxiliary hormonal defence mechanism against leafy gall syndrome in *Arabidopsis thaliana*. J Exp Bot 66(16):5123–5134
Stes E, Francis I, Pertry I, Dolzblasz A, Depuydt S, Vereecke D (2013) The leafy gall syndrome induced by *Rhodococcus fascians*. FEMS Microbiol Lett 342:187–194
Stes E, Prinsen E, Holsters M, Vereecke D (2012) Plant-derived auxin plays an accessory role in symptom development upon *Rhodococcus fascians* infection. Plant J 70:513–527
Stes E, Vandeputte OM, El Jaziri ME, Holsters M, Vereecke D (2011b) A successful bacterial coup d'état: how *Rhodococcus fascians* redirects plant development. Annu Rev Phytopathol 49:69–86
Su YH, Liu YB, Zhang XS (2011) Auxin-cytokinin interaction regulates meristem development. Mol Plant 4(4):616–625
Temmerman W, Vereecke D, Dreesen R, van Montagu M, Holsters M, Goethals K (2000) Leafy gall formation is controlled by *fasR*, an AraC-type regulatory gene in *Rhodococcus fascians*. J Bacteriol 182(20):5832–5840
Thimann KV, Sachs T (1966) The role of cytokinins in the "fasciation" disease caused by *Corynebacterium fascians*. Am J Bot 53:731–739
Tilford PE (1936) Fasciation of sweet peas caused by *Phytomonas fascians* n. sp. J Agric Res 53:383–394
Toussaint JP, Pham TTM, Barriault D, Sylvestre M (2012) Plant exudates promote PCB degradation by a rhodococcal rhizobacteria. Appl Microbiol Biotechnol 95:1589–1603
Traw MB, Kniskern JM, Bergelson J (2007) SAR increases fitness of *Arabidopsis thaliana* in the presence of natural bacterial populations. Evolution 61(10):2444–2449
Trivedi P, Pandey A, Sa T (2007) Chromate reducing and plant growth promoting activities of psychotropic *Rhodococcus erythropolis* MtCC 7905. J Basic Microbiol 47:513–517
van der Geize R, Dijkhuizen L (2004) Harnessing the catabolic diversity of rhodococci for environmental and biotechnological applications. Curr Opin Microbiol 7:255–261
Vereecke D (2018) Comment on "Evolutionary transitions between beneficial and phytopathogenic *Rhodococcus* challenge disease management". eLife 7:e35238
Vereecke D, Cornelis K, Temmerman W, Jaziri M, Van Montagu M, Holsters M, Goethals K (2002) Chromosomal locus that affects the pathogenicity of *Rhodococcus fascians*. J Bacteriol 184:1112–1120

von Bargen K, Haas A (2009) Molecular and infection biology of the horse pathogen *Rhodococcus equi*. FEMS Microbiol Rev 33:870–891

Weinthal D, Barash I, Panijel M, Valinsky L, Gaba V, Manulis-Sasson S (2007) Distribution and replication of the pathogenicity plasmid pPATH in diverse populations of the gall-forming bacterium *Pantoea agglomerans*. Appl Environ Microbiol 73:7552–7561

Weyens N, Beckers B, Schellingen K, Ceulemans R, Croes S, Janssen J, Haenen S, Witters N, Vangronsveld J (2013) Plant-associated bacteria and their role in the success or failure of metal phytoextraction projects: first observations of a field-related experiment. Microbiol Biotechnol 6 (3):288–299

Zhang Y, Bignell DR, Zuo R, Fan Q, Huguet-Tapia JC, Ding Y, Loria R (2016) Promiscuous pathogenicity islands and phylogeny of pathogenic *Streptomyces* spp. Mol Plant Microbe Interact 29:640–650

Zhang Y, Loria R (2017) Emergence of novel pathogenic *Streptomyces* species by site-specific accretion and cis-mobilization of pathogenicity islands. Mol Plant Microbe Interact 30:72–82

Zhao GZ, Li J, Zhu WY, Tian SZ, Zhao LX, Yang LL, Xu LH, Li WJ (2012) *Rhodococcus artemisiae* sp. nov., an endophytic actinobacterium isolated from the pharmaceutical plant *Artemisia annua* L. Int J Syst Evol Microbiol 62:900–905

Index

H. M. Alvarez (ed.), *Biology of Rhodococcus*, Microbiology Monographs 16,
https://doi.org/10.1007/978-3-030-11461-9

W

X

The manufacturer's authorised representative in the EU is Springer Nature Customer Service Centre GmbH, Europaplatz 3, 69115 Heidelberg, Germany. If you have any concerns regarding our products, please contact ProductSafety@springernature.com

Printed and bound by CPI Group (UK) Ltd, Croydon, CR0 4YY
22/09/2026
02221050-0002

Trends in Logic

Volume 52

TRENDS IN LOGIC
Studia Logica Library

VOLUME 52

The book series Trends in Logic covers essentially the same areas as the journal Studia Logica, that is, contemporary formal logic and its applications and relations to other disciplines. The series aims at publishing monographs and thematically coherent volumes dealing with important developments in logic and presenting significant contributions to logical research.

Volumes of Trends in Logic may range from highly focused studies to presentations that make a subject accessible to a broader scientific community or offer new perspectives for research. The series is open to contributions devoted to topics ranging from algebraic logic, model theory, proof theory, philosophical logic, non-classical logic, and logic in computer science to mathematical linguistics and formal epistemology. This thematic spectrum is also reflected in the editorial board of Trends in Logic. Volumes may be devoted to specific logical systems, particular methods and techniques, fundamental concepts, challenging open problems, different approaches to logical consequence, combinations of logics, classes of algebras or other structures, or interconnections between various logic-related domains. Authors interested in proposing a completed book or a manuscript in progress or in conception can contact either christi.lue@springer.com or one of the Editors of the Series.

More information about this series at http://www.springer.com/series/6645

Adam Rieger · Gareth Young
Editors

Dialetheism and its Applications

Springer

Editors
Adam Rieger
University of Glasgow
Glasgow, UK

Gareth Young
University of Glasgow
Glasgow, UK

ISSN 1572-6126 ISSN 2212-7313 (electronic)
Trends in Logic
ISBN 978-3-030-30220-7 ISBN 978-3-030-30221-4 (eBook)
https://doi.org/10.1007/978-3-030-30221-4

This Springer imprint is published by the registered company Springer Nature Switzerland AG
The registered company address is: Gewerbestrasse 11, 6330 Cham, Switzerland

Introduction

Dialetheism is the view that there are true contradictions. For most of the history of Western philosophy at least, it has been regarded as absurd, though a small number of figures, such as Heraclitus and Hegel, have dissented. Logical developments in the twentieth century, traceable back at least to a 1954 Argentinian PhD thesis,[1] have made possible a more systematic investigation of the position, which has slowly built into a significant level of activity in recent decades. Undoubtedly, the person who deserves the largest share of the credit (or blame) for this is Graham Priest; starting with the 1979 paper 'The Logic of Paradox',[2] through the comprehensive *In Contradiction*[3] and numerous other publications, Priest has indefatigably nurtured the topic from the 'lunatic fringe' into what is now a flourishing research area. The current volume arose from a conference held at the University of Glasgow in December 2012 to celebrate 25 years since the publication of the first edition of *In Contradiction*. Some of the papers are versions or descendants of material presented there; others are invited contributions from scholars who were unable to attend the conference.

Whilst early work on dialetheism tended to focus on whether it is even coherent, and its application to the semantic paradoxes, increasingly dialetheism is finding applications in wider areas of philosophy. The potential here, indeed, seems very broad: in a typical philosophical dispute, we may have an apparently good argument for P and another one for not-P. The usual approach is to try to locate where one argument is at fault, perhaps at the cost of giving up something highly plausible, but dialetheism holds out the prospect of maintaining that both the arguments are not just apparently, but actually, sound. Of course, it is not as easy as this. There is a cost to a dialetheist solution as well, and every case must be judged on its merits. By no means every author in this collection is a supporter of dialetheism;

[1]Asenjo, F.G. (1954). La idea de un calculo de antinomias. Seminario Mathemático, Universidad de La Plata.

[2]Journal of Philosophical Logic, 8(1), 219–241.

[3]Dordrecht: Martinus Nijhoff, 1987. 2nd expanded edition, Oxford: Oxford University Press, 2006.

but they have in common that they think it is worth considering and engaging with; where it is rejected, the rejection must be argued for.

The first two papers in the volume are concerned with dialetheism's most familiar application so far, in solving logical paradoxes. Curry paradoxes—such as 'If this sentence is true then the moon is made of green cheese'—are awkward for the dialetheist. The classic set-theoretic and semantic paradoxes arise from apparently sound reasoning; dialetheism has the attractive feature of endorsing this reasoning, whilst avoiding trivialism by adopting a *paraconsistent* logic, which abandons *ex contradictione quodlibet*, a principle the intuitive validity of which is, in any case, debatable. In contrast, the Curry paradox trivialises *directly*, without passing through a contradiction (we can clearly replace the consequent of a Curry conditional by any sentence we choose). The paradox relies only on simple and apparently compelling principles of reasoning involving conditionals, and the dialetheist has to pay the price of giving at least one of these up.

In their paper 'Liars with Curry: dialetheism and the prospects of a uniform solution', Ben Burgis and Otávio Bueno consider in particular the implications of Curry paradoxes for another claimed advantage of dialetheism, its ability (unlike conventional solutions of the paradoxes) to obey the 'Principle of Uniform Solution'—the idea that paradoxes of similar structure should be given similar solutions. Burgis and Bueno, however, argue that the Curry paradox casts severe doubt on whether dialetheism really has this claimed advantage; whichever of a number of options the dialetheist chooses, the cost of preserving the principle will be prohibitively high.

A further advantage that has been claimed on behalf of a dialetheic solution to the paradoxes is immunity to so-called *revenge* problems. These concern a common phenomenon, whereby an attempt to classify a liar sentence into some semantic category merely results in a renewed paradox involving that category. In the simplest case, attempting to classify 'This sentence is false' as, say, *defective*, where defective sentences are neither true nor false, runs into problems with 'This sentence is either false or defective'. Some vital semantic notion must remain inexpressible in the object language or inconsistency results.

By facing contradiction head-on, dialetheism has been claimed to avoid revenge problems and allow for genuinely semantically closed languages. But Gareth Young, in his paper 'A revenge problem for dialetheism', argues that this claim is ungrounded. Taking the dialetheism of Priest as the most promising revenge-free solution, Young argues that dialetheists of this sort must accept as contradictory, not only truth, but validity: in other words, some inference principles are both valid and invalid. He argues that, depending on the details of the theory, all, or almost all, inference principles can be shown to be dialetheically invalid. This, he claims, makes the notion of invalidity inexpressible for the dialetheist and deprives them of the ability to express crucial semantic claims about their theory. Hence, they too fall foul of a revenge problem.

The two following papers discuss the prospects for applying dialetheism in new areas, somewhat distant from its traditional hunting ground. In his paper 'It was so revolting I couldn't take my eyes off it', Graham Priest introduces what he argues is a new form of dialetheia. Before doing so, he briefly surveys the main areas in which dialetheias have previously been thought to occur. Because the view is so closely associated with the paradoxes, it is interesting that Priest writes that he has 'never thought of the paradoxes as the most transparent example'. Rather, he thinks cases of conflicting laws provide this, say when one law stipulates of a person, Emily, that she is permitted to vote, and another law that she is forbidden.

Priest's new examples concern psychological dialetheias, in particular cases where one is simultaneously attracted and repelled by something. Priest argues that in such cases one is both attracted and not attracted; this, he claims, is not simply a matter of having contradictory attitudes (as in both wanting and wanting not to do something), but a genuine dialetheia. He defends this claim against various attempts to defuse the contradiction.

Sebastiano Moruzzi and Annalisa Coliva's paper 'Dialetheism in action: a new strategy for solving the equal validity paradox?' explores the prospects for giving a dialetheist account of cases of 'faultless disagreement', such as when one speaker asserts 'ginger is tasty' and another 'ginger is not tasty'. Standard examples such as this concern judgments of taste, though, depending on one's views, the phenomenon may occur more widely, for example, in aesthetics or ethics. The puzzle is that, whilst intuitively such cases seem possible (or so many have thought), it is hard to analyse precisely what is going on in them whilst simultaneously vindicating both that they are cases of genuine *disagreement* and that the parties are *faultless*.

After exploring various alternative solutions to their version of the paradox, Moruzzi and Coliva investigate the possibility of a dialetheic solution, in which the law of non-contradiction is abandoned. The dialetheist, they argue, has some difficulty in explaining why these cases involve genuine disagreement, since in a dialetheic framework the two parties could, it seems, each admits the truth of their opponent's statement as well as their own. After considering dialetheic responses to this problem, they investigate the relationship between the dialetheic solution and trivialism (the view that every sentence is both true and false). This is a threat in this context, since, in contrast to the usual semantic and set-theoretic paradoxes, the dialetheias may be universal, at least in those domains where faultless disagreement can arise.

The following three papers all explore, in different ways, how dialetheism can be helpful in reasoning about impossibilities, even if one is not convinced that these occur in the actual world. Ira Kiourti's paper 'An excess of dialetheias: in defence of genuine impossible worlds' examines the question of whether impossible worlds can be reconciled with a Lewis-style modal realism. There are some attractions to admitting impossible worlds, for example, to give plausible semantics to counterfactuals with impossible antecedents (these all come out trivially true on a standard Lewis–Stalnaker theory). But Lewis himself thought impossible worlds should be rejected. Kiourti proposes two ways of resisting Lewis's conclusion. The more moderate is to reject Lewis's principle governing negation, that for A to be true at world w is simply for A to fail to be true at w. But this principle, Kiourti argues, begs the question against impossible worlds and can be rejected whilst retaining classical logic. Alternatively, and more radically, one can retain Lewis's semantics for negation but abandon classical logic and embrace dialetheias. Kiourti defends this approach against the charge that it trivialises, by making everything true at every world.

Lewis's scepticism about reasoning with impossibility is also a starting point for Frederick Kroon's paper, 'Game of Truth: truth, fictionalism, and semantic paradox'. Suppose the moral that one draws from the semantic paradoxes is that our naive concept of truth is inconsistent, as has been a popular (though by no means universal) reaction. One option is to accept the inconsistencies and become a dialetheist. Another is to replace the naive concept by a new one, salvaging as much as possible from the naive conception whilst avoiding inconsistency. That might seem to exhaust the possible responses, but Kroon's paper is an exploration of a third option: fictionalism about truth. The paradoxes show that there is really no property of truth; nevertheless, argues Kroon, it is extremely useful to pretend that there is one. Just as on Field's view of mathematics, or Joyce's view of morality, there are no such things as numbers or the property of goodness, but it is a convenient practice to behave as though there are, Kroon thinks the existence of truth is a useful fiction. Because of the paradoxes, the fiction, in this case, will be inconsistent—like Priest's story of Sylvan's box which is both empty and not-empty.[4] The logic of the fiction will have to be paraconsistent, to avoid triviality. But the contradictions will not, as Kroon puts it, 'leach out into the class of ordinary statements about reality'.

Chris Mortensen's short paper, 'Dialetheism and impossible figures', has some ideas in common with Kroon's. The moral we should draw from impossible figures such as the Penrose triangle, he suggests, is that we can imagine contradictory situations and reason about them; such reasoning is non-trivial, and so requires paraconsistent logic. (Mortensen also mentions Priest's story of Sylvan's box in this context, as achieving a similar effect by literary, rather than visual, means.) Since the objects depicted need not exist, the argument yields only weak rather than strong paraconsistency—that is, it does not establish that there are dialetheias in the world.

The final two papers are more technical in nature. Amongst solutions to the liar paradox which preserve the naive T-rules in full but change the logic in order to avoid catastrophe, a natural desideratum is *classical recapture*: the principle that discourse not involving the problematic semantic notions such as truth should continue to be governed by classical logic. Andrew Tedder and Stewart Shapiro's paper 'Making truth safe for intuitionists' considers the prospects of this kind of recapture if we start with *intuitionist* logic. The usual paradoxical reasoning is intuitionistically valid, and intuitionists accept *ex contradictione quodlibet*, so the paradoxes are as potentially troubling for them as they are for a classicist. Tedder and Shapiro consider the prospects for 'intuitionist recapture' for a variety of logics, some with a dialetheic interpretation and some not. As well as demonstrating some

[4]Priest, G. (1997). Sylvan's Box: A Short Story and Ten Morals. Notre Dame Journal of Formal Logic, 38(4), 573–582.

results, they sketch out a terrain which promises to be fertile ground for future work.

Inconsistent mathematics looks likely to be a growth area in the coming years. A natural idea is to preserve naive set theory, preventing triviality by adopting a paraconsistent logic. But which logic? It must be weak enough to avoid paradox (including Curry), but strong enough to be usable in proving mathematical theorems. Guillermo Badia and Zach Weber's paper 'A substructural logic for inconsistent mathematics' presents a logic designed for this purpose. The various desiderata prove impossible to satisfy with a standard set of connectives; their logic, therefore, has two conditionals and two negations. As an illustration of what the logic can do, they argue that it justifies induction in naive set theory. As with the Shapiro and Tedder paper, and indeed the collection as a whole, one is left impressed not only with what has already been achieved, but how much interesting work remains to be done.

Adam Rieger
Gareth Young

Contents

Contributors

Guillermo Badia School of Historical and Philosophical Inquiry, University of Queensland, Brisbane, Australia

Otávio Bueno Department of Philosophy, University of Miami, Coral Gables, FL, USA

Ben Burgis Perimeter College, Georgia State University, Alpharetta, GA, USA

Annalisa Coliva University of California, Irvine, CA, USA

Ira Georgia Kiourti Independent researcher, London, UK

Frederick Kroon Department of Philosophy, University of Auckland, Auckland, New Zealand

Chris Mortensen University of Adelaide, Adelaide, Australia

Sebastiano Moruzzi Department of Philosophy and Communication Studies, University of Bologna, Bologna, Italy

Graham Priest Departments of Philosophy, CUNY Graduate Center and the University of Melbourne, Melbourne, Australia

Stewart Shapiro Ohio State University, Columbus, OH, USA

Andrew Tedder Institute of Computer Science, The Czech Academy of Sciences, Prague, Czech Republic

Zach Weber Department of Philosophy, University of Otago, Dunedin, New Zealand

Gareth Young University of Glasgow, Glasgow, Scotland, UK

Liars with Curry: Dialetheism and the Prospects for a Uniform Solution

Ben Burgis and Otávio Bueno

Abstract Some dialetheists have claimed that one of the central advantages of their approach to the Paradoxes of Self-Reference is that they are able to offer a unified solution to structurally similar paradoxes that arise in the semantic and set-theoretic realms (Priest in Mind 103:25–34, [12] and Beyond the limits of thought. Clarendon Press, Oxford, [16]). They argue that since the structures of all of these paradoxes conform with the Inclosure Schema (IS), the Principle of Uniform Solution (PUS) dictates that we should solve them all the same way. But the dialetheist's approach to PUS collapses when it comes to the Curry Paradox, to which any solution based on dialetheism seems inapplicable. We show that a particular version of a 'paracomplete' theory (inspired by Field in Saving truth from paradox. Oxford University Press, Oxford, [8]) has available to it a way of avoiding these problems that the dialetheist cannot mirror without losing *Modus Ponens*. JC Beall has suggested a way of minimizing this loss, but Beall's strategy runs up against a further difficulty about the PUS. This one involves the Irrationalist's Paradox. We conclude with a dilemma: The dialetheist can either reject the principles that are used to accommodate ordinary reasoning when *Modus Ponens* or Disjunctive Syllogism fail, and thus live without the ability to mimic these inferences, or they can sacrifice the Principle of Uniform Solution by solving an extremely Liar-like semantic paradox in a way that has nothing to do with their solution to the Liar and Curry's Paradox. In either case, the prospects for a plausible and truly uniform dialetheic solution to the paradoxes of self-reference are grim.

Keywords Dialetheism · Liar paradox · Curry's paradox · Graham Priest · Hartry Field · JC Beall

B. Burgis (✉)
Perimeter College, Georgia State University, Alpharetta, GA 30022, USA
e-mail: benburgis@gmail.com

O. Bueno
Department of Philosophy, University of Miami, Coral Gables, FL 33124, USA
e-mail: otaviobueno@mac.com

A. Rieger and G. Young (eds.), *Dialetheism and its Applications*, Trends in Logic 52,
https://doi.org/10.1007/978-3-030-30221-4_1

1 The Principle of Uniform Solution

Graham Priest argues for a "dialetheic" solution to several related paradoxes, in each case affirming that the resulting contradiction is actually true and using paraconsistent logic, in which some contradictions can be true without *everything* being true, to limit the damage. Thus, 'Liar sentences' like (1) below are both true and untrue.

(1) This sentence is untrue.

So is every sentence in the infinite series of sentences in Yablo's Paradox, each of which says that all the ones that come after it in the series are untrue. When it comes to Russell's Paradox, the set of all sets that are not members of themselves both is and is not a member of itself.

Priest has claimed in several papers that an advantage of his dialetheist approach over more standard views is that he can offer *unified* solutions to all of these paradoxes:

> The Principle of Uniform Solution (PUS) counsels: Same kind of paradox, same kind of solution. Since standard solutions to the set-theoretic paradoxes do nothing to address the semantic paradoxes, and vice versa, this shows that each is inadequate [15, p. 123].

One might wonder about which paradoxes are to be counted as being the "same kind" as one another for the purposes of the PUS, but Priest has an answer for this as well. He takes the family of paradoxes, semantic and set-theoretic—sometimes referred to together as "the paradoxes of self-reference"—to all be of the "same kind" because they all satisfy the Inclosure Schema (IS). The technical details of the IS need not concern us here, except to note that it involves three elements, Existence, Closure and Transcendence, and that when the dialetheist affirms, for example, that there is a set of all sets that are not members of themselves, that it is a member of itself, and that it is not a member of itself, they are affirming all three.[1]

In *Beyond the Limits of Thought*, Priest insists on this point:

> All the traditional paradoxes of self-reference are inclosure contradictions. That is, the structure described in the Inclosure Schema explains all of these contradictions. Hence, by the PUS, one should be satisfied with nothing less than a unified solution to the family. Does any extant solution provide this? [16, p. 167].

According to Priest, the answer is a loud and clear *No*—with the exception of dialetheism:

> The only satisfactory uniform approach to all these paradoxes is the dialetheic one, which takes the paradoxical contradictions to be exactly what they appear to be. The limits of thought which are the inclosures are truly contradictory objects [16, p. 169].

[1] See Priest [16, Chaps. 9–11], for details. The crucial point brought out by Priest's set-theoretic model of this structure is that in all Inclosure paradoxes, Closure and Transcendence are a 'package deal'. Objects transcend the relevant categories *because* they belong to them, and they belong to them *because* they transcend them.

To be fair, though, a ‘paracomplete’ theorist, who solves the semantic paradoxes not by accepting certain contradictions ($\alpha \wedge \neg\alpha$) but by rejecting certain instances of the Law of the Excluded Middle (such that for some sentences α, the paracompletist rejects α *and* rejects $\neg\alpha$) would be well-placed to be able to offer a solution that Priest could accept as being as uniform as his. (Thus, by his lights, one would have to use some other criteria to decide between the two solutions.) Just as paracompletist rejects the claim that Liar sentences are true *and* rejects the claim that they are untrue (but they accept that sentences that say of themselves that they are untrue exist, that such sentences are meaningful, that they have the truth conditions they seem to, and so on), they could accept the existence of the Russell Set, but reject both the claim that it *is* a member of itself *and* the claim that it is *not* a member of itself.

As it happens, the primary advocate of the paracomplete approach, Hartry Field, is committed to mathematical orthodoxy on set-theoretic matters. In the most rhetorically amusing (if not, perhaps, the most philosophically compelling) passage of his book *Saving Truth from Paradox*, Field dismisses out of hand Priest’s willingness to accept logically revisionary conclusions on such well-established mathematical subjects:

> I think this conclusion that there are dialetheias within arithmetic defies belief, and I don’t think insisting on it will bring many converts to dialetheism. (As John Lennon put it: “If you go carrying pictures of Chairman Mao, you ain’t going to make it with anyone anyhow”.) [8, p. 377].

A bolder sort of paracomplete theorist, though—one less moved by the lesson of Lennon’s lyric—could ape all of Priest’s moves and arrive at an exactly equally uniform solution to the various paradoxes. Where Priest accepts Existence, Closure and Transcendence for each of the paradoxes—semantic and set-theoretic—our hypothetical bold paracompletist could accept Existence in every case and, in every case, deny both Closure and Transcendence.[2]

By contrast, a more orthodox sort of paradox-solver, still wedded to classical logic, who (a) handled Russell’s Paradox by rejecting naïve set theory in favor of the view that the only sets that exist are those found in the cumulative hierarchy of orthodox contemporary set theory, and (b) solved the Liar Paradox by arguing that there are exceptions to the T-Schema, would be (by making move (a)) evading the contradiction about the Russell Set’s self-membership status by denying Existence, and (by making move (b)) evading the contradiction about the Liar’s truth status by accepting Existence but denying both Closure and Transcendence. This, Priest claims, is unacceptably disunified.

---

[2]Oddly, in Priest’s comparison of “Field’s approach” and dialetheism, he uses the PUS to critique Field’s views on set theory without acknowledging that Field’s views on the semantic paradoxes could easily be extended into the set-theoretic realm [17, pp. 48–51].

2 Begging the Question? (I): Dialetheism and Classical Logic

At this point, there might be room for an (apparently quite strong) objection to Priest's reasoning. One could happily grant that the PUS is a *prima facie* norm of good reasoning, but still insist that contradiction-avoidance is a far more basic and important norm. Thus, given the choice between a unified-but-inconsistent solution and a consistent-but-disunified one, the latter might seem vastly rationally preferable.

Priest has a standard response to this sort of objection. In his book *In Contradiction*, where he lays out the most exhaustive version of his case for a dialetheic solution to the paradoxes, he frequently says things like this:

> The inconsistency of our linguistic principles is the very thesis that I am affirming. Hence [...] consistency cannot be invoked as a regulative principle without begging the question against me [11, p. 5].
>
> Any appeal to this principle [that from a contradiction, anything follows] or to classical logic in general, can only, therefore, beg the question [11, p. 6].
>
> [...] what is at issue here is the very correctness of classical logic. Hence the issue cannot ([even] partially) be settled in this way without thoroughly begging the question [11, p. 160].

He claims that "the semantic [and set-theoretic] paradoxes are *bona fide* sound arguments" for the claim that the truth is inconsistent [11, p. 10]. Thus, to elevate above all else the assumption that the truth *must* be consistent, as we are doing when we say that if the best consistent solution is disunified, it is still preferable to a unified but inconsistent solution, is simply to assume the very thing that's at issue in the debate. We should keep this principle in mind—*question-begging should be avoided no matter how unorthodox or obviously false the position being argued against may seem*—since we will see in due course that Priest himself rather flagrantly violates it.

3 Liars and Curry Sentences

A more accommodating critic of dialetheism, one who was willing to accept the principle just mentioned, might still argue that the orthodox solutions to the semantic and set-theoretic paradoxes are unified after all. On the same level of abstraction that we can say that Priest's solutions to the various paradoxes all 'affirm all elements of the Schema', we can say of the orthodox solutions that they are unified in the sense that they all 'circumvent the Schema'. (Indeed, this is precisely what Nicholas Smith argues; see [21].)

Priest will have none of this. If two paradoxes are obviously instances of the same phenomenon, they should be treated alike.

> The Liar Paradox comes in many varieties. In some of them, self-reference is achieved by means of a demonstrative ("this sentence", etc.); in others, it is attained by a description ("the first sentence on such-and-such a page", etc.); in still others, it is achieved by some sort of diagonal argument employing Gödel numbers as names. At some level of abstraction, these paradoxes are, presumably, of different kinds. The first depends on context-dependent features of utterance; the second presupposes some account of the way that descriptions work. The third employs numbers. Suppose that I propose three corresponding kinds of solution. The first kind of paradox is solved by some theory of speech acts; the second is solved by some theory of the semantics of description; the third is solved by an appeal to mathematical nominalism. Even if all the solutions were to work for their own kind, it appears to me undeniable that, collectively, there is something wrong with them. [...] The solutions may avoid the paradoxes, but they do so by attacking peripherals, not essentials [15, p. 124].

We share Priest's intuition here. When deciding how to group paradoxes into types for the purposes of applying the PUS, we can't stray too far from the obvious facts of the case. The Liar Paradox is a single phenomenon. We would, however, take this one step further. What about Curry's Paradox, which revolves around sentences like (2)?

(2) If this sentence is true, *everything* is true.

By a familiar series of steps, we can plug (2) into Tarski's Biconditional Truth Schema ($Tr\langle\alpha\rangle \leftrightarrow \alpha$), and, after applying a few otherwise unobjectionable logical rules—notably Contraction, which allows us to obtain $\alpha \rightarrow \beta$ from $\alpha \rightarrow (\alpha \rightarrow \beta)$—derive from the mere statement of (2)'s truth conditions the conclusion that *everything is true*...for example, it is true that 9/11 was an inside job, true that Obama was born in Kenya, true that the earth is flat, true that the moon is made of green cheese, *and*, as Slavoj Zizek might put it, *so on and so on and so on*.

A crucial difference between Curry and the Liar is that classifying sentences like (2) as dialetheias would be a disaster, even given paraconsistency. If (2) is both true and false, it is true, and the moon is made of green cheese. As such, it looks like the way that Priest and other dialetheists solve the Liar Paradox *can't* be applied to Curry. How big a problem is this for Priest's claim that dialetheists can offer a uniform solution to the paradoxes of self-reference?

The suggestion that the standard semantic and set-theoretic paradoxes are 'of the same kind' is interesting and somewhat plausible. That Curry and the Liar Paradox are of a kind, however, is painfully obvious in just the way that it is intuitively undeniable that Liar-type sentences that refer to themselves by means of demonstratives and descriptions are of a kind. (1) and (2) are both paradoxical sentences whose paradoxicalness relies on self-referential truth talk—they each assert something about their own truth value. Once fed into the Biconditional Truth Schema, both sentences—given the rules of classical logic—generate triviality, leading to the conclusion that *absolutely everything* is true. One does so by means of the inference rules of classical conditionals, while the other does so by means of the classical principle *ex contradiction quodlibet*—from a contradiction, anything follows—but to hang much on this distinction is to attack peripherals, not essentials.

Strangely, though, this is precisely what Priest does. He makes the conditionals in his favored logic "Curry-proof" by rejecting Contraction, so that (according to him) $\alpha \rightarrow (\alpha \rightarrow \beta)$ can be true without $\alpha \rightarrow \beta$ being true.

Of course, Priest rejects various other rules of classical logic as well. For example, for obvious reasons, he rejects *ex contradiction quodlibet.* (As he points out, since not everything is true, every true contradiction would constitute a counter-example to the rule [11, pp. 5–6]!) For similarly straightforward reasons, it looks like any dialetheist would have to reject *reductio ad absurdum.*[3] (After all, it would be very odd if, for example, a dialetheist solution to Russell's Paradox had to be combined with denying the existence of the set of all sets that aren't members of themselves!) In these cases, though, he makes a good case for the claim that our ordinary assumption that such rules are universally truth-preserving is inseparable from our assumption that contradictions are never true.

The problem is that Contraction does *not* in any way rely on the latter assumption. The mere existence of true contradictions would not provide us with counter-examples to Contraction. Thus, Priest's solutions to the Liar and Curry's Paradox are flagrantly disunified.

4 Curry and Inclosure (I): Priest's First Attempt

Does the structure of Curry fall under the IS? In the past, Priest has moved very delicately on that front.

> [...] whether or not Curry paradoxes fit the Schema depends on how the conditional involved is interpreted. If it is interpreted as a material conditional, Curry paradoxes fit the Schema. If not, they do not. Such paradoxes are, therefore, of a different kind ([14, p. 387]; see also [16, pp. 168–169]).

This is fairly weak sauce. After all, the sort of classical logician we considered above—the one who evaded the set-theoretic paradoxes in the mathematically orthodox way, and the semantic paradoxes by postulating exceptions to the Biconditional Truth Schema—could tell a precisely similar story. *Are semantic paradoxes like the Liar Inclosure Schema paradoxes? Well, if their truth conditions are given by the Biconditional Truth Schema, they are, and if not, not. Thus, they're of a different kind than the set-theoretic paradoxes, which are definitely IS paradoxes. We can solve all real IS paradoxes in a completely unified way!*

Indeed, even the hypothetical-paradox solver Priest considers above, the one who solves different versions of the Liar differently depending on the means of securing

[3]Alan Musgrave reports being shocked that Priest told him that *reductio* is "not always" valid. "Logicians", Musgrave complains, "used to think an argument form was *valid*, full-stop, or *invalid*, full-stop, not valid here and invalid there" [9, p. 338]. Given the assumption that there are true contradictions, though, it's not obvious why it should be so absurd to say that argument forms that preserve truth in contexts where no contradictions are true should fail to do so in inconsistent domains.

self-reference, could offer up a precisely similar explanation for how they solved all real Inclosure Schema paradoxes in a uniform manner. *If such-and-such theory of speech acts were right, the version of the Liar that started with the words "this sentence" would be an Inclosure Schema paradox. If not, not. If a certain picture of how descriptions secured reference were right, the version of the Liar that secured self-reference with a definite description would be a Schema paradox. If not, not...*

This method of distancing Curry from the IS makes the IS analytically worthless as a tool for distinguishing uniform from non-uniform solutions. Any paradox-solver who wants to (a) solve the Liar Paradox by regarding Liar sentences as both true and false and rejecting *ex contradiction quodlibet*, (b) solve Curry's Paradox by regarding Curry sentences as simply false and rejecting Contraction, (c) honor the PUS, and (d) use the IS to decide what counts as the "same kind of paradox" must do better.

5 Curry and Inclosure (II): Beall Versus WRPHC

More recently, Priest and a phalanx of co-authors—Zach Weber, David Ripley, Dominic Hyde, and Mark Colyvan—have offered what looks like just such a superior alternate strategy for making sense of (a)–(d). They grant that Curry *does* fall under Inclosure, *if* "Curry's paradox" means the paradox about sentences with consequents like the consequent of (2) above—that *everything* is true. (Conventionally, this consequent is denoted by the triviality symbol $\perp$.) *Everything* includes (2)'s negation. As Beall [3] argues, this gives us both Closure and Trascendence in pretty much the same way we get those elements in the formal structure of the Liar Paradox. Weber, Ripley, Priest, Hyde, and Colyvan—henceforth, following Beall's usage in the exchange, we'll call them WRPHC—don't dispute Beall's technical point.

> So Curry's paradox fits the schema? No; this is not Curry's paradox! The problem posed by Curry-style reasoning is that it allows us to establish an *arbitrary* sentence, not just $\perp$. If A is any sentence and s is the sentence $Ts \rightarrow A$, the Curry reasoning allows us to establish A. This is problematic whether A is true or false [23, p. 823].

To see their point, let's consider a few other Curry sentences:

(3) If this sentence is true, the moon is made of green cheese.
(4) If this sentence is true, Hillary Clinton is the 45th President of the United States.
(5) If this sentence is true, Donald Trump is the 45th President of the United States.
(6) If this sentence is true, Canberra is the capital of Australia.

Assuming the validity of Contraction, the T-Schema truth conditions of (3) entail that the moon is made of green cheese. The truth conditions of (4) entail that Hillary Clinton is the President of the United States. Since astronomers tell us that the moon is not made of green cheese, and anyone who reads a newspaper knows the American Presidential election went the other way, it's obvious why these are paradoxical results... But what about (5) and (6)? "Donald Trump is the 45th President of the United States" and "Canberra is the capital of Australia" aren't paradoxical results.

They aren't even false. Nevertheless, there does seem to be something paradoxical about the idea that they can be derived from a mere statement of the *truth conditions* of (5) and (6). As WRPHC put it, focusing on (6), there's "something crazy about thinking we can establish that Canberra is the capital of Australia in *this way* [...]" [23, p. 823, our emphasis].

Beall's formal analysis applies to (3) and (4). Since Hillary Clinton is not the President and the moon is not made of green cheese, in each case we can use Contraposition and *Modus Ponens* to derive the negation of the antecedent and hence the negation of the whole conditional. But since the consequents of (5) and (6) are *true*, there's no route through Contraposition to fitting these sentences into the formal structure of the IS.

> Beall is attempting to appeal to the principle of uniform solution: same kind of paradox, same kind of solution [16, Sects 11.5, 17.6]. What is at issue, then, is the question of whether the Curry paradox is an inclosure paradox. All the instances of the Curry paradox—whatever the A is—are clearly of the same kind. And it is not an inclosure paradox, since instances of it do not fit the schema. By the uniform solution principle, the Curry paradoxes all require the same treatment, but not (necessarily) the same as that of the Liar [...] [23, p. 823].

Beall's reply to WRPHC on this point takes the form of a dilemma. What, he asks, is the IS supposed to apply to—entire sentential schemes, like ($s = (Ts \rightarrow A)$), or individual instances of that scheme, like (2)–(4)?

Here, we [...] branch into two cases.

> Super-paradoxical case: a sentential scheme is an inclosure paradox iff *all* instances are (inclosure) paradoxical.
>
> Sub-paradoxical case: a sentential scheme is an inclosure paradox iff *some* instances are (inclosure paradoxical) [4, p. 843].

Consider (7), below, which is recognizably a version of the Liar Paradox—in fact, it's just the sort of sentence that might be brought up as a revenge Liar to frustrate certain non-dialetheist solutions to the paradox:

(7) Either this sentence is false or Hillary Clinton is the 45th President of the United States.

Given that the second disjunct is contingently false, (7) is true iff the first disjunct (and hence the whole sentence) is false. In other words, it is true iff it is false like any other Liar sentence, and hence it fits the IS. So far, so good. Now consider (8):

(8) Either this sentence is false or Donald Trump is the 45th President of the United States.

Given the contingent truth of the second disjunct, (8) can't be made true by being false. Hence, the IS doesn't apply. But (7) and (8) are both instances of the sentential scheme ($s = (Fs \vee A)$)! Thus, Beall concludes, if a sentential scheme must be super-paradoxical to be a schema paradox, even the Liar Paradox is not a schema paradox. If, on the other hand, being sub-paradoxical is enough, Curry is a schema paradox too. Either both paradoxes or neither of them fit IS.

6 Priest on Individuating Arguments

Priest has responded (this time without the collaboration of R, P, H, and C).

> [Beall] asks whether a paradox is an individual sentence (such as a sentence like $T(C) \Rightarrow \bot$) or a sentential schema (such as the schematic $T(C) \Rightarrow A$). The answer is: neither. It is an argument. It would be wrong to identify a paradox with the sentence at the end of the argument: it would be silly, for example, to take Zeno's Arrow Paradox to be that the arrow does not move—or even that it does and does not move. What is important about the paradox is the argument for the conclusion. [20, pp. 115–116]

Whether Priest has succeeded in escaping Beall's dilemma now depends on the sticky question of how to individuate arguments. Priest is surely right that multiple arguments can have the same conclusion. The Teleological Argument and the Ontological Argument, for example, share a conclusion, but they're clearly distinct arguments.

One could demand that two argument-instances share not only a conclusion but *all* of their premises and *all* the same inferential steps in order to count as two instances of the same argument, but (as Priest quite correctly recognizes) this would be too demanding.

> There are *some* details of a proof that are "parametric". They can be filled in in a number of different ways. And how one does so is irrelevant to what is going on. Thus, in the case of the first proof of Gödel's theorem, the precise details of the axiomatization of PA, or of how the Gödel coding is done, are irrelevant to the proof. [20, p. 116]

So far, so good. The question is now *which* details are parametric. He gives two examples—one about argument *forms* and one about arguments themselves.

> John says 'The Bible says that one should take an eye for an eye and a tooth for a tooth, so one should take an eye for an eye and a tooth for a tooth'. Mary says 'The Bible says that you should turn the other cheek, so you should turn the other cheek'. They are both using the same form of argument: 'The Bible says that A; ergo *A*'. The *A* here is a parameter. [20, pp. 116–117]
>
> Consider an argument for some conclusion which uses classical *reductio*, and so is intuitionistically invalid. Consider, also, an intuitionistically valid argument for the same conclusion. Most would agree that these are different arguments, since they use different form of inference. [20, pp. 121–122]

Priest is clearly right that there's a quite strong sense of 'form' in which the two Bible-based arguments have the same form ('Appeal to the Bible'). It's far less clear that he's to say any two argument-instances that use different forms in the sense at issue in the second passage would be 'different arguments'—or that he's right that 'most' would share his intuition. Take 01 and 02:

(01) Assume for the sake of argument that the most perfect being we can imagine does not exist. We can then imagine a more perfect being—one just like the most perfect imaginable being except that it has the additional perfection of existing. But this is a contradiction—we have now, by hypothesis, imagined a being more perfect than the most perfect being we can imagine. Therefore, such a being can't fail to exist. There is a God.

(02) God is by definition the most perfect being we can imagine, and existence is contained within the idea of perfection. If such a being exists as an object of thought, it exists in reality. It clearly *does* exist as an object of thought. As such, it exists in reality.

Are (01) and (02) two arguments or two merely parametrically distinct instances of one argument—the Ontological Argument? Well, (01) relies on *reductio* and Double Negation and is intuitionistically invalid, while (02) relies on *Modus Ponens* and is intuitionistically unobjectionable. Nevertheless, we suspect that most would regard these as two merely parametrically distinct instances of the Ontological Argument, and that most would have a point.

Whatever one makes of this issue, there's a more basic problem here for Priest. His two examples point in opposite directions. If all that matters is that two arguments have the same *form* at the exact level of precision at which the two Appeals to the Bible have the same form, then all argument-instances in which a conclusion A from the T-Schema truth conditions of $T(C) \Rightarrow A$ based on Contraction and *Modus Ponens are* instances of the same argument, regardless of whether the content of A has anything to do with C. On the other hand, if (as suggested by the second example) two argument-instances that get to the same conclusion from exactly the same premises count as instances of different arguments if they use different inferential steps, then not even all instances of the Liar Paradox are instances of the same argument! (There are at least half a dozen different ways of getting from the T-Schema truth conditions of a standard Liar sentence to the conclusion that the sentence in question is both true and untrue.) If Priest wants to individuate arguments in a way that makes the Liar Paradox one argument and *some* instances of Curry's Paradox an argument of the same type and others an argument of a different type, neither of his examples are going to provide any useful guidance about how the individuation should proceed.

Things only get more confusing when one considers a complication about the structure of Curry's Paradox (see [6]). So far we've been discussing the standard conditional-Curry, but there's also a version of Curry—the validity-Curry—which concerns the validity predicate rather than a conditional. The validity-Curry extends the conditional-Curry in such a way that the former is not automatically solved when the latter is. While one may resolve the conditional-Curry by invoking a non-classical—in particular, a contraction-free—logic, this may not be sufficient to solve the validity-Curry [6], but see [22]. Some theorists believe that the validity-Curry can only be solved by looking at structural rules like Cut and Structural Contraction. If so, so much the worst for the prospects of a uniform dialetheist solution to the Liar and to all forms of Curry. One way or the other, the crucial point remains: both versions of Curry's Paradox seem to be typical instances of Liar-like paradoxes of self-reference. The fact that Priest's classification does not take them to be so seriously questions the adequacy of the proposed classification in the first place.

7 Curry and Inclosure (III): A More Basic Dilemma

So, who's right about all of this—Beall or Priest? The answer is that it doesn't matter. If Beall is right, so much the worse for Priest's claim to be offering a unified solution to all the IS paradoxes. If Priest is right and Curry is *not* an IS paradox, so much the worse for the Inclosure Schema as a plausible dividing line between types of paradoxes. The various types of sentences that make paradoxical claims about their own truth-values, such that the instances of the Biconditional Truth Schema for these claims generate triviality (given the rules of classical logic), aren't just different members (as the Liar and Russell's Paradox may be) of the same broad 'family' of paradoxes. They're more like Siamese twins. Given that Curry and the Liar share precisely the same paradoxical traits, if one of them falls under the IS and the other does not, this seriously undermines the plausibility of the IS as a delineator of which paradoxes fall into the same type and thus require the same solution.

It may be objected that Curry's Paradox can be formulated using sentential schemes different than the one considered by Beall and WRPHC above. Some of these alternate formulations don't involve self-reference, and thus look far less like typical versions of the Liar Paradox. While this is true, it is beside the point for two reasons. The first reason is that the existence of Curry sentences of the more familiar type considered above is enough to generate the dilemma sketched out in the previous paragraph. Interestingly enough, it is precisely these self-referential versions of Curry's Paradox that Priest himself considers "the strongest form of it [Curry's Paradox]" that he knows [18], p. 83. As we've established, for these self-referential versions, whether they fit the Inclosure Schema or not, problems emerge for the dialetheist. The second is that the existence of non-self-referential versions of Curry is not a disanalogy between Curry and the Liar. There are Liar-like sentences—most obviously the sentences in the Yablo series—that don't appear to be even indirectly self-referential.[4] In the language of Kripke's and similar solutions to the paradoxes, all such sentences are at least "ungrounded", even if they aren't technically self-referential. So are non-self-referential Curry-like sentences. In other words, when it comes to both Curry and the Liar, we have canonical versions of the paradoxes that revolve around self-referential sentences, and in both cases, versions can be generated that at least apparently avoid self-reference. When it comes to all versions of both paradoxes, the problem revolves around ungrounded truth talk that leads to triviality when it is fed into the Biconditional Truth Schema.

We can even generate sentences like (9) below that intuitively count as both Liars *and* Curry sentences!

(9) If this sentence is true, it is false.

Perhaps the dialetheist could simply grant that the Inclosure Schema does not provide an adequate classification of paradoxes, and just move on. But there is a significant cost associated with this move. As we saw above, a crucial argument in favor

[4]Whether Yablo's Paradox involves some element of disguised self-reference is controversial. See Priest [13] and Bueno and Colyvan [7].

of a dialetheic treatment of the paradoxes of self-reference is the resulting unified approach to them. Once Curry's Paradox and the Liar, despite their fine-grained similarity, are nevertheless treated as belonging to distinct categories, the dialetheist's defense of a uniform approach to self-reference paradoxes becomes untenable. A key argument in favor of dialetheism is then gone, together with the claim that "the only satisfactory uniform approach to all these [self-reference] paradoxes is the dialetheic one" [16, p. 169].

8 Begging the Question? (II): Dialetheism and Trivialism

If, as Priest thinks, the Liar and the set-theoretic paradoxes can be seen as *prima facie* sound arguments for dialetheism—such that the challenge for non-dialetheists is to find a mistaken premise or a mistake in the reasoning connecting those premises to the resulting contradiction—Curry's Paradox can be seen as a *prima facie* sound argument for *trivialism,* the claim that *absolutely everything* is true.

Seen in this light, Priest's approach to Curry—rejecting an otherwise plausible logical rule in order to avoid the paradoxical conclusion—seems question-begging in precisely the sense that objections to Priest's dialetheic solution based on pure contradiction-avoidance are (according to Priest) question-begging. What's the difference?

One could argue that the difference is just that it is impossible to beg a question against the trivialist, because the trivialist accepts everything, necessarily including the wrongness of trivialism. You can't illegitimately assume what's in dispute, because nothing is really 'disputed' by the trivialist.

This would, however, be entirely too hasty. The epistemic reasons that begging the question is barred from rational debate don't cease to apply in this case. Priest himself has some very sensible things to say about precisely this subject in *Doubt Truth to Be a Liar* [19, Chap. 3] and when looking at how he fares on these grounds, it will be sufficient to compare his actual performance with the standards he himself lays out in that discussion.

What he asks us to do in that book is to imagine, when arguing against the trivialist, that we are not trying to convince the trivialist (who, after all, is already convinced of everything!), but rather that we are trying to convince a neutral arbitrator who has not yet made up her mind about the issue of trivialism, and who will (using normal standards of evidence) accept or reject arguments brought to her by both sides. Convincing the trivialist of the wrongness of trivialism is a trivial matter. Convincing the imaginary arbitrator is not. It is the latter, harder standard to which we should set ourselves.

> Convincing the arbitrator is [...] not as easy as it might first appear. [...] For example, an obvious argument to use is that the trivialist's position is inconsistent, and so ought not to be believed. (Naturally, a dialetheist is not going to make this objection; but a classical logician might.) It is clear that this argument fails, though, since it begs the question. The trivialist

> affirms [...] all contradictions. Simply to claim that these cannot be true is to take for granted part of what is at issue [18, p. 66].

Things are not so hopeless as they may appear, however. In fact, there is excellent reason to be confident that a nontrivialist with normal beliefs about the world could make a lot of good, convincing anti-trivialist arguments of the kind that the hypothetical neutral arbitrator *would* allow.

> The trivialist believes many strange things. They believe, for example, that you are a scrambled egg. This is objectionable since there is just no evidence to justify it. The rational person should not believe anything for which there is no reason. The arbitrator must agree. And if the trivialist argues that there is a reason, namely that it follows from trivialism, then it is they who now beg the question. The trivialist will also claim that there is independent evidence. (They will claim everything.) But claiming does not of course make it so. The evidence needs to be produced for the arbitrator. [...] [18, p. 66].

This should be enough to get across Priest's general idea about how to argue against trivialism.[5] We agree that this is a reasonable standard for evaluating anti-trivialist arguments.

What will happen, though, when the trivialist presents the neutral arbitrator with Curry's Paradox, which seems to show that trivialism is true in just the way that (according to both Priest and the trivialist) the Liar Paradox shows that dialetheism is true? Oddly, Priest nowhere mentions Curry in his discussion of arguing with trivialists in *Doubt Truth to Be a Liar* [19, Chap. 3]. Elsewhere in the book [19, pp. 99–101], however, he mentions the paradox to justify the rejection of Contraction in his favored logic LP (Logic of Paradox). A footnote directs the reader to a longer discussion of Contraction in *In Contradiction* [18, Chap. 6]. There, he explicitly frames the entire discussion in terms of "Curry-proofing" his conditionals.

If this were Priest's defense against the trivialist's argument-from-Curry, the neutral arbitrator could hardly be impressed.

9 Keeping Score: Dialetheism, Paracompleteness, and Uniform Solutions

Of course, while Priest frames his discussion in *In Contradiction* and even in *Doubt Truth to Be a Liar* in terms that the arbitrator would have to reject as question-begging in the context being considered—"Curry-proofing" his conditionals—he could argue that there are independent reasons to prefer non-classical conditionals. Indeed, in some discussions of exactly which kinds of conditionals to substitute for

---

[5]Once this framework is established, Priest goes on to introduce various complexities to the scrambled egg argument, and to make other arguments against trivialism. Whether or not all of his arguments go through is beside the point for our current purposes. The important part of the discussion from *Doubt Truth to Be a Liar* for our purposes here is not Priest's anti-trivialist argumentation, but his conceptual framework for how anti-trivialist argumentation should proceed.

the regular classical kind, he has gestured at such considerations, such as familiar 'paradoxes of material implication'.

> Let A be 'you will not be harmed tomorrow'. This, let us pray, is true. Let B be 'you jump off the Empire State Building tomorrow'. Then from the principle in question [which is that A entails $(B \rightarrow A)$] 'if you jump off the top of the Empire State Building tomorrow you will not be harmed'—which, it would seem, is patently false [17, p. 49].

Charitably, let's assume that—although he never hints at such a move in his discussions of trivialism—Priest would, confronted with the hypothetical trivialist pushing an argument from Curry's Paradox and the hypothetical neutral arbitrator, make *this* argument against standard contracting conditionals, rather than (as he usually does) use the paradox itself as his primary reason for rejecting Contraction. What would the arbitrator say (assuming that she was perfectly rational)?

This is, of course, a hard question to answer, as the case is notoriously controversial. Some classical logicians think that these sorts of 'paradoxes' can be solved through disambiguation of different sorts of conditionals, or an analysis in terms of conversational implicatures. Other theorists strongly disagree, and think such moves fall short in various ways. Let's assume for the sake of argument that Priest really does have a good case here. Does it matter?

Recall the standard he laid down when criticizing the hypothetical paradox-solver who wanted to handle different versions of the Liar Paradox differently.

> Even if all the solutions were to work for their own kind, it appears to me undeniable that, collectively, there is something wrong with them [...] [15, p. 124].

In other words, even if the paradox-solver in question had a good case to make on speech acts, and a good case to make on definite descriptions, and so on, it wouldn't matter, because the PUS would be in effect. The different Liars are obviously examples of the same phenomenon, so even if the disjoint solutions otherwise "work", they must be inadequate. By this standard, no matter how good Priest's non-dialetheic arguments for non-classical conditionals may be, it simply doesn't matter. If he solves the Liar with dialetheism, and Curry with a rejection of Contraction—and the two moves are, as they seem to be, entirely unrelated—his solutions still fail his own test of uniformity.

One might be tempted to conclude at this point that the goal of a unified solution to the Liar, Curry, and the various set-theoretic paradoxes is unachievable. Various classical solutions—e.g. those that allow exceptions to the Biconditional Truth Schema—might spike Curry as easily as the Liar, but they can't be extended to the paradoxes of naïve set theory. Priest's dialetheic approach can handle the Liar and the set-theoretic paradoxes, but not Curry. Uniformity seems out of reach in either case.

Remember, though, that the paracomplete approach to the Liar *can* be extended to the set-theoretic cases. How does the paracompletist fare with regard to Curry? If they reject Contraction, is this move just as disconnected from their paracompleteness as Priest's rejection of Contraction is from his dialetheism?

To answer that, start by noticing that Contraction is simply a short cut for the inferential work that can be done more slowly with the combination of Conditional Proof and *Modus Ponens*. (If we have $\alpha \rightarrow (\alpha \rightarrow \beta)$ and we assume α for Conditionalization purposes, we can get to β with two applications of *Modus Ponens*.) Nothing about a paracomplete solution to the paradoxes gives the paracompletist a special reason to reject *Modus Ponens*, but there *is* a way to get from paracompletness to a rejection of Conditionalization.

To see how this works, first assume that conditionals work the way they do in classical logic, such that $\alpha \rightarrow \beta$ is logically equivalent to $(\neg\alpha \vee \beta)$. It's important to note that Field himself doesn't make this move. He shares Priest's commitment to non-classical conditionals, though of course the details of his account are different.[6] We can perhaps call the hypothetical version of paracompleteness that treats conditionals classically "semi-classical completeness".

Secondly, note a related advantage that Field has over Priest when it comes to the Liar and Curry. Whether or not he can give them a uniform *solution*, he can at least give a uniform *diagnosis* of both paradoxes, classifying the self-referential (or at least ungrounded) sentences about truth at the heart of all versions of both paradoxes as exceptions to Excluded Middle. He thinks we should reject both Curry sentences and their negations, just as (according to him) we should reject both Liar sentences and their negations. As discussed in Sect. 3, above, Priest can't make a parallel move and accept both Curry sentences and the negations of those sentences without either being committed by *Modus Ponens* to trivialism.[7]

This difference leads to a surprisingly simple route from semi-classical paracompleteness to a principled rejection of Conditionalization. (Field himself suggested this route in conversation.) If Conditionalization is universally truth-preserving, all sentences of the form $\alpha \rightarrow \alpha$ must be true. (The proof is two steps long. Assume α for Conditionalization. Conclude α. QED.) But wait! If the conditional is material, $\alpha \rightarrow \alpha$ is equivalent to $\neg\alpha \vee \alpha$—which is just an instance of Excluded Middle!

10 Beall and Semi-classical Dialetheism

This comparative advantage of paracompleteness over dialetheism in satisfying the PUS hinges on the fact that, unlike dialetheists (even hypothetical dialetheists who accept the classical material conditional), material-conditional-embracing paracompletists have available to them a principled reason, independent of "Curry-proofing",

[6]The 'paradoxes of material implication' play no role in motivating Field's conditionals, which work just like classical conditionals in contexts where Excluded Middle holds. Crucially, though, they behave very differently in paracomplete contexts. This in term allows Field to generate a transfinite hierarchy of determinateness operators, such that he can say of any given paradoxical sentence that it is (in some sense or another) not determinately true. The hypothetical version of paracomplete theory being proposed would thus have costs in terms of expressive completeness.

[7]Can dialetheists simply reject *Modus Ponens* as a way out? We will consider this option, and note its significant costs, in the next sections.

to reject Conditionalization. However, one doesn't need a principled reason to reject Conditionalization to have a principled reason to reject Contraction. One needs *either* a reason to reject Conditionalization *or* a reason to reject *Modus Ponens*. Dialetheists *do* have the latter available to them.

As dialetheists have known since the early days of the paraconsistent project, the combination of dialetheism and material conditionals makes *Modus Ponens* invalid. After all, given Double Negation, instances of *Modus Ponens* ($\alpha \rightarrow \beta$, α // β) are equivalent to instances of Disjunctive Syllogism ($\neg\alpha \vee \beta$, $\neg\neg\alpha$ // β), which is paraconsistently invalid.

Traditionally, dialetheists have seen this as a major problem, and tried to get around it by extending their favored paraconsistent logic with a conditional in which this equivalence fails, so that Disjunctive Syllogism can be regarded as invalid without *Modus Ponens* being sacrificed. A "semi-classical dialetheist", however, who was content to let this equivalence stand, *would* be able to offer both a uniform diagnosis of Liar and Curry sentences and a uniform solution to the two paradoxes. Without *Modus Ponens,* (2) above can be true and false (and hence true) without *everything* being true.

Unlike the semi-classical paracompletist, the semi-classical dialetheist is not a hypothetical creature. JC Beall advocates precisely this position in [2]. He argues that regarding *Modus Ponens* as invalid isn't as counterintuitive a move as it may initially seem. We still have a way of reasoning away from, for example, the premises "Lucy is a miniature schnauzer" and "If Lucy is a miniature schnauzer, then Lucy is a dog" to the conclusion "Lucy is a dog". We just need an extra step.

> Many have thought that when we 'do modus ponens' or 'follow modus ponens', we are simply inferring (say, accepting something new) *in accord with logic*, that is, *in accord with what our theory logically implies*. Since $\{A, A \supset B\}$ doesn't imply B, we can't be 'in accord with logic' when we infer B from $\{A, A \supset B\}$, when we accept B into our expanded theory. We are close [2, p. 10].

The 'close' inference that Beall still takes to be valid is the inference from all of the elements of the set $\{A, B\}$ to a choice of between the two elements of the set $\{B, A \wedge \neg A\}$. Since Beall is committed to the principle that it is rational to reject contradictions whenever we can plausibly do so, in any case where A and B are ordinary, non-paradoxical sentences, it's rational to reject $A \wedge \neg A$ and hence to accept B.

The formal theory he uses to make sense of all this is LP+ a multi-conclusion version of Priest's Logic of Paradox. Crucially, though, he accepts—granting a point to which he credits an anonymous referee—that, "given background (so-called structural) features of LP+ together with the clauses for disjunctions", his strategy for explaining how we can rationally 'do' what looks like *Modus Ponens* without *Modus Ponens* being valid could just as easily be run with the *single*-conclusion inference from $\{A, B\}$ to $(B \vee (A \wedge \neg A))$. He favors the multi-conclusion presentation for two reasons. The first is that the single-conclusion version "need not always hold for non-classical logics" and may not hold even for more complicated variations of his favored framework. The second is that "the 'choices' idea [...] seems to be more intuitive when presented in the more general multi-conclusion guise [...]" [2, p. 11].

Crucially, though, Beall seems to accept that the distinction between rationally rejecting one element of a set of possible conclusions and thus rationally accepting the other and rationally rejecting one disjunct of a disjunction and thus rationally accepting the other doesn't make a difference. In principle, however the choice is formally represented, if we have a choice between accepting two claims and a rational reason to reject one, we have a rational reason to accept the other.

The reason this point is worth belaboring is that it brings something odd and important into sharp relief. Beall's "choice" idea circa 2013 is just a disguised version of a strategy for "doing" Disjunctive Syllogism that Priest proposed in the first edition of *In Contradiction* in 1987 and then abandoned (for several excellent reasons) when the second edition came out in 2006.

11 Why Priest Now Rejects Principle R

Like *Modus Ponens*, Disjunctive Syllogism (DS) is one of the classical inference rules that seems to be most intimately tied to our ordinary reasoning practices. The dialetheist thus needs an explanation of what's going on when we infer from, for example, "Gareth is at home or at the pub" and "Gareth is not at home" to "Gareth is at the pub". If we're not engaged in a straightforward, deductively valid inference from the premises to the conclusion, do we have any rational reason to add "Gatheth is at the pub" to our expanded theory of the world?

Various dialetheists have offered various strategies for avoiding the counterintuitive conclusion that ordinary reasoning gets this case wrong. Priest himself has offered multiple strategies. The first one he gives in the first edition of *In Contradiction* involves Principle R:

> **Principle R**
>
> If a disjunction is rationally acceptable and one of the disjuncts is rationally rejectable, then the other is rationally acceptable.
>
> Suppose that someone accepts a disjunction, $\alpha \vee \beta$. Then nothing, as yet, force them to accept either disjunct: they may be agnostic about both. Let us suppose that they now come to reject one of the disjuncts, say α, while continuing to accept the disjunction. Then the person is rationally committed to accepting the other, β. The rationale for this is precisely the truth condition for disjunction: $\alpha \vee \beta$ is true iff α is true or β is true. So if one limb of the disjunction is not on, the other must be.
>
> It is important to see that this is not an application of DS. The argument is not: $\alpha \vee \beta$ is true but α is not true; therefore β is true. The justification is not a formal one but a pragmatic one: $\alpha \vee \beta$ is rationally acceptable; β is rationally rejectable; hence β is rationally acceptable [11, p. 113].

While it's possible (and, according to Priest, sometimes rational) to accept both a sentence and its negation, he takes it to be psychologically impossible to both accept and reject the same sentence at the same time. Thus, Priest then thought, Principle R survives the dialetheist's sacrifice of Disjunctive Syllogism.

The problem, as he realized between the first and second editions, is that it doesn't follow from the impossibility of *actually* accepting and rejecting the same sentence at the same time that we couldn't have a rational *obligation* to do just that. There could be rational dilemmas—cases where anything that we can do is irrational.

> If there are rational dilemmas [...] this principle [Principle R] cannot be correct. Let α be something that is both rationally acceptable and rejectable. For any β, α entails $\alpha \vee \beta$, so this is rationally acceptable too. By Principle R, β is rationally acceptable too [19, p. 276].

Priest gives several arguments for the existence of rational dilemmas. He further argues that it's particularly hard for *dialetheists* to avoid such dilemmas.[8] Let Ground A be whatever grounds the acceptability of rationally acceptable sentences. Let Ground B be whatever it is about rejectable sentences that makes it rational to reject them.

> A dialetheist cannot rule out *a priori* the occurrence of Ground A and Ground B, and so of the dilemma generated. [18, p. 110 n. 15][9]

Perhaps, though, the difference between Priest's relatively expansive version of dialetheism and Beall's more constrained view is relevant here. Beall has said that a "largely philosophical" difference between the two views is that Priest starts from a recognition of the possibility of "gluts in general" [1, p. 130]. By contrast, Beall himself believes that only "the standard 'spandrels of truth'—truth-theoretic paradoxes and other semantic paradoxes—deliver gluts".

> But these are odd and rare beasts, tangled up with truth and falsity on their surface—for example, 'I am not true'. It takes substantial argument to extend the lesson of gluts beyond such a limited realm [3, p. 831].

As it turns out, though, we don't *have* to extend the lesson of gluts beyond the limited realm of paradoxical sentences in order to create problems for Beall here. One of several reasons that Priest now rejects Principle R concerns precisely one of those "odd and rare beasts". In this case, rational acceptability and rational rejectability are the concepts tangled up on its surface.

> Consider a claim of the form, 'It is irrational to believe this claim', i.e. something of the form:
>
> It is irrational to believe p.
>
> where it, itself, is p. Suppose you believe p. Then you believe something, and at the same time believe that it is irrational to believe it. This, presumably, is irrational. Hence, you ought not to believe p: $O\neg Bp$. But we have just shown that p is true! Hence you ought to believe p [...] This is a version of the Irrationalist's Paradox [19, p. 111].

---

[8]These arguments take up four sections—6.5–6.9—in [18]. When he renounces Principle R in [19], he refers the reader to that discussion.

[9]A desperate way to avoid this might be to deny that there's any feature that sentence can have that *always* make it rationally rejectable. The problem with this move is that it makes the whole notion of rational rejection rather mysterious. If there's no feature of sentences that makes them rationally rejectable, can we say anything about them except that they're rejectable because they're rejectable?

Of course, if Beall solved this paradox by claiming that self-referential sentences about irrationality were meaningless, or perhaps by claiming that we should postulate an infinite hierarchy of rational rejectabilities, such that a rank-n rejectability operator can only be used to reject rank-n-minus-one sentences, he could avoid having to grant the existence of rational dilemmas and thus hold onto Principle R.[10] Once *this* odd and rare beast has been treated that way, however, why not make the parallel move for sentences like (1) and (2) above and keep not only Principle R, but Disjunctive Syllogism and *Modus Ponens*?

12 Conclusion

One of the central advantages of a dialetheic approach to the paradoxes of self-reference is supposed to be that such an approach allows us to handle all of the paradoxes in the same way. *Whether or not* Curry's Paradox fits the Inclosure Schema, no approach that treats Liars, but not Curry sentences, as dialetheias can claim this advantage.

Conversely, any (non-trivial) dialetheic approach that treats *all* of truth's paradoxical spandrels dialethically does so at the expense of *Modus Ponens*. If Principle R were available to dialetheist paradox-solvers, the sacrifice of *Modus Ponens* might turn out to be a manageable cost. Given the existence of the Irrationalist's Paradox, however, a further dilemma presents itself. The dialetheist can either reject Principle R, and thus live without the ability to "do" *Modus Ponens* even in an indirect way, or they can sacrifice the Principle of Uniform Solution in a different way by solving an extremely Liar-like semantic paradox in a way that has nothing to do with their solution to the Liar and Curry's Paradox. Regardless of which of these dismal choices any particular dialetheist adopts, the prospects for a plausible and truly uniform dialetheic solution to the paradoxes of self-reference are grim.

[10] Actually, it's not clear that Beall *could* endorse a hierarchical solution here without giving up on his transparent theory of truth. As he points out in his discussion of Field's solution to the paradoxes in *Spandrels of Truth* [1] one of the things transparent truth predicates are good for is referring to every level of an infinite hierarchy at the same time. Thus, we could have sentence (10):

(10) It is irrational in at least one of the infinitely many senses of *irrational* to believe this sentence.

References

1. Beall, J. C. (2009). *Spandrels of truth*. Oxford: Oxford University Press.
2. Beall, J. C. (2013). Free of detachment: Logic, rationality and gluts. *Noûs, 49*(2), 410–423.
3. Beall, J. C. (2014). End of inclosure. *Mind, 123*, 829–849.
4. Beall, J. C. (2014). Finding tolerance without gluts. *Mind, 123*, 791–811.
5. Beall, J. C., & Armour-Garb, B. (Eds.). (2005). *Deflationism and paradox*. Oxford: Clarendon Press.
6. Beall, J. C., & Murzi, J. (2013). Two flavors of curry's paradox. *Journal of Philosophy*, 143–165.
7. Bueno, O., & Colyvan, M. (2003). Paradox without satisfaction. *Analysis, 63*, 152–156.
8. Field, H. (2008). *Saving truth from paradox*. Oxford: Oxford University Press.
9. Musgrave, A. (1999). Critical rationalism and logical omniscience. In Musgrave [10].
10. Musgrave, A. (1999). *Essays on realism and rationalism*. Amsterdam: Rodopi.
11. Priest, G. (1987). *In Contradiction*. Dordrecht: Kluwer.
12. Priest, G. (1994). The structure of the paradoxes of self-reference. *Mind, 103*, 25–34.
13. Priest, G. (1997). Yablo's paradox. *Analysis, 57*, 236–242.
14. Priest, G. (1998). The import of inclosure: Some comments on Grattan-guiness. *Mind, 107*, 835–840.
15. Priest, G. (2000). On the principle of uniform solution: A reply to Smith. *Mind, 109*, 123–126.
16. Priest, G. (2002). *Beyond the limits of thought* (2nd ed.). Oxford: Clarendon Press.
17. Priest, G. (2005). Spiking the field artillery. In Beall & Armour-Garb (Eds.).
18. Priest, G. (2006). *In Contradiction* (2nd ed.). Oxford: Clarendon Press.
19. Priest, G. (2006). *Doubt truth to be a liar*. Oxford: Oxford Clarendon Press.
20. Priest, G. (2017). What *If*? The exploration of an idea. *Australasian Journal of Logic, 14*, 54–127.
21. Smith, N. J. J. (2000). The principle of uniform solution (of the paradoxes of self-reference). *Mind, 109*, 117–122.
22. Wansing, H., & Priest, G. (2015). External curries. *Journal of Philosophical Logic, 44*, 453–471.
23. Weber, Z., Ripley, D., Priest, G., Hyde, D., & Colyvan, M. (2014). Tolerating gluts. *Mind, 123*, 813–828.

A Revenge Problem for Dialetheism

Gareth Young

Abstract Perhaps the most powerful argument that has been made in favour of the view that some contradictions are true (dialetheism) is that it allows for a solution to the logical paradoxes which is immune to the well-known problem of revenge. The version of the view which would seem to have the best chance of avoiding the problem is a particularly thoroughgoing dialetheism, most prominently defended by Graham Priest, which takes paraconsistent set theory as its working metatheory. The purpose of this paper is to characterise a revenge problem for this thoroughgoing dialetheism, involving the notion of invalidity. I argue that the inconsistency of the metatheory commits dialetheists of this sort to accepting as contradictory, not only truth, but validity: in other words, some inference principles are both valid and invalid. I show that, depending on the details of the theory, all, or 'almost all' (in a sense to be explained), inference principles can be shown to be dialetheically invalid. I argue that this gives rise to a revenge problem for dialetheism, since it makes the notion of invalidity inexpressible for the dialetheist and deprives them of the ability to express crucial semantic claims about their theory.

1 Introduction

Perhaps the most powerful argument in favour of dialetheism (the view that some contradictions are true) concerns the logical paradoxes and proceeds by elimination. In a nutshell, the argument is this: every other attempt to solve the logical paradoxes suffers (indeed *must* suffer) from the revenge problem, which is so serious as to refute any theory subject to it; dialetheism does not suffer from the revenge problem, so we should endorse dialetheism. Graham Priest, Dialetheism's most prominent defender, argues like this (though more subtly and at greater length, of course) in a number of places, but especially in his [10], where considerations of revenge are invoked against alternative theories frequently.

G. Young (✉)
University of Glasgow, Glasgow, Scotland, UK
e-mail: gareth.young@glasgow.ac.uk

A. Rieger and G. Young (eds.), *Dialetheism and its Applications*, Trends in Logic 52,
https://doi.org/10.1007/978-3-030-30221-4_2

Responses to this situation from opponents of dialetheism have mostly come in two forms: the first is to attempt to characterise alternatives to dialetheism which do not suffer from the revenge problem (for a recent attempt at this, see Field [5]). The second is to attempt to show that dialetheism, too, has a revenge problem. The point of the second response is not to vindicate any particular theory of the paradoxes, but simply to show that dialetheists are at no advantage when it comes to revenge, and so to undermine the argument for dialetheism which proceeds by appeal to its purported immunity to revenge.

This paper is a contribution to the second of these projects: my aim is to characterise a novel revenge problem for dialetheism, and so to undercut a significant part of the motivation for the view. Dialetheism comes in, broadly, two forms. The first, whose most prominent defender is Beall (see, especially, his [2]), takes, as its metatheory, classical ZF set theory. I won't argue for this claim here, but this view seems to me to be subject to the standard revenge problems levelled at non-dialetheist theories by Priest, so I won't discuss this theory further. The target of this paper is the more interesting case (from the point-of-view of revenge), which is dialetheism of the sort which takes para consistent is set theory as its working metatheory. Dialetheists such as Priest may not like this way of describing the view. One feature of dialetheism of this sort is that it allows for the identification of object and metatheory, and so, in Priest's view, collapses the distinction between object and metatheory: we simply have a single, dialetheic theory which is able to provide its own semantics. Nonetheless, it seems easier to me, for explanatory purposes, to continue to talk of dialetheism's object theory and metatheory. This will do no harm, so long as we remember that, ultimately, the aim of the dialetheist is to identify them.

The revenge problem I argue is faced by this theory concerns the notion of invalidity. The inconsistency of the thoroughgoing dialetheist's metatheory commits them to the inconsistency, not just of truth, but of validity. In other words, it commits them to some inference principles being both valid and invalid. The extent of this overlap, between the valid and the invalid inference principles, is surprisingly great. On certain versions of the view, one can show, reasonably straightforwardly, the invalidity of every inference principle there is. Things are trickier on the version of the view defended by Priest, but even in this case, I show the invalidity of every inference principle whose conclusion lacks a conditional. I also show, in a piecemeal fashion, that a number of inference principles with a conditional in the conclusion are also invalid. Whether this can be extended to the general result that every inference principle is invalid, on Priest's view, is unclear at this point. Even if this result cannot be obtained, the weaker result, that every inference principle with a conditional-free conclusion, as well as a number of principles with conditional conclusions, is invalid, is still significant. Every inference principle anyone is ever likely to use, and every inference principle to be found in, say, a logic textbook, is invalid.

This result is surprising and important for our understanding of dialetheism, but it also, I argue, gives rise to a revenge problem. In classical logic, the validity of the principle of explosion means that contradictions are trivialising: a single contradiction entails the truth (and falsity) of every sentence. So dialetheists, if they want to avoid this absurd result, must endorse some non-classical logic which eschews the principle of explosion and those principles which give rise to explosion, such as disjunctive syllogism, which would trivialise the theory, were they valid. For this reason, it is an absolutely vital fact about dialetheism that, for example, disjunctive syllogism is invalid. But if dialetheists try to express this fact in the obvious way by saying "Disjunctive syllogism is invalid", they say something vacuous, which either applies to every inference principle whatever, or, at the least, to every inference principle anyone is likely to use, such as *modus ponens*, conjunction-elimination, disjunction-introduction, and so on. I argue that the effect of this is to rob dialetheists of the ability to express the notion of invalidity, and so to prevent them from making crucial semantic claims about their view. If I am correct, this would seem to be a revenge problem every bit as serious as those which dialetheists level at their opponents. I will call this problem the "invalidity revenge problem".

Other attempts have been made to get revenge on dialetheists, some of which are related to my own (especially those involving expressions like "just false" and "non-dialetheia"). I begin, in Sect. 2, by discussing some of these, especially as defended in Shapiro [12], partly because they will help to illustrate my own revenge objection, and partly because I wish to distinguish mine from theirs. I then, in Sect. 3, present arguments that dialetheists are committed to the invalidity of a great many valid inference principles, where the exact route we take to establish this varies with the dialetheists specific views. In Sect. 4, I discuss the revenge problem I claim arises from these result and I compare the invalidity revenge problem to the just false problem, arguing that the former is importantly distinct from, and more serious than, the latter. I also discuss some potential responses a dialetheist might offer to the invalidity revenge problem, and find each wanting.

2 Dialetheism, Revenge and the 'Just False' Problem

Dialetheists admit that there exist true contradictions but, as non-trivialists, they take these dialetheia to be the exception: most sentences are not like this. But dialetheists seem to run into expressive difficulties when trying to classify the "normal" sentences as such. The reason for this is that the cluster of notions one would wish to employ in these characterisations turn out to be inconsistent. This has been developed into an objection to dialetheism by a number of philosophers (for example, Batens [1], Parsons [8], Littman and Simmons [7], but the objection is best presented, it seems to me, by Shapiro [12], and so I draw on Shapiro's characterisation of the problem in what follows.

Shapiro argues that dialetheists cannot express the notions 'just true' (though he calls this 'simple truth'), 'just false' (though he calls this 'simple falsity') 'non-dialetheia', consistency and, relatedly, they cannot express disagreement. The reasons for each are, in essence, the same: it seems part of the meaning of these notions that they behave consistently, or, at the least, not too inconsistently. In this section, I discuss two of these notions, 'just false' and 'non-dialetheia', the reasons we might think they are inexpressible for dialetheists and some dialetheist responses to the problems. I discuss these two only since they relate most closely to my own revenge problem, and so best help illustrate it, and because the arguments concerning the others are similar enough that these should be sufficient to illustrate the general problem.

One very simple way of characterising dialetheism (of the non-trivialist variety) is as the view that some statements are both true and false, some are just true and some are just false. Claims of this kind appear with great frequency in descriptions of dialetheism, by dialetheists as well as others. So dialetheists ought to be able to express the notions involved in such claims; in particular, they should be able to express 'just true' and 'just false'. As I have said, I focus here on the latter notion (though much the same things are true of the former). We take the obvious definition of 'just false' as 'false and not true' as Priest does (for example in his [11, p. 137], though he endorses this in a number of other places too).

We can see that the notion of 'just false', defined in the obvious way, is inconsistent by considering the following sentence, Σ, equivalent to $F <\Sigma> \wedge \neg T <\Sigma>$.

Intuitively, this sentence says of itself that it is just false. We can derive contradiction from Σ in much the same way as with the strengthened liar sentence:

(1) $T<\Sigma>$	(suppose for reductio)
(2) Σ	(1, T-scheme)
(3) $F<\Sigma> \wedge \neg T<\Sigma>$	(2, substituting equivalents)
(4) $\neg T<\Sigma>$	(3, $\wedge$-elimination)
(5) $\neg T<\Sigma>$	(1, 4, reductio)
(6) $F<\Sigma>$	(5, exhaustion)
(7) $F<\Sigma> \wedge \neg T<\Sigma>$	(5, 6, $\wedge$-introduction)
(8) Σ	(7, substituting equivalents)
(9) $T<\Sigma>$	(8, T-scheme)
(10) $T<\Sigma> \wedge (F<\Sigma> \wedge \neg T<\Sigma>)$	(7, 9, $\wedge$-introduction)

So Σ is both true and just false, and so the notion 'just false', defined in this way, is inconsistent. In fact, matters are simpler than this. Though it is not Priest's view, a number of other philosophers (for example, Field [5] and Beall [2]) have endorsed the view that the T-scheme should be fully transparent and, hence, contraposable. So, on these views, (on the assumption that falsity is defined as truth of negation) $F <\alpha> \leftrightarrow \neg T <\alpha>$ holds. On this view, take a sentence, α, and suppose it is false, so we have $F <\alpha>$. Given the properties of the T-scheme just mentioned (in particular, what Priest calls "exclusion"), we have $\neg T <\alpha>$, and so $F <\alpha> \wedge \neg T <\alpha>$. So, on the assumption that α is false, α is *just* false. Since, obviously, the converse also follows, 'false' and 'just false' are equivalent, on this view. So every falsehood is just false, including every dialetheia. So, paraphrasing Shapiro [12], saying that something is just false says no more than that it is false: the 'just' comes for free and adds nothing.

So, if a dialetheist endorses the transparency of the T-scheme, 'just false' makes no distinction between dialetheia and the non-dialetheia we wish to pick out: it fails to capture what we want when we utter expressions like 'Some sentences are both true and false, some are just true and some are just false'. For all the distinction drawn, we may as well have said 'Some sentences are both true and false, some are true and some are false', which is not at all what's intended.

Priest responds on behalf of this variety of dialetheism that "…this objection is entirely question-begging. If one does subscribe to the exclusion scheme, there *is* no distinction—nor, therefore any work to be done to draw it" [10, pp. 292–293]. I disagree that the objection is question-begging. There certainly seems to be a difference between 'false' and 'just false' captured by the difference between sentences like 'Some sentences are false, but some are just false', on the one hand, and 'Some sentences are false, but some are false', on the other. These, and sentences like them, again, are uttered by dialetheists frequently, and it seems to me that we clearly understand what is intended. To pick an arbitrary example, Priest and Berto (1998) characterise an historical form of dialetheism as follows:

> In ancient Indian logic/metaphysics, there were standardly four possibilities to be considered on any statement at issue: that it is true (only), false (only), neither true nor false, or both true and false. Buddhist logicians sometimes added a fifth possibility: none of these.

For another example, just before Beall defends the view that 'true' and 'just true' are equivalent (since his metatheory is consistent, I don't discuss his view here) he says:

> Rational dialetheists maintain that some (actually, many) [truths] are *just true*; they reject that all or even most claims are gluts. Indeed, on my account, it is only the spandrels of [truth] (or related notions) that are gluts; the rest are 'just true'." [2, p. 48]

The 'only' used in the first passage adds something to the characterisation of the view described above (why else would the authors have written it?) Similarly, though perhaps Beall adds "scare quotes" to indicate that the notion he is discussing is problematic, I suspect most readers understand exactly what these sentences mean, and so grasp a difference between "true" and "just true" (and, correspondingly between 'false' and 'just false').

If it were a straightforward consequence of dialetheism generally that this distinction could not be accounted for, this would be a problem for dialetheism, and it would not be question-begging to says so. But the equivalence of falsity and just-falsity is not an immediate consequence of dialetheism, and so the charge of question-begging seems to me to have even less plausibility.

Though many philosophers have endorsed the account of the T-scheme on which the preceding remarks are based, as I have said, this is not Priest's view. Priest rejects the principle of exclusion according to which falsity entails untruth: $F <\alpha> \rightarrow \neg T <\alpha>$. This means that the situation for Priest, with regards to the just false problem, is different from that sketched immediately above. In particular, since falsity does not entails untruth, we cannot in general show that an arbitrary falsehood is thereby untrue and, hence, just false.

The notion of just falsity, though, is still inconsistent for Priest, since he still accepts the proof above. Moreover, we can show, since Priest accepts exhaustion, $\neg T <\alpha> \rightarrow F <\alpha>$, that any sentence, α, which is untrue, is thereby false and, hence, just false. So the strengthened liar, λ, equivalent to $\neg T <\lambda>$, since it can be shown to be both true and untrue, is both true and just false. The same holds for every untrue dialetheia, of which there are infinitely many.

Part of the point of notions like just false certainly seems, on the face of it, to involve those notions behaving consistently: we want to pick out the sentences which are *just* false. Still, it might be thought, it is unfair to expect these notions to behave consistently for a dialetheist, since they are semantic notions which can be employed in the construction of liar sentences, which are the very sorts of thing dialetheists urge us to treat inconsistently. But, Shapiro claims [12, p. 342], the point of introducing the notion of just false in the first place was to distinguish falsehoods from at least paradigmatic dialetheia, which the strengthened liar is. So, even if we don't demand the notion behave consistently, if the strengthened liar, as well as infinitely many other untrue dialetheia, are just false, we might think the notion too inconsistent to be of use.

In response to this, Priest says "I am not sure what the first place was in this case, nor, therefore, what its point was. But since F $<\alpha>$ and $F<\alpha> \wedge \neg T<\alpha>$ are not logically equivalent, there is a distinction between being false and being simply false. The fact that some sentences (be they paradigm dialetheias or anything else) may be in both camps is just one of those contradictory facts of life that populate the dialetheic landscape [The strengthened liar] is simply false—and it is true as well" [9, p. 294].

So Priest's response to the problem is to simply accept the inconsistency of 'just false' and deny that this prevents it from meaning just false, regardless of its inconsistency in the case of 'paradigm' dialetheia, because we still have a distinction (because of the failure of exclusion) between being false and being just false. My sympathies lie with Shapiro on this issue. It seems to me that the point of expressions such as 'just false' is to consistently pick out those sentences which are false only. Having said that, it is difficult to see what argument can be given to this effect whose premises a dialetheist like Priest would be happy to accept. This is not, of course, to concede that Priest is right; rather the situation seems to have reached a stalemate. It is worth re-iterating, though, that this response from Priest is only available to him because of his rejection of exclusion. If one accepts the equivalency of falsity with untruth, the distinction between falsity and just-falsity is collapsed.

A closely related problem concerns the notion 'non-dialetheia'. Shapiro points out [12, pp. 342–343], attributing the original complaint to Parsons [8], that there are a number of ways of expressing that something, α, is a dialetheia. We might say, though this may not exhaust the possibilities, $\alpha \wedge \neg\alpha$, or $T<\alpha> \wedge F<\alpha>$, or perhaps $T<\alpha> \wedge \neg T<\alpha>$. Corresponding to each of these is a way of expressing what it is to be a non-dialetheia: $\neg(\alpha \wedge \neg\alpha)$, $\neg(T<\alpha> \wedge F<\alpha>)$ and $\neg(T<\alpha> \wedge \neg T<\alpha>)$, respectively. Each of these is inconsistent, in the sense that each holds of some dialetheia (so some dialetheia are non-dialetheia, on each of these definitions). The first, $\neg(\alpha \wedge \neg\alpha)$, is a dialetheic logical truth, and so holds for any sentence, α, including every dialetheia. So this option is inadequate. The third option, $\neg(T<\alpha> \wedge \neg T<\alpha>)$, is also true of every sentence, α, as Priest has shown [10, pp. 90–91). The second option, $\neg(T<\alpha> \wedge F<\alpha>)$, differs only if exclusion fails. If both exclusion and exhaustion hold, then falsity and untruth are equivalent and, hence, $\neg(T<\alpha>$

$\wedge \neg T <\alpha>)$ and $\neg(T <\alpha> \wedge F <\alpha>)$ are equivalent. This would mean the second option is also true of any α. Since Priest rejects exclusion, things are different for him.

The situation is similar to the one described above for 'just false'. Since it does not follow from the falsity of some sentence, α, that α is untrue, it does not follow from the falsity of α that α is not both true and false. So, for Priest, this definition of non-dialetheia, though inconsistent (some dialetheia are non-dialetheia), is not vacuous (not all dialetheia are non-dialetheia).

Unsurprisingly, Priest chooses the option which is not vacuous: α is a dialetheia just if $T <\alpha> \wedge F <\alpha>$ and so a non-dialetheia just if $\neg(T <\alpha> \wedge F <\alpha>)$. Though the definition is not vacuous, Shapiro still thinks it is problematic. Any sentence which is untrue is thereby *not* both true and false, and so not a dialetheia, on this definition. This includes paradigm dialetheia like the strengthened liar. He says:

> It is surely reasonable to demand that our definitions be non-vacuous. If we have to say that *every* sentence is a non-dialetheia, then the notion is useless. And the notion is all but useless if we have to say that every untruth is also a non-dialetheia, including the Original Liar and nearly all the dialetheias we run across in the course of thinking about this stuff—every non-truth and every non-falsehood are also non-dialetheias. I would have thought that a dialetheist, like Priest, would *deny* that the Original Liar is a non-dialetheia, rather than asserting that it is one. [12, p. 344]

Priest's response is to down-play the overlap between the dialetheia and the non-dialetheia:

> Shapiro overplays his hand here. There is indeed an overlap in the categories of being and not being a dialetheia. But the only denizen of the overlap we have on the table is $\xi(\neg \mathrm{T} <\xi>)$. And this is very special. By its particular properties, it is true and it is not true—and so not (true and false). But the same is not the case for any other of the standard paradoxes of self-reference (including the liar in the form 'this sentence is false')...And I am quite happy to assert that ξ is a non-dialetheia, provided that I can add that it is as well. (2006, p. 294)

The strengthened liar may be the only denizen of the overlap 'on the table', but it is certainly not the only denizen of the overlap there is. For one, once we have one dialetheia of this kind, it follows that there are, by similar constructions, infinitely many. So at least in terms of cardinality, there is not only one such contradiction. Still, a more charitable way of reading Priest is as claiming that the strengthened liar is the only *kind* (in some sense) of paradox which is both true and untrue. But this isn't right either. Many of the other paradoxes of self-reference can be 'strengthened' in a similar way to the liar.

For example, Grelling's paradox is treated dialetheically by Priest. Grelling's paradox arises from the predicate 'heterological', which is defined as being true of a predicate when that predicate does not apply to itself. For example, 'monosyllabic' is heterological, since it is not monosyllabic. The contradiction is produced by applying the predicate to itself: that is, by asking whether 'heterological' is heterological. But we can define heterologicality differently to obtain a strengthened version of the paradox: define a predicate to be heterological when the result of applying it to itself is, not false, but untrue. The dialetheia generated by this version of the paradox is both true and untrue and, hence, a non-dialetheia.

A little more precisely, if we introduce our heterologicality predicate, *Het*, defined such that it holds of a predicate just if it is untrue that the predicate applies to itself, then we can derive contradiction as follows:

(1) $T{<}Het{<}Het{>}{>}$ (suppose for *reductio*)
(2) $Het{<}Het{>}$ (1, T-scheme)
(3) $\neg T{<}Het{<}Het{>}{>}$ (Definition of Het)
(4) $\neg T{<}Het{<}Het{>}{>}$ (1, 3, *reductio*)
(5) $Het{<}Het{>}$ (4, definition of Het)
(6) $T{<}Het{<}Het{>}{>}$ (5, T-scheme)

Another example is Russell's paradox, which can be strengthened by introducing a satisfaction predicate, along with the naïve principles for satisfaction. We take the Russell set to be $\{x: \neg Sat(x, v\ v\}$ and prove a contradiction as follows:

(1) $R \in R$ (suppose for *reductio*)
(2) $\neg Sat(R, {<}v \in v{>})$ (1, comprehension)
(3) $Sat(R, {<}v \in v{>})$ (2, exhaustion)
(4) $R \notin R$ (3, definition of satisfaction)
(5) $R \notin R$ (1, 4, *reductio*)
(6) $Sat(R, {<}v \in v{>})$ (comprehension, contraposition, DNE)
(7) $R \in R$ (6, definition of satisfaction)
(8) $\neg Sat(R, {<}v \in v{>})$ (7, comprehension)

Since for x to satisfy φ is for φ to be true of x, it is both true and untrue of this Russell set that it is self-membered, and we have another denizen of the overlap. Similarly, we can strengthen Quine's paradox (modifying the crucial expression to "Yields an untruth when appended to its own quotation"), and Yablo's (by taking each sentence in the relevant sequence to predicate untruth of all later sentences).[1] So the strengthened liar does not seem to be as singular as Priest's remarks suggest. Many of the standard paradoxes of self-reference can be strengthened to yield versions which Priest must treat as both true and untrue, and, hence as non-dialetheia. One might think this lends some plausibility to Shapiro's claim that the overlap between the dialetheia and the non-dialetheia is too great for 'non-dialetheia' to be of much use.

On the other hand, the notion is still not vacuous: not every dialetheia is a non-dialetheia. In particular, Priest may still think that the fact that a significant number of the self-referential paradoxes, especially those characterised using falsehood rather than untruth, are both true and false but *not* untrue, is sufficient for the term 'non-dialetheia' to preserve its meaning and usefulness.

[1] In fact, it is very common in the literature for Yablo's paradox to be expressed in this, strengthened form, employing untruth instead of falsity. Indeed, Yablo's original formulation of the paradox in his [13] was characterised in this way.

My sympathies are, again, with Shapiro here, that non-dialetheia ought to behave consistently, or at least, less inconsistently than it does, for it to mean what's needed. But as with the just false problem, it is difficult to see how, if Priest takes the view described above, one can argue for this conclusion from premises Priest is likely to accept. Again, though, this is only because Priest rejects exclusion: if this principle is accepted, all the obvious definitions of non-dialetheia collapse into vacuity.

The problems discussed, of the apparent expressive limitations of dialetheism, arise from the obvious definitions of 'just false' and 'non-dialetheia' respectively. One avenue of response is to define a new, unobvious, definition of 'just false' and, correspondingly, of 'non-dialetheia', which behaves consistently.

The most obvious problem facing such attempts is the threat of revenge paradoxes employing the newly defined notion. So, for any newly-defined notion *just false*, we construct a sentence:

$$(*) \text{ The sentence marked } (*) \text{ is } just\ false$$

From this sentence, little is required to derive the conclusion that (*) is both true and *just false*. This obstacle must be avoided if the notion to do its job, which is no easy task.

Still, attempts have been made to solve the problem in a way which avoids this. Some options which have been considered are a form of primitive exclusion [4], a form of negation called 'arrow falsum' as well as non-logical 'shriek rules' (see Beall [3] for a discussion of the second two, and Young [14] for a critical discussion of Beall's proposals).

I don't discuss these options here for a few reasons. The main reason is that each of the attempts at solving this problem with a new definition of 'just false' has been carried out in a setting where the metatheory is taken to be consistent. The views under discussion in this paper are ones which take the metatheory to be inconsistent. It is not obvious how these solutions behave in this different setting, so I set them aside here.

A second reason is that to take devices like arrow-falsum, shriek rules or primitive exclusion as putative solutions to the just false problem is to conflate the just false problem with the, admittedly related, problem of exclusion. For a discussion of the distinction between the exclusion problem and the just false problem, and a critical discussion of Beall's [3], see Young [14].

So, although it may be that some new, consistent notion of 'just false' can be defined for metatheoretically inconsistent dialetheism, no such definition has, to my knowledge, been offered. The principal reason for this, I suspect, is that if such notions are strong enough to capture anything like 'just false' a revenge liar will demonstrate them to be inconsistent, defeating their purpose. So Priest, the most prominent proponent of this version of dialetheism, simply settles for the original definitions and accepts their inconsistency. This response is, as I will argue, far less plausible in the case of the invalidity revenge problem presented in this paper.

3 Inconsistent Validity and Revenge

Dialetheism which takes classical *ZF* set theory as its metatheory, of course, allows some sentences to be both true and false: that is, some sentences are assigned both 1 and 0 in the same model. But this inconsistency infects the metatheory no more than my views become inconsistent when I report that, according to Graham Priest, some sentences are both true and false. But if the metatheory is inconsistent, we should expect the assignments of truth-values to sentences themselves to become inconsistent: that is, we should expect some sentences both to be assigned 1 and not assigned 1 in the same model, and some sentences both to be assigned 0 and not assigned 0 in the same model.

As I show in this section, we do indeed find that metatheoretically inconsistent dialetheism makes inconsistent assignments of truth values to sentences. One significant effect this has is to make, not just truth, but *validity* dialetheic. This fact is significant and interesting on its own, but it has greater significance because it leads to a revenge problem: we can show that so many inference principles are invalid that dialetheism becomes unable to express the notion of invalidity. This is a very serious expressive limitation, since there are crucial semantic facts about the theory, for example the invalidity of the principle of explosion ($A, \neg A \vDash B$), which the dialetheist must use the notion to express.

The ease with which, and the extent to which, it may be shown that inference principles are invalid on this view varies with respect to a number of factors, especially the behaviour of the truth-predicate. So I take a number of options in turn and show the extent of the overlap between validity and invalidity for each option. On the first two options, it is straightforward to show the invalidity of every inference principle there is. On the second two options, because of the behaviour of Priest's truth predicate (in particular, the failure of the principle Priest calls "exclusion" [10, pp. 70–72], things are more difficult. I begin by showing that every inference principle capable of taking an atomic sentence as its conclusion is invalid. I then extend this, using a result by Heck (2013), to every inference principle whose conclusion contains only extensional connectives.

Next, I address principles whose conclusion contains a conditional. I show, in a piecemeal way, that a number of these are invalid. I do not, at the present, have a general result that every such principle is invalid, though I suspect there is one to be had. I discuss a number of avenues of response to this problem and argue against them. Throughout most of this chapter, I assume that the dialetheic metatheory under discussion is carried out in paraconsistent set theory, where the underlying logic is *LP* with the addition of a suitable, relevant conditional. I first discuss the view which takes truth to be transparent, and so which takes the biconditional in the T-scheme to contrapose. I don't take a view on whether the conditional added to *LP* ought itself to contrapose since, if it does not, the same effect may be achieved (as regards the T-scheme) by stipulating, in addition to $T <\alpha> \leftrightarrow \alpha$, $\neg T <\alpha> \leftrightarrow \neg\alpha$. I conclude the chapter with a brief discussion of the prospects of developing my revenge problem on Priest's model-theoretic strategy. I give arguments which seem to show that there must be metatheoretically inconsistent assignments of truth values to sentences (some both get 1 and don't), but point out the difficulties involved in showing how these might invalidate valid inference principles.

Validity is understood throughout as necessary preservation of truth, as defined by *Val*, such that

$$A \vDash B \text{iff } (Val) \text{ for every interpretation } I,\ (I \Vdash A) \rightarrow (I \Vdash B)$$

I assume throughout, as Priest accepts [10, p. 8] that the conditional validates the principle $A \wedge \neg B \vDash \neg(A \rightarrow B)$ and so any interpretation I in which A is assigned 1 and B is not assigned 1 will invalidate $A \vDash B$.

3.1 Inconsistent Validity with Transparent Truth

Though it is not Priest's view that truth is transparent, a number of philosophers have found this view attractive. If the adoption of metatheoretically paraconsistent dialetheism were to become more widespread, it seems reasonable to suppose that a number of dialetheists would want the T-scheme to behave in this way, and so I discuss some possibilities in this section. For those more interested in the views of Priest himself, some of the subtleties of this section can safely be skipped.

Defenders of this view accept the T-scheme in full generality. They also accept both of:

$$(Exclusion)\quad T<\neg\alpha> \rightarrow \neg T<\alpha>$$
$$(Exhaustion)\quad \neg T<\alpha> \rightarrow T<\neg\alpha>$$

One would expect them to accept similar principles regarding truth-in-a-model (TIAM) so, in Priest"s relational semantics, for any interpretation ρ:

$$(TIAM\text{-}Exclusion)\ <\neg\alpha>\rho 1 \rightarrow \neg<\alpha>\rho 1$$
$$(TIAM\text{-}Exhaustion)\ \neg<\alpha>\rho 1 \rightarrow <\neg\alpha>\rho 1$$

In fact, Priest suggests that they may not need to endorse these principles. He says, discussing a claim from Shapiro about the exclusion principle, that "[o]ne should note, here, that we are talking about truth/falsity *simpliciter*, not truth/falsity in an interpretation. Truth/falsity in an interpretation is quite different. Even if one endorses the exclusion schema, this does not entail that, for any given interpretation, ρ, $<\alpha>\rho 0 \rightarrow \neg <\alpha>\rho 1$" [10, p. 293].

If one accepts Exclusion, it's unclear why one would reject *TIAM-Exclusion*. To accept an asymmetry of this kind between truth/negation *simpliciter* and truth-in-a-model/$\neg$ is to make the semantics of the theory ad hoc and unnatural. After all, the semantics are meant to be characterising the features of actual negation, and, via truth-in-a-model, actual truth. But if *Exclusion* holds of truth, though the corresponding principle fails of truth in a model, then there is an important asymmetry between the real notions and the notions as they appear in the semantics.

Shapiro argues that Priest's semantics are artificial, since he denies *TIAM-Exclusion*, and is thereby denied a homophonic semantics [12, p. 352]. Priest responds by denying that a homophonic semantics is desirable, and claims that the objection amounts to nothing more than the question-begging insistence that dialetheic negation is not negation [10, pp. 294-295). Moreover, he thinks, once we admit dialetheia, homophonic semantics are unappealing since, if α is a dialetheia, $\neg\alpha$ holds, but then so does α. This point is not persuasive, however, since, on the assumption that dialetheism is true, there is nothing problematic about accepting both a and $\neg a$. Having said that, if Priest is right to reject *Exclusion* in the case of truth, he ought also to reject *TIAM-Exclusion* for truth-in-a-model, for the same reason that those accepting *Exclusion* of truth ought to accept *TIAM-Exclusion* for truth-in-a-model: if the semantics are otherwise there is a problematic asymmetry between the notions occurring in the model theory and the real semantic notions we are trying to model. So one ought to accept *TIAM-Exclusion/TIAM-Exhaustion* if and only if one accepts the corresponding principle about truth *simpliciter* i.e. *Exclusion/Exhaustion*.

So, on the view under discussion, the dialetheist has accepted both *Exclusion* and *Exhaustion*, as well as their truth-in-a-model counterparts, *TIAM-Exclusion* and *TIAM-Exhaustion*. Given these principles, the invalidity of every inference principle there is follows straightforwardly from the existence of a trivial model of the logic *LP*. In the trivial model, every atomic sentence receives the values 1 and 0. By *TIAM-Exclusion*, then, every atomic sentence receives 1 and fails to receive 1. By *TIAM-Exclusion, TIAM-Exhaustion*, and the recursive clauses for the extensional connectives, every conditional-free complex sentence receives 1 and fails to receive 1. We extend this to sentences containing a conditional by considering, in Priest's tertiary semantics, the instance of the relation $Rw\bot w\bot w\bot$ (that is, the instance which relates the trivial model, or rather, in this case, the trivial world, to itself). This makes every conditional false at the trivial world and, by *TIAM-Exclusion*, untrue at that world. This makes every inference principle invalid since each has an instance, at the trivial world, at which the premises are true and the conclusion fails to be true at that world.

We can illustrate with a couple of examples. To take the simplest, consider

$$(\textit{Reflexivity})\, R \vDash R$$

Take some atomic sentence, φ, and substitute for R to obtain the following instance of *Reflexivity*

$$\varphi \vDash \varphi$$

Since φ both relates to 1 and fails to relate to 1 in the trivial model, in this instance of *Reflexivity*, the premise relates to 1 and the conclusion fails to. So *Reflexivity* is invalid. To take another example, $\wedge$-introduction:

$$(\wedge\text{-}Intro)\ A, B \vDash A \wedge B$$

Take the conjunction of sentences $\varphi \wedge \psi$, which receives 1 and fails to receive 1 in the trivial model (as do its conjuncts). Substituting, we obtain the following instance of $\wedge$-*Intro*:

$$\varphi, \psi \vDash \varphi \wedge \psi$$

Again, since φ, ψ and $\varphi \wedge \psi$ receive 1 and fail to receive 1 in the trivial model, the premises in this instance of $\wedge$-*Intro* receive 1, the conclusion fails to receive 1 and so the inference is invalid. This strategy invalidates every inference principle there is.

Though I have argued that dialetheists should accept both *TIAM-Exclusion* and *TIAM-Exhaustion* for truth-in-a-model so long as they accept their counterparts for truth *simpliciter*, one might wonder what effect it would have on invalidity to reject *TIAM-Exclusion* whilst accepting *Exclusion*. The answer, I think, is nothing: every inference is still invalid, though our strategy for demonstrating this will have to be slightly different, and more akin to the strategy to be employed in the next section against Priest's view. So, in addition to being an unattractive feature of the semantics for independent reasons, it does not help, on its own, with this revenge problem.

On this strategy, we construct special, model-theoretic liar sentences, such as L equivalent to:

$$(L)\, \neg(M \vDash L).$$

This sentence, informally, says of itself that it is not true in the model *M*. If we try to prove a contradiction from this, following similar reasoning to the strengthened liar, we find that the proof breaks down at the point at which the T-scheme would normally be invoked, since we have no equivalent to the T-scheme for truth-in-a-model. What we must do instead is to select *M* carefully, such that something like the T-scheme holds for it. The model we are interested in is the actual model where truth-in-a-model matches exactly truth simpliciter. It is a great virtue of metatheoretically inconsistent dialetheism allows for the existence of such a model, since otherwise, arguably, truth-in-a-model fails to characterise truth simpliciter.

The obvious condition to stipulate in picking out this model, *M*, is that, for any sentence α,

$$(*)\ M \Vdash \alpha \leftrightarrow T <\alpha>$$

One question arising here is what sort of conditional we should take (*) to contain. I suggest that it ought to contrapose (and so, if "$\rightarrow$" is not taken to contrapose, (*)

is really $M \Vdash \alpha \leftrightarrow T <\alpha> \wedge \neg M \Vdash \alpha \leftrightarrow \neg T <\alpha>$). The reason for this is that the conditional must contrapose if truth in M is to exactly match (both the positive and negative extensions of) truth simpliciter; if it does not, we are left with the possibility, for any α, of having $\neg M \Vdash \alpha$ but failing to have $\neg T <\alpha>$ and vice versa. In fact, if the conditional does not contrapose, (*) does not specify a unique model: for there are infinitely many models, each satisfying non-contraposable (*), but differing in their assignments of $\neg M \Vdash \alpha$ and $\neg T <\alpha>$, respectively.

In any case, I see nothing dialetheically objectionable about the conditional in (*) being contraposable. Certainly there is nothing dialetheically problematic about the mere existence of a model satisfying (*) so-construed. This being so, it seems reasonable that we be able to pick it out as above. It's worth pointing out that Priest's reasons for rejecting the contraposability of the T-scheme do not apply here: one can perfectly well accept the contraposability of (*) and reject the contraposability of the T-scheme.

So there is nothing wrong with us taking (*) in its strongest, contraposable, form. Even if this is rejected, however, this does not avoid the revenge problem. It does, however, collapse the difference, from the point-of-view of this revenge problem, between the view currently under discussion and the view of Priest's discussed in the next section. So, if a dialetheist accepts the transparency of truth *simpliciter*, but rejects *TIAM-Exclusion* as well as the contraposed form of (*), then they will fare exactly as Priest's view does there, though for the reasons given, their view will, additionally be unpleasantly ad hoc.

To return to the case where (*) is accepted as contraposable, the proof of contradiction from L proceeds as follows:

(1) $M \Vdash L$	(suppose for *reductio*)
(2) $T{<}L{>}$	(1, (*))
(3) L	(2, T-scheme)
(4) $\neg(M \Vdash L)$	(3, substituting equivalents)
(5) $\neg(M \Vdash L)$	(1, 4, *reductio*)
(6) L	(5, substituting equivalents)
(7) $T{<}L{>}$	(6, T-scheme)
(8) $M \Vdash L$	(7, (*))
(9) $M \Vdash L \wedge \neg(M \Vdash L)$	(5, 8, $\wedge$-Intro)

So L both receives 1 in M and fails to receive 1 in M. We can now demonstrate the invalidity of *Reflexivity* by substituting L to obtain the instance $L \vDash L$ in which the premises receive 1 and the conclusion fails to receive 1. In fact, since, on the view under consideration, the T-scheme is transparent, and so *Exclusion* holds, the effect of (*) understood contraposably, is to reinstate *TIAM-Exclusion* in the restricted case of M. So, for any complex sentence, A, built from L using the recursive clauses for the connectives, A both receives 1 in M and fails to. By (*), then, every sentence containing a conditional fails to receive 1 in M and so every inference principle whose conclusion contains a conditional is invalid.

So, if the dialetheist accepts that truth is transparent, and accepts *TIAM-Exclusion*, it follows from the existence of the trivial model (or world) that every inference principle is invalid (so long as one accepts Priest's account of the conditional). If, problematically, I have argued, they reject *TIAM-Exclusion* (despite accepting *Exclusion*), then so long as they accept (*) in its contraposable form, it follows from the universal model M and the existence of a liar sentence L, that every inference principle is invalid. If (*) is accepted only in its non-contraposable form, then the situation is the same as is described in the Sect. 3.2.

3.2 Invalidity with a Non-contraposable T-Scheme

If the conditional in the T-scheme does not contrapose, the situation is different to that described above. Though it is true that every sentence in the trivial model receives both 1 and 0, in the absence of *TIAM-Exclusion*, we cannot demonstrate that any such sentence additionally fails to receive 1. So the straightforward strategy of employing the trivial model will not work in this case. Instead, we employ the strategy which invokes the universal model M. The proof of a the contradiction $M \Vdash L \wedge \neg(M \Vdash L)$ proceeds exactly as in the previous section, since it does not depend on the contraposability of the T-scheme. So we may use L to demonstrate the invalidity of any inference principle whose conclusion is a sentence letter, for example: $\wedge$-elimination, $\vee$-elimination (reasoning by cases), *modus ponens*, reflexivity and double negation elimination.

The relevant instances being:

$$(\wedge\text{-}Elim)\ L \wedge L \vDash L$$
$$(\vee\text{-}Elim)\ L \vee L, (L \rightarrow L), (L \rightarrow L) \vDash L$$
$$(MP)\ L, L \rightarrow L \vDash L$$
$$(Reflexivity)\ L \vDash L$$
$$(\text{DNE})\ \neg\neg L \vDash L$$

In each case, since L both receives 1 and fails to receive 1, the premises in each instance receive 1 while the conclusion fails to do so, and each inference principle is invalid.

One might suppose that L could be used to invalidate inference principles with more complex conclusions. For example, one might think the following instance would do to eliminate $\wedge$-introduction:

$$(\wedge\text{-Intro})\ L, L \vDash L \wedge L$$

Unfortunately, this does not work. The reason is that the failure of *Exclusion* appears to leave the behaviour of untruth somewhat underdetermined. One would think that, if a sentence, A, is untrue that each of $A \wedge A$, $A \vee A$, $\neg\neg A$ would also be untrue (in fact, one would think it *obvious* that this is so). However, given the failure of *Exclusion*, there is no obvious way of showing that this is the case. I do not think this an advantage to Priest's view. In fact, it seems highly undesirable that untruth behaves in this way, though it does make getting revenge on Priest's view slightly more difficult.

What we must do instead, for inference principles with more complex conclusions, is to tailor our liar sentence to the particular inference principle. For example, to invalidate $\wedge$-introduction, we use the sentence A equivalent to:

$$(A)\, \neg(M \Vdash A \wedge A)$$

Intuitively, A says that the result of conjoining it with itself is untrue in M. We prove a contradiction from A as follows:

(1) $M \Vdash A \wedge A$	(suppose for *reductio*)
(2) $T{<}A \wedge A{>}$	(1, (*))
(3) $A \wedge A$	(2, T-scheme)
(4) A	(3, $\wedge$-elim)
(5) $\neg(M \Vdash A \wedge A)$	(3, substituting equivalents)
(6) $\neg(M \Vdash A \wedge A)$	(4, *reductio*)
(7) A	(5, substituting equivalents)
(8) $A \wedge A$	(6, $\wedge$-introduction)
(9) $T < A \wedge A >$	(7, T-scheme)
(10) $M \Vdash A \wedge A$	(8, (*))
(11) $(M \Vdash A \wedge A) \wedge \neg(M \Vdash A \wedge A)$	(5, 9, $\wedge$-introduction)]

So, in M, $A \wedge A$ both receives 1 and fails to receive 1. We can then substitute A to obtain the instance of $\wedge$-introduction:

$$A, A \Vdash A \wedge A$$

So the premises will receive 1 in M and the conclusion will fail to receive 1 and so $\wedge$-introduction is invalid. The case establishing the invalidity of $\vee$-introduction, $A \vDash A \vee B$, employing the sentence B, equivalent to $\neg(M \Vdash B \vee B)$ mirrors that of $\wedge$-introduction almost exactly. For double-negation-introduction involves the sentence N, equivalent to $\neg(M \Vdash \neg\neg N)$, from which a contradiction is proved thus:

(1) $M \Vdash \neg\neg N$	(suppose for *reductio*)
(2) $T{<}\neg\neg N{>}$	(1, (∗))
(3) $\neg\neg N$	(2, T-scheme)
(4) N	(3, double-negation-elimination)
(5) $\neg(M \Vdash \neg\neg N)$	(4, substituting equivalents)
(6) $\neg(M \Vdash \neg\neg N)$	(5, *reductio*)
(7) N	(6, substituting equivalents)
(8) $\neg\neg N$	(7, double-negation-introduction)
(9) $T{<}\neg\neg N{>}$	(8, T-scheme)
(10) $M \Vdash \neg\neg N$	(9, (∗))
(11) $(M \Vdash \neg\neg N) \wedge \neg(M \Vdash \neg\neg N)$	(6, 10, $\wedge$-introduction)

Again, we now substitute N into double-negation-introduction to obtain the instance $N \vDash \neg\neg N$ in which the premise relates to 1 and the conclusion fails to and, thus, double-negation introduction is invalid. N will also serve to invalidate Priest's *reductio* rule, $\alpha \to \neg\alpha \vDash \neg\alpha$ by substituting $\neg N$ for α.

One question is whether this piecemeal strategy can be developed into a general result. I think, by appeal to a result by Heck [6], that it can. In his paper, Heck analyses the extent to which genuine self-reference is possible in certain expanded languages for arithmetic. The result in which I am interested is his proof of the *Structural Diagonal Lemma*. In the language *PAs*, which is $\{0, S, +, \times\}$ extended by the addition of a truth-predicate T.

He states the Structural Diagonal Lemma as follows [6, p. 9]:

Structural Diagonal Lemma: Let P be a truth-functional schema in (distinct) sentence letters $p1,\ldots, pn$. Let $A(x)$ be the substitution instance of P in which each of pi has been replaced by a corresponding formula $Ai(x)$ containing just x free. Then there is a formula G such that:

(1) G is the substitution instance of P in which each pi has been replaced by a corresponding formula Gi;
(2) $PAs \vdash Gi \equiv Ai({<}G{>})$
(3) $PAs \vdash G \equiv A({<}G{>})$

I refer the reader to Heck's paper [6, pp. 10–12] for the proof, of which Heck says "nothing in the proof actually requires the assumption that $A(x)$ contain *only* x free, so [the proof] easily adapts to a proof of the form that allows parameters" [6, p. 10, n. 11]. There is no machinery here which is dialetheically unacceptable and so no reason, that I can see, for a dialetheist to reject the Structural Diagonal Lemma.

This establishes that, for any inference principle whose conclusion contains only extensional connectives, there is a sentence G, equivalent to $\neg(M \Vdash G)$, such that the G has the same logical form as the conclusion of the principle. We can then reason in the standard way to the conclusion that $M \Vdash G$ and $\neg(M \Vdash G)$, which invalidates the principle. This delivers the result that any inference principle whose conclusion does not contain a conditional is invalid.

This leaves the inference principles which contain in their conclusion a conditional. Since Priest's conditional is not truth-functional, it is not covered by Heck's result. One way one might expect to demonstrate the invalidity of such principles, following the piecemeal strategy above, would be as follows. Consider some inference principle whose conclusion is $P \rightarrow Q$ and take the sentence β, equivalent to $\neg(M \Vdash \beta \rightarrow \beta)$. Unfortunately, this does not work, since β is not a liar sentence: $\beta \rightarrow \beta$ is logically true (only), and so β, which says of $\beta \rightarrow \beta$ that it is untrue, is false (only).

What we might do instead is use the sentence κ, equivalent to $\neg(M \Vdash t \rightarrow \kappa)$, where "$t$" is a truth constant; most simply, we can simply take "t" to be the conjunction of all truths. Two useful features of t are that $t \rightarrow \alpha \vDash \alpha$, and $\alpha \vDash t \rightarrow \alpha$ both hold, which I will call "t-elim" and "t-intro", respectively. Given these features, contradiction is proved from κ as follows:

(1) $M \Vdash t \rightarrow \kappa$	(suppose for reductio)
(2) $T\langle t \rightarrow \kappa\rangle$	(1, (*))
(3) $t \rightarrow \kappa$	(2, T-scheme)
(4) κ	(3, t-elim)
(5) $\neg(M \Vdash t \rightarrow \kappa)$	(5, substituting equivalents)
(6) $\neg(M \Vdash t \rightarrow \kappa)$	(1, 5, reductio)
(7) κ	(6, substituting equivalents)
(8) $t \rightarrow \kappa$	(7, t-intro)
(9) $T\langle t \rightarrow \kappa\rangle$	(8, T-scheme)
(10) $M \Vdash t \rightarrow \kappa$	(9, (*))
(11) $(M \Vdash t \rightarrow \kappa) \wedge \neg(M \Vdash t \rightarrow \kappa)$	(6, 10, $\wedge$-introduction)

As before, we now have a counter example to any inference principle whose conclusion is an arrow flanked by two sentence letters. This strategy can be applied in a piecemeal way to a number of inference principles whose conclusions contain conditionals. For a final example, I show how to invalidate the principle $\alpha \rightarrow \beta \vDash (\gamma \rightarrow \alpha) \rightarrow (\gamma \rightarrow \beta)$, which Priest has shown holds for his conditional [10, pp. 92–93]. This argument takes advantage of the fact that, since $\vDash t$ and $\vDash t \rightarrow t$, we have $t \vDash t \rightarrow t$ and $t \rightarrow t \vDash t$, so t and $t \rightarrow t$ are equivalent. We employ the sentence γ, equivalent to $\neg(M \Vdash ((t \rightarrow t) \rightarrow (t \rightarrow \gamma))$ and demonstrate that contradiction follows thus:

(1) $M \Vdash ((t \to t) \to (t \to \gamma))$	(suppose for *reductio*)
(2) $T{<}(t \to t) \to (t \to \gamma){>}$	(1, (∗))
(3) $T{<}(t \to t) \to (t \to \gamma){>}$	(2, T-scheme)
(4) $t \to (t \to \gamma)$	(3, substituting equivalents)
(5) $t \to \gamma$	(4, t-elim)
(6) γ	(5, t-elim)
(7) $\neg(M \Vdash ((t \to t) \to (t \to \gamma))$	(6, substituting equivalents)
(8) $\neg(M \Vdash ((t \to t) \to (t \to \gamma))$	(1, 7, *reductio*)
(9) γ	(8, substituting equivalents)
(10) $t \to \gamma$	(9, t-intro)
(11) $t \to (t \to \gamma)$	(10, t-intro)
(12) $(t \to t) \to (t \to \gamma)$	(11, substituting equivalents)
(13) $T{<}(t \to t) \to (t \to \gamma){>}$	(12, T-scheme)
(14) $M \Vdash ((t \to t) \to (t \to \gamma))$	(13, (∗))
(15) $M \Vdash ((t \to t) \to (t \to \gamma)) \wedge \neg(M \Vdash ((t \to t) \to (t \to \gamma)))$	(8, 14, $\wedge$-introduction)

As before, we simply substitute γ into the principle in the obvious way, and so have a counterexample to its validity. It is unclear whether something like this strategy can be extended to a general result for every inference principle with a conditional conclusion; at any rate, I do not have a general result at the moment.

So, on Priest's strategy, both *Exclusion* and *TIAM-Exclusion* fail. This rules-out the strategy discussed in the previous section which employed the trivial model. Instead, we used the strategy which employs the universal model M and constructs "bespoke" liar sentences for each inference principle we wish to invalidate. Heck's result allows us to show that this delivers a strategy which invalidates every inference principle whose conclusion is extensional. Finally, any principle with a conditional conclusion, able to take atomic sentences as antecedent and consequent, is also invalid. This can be extended in a piecemeal fashion to some principles with more complex conditional conclusions, but, as yet, no more general result has been demonstrated.

Whether, in the end, Priest's view, by rejecting *Exclusion* and *TIAM-Exclusion*, fares any better here than versions of dialetheism accepting those principles depends on whether the stronger result, that every inference principle whose conclusion contains a conditional is invalid, can be established. If this result can't be established, it seems to me that Priest's view is not *much* better off. It is still the case that a *great* many inference principles are, on his view, invalid, including all the introduction and elimination rules for the logical constants, and all of the principles we are likely to find in a logic textbook, or use in everyday reasoning.

4 The Invalidity Revenge Problem

The fact that dialetheists are committed to the inconsistency of validity (at least for versions of dialetheism with a paraconsistent metatheory) is interesting and important, all the more so because of the extent of its inconsistency. At worst, every inference principle is invalid, and at best, almost every inference principle one is likely to encounter is invalid. The revenge problem this generates is that is renders dialetheists unable to express invalidity. One of the crucial features of dialetheism is that it is, we think, non-trivial: dialetheists are able to accept inconsistency, but also contain inconsistency. So it is a crucial semantic fact about dialetheism that principles such as explosion and disjunctive syllogism, which would otherwise trivialise the theory, are invalid. If dialetheists cannot express this fact, then their view suffers from expressive limitations of the very kind they argue that non-dialetheists fact: that is, they have a revenge problem.

Consider, first, the case in which every inference principle is invalid. This is the position of metatheoretically inconsistent dialetheism in which *TIAM-Exclusion* holds, and perhaps also Priest's view, pending some more definitive result concerning inference principles with conditional conclusions. In this case, "invalid inference principle" and "inference principle" are logically equivalent notions. The "invalid" comes for free and adds nothing. So, for all the distinction the notion draws, instead of 'Disjunctive syllogism is invalid', we may as well have said "Disjunctive syllogism is an inference principle".

Just as, with the introduction of dialetheism about truth, it became necessary to introduce notions like 'just true' and 'just false' to characterise the sentences which are not dialetheia, it seems reasonable to want similar notions once we recognise that validity is dialetheic. So we may wish to introduce notions such as 'just valid' and 'just invalid'. In fact, the former notion is redundant, in this case, since no inference principle is just valid. Still, though every principle is invalid, what is special and bad about principles like disjunctive syllogism and explosion is that they are *just* invalid, whereas the legitimate principles, like reflexivity and *modus ponens*, are both valid and invalid.

The obvious definition of 'just invalid' of course, is 'invalid and not valid' Unfortunately, since 'invalid' and 'not valid' are equivalent notions, 'invalid' and 'just invalid' are also equivalent. So, on this obvious definition, all inference principles are both invalid and just invalid. So, on this version of dialetheism, both invalidity and just-invalidity are vacuous notions which fail to pick out the feature of disjunctive syllogism that we are interested in. So, if dialetheists accept *TIAM-Exclusion*, they have a revenge problem. Specifically, they cannot express the crucial semantic facts about their theory which require, for their expressibility, non-vacuous notions of invalidity and just-invalidity.

Assuming that we cannot establish the stronger result mentioned above, Priest's view (or, more generally, views in which *TIAM-Exclusion* fails) is different, but only slightly. The problematic overlap, however, is still enormous. The notion being employed in sentences such as 'Disjunctive syllogism is invalid' still applies to every inference principle whose conclusion lacks a conditional, and a number of others, including reflexivity, *modus ponens*, $\wedge$-introduction, and almost every other inference principle anyone ever uses. It is still the case, on this version of dialetheism, that 'just invalid' is equivalent to 'invalid' and so each of these principles is also *just* invalid. So, even if dialetheists reject *TIAM-Exclusion*, and assuming the stronger result cannot be established for these theories, they still have a revenge problem. Both 'invalid' and 'just invalid' are equivalent, so these theories cannot express the latter as distinct from the former. Moreover, the overlap between these notions and validity is so significant that they are deprived of their usefulness and so cannot be used to express the crucial semantic facts about dialetheism which require either consistent, or at least less inconsistent, notions of invalidity.

4.1 Invalidity Revenge and Just False Revenge

It is worth briefly discussing the differences between the 'just false' problem already extant in the literature and the invalidity revenge problem I have just introduced, in case it be thought that nothing new has been added.

The first difference is that the 'just false' problem concerns only the notion 'just false' leaving 'false' alone, and not threatening any further, unpalatable properties of 'false' that the dialetheist has not already happily accepted. The invalidity revenge problem, on the other hand, concerns both invalidity and just-invalidity, demonstrating that the notions are both equivalent and vacuous (or near enough).

A further difference is that the problematic overlap in the case of the invalidity revenge problem is far greater than in the case of the just false problem. In the case of the just false problem, for dialetheists of Priest's sort at least, the cases where 'just false' and 'true' overlap seem to be restricted to strange paradoxical sentences like the strengthened liar. Priest's defence of his view against the charge that it does not allow for the expressibility of 'just false' appeals to this very fact. But in the case of the invalidity problem, all the valid inference principles one is likely to encounter are also both invalid and just invalid.

4.2 Avenues of Reponse

I now turn to how a dialetheist might respond to the invalidity problem I have characterised. The first objection I will consider concerns an alternative notion of validity proposed by Priest (which he confirms in correspondence was given to head-off possible objections of the sort given in the present paper). The definition in question appears definition *In Contradiction*, 5.2:

$$(Val^{*})\quad A \vDash B \text{ iff for every interpretation } I, T <(I \Vdash A \rightarrow I \Vdash B)>$$

Which we can contrast with the more standard definition I used in the above:

$$(Val) \text{ for every interpretation } I,\ (I \Vdash A) \rightarrow (I \Vdash B)$$

(*Val**) delivers the same validities as (*Val*), though not the same invalidities because, for Priest, the conditional in the T-scheme does not contrapose. In effect, it becomes harder to show that an inference principle is invalid, on (*Val**). This is because, in the relevant case, one can't just find a counter-example in the usual way—we must take the further, tricky step of showing that $I \Vdash A \rightarrow I \Vdash B$ is not merely false but untrue.

However, I don't think this move helps Priest achieve his goal, since it's now difficult to see how he could show any inference principle at all to be invalid. In fact, his own proofs of the invalidity of disjunctive syllogism don't go through, on this definition, since they don't take the further step of showing the relevant untruth (for example, see Priest's proof of Fact 3, Sect. 5.5, [10]). Without further work to show that the desirable invalidities can be maintained without the undesirable ones, this suggestion seems like a non-starter.

One alternative response would be to accept that invalidity is vacuous and to be given up, but to define 'just invalid' in some new way to express what is wrong with inferences like disjunctive syllogism. This will, I think, be parasitic on some more general, consistent redefinition of 'just false' (or perhaps, 'just untrue'). The reason for this is that every counterexample to a valid inference principle described in the previous section consisted in finding some sentence which, in some model, was both true and untrue and inserting it into the inference principle in question such that the premises were true, but the conclusion untrue, in that model. In each case, because the inferences were valid too, the conclusion, in addition to being untrue, was also true (again, in the model). So the only time we have counterexamples to valid inference principles is when we construct cases in which the premises are true while the conclusion is both true and untrue. This is different from inference principles which are *just* invalid, because in those cases we have counter-instances in which the premises are true and the conclusion false only (or untrue only).

So what the dialetheist wants here is some notion, some form of negation, which expresses that a sentence fails to be true, where this must be understood consistently (i.e. it cannot be such that something is untrue in this sense, but also true). We call this new notion *NEG* and define an inference to be invalid, $A \nvDash B$ just if there is some model I such that $I \Vdash A \wedge NEG\text{-}(I \Vdash B)$. We can also define a sentence, α, to be just false if and only if $F <\alpha> \wedge \neg T <\alpha>$. But, of course, we can also construct a sentence, β, equivalent to: (β) $NEG\text{-}T <\beta>$.

From which we require very little to deduce $T <\beta> \wedge NEG\text{-}T <\beta>$, defeating the purpose of introducing *NEG* in the first place. An argument for this from β is as follows:

(1) $T{<}\beta{>}$	(suppose for *reductio*)
(2) β	(1, T-scheme)
(3) $NEG\text{-}T{<}\beta{>}$	(2, substituting equivalents)
(4) $NEG\text{-}T{<}\beta{>}$	(3, *NEG-reductio*)
(5) β	(4, substituting equivalents)
(6) $T{<}\beta{>}$	(5, T-scheme)
(7) $T{<}\beta{>} \wedge NEG\text{-}T{<}\beta{>}$	(4, 6, $\wedge$-introduction)

All that is required for this argument, beyond principles already accepted by dialetheists is that a *reductio* rule is valid for *NEG*. This need only be the weak *reductio* rule Priest already accepts for negation, $\alpha \rightarrow NEG\text{-}\alpha \vDash NEG\text{-}\alpha$. It is difficult to see how the failure of such a rule could be motivated (at least, in the present context), but there are other arguments for $T <\beta> \wedge \neg T <\beta>$ which don't involve it anyway. For example:

(1) $T <\beta> \vee NEG\text{-}T <\beta>$ (*NEG-LEM*)

Case 1:

(1a) $T <\beta>$
(1b) β (1a, T-scheme)
(1c) $NEG\text{-}T <\beta>$ (1b, substituting equivalents)
(1d) $T <\beta> \wedge NEG\text{-}T <\beta>$ (1a, 1c, $\wedge$-introduction)

Case 2:

(2a) $NEG\text{–}T<\beta>$
(2b) β (2a, substituting equivalents)
(2c) $T <\beta>$ (2b, T-scheme)
(2d) $T <\beta> \wedge NEG\text{-}T <\beta>$ (2a, 2c, $\wedge$-introduction)

(2) $T <\beta> \wedge NEG\text{-}T <\beta>$ (1, 1d, 2d, reasoning by cases)

This argument, too, might be blocked, perhaps by denying that *NEG* obeys the law of excluded middle, or that reasoning by cases is valid, but again, there doesn't seem much motivation for this, and the more principles of this kind which fail for *NEG*, the less it looks like a form of negation at all. Perhaps there is some way to characterise a consistent form of negation for dialetheists whose meta-theory is paraconsistent, but it is not at all obvious how this would be achieved, and it has not, to my knowledge, been seriously attempted thus far.

One attempt which has been made for dialetheists with a consistent meta-theory is worth discussing for illustrative purposes. Franz Berto, in, for example, his [4] has developed a primitive notion of metaphysical exclusion, characterised as a predicative functor, holding between properties which metaphysically exclude one another. Though the notion is primitive, and so has no explicit definition, there are still formal principles which hold of it. One such is the principle in the meta-theory that a property excludes another if and only if their extensions are (necessarily) disjoint. Since Berto's meta-theory is consistent, this principle does not make exclusion inconsistent in the object theory. It is difficult to see how anything like this principle could hold in the present, metatheoretically inconsistent context (where extensions are allowed to be both disjoint and not) without thereby introducing inconsistency into the notion of exclusion. It is not required, of course, that a *definition* be given of a notion which is supposed to be primitive, but we should still expect some formal constraints to hold of exclusion, and it is unclear what these could be such that the notion doesn't turn out inconsistent.

Still, supposing some constraints could be given to make the notion sufficiently wieldy, we might then hope to characterise an invalid inference, $A \nvDash B$ as being such that there is some model I such that A is true in I and A's truth in I excludes B's truth in I.

This might work for some invalid inference principles, for example $A \vDash B$; presumably, so long as the truth of some sentences excludes the truth of some others, there is some instance of this such that the truth of the premise excludes the truth of the conclusion. The notion is less helpful in other, more interesting cases, however. The reason is that the obvious counterexamples to the just-invalid inference principles in which we are interested, for example, disjunctive syllogism or explosion, involve necessarily dialetheic sentences arising from semantic paradoxes like the liar. For example, a counter-instance to disjunctive syllogism might be $\neg$L, L $\vee$ A $\vDash$ A, where L is some liar sentence and A is some sole-falsehood, such as 'Graham Priest is 10-feet-tall'. But since the liar sentence L is, presumably, necessarily true and false, it is difficult to see how it could metaphysically exclude anything and, in particular, how it could be thought of as excluding my being 10-feet-tall. All the paradigmatic dialetheia are of this necessary sort, so it does not appear that metaphysical exclusion could help with this problem, even if a satisfactory account of the notion could be made in the metatheoretically paraconsistent context.

Another strategy dialetheists might wish to pursue in response to the invalidity revenge problem appeals to triviality. One problematic feature of inference principles like disjunctive syllogism and explosion, and the main reason we want to be able to express their invalidity, is that, were they valid, the theory would trivialise. So, it seems reasonable to wish to invoke this fact in our response to the problem. The response would be to give up on notions like invalidity and just-invalidity, but to express that a certain inference principle is bad by pointing to the fact that, were it added to the theory, the theory would entail everything.

One might think that, rather than answering the objection, this simply underlines it. It is of course correct that these principles would be trivial were they valid, but this is the very reason we want to be able to say they are *invalid*. If I notice, when first introduced to dialetheism, that there is a straightforward argument from dialetheism to triviality via disjunctive syllogism, and inquire as to the principle's validity, it will hardly assuage my concerns if the dialetheist simply confirms that the principle does indeed entail triviality.

In any case, this response will not work, since not all invalid inferences would trivialise if added to dialetheism. To take Priest's theory as an example, not all inferences he rejects are rejected because they would, if he accepted them, entail everything. For example, for all α, $T <\neg\alpha> \vDash \neg T <\alpha>$, which is a rule form of what Priest calls 'Exclusion', is rejected by appeal to his teleological account of truth. It is, for Priest, invalid, but it would not, so far as I can see, trivialise his theory if he accepted it. For another example, related to the previous , Priest rejects the contraposability of the conditional: $P \to Q \vDash \neg Q \to \neg P$ is, for him, invalid. But the reasons for this, again, are not the threat of triviality, but philosophical concerns about what conditional ought to appear in the T-scheme.

For a final example, one obviously bad inference principle is $A \vDash$ Priest is a fried egg. This would, if valid, allow us to conclude that Priest is a fried egg from any sentence, but it would not, so far as I can tell, entail *everything*. It might be objected that since the conclusion of the principle is an interpreted sentence that it is not an inference principle, but this doesn't matter. It is still a principle by which one might reason and doing so would be bad. This fact ought to be able to be expressed dialetheically but cannot be if the only way we can express what's bad about invalid inference principles is to point to their triviality.

References

1. Batens, D. (1990). Against global paraconsistency. *Studies in Soviet Thought, 39*(3–4), 209–229.
2. Beall, J. C. (2009). *Spandrels of truth*. Oxford University Press.
3. Beall, J. C. (2013). Shrieking against gluts: the solution to the 'just true' problem. *Analysis, 73*(3), 438–445.
4. Berto, F. (2014). Absolute contradiction, dialetheism, and revenge. *Review of Symbolic Logic, 7*(2), 193–207.
5. Field, H. (2008). *Saving truth from paradox*. Oxford University Press.
6. Heck, R. (2007). Self-reference and the languages of arithmetic. *Philosophia Mathematica, 15*(1), 1–29.
7. Littman, G., & Simmons, K. (2004). A critique of dialetheism. In G. Priest, J. C. Beall, & B. Armour-Garb (Eds.), *The law of non-contradiction*. Oxford University Press.
8. Parsons, T. (1990). True contradictions. *Canadian Journal of Philosophy, 20*(3), 335–353.
9. Priest, G. (2006). *In contradiction: A study of the transconsistent*. Oxford University Press.
10. Priest, G. (2010). Hopes fade for saving truth. *Philosophy, 85*(1), 109–140.
11. Priest, G., & Berto, F. (2017, Spring). Dialetheism. In E. N. Zalta (Ed.), *The stanford encyclopedia of philosophy*. https://plato.stanford.edu/archives/spr2017/entries/dialetheism/.
12. Shapiro, S. (2004). Simple truth, contradiction, and consistency. In *The law of non- contradiction*. Oxford University Press.
13. Yablo, S. (1985). Truth and reflection. *Journal of Philosophical Logic., 14*, 279.
14. Young, G. (2015). Shrieking, just false and exclusion. *Thought: A Journal of Philosophy, 4*(4), 269–276 (2015).

It Was so Revolting I Couldn't Take My Eyes Off It

Graham Priest

Abstract Dialetheism is the view that some contradictions are true. One might naturally ask for examples. This paper offers a new one. There is a well known psychological phenomenon (noted, for example, by Plato in the *Republic*) in which something is so repulsive that one is compelled to look at it. One is attracted and repelled. *Prima facie*, that is a contradiction, and, given the context, a true one. I argue that is exactly what it is. A brief discussion of dialetheism frames the topic.

1 Introduction: Dialetheism

A dialetheia[1] is a pair of statements of the form A and $\neg A$ which are both true—or, assuming that conjunction behaves as one would expect, a true statement of the form $A \wedge \neg A$. Dialetheism is the view that some things are dialetheias. Dialetheism is a view that has been endorsed by a few thinkers in the history of philosophy—perhaps, most notably, Hegel.[2] Since Aristotle's somewhat confused and ultimately ineffectual defence of the Principle of Non-contradiction, though, it has been a highly unorthodox view in Western philosophy.[3] Dialetheism has resurfaced again in the last 40 years, however; and this time defended with all the resources of modern formal logic.[4]

---

[1] Talks based on versions of this paper were given at Clemson University, the University of Adelaide, the Central European University (Budapest), as the 2017 Kretzmann lecture at Cornell University, and at UNILOG2018 in Vichy. My thanks go to many members of the audiences in those places for their very helpful comments. Many thanks, too, go to Don Baxter, for setting me straight on some aspects of his view.

[2] See Sect. 3 of Priest [12].

[3] For an analysis of Aristotle's defence, see Chap. 1 of Priest [11].

[4] Priest [9] might be taken as crystalising the beginnings of the new movement.

G. Priest (✉)
Departments of Philosophy, CUNY Graduate Center and the University of Melbourne, Melbourne, Australia
e-mail: priest.graham@gmail.com

A. Rieger and G. Young (eds.), *Dialetheism and its Applications*, Trends in Logic 52,
https://doi.org/10.1007/978-3-030-30221-4_3

If someone claims that dialetheism is true, one might fairly ask for examples of dialetheias. All examples are philosophically contentious. However, perhaps everyone's favourites are the paradoxes of self-reference—with a myopic fixation on the liar paradox. This is the sentence:

- this very sentence is false

If it is true, it is false; if it is false, it is true. So it seems to be both. One might say that this conclusion can be avoided if it is neither true nor false; and so it can; but then we are faced with the "extended" paradox:

- this very sentence is either false or (neither true nor false)

If it is true it is either false or neither; if it is false, it is either false or neither, and so true; and if it is neither, it is either false or neither, and so true again. We are back with contradiction. There is much discussion of these matters in the contemporary literature; but this is not the place to go into it.[5] Here, my point is simply to give a putative example of a dialetheia.

Discussions concerning the paradoxes of self-reference have so much held center-stage in contemporary discussions of dialetheism, that some seem to think of it simply as a view about the paradoxes. It is not: it is a quite general metaphysical/semantic view about truth and negation. The paradoxes are but a putative example. And for my part, I have never thought of the paradoxes as the most transparent example. What I take to be so, concerns the philosophy of law.

There are many things that a duly constituted legislature cannot make true. For example, it cannot make true the statement that the Moon is 40 km from the Earth. Yet things of some kind they certainly can make true, notably statements to the effect that certain people have or do not have legal rights (in the jurisdiction in question). Thus, suppose that a duly constituted legislature passes a piece of legislation containing the following two clauses[6]:

- Every property holder may vote.
- No woman may vote.

One may imagine that at the time the legislation was passed, it was just not a serious possibility (legal or factual) that a woman could hold property. The law, then, generated no contradiction; and if no woman ever comes to own property, it remains so. However, the times changes, and we may suppose that women do come own property. Perhaps this was first a *de facto* matter, then a *de jure* one. Eventually, a woman who owns property turns up at the polling booth wanting to vote. Let us call her Emily. Then, patently, Emily, being a property holder, has the right to vote; and, being a woman, she does not. The law plus contingent circumstances have generated a dialetheia.

One might reply that once this situation arose, the law would be changed, either by new legislation being passed, or more likely, by a judge making an appropriate

[5]One point of entry into the literature is Beall [2].

[6]The example is a toy one; real life example are more complex and arguable. The example makes the point, however.

ruling. This is true: the law is meant to be functional, and this contradiction is clearly dysfunctional. But this is beside the point. The reason that the law needed to be changed was exactly that it was contradictory. So after the change in law there is no longer a dialetheia; but before there certainly was.[7]

Let us turn to a third putative example of a dialetheia. Many philosophers have held that there are things which are beyond the limits of our language/concepts. Indeed, they argue that there are such things. Of course, to argue that there are such things one has to talk about them. So they are not beyond those limits. Many philosophers have found themselves in this situation.[8]

- In the *Critic of Pure Reason* Kant argues, thereby applying categories, that we must suppose that there are some noumenal entities (such as God). The categories cannot apply to such entities, however, since the criteria for their application are spatio/temporal, and such entities are outside space and time.
- Wittgenstein in the *Tractatus* gives us a theory of the relationship between language and the world. Propositions are arrangements of names; states of affairs are arrangements of objects; and a proposition describes a state of affairs if the names in it are arranged with the same form as the objects in the state of affairs which they name. But form cannot be an object: it is the *way* that names are arranged in a proposition or objects are arranged in a state of affairs. So one cannot talk about form. The *Tractatus* is, of course, replete with such talk.
- At the very beginning of *Sein und Zeit* Heidegger famously asked the question of being (*Seinsfrage*): what is being, what is it that makes beings be? And he tells us immediately that there is one central mistake to avoid. Being is not itself *a* being. It follows that one cannot answer the *Seinsfrage*. To do so one would have to say 'Being is such and such', and hence to treat it as a being. Indeed, one cannot even ask it. For even to say 'What is being?' is to treat it as a being (cf., What is the United Nations?).

The three philosophers just mentioned were well aware of the putative contradiction involved in speaking of the ineffable, and tried to wriggle out of it.[9] Their attempts were singularly without success, though this is not the place to go into the matter.[10]

Of course, one may not buy into the projects of Kant, Wittgenstein, or Heidegger. One might even take the contradiction at issue here as showing that there are no things beyond our language/concepts. But if one does buy into any of these projects, or ones like them,[11] then one would appear to be stuck with the fact that there is something that is both effable and ineffable. This is our third example of a dialetheia.

Our three examples of dialetheias do not exhaust the possibilities. There are others, concerning change, motion, time, and further matters.[12] But this will suffice here.

---

[7] For further discussion, see Priest [9], Chap. 13.

[8] On the following, see Priest [10], Chaps. 5, 12, and 15.

[9] With the exception of the later Heidegger. See Casati [3].

[10] See the discussion in Priest [15].

[11] As I do. See Priest [13], 13.11.

[12] On which, see Priest [9, 10].

For the present section is just a—somewhat lengthy—prologomenon to what is to follow. In this, I wish to offer a new candidate for a dialetheia. Perhaps it is not a candidate that will, on its own, convince someone of the truth of dialetheism. But once the possibility of dialetheism is taken seriously, it certainly looks like a good candidate.

2 The Phenomenon in Question

By a *psychological dialetheia* I mean a dialetheia which describes some agent's mental state. Normally, our mental states are, it would seem, quite consistent. If I am thinking of the Taj Mahal, I am not also not thinking of it. If I see a polar bear, I do not also not see it. But one should not be taken in by an inadequate diet of examples—as Wittgenstein put it.[13] Unusual things may happen in unusual situations; and arguably some odd sorts of situations may give rise to psychological dialetheias. Thus, take the phenomenon of self-deception. In this, one might argue, an agent both believes and does not believe something. (Note that this is different from believing something and believing its negation. That mental state is quite consistent.) That is not the phenomenon I wish to discuss here, though.[14] I wish to discuss a quite different sort of unusual situation.

This was well illustrated by Plato in the *Republic* 439a[15]:

> Leontius, the son of Aglaion, was going up from Piraeus along the outside of the North Wall when he saw some corpses lying at the executioner's feet. He had an appetite to look at them but at the same time he was disgusted and turned away. For a time he struggled with himself and covered his face, but, finally, overpowered by the appetite, he pushed his eyes wide open and rushed towards the corpses saying, "Look for yourselves, you evil wretches, take your fill of this beautiful sight!"

Note that this is not simply a case of wanting to look and wanting not to look,[16] which is quite consistent. Being repelled by something is not simply wanting not to look. For example, I might want not to look at something (for example, pornography) because my mother told me not to, and I wish to respect here wishes. I may not be repelled by it at all, however. As far as I can see, wanting not to do something does not have any very significant phenomenology. By contrast, mental revulsion has a very powerful phenomenology.

This kind of situation on view here is one, I think, that many, perhaps most, of us have occasionally experienced: a situation where something is so revolting that one cannot take one's eyes off it. For example, one may sometimes go past a gruesome traffic accident, where mangled people and bodily parts are strewn round a crash-site. It is revolting, but one cannot help but look. Another example: many of the portraits

[13] *Philosophical Investigations*, section 593.

[14] It is discussed in Priest [14], Chap. 16.

[15] The translation is taken from Grube [7].

[16] Though it may be that as well.

of the British artist Francis Bacon are of faces horribly distorted. They are hideously revolting, but riveting. In situations like this, one is phenomenologically attracted to and repelled (or repulsed) by the same thing.[17]

This is the phenomenon that I want to examine.

3 Attraction and Repulsion

Let us write Ax for 'The agent in question is attracted by x', and Rx for 'The agent in question is repelled by x'. Then in such situations, for some x, we have $Ax \wedge Rx$. However, *prima facie* at least, *being attracted by* and *being repelled by* are contraries. That is: $\neg(Ax \wedge Rx)$, and so $\neg Ax \vee \neg Rx$. So we have $(Ax \wedge Rx) \wedge (\neg Ax \vee \neg Rx)$, from which it follows that $(Ax \wedge \neg Ax) \vee (Rx \wedge \neg Rx)$. Whichever disjunct is true, we have a psychological dialetheia—and given the symmetry of the situation, presumably two of them.

A natural thought at this point is that A and R are not really contraries. After all, in the kind of situation in question, *something* is both. But not so fast.

Take, first, not mental attraction and repulsion, but physical attraction and repulsion. Moving towards something and moving away from it (in Euclidean space) are certainly contraries. So, for an object to be *forced* to move towards something and *forced* to move away from it by the nett force in action are also contraries. Now come back to psychological attraction and repulsion. These are attraction and repulsion, not for a body, but for a mind; but the point is the same. The mind is being forced towards something and away from it simultaneously. These are contraries for the same reason. The point is nicely made by Don Baxter[18]:

> To have an internal conflict like Medea's is like trying to move in opposite directions.

Of course, it is not a contradiction to say in the physical case that one *component* (say electro-magnetic) of the nett force is attracting, and another (say, gravitational) is repelling. The analogue of the distinction between a nett force and its components in the mental case is not entirely clear. However, there is no conflict in a body that moves under the influence of component forces: it simply moves according to the their resultant. Yet there is a distinct tension in the kind of mental state we are talking about; indeed, the agent is *torn* both ways. So whatever the correct analogy is, it would seem to be the analogue of the nett physical force that is at issue.

---

[17]The phenomenon may well be present in non-human animals as well. See Eibl-Eibsfeldt [5], Chap. 10.

[18]Baxter [1, p. 3]. In Greek mythology Medea was a character who was torn as to whether she should kill her children.

4 Distinguo

When one meets an apparent contradiction, a natural way to try to avoid it is to draw an appropriate distinction. To say that it is 3 and 8 pm, is a contradiction. The contradiction disappears once we note that it is 3 pm in New York, and 8 pm in London. So if it appears that *s* is attracted and repelled by *o*, one might attempt to draw a contradiction-defusing distinction in the same way. There are two obvious loci for such a distinction. One concerns the subject, *s*; the other concerns the object, *o*.

Take the object first: one might say that it is not *o* itself that attracts/repels. It is a certain aspect of *o* which attracts/repels. Things like this can certainly sometimes be the case: one might be attracted by someone's beauty, but repelled by their political views. However, this move does not work here. In the case we are dealing with, it is one and the same object that is attractive and repulsive: the car-crash or the painting. Indeed, it is precisely the repulsiveness of an object which is attractive. Thus, it is the horror of the car crash which is both revolting and fixating. It is one and the same thing that attracts and repels.

So let us move to distinctions on the subject side. One might say that it is *part of s* which is being attracted, and *part of s* which is being repelled. After all, it is no contradiction for part of an object to be physically forced towards something, and part of it to be forced away from it. Being forced towards and forced away are contraries only when it is the whole which is so forced.

At this point let us return to Plato. The passage I quoted is taken from Plato's argument that the soul has different parts—roughly, the rational, the moral, and the appetitive. He has just enunciated a version of the Principle of Non-contradiction. Given this, it cannot be the same thing that is attracted and repelled. It is the moral part which is repelled, and the appetitive part which is attracted. The moral part creates a desire to avert one's gaze; the appetitive part creates the desire too gaze.

This is not the place to take on Plato's moral psychology, but I note that one cannot appeal to the Principle of Non-contradiction in the present context. That is exactly what is at issue. Moreover, even if it is different mental mechanisms which generate the conflict, this is actually irrelevant. What is at issue here is the phenomenological state in which the agent finds themself, and this is a unity. It does not fall apart cleanly into two neat fragments. Even if it is produced by different parts of the soul, it, itself, is a single conflicted state of a single agent.

Quite generally, and the details of Plato's moral psychology aside, the attempt to defuse the contradiction by appealing to different parts of the agent fails for similar reasons. What we are concerned with is not the different aspects of the agent's cognitive psychology which generate the phenomenological state: it is the phenomenological state itself. And it is one and the same phenomenological state which is torn. As Baxter, again, puts it[19]:

> ... it is going too far to conclude that there are distinct parts of the self or distinct co-habiting selves. Such a conclusion neglects the unitariness of the conscious self. It is one oneself who tries to move in each opposite direction.

[19]Baxter [1, p. 4].

5 Baxter's Aspects

I have quoted Baxter with approval a couple of times above. In the paper from which the quotations come, he is very much concerned with the kind of phenomenon I am discussing here.[20] He also advances a variant of the strategy of making a distinction in the subject, but of a very different kind.

Baxter postulates a kind of entity he calls *aspects*. These are neither properties nor objects, but a *sui generis* kind of entity. In the sort of situation when a person, say Emily, e, is attracted (A) by on object, and repelled (R) by it, then Emily has aspects $e_x[Ax]$ and $e_y[Ry]$, such that it is the former of which one may truly predicate attraction, and the latter of which one may truly predicate repulsion: $A(e_x[Ax])$ and $R(e_y[Ry])$. The aspects are not parts of Emily, however. Indeed, they are identical with her! So $e_x[Ax] = e = e_y[Ry]$. The unity of the unfortunate Emily is, therefore, not threatened.

There are a number of things to note about Baxter's account. First, there would seem to be no independent grounds to posit aspects. They are simply a way of protecting the Principle of Non-contradiction. In this way the postulation is *ad hoc*. It is quite unlike the situation concerning 3 and 8 pm: we have quite independent reasons for believing in different time zones. Baxter claims that one may experience these aspects. But what we experience is the torn state. To say that we experience aspects is precisely to posit them.

Next, it would appear that we are going to violate the principle of the Indiscernibility of Identicals, since $A(e_x[Ax])$ and $e_x[Ax] = e_y[Ry]$, but to infer that $A(e_y[Ry])$ would restore the contradiction we are trying to get rid of. Baxter is well aware of this. He suggests that the correct way to formulate the principle is as applying only to objects. Now, whether it is correct or not, I see little reason for this formulation, except to preserves the Principle of Non-contradiction. Aspects are the kinds of thing which can have properties, and can be identical with other things (objects and aspects). At the very least, then, this seems to be a very counter-intuitive way to formulate the principle.

Thirdly, the theory tells us that $A(e_x[Ax])$ and $R(e_y[Ry])$, but what are we to say of $A(e)$ and $R(e)$ themselves? Baxter's theory is silent on this matter, though both cannot be true. At the very least, this is an incompleteness in the theory. But worse, despite the move to retain the unity of the subject in question by endorsing the identity of Emily and her aspects, in the end the theory appears not to do so. After all, it is *Emily* who is attracted and repelled: it is *she* who is torn. But this is exactly what we do not have.

Better, it seems to me, just to accept the contradiction at face-value, and so avoid all the fast footwork, clever though it is.

---

[20] And the paper may be consulted for an illuminating discussion of it. In the final part of the paper, he extends his analysis to non-mental matters. I would not follow him down this path.

6 Content and State

I have emphasised, as does Baxter, the unity of the *phenomenological* state at issue here. This invites another objection. The phenomenological content is, let us agree, contradictory. It does not follow that the agent's mental state is itself contradictory. Let me explain.

Consider the waterfall illusion.[21] In this, a subject's visual system is conditioned by constant motion in one direction. After an appropriate time, the conditioning is stopped, and a negative after-image is perceived. The visual field will appear to be moving in the opposite direction. But if the agent focuses their attention on a single point in the field, say a black dot, *o*, it will appear to be stationary. *o* appears to be both moving and stationary, as subjects naturally report matters. The content of the visual field is contradictory: *o* is both moving and stationary. But this is an illusion: in actuality, *o* is quite consistently stationary. So, it may now be suggested, if *o* is the attractive/repulsive object, the phenomenological content of the agent's mental state is contradictory: *o* is both attractive and repulsive. It does not follow that the agent's mental state is contradictory: that *s* is both attracted and repelled.

The distinction between the phenomenological content of an agent's state, and the state itself is a good one. From the fact that *o* appears to be both moving and stationary, it does not follow that it *is* both moving and stationary. In the present case, however, the distinction collapses. In the motion case, to get from appearances to reality, we would need the bridge principles:

- if an object, *o*, appears to be moving to subject *s*, *o* is moving
- if an object, *o*, appears to be stationary to subject *s*, *o* is stationary

which principles appear to have little to recommend them. By contrast, in the attraction/repulsion case, we require the principles:

- if an object, *o*, appears attractive to subject *s*, *s* is attracted by *o*
- if an object, *o*, appears repulsive to subject *s*, *s* is repelled by *o*

These principles are hard to gainsay. There seems to be little more than an object being attractive/repulsive to a subject than it appearing to them in that way. Unlike the case of perceived motion, then, the contradiction of the subject's phenomenological content and the objective contradiction go hand in hand.[22]

---

[21] See Priest [11, Sect. 3.3].

[22] There is an interesting intermediate case. There is an auditory phenomenon in which a note sounds to be continually ascending, even though it also appears to stay put (See Shepard [16]). The corresponding bridge principles required to generate a contradiction are:

- if a sound, *o*, appears to be rising to subject *s*, *o* is rising
- if a sound, *o*, appears to be stationary to subject *s*, *o* is stationary

The status of these principles is moot, simply because the notion of a sound balances precariously between the subjective and the objective.

7 The Physical Basis

At this point one might essay another objection. If the agent's mental state is contradictory, this means that its physiological base must be contradictory too. So physics itself must be inconsistent. Surely that cannot be right?

The objection, however, presupposes an over-simple reductionism. Whilst mental states may well supervene on brain states, it does not follow that a contradictory mental state entails a contradictory physical state. Attraction to o might be encoded in some neurological structure, s_A. Repulsion from o might be encoded in some other neurological state, s_R. It may be that normally the states s_A and s_R inhibit each other, but that in odd circumstances they do not do so.

As an analogy (not to be pushed too far), if someone is asked a question of a certain kind, they may answer it verbally or by pointing; and they will normally give what amount to the same answers. But in the case of a "split brain" patient (whose *corpus callosum* has been cut), this may not be so.[23] For example, suppose that a cup is shown to only the lefthand side of such a person's visual field. If the person is asked what they see, they may say that they see nothing. The visual input goes to the righthand side of the brain, and does not reach the speech centres, which are on the lefthand side. If asked to *point* to what they see, however, they may point to a picture of a cup with their hand. The visual input does reach the motor centres that control this. Normally the visual input goes into both sides of the visual field, so this does not happen; but in the experimentally contrived situation of the kind described, it does.

I note that different parts of the brain do, in fact, seem to be involved in affective states such as attraction and repulsion. Positive affect seems to be correlated with activity in the left frontal cortex, whilst negative affect is correlated with activity in the right frontal cortex.[24]

8 Conclusion

The phenomenon of being attracted and repelled by one and the same thing is, indeed, a very strange one. The phenomenological state is one, but intrinsically conflicted. This would seem to give rise to a psychological dialetheia. And that is exactly what, I have argued, it does.[25]

---

[23] See Gazzaniga [6].

[24] See Davidson [4]. The matter is, however, a complex one. See Harmon-Jones et al. [8].

[25] The editors of this volume asked me to respond to a comment made by a referee, who said (their italics): 'What is really remarkable is that Priest decided to endorse once and for all the view according to which contradictions are *not* ontological, in the sense of an object a and a property P not related to or independent from human mind (*sic*). Indeed, a central issue on the dialetheist view on paraconsistency used to be examples of real contradictions that would succeed as examples of dialethias'. Now, first, there is absolutely nothing I say in this paper to suggest that all dialethias

References

1. Baxter, D. (2017). Self-differing, aspects, and Leibniz's Law. *Noûs, 51*, 1–21.
2. Beall, J. C. (2016). The liar paradox. In E. Zalta (Ed.), *Stanford encyclopedia of philosophy*. Retrieved from https://plato.stanford.edu/entries/liar-paradox/.
3. Casati, F. (2016). Being. A dialetheic interpretation of the late heidegger. Ph.D. thesis, University of St Andrews.
4. Davidson, R. (1992). Emotion and affective style: Hemispheric substrates. *Psychological Science, 3*, 39–43.
5. Eibl-Eibsfeldt, I. (1975). *Ethnology*. New York, NY: Reinhart and Winston.
6. Gazzaniga, M. S. (1998, July). The split brain revisited. *Scientific American*, 51–55.
7. Grube, G. M. A. (1992). *Plato: Republic*. Indianapolis, IN: Hackett Publishing Co.
8. Harmon-Jones, E., Gable, P., & Peterson, C. (2009). The role of asymmetric frontal cortical activity in emotion-related phenomena: A review and update. *Biological Psychology, 84*, 451–462.
9. Priest, G. (1987). *In contradiction* (2nd ed., 2006). The Hague: Martinus Nijhoff and Oxford: Oxford University Press.
10. Priest, G. (2002). *Beyond the limits of thought* (2nd ed.). Oxford: Oxford University Press.
11. Priest, G. (2006). *Doubt truth to be a liar*. Oxford: Oxford University Press.
12. Priest, G. (2007). Paraconsistency and dialetheism. In D. Gabbay & J. Woods (Eds.), *Handbook of the history of logic* (pp. 129–204), North Holland.
13. Priest, G. (2014). *One*. Oxford: Oxford University Press.
14. Priest, G. (2016). *Towards non-being* (2nd ed.). Oxford: Oxford University Press.
15. Priest, G. (2019). Objects that are not objects. In M. Szatkowski (Ed.), *Quo vadis, metaphysics? Essays in honor of Peter van Inwagen*. Berlin: De Gruyter Studium.
16. Shepard, R. N. (1964). Circularity in judgements of relative pitch. *Journal of the Acoustical Society of America, 36*, 2346–2353.

---

are mind-dependent. I am offering one which obviously is, in a certain sense. I am simply *adding* this example to the old list. Moreover, by the definition with which I started this essay, a dialetheia is simply a true statement of the form $A \wedge \neg A$. The referee seems to think that this is not sufficient for me: the contradiction has to be 'real'. I have no idea what they mean by this; nor is it something of a kind I have ever said.

Dialetheism in Action: A New Strategy for Solving the Equal Validity Paradox?

Sebastiano Moruzzi and Annalisa Coliva

Abstract This paper starts from the Equal Validity Paradox, a paradoxical argument connected to the so-called phenomenon of faultless disagreement. It is argued that there are at least six strategies for solving the paradox. After presenting the first five strategies and their main problems, the paper focuses on the sixth strategy which rejects the assumption that every proposition cannot be both true and false. Dialetheism is the natural candidate for developing strategy six. After presenting strategy six in detail, we formulate a normative problem for the dialetheist and offer a tentative solution to it. We then elaborate further considerations connecting strategy six to pluralism about truth and logic. Even if strategy six is a hard path to take, its scrutiny highlights some important points on truth, logic and the norms for acceptance and rejection.

1 Relativism and the Equal Validity Paradox

A well known problem, perhaps the main difficulty, for any relativistic reading of the concept of faultless disagreement[1] is that a genuine dispute seems incompatible with the idea of equal validity—namely with the idea that the opinions involved in the dispute are equally valid. Suppose one party maintains that p and another one

[1] The expression "faultless disagreement" has been introduced in the literature by Kölbel [27]. See [18] for a useful normative analysis of the phenomenon.

This paper has been developed along a considerable amount of time. The general strategy of considering the prospects of dialetheism for making sense of relativism was first sketched in [11]; Sects. 1–4.2 have been elaborated jointly, while part of Sects. 5 and 5.1–6 have been developed by Sebastiano Moruzzi.

S. Moruzzi (✉)
Department of Philosophy and Communication Studies, University of Bologna, Bologna, Italy
e-mail: sebastiano.moruzzi@unibo.it

A. Coliva
University of California, Irvine, CA, USA
e-mail: a.coliva@uci.edu

A. Rieger and G. Young (eds.), *Dialetheism and its Applications*, Trends in Logic 52,
https://doi.org/10.1007/978-3-030-30221-4_4

that not-p.[2] The Law of Non-Contradiction (henceforth "LNC") tells us that they can't both be right (or wrong) hence relativism on faultless disagreement cannot be coherently formulated. Let's call this problem *Equal Validity Paradox*.

The present work intends to explore a strategy for blocking the Equal Validity Paradox that targets LNC. So, before proceeding it is important to be clear about the present interpretation of LNC. It is well known that there are several ways to formulate LNC, in fact several aspects have to be fixed in order to state this principle properly. First, it must determined the nature of the law: whether syntactic, pragmatic, semantic or metaphysical; secondly, it must be clarified the type of contradiction involved: whether explicit or implicit; thirdly it must be decided what are the objects of contradiction: whether they are sentences (token,types), statements, claims, propositions, or states of affairs; finally it must be said which type of negation is involved: whether classical negation or non-classical negation. These choices give rise to a potentially large number of formulations of LNC.[3] Given the orthodox assumption that propositions are the objects of our attitudes and since we are primarily interested in the attitudes of disputing subjects, we will use the following semantic formulation[4]:

LNC (semantic reading) propositions p and *not-p* cannot be both true in the same circumstances.

Where the notion of circumstance is intuitively taken to represent those aspects of reality against which the proposition is evaluated (possible worlds are the standard case).

Now that we have specified the reading of LNC that we will be using, it is time to get clear about the structure of the paradox by using as a working example a dispute on matters of basic taste. Let's illustrate it in details by means of the following utterances:

Mary: "Ginger is tasty"

Jane: "Ginger is not tasty"

Call these *yes-no utterances*, and let's conceive an utterance as an ordered couple $<s, c>$ where s is the uttered sentence and c the context of use of the sentence relevant for the utterance. Intuitions related to the subjective character of taste discourse suggest that yes-no utterances must be equally valid. Suppose their equal validity involves that they their semantic status is on a par. If so, they must have an equal semantic status that does justice to their equal validity. The most natural hypothesis is thus that they are both true. Given an equivalence schema for utterance truth—if an utterance $<s, c>$ says that p then 's' is true in c iff p—and an equivalence schema for propositional truth—it is true that p iff p—, and given what yes-no utterances say, we have that, in the same world, the proposition that ginger is tasty and the proposition that ginger is not tasty are both true. By LNC these two latter propositions cannot both be true in the same world, hence, by *reductio*, we conclude that, despite the equal

---

[2]We use italics for referring to propositions.

[3]Grim [21] has in fact calculated that there at least 240 possible formulations of LNC!

[4]See [10, pp. 166–69] for a discussion of the semantic formulation.

validity intuition, yes-no utterances cannot both be true. Thus yes-no utterances must differ in truth-value. Conclusion: the equal validity intuition cannot find a coherent formulation in terms of equality of truth of the utterances involved in a dispute of inclination.

Here is a semi-formal presentation of the argument:[5]

1	(1)	<s,c> and <s',c'> are equally valid	(Equal Validity)
2	(2)	If <s,c> and <s',c'> are equally valid, then <s,c> and <s',c'> are both true wrt to the same circumstances	(Alethic Equal Validity)
3	(3)	if <s,c> expresses <p>, then (<s,c> is true only if <p> is true in circ(<s,c>).	(Utterance-Proposition Truth Schema)
4	(4)	if <s',c'> expresses <q> then (<s',c'> is true only if <q> is true in circ(<s',c'>).	(Utterance-Proposition Truth Schema)
5	(5)	<s,c> expresses <p>	(Propositionality)
6	(6)	<s',c'> expresses <q>	(Propositionality)
1, 2	(7)	<s,c> and <s',c'> are both true wrt to the same circumstances	(Modus ponens: 1, 2)
3, 5	(8)	<s,c> is true only if <p> is true in circ(<s,c>)	(Modus ponens: 3, 5)
4, 6	(9)	<s',c'> is true only if <q> is true in circ(<s',c'>)	(Modus ponens: 4, 6)
1, 2, 3, 5	(10)	<p> is true in circ(<s,c>)	(Modus ponens: 7, 8)
1, 2, 4, 6	(11)	<q> is true in circ(<s',c'>)	(Modus ponens: 7, 9)
1, 2	(12)	circ(<s,c>) = circ(<s,'c'>)	(7)
13	(13)	<q> = <not-p>	(Contradiction)
1, 2, 4, 6, 13	(14)	<not p> is true in circ(<s,c>)	(11, 12, 13)
1, 2, 3, 4, 5, 6, 13	(15)	<not p> and <p> are true in circ(<s,c>)	(Introduction of conjunction: 10, 14)
16	(16)	∀P,C(<P> and <not-P> are not both true in C)	(LNC)
1, 2, 3, 4, 5, 6, 13, 16	(17)	Contradiction	(15, 16)

The Equal Validity Paradox is generated by the following theses: (1) yes-no utterances are equally valid (Equal Validity); (2) if yes-no utterances are equally valid then they are both true with respect to the same circumstances (Alethic Equal Validity); (3) yes-no utterances both express propositions (Propositionality); (4) these propositions are contradictory (Contradiction)[6]; (5) the alethic profile of the truth property for these utterances and the propositions expressed is governed by the utterance-

[5]We use '<p>' to name the proposition that p, and 'circ(<s,c>)' to name the circumstances of evaluation relative to the utterance of 's' in context c.

[6]We are appealing to contradiction because in the example in question are involved sentences of the from "p" and "not-p". The more general relation should be the one of incompatibility which can involve also being contraries.

proposition truth-schema (Utterance-Proposition Truth Schema); and, finally, (6) these propositions cannot both be true in the same circumstances (LNC).[7]

2 Six Ways Out of the Paradox

A way to object to the paradox is to criticise the reasoning on the basis that some of inferences are invalid. We will set aside this solution since we do not think that objecting to modus ponens or conjunction introduction in relation to the inferences involved in the paradox is an helpful strategy for explaining of the nature of the paradox.

Given the structure of the paradox there are at least six solutions to the Equal Validity Paradox: (i) to deny Equal Validity; (ii) to deny Alethic Equal Validity; (iii) to deny Propositionality; (iv) to deny Contradiction; (v) to deny Truth Equivalences; and, finally, (vi) to deny LNC. In this section we will sketch these strategies.

2.1 Strategy I: Revisionary Versus Descriptive Approaches

Strategy (i)—i.e. the denial Equal Validity—consists in a revisionary approach to the problem.[8] According to the revisionary approach disputes of inclination depend on a misguided view on the subject matter: when confronted with disputes of taste, for example, we wrongly think that both views are legitimate, whereas there is no sense of entitlement for these views capable of making sense of both views as rationally responsible attitudes—the only rational attitude is that of abstaining from believing and disputing. Such a radical stance on the problem denies the phenomenon we wish to account for—i.e. the intuition that subjective discourse has its own specificity. Of course, such a radical departure from our practice needs an articulated motivation, for the challenge for any revisionary approach is the possibility of elaborating a theory that offers a descriptive account of the phenomenon while at the same time solving the Equal Validity Paradox. Strategies (ii)–(vi) can then be seen as attempts to solve the paradox and to offer at the same time a descriptive account of faultless disagreement, whereas strategy (i) can be seen as the fallback strategy in case no other strategy is viable.

Before moving to the remaining solutions to the paradox it is important to stress the dialectical role of the Equal Validity Paradox with respect to the problem of faultless disagreement. Going revisionary means that the allegedly Equal Validity Paradox shows that our practice of judging and disputing about subjective domains is intrinsically misguided: no positive account can make sense of this practice as a rational practice, the paradox is thus taken as the symptom of the defectiveness of the

[7]For a more detailed and slightly different reconstruction of the Equal Validity Paradox see [13].

[8]See for example [24] for the thesis that there is no faultless disagreement.

practice.[9] On the contrary, a descriptive account saves the phenomenon by rejecting some theoretical theses—i.e. theses 2–5—involved in the paradox.

The challenge to any descriptive solution to the Equal Validity Paradox is thus the request to show that the theoretical revision invoked can indeed avoid a revisionary approach to the phenomenon—in other words the challenge amounts to the request of showing that the rejection of any of the theses 2–4 is consistent with the acceptance of thesis 1.

In the following, five descriptive solutions to the Equal Validity Paradox will be outlined. These strategies can be divided in two camps: *semantic strategies* (Sect. 3) and *alethic strategies* (Sect. 4). Semantic strategies appeal to the workings of the semantics of natural languages in order to block the paradox, whereas alethic strategies invoke some truth-property to show where the Equal Validity Paradox goes wrong. The paper will focus then (Sect. 4.2) on the last descriptive strategy—strategy (vi)—which is based on a dialetheist rejection of LNC. In Sect. 5 a normative problem for strategy (vi) is developed and in Sect. 5.1 a possible solution to it is sketched. Finally, in Sect. 6 some connections between strategy six and pluralism on logic and truth are explored.

3 Semantic Strategies

Semantic strategies include at least three strategies—i.e. rejection of either theses 2, or 3 or 4—which are well known in the literature. Yet for each of these routes there are well-founded doubts that they can actually avoid a revisionary approach to faultless disagreement. In this section we will sketch these strategies and will rehearse some worries on the feasibility of these options.[10]

3.1 Strategy (ii): The Denial of Alethic Equal Validity—Realism, Contextualism and Relativism, Indeterminism

Strategy (ii) is the strategy shared by truth-relativists, non-indexical contextualists and realists. Alethic equal validity is rejected by realists who account for equal validity in epistemic terms, truth-relativist and non-indexical contextualist deny altogether Alethic Equal Validity by relativizing propositional truth to non-standard circumstances of evaluation.[11]

[9]Following [53], this solution is an "unhappy face" solution to the paradox.

[10]See also [13].

[11]For realism see [4, 52]; for assessment-sensitive relativism see [29, 30, 32], for non-indexical contextualism (aka truth relativism) see [28].

It is well-known what the difficulties for a realist treatment are: the price of preserving classical semantics is to make room for the existence of a fact of the matter on subjective questions—such as whether or not a piece of food is tasty—where it is hard to imagine what else, if not our judgements, could determine what the correct answer is.[12] More importantly: if truth is so remotely connected to our practices of subjective judgements, it is difficult to describe these practices as rationally sustainable when it comes to the activity of disputing.

As for truth-relativism and non-indexical contextualism, these positions share the main difficulty in making sense of the idea of faultless disagreement when the relativisation of propositional truth is in place: it becomes unclear, how according to non-indexical contextualism, two utterances, which are true at different circumstances, involve any incompatibility of their respective assertions; whereas truth-relativism does not seem to have a non-metalinguistic way to express the idea of faultlessness since from each perspective each subject is right is criticising the opponent's view.[13]

Another implementation of strategy (ii) is to argue that disputes of inclination operate in areas of discourse where there is no fact of the matter as to whether an utterance or a proposition is true. Call this strategy *indeterminism*. The main gist of indeterminism is that there is an "absence of mistake-makers" when yes-no utterances are involved [7, p. 64], and this absence is due to the absence of any fact of the matter that would otherwise determine one of the utterances as incorrect. A natural way to implement indeterminism is to argue that yes-no utterances have a third semantic status different—and incompatible—with truth. A well known way to model this option is to adopt a strong Kleene semantics and hold that yes-no utterances have a gappy semantic status as they are neither true nor false.

The problem with indeterminism lies in the tension between the thesis of semantic under-determination as a third semantic status other than truth and falsity and the thesis of equal validity.

If we adopt the rather weak truth-norm:

Norm it is permitted to assert p only if p is true

yes-no utterances (and their respective beliefs) would count as incorrect.[14] Could we reasonably modify the norm for allowing that a gappy proposition is assertable?

Norm* it is permitted to assert p only if p is true or gappy.

Being a gappy proposition would thus par with truth in being a semantic value compatible with the correctness of an assertion and a belief.

However Norm* cannot capture the normative role of a gappy proposition. As [15] noticed, the normative role of a semantic value has a close connection with the

---

[12]See [61].

[13]The critical literature on MacFarlane's relativism is huge. For some recent critical points see [19, 20, 34, 35, 49, 55, 58, 60].

[14]The debate on the normativity of truth is wide and complex. We are here assuming a rather weak norm. See Ferrari 2016 and ms for an analysis on the different varieties of normativity that could exert on truth.

notion of logical consequence: if a semantic value plays a central positive role for assertion, then it has a semantic status that plays a positive role for the practice of inferring deductively. Truth is such a semantic value in classical logic, where the only other semantic value is the false. Generalising from classical logic, Dummett pointed out that a *designated value* is a semantic value that plays the same role for logical consequence as the role played by truth in classical logic. Thus, on the model theoretic approach to logical consequence, being a designated value is being one of the values that are preserved in a deductively valid inference: in many-valued logic, validity is preservation of these designated values.

Now, if we grant these considerations, Norm* determines that the status of being gappy is a designated value, but that can't be right since in strong Kleene semantics the third value is interpreted as undesignated.

3.2 *Strategy (iii): The Denial of Propositionality—Expressivism, Second-Order Indeterminacy*

Whereas the latter three positions agree on the idea that in disputes of inclination the discourse involves genuine propositions and hence it is truth-apt, strategy (iii) rejects the idea that yes-no utterances express genuine propositions; one way to make sense of this option is to go expressivist.[15] Expressivism has been historically the preferred route for escaping the problem: given that no genuine proposition is involved, there is no point in making sense of the truth-conditions of yes-no utterances. However, well known problems afflict the expressivist route, first and foremost the so-called Frege-Geach problem.[16] Whether or not this daunting problem is genuine is beyond the scope of this paper. Suffice here to say that [54] has thoroughly elaborated the expressivist position concluding that the Frege-Geach problem is answerable only by incurring in heavy costs that compromise compositional semantics.

Another way to implement strategy (iii) is to hold that a yes-no utterance expresses one proposition but that it is indeterminate which one it actually expresses. Norm and this indeterminacy thesis jointly entail that it is indeterminate whether yes-no utterances are permitted. Note that this indeterminacy strategy is different from the gappy indeterminism mentioned in relation to strategy (ii). Whereas gappy indeterminism claims an indeterminacy regarding the truth-value of the proposition expressed (first-order indeterminacy), this indeterminacy thesis concerns which proposition is expressed by a yes-no utterance. Borrowing the terminology of Matti Eklund,[17] we can call this indeterminacy *second-level indeterminacy* and the indeterminacy of gappy indeterminism first-order indeterminacy.

[15] Ayer [2] is the first example of this strategy in relation to aesthetics and morality. For contemporary examples of this strategy see [22, 23, 50].

[16] See [57].

[17] See [17].

The basic worry for this second-order indeterminacy strategy is that the indeterminacy in the content expressed casts doubt on the existence of a genuine disagreement: if Mary's utterance does not determinately express that ginger is tasty, then Jane's utterance does not determinately contradict what Mary has said. A further worry is related to the appropriate stance related to second-order indeterminacy. If Mary fails to express a proposition determinately, it seems that the appropriate stance to have in this case is to withhold belief from any of the propositions that her utterance indeterminately expresses. Agnosticism seems thus the mandated attitude in disputes of inclination, and thus the proposal seems to fall back onto a form of revisionism.

3.3 Strategy (iv): The Denial of Incompatibility—Indexical Contextualism

As to the strategy (iv)—the denial of Incompatibility—indexical contextualism is the case in point: indexical contextualists deny that the same proposition is involved in disputes of inclination; according to them the semantic content of evaluative expressions such as "tasty" is sensitive to the context of use. Whether or not the contextualist semantic story is credible, the most pressing problem for this strategy is the so-called lost disagreement problem: the difficulty of making sense of the idea of disagreement if yes-no utterances do not express incompatible contents.[18]

None of the previous considerations is of course intended to provide a knock-down argument against these proposals, nor it is the aim of this work to do so. What matters here is rather to show that there is a well-grounded suspicion that these solutions to the Equal Validity Paradox cannot but end up in being revisionary with respect to the phenomenon of faultless disagreement.

4 Alethic Strategies

Whereas the former strategies and their respective problems have been widely explored in the recent literature on faultless disagreement, the remaining two strategies—strategies (v) and (vi)—haven't received the same extensive treatment. Option (v) denies equivalence schema either for utterance truth or for propositional truth, whereas option (vi) rejects LNC. Since strategy (vi) is here considered via the adoption of dialetheism, this strategy can be conceptualised as the adoption of thesis that truth is not a consistent property. Since these two last options require the abandonment of some orthodox principles about truth, they are labeled *alethic strategies*.

[18]For the basic worry of the lost disagreement problem see [12]. For sophisticated forms of contextualism see [14, 56, 63].

4.1 Strategy (v): The Denial of Utterance-Proposition Truth Schema

According to this strategy a yes-no utterance expresses a proposition but the truth of the utterance does not entail the truth of the proposition expressed:

<s,c> expresses <p> and <s,c> is true but <p> is not true in circ(<s,c>)

Utterance-Proposition Truth Schema would thus fail. It is hard to find a principled motivation for this strategy, nor we do know of any proposal in the literature implementing this strategy. Few remarks will suffice for the purposes of this work. Given that yes-no utterances are true they are correct as speech acts. On the other hand, given that the propositions expressed by these utterances are not true in the relevant circumstances, believing these propositions would be incorrect.[19] It would follow that the correctness conditions belief and assertion potentially diverge. When Mary utters "Ginger is tasty" her linguistic move is correct, while her internal representation of the taste of ginger is mistaken. One way to make sense of this option is to hold a pragmatic story explaining why assertions about taste such as Mary's are correct, while at the same holding an error theory about taste-properties: there is no genuine taste property, but only physiological properties that play a different role from the evaluative role that in our ordinary talk we attribute to taste-vocabulary.

The basic problem with this way of implementing strategy (v) is that it is clearly committed to a revisionary stance: we are systematically mistaken in having beliefs about what is tasty.[20]

4.2 Strategy (vi): The Denial of LNC—Dialetheism

The last option is to deny LNC. This can be done by claiming that yes-no utterances express propositions that are both true and false. From a logical point of view, being true and false is here treated as a the third semantic status which is designated. This third status is the dual status of a truth-value gap: a truth-value glut. The most popular logic for modeling this view is LP—the Logic of Paradox—which is exactly

[19] Notice that in order to avoid violation of LNC, it should claimed that either (a) one proposition is true and the other is false; or (b) the propositions expressed are not contradictory; or (c) that <p> is untrue (but not false) and that <not-p> is false. As for option (a), it seems tantamount to realism, so option (a) collapses into strategy (ii). Analogously, option (b) seems just indexical contextualism, so we have a collapse into strategy (iv). Finally, option (c) requires to adopt some version of three-valued semantics with a value (call it "untruth only") that is excluded by truth of a negation. This latter option seems the most promising option for avoiding the collapse into other strategies.

[20] Assuming that propositions are objects of belief.

like strong Kleene semantics except that the third value is designated.[21] This brings us to see option vi a more promising solution than the gappy theory: *dialetheism*.

According to dialetheism, LNC fails[22] since a yes-no utterance expresses a proposition that is a dialetheia—i.e. a proposition that is both true and false. This latter solution seems more promising: since a dialetheia is both true and false, it is also false but its falsity does not imply its untruth and thus Norm* does not evaluate yes-no utterances as incorrect.

Whether or not a contradiction is acceptable is, for a dialetheist, a question to decide case by case. According to Priest, there seems to be no general criterion for deciding whether or not a contradiction is rationally acceptable.[23] The phenomenon of equal validity *could* then be a reason to accept a contradiction as true.

This dialetheist solution to the Equal Validity Paradox allows one to say that in a dispute where Mary asserts that ginger is tasty and where Jane denies it, Mary's and Jane's opposite judgements are actually contradictory, so no hidden or extra parameters are invoked for relativizing yes-no utterances' truth (as opposed to contextualism or truth-relativism); moreover it allows one to maintain that Mary's and Jane's judgements are equally valid because of they are value gluts.

Note that this solution can be seen as very similar to a relativistic solution in the following sense. Though it is difficult to state precisely what a fully general relativistic doctrine is, we can try to gesture at least at the *spirit* that has motivated relativistic proposals. Relativism finds its inspiration in the basic and rough idea that there are areas of discourse where we cannot describe reality in absolute terms without falling into contradiction. This thought can be developed in two ways: (i) in a metaphysical way by claiming that reality is somehow richer than what is described by any individual consistent description; (ii) alternatively, it can be developed in terms of representation by claiming that our ways of describing reality are expressively inca-

---

[21] The idea of using dialetheism for accounting for the phenomenon of faultless disagreement is sketched in [7] where it is claimed that analetheism—which interprets gappy propositions as designated—is a better option. See [9] for a presentation of analetheism. Beall [7] prefers this strategy to dialetheism in relation to taste on the grounds that analetheism makes sense of the idea that there is no fact of the matter in matters of sense (taste propositions are neither true nor false). We agree with [1] in thinking that analetheism does not do justice to the truth-norm for the same problems we have raised before in relation to the indeterminist gappy theory (see *supra* Sect. 3.1). Insofar as truth is the norm for assertion, we do not think that this model of indeterminacy is sufficient for accounting for the rational sustainability of a dispute. We will thus focus on dialetheism.

[22] Recall that we are using a semantic reading of LNC. In LP the formula $\neg(p \wedge \neg p)$ is never false only since, when p is dialetheia, the formula itself is a dialetheia.

[23] "I am frequently asked for a criterion as to when contradictions are acceptable and when they are not. It would be nice if there were a substantial answer to this question—or even if one could give a partial answer, in the form of some algorithm to demonstrate that an area of discourse is contradiction free. But I doubt that this is possible. Nor is this a matter for surprise. Few would now seriously suppose that one can give an algorithm—or any other informative criterion—to determine when it is rational to accept something. There is no reason why the fact that something has a certain syntactic form—be it $p \wedge \neg p$ or anything else—should change this. One can determine the acceptability of any given contradiction, as of anything else, only on its individual merits" [45, p. 423].

pable of providing an absolute and coherent description of it.[24] Now, historically this basic thought has been developed by introducing some mechanism of relativisation because the idea of embracing a contradiction was abhorred.[25] However, if this rough description of the basic relativistic thought is right, it is perfectly consistent with the relativistic spirit to go dialetheist by claiming that there is no consistent absolute description of reality because in some areas of discourse contradictions are true. In analogy with the metaphysical and representational version of the basic relativistic thought, dialetheist relativism can mirror these two versions by adopting, respectively, either metaphysical dialetheism (e.g. there are inconsistent facts in the subjective domain) or semantic dialetheism (e.g. natural languages are intrinsically inconsistent when it comes to areas of subjective discourse; see [33] for the distinction).

To sum up, dialetheism has a way of dealing with the contradiction involved in a dispute that gives rise to faultless disagreement. It also seems capable of avoiding the problems that contextualist proposals usually face—namely the disappearance of disagreement. At the same time it distances itself from truth-relativism in making sense of equal validity, while truth-relativists have no way to make sense of the idea that relatively to the same parameter, to which truth is relativized, both opinions are equally valid.

In the following sections strategy (vi) will be scrutinised in detail in order to assess its stability.

5 The Normative Problem for Strategy (vi)

Consider again our yes-utterances: Mary utters "Ginger is tasty" and Jane utters "Ginger is not tasty". Suppose that basic taste discourse exhibits faultless disagreement and that dialetheism should be applied to it. Incompatibility between the utterances would fail: both parties should admit that also their opponent is right, not just from her own point of view, but *tout court*. This consequence is itself a virtue of this strategy since it satisfies what Crispin Wright has dubbed *Parity*:

> [...]it is the requirement that faultlessness be appreciable, and endorsable, from the point of view not just of neutrals but of the committed parties in a dispute [59, p. 439]

In addition to Parity, [59] lists three other features of a dispute involving faultless disagreement. Two of these features are mentioned in the Equal Validity Paradox: Contradiction (equivalent to the principle Incompatibility) and Faultlessness (equivalent to Equal Validity). To these features Wright adds Sustainability: that is perfectly rational for the disputants to hold on disputing.

In the dialetheist setting Parity seems to jeopardise the sustainability of the dispute: if Mary thinks that Jane is not faulty in thinking that ginger is *not* tasty because Mary

---

[24] See [3, 11].

[25] Aristotle (*Metaphysics*, Γ 4) famously held that LNC cannot admit exceptions. See [44] for a discussion of Aristotle on LNC.

recognises that it is true that ginger is not tasty, and if Jane recognises that Mary is not faulty in thinking that ginger *is* tasty, then given they respective beliefs, they seem to have no real disagreement insofar as the latter requires a belief in the incompatibility concerning the truth of the contents believed. They should therefore cease to quarrel since they should both recognise that also the other party is right. The normative significance of holding contradictory contents evaporates since it does not involve any genuinely sustainable disagreement. Strategy (vi) thus collapses into a form of *revisionism* instead of being a plausible way of pursuing a *descriptive* project: instead of delivering a description of a targeted area of investigation that matches with our ordinary representation of it, the dialetheist delivers a description of the area of investigation that does not match with our ordinary representation of it.[26]

Assume the previous norm of assertion is in place:

Norm it is permitted to assert p only if p is true.

The latter considerations can then be presented as a normative elaboration of a familiar problem for dialetheism in making sense of an exclusive notion of negation [5, 10, 38]:

ARGUMENT 1:

Suppose p is a dialetheia and that A and B have evidence for this;
Suppose A asserts p;
Suppose B's typical of way of disputing with A is by asserting not-p;
B's assertion of not-p does not prevent her from also accepting p (since p is a dialetheia Norm is respected);
Incompatibility between A and B's assertions is lost;
The dispute between A and B on p is not rationally sustainable.

In a nutshell, the problem stems from the nature of the dialetheia: if a proposition is a dialetheia, the mutual correctness of accepting the proposition and its negation does not seem to leave space for any substantial disagreement: both accepting p and rejecting p—intended as accepting not-p[27]—are correct. So opposite attitudes over p do not seem to motivate any rational dispute.

A possible way out of this problem is to distinguish four notions in play here [47, pp. 96–99], [48, p. 103]:

[26] This normative problem in relation to faultless disagreement is outlined in [11, pp. 151–152]. See also [59, p. 438].

[27] The equivalence between acceptance of not-P and rejection (or denial) of P is known as "denial equivalence" in [51] and as "classical denial" in [36]. Also gappy theorists reject this equivalence—see [37].

Acceptance of *p* = mental state of believing *p*;
Assertion of *p* = the speech act of expressing the act of accepting *p*;
Rejection of *p* = the mental state of refusing to believe *p*[28];
Denial of *p* = the speech act of expressing the act of rejecting *p*.

Here is how Priest explains these distinctions:

> Someone who rejects A cannot simultaneously accept it any more than a person can simultaneously catch a bus and miss it, or win a game of chess and lose it. If a person is asked whether or not A, he can of course say 'Yes and no'. However this does not show that he both accepts and rejects A. It means that he accepts both A and its negation. Moreover a person can alternate between accepting and rejecting a claim. He can also be undecided as to which to do. But do both he can not [43, p. 618].

Hence acceptance and rejection are exclusive (but not exhaustive, agnosticism is always a third possible stance). Familiar examples in which these distinctions are applied are gappy sentences (given that a sentence can be untrue without being false, having ground for the denial of *p* and not-*p* does not ground, respectively, the acceptance of not-*p* and of *p*) and, crucially here, glutty sentences: given that a sentence can be false without being untrue, having ground for the acceptance of *p* and acceptance of not-*p* does not ground, respectively, denial of not-*p* and of *p*. Priest's point (Vs Frege and Geach) is thus that denying *p* does not always involve asserting not-*p*.

Now let's assume this notion of rejection, at least for the sake of the argument, though it is unclear what information is conveyed by a subject's denial.[29] A way to express incompatibility between attitudes can then be regained: the correctness of A's acceptance of *p* excludes the correctness of B's rejection of *p*.[30]

Given this distinction, [48, p. 110] states three different normative principles for rationality involving acceptance and rejection:

(Accept-p) One ought rationally to accept *p* if there is good evidence for the truth of *p*.
(Accept-Not-p) One ought rationally to accept not-*p* if there is good evidence for the falsity of *p*.
(Reject-p) One ought rationally to reject *p* if there is good evidence for the untruth of *p*.

First, these principles are epistemic norms and not alethic norms—principles about what is rational to believe on the basis of evidence for the truth of a proposition

[28] We can interpret acceptance of *p* and rejection of *p* as "yes-or-no questions, respectively reading A? Yes! and A? No!" [36, §2].

[29] See [10, 21]. Berto [10, pp. 174–178] states some worries regarding the idea that acceptance and rejection are primitively exclusive mental states. Berto argues that it is better to explain this incompatibility between attitudes through a notion of content exclusion. We agree with Berto's point, however this issue will not be crucial for the following discussion.

[30] In other words the exclusivity relation between acceptance and rejection involve that accepting *p* and rejecting *p* cannot be jointly correct attitudes. For the notion of joint correctness and its relation to disagreement see [32].

and not principles about what is permitted to believe in relation to the truth-value of a proposition. Second, note the Reject-p is not an exclusive norm for rejection since accepting p and rejecting p can be jointly correct attitudes. An exclusive norm would be

(Reject-p-Ex) One ought to reject p if p is untrue.

However, [36] have argued that an exclusive denial norm—plus the assumption that denial is expressible in the language—gives rise to a paradox once very minimal assumptions are in place.[31]

Contrary to the norms for exclusive denial, Reject-p does not make assertion and rejection normatively incompatible. However, according to Priest, these acts cannot be performed at the same time and hence, though they are not normatively incompatible, they are nonetheless metaphysically incompatible: given any proposition p a subject cannot perform both the act of accepting p and the act of rejecting p. Murzi and Carrara object against the metaphysical impossibility [36, sect. 32].

In order to recapture the normative incompatibility between acceptance and rejection, we could modify the rejection rule as follows:

(Reject-p*) One is permitted to reject P if there is good evidence for Ps untruth, unless there is also good evidence for its truth.[32]

According to Reject-P* one may deny P if one has good reasons for thinking that P is untrue *only*.

Carrara and Murzi criticise this norm on the basis of its asymmetry with the acceptance norm:

> Unlike assertion, any denial may later turn out to be incorrect, since any false sentence can in principle be discovered to be a glut. Thus, you can disagree with my assertion that $0 \neq 0$, and thus deny $0 \neq 0$. But, even if you can prove $0 = 0$, and hence disprove $0 \neq 0$, you can never be fully confident that your denial is correct: a proof of $0 \neq 0$ may always turn up. By contrast, if you have proved $0 = 0$ and thereby assert it, you can be fully confident that your assertion is correct.We find this asymmetry problematic: nothing in our practice of asserting and denying things, it seems to us, suggests that assertion can be indefeasible in a way that denial is not [36].

In the quoted passage it seems crucial that evidence for accepting $0 = 0$ is a proof, and thus an indefeasible evidence. At the same time the evidence sufficient for rejecting p is defeasible: once it is discovered that p is a dialetheia its rejection becomes unpermitted for the subject. But this argument assumes that assertion always requires indefeasible evidence. However in many other (non mathematical) cases the evidence for a proposition is feasible: I believe that this fish is a mackerel but I am

---

[31] "If denial is to serve as a means to express disagreement, it must be rationally impermissible to both assert and deny A. Yet, in view of the Paradox of Deniability, no comprehensive set of norms for exclusive denial can be formulated in the glut-theorist's language" [36]. See also [51] for similar worries.

[32] This a modified version considered by Murzi and Carrara [36] under the label "Deny(U)*".

wrong due to bad lighting conditions.[33, 34] If we bracket this issue we nevertheless have the following problem:

ARGUMENT 2:

Suppose p is a dialetheia and that A and B have evidence for this.
Suppose A asserts p.
Suppose B's way of disputing with A is rejecting p.
B's rejection of p prevents her from also by accepting p (Incompatibility satisfied).
But since p is a dialetheia and B has evidence of it, B is not permitted to reject p if B has evidence for the truth of p (Reject-p*).
B ought not rationally to reject p (ditto for A wrt not-p).
The dispute between A and B on p is neither faultless nor rationally sustainable.

The conclusion of Argument 2 is thus worse than the conclusion of argument of 1 for the dialetheist solution to the Equal Validity Paradox: assuming that both subjects have evidence that p is a dialetheia, not only can we make no sense of the rationality of the dispute, but B's rejection of p is also incorrect. Hence,faultlessness is lost. Therefore, a dispute on matters of inclination cannot be presented as involving the norm of rejection. However, the mere acceptance of a proposition and of its negation does not justify any sense of genuine disagreement between the parties to the disputes.

In principle there are several reasons why a subject could start disputing over a proposition that has been recognised a dialetheia. She could, for example, be interested in persuading the other subject to endorse some course of action instead of a different one. The point of Argument 2 is that there are no *epistemic* reasons (reasons for believing) that could motivate a dispute once a dialetheia is recognised. We think that precisely for this reason a dialetheist strategy (vi) must go revisionary in the end.

In conclusion, Arguments 1 and 2 set up a dilemma for a dialetheist solution to the Equal Validity Paradox: either a dispute between subjects A and B involves A's acceptance of a proposition and B's mere acceptance of its negation; or else, it involves A's acceptance of a proposition opposed to B's rejection of it; if the former, incompatibility between the correctness of subjects' attitude is lost and hence, if the dispute is rational, ignorance of the fact that the proposition is a dialetheia must be then imputed to subjects (they mistakenly take acceptance of the negation of the proposition to be equivalent to the rejection of it). If, on the other hand, the dispute between subjects A and B involves A's acceptance and B's rejection of a proposition, B's rejection could be rational only insofar as she ignores that the proposition is a dialetheia, for recognition that the proposition p is a dialetheia involves recognition

---

[33] It is also doubtful that in mathematics the evidence for an assertion must always be a proof.

[34] The other option Carrara and Murzi consider is [6]'s proposal according to which we can add the so-called shriek-rules to express that a sentence is true-only. For each sentence A which is not glutty it is added a non-logical rule of the form: $A, \neg A \vdash \bot$, where '$\bot$' is an operator that implies triviality. By means of the shriek-rules we can then express that a theory is consistent by stating shriek-rules for all the predicates—a shrieked theory. A shrieked-theory is thus either consistent or trivial. Notice, however, that this proposal will not work if the dialetheist position for the Equal Validity Paradox entails local triviality as argued below—see infra Sect. 6.

that both p and not-p are true and hence that rejection of each is incorrect. In both cases the idea that the dispute is rationally conducted and sustainable is at odds with the assumption that subjects have correct beliefs about the dialetheic status of the proposition. Hence strategy (vi) is committed to a form of revisionism and it no longer seems a promising strategy for making sense of a descriptive solution to the Equal Validity Paradox.

5.1 The Epistemology of Dialetheias

A natural answer to the normative problem raised so far is to deny the assumption that both subjects have evidence of the dialetheic status of p. Perhaps in this way the rationality of the dispute could be regained.

The assumption that both subjects have evidence that the relevant proposition is a dialetheia needs more careful scrutiny. The crucial question is: how can subjects acquire evidence that a proposition on matters of taste is a dialetheia? Consider again the proposition that ginger is tasty. Mary tastes ginger finding it tasty, whereas Jane (tasting the same piece of ginger) finds it disgusting. At the time of tasting neither of them has first-order evidence—i.e. evidence coming from their senses—that ginger is different from how their respective tastes presently inform them it is. After expressing their own views, they have nonetheless higher-order evidence—evidence coming from what the other believes—that ginger is different from how each of them finds it. Suppose they are open to the view that the proposition that ginger is tasty is a dialetheia: each subject would both have (first-order) evidence for the proposition she has asserted and (second-order) evidence for the opposite proposition. Assuming that Mary and Jane are epistemic peers with respect to tasting ginger, they would thus have evidence that the proposition that ginger is tasty is a dialetheia. Would they be irrational in disputing given this epistemic situation?

The standard problem in the epistemology of disagreement is how it can be rational to disagree with an epistemic peer: the root of the problem comes from the intuition that if evidence is knowingly shared by subjects with the same skills and if it is agreed that only one view can be right, then the responsible attitude to take, once disagreement comes into light, is agnosticism (or, at any rate, a downgrading of one's belief/confidence that ginger is tasty (ditto for "Ginger isn't tast"). If I believe a proposition based on some evidence, further evidence that a peer takes a different and incompatible view is in its turn evidence against that proposition. Such evidence undermines the rationality of my belief. However, once we drop the belief that the propositions in play are incompatible, the problem vanishes: the evidence for the opposite view does not automatically speak against my own view. So the question of the rationality of disagreeing between epistemic peers in the case of a dialetheia cannot be framed as the standard problem of the epistemology of disagreement.

The problem for the dialetheist strategy to the Equal Validity Paradox is in fact different, it is the problem of making sense of why Mary and Jane, say, should take one view *and not both* if they don't believe that their views are incompatible. To put the problem in a nutshell: if they have evidence that the proposition that ginger is tasty is a dialetheia, why should not they believe this proposition and its negation? But if they believed both propositions, then any sense of there being a rationally sustainable dispute would vanish on the assumption that both subjects have evidence that the proposition is a dialetheia.

Here is one way to answer this epistemological worry. Notice that each subject has access to different pieces of evidence for the two views: one piece is first-order (coming from the senses) for each view, the other is second-order (coming from the acknowledgement of the view of the other) for the opposite view. Following a suggestion made by [26], there is a problem in the role that second-order evidence plays for deciding what to believe when this evidence is grounded in a belief based on the same body of evidence as the one on which one's belief is based. More specifically: If Mary credits the second-order evidence that Jane believes that ginger is not tasty as a *further* evidence relevant to deciding what to believe, she then gives *additional* weight to Jane's belief that ginger is not tasty is true when she has already considered (to the best of her ability) all first-order elements on which Jane's belief is based. Kelly's point is that this additional weight is suspicious.

Kelly's point can be developed in a slightly different direction: the second-order evidence coming from the opponent's view cannot be as strong as my first-order evidence, in particular not strong enough to justify a belief in the opposite view. Nonetheless, this second-order evidence is evidence that the proposition that ginger is tasty is a dialetheia. So here there might be an explanation of why it is rational to have different beliefs while at the same time having evidence that a proposition is a dialetheia: I do not believe that the propositions that ginger is tasty and that ginger is not tasty are incompatible, for all I know I am open to believe that they can both be true. I taste ginger and, because of my subjective physiological reactions, I feel a pleasant flavour and, because of this evidence, I form the belief that ginger is tasty. Later on, a subject that I consider an epistemic peer forms the opposite belief, while tasting the same piece of ginger. However, I do not thereby form the belief that ginger is not tasty. Not because I exclude the presence of a dialetheia, but because the considerations of the opponent are based on the same evidence that has led me to believe that ginger is tasty. Nonetheless the fact a peer has an opposite opinion provides evidence for the view that ginger is not tasty. Still, this evidence does not play a role in forming or maintaining a belief that ginger is not tasty, but it plays the role of making me suspicious about the *uniqueness* of the correctness of my view, that is why I am open to discussing the matter further.

The former explanation, if successful, can, at best, explain why it is rational to take different views given the asymmetry of the type of evidence for the propositions forming a dialetheia. It remains however an open problem how to justify the rationality of sustaining different views once subjects have the same type of evidence for the truth of the propositions forming the dialetheia. Thus, it seems that if we grant knowledge of the dialetheist solution to the Equal Validity Paradox to subjects involved in a dispute of taste, they should not be prevented from believing both propositions, and thus any sense of disagreement would vanish once again.

6 Dialetheism, Trivialism and Pluralism

Call a *domain of discourse* a set of propositions that concern a certain subject matter.[35] It is not easy to define what a subject matter is, but we have an intuitive grasp of this notion when we say that the domain of morality is made of propositions that concern what is good and bad, that the domain of mathematics is the set of propositions that concern abstract properties such as being a set or being the sum of two numbers etc. Now the Equal Validity Paradox does not just arise for scattered couples of propositions that are related to yes-no utterances, rather it arises for a whole set of propositions that belong to a domain of discourse. Thus Mary and Jane's yes-no utterances concern the subject matter of basic taste: what is tasty, disgusting, delicious and so on. Hence, if the Equal Validity Paradox arises for Mary and Jane's yes-no utterances, then it quite plausibly arises for the whole domain of basic taste.

If the Equal Validity Paradox targets at least an entire domain, then strategy (vi) involves a form of "local atomic trivialisation" for domains of discourse: whereas paradoxical statements are a limited class of sentences, applying dialetheism to all domains of discourse targeted by the Equal Validity Paradox would imply that at least all atomic sentences belonging to those domains would be true and false. Not just isolated true contradictions, but a lot of our statements would present truth-value gluts.

Now it is well known that if all the atomic sentences of a language whose logic is given by LP are dialetheias, then this entails the existence of a trivial model for the language: in fact it is a property of this logic that if all atomic sentences are both true and false, then all sentences are true and false.[36]

There are no logical reasons against trivialism. It has been argued that trivialism cannot be easily dismissed as logically absurd [46, Chap. 3], [25, 47]. Some have argued against it on the basis of its theoretical uselessness [8, p. 34], others [42] that there are statistical reasons for thinking that it is false.

Though global trivialism has been defended [25], it is a very radical position to accept for it is unclear what would constitute a general reason for accepting all kinds of contradictions and it is very difficult to provide different justifications for every domain of discourse. In fact, the standard motivations for holding that a proposition is a dialetheia—e.g. reasons ranging from semantic paradoxes such as the Liar, to set-theoretic paradoxes such as Russell's paradox, or to metaphysics such the problem of change etc.—do not seem to be systemic enough to be generalisable to all atomic sentences of a domain. A fortiori they do not seem a sufficient motivation for considering all atomic sentences in all domains as dialetheias.

The previous considerations entail that if you assume a logically monist stance on dialetheism—i.e. LP is the only correct logic—, then global trivialisation will arise only insofar as you think that the Equal Validity Paradox arises for atomic proposi-

---

[35]For issues related to domains of discourse see [62].

[36]"LP has its so-called trivial model: if all atomic sentences of the relevant language are both true and false (or both true and false at the base world where truth is evaluated, if we have a worlds semantics for LP plus arrow), then all sentences are true and false." [10, p. 170].

tions belonging to *all* domains of discourse. The idea that the equal validity paradox arises for all atomic propositions can be seen as an extreme form of relativism—let's call this position *global atomic equal validity*. If we conceptualise a judgement directed towards an atomic proposition as a judgement attributing properties or relations to objects—i.e. as representing that certain things are related in such and such way—then the thought of global atomic equal validity is that it is always equally correct to hold that objects are related so and so and to hold the contrary opinion. Global atomic equal validity is a difficult doctrine to accept. Can this doctrine be stabilised in order to avoid the collapse into global triviality—i.e trivialism for all domains?

The trivial model property for LP entails that if you hold global atomic equal validity and you think that dialetheism is the right answer to the Equal Validity Paradox for all domains, then you get full triviality for all domains. Given full triviality, everything is true and hence equal validity holds for all propositions. Thus, to conclude, global atomic equal validity, logical monism and dialetheism constitute a cocktail for an extreme form of global relativism: global trivialism. Hence, if you don't like global trivialism, you had better give up either global atomic equal validity or logical monism or dialetheism.

Thus a moderate implementation of strategy (vi) is to hold that the logic of dialetheism is the logic of all domains of discourse—i.e. to be a dialetheist logical monist—and to deny global atomic equal validity—i.e. only in some domains such as the domain of basic taste all atomic propositions[37] are dialetheias. Assuming that the dialetheist semantics is given for all domains, no trivial model would then follow.

Another moderate implementation of strategy six is to hold on to dialetheism, to deny global atomic equal validity because you think that equal validity holds only in some domain of discourse and to have a pluralist stance on logic.[38] In fact if there is more than one relation of logical consequence, ad if logical pluralism is domain-based—i.e. each domain has its own correct logic—then global trivialism is avoided. Notice, however, that if the dialetheist is a domain-based logical pluralist and if she holds that dialetheism is the solution to the Equal Validity Paradox, then in all domains targeted by the Equal Validity Paradox you will end up with local atomic trivialisation. Such a position would then be committed not to global trivialism—i.e. all propositions are true—, but to a local form of trivialism: all propositions of a domains—such as the domain of basic taste—are true. Whether or not this result is problematic, it can be shown to be an interesting consequence in relation to the theory of truth. In fact, local atomic trivialisation can be seen as the idea that truth has the property of being inconsistent in certain domains. Would this be a new form of alethic pluralism?[39] Only if we can stabilise the idea of local trivialism and we

---

[37]Together with all the other propositions targeted by the dialetheist such as the liar, Russell's paradox, sentences expressing change etc.

[38]For domain-based logical pluralism see [31, Chap. 5], [39, 40].

[39]For alethic pluralism see [31] and the collection [41].

can make the case that the property of being inconsistent fits the alethic pluralist framework as a property that can play the truth role.

Is local trivialism untenable? Surely it is an extreme position. But notice that the arguments given by Priest against global trivialism do not apply to a form of local trivialism motivated by the faultless disagreement in the domain of basic taste. In fact, [47] argues against three possible arguments for (global) trivialism. First, he considers empirically-based evidence for trivialism. One way to argue from empirical evidence to trivialism is to hold that optical illusions related to colours are evidence for the more general thesis that all states of affairs are inconsistent:

> How do I know that the room does not go red and green all over when I put on the glasses? Again, the facts that only I see it in this way, that the colouring comes and goes with the glasses, etc., clearly make the hypothesis that the colouring is subjective a superior one. [47, Sect. 3.5]

But this line of argument is not effective against an empirically based evidence related to the phenomenon of faultless disagreement in relation to basic taste. If we take seriously the idea of disagreement, then faultless disagreement in the discourse of basic taste might be taken as empirical evidence for the existence of dialetheias.

Priest's second argument against trivialism is centred on the idea that assuming that the acquisition of meaning requires contrastive applications (correct and incorrect applications of expressions), trivialism implies that language is meaningless since there are correct applications only. Priest's himself recognises that this argument is not effective against trivialism:

"They may not be able to assert anything, but that does not stop them [from] uttering things. They may take themselves to be just babbling—to use a helpful phrase of Peter Unger. But we don't; and we still have to take into account the force of what they say, or at least, of what we take to be that force." [47, Sect. 3.8]

Moreover the objection is less forceful against local trivialism, since the local trivialist might argue that the acquisition of the language of the domain in question—e.g. taste predicates—is secondary to the acquisition of the language of other non trivial domains.

The third argument against trivialism is a sort of "transcendental deduction from certain features of consciousness to the impossibility of being a trivialist" [47, Sect. 3.9]: the argument is based on the idea that choosing is intending to bring about a state of affairs, but "One cannot intend to act in such a way as to bring about some state of affairs, s, if one believes s already to hold". The trivialist believes that everything already holds, hence the trivialist cannot choose. But is it part of our phenomenological life that we (including the trivialist?) ought to choose. Thus, "We cannot, therefore believe everything to be the case about this world". Hence, we have a sort of transcendental argument against trivialism.

However suppose you have trivialism for the domain of basic taste. What is intending to act for basic taste propositions? Does Mary intend to act to bring about that ginger is tasty? Mary intends to act for tasting ginger, but the latter is not a basic taste proposition. Mary tastes ginger and then she judges that ginger is tasty. Her purpose was to taste ginger, once the purpose has been reached she has deliberated that ginger is tasty. However Mary did not act to bring about that ginger is tasty, nor to bring about her deliberation that ginger is tasty. It could be replied that if May intended to eat ginger she would be unable to choose what to eat in the future, assuming, as seems plausible, that in general we choose to eat tasty food and to avoid food we don't consider tasty. The objection would be effective if Mary's evidence for thinking that ginger is tasty were the same as her evidence for thinking that ginger is not tasty. However, crucially, Mary's evidence for thinking that ginger is tasty is stronger than her evidence for thinking the contrary as we have argued before (see *supra* 5.1). Hence, Mary can rationally choose what to eat in the future even if basic taste propositions are dialetheias.

Priest's objections to global trivialism thus do not apply to local trivialism. Whether or not local trivialism is acceptable is not easy to answer. On the one hand, it offers a dialetheist solution to the Equal Validity Paradox leaving open the possibility that truth is realised by different properties in other domains. On the other hand, it endangers to destabilize the intelligibility of the targeted domains. Thus, whether or not a dialetheist solution to the Equal Validity Paradox should embrace local trivialism depends on the reasons we have for being pluralist. If the explanatory advantages gained with pluralism overcome the disadvantages of local trivialism, then local trivialism can be overall acceptable. Otherwise, better to pursue strategy (vi) in a monist framework.

7 Conclusions

Of the five descriptive strategies to solve the Equal Validity Paradox, the dialetheist implementation of strategy (vi) is in danger of not doing justice to the rational sustainability of disputes related to the phenomenon of faultless disagreement. Strategy (vi) thus risks to fall back onto strategy one—the revisionary strategy—, a danger shared by all other descriptive strategies. In this work it has been argued that strategy (vi) has some space for resisting the revisionary challenge and that pursuing strategy six might have interesting consequences for pluralism about truth. In fact, if strategy (vi) is applied to some domains only—like the domain of basic taste—, it provides new motivations for claiming that in these domains truth is inconsistent. If in other domains truth turns out to be realised by truth-properties such as coherence, correspondence or super-assertability that are not inconsistent, we would have a new case for alethic pluralism.

Whether or not strategy (vi) amounts to the most stable solution to the Equal Validity Paradox is still an open question, but a question that is worth enquiring.

References

1. Armour-Garb, B., & Priest, G. (2005). Analetheism: A pyrrhic victory. *Analysis*, *65*(286), 167–173.
2. Ayer, A. J. (1936). *Language, truth and logic*, London: V. Gollancz.
3. Baghramian, M., & Coliva, A. (forthcoming). *Relativism*. London: Routledge.
4. Baker, C., & Robson, J. (2017). An absolutist theory of faultless disagreement in aesthetics. *Pacific Philosophical Quarterly*, *98*(3).
5. Batens, D. (1990). Against global paraconsistency. *Studies in East European Thought*, *39*(3–4), 209–229.
6. Beall, J. (2013). A simple approach towards recapturing consistent theories in paraconsistent settings. *Review of Symbolic Logic*, *6*(4), 755–764.
7. Beall, J. C. (2006). Modelling the 'ordinary view'. In P. Greenough & M. Lynch (Eds.), *Truth and relativism* (pp. 61–76). Clarendon Press.
8. Beall, J. C. (2009). *Spandrels of truth*. Oxford, UK: Oxford University Press.
9. Beall, J., & Ripley, D. (2004). Analetheism and dialetheism. *Analysis*, *64*(281), 30–35.
10. Berto, F. (2008). Adynaton and material exclusion. *Australasian Journal of Philosophy*, *86*(2), 165–190.
11. Coliva, A. (2009). *I modi del relativismo*. Roma-Bari: Laterza.
12. Coliva, A., & Moruzzi, S. (2014). Basic disagreement, basic contextualism and basic relativism. *Iride*, *27*(2), 537–554.
13. Coliva, A., & Moruzzi, S. (2014). Faultless disagreement and the equal validity paradox. In S. Caputo, F. Bacchini, & M. Dell'Utri (Eds.), *New frontiers in truth*. Newcastle upon Tyne: Cambridge Scholars Publishing.
14. de Sa, D. L. (2015). Expressing disagreement: A presuppositional indexical contextualist relativist account. *Erkenntnis*, *80*(1), 153–165.
15. Dummett, M. (1959). Truth. *Proceedings of the Aristotelian Society*, *59*. Reprinted with a post-scriptum in Dummett (pp. 1–24) (1978).
16. Dummett, M. (1978). *Truth and others enigmas*. Duckworth.
17. Eklund, M. (2010). Vagueness and second-level indeterminacy. In R. Dietz & S. Moruzzi (Eds.), *Cuts and clouds: Vaguenesss, its nature and its logic*. Oxford University Press.
18. Ferrari, F. (2016). Disagreement about taste and alethic suberogation. *The Philosophical Quarterly*, *66*(264), 516–535. http://pq.oxfordjournals.org/content/66/264/516.abstract.
19. Ferrari, F. (2016). Relativism, faultlessness and parity: Why we should be pluralists about truths normative function. *Argumenta*, *2*(1), 77–94.
20. Ferrari, F., & Wright, C. (2017). Talking with vultures, *Mind 126*(503), 911–936. http://dx.doi.org/10.1093/mind/fzw066.
21. Grim, P. (2004). What is a contradiction? In G. Priest, J. Beall, & B. P. Armour-Garb (Eds.), *The law of non-contradiction: New philosophical essays* (pp. 49–72). Oxford University Press.
22. Huvenes, T. T. (2012). Varieties of disagreement and predicates of taste. *Australasian Journal of Philosophy*, *90*(1), 167–181.
23. Huvenes, T. T. (2014). Disagreement without error. *Erkenntnis*, *79*(1), 143–154.
24. Iacona, A. (2007). Faultless or disagreement. In M. Kölbel & M. García-Carpintero (Eds.), *Relative truth*. Oxford University Press.
25. Kabay, P. (2010). *On the plenitude of truth: A defense of trivialism*. Saarbrücken: Lambert Academic Publishing.
26. Kelly, T. (2005). The epistemic significance of disagreement. In J. Hawthorne & T. Gendler (Eds.), *Oxford studies in epistemology* (Vol. 1, pp. 167–196). Oxford University Press.
27. Kölbel, M. (2003). Faultless Disagreement. *Proceedings of the Aristotelian Society*, *54*(1), 53–73.
28. Kölbel, M. (2004). Indexical relativism versus genuine relativism. *International Journal of Philosophical Studies*, 297–313.
29. Lasersohn, P. (2005). Context dependence. *Disagreement, and Predicates of Personal Taste, Linguisticis and Philosophy*, *28*(6), 643–686.

30. Lasersohn, P. (2013). Non-world indices and assessment-sensitivity, Inquiry: An Interdisciplinary. *Journal of Philosophy*, *56*(2–3), 122–148.
31. Lynch, M. P. (2009). *Truth as one and many*. Oxford University Press.
32. MacFarlane, J. (2014). *Assessment sensitivity: Relative truth and its applications*. Oxford: Oxford University Press.
33. Mares, E. D. (2004). Semantic dialetheism. In G. Priest, J. C. Beall, & B. Armour-Garb (Eds.), *The law of non-contradiction*. Clarendon Press.
34. Marques, T. (2014). Relative correctness. *Philosophical Studies*, *167*(2), 361–373. http://dx.doi.org/10.1007/s11098-013-0100-3.
35. Moruzzi, S. (2008). Assertion, belief and disagreement: A problem for truth-relativism. In M. Kölbel & M. García-Carpintero (Eds.), *Relative truth*. Oxford University Press. In corso di pubblicazione.
36. Murzi, J., & Carrara, M. (2014). Denial and disagreement. *Topoi*, *34*(1), 109–119.
37. Parsons, T. (1984). Assertion, denial, and the liar paradox. *Journal of Philosophical Logic*, *13*(2), 137–152.
38. Parsons, T. (1990). True contradictions. *Canadian Journal of Philosophy*, *20*(3), 335–353.
39. Pedersen, N. (n.d.). On moderate pluralism about truth and logic: Truth and logic as one, quasi-truth and quasi-logic as many.
40. Pedersen, N. J. L. L. (2014). Pluralism x 3: Truth, logic, metaphysics. *Erkenntnis*, *79*(2), 259–277.
41. Pedersen, N. J. L. L., & Wright, C. D. (2013). *Truth and pluralism: Current debates*. Oxford University Press.
42. Plebani, M. (2015). Could everything be true? Probably not. *Philosophia*, *43*(2), 499–504.
43. Priest, G. (1989). Reductio ad absurdum et modus tollendo ponens. In G. Priest, R. Routley, & J. Norman (Eds.), *Paraconsistent logic. Essays on the inconsistent* (pp. 613–626). Philosophia Verlag, München.
44. Priest, G. (1998). To be and not to be-that is the answer. On Aristotle on the law of non-contradiction. *Logical Analysis and History of Philosophy*, *1*.
45. Priest, G. (1998). What is so bad about contradictions? *Journal of Philosophy*, *95*(8), 410–426.
46. Priest, G. (2000). Could everything be true? *Australasian Journal of Philosophy*, *78*(2), 189–195.
47. Priest, G. (2005). *Doubt truth to be a liar*. Oxford University Press.
48. Priest, G. (2006). *In contradiction: a study of the transconsistent*. Oxford University Press.
49. Raffman, D. (2016). Relativism, retraction, and evidence. *Philosophy and Phenomenological Research*, *92*(1), 171–178.
50. Richard, M. (2008). *When truth gives out*. Oxford University Press, UK.
51. Ripley, D. (2015). Embedding denial. In C. Caret & O. Hjortland (Eds.), *Foundations of logical consequence* (pp. 289–309). Oxford University Press.
52. Schafer, K. (2011). Faultless disagreement and aesthetic realism. *Philosophy and Phenomenological Research*, *82*(2), 265–286.
53. Schiffer, S. (2003). *The things we mean*. Oxford: Oxford University Press.
54. Schroeder, M. A. (2008). *Being for: Evaluating the semantic program of expressivism*. Oxford University Press.
55. Stanley, J. (2016). On a case for truthrelativism. *Philosophy and Phenomenological Research*, *92*(1), 179–188.
56. Sundell, T. (2011). Disagreements about taste. *Philosophical Studies*, *155*(2), 267–288.
57. Woods, J. (forthcoming). The frege-geach problem. In D. Plunkett & T. McPherson (Eds.), *The Routledge handbook of metaethics*. Routledge.
58. Wright, C. (2007). Relativism about Truth Itself: Further thoughts about the Very Idea. In M. Kölbel & M. García-Carpintero (Eds.), *Relative truth*. Oxford University Press.
59. Wright, C. (2012). Mind, meaning and knowledge: Themes from the Philosophy of Crispin Wright. In *Replies part III: Truth, objectivity, realism and relativism* (pp. 418–450). Oxford University Press, Oxford.

60. Wright, C. (2016). Assessment sensitivity: The manifestation challenge. *Philosophy and Phenomenological Research*, *92*(1), 189–196.
61. Wright, C. J. G. (2001). On being in a quandary. Relativism vagueness logical revisionism. *Mind*, *110*(437), 45–97.
62. Wyatt, J. (2013). Domains, plural truth, and mixed atomic propositions. *Philosophical Studies*, *166*(1), 225–236.
63. Wyatt, J. (forthcoming). Absolutely tasty: An examination of predicates of personal taste and faultless disagreement. *Inquiry*.

An Excess of Dialetheias: In Defence of Genuine Impossible Worlds

Ira Georgia Kiourti

Abstract David Lewis famously dismisses genuine impossible worlds on the basis that a contradiction bound within the scope of his modifier 'at w' amounts to a contradiction *tout court*—an unacceptable consequence. Motivated by the rising demand for impossible worlds in philosophical theorising, this paper examines whether anything coherent can be said about an extension of Lewis' theory of genuine, concrete possible worlds into genuine, concrete impossible worlds. Lewis' reasoning reveals two ways to carve out conceptual space for the genuinely impossible. The first is to abandon Lewis' classical translation schema for negation, on the basis that it begs the question against incomplete and inconsistent worlds. I argue that, whilst this option incurs some loss in the semantics, it preserves the core spirit of Lewis' metaphysics. The alternative is to bite the bullet, abandon classical logic and embrace true contradictions. The key challenge with this strategy is that the resulting theory seems committed to a particularly strong kind of dialethism—one that even dialethists would be reluctant to accept. I motivate such a dialethic account of genuine impossibilia using Lewis' own methodology and defend it against triviality objections. I close with a few comments on why impossible worlds should not be reduced to set theoretic constructs out of possible worlds.

1 Introduction

Realism about possible worlds has proven useful in a number of philosophical endeavours. The force behind impossible worlds stems from similar considerations of utility. If possible worlds permit the individuation of propositions and properties via sets of worlds and their parts, impossible worlds extend this treatment to necessary and impossible such intensions. If possible worlds allow for the meaningful evaluation of counterfactual conditionals, impossible worlds extend that treatment to subjunctives with impossible antecedents, that is, counterpossibles. But while there are a number of actualist impossible world proposals in the literature, the prevailing

---

I. G. Kiourti (✉)
Independent researcher, London, UK
e-mail: irakiou@gmail.com

A. Rieger and G. Young (eds.), *Dialetheism and its Applications*, Trends in Logic 52,
https://doi.org/10.1007/978-3-030-30221-4_5

view is that David Lewis' genuine realism about possible worlds cannot embrace impossible worlds on a par with possibilia.[1] Indeed, Lewis [18] is the first to dismiss such an extension of his theory. But impossible worlds are becoming increasingly mainstream in philosophical theorising and this might be considered a rising cost for genuine realism. As Stalnaker's "Will" puts it "..if…modal realism can find no room for impossible worlds, so much the worse for…modal realism." [41, p. 199].[2] The aim of this paper is to carve out some conceptual space for genuine impossible worlds by addressing Lewis' own argument against them. The remainder of this section presents some motivations for impossible worlds followed by Lewis' objection. I then proceed to examine two alternative responses to Lewis.

1.1 Impossible Worlds

A key driver behind impossible worlds stems from what is often called 'the granularity problem'. Insofar as possible worlds help us to individuate distinct propositions via distinct sets of possible worlds, they offer too coarse-grained an account to adequately distinguish all intuitively distinct thought contents. Consider: Mathematicians have tried to square the circle for centuries before it was proven impossible. In doing so they entertained an impossible proposition, which, in turn, is identified with the null set of worlds (for it is true at no possible world). But then, to entertain the idea of squaring the circle is to entertain each and every impossible proposition—'1 + 1=3', 'red is green', 'the law of excluded middle is false', 'Tim (the time traveller) is dead and not dead' and so on, also identified with the null set. But, surely, these are distinct thoughts for one can entertain, believe or desire one without thereby also entertaining, believing or desiring every other. The situation is similar when it comes to necessary propositions, usually identified with the set of all worlds. One can presumably know that one plus one equals two but not thereby know that Hesperus is Phosphorus or that if all men are mortal and Socrates is a man, then Socrates is mortal. Venturing into worlds that verify impossibilities (and falsify necessities) allows us to differentiate all kinds of intuitively distinct propositions.[3]

Another key motivation behind impossible worlds is the desire to build a semantics for counterpossible conditionals that does not render them all vacuously true. According to the standard Lewis-Stalnaker semantics for counterfactuals, a counterfactual is true just when all worlds closest (or more relevantly similar) to ours that verify the antecedent also verify the consequent (or just when any worlds that do

[1]C.f. Vander Laan [13], Varzi [42], Zalta [46], Restall [37], Nolan [25], Mares [21] and Barwise [1], more recently Mark Jago [7–10]. Impossible worlds have even been employed toward a *reductio* against genuine realism, e.g., by Naylor [24]—see Divers [6], Perszyk [27] for a defence.

[2]See also the real-life William Lycan [20].

[3]C.f. e.g. Barwise [1], Berto [2] Jago [9], Priest [34].

are closer than any that don't).[4] But, without impossible worlds, the same treatment does not extend to counterpossibles, since antecedents of such conditionals are false at all worlds, trivially satisfying these truth conditions.[5] Now, if my table suddenly sprouted wings and flew away I would sit up and take notice. Ditto if my table suddenly became a square circle, showing me how this is done. By the same token, it's false that if my table suddenly sprouted wings I wouldn't bat an eyelid; and false that if my table revealed to me the one true way of squaring the circle, I wouldn't be surprised. But the latter comes out true on the standard account.[6]

Note that both impossible propositional content and counterpossible reasoning strike deeply at philosophical theorising. We seem to entertain impossibilities and, as Daniel Nolan [25, pp. 544–546] points out, reason from such hypotheses in logic and metaphysics rather frequently. We regularly evaluate mutually exclusive metaphysical theories or logical systems to assess their consequences. But whilst these are epistemically possible, if one such theory is true, then, presumably, it is necessarily true and hence its rivals impossible. Yet the content of these theories is not trivial. Nor is it that any odd thing is true under such hypotheses. It seems unlikely that two philosophers may strongly disagree about, say, the behaviour of negation or about the nature of properties, yet that everything at least one of them says is trivially true. A more reasonable interpretation would be to say that philosophers, mathematicians, logicians and the like are capable of having meaningful thoughts and making both true and false assertions. As Berit Brogaard and Joe Salerno put it: "[p]erhaps much of philosophy is vacuous, uninformative and fallacious. But if it is, it is not for systematic misuse of the counterfactual" [4, p. 644]. Impossible worlds allow for contentful philosophical hypotheses and elegantly extend the standard treatment of counterfactuals to cover subjunctives of all kinds.[7]

1.2 Genuine Realism and Impossible Worlds

David Lewis argues in a footnote of his *Plurality* [18] that drawing the line at impossible worlds is a consequence of his overall logical and metaphysical framework. For Lewis, "whatever way a world might be is a way that some world is" [18, p. 75] and worlds are *bona fide* genuine (or concrete) totalities that do not differ from our world in kind. The resulting framework involves a basic ontology of individuals, sets and

[4]C.f. Lewis [16], Stalnaker [40].

[5]The same goes for conditionals with necessary consequents, which are true at all worlds, rendering every such conditional trivial.

[6]Williamson [43] denies the intuitive falsehood of such conditionals and presents a series of objections to non-trivial counterpossibles. See e.g. Brogaard and Salerno [4] and Berto et al. [3] in response. Kment [12], Nolan [25] and Vander Laan [14] also defend counterpossibles.

[7]Nolan [26], Kment [12] and Brogaard and Salerno [4] take this a step further and argue that counterpossibles can also serve in the analysis of metaphysical concepts such as modality, essence, intrinsicality, dispositions.

combinations thereof, all couched in classical logic, and the argument against impossible worlds therein turns on the theory's alethic nature of representation. Namely, for something to be 'true at' a Lewis-world is for it to be literally true when quantifying over everything in that world.[8] Here's Lewis:

> [The] discussion of restricting modifiers enables me to say why I have no use for impossible worlds, on a par with the possible worlds. For comparison, suppose travellers told of a place in this world – a marvellous mountain, far away in the bush – where contradictions are true. Allegedly we have truths of the form 'On the mountain both P and not P'. But if 'on the mountain' is a restricting modifier, which works by limiting domains of implicit and explicit quantification to a certain part of all that there is, then it has no effect on the truth-functional connectives. Then the order of modifier and connectives makes no difference. So 'On the mountain both P and Q' is equivalent to 'On the mountain P and on the mountain Q'; likewise 'On the mountain not P' is equivalent to 'Not on the mountain P'; putting these together, the alleged truth 'On the mountain P and not P' is equivalent to the overt contradiction 'On the mountain P, and not: on the mountain P'. That is, there is no difference between a contradiction within the scope of a modifier and a plain contradiction that has the modifier within it. So to tell the alleged truth about the marvellous contradictory things that happen on the mountain is no different from contradicting yourself. But there is no subject matter, however marvellous, about which you can tell the truth by contradicting yourself. Therefore there is no mountain where contradictions are true. (Lewis [18], p. 7, fn 3)

In short, given how genuine worlds represent, the modifier 'at *w*' is logically transparent. In particular, following Lewis [15] where the translation schema for the truth-functional connectives is laid out in full, the truth-at-*w* conditions for conjunction and negation are:

[AND] at w $(A \wedge B)$ *iff* $[(\text{at } w\, A) \wedge (\text{at } w\, B)]$[9]
[NOT] at w $(\neg A)$ *iff* $\neg(\text{at } w\, A)$

These give us premise:

(1) $[\text{at } w\, (A \wedge \neg A)]$ *iff* $[(\text{at } w\, A) \wedge \neg(\text{at } w\, A)]$

Plug in the assumption of a contradictory world:

(2) at w $(A \wedge \neg A)$

and we get an outright contradiction:

(3) $(\text{at } w\, A) \wedge \neg(\text{at } w\, A)$

Drawing a *reductio,* Lewis rejects (2).

While there is no questioning the classical validity of this argument, one wonders if the hypothesis of impossible worlds has been given a chance. Can this be done? The obvious way to challenge this argument is at premise (1). I do so in Sect. 2. The alternative is to motivate a dialethic theory of genuine impossible worlds. I do so in Sect. 3. In Sect. 4, I say a few words against an alternative solution proposed by Berto [2].[10]

[8] C.f. also Vander Laan [13] for a related argument against genuine impossible worlds.
[9] I use (unquoted) letters *A*, *B* throughout to denote propositions, unless stated otherwise.
[10] C.f. also Kiourti [11, Sect. 3].

A few caveats before we proceed.

This paper is not meant as a full defence of genuine impossible worlds, save to address Lewis' own objection. One of the key benefits of genuine realism is its reductive analysis of modality:

(P) Possibly *A iff* there is a world, *w*, and *at w A*

This biconditional is reductive in that it employs no modal terms on the right-hand side. The term 'world' is defined non-modally as an isolated sum of all and only spatiotemporally related individuals and truth at a world is nothing more than truth when quantifying over that sum's domain. But add worlds with round squares and inconsistent time travellers and the analysis breaks down, right to left. For then there will be a world where *A* is true, but *A* isn't possible. But amending (P)'s right hand side to read 'possible world' to avoid the problem introduces circularity. And so, it looks like genuine impossible worlds can't even get off the ground.

I'll say two things in response. First, Lewis' reduction of modality does not constitute the full extent of Lewis' reductive program. Unlike actualist theories for whom intensions like propositions and properties are primitives (or used to individuate worlds just as much as worlds are used to individuate such intensions), Lewis' is the only fully reductive framework around, where intensions are reduced to their extensions. And so, I believe that there's sufficient reason to look at impossible worlds within Lewis' reductive programme, irrespective of Lewis' reduction of modality in particular. Second, it's not clear that we lose the latter. As Lewis admits, most often "modality is restricted quantification…by means of so called accessibility relations" [18, p. 7]. These, in turn, are non-modal relations of "comparative similarity" [18, p. 234]. So, we could potentially amend (P) to:

(P^a) Possibly *A iff* there is an *accessible* world, *w*, and *at w A*.[11]

Fixing the relevant accessibility relation, of course, is far from trivial. Most likely, it will involve similarity with respect to our best systematisation of logical and analytical facts in the epistemically ideal case—or as Jon Barwise puts it "those regularities that fall out of the way that humans individuate objects, properties and relations" [1, p. 496]. But this is a topic for another occasion.[12] Let's agree for now that there's at a least *prima facie* reason to proceed with examining Lewis' own argument against genuine impossible worlds.

There are as many senses of 'impossible' as there are kinds of possibilities. For present purposes, I need only assume the existence of logically impossible worlds and more narrowly, contradictory worlds, since Lewis' reasoning turns on that point. Needless to say, limiting the discussion in this way does nothing to give us a plenitude of impossibilities. I leave that, too, for another occasion.[13] Lastly, I assume classical logic as the default, since Lewis' theory is laid out in those terms.

---

[11] In this case any notion of absolute possibility becomes trivial. See Mortensen [23] for a defence of the idea that anything is possible.

[12] This takes a leaf from Lewis' [16] regularity theory of natural laws. C.f. Kiourti [11, Sect. 5].

[13] c.f. Kiourti [11, Sect. 6]. See Reinert [36] for a critique.

2 A Classical Translation Schema for Negation?

The most obvious point at which to challenge Lewis' argument is premise (1). But to do so is to question Lewis' truth definition for the logical connectives, in particular conjunction and negation. Based on this truth definition, Lewis explains that there's no room for impossible worlds within his framework. But upon closer inspection it's not clear that

[NOT] at w $(\neg A)$ *iff* $\neg$(at w A)

can be held onto when entertaining the hypothesis of impossible worlds. For [NOT] is nothing other than the classical truth conditions for negation applied to truth-at-w. As such it begs the question against the very hypothesis of inconsistent and incomplete worlds, genuine or otherwise.[14] [NOT] states that something is false at a world w just when it's not (also) true there, i.e. that worlds are consistent and complete. And it's clear that this cannot be maintained in the face of impossible worlds, which *ex hypothesi* are, at the very least, inconsistent. As Lewis points out, [NOT] doesn't hold if worlds are more like stories:

> 'Fred says that not P' and 'Not: Fred says that P' are independent: both, either, or neither might be true. If worlds were like stories…there would indeed be room for worlds according to which contradictions are true. The sad truth about the prevarications of these worlds would not itself be contradictory. (Lewis [18], p. 7. fn 3)

But while it is true that Lewis-worlds are concrete, genuine bits of reality that we can quantify over, it is not immediately obvious why this sanctifies [NOT].[15] Consider [FRED]:

[FRED] Fred says $(\neg A)$ *iff* $\neg$(Fred says A)

What makes [FRED] false? Well, it is false if the left- and right-hand-side truth values do not co-vary—that is if both, either or neither could be true. How can this happen? Well, 'Fred says $(\neg A)$' can be true and '$\neg$(Fred says A)' false if Fred contradicts himself. Similarly, '$\neg$(Fred says A)' can be true, while 'Fred says $(\neg A)$' false when Fred fails to utter or deny A. In short, [FRED] fails because what Fred says may be inconsistent or incomplete. But isn't it, *ex hypothesi*, exactly the same with impossible worlds? When it comes to such worlds, [NOT] ought to fail for the same reasons [FRED] does. Maintaining [NOT] in deriving premise (1) begs the question against the very hypothesis of genuine impossible worlds.

Now, [NOT] states Lewis' own definition of how negation behaves at a world. Isn't this sufficient grounds to hold onto [NOT] and reject anything that requires its revision? But while it is important to preserve the core tenets of genuine realism, it is natural to allow that any extension of a classically based theory into impossible

[14]C.f. Lycan [20], Routley [38, Sect. 1.10], Stalnaker [41]. Priest [29].

[15]It may be easier to imagine what it would be for a story, rather than a genuine world, to be incomplete or inconsistent but that's beside the point. We are only asked to assume for the sake of argument that some worlds are so. This is impossible, moreover, so any imaginative limitations are arguably irrelevant.

worlds will require *some* logical revision. [NOT] is a prime candidate for rejection since it essentially states that all worlds behave classically.

One may object that to deny [NOT] is to abandon classical logic. But this isn't so. All we're abandoning is the classical translation schema for negation at a world, not negation *simpliciter*.[16] What we need in order to maintain classical logic in the overall theory is to keep negation *simpliciter* classical, i.e. preserve:

[NEG] $\neg B$ is true *iff* B is not true.[17]

And clearly [NOT] must be rejected for [NEG] to remain unaffected. For [NOT] can be read as an equivalence between the falsehood-at-w of a proposition A and the falsehood *simpliciter* of another proposition B about w, where $B = at\ w\ A$. Then, if at w A and at w $\neg A$, for some w, [NOT] gives us (at w A) and $\neg$(at w A), i.e. $B \wedge \neg B$ *simpliciter*, violating [NEG]. Rejecting [NOT] blocks this step and is precisely what allows the theory's home-language, and truth *simpliciter*, to remain classical.

2.1 Negation and Representation

Does Lewis have independently good reasons for holding onto [NOT]? The strongest argument for the preservation of [NOT] rests on Lewis' overall extensional metaphysical framework and, in particular, its alethic nature of representation as truth *about* a particular domain. A genuine world represents a proposition as 'true at' it by having that proposition be a truth *about* it, i.e. literally *describe* that world. In that sense propositions are a kind of property for Lewis. He says as much: "I identify propositions with certain properties—namely with those that are instantiated only by entire possible worlds" [18, p. 53].[18] And so we could say that representation, i.e. truth at w for Lewis, amounts to a kind of predication—the predication *simpliciter* of a particular proposition A of a world w[19]:

(R) at w A *iff* Aw

(R) says that A is true at w just when A truly describes w. By the same token, 'A is false at w' becomes:

(R$\neg$) at w $(\neg A)$ *iff* $\neg(Aw)$

---

[16] Talk of truth '*simpliciter*' refers to truth in the home language as opposed to the truth-at-w of possible world semantics. I am not concerned with any particular theory of truth here.

[17] C.f. Priest [29].

[18] One could ironically therefore argue that truth-at-w amounts to set-membership for Lewis just as much as it does for ersatzers. But it isn't quite so. A proposition true at a Lewis-world truly characterises that world in a sense that propositions true at ersatz-worlds typically do not, something which e.g. Zalta [46] finds objectionable about ersatz-theories.

[19] In setting forth counterpart theory, Lewis [15] uses a similar notation 'φ^w' to signify that sentence φ holds at w. Since we're concerned with the metaphysics of representation here, let's make explicit the idea that representation is alethic for Lewis by using plain subject-predicate notation, referring directly to a proposition A and world w that it literally describes.

That is, $\neg A$ is true at w when it is not the case that A holds of w. And this is a restatement of

[NOT] at w $(\neg A)$ *iff* $\neg$(at w A)

And so, the argument goes, [NOT] is an indispensable part of genuine realism, leaving no room for genuine impossible worlds.

But strictly speaking, the notion of alethic representation captured by (R)—as literal, true predication of (or property instantiation by) a world—does not give us (R$\neg$). (R), by itself, says nothing about falsehood, or about the rest of the truth-functional connectives for that matter. Arguably the truth-at-w-condition for negation that strictly falls out of (R) fails to distinguish between (R$\neg$) and a weaker principle:

(WR$\neg$) *at* w $(\neg A)$ *iff* $[\neg A]w$

This opaque, narrow scope reading of negation is still compatible with Lewis' core extensional metaphysics of truth at w as alethic, captured in (R). (WR$\neg$) states that the predicate $[\neg A]$ can be truly ascribed to world w. Switching from (R$\neg$) to (WR$\neg$) makes room for genuine impossible worlds without contradiction spilling into the home language.[20]

That the notion of truth at a world as literal predication alone does not give us Lewis' full translation schema can readily be seen in the case of the other logical connectives. Let's take '$\wedge$'. The weaker schema for '$\wedge$' that drops out of (R), gives us:

[WR$\wedge$] *at* w $(A \wedge B)$ *iff* $[A \wedge B]w$

To get the full translation schema for conjunction ([AND]) we need one more step, namely to export the 'and' and parse the sentence. But Lewis' core metaphysics of representation as simply truth about a particular domain, or as the literal ascription of properties to worlds, does not strictly necessitate this step. Representation as true description gives us, at most, that for any formula φ: $<$*at-w-*φ *iff* $\varphi w>$, it does not tell us how to parse φ. We need to also grant extensionality for the logical connectives to go that far, i.e. decree that they behave classically at all worlds. But *ex hypothesi* they don't. We can extend these opaque truth conditions to the rest of the truth functional connectives. The upside is that we can thus accommodate all kinds of impossible worlds, for instance worlds where even the simplest logical truths fail – such as a world where $A \wedge B$ is true but A is not and other such logical peculiarities—given that we have imposed no logical structure on such worlds.[21] Where a world is consistent and complete, we revert to Lewis' original truth conditions.[22]

---

[20] We can think of (WR$\neg$) as involving predicate negation. Routley [38] also discusses this option to give Meinong's ontology of possible and impossible objects a consistent reading [38, pp. 86–89 and pp. 92–96]. No contradiction ensues since the conjunction $Aw \wedge [\neg A]w$ simply amounts to a conjunction of two logically unrelated sentences. Similarly, the property *qua* set of things $|[\neg A]|$ is not identical to $|\neg A|$, the complement of $|A|$.

[21] Some of these worlds might fit the models of different non-classical logics, others might be unruly or 'open', in the spirit of Priest [34, Sect. 1].

[22] One might object that the contradictions we sought to eliminate crop up at a higher level. What about a world w such that $[\neg A]$ at w and also $\neg[\neg A]$ at w? But obviously, given (R) and (WR$\neg$),

The idea of logically opaque, non-parsable genuine impossible worlds, *qua* domains of quantification, is not unintuitive. If I were to open my fridge and see that there both is and isn't beer in it, would I immediately conclude that there's beer in my fridge? Would I conclude that there isn't? I might well be reluctant to draw the usual conclusions. Can I offer you some beer? Yes and no. I couldn't confidently give you a singular answer. Impossible worlds are strange places after all. It's only natural that we can't reduce truths about them to their individual components (and certainly not using classical rules).

2.2 Some Objections

One may object that the present proposal doesn't really give us contradictions true at impossible worlds at all, but some sort of 'quasi-contradictions'—worlds where negation merely means something different since it does not answer to the classical translation schema.[23] Now whether the notion of a true (or true at w) contradiction changes the meaning of negation per se is neither here nor there in the present context. The hypothesis is that some worlds *are* inconsistent. What we're concerned with is the translation of said contradiction into the home language. The proposed translation no more prevents contradictions from being true at impossible words than a putative translation of 'has four legs and a tail at w' to mean 'has five legs' in the home language means that dogs, horses and the like in w really have five legs. What we have lost is some transparency on a semantic level. A contradiction true *at* w no longer is a contradiction true *about* w. We don't have the expressive resources to say what exactly *is* true *about* w other than to say that it *would* amount to a contradiction *were* we able to parse it.

Have we multiplied properties unnecessarily? For an impossible world to represent a falsehood $\neg A$ is no longer for that world to straightforwardly falsify A, but instead for it to verify some other mysterious, ad hoc property $[\neg A]$. But the relevant properties are not arbitrary since there is a principled method for their introduction. We arrive at them by recognising that (a) Lewis' classical truth conditions for the logical connectives are not strictly necessitated by the idea of representation as alethic and (b) these are a natural candidate for rejection under the impossible worlds hypothesis. We still have simplicity in kind, that is, a purely extensional metaphysics of worlds, individuals and sets thereof and truth at w still amounts to truth when quantifying over w. It's just that the domains of impossible worlds are logically opaque. Now, is the worry that the complex truths at (properties of) impossible worlds do not supervene on particular matters of fact there and so that they go against Humean supervenience? If so, even Lewis [19, pp. x–xi] admits that Humean supervenience

we get at most: $[\neg A]w$ *and* $[\neg[\neg A]]w$, that is we cannot export negation (nor collapse it unless we know more about how negation behaves at said world). No contradiction ensues in the home language.

[23] C.f. Stalnaker [41] in a similar spirit.

might hold only contingently. It certainly need not be built into any conception of impossible worlds.[24]

In sum, rejecting Lewis' classical translation schema for the logical connectives is a justified move when it comes to impossible worlds, since these beg the question against such worlds. Doing so might constitute some departure from the original theory but it preserves Lewis' core spirit of alethic representation as literal ascription of a property to a world and, in that sense, a fully extensional metaphysics. The same cannot be said about the semantics.[25] What's the price of full semantic and metaphysical transparency? Dialethism, to which I now turn.

3 Impossibilist Dialethism

A bolder response to Lewis' reasoning to is to bite the bullet, retain Lewis' extensional semantics and go dialethic—that is, embrace true contradictions in the home language. This move is not unfounded. As Takashi Yagisawa puts it:

> When you contradict yourself, what you are saying could not possibly be true. That is indeed a good reason for the conclusion that you cannot tell the truth about anything *possible* by contradicting yourself. But it is hardly a good reason against impossibilia. [44, p. 203][26]

But, as John Divers [6, pp. 76–77] counters, abandoning classical logic merely to make room for impossible worlds (and avoid triviality) seems like a bad motivation. Commitment to logically impossible worlds should no more force us to amend our notion of logical consequence than commitment to physically impossible worlds ought to force us to change our physical laws. Moreover, this commits the resulting theory to a very strong form of dialethism—not that *some* contradictions are true, as most dialethists would argue, but that *every* contradiction is true, albeit at some impossible world. And it says precious little, besides, about the kind of logic a theory of genuine impossible worlds can adopt to reason from its hypotheses. I propose a divide and conquer approach to these matters, drawing from Lewis' own methodology.

---

[24]This admission by Lewis (that core aspects of his metaphysics may hold only continently) can also potentially allow for exotic impossible worlds containing e.g. the entire platonic universe. This is arguably fine as long as such entities play no fundamental role in theory construction.

[25]Can these opaque, intensional connectives be given an extensional world semantics? Perhaps. But any choice of semantics will colour both the metaphysics and what sorts of things impossible worlds can represent. For instance, a Routley semantics for negation [39] decrees that for a world to render true $\neg A$ is for A to be false at its mirror world w^* and this imposes particular relationships between distinct impossible worlds. It also decrees that whenever $\neg\neg A$ is true at w, A also is, which is something we may not want at all impossible worlds. So I'd rather refrain from choosing a particular semantics for intensional operators. One could reason about such worlds instead using counterpossible conditionals. (c.f. [25].

[26]Yagisawa [45] has since developed a distinct type of genuine realism about possible and impossible worlds but does not seem to have amended his [44] view on this matter.

3.1 A Pragmatic View

For all we have said, nothing contradictory is—or could ever be—true about our world.[27] And so, there is a sense in which commitment to logically impossible worlds no more alters our notion of logical consequence than commitment to nomologically impossible worlds alters our natural laws. When we say that nothing travels faster than light, we quantify over everything in our world. It is actually true and necessarily so, let's suppose, a law (true at all nomologically accessible worlds) that nothing travels faster than light. There may exist, unrestrictedly, tachyon-fish in some exotic worlds but they do not concern actual physical theory construction. Similarly, when we say that there are no true contradictions, we need only quantify over everything actual.[28] It's actually true and necessarily so, let's suppose, a law (true at all logically accessible worlds) that there are no true contradictions. There may, unrestrictedly, be true contradictions in some exotic worlds but they do not concern actual logical theory construction. In both cases, the respective laws hold because they pick out some property of our world and worlds similar to it. So even if impossible worlds violate the law of non-contradiction, that law holds actually, and necessarily so. And so, there is a very real sense in which we have not actually changed our default logic, nor embraced dialethism, to accommodate impossible worlds.

Now, this pragmatic response can be justified using Lewis' own methodology. Unlike actualists, according to whom all there is, is all there is actually, it is a well-known matter of contention that Lewis embraces *non-actualia*. That is, he rejects:

(A) *Actually* $A \equiv A$

Let A be 'there are no talking donkeys'. Whilst there are actually no talking donkeys—that is, with quantifiers restricted to all things actual—there are talking donkeys *simpliciter*—that is, with quantifiers unrestricted. Talking donkeys seem perfectly possible and so, for Lewis, these are denizens of some world. Now, the claim that unrestrictedly there are talking donkeys is usually met with incredulous stares. Lewis [18, p. 133] softens these stares by running something like this: Received opinion (common sense) says that there are no talking donkeys. But this typically doesn't distinguish between actual truth—quantifiers restricted—and truth *simpliciter*. Once these come apart the incredulity ought to be defused. Received opinion is respected in that there actually are no talking donkeys and we only have to stretch that extra mile to accept the existence of (actually non-existent) talking donkeys *simpliciter*, a hypothesis about which we ought no longer to have the same intuitions that we associate with what is actually true. Moreover, we can go on simply ignoring the unrestricted existence of talking donkeys and go about our business as if they did not exist at all, given that actually, and so to all intents and purposes, they don't.

Now, Lewis' argument for softening incredulity regarding talking donkeys can equally be applied to soften incredulity regarding true contradictions. For while Lewis

---

[27] See in the same spirit Routley [38, pp. 87–88].

[28] Routley [38, pp. 894] and Priest [33, Sect. 8] similarly propose restricted domains for classical inferences.

rejects (A), he typically doesn't distinguish between absolute or logical necessity and truth *simpliciter*; i.e. between propositions true at every possible world, and propositions simply true, or true "no matter what" [18, p. 7]. That is, Lewis embraces the truth of:

(N) *Necessarily* $A \equiv A$[29]

But add impossible worlds into this picture, and (N) fails. Let *A* be 'there are no true contradictions'. While necessarily there are no true contradictions—that is with quantifiers restricted to all possible (accessible) worlds—there are true contradictions *simpliciter*. Yet more incredulous stares. Here, we can once more apply some Lewis-balm. Received opinion (common sense?) says that there are no true contradictions. But this doesn't typically distinguish between necessary truth—quantifiers restricted—and truth *simpliciter*. Once these come apart, the incredulity ought to be defused: received opinion is respected in that no contradiction is ever *possibly* true. We only have to stretch that extra mile to accept that *per impossibile* there are some (impossibly) true contradictions *simpliciter*, a hypothesis about which we ought no longer to have the same intuitions that we associate with what is necessarily true. Moreover, we can go on simply ignoring these cases, since classical logic actually holds and necessarily so. Upshot? We can justify (and banish) dialetheias that result from venturing into genuine impossible worlds using tools from within Lewis' own methodological arsenal.

3.2 Some Evaluative Remarks

So far so good. I have argued that we have not actually embraced dialethism, that we can keep reasoning using classical logic for all practical purposes and that we are just as justified in doing so as we are in believing that no donkey talks.

An immediate objection here is that classical reasoning has been bought back at the cost of compromising its generality.[30] A key feature of logic proper is that it is topic-neutral, its scope of application universal and unconstrained by subject-matter. By localising the applicability of classical logic to parts of reality, the proposal sacrifices its topic-neutrality and so its status as a logic proper. Worse, talk of actual truth preservation subsumes logic proper under the possible world semantics interpretation of quantified modal logic—effectively a semantics for reasoning with boxes and diamonds. This lacks, again, the required generality.

[29]Lewis has this in mind when he says: "Necessarily all swans are birds iff, for any world W, quantifying over parts of W, all swans are birds...iff all swans, no matter what world they are part of, are birds." [18] A caveat. One night argue that (N) breaks down (right to left) when we consider certain *simpliciter* truths like 'there is a talking donkey.' This is not true at all worlds and so not necessarily true properly speaking. But arguably, these are necessary truths nonetheless derived from Lewis' overall metaphysics. In any case, the argument can go through with (N) expressed as a left to right conditional.

[30]I thank John Divers for this critique.

The reply is that while classical logic is no longer general in the sense of being applicable irrespective of subject-matter, it is general enough in that (for all we have said) there is actually nothing to which it does not apply; no actual reinterpretation of the non-logical terms that can issue a counterexample. In that sense, classical logic is no less general for a genuine impossibilist than for anyone who does not believe in the existence of concrete worlds besides this one. Arguably, the notion of 'actual' here is not dependent on some operator '@' in possible world semantics but merely serves to pick out a particular domain of discourse. At least in that sense, the restricted validity of classical consequence can be detached from reasoning with boxes and diamonds.

Now, what about truth *simpliciter*? Is there any justification for abandoning classical discourse there? Here we are back with Yagisawa [44]. Why can't you tell the truth about the impossible by contradicting yourself? Possible worlds already furnish us with some strange existents. What's so bad about contradictory ones?[31] A Quinan might retort that to change the logic is to change the subject.[32] We started talking about the theoretical utility of impossible worlds, now we're talking about the behaviour of the logical connectives. But, arguably, the subject changed the minute we started entertaining the genuine impossible. The inconsistency we encounter is not spurious but explicitly built into the hypothesis under review. In that context, the charitable interpretation might be that the hypothesis is not trivial, and that classical discourse may not be the best framework for its evaluation.

Now, should we go on and embrace such a theory? This requires further examination, most likely on the basis of how well it fares on a cost-benefit comparison against its rivals. For now, let's just note that while the resulting theory embraces an abundance of dialetheias, it is not trivial. While every contradiction is true at some impossible world, it does not follow that every contradiction is true *simpliciter*. For suppose $A \wedge \neg A$ is true at w_1. Then, in the home language, we get $Aw_1 \wedge \neg Aw_1$ *simpliciter*. Not $A \wedge \neg A$. Take some other world, w_2, at which that particular contradiction $A \wedge \neg A$ is false. Then the contradiction $Aw_2 \wedge \neg Aw_2$ will be false *simpliciter*. So not every contradiction is true.

3.3 *Triviality Further Down the Line?*

Mark Jago repeatedly charges genuine conceptions of impossible worlds with triviality.[33] Here's one example:

> If we can export contradictions from impossible worlds, however, every sentence will be true (simpliciter). Consider any sentence 'A' and some impossibility 'Fx'. Then there is an impossible world such as $A \wedge Fx$. If that world is genuine, we can export a genuine x which is

[31] Priest, e.g. [31–33] goes a long way toward demystifying the notion.

[32] C.f. Quine [35, p. 81].

[33] Jago [7–10].

such that $A \wedge Fx$, which (classically or paraconsistently) entails that 'A' is true (simpliciter). Hence, absurdly, any 'A' whatsoever is true. [10, p. 716]

But, something seems to go awry with this style of argument. Let's say that some world w renders true '$A \wedge Fx$', so that ($A \wedge Fx$)-at-w. Then, Lewis' [15] translation into the home language (using letter 'w' to range only over worlds and 'Ixw' to mean 'x is in w') gives us:

$\exists w\ \exists x\ (Aw \wedge Ixw \wedge Fx)$

So, at most, we get $\exists w(Aw)$ *simpliciter*. Even if we reduce everything to particulars, for instance, world a and individual b (in a) satisfying A and F respectively, at most we arrive at $Aa \wedge Fb$ *simpliciter*. At no point do we arrive at $A \wedge Fx$, hence 'A' being true *simpliciter*.[34]

Am I missing something? Clearly, there's no puzzle where sentence 'A' contains variables. Presumably some worlds will contain individuals that satisfy the relevant properties rendering 'A' true and others won't. So, the proposition expressed by 'A' will hold of (i.e. 'at') some worlds $w_1, w_2 \ldots w_n$ and not others $w'_1, w'_2, \ldots w'_n$. So, for instance, Aw_1 will be true but Aw'_1 won't be. So not any sentence 'A' will be true *simpliciter*. Now, granted, when something is F in w, that very thing is F *simpliciter*. The ascription of properties to individuals is absolute. But this does not give us trivialism. Let's say 'A' is some such sentence ascribing a property to an individual absolutely: Fa. Then 'A' will be true in the particular world containing the named individual and it will also be true *simpliciter*—that is quantifying over everything there is. But that is beside the point since this is not true for any 'A'. To see this, suppose 'A' refers to (part of) a consistent world w (there are lots of them around). Then '$\neg A$' will not also be true at w, hence not true *simpliciter*. So not any 'A' is true *simpliciter*. Even if we grant that for every property F, possible or impossible, some *thing* (world or part thereof) in the plurality satisfies it, we still don't get triviality. For *every* property F won't hold of *every* object and so not every sentence 'Fa', 'Fb', etc. and their negations will be true *simpliciter*. Once more we don't have triviality.[35] Finally, let's ask: does 'A' pick out a proposition, say, about sets or numbers, supposed to be true "*from the standpoint of* a world"? [17, p. 40]. Even that wouldn't give us triviality, bar perhaps about sets or numbers. And even then, for a theory of genuine impossible worlds to blossom, truths about sets, numbers and the like need to vary at impossible worlds, to properly be truths *about* those worlds, never mind how.[36]

[34] See Priest [34, pp. 83–85] for a related discussion.

[35] Now, what if *per impossibile,* every F did hold of every object, *simpliciter*? That is a higher order theoretical impossibility and as such, we need to devise some way to represent it, drawing from existing tools within genuine realism. Note that this is an impossibility *de re* about every object *per impossible* being every which way. Were this a possibility it would be represented using counterparts. Let's adopt this idea here and let some totality of objects (say in a trivial world) stand in for the plurality itself. See also footnote 37.

[36] Lewis allows that not all sets exist from the standpoint of all worlds [17, p. 40]. In saying this, he has unit sets of particulars in mind, but there's nothing to stop genuine impossibilism from decreeing that arithmetical truths and the like are also true from the perspective of a subset of worlds only. This is also the natural way to respond to e.g. Jago's [7, p. 64] impossible real numbers or Yagisawa's

And so not just anything will be true *simpliciter*. I submit that this objection fails to take into account that a truth *at* a genuine world, possible or impossible—*A*—and an exported truth *about* it—*Aw*—are not identical. So even if any '*A*' is rendered true at some world, not any '*A*' is true *simpliciter*.

The opponent of genuine impossibilia can now pull out a trump card. This is the Church false constant (CFC), a well-defined special atomic sentence $\bot$ in paraconsistent logic that entails any arbitrary sentence *A* (c.f. [7, p. 64]). This is presumably conceived of as capturing some impossible proposition or property and, given genuine realism and a generous comprehension principle on worlds, it can thereby be truly ascribed of (something at) some impossible world. But if something instantiates $\bot$ at *w*, it instantiates $\bot$ *simpliciter*. It is such that 'everything is true'. We're back with triviality.

One can only reply here by returning to basics. The whole purpose of impossible worlds, genuine or not, is not to render impossibilities true. It is to represent them. Granted, genuine worlds do so by exemplifying those impossibilities. But there's still room for manoeuvre. What's the impossibility in question? By all accounts it should be this: that the CFC is (actually) impossible, i.e. that the world could not be such that the CFC is true. And what is needed to represent it? A world exemplifying said impossibility, namely a world where the CFC *is* true; a world that *is* as the CFC describes. We have such a world: the trivial world. It represents the impossibility that CFC actually obtains; that our world could not have been like *that*. Sure, the trivial world instantiates CFC *simpliciter*, absolutely, not relative to anything. It really *is* that way, *simpliciter*. What way? Trivial. The CFC refers to 'every' sentence. It's those quantifiers we restrict to the world that does the representing: everything is true at that world. If the opponent insists that, no, this is not good enough, we've reached a dead end. CFC describes a trivial world (one that could not have been ours). And that we have given. No more is required.

Now, there's another, higher order, theoretical impossibility that can be invoked here and that's that the entire body of true propositions, *simpliciter*, could not have been trivial. Can genuine impossibilism represent *that*? Arguably yes. Again, a trivial world (or collection thereof) can do the job. As far as I can see, there's no methodological reason why genuine realism cannot lie when asked to represent such higher order theoretical impossibilities about the plurality itself by (falsely) restricting its quantifiers to a few relevantly useful objects.[37]

I conclude that a dialethic theory of genuine impossible worlds is not trivial.

[44, p. 195] trouble with property identity. I here leave aside exactly how worlds are to represent such impossibilities or how to carve up (il)logical space into more useful components. For some discussion see Kiourti [11, Sect. 5].

[37]C.f. also, in that spirit, Cameron [5], who presents, *per impossibile*, a stipulated world such that if it exists, then any arbitrary *A* is true *simpliciter.* But surely, there's no reason why the proponent of genuine impossibilism need worry about objections that render the theory trivial by stipulation. I propose that higher order theoretical impossibilities can all be (mis)represented in a uniformly principled manner by restricting the relevant universal quantifiers to a subset of the plurality for representational purposes. This point bears further discussion, which I won't here carry out in full. Lewis' [15] counterpart theoretical language—elegantly expressed in the idea that propositions are in effect properties of worlds [18]—makes room for the idea that genuine realism

3.4 No Universal Logic

What, if any, logical principles can a dialethist about genuine impossible worlds use to reason from their hypotheses? Well, this depends on how permissive a comprehension principle on impossible worlds we want. The worry, as Daniel Nolan notes, is that

> …if the motivation is to provide a logic, which applies to every situation, possible or not, then that logic will have few principles indeed… [for]..if there is an impossible situation for every way things cannot be, there will be impossible situations where even the principles of subclassical logics fail… [25][38]

Suppose we adopt a weak four-valued paraconsistent logic to reason from the ontological hypotheses of the extended theory. One such candidate, seemingly weak enough, is First Degree Entailment (FDE), with the material conditional $A \supset B$ defined as $\neg A \vee B$. Under FDE there are no logical truths. So, we can allow worlds where any rule—even identity—expressed as a material conditional fails, while staying within the logical confines of the theory. But even under FDE, simplification holds, that is, from $A \wedge B$ you can derive $A(B)$, even if $A(B)$ is also false. So, simplification holds necessarily. But then, given a generous comprehension principle on impossible worlds, there should be a world where simplification fails. Hence simplification is not universally valid. Generalising, from the supposition that any rule R is general enough to allow us to reason about anything in the extended plurality, it follows that R ought to fail with regard to some domain of the extended plurality.

One way to respond to these worries is along the traditional paraconsistentist lines: Any rule stated as a conditional can be true—or false—at a world, without this compromising the logical rules governing that world. Following Priest [28, pp. 295–297], we can represent the failure of a logical law at an impossible world by having the corresponding conditional be false there. If so, we can adopt simplification to reason from the theory's hypotheses and still allow for worlds where the conditional $(A \wedge B) \rightarrow A$ fails, extending FDE into a proper relevant logic with which to reason from the overall theory, with a stronger conditional '$\rightarrow$'. This still allows for any logical truth in the form of a conditional $A \rightarrow B$ to fail at impossible worlds, whilst allowing us to adopt a universal logic to reason about the plurality.[39] This proposal is generous enough in that, although sentences do not vary arbitrarily in truth value

---

is ultimately a theory of objects and properties first, and sentences and truths second. In that sense all *de dicto* modalities are ultimately modalities *de re* about a special kind of composite objects, i.e. worlds. If so, impossibilist genuine realism can 'lie' after a fashion: objects can stand in for other objects, vicariously representing impossibilities about them just like Humphrey's counterparts can stand in for Humphrey, vicariously representing possibilities about Humphrey. Now what about an impossibility that concerns the entire plurality—*everything*? It seems in keeping with the spirit of genuine realism, especially when it comes to such higher order theoretical impossibilities, to allow smaller collections of such objects to stand in for, and misrepresent, the plurality itself.

[38]See also [6].

[39]This can be done by having worlds where the usual truth-conditions for conditionals do not apply [28], or by building it into the theory that logical truths (expressed as conditionals) do not hold at all worlds [22].

at such worlds, i.e. as Priest [29, p. 485] notes, we still get, for any sentence, worlds where it is true and worlds where it is false.[40]

One might object to this strategy on the grounds that it still puts artificial limits on what kinds of impossible worlds there are. Nolan [25, p. 569, fn 15] complains that such "dialethic compromises" leave something to be desired. What we want is a world about which, from $A \wedge B$ we cannot infer A; not a world where $(A \wedge B) \rightarrow A$ is false but where we can infer A, even if $\neg A$ also holds. To have such a permissive conception of impossible worlds we have to abandon any attempt at a logic general enough to reason about the entire pluriverse. An argument for not dismissing this option out of hand is that if we grant the existence of logically impossible worlds, it's only natural to suppose there is no one logic to rule them all. We can still partly systematise the resulting theory and ignore some 'open worlds', as Priest [34] calls them, or 'non-deductive situations', as Routley [38] calls them, as being outside any logical bounds.[41] When it comes to parts of reality outside the theory's logical bounds, we can use a subjunctive conditional as per Nolan [25]. This would obey no rules without exception (thus would not systematise the theory), but it would allow us to reason counterfactually based on rules of thumb and pragmatic considerations.[42]

In sum, I submit that one can motivate a dialethic theory of genuine impossible worlds, justify it using Lewis' own methodology and defend it against triviality and related objections.

4 Much Ado About Nothing?

Here's a very different objection to the present enterprise: do we need to go this far to reap the benefits of impossible worlds for genuine realism? Berto [2] argues that a genuine realist can venture into ersatz impossible worlds with zero cost, by taking them to be set-theoretic constructions out of possible worlds and their parts.[43] He argues that this is the best way forward because it (a) preserves Lewis' reductive analysis of modality, which employs genuine worlds alone and (b) avoids Lewis' *reductio*. Even Lewis himself comments that "...there is no reason not to reduce [impossible worlds] to something less objectionable, such as sets of propositions" [16, p. 16].[44] Much like an explorer cutting a path through the thicket right next to a highway, this could render the current enterprise pointless.

---

[40]For sentences to be able to take on arbitrary truth values at worlds, we need deductively 'open' worlds along the lines of Priest [34].

[41]Routley [38, p. 895 and 898] distinguishes between two kinds of logically impossible situations (or worlds, in our case): situations where logical rules apply but the relevant conditional statements of these rules fail; and 'non-deductive' situations, where even the principles of a general logic fail to hold, remarking that it is only to be expected that even a universal logic will not apply to non-deductive situations. C.f. also Priest [34, Sect. 1].

[42]C.f. Berto et al [3] for a formalisation of counterpossible reasoning.

[43]C.f., in the same spirit, Kiourti [11, Sect. 3].

[44]This is effectively a proposal that genuine realism can do without impossible worlds.

But I believe that exploring genuine impossibilia is a worthwhile enterprise. Firstly, as Priest [30] argues, it's not obvious that we can pose a fundamental ontological distinction between possible and impossible worlds any more so than we can pose a fundamental ontological distinction between those worlds that are physically possible and those that are physically impossible.[45] Secondly, set theoretic constructions do not give us a rich enough tapestry of impossible worlds and fail altogether to deliver impossible individuals. These latter would normally serve to differentiate impossible properties as well as necessarily coextensive yet distinct properties, such as triangularity and trilaterality. Berto [2] recognises this and proposes that we differentiate, say, triangularity and trilaterality as per original Lewis [18, p. 56], namely by structural means, identifying these as higher order structured properties <T, A> and <T, S>. Berto suggests we can then use such constructions to build set-theoretic ersatz impossible individuals. Sure we can. But these will play no (or limited) theoretical role since the individuation of properties happens at a lower structural level. And the same piecemeal process is required for impossibilities not instantiated by any genuine world (and so not true at any set of sets of such worlds), such as '*a* is not self-identical', 'nothing exists' or 'some things are spatiotemporally distinct from their parts'. The same goes for other coextensive propositions, not differentiated at the level of possible worlds, like 'Phosphorus is *F*' and 'Hesperus is *F*, as noted by Jago [7, p. 65] who also charges Berto's hybrid account with failing to deliver a fine enough grain of impossibilities.

In sum, there is at least a *prima facie* reason to explore alternatives. Going hyperextensional, by contrast, and adopting a genuine theory of impossible worlds and individuals offers rich enough resources for a uniform extension of Lewis' program right at the outset. Perhaps too rich for some. But that is a matter of opinion.

5 Conclusion

My aim was to carve out conceptual space for the genuinely impossible and this I have done. One way to do so is to abandon Lewis' classical translation schema for the truth functional connectives. This is a principled move in the face of genuine impossible worlds and, while costing the resulting account its semantic transparency, it doesn't thereby compromise genuine realism's core extensional metaphysics or its alethic nature of representation as property instantiation. To keep both metaphysics and semantics fully extensional, by contrast, is to espouse a dialethic theory of genuine impossible worlds. To make room for such a theory, I have argued that the resultant dialethism is far from trivial, that it can be justified using Lewis' own methodology and that it can ultimately be ignored for all practical purposes.[46]

[45] Priest [30, p. 582] argues, in particular, that there's no non-question-begging argument to establish such a distinction.

[46] For useful discussions and comments, I thank Graham Priest, Katherine Hawley, Stephen Read, Daniel Nolan, Patrick Greenough, John Divers, Simon Prosser, Colin Caret and an anonymous

References

1. Barwise, J. (1997). Information and impossibilities. *Notre Dame Journal of Formal Logic, 38*(4), 488–515.
2. Berto, F. (2010). Impossible worlds and propositions: Against the parity thesis. *Philosophical Quarterly, 60,* 471–486.
3. Berto, F., French, R., Priest, G., & Ripley, D. (2018). Williamson on counterpossibles. *Journal of Philosophical Logic, 47*(4), 693–713.
4. Brogaard, B., & Salerno, J. (2013). Remarks on counterpossibles. *Synthese, 190*(4), 639–660.
5. Cameron, R. (2010). Worlds and individuals, possible and otherwise. *Analysis, 70*(4), 783–792.
6. Divers, J. (2002). *Possible worlds*. London: Routledge.
7. Jago, M. (2012). Constructing worlds. *Synthese, 189*(1), 59–74.
8. Jago, M. (2013). Against Yagisawa's modal realism. *Analysis, 73,* 10–17.
9. Jago, M. (2014). *The impossible: an essay on hyperintensionality*. Oxford: Oxford University Press.
10. Jago, M. (2015). Impossible worlds. *Noûs, 49*(4), 713–728.
11. Kiourti, I. G. (2010). *Real impossible worlds: the bounds of possibility*. Ph.D. Dissertation. University of St Andrews. https://research-repository.st-andrews.ac.uk/handle/10023/924.
12. Kment, B. (2014). *Modality and explanatory reasoning*. Oxford: Oxford University Press.
13. Laan, V., & David, A. (1997). The ontology of impossible worlds. *Notre Dame Journal of Formal Logic, 38*(4), 597–620.
14. Laan, Vander D. (2004). Counterpossibles and similarity. In F Jackson & G Priest (Eds.) *Lewisian themes: The philosophy of David K. Lewis*, (pp. 258–275). Oxford: Clarendon.
15. Lewis, D. K. (1968). Counterpart theory and quantified modal logic. *Journal of Philosophy, 66,* 113–126.
16. Lewis, D. K. (1973). *Counterfactuals*. Oxford: Blackwell.
17. Lewis, D. K. (1983). *Philosophical papers* (Vol. I). Oxford: Oxford University Press.
18. Lewis, D. K. (1986). *On the plurality of worlds*. Oxford: Blackwell.
19. Lewis, D. K. (1986). *Philosophical papers* (Vol. II). Oxford: Oxford University Press.
20. Lycan, W. G. (1994). *Modality and meaning*. Dordrecht: Kluwer.
21. Mares, E. D. (1997). Who's afraid of impossible worlds? *Notre Dame Journal of Formal Logic, 38*(4), 516–526.
22. Mares, E. D. (2004). *Relevant logic: a philosophical interpretation*. Cambridge: Cambridge University Press.
23. Mortensen, C. (1989). Anything is possible. *Erkenntnis, 30,* 319–337.
24. Naylor, M. B. (1986). A note on David Lewis's realism about possible worlds. *Analysis, 46,* 28–29.
25. Nolan, D. P. (1997). Impossible worlds: A modest approach. *Notre Dame Journal of Formal Logic, 38*(4), 535–572.
26. Nolan, D. P. (2014). Hyperintensional metaphysics. *Philosophical Studies, 171*(1), 149–160.
27. Perszyk, K. (1993). Against extended modal realism. *Journal of Philosophical Logic, 22,* 205–214.
28. Priest, G. G. (1992). What is a non-normal world? *Logique Et Analyse, 35,* 291–302.
29. Priest, G. G. (1997). Editor's introduction. *Notre Dame Journal of Formal Logic, 38*(4), 481–487.

---

referee, as well as friends, colleagues and audiences at the University of St. Andrews Philosophy Department and the Arché Research Centre, the Sheffield 2008 Graduate Conference, the Fourth World Congress of Paraconsistency at the University of Melbourne, the Institute of Philosophy and the LSE Choice Group in London, the University of Nottingham philosophy research seminar and the Philosophy Departments at Kyoto University and Ruhr University Bochum. I am also grateful to the Arts and Humanities Research Council and the University of St. Andrews for their financial support.

30. Priest, G. G. (1997). Sylvan's box: A short story and ten morals. *Notre Dame Journal of Formal Logic, 38*(4), 573–582.
31. Priest, G. G. (1998). What is so bad about contradictions? *The Journal of Philosophy, 95*(8), 410–426.
32. Priest, G. G. (1999). Perceiving contradictions. *Australasian Journal of Philosophy, 77*(4), 439–446.
33. Priest, G. G. (2006). *In contradiction: a study of the transconsistent* (2nd ed.). Oxford: Oxford University Press.
34. Priest, G. G. (2016). *Towards non-being: the logic and metaphysics of intentionality* (2nd ed.). Oxford: Oxford University Press.
35. Quine, W. V. O. (1986). *Philosophy of logic* (2nd ed.). Cambridge, MA: Harvard University Press.
36. Reinert, J. (2018). The truth about impossibility (and how not to tell it). *Philosophical Quarterly, 68*(271), 307–327.
37. Restall, G. (1997). Ways things can't be. *Notre Dame Journal of Formal Logic, 38*(4), 583–596.
38. Routley, R. (1980). *Exploring Meinong's jungle and beyond*. Departmental Monograph, Research School of Social Sciences, Australian National University, Canberra.
39. Routley, R., & Routley, V. (1972). The semantics of first degree entailment. *Noûs, 6,* 335–359.
40. Stalnaker, R. C. (1968). A theory of conditionals. In: N. Rescher (Ed.), *Studies in logical theory, American philosophical quarterly monograph series 2* (pp. 98–112). Oxford: Blackwell.
41. Stalnaker, R. C. (1996). Impossibilities. *Philosophical Topics, 24,* 193–204.
42. Varzi, A. C. (1997). Inconsistency without contradiction. *Notre Dame Journal of Formal Logic, 38*(4), 621–639.
43. Williamson, T. (2007). *The philosophy of philosophy*. Oxford: Blackwell.
44. Yagisawa, T. (1988). Beyond possible worlds. *Philosophical Studies, 53,* 175–204.
45. Yagisawa, T. (2010). *Worlds and individuals, possible and otherwise*. Oxford: Oxford University Press.
46. Zalta, E. N. (1997). A classically-based theory of impossible worlds. *Notre Dame Journal of Formal Logic, 38*(4), 640–660.

Game of Truth: Truth, Fictionalism, and Semantic Paradox

Frederick Kroon

Abstract According to dialetheists, there are true contradictions. Anti-dialetheists deny this. David Lewis famously thought that the dialetheist's toleration of contradiction was beyond the pale and made dialogue between the two sides impossible. But this sceptical view presents him with at least two problems. First, what do we do about the apparent appearance of contradiction when we reason about certain topics such as truth? Secondly, contrary to any summary dismissal of contradiction, don't we often at least contemplate and imagine things that are impossible? Lewis responded that we sometimes make believe that impossibilities are possible, "subtle ones at least", but that these cases seem to require a distinction between the subtle ones and the blatant ones, a distinction that he for one didn't know how to draw. In this paper, I argue that there is a reading of Lewis's remarks on which they not only suggest an attractive account of the imaginability of the impossible, but also a potentially compelling solution to the kind of paradoxes that drive dialetheism, as well as an equally natural caution about how far such a solution can be taken.

1 Introduction

According to dialetheists, there are true contradictions. Anti-dialetheists deny this. They might say more, of course. They might give reasons for the claim that contradictions can't be true. That might be a mistake, since dialetheists have a way of agreeing with this claim. After all, they think that these are *contradictions*, hence *impossibilities*. An alternative anti-dialetheist strategy is to cite the Law of Non-Contradiction and declare the law to be inviolable: nothing contravening the law should be accepted

F. Kroon (✉)
Department of Philosophy, University of Auckland, Auckland, New Zealand
e-mail: f.kroon@auckland.ac.nz

A. Rieger and G. Young (eds.), *Dialetheism and its Applications*, Trends in Logic 52,
https://doi.org/10.1007/978-3-030-30221-4_6

as true.[1] But how would we then justify the (peculiar) elevation we give to this one law? Shouldn't the law's inviolability be argued rather than simply asserted (especially as the other side does its side of the debate by providing detailed reasons why we should accept that there are true contradictions)?

We seem to be at the start of a dialectical game of snakes and ladders. Some have argued that there are impasses at nearly every turn, a situation not helped by the many versions of the Law of Non-Contradiction in existence.[2] Others are confident that there is a way of achieving resolution.[3] David Lewis was not one of those. In response to an invitation by Graham Priest and JC Beall to contribute to a collection on the Law of Non-Contradiction,[4] he declined the invitation, and had this to say (in a letter to Priest and Beall on 21 July 1991):

> To conduct a debate, one needs common ground; principles in dispute cannot of course fairly be used as common ground; and in this case, the principles *not* in dispute are so very much less certain than non-contradiction itself that it mattes very little whether or not a successful defence of non-contradiction could be based on them. [27, p. 176]

For Lewis, then, the dialetheist's toleration of contradiction is beyond the pale.[5] In this paper I am going to assume that Lewis is right about this.[6] But this sceptical view presents him with at least two problems. First, what, then, do we do about the apparent appearance of contradiction when we reason about certain topics such as truth? Secondly, contrary to any summary dismissal of contradiction, don't we often at least contemplate and imagine things that are impossible, for example stories in which people travel back in time and change the past, or the antecedents of counterpossibles like 'If all magnitudes were rational, then Pythagoreanism would be vindicated'? So how can there be a blanket intellectual ban on the very thought that there can be contraventions of the Law of Non-Contradiction?

Given this second objection from the work of Priest in particular, Lewis added the following (in a further letter to Priest on 9 January 2001):

> That doesn't stop me from sometimes making believe that impossibilities are possible, subtle ones at least. I agree with you about the many uses to which we could put make-believedly possible impossibilities, if we are willing to use them. The trouble is that all these uses seem to require a distinction between the subtle ones and the blatant ones (very likely context-dependent, very likely a matter of degree) and that just what I don't understand. [27, p. 177]

---

[1]Here too we should be careful. The first and second clauses may come apart, since the law may be valid—true in all interpretations—and yet allow some contradictions to be true. It is valid in Priest's LP, for example.

[2]See especially Patrick Grim's excellent discussion in Grim [16].

[3]Grim [16] is slightly hopeful. Bueno and Colyvan [6] and especially Priest [33] are far more confident.

[4]Published in 2004 as *The Law of Non-Contradiction; New Philosophical Essays* [35].

[5]See also Lewis [25].

[6]See Bueno and Colyvan [6] for a direct response to Lewis's argument.

While it is not entirely clear how we ought to interpret Lewis's remarks,[7] I think there is a reading on which they not only suggest an attractive account of the imaginability of the impossible, but also a potentially compelling solution to the kind of paradoxes that drive dialetheism (thus taking care of the first problem in the preceding paragraph), as well as an equally natural caution about how far such a solution can be taken. The present paper is an attempt to say more about this kind of solution, and also to suggest that Lewis's apparent doubts about how far the solution can be taken are misplaced. Because semantic paradox remains for many the most obvious and most compelling entry into dialetheism, the paper will be restricted to a discussion of the way all this impacts on the notion of truth in particular.

The structure of the paper is as follows. In Sect. 1 I discuss the view of a number of classically minded "inconsistency theorists": those who agree with dialetheists that the concept of truth is inconsistent, but who want to retain the Law of Non-Contradiction and who argue instead for modifying the concept of truth in a limited way or replacing it. After giving reasons for rejecting this approach, I turn to the idea, suggested in the quote from Lewis's second letter, that we might invoke make-believe to deal with the apparent inconsistency of the concept of truth. Section 3 describes and rejects one version of this idea, a way that aims to show that the inconsistency is *only* apparent. In Sect. 4 I describe what I take to be a better version: one on which truth gives rise to contradiction, but the contradiction is confined to the make-believe. Section 5 discusses Lewis's objection that many of the ensuing contradictions are "blatant" and hence unimaginable and not the proper subject of make-believe. The concluding section summarises the argument of the paper.

2 The Inconsistency of the Concept of Truth

Dialetheists—those who think there are true contradictions, in defiance of the Law of Non-Contradiction—think that their rejection of *ex contradictione quodlibet* means that they can't be saddled with the claim that anything whatsoever is true (as well false). They believe that there are true contradictions aplenty, but that such true contradictions can be pretty well insulated from the rest of our beliefs. Probably the most familiar examples of allegedly true contradictions of this kind are sentences that generate semantic paradox, such as the Liar ('This sentence is false', or, in strengthened form, 'This sentence is not true'; (*) 'The first starred sentence on this page is false', and so on).

But dialetheists are not the only ones to have been persuaded of the inconsistency of the concept of truth (although, to be fair, their work was instrumental in persuading others). An increasing number of philosophers now agree that the concept of truth is indeed inconsistent, and so have also become "inconsistency theorists", but refuse

[7] In part, because it is not entirely clear how to square what he says with his well-known remarks on inconsistent fiction in 'Truth in Fiction' [24].

to accept dialetheism's take on this inconsistency.[8] In particular, since they deny that there really can be true contradictions, they think the fault lies in the *concept* of truth.

The question, then, becomes what to do about the inconsistency, since it seems to follow that we and our theories are in considerable intellectual trouble if we continue using the concept. Although his focus wasn't semantic paradox, David Chalmers offers a useful methodological directive [9]. He thinks that philosophers, like scientists, should be asking themselves whether existing concepts, or the properties and relations they stand for, are the best tools for understanding the world, and, if not, whether we should look for suitable replacements. If the concept of truth really is inconsistent, Chalmers' methodological directive suggests we have an iron-clad case for replacement.[9]

What might be an example of something that might replace the ordinary notion of truth? Tarski, of course, did think that the ordinary concept of truth was inconsistent, and offered us his hierarchy of truth-predicates. Others like Kripke and revision theorists have produced alternative theories of truth, describing competing concepts of truth that might again serve as replacements for the ordinary inconsistent concept (but of course the rhetoric of these theorists wasn't replacement; they presumably thought they were formally explicating an intuitive concept). More recently, Kevin Scharp has explicitly proposed two replacement concepts for the ordinary inconsistent concept of truth, based on two rules—the descending truth rule and the ascending truth rule—that are constitutive of the ordinary concept and that when operating in concert (together with principles governing logical concepts)[10] permit the derivation of semantic paradoxes. These rules are: 'True <p> → p' and 'p → True <p>', and the replacement concepts are the concept of descending truth, or D-truth (where D-true <p> → p) and ascending-truth or A-truth (where p → A-true <p>) [36, 37].

Tarski was famously dismissive of the concept of truth in play in ordinary language, and just wanted a concept or concepts of truth for formalised languages, like the language of mathematics. By contrast, Scharp is very clear that his interest is in the ordinary concept of truth, and that it is the ordinary concept that is inconsistent and needs replacing. His overall programme is complicated by a desire to blend the idea of replacement with the ambition to offer a descriptive theory of the predicate 'true' that is sensitive to the need for replacement, a theory that turns out to be both indexical and assessment-relative. The details don't concern us here. For my present purposes, it is enough to point out how complex assessments of truth and falsity have become under such a scheme. It is hard to believe that anything like this reflects our ordinary understanding of the way 'true' functions.

Consider two worries in particular. The first involves Kripke's point that semantic evaluations for sentences that use the word 'true' can be risky—that there is no a priori

[8]E.g., Azzouni [1], Barker [4], Burgess and Burgess [7], Burgess [8], Eklund [10, 11], Johnston [17], Patterson [29], and Scharp [36, 37].

[9]See Eklund [11].

[10]The inconsistency of the concept of truth should of course be seen as a joint affair: logic is also implicated in the inconsistency, together with the semantic rules that allow for self-reference. But the standard inconsistent theorist's line on this is that the latter are harmless until the concept of truth is deployed, hence the latter should be singled out.

route to telling when they lead to paradox [19]. This point seems to me enormously valuable. But on Scharp's replacement strategy it seems we don't have any risk unless and until we combine the rules for A-true and D-true; we can do this without risk when the context is what Scharp calls safe but not otherwise. That seems to me to put things the wrong way around. It hides the fact that risk is a natural consequence of using the concept of truth, where the concept we use is a unitary concept rather than two artificially separated concepts used in concert.

Scharp's way out faces other difficulties, some of them brought out in a recent paper by Eklund [11]. Consider the familiar justification for the use of truth in semantics. By stating what speakers have tacit knowledge of, truth-conditional semantics provides a psychological explanation of how our knowledge of language allows us to gather and communicate information about the world, thereby providing a generally successful account of linguistic activity. Call this the *cognitivist* justification for using the concept of truth in semantics. Eklund concludes that on this justification what justifies the use of truth in semantics is that speakers have tacit knowledge of contents that involve this concept:

The concept of truth employed in semantics must be the concept of truth that occurs in the contents that speakers have tacit knowledge of. Speakers must have this concept of truth. Given the cognitivist justification for theoretical employment of the truth predicate, there is then a problem for the idea that the ordinary concept of truth ought to be replaced for theoretical purposes [11, pp. 295–296].

As Eklund points out, this concern rests on a psychologically realist way of understanding the role of truth in truth-theoretic semantics, and the concern may prove to be misplaced on a deflationary understanding of this role [11, pp. 297–298].[11] The same cannot be said of the final, and perhaps most important, concern. Even if the concept of truth is inconsistent, that is no reason to think that the concept of truth is not good enough for ordinary everyday purposes. (For Eklund, "[i]f any version of the inconsistency view contradicted this, that would be a good reason to abandon the view" [303].) Working with a replacement concept is likely to be far more taxing than working with the ordinary concept and using our ordinary rules of inference—something that seldom gets us into trouble.

3 Making Believe in Truth I

I have argued, all too briefly, against replacement responses to the inconsistency of the concept of truth. The argument suggests that not relying on the concept, if that is even possible, will incur heavy practical and theoretical costs. So what should we do?

In this kind of situation, it is best to adopt a very different approach. We should keep on using the concept of truth while recognizing that taking the concept seriously

[11] Although in the absence of decisive reasons favouring the latter it seems important to pay the concern some heed.

as a guide to reality would be disastrous. In other words, we should treat it as a kind of fiction. And we should do this because of the enormous benefits of doing so—because of what we would lose if we were to give up on the concept. That is, we should do what many philosophers recommend we do for a fragment of discourse that has proved troublesome in some way but that is eminently worth preserving despite this: adopt a fictionalist stance towards it, where fictionalism with respect to some discourse D holds that (i) the sentences of D systematically misrepresent the world if construed at face-value (generally, because crucial terms in D fail to latch on to genuine individuals or properties); (ii) there is nonetheless utility in continuing to use D; and (iii) treating sentences of D as if they were part of a fiction is what allows us to explain how it can be useful to continue using D.[12]

Before we return to discourse involving the word 'true', here is a quick reminder about how some other well-known fictionalisms satisfy this rubric. Consider first Hartry Field's version of mathematical fictionalism. This holds that the sentences of mathematics are false since there are no mathematical entities, but that principles of mathematics like the Peano axioms are at least true *in the story of mathematics*, a story whose use in deriving consequences from physical theories allows us easy access to what these theories tell us about physical reality [12, 13]. Another well-known example is moral fictionalism, of the kind defended by Richard Joyce [18]. Joyce holds that moral claims are all false since there are no such properties as goodness and being obligated (these are what Mackie called *queer* properties, and they have no place in a modern world-view). Nonetheless, engagement in moral deliberation and discourse is useful for streamlining and policing our decision-making, and because of that we should *pretend* that many familiar moral precepts are true. Such precepts are fictions, or like fictions in crucial respects. We accept that torture is wrong by *pretending* we believe it, just as we do with the claim from *Peter Pan* that a crocodile bit off Hook's arm.

Note that these two fictionalisms satisfy condition (iii) in rather different ways. Field's is what is known as a prefixing form of fictionalism, because of the way it relies on an 'in-the-story/fiction' prefix. By contrast, Joyce's fictionalism is a prefacing form of fictionalism (the prefix/preface contrast is found in [28]); on such an account of morality, the fictionalist implicitly prefaces whatever is said with a special kind of disclaimer: let's do as if the following is in fact the case, even though it isn't. Given the close association between fiction and pretense on modern theories of fiction, this is a second way in which the notion of fiction can be used to ground the usefulness of engaging in certain fragments of discourse. In the current jargon, both are *revolutionary* forms of fictionalism: they underwrite a revolutionary directive that we treat mathematics and morality this way; they do not say that mathematics and morality as ordinarily understood are to be construed in such a way (the latter view would make them *hermeneutic* forms of fictionalism).

Now return to the case of 'true'. For the fictionalist about truth-talk there is no such property as *truth*. The kind of inconsistency theorist we have been talking about

[12]Here I am using the account of philosophical fictionalism defended in Armour-Garb and Woodbridge [2], easily the clearest and best motivated account of the approach to date.

has a good reason for taking this line: the concept of truth is inconsistent, and so there is no property it picks out. (Depending on the way we count constitutive conditions, there might of course be properties that satisfy a reasonable number of the conditions constitutive of the concept of truth: Kripke's grounded truth, say, or even Scharp's D-truth and A-truth. But I take it we shouldn't count these as the designation or semantic value of 'true'.)[13] But even though there is no such property, the fictionalist thinks it will be useful to pretend or make believe that there is, since this will allow us to retain the full range of benefits of using the concept of truth where using it is unproblematic, while not incurring the costs that its use would normally incur in problematic contexts. (In particular, the paradoxical consequences of certain uses of 'true', for example in the Liar, will be confined to the pretense or make-believe; the contradictions will not leach out into the class of ordinary statements about reality.)

The above motivation for a fictionalist account of truth has focused on the inconsistency of our concept of truth. But this is not the only motivation found in the literature. The most developed fictionalist account of truth-talk, and one that deliberately tries to deal with the problem of semantic paradox along the way, is the account advanced by the philosophers of language Bradley Armour-Garb and James Woodbridge, most recently in their book *Pathology and Pretense*. Unlike the inconsistency-theoretic line taken above, they take their lead from their deflationism about truth. They argue that if one is a deflationist about truth—and they think that one ought to be, since there are problems with the two alternatives of eliminativism about truth-talk and holding that there is a real, robust, perhaps explanatorily important, property of truth—then one should be a fictionalist about truth-talk.[14]

Armour-Garb and Woodbridge propose the following account of fictionalism for various fragments of discourse, including truth-talk. What they call semantic pretense-involving fictionalism, or SPIF, about some fragment of discourse D holds that there is semantic infelicity in an utterance of a sentence from that fragment, often because one of the terms fails to refer. Consider negative existential statements, for example. 'Frodo doesn't exist' expresses no proposition if taken at face value since 'Frodo' lacks a bearer. SPIF's account of how to understand the utterance is that it involves the *pretense* that the name 'Frodo' has a bearer and that 'exists' is a genuinely descriptive predicate. The rules for using this predicate in the game of make-believe associated with negative existentials imply that it is correct to imagine that Frodo has the descriptive property of non-existence just if there is no individual

---

[13]This is controversial. A number of inconsistency theorists such as Azzouni and Patterson hold that the term doesn't stand for any property, while more moderate inconsistency theorists like Eklund and Scharp deny this, or at any rate deny that this follows from the inconsistency of the concept of truth. Eklund, for example, thinks that there might be close enough deservers of the term [11]. In my view, the inconsistency goes so deep that there is no reasonable prospect of a designation of 'true' (which is not to deny that some properties have many of the most desirable features that we would want in a property of truth).

[14]They argue this on the basis of a certain style of argument put forward by Stephen Yablo for the case of mathematical fictionalism. The details don't concern us. (See [2], Chap. 4).

picked out by the name 'Frodo' as the speaker understands it.[15] The latter claim captures the serious content to which an utterance of the sentence redirects its hearers. It is the ability to talk *through* the pretense in this way, affirming some genuine claim by pretending to affirm another claim, that shows how the original fragment of discourse remains useful despite its sentences being false or contentless.

Now consider truth-talk. Armour-Garb and Woodbridge think that when we use truth-talk our talk should be understood as involving the overarching pretense that 'true' and 'false' stand for genuine, substantive properties.[16] The various rules of generation for this pretense then specify under what circumstances we are to imagine or pretend that p is true/p is false. Thus, one set of rules says that the pretense displayed in an utterance of '[The proposition] that *p* is true' (alternatively, '... is false') is prescribed just if *p* (alternatively, not-p) is the case. Thus, from the perspective of the game it is correct to say 'The proposition that Obama was born in the U.S. is true' (despite there being no such property as truth) since Obama was indeed born in the U.S. Similarly, it is correct from the perspective of the game to say 'The sentence 'Obama was born in the U.S.' is true'.

How does all this help with semantic paradox? As in the case of statements involving the predicate 'exists', Armour-Garb and Woodbridge maintain that statements involving 'true' and 'false' express contents indirectly, through the pretenses involved in such statements. What typical instances of truth-talk indirectly specify are real-world conditions—M-conditions, as Armour-Garb and Woodbridge call them—that articulate what the world has to be like for the pretenses in question to be prescribed. For example, the M-condition for 'Snow is white' is simply the worldly condition that snow is white, and the fact that this condition obtains is what makes it correct from the perspective of the pretense to utter 'The proposition that snow is white is true'. What goes wrong in cases of apparent semantic pathology like 'This very sentence is false' is that the sentence doesn't direct us to any M-conditions that mandate either the pretense that the sentence is true or that it is false, so that the sentence is strictly contentless. (Their term for this status is s-defective.) The same is true, they argue, in the case of apparent pathology involving other well-known paradoxical sentences, such as Curry paradoxical sentences and the sentences in a Yablo chain.[17]

Armour-Garb and Woodbridge's discussion is deep and wide-ranging, but an inconsistency theorist is bound to wonder whether contradiction can indeed be avoided in this way. Two problems loom. The first is that the concept of an M-condition is a notion of *content*, so a semantic notion, and as such no less prone to paradox than the concept of truth itself. To tie the paradoxicality of the notion of truth to the absence of M-conditions raises the worry that the possible paradoxicality of the latter will mean that the task is left incomplete. Armour-Garb and Woodbridge

---

[15]The idea of rules of generation that specify what we are to imagine when we engage in a particular game of make-believe is Walton's. See especially Walton [39].

[16]For Armour-Garb and Woodbridge, such pretenses need not involve speakers' actually pretending or even alluding to pretense, a complication that doesn't affect what I have to say about their view.

[17]They also apply the strategy to truth-teller variants, like 'This very sentence is true', where the symptom of pathology is indeterminacy rather than inconsistency, as well as to cases that display both symptoms (what they call symmetrical and asymmetrical naysayers).

discuss this worry and try to allay it by considering a number of possibly problematic sentences that use the concept of an M-condition and then defusing the apparent paradoxes. I doubt, however, that such an approach is able to allay the worry in its full generality: even if their responses work at the level of individual sentences, how can we be sure that the notion of an M-condition is paradox-free?[18]

The second problem is this. Armour-Garb and Woodbridge take what is assertible from within the fiction of truth to be determined purely by what is the case outside of this fiction. 'The proposition that snow is white is true' is assertible from the perspective of this fiction or pretence (and so is itself quasi-assertible) just if snow is indeed white. But that is not how most fictions work. Take the Sherlock Holmes stories. It is quasi-assertible that Holmes lived in the capital of Great Britain, even though nothing in the stories explicitly says this. But it would of course be wrong to claim that the pretense portrayed in an utterance of 'Holmes lived in the capital of Great Britain' is prescribed solely in virtue of London's in fact being the capital of Great Britain. Playing a pivotal role are descriptions from the fiction that deliver constitutive facts about Holmes, in this case that Holmes lived in London. In the same way, it is arguable that to be the kind of fictionalist about truth-talk who is properly attentive to the power of semantic paradox we should take the fiction of truth to contain descriptions of truth that are constitutive of the concept. Any other fiction simply fails to do justice to the concept.

Another way to put the point is as follows. Armour-Garb and Woodbridge have a very particular conception of their fictionalist project. They are deflationists about truth first and foremost, and for deflationists what is crucial about the terms 'true' and 'false' is that they do not pick out genuinely substantive properties. Instead, saying that sentence '*s*' is true, or that it is true that s, is just an indirect way of saying that *s*. Armour-Garb and Woodbridge capture this insight in terms of rules of generation for pretenses centred on the concept of truth. What inconsistency theorists regard as conceptual facts about the concept of truth are for Armour-Garb and Woodbridge generalizations to be tested once we see the workings of a fictionalism that is driven by such a deflationary conception of truth. And they think that on such an understanding of the concept of truth, any attempted inference to semantic paradox will simply dissolve, the result of a faulty understanding of how attributions of truth and falsity arise.

My own view is that this is a sanguine hope. Earlier I suggested that Armour-Garb and Woodbridge's framework may well not be able to escape inconsistency, but above I also argued that the inconsistency of the concept of truth runs too deep, the result of the constitutive role played by the T-schema and the associated ascending and descending truth rules rather than being a hazardous generalisation. In my view, we should accept both the inconsistency of our concept of truth and the apparently deflationary behaviour of truth as captured by the T-schema. Given the practical indispensability of the concept, that argues for a version of fictionalism that looks rather different from that of Armour-Garb and Woodbridge. It won't be driven in the

[18] Kroon [21] argues that for the existence of revenge paradoxes. Armour-Garb and Woodbridge [3] give a response.

same way by deflationism (although it is consistent with deflationism, on a certain understanding of that perspective) but by the manifest inconsistency of the concept and so the absence of even a purely *logical* property of truth. This means that we can at most *pretend* that there is such a property of truth (on a prefacing view of fictionalism), or that it is only in the *fiction* of truth that there is such a property (on the prefix view of fictionalism). I say more about what such a fictionalism might look like in the next section.[19]

4 Making Believe in Truth II

If inconsistency theorists are right in their account of the conditions under which to apply this word, then 'true' doesn't have what it takes to classify sentences coherently in terms of their ability to represent the world. Note that there is nothing particularly odd in the idea that words that appear semantically and syntactically healthy don't fulfil their expected semantic functions. If, as direct reference theorists believe, a proper name is a word whose semantic function is simply to stand for an object, then 'Frodo' and 'phlogiston' both fail to fulfil their expected semantic functions. Similarly, we may have predicates that don't fulfil the semantic function of marking out a genuine property. Alongside fictional names such as 'Frodo' are fictional predicates such as 'unicorn' and 'hobbit', and alongside non-fictional empty names such as 'phlogiston' are predicates that somehow involve reference to what these names unsuccessfully denote (for instance, 'dephlogisticated'). Neither type of predicate marks out genuine properties.

The reasons for believing that this or that term doesn't mark out a genuine individual or property can vary substantially. Sometimes the reason has to do with the term's origin in the telling of an invented story, as is the case with 'Frodo' and 'hobbit'. Sometimes the reason has to do with empirical or even logical evidence that the term marks out no individual or property, as is the case with 'phlogiston' and 'dephlogisticated'.[20] There is a stark difference between the two sorts of reasons. The terms 'Frodo' and 'hobbit' occur in works of fiction, and don't disappear from language despite their failure to mark out genuine individuals or properties. They continue to have a (pretend-) use in so far as readers and movie watchers imaginatively engage with these works. That is not so in the case of 'phlogiston' and 'dephlogisticated'.

Armour-Garb and Woodbridge suggest that 'true' is like 'hobbit' rather than 'phlogiston'; it is a term that is used in the context of the pretense that the term stands

[19]An earlier foray into the kind of revolutionary fictionalism I favor is found in Kroon [20]. See also Burgess [8].

[20]Sometimes the conviction arises from logic. Consider, for instance, the following name introduction: Let 'Sylvo' name the famous Australian dialetheist who proved that $0 = 1$. Similarly, we might let a Sylvon be a follower of Sylvo, so that 'Sylvon' names a relational property if it names a property at all. Logic tells us that there is no such person as Sylvo, and hence 'Sylvo' and 'Sylvon' don't fulfil their semantic function of marking out an individual and a corresponding relational property.

for a substantive property, and it continues to be so used because of the expressive usefulness of truth-talk. As they are at pains to point out, the resulting fictionalism about truth is a form of hermeneutic fictionalism. I prefer a form of revisionary or revolutionary fictionalism on which we, or at least enlightened semantic theorists, understand that the concepts of truth and falsity are incoherent, with the result that 'true' and 'false' don't fulfil their expected semantic functions of determining genuine properties, not even genuine *logical* properties. But I agree with their emphasis on the usefulness of truth-talk. For the revolutionary fictionalist about truth, this means that it is incumbent upon us to adopt the *fiction* that there are such properties (take this to be neutral between a prefix view of fictionalism and a pretense or preface view).

Before we say more about what gets included in this "fiction of truth", consider the reasons why truth-talk is worth retaining as a fiction. If you are a correspondence theorist about truth but impressed by the arguments for truth's inconsistency, one reason would be the way ascribing truth or falsity allows one to indicate that correspondence relations hold between 'true'-free sentences (or the propositions they express) and the world. Deflationists won't be impressed by this consideration, but they can accept other considerations. In particular, they can accept more familiar accounts of the expressive usefulness of truth-talk. They can say that when coupled with other resources (e.g., quantifiers, descriptions, etc.) the truth-predicate enables its users to express agreement with (or perhaps indirectly co-assert) a certain body of claims even when they don't have direct access to these claims. (Consider, in particular, the truth-predicate's *generalizing* role, which permits speakers to endorse a potentially infinite collection of claims gathered together by a quantifier, as in 'Everything Sally believes is true', even though they lack the capacity to state infinite conjunctions and disjunctions.)[21] One doesn't need to be a deflationist to be impressed with such reasons for retaining truth-talk.

In short, despite truth's inconsistency and (hence) the non-existence of a property of truth, there are compelling reasons for retaining truth-talk, much as Hartry Field took himself to have compelling reasons for retaining number-talk despite the non-existence of numbers. But if it is to reflect such reasons, what will a suitable fiction of truth be like? For one thing, it will be internally inconsistent since the presence of the T-schema means that sentences like the Liar sentence will turn out to be both true and not true, and once we adopt it is therefore incumbent upon us to make sure that it is not deductively closed with respect to classical rules of inference (else, everything would be the case in the fiction, and such a trivialist fiction would no longer be useful). Since there appears to be no other classically acceptable way to deal with such contradictions, for example by using what Lewis calls the methods of intersection and union [24],[22] and since we certainly want the fiction to contain

[21] For an excellent account of the expressive usefulness of truth-talk from a deflationist perspective, and its capture in a version of hermeneutic fictionalism, see Chap. 4 of Armour-Garb and Woodbridge [2].

[22] Lewis used these accounts to deal with venial contradictions in fictions, such as Dr Watson's shifting war-wound in the Holmes stories. He denied that such accounts could work with blatantly inconsistent stories [24].

all *reasonable* logical consequences of the sentences contained in the fiction, the underlying logic of the fiction will need to be a paraconsistent logic like Priest's LP.

This does not mean that in using the fiction of truth we are bound by an unwieldy logic. That would nullify an argument used at the end of Sect. 2 against replacement strategies for the inconsistent concept of truth: there we concluded that we should respect the fact that combining the intuitive concept of truth and our ordinary classical patterns of inference for the most part doesn't get us into trouble. But that observation is consistent with employing a fiction of truth whose underlying logic is paraconsistent, since in using this fiction nothing forbids us from using classical patterns of inference when things are demonstrably safe, as they typically are. Even Priest thinks it is fine to rely on what he calls "quasi-valid" inferences when things are safe; that is, inferences that are truth-preserving provided all the truth-values involved are classical [30, pp. 231–237]; cf. especially methodological maxim MM). His reasons apply with equal force to the fiction of truth.

More needs to be said, of course. We need to fill out the fiction, since we want to be able to say what becomes quasi-assertible where ascriptions of truth and falsity are concerned (that is, what sentences involving these semantic predicates hold in the fiction of truth), in a way that delivers on the positive reasons mentioned above for retaining truth-talk. The justification of what gets included won't be the same as that offered by Armour-Garb and Woodbridge, since the present approach needs to rely on rules that are constitutive of the concept of truth and not simply on rules of generation. The first thing to say, then, is that the fiction should be closed under the both the descending- and ascending-truth rules. But that is not enough, since we need "true" sentences containing the predicate 'true' to belong to the fiction, sentences like 'It is true that snow is white': roughly speaking, the fiction of truth should contain all sentences that say what would be the case if there was such a property as truth, and that will certainly include factual claims like 'It is true that snow is white', 'It is true that it is true that snow is white and grass is green' and so on. So at a minimum we need the following entry rule: if S is the case (i.e., is genuinely the case, not just the case in a fiction) then S has the property of truth in the fiction of truth. Given that the fiction of truth is closed under the descending- and ascending-truth rules we can infer that if S is the case then not only does S have the property of truth in the fiction of truth, but S itself holds in the fiction of truth.

In addition, we should expect the fiction of truth to be coherent in at least the following sense. Let S be a sentence that makes no reference to semantic or other notions that only belong in a fiction; if S is genuinely and not merely fictionally the case, then not-S, unlike S, does *not* hold in the fiction of truth. (Similarly with 'not-S' and 'S' in place of 'S' and 'not-S'.) So we need to insist on an exit rule: if T holds in the fiction of truth, where T is a sentence that makes no reference to semantic or other notions that only belong in a fiction, then T holds outside the fiction. (As before, I am assuming that classical logic holds sway outside the fiction of truth.) This would show why we can legitimately infer that snow is white from, say, 'Everything Sally believes is true' (suppose this holds in the fiction) and 'Sally believes that snow is white'.[23]

This still cannot be the full story, of course. Among other things, we need to know what to do about truth-teller variants of the Liar paradox. And the approach needs to be adapted to deal with other truth-involving paradoxes like Curry's paradox. To the extent that the above fictionalism presents us with a solution to semantic paradox, it is at best a partial solution.[24]

5 Lewis's Scepticism About Impossibilities

At the start of this paper I gave a brief description of Lewis's attitude to dialetheism. Now that we have a fictionalist proposal on the table, one that has resonances of the appeal to make-believe also found in Lewis's comments, it would be interesting to see how it copes with Lewis's concerns about dialetheism. Here is a slightly expanded take on those concerns. First, he thinks there is no debating those who think they have provided good reasons for thinking there are true contradictions. When one side uses B to contest A, and so thinks A to be no less contestable than B, and the other side thinks A is absolutely uncontestable, with any moves to discredit it (such as reliance on B) far more subject to doubt, there simply is no common ground to make debate possible or fruitful. That is precisely what is the case when A is the Law of Non-Contradiction.[25] And this remains so even if we are faced with seemingly insoluble semantic and set theoretic paradoxes. As Lewis had put the point many years earlier:

> No truth does have, and no truth could have, a true negation. Nothing is, and nothing could be, literally both true and false. We know this for certain, and a priori, and without any exception for particularly perplexing subject matters. [24, p. 101]

Lewis also makes it clear in his second letter to Priest that none of this is supposed to show that we can't reason about impossibilities, in the sense of seriously considering the thought that some impossibility is true or at least possible. But he thinks that we only do this when the impossibility is subtle, not blatant. And he thinks we can't use the fact that we can do this to infer that such impossibilities must therefore be possible after all, and so must hold in some world—an impossible world. That

[23]This condition amounts to a kind of conservativeness condition. As Zach Weber reminds me, this kind of conservativeness condition is ensured in a logic like LP through the monotonicity of valuations on LP: classical values stay classical (see Priest [34, 352–353]).

[24]Thanks to Brad Armour-Garb for useful discussion on this issue.

[25]Note that Priest doesn't think the inference to a contradiction from the Liar sentence involves steps that *can't* be contested. He adopts a kind of coherentist approach on which no principle or rule of inference should be designated as uncontestable but overall simplicity and systematicity determines what stays and what goes [33]. The point remains, however, that for Lewis no such contest correctly represents the burden of proof: the Law of Non-Contradiction is the gold standard to which there can be no exceptions.

would be a step too far. To the extent that we reason about impossibilities in this way we do so by *making believe* that certain propositions that we suspect, or even know to be impossible, are possible. We don't do it by believing that they are possible.

In this paper I have sided with those who think the Law of Non-contradiction does have this kind of status. But I have also sided with those who think that the concept of truth is inconsistent, and have argued for retaining the concept of truth despite the inconsistency by treating truth as a fiction. This fiction requires us to say that the Liar is both true and false, so not true. But for p and ¬p to be both the case is not just impossible; it is surely utterly unimaginable. So it looks as if Lewis's doubts are appropriate: there is no make-believing such a fiction, which seems to spell doom for the fictionalist project. And it does so on both a prefacing and prefixing understanding of such a fiction: coherence of a fiction requires its content to be minimally imaginable, even if it can ultimately be shown not to be possible.[26]

Recall that Lewis himself expresses puzzlement about aspects of his view. In his second letter he writes that conceding, as he now does, that we can reason about impossibilities requires a distinction between the subtle ones and the blatant ones—"very likely context-dependent, very likely a matter of degree"—and that he doesn't understand how such a distinction is to be made. But one thing he remains utterly clear about is that the blatant contradictions, such as the Liar and its negation, are beyond the pale: they are impossibilities of a kind we cannot reason about—they are not even make-believe impossibilities.

How should we respond? I think by allowing that there are grades of imaginability, and that depending on the issue at hand it may be wrong to insist on the highest grade. The topic deserves a lot more space than I am able to give it here, but here is a sketch of a response to Lewis's worries. To begin with, it is important to remember that for Lewis there is no close link between imaginability and possibility: as he put it,

> [I]maginability is a poor criterion of possibility. We can imagine what is impossible, *provided that we do not imagine it in perfect detail and all at once*.[27] [26, p. 56; my italics]

The italicised clause is important, and will play a role in my response. It reminds us that even Lewis allows that we can imagine the impossible, but that he must think aspects of the scenario are in that case hidden from us. Lewis's point in his letter to Priest about being able to make believe subtle impossibilities is, I take it, a generalization of this point.

But what makes an impossibility "subtle"? Consider Priest's *Sylvan's Box*, a fictional story of how he found a box among Richard Sylvan's belongings that was both

---

[26]I take this to be a lesson of Lewis's letter to Priest, read in conjunction with Lewis [24]. Note that Lewis also holds that some fictions contain merely "venial" contradictions; their impossibility/unimaginability can be dealt with by taking them to be sums of two or more consistent fictions, and so are no further cause for concern (Lewis [24], 274 ff).

[27]Lewis adds: "It is impossible to construct a regular polygon of nineteen sides with ruler and compass; it is possible but very complicated to construct one of seventeen sides. In whatever sense I can imagine the possible construction, I can imagine the impossible construction just as well." This suggests he has in mind visual imagery, but in the discussion that follows these remarks he shows he has much bigger fish to fry.

empty and non-empty [32]. Priest takes this to show not just that it is true in the story that the box is both empty and non-empty but that we can *imagine* a situation in which the box is both empty and non-empty. Lewis, as we saw earlier, was inclined to agree, but that is surely surprising since one of the morals of Priest's thought experiment is that in the story *Sylvan's Box* it is true that the box is both empty and not empty (other hypotheses are discarded in the story), and that readers *imagine* this to be true.

How can this be? The beginnings of an answer are found in Peter Kung's recent work on the topic (see Kung [22, 23]). Consider my favorite great-uncle Jo, a bachelor to the end. Even though I now have only a vague memory of what he looked like, and even if that vague image of him most closely matches a different person whom I have never met or heard of (call him Mr X), when I imagine great-uncle Jo as a committed polygamist I am imagining something about great-uncle Jo, not about Mr X. (I could even imagine great-uncle Jo having had very different parents from the one he actually had, something that in Kripke's view would be to imagine a necessary falsehood.) Similarly, I can imagine water being XYZ even if water is necessarily H_2O. Such an imagining is not focused on stuff that merely has the outward appearance of water, contrary to what Kripke seems to have thought [23]; it is focused on water in just the way my imagining was focused on great-uncle Jo in the earlier case.

Kung puts it this way: imagination should be construed as

> *constructing* scenarios out of existing materials rather than as a telescope through which we survey genuine possibilities. When we think about imagination as constructive rather than telescopic, it becomes clearer how we can imaginatively construct impossible scenarios: we combine ingredients that together amount to an impossible situation [23, p. 95].

Such a *construction model* "respects commonsense about imagination's flexibility and power, including the power to imagine the impossible" (ibid.). Notice how it does this. First of all, the fact that I am imagining things about great-uncle Jo and water shows how some of the content of what we imagine is *assigned* content. It is as if the pictured items carry labels [22]. But not always. Some of the content is merely stipulated. If I imagine my great-uncle Jo as a polygamist but despite this still being my favourite great-uncle, these are things I stipulate, and so we might talk about stipulative content [23, pp. 107–108]. More generally and more simply, we might say that imagination is stipulative in a broad sense: the imagination stipulates not only what things there are but also how they are in the imagined situation. So construed, we can now apply the construction model to the case of our engaging with fictional stories, situations where we are involved in extended stretches of imaginative activity.

Once again, there seems to be no bar to making believe impossibilities. Take the case of *Sylvan's Box*. Even Lewis seems to have agreed he could imagine or make believe the story's plot. But how, if it contains an explicit contradiction? The answer is that the contradiction is a central part of the story, and the imagination stipulates that the story correctly describes the world. As we imagine this world we notice and go along with its weirdness, even agreeing that the simultaneous emptiness and non-emptiness of the box (rather than some weird quantum-mechanical event) constitutes the best explanation of what Priest and Griffin see. We have to work to get to the point where we agree with this. Had we been presented with an arbitrary contradiction without a backstory it is doubtful that the imaginative exercise could have worked. (Recall again what Lewis says about imagining what is impossible, *provided that we do not imagine it in perfect detail and all at once*. The story makes the lack of "perfect detail" part of the story, even though it is perfectly clear that in the story the box is both empty and not empty. A story that simply instructs you to imagine both p and ¬p, without any backstory to help and guide you, is likely to meet imaginative resistance.)[28]

It should be clear where this is heading. The fiction of truth described in Sect. 3 does *not* simply invite readers to imagine a series of contradictions. Once again, any such fiction would be likely to meet imaginative resistance. Instead, the fiction asks us to make believe that there is a genuine property of truth (the first stipulation), one that is regulated by the unrestricted T-schema (a second stipulation): stipulations that should not meet imaginative resistance since they correspond to what we unreflectively believe anyway. There will then be numerous propositions p such that p is true and p is false/not true in the fiction. But these are consequences of something that is stipulated to be so in the imaginative game we play, just as the fact that Sylvan's box is both empty and not empty is something we have to imagine on the basis of imagining the rest of the story. And as in the case of *Sylvan's Box* the story makes it clear that we are not asked to imagine this "in perfect detail and all at once". We are asked to imagine this against a backstory that should make the contradictions in this kind of case seem acceptable and tolerable where other contradictions are not. In particular, they do not involve imagining, of some proposition p that doesn't involve semantic notions, that p both holds and doesn't hold (in this sense it seems to require less than *Sylvan's Box*). Instead, they involve imagining of certain propositions p that can only be characterized in terms of a special fictional property of truth—a property that couldn't possibly exist—that p both holds and doesn't hold. Nothing in this requires a supremely difficult (content-wise) imaginative task in which we imagine, "in perfect detail and all at once", contradiction that infects the *world*.[29]

But don't we also want to say that no proposition p and its negation can both be the case? Nothing in what I have said denies this. We can even agree that there is a good sense in which we can't imagine both a proposition and its negation "in perfect detail and all at once", which may reflect the fact that the Law of Non-Contradiction is a priori known to have no exceptions. More generally, we should distinguish between

[28]For a survey of philosophical work on the imagination, including the phenomenon of imaginative resistance, see Gendler [15]. (This is the phenomenon that occurs when a subject finds it difficult or problematic to engage in some sort of prompted imaginative activity, such imagining that horrendous acts of cruelty are morally innocuous.) Gendler herself doubts that logical inconsistency provokes imaginative resistance. In Gendler [14] she offers the story of the Tower of Goldbach in which 7 + 5 is both equal to and distinct from 12. Others, eg Stock [38] finds that same scenario completely impossible to imagine. I side with Gendler, and think that doubters adopt the standard of trying to imagine all this "in perfect detail and all at once"—an inappropriate standard in my view.

[29]Something similar about such consequences is said by Beall [5]. Beall takes semantically paradoxical sentences to be just spandrels: unintended but unavoidable by-products of adding the transparent truth predicate to our language, and so not something that should surprise or worry.

succeeding in imagining a certain scenario and having evidence that the scenario is genuinely possible [22]. In the present case, the distinction has all the more bite because on the proposed fictionalism about truth there is no property of truth outside of the fiction, so the propositions that cause trouble and yet are imagined to hold when we engage with the fiction of truth are only *pretend*-propositions. Nothing so simple is the case with the (un)usual staple of contradictions that can appear in fictional stories: people going back in time who succeed in killing their grandfathers, boxes that are both empty and not empty at the same time, and so on.

6 Conclusion

The view that we can be inconsistency theorists about truth without being dialetheists has now become relatively popular. What I have suggested in this chapter is that the inconsistency theories currently on the market face some uncomfortable problems. In particular, replacement of the notion of truth, in part or in whole, just does not seem like a live option. The arguments, admittedly sketchy, that I used to press this point suggest instead that the concept of truth should be retained, despite the inconsistency, because of the practical usefulness of the concept—a revolutionary fictionalist perspective on truth.

Early on in the paper we met Lewis's scepticism about making believe that contradictions can be true. The final section addressed that worry head on. No doubt a great deal more should be said about the idea that we *can* make believe impossibilities, including blatant ones like the Liar, but my hope has been to at draw some of the sting of Lewis's complaint, and thereby help to make intellectual room (or perhaps *add* to the intellectual room) for a revolutionary, semantic-paradox-busting, form of fictionalism about truth.[30]

References

1. Azzouni, J. (2007). The inconsistency of natural languages: How we live with it. *Inquiry, 50*, 590–605.
2. Armour-Garb, B., & Woodbridge, J. (2015). *Pretense and pathology: Philosophical fictionalism and its applications*. Cambridge: Cambridge University Press.
3. Armour-Garb, B., & Woodbridge, J. (2018). Replies. *Analysis, 78*(4), 718–736.
4. Barker, J. (1998). *The inconsistency theory of truth*. Unpublished Ph.D. Dissertation. Princeton University.
5. Beall, J. C. (2009). *Spandrels of truth*. Oxford: Oxford University Press.
6. Bueno, O., & Colyvan, M. (2004). Logical non-apriorism and the law of non-contradiction. In: Priest et al. (pp. 156–175).
7. Burgess, A., & Burgess, J. P. (2011). *Truth*. Princeton, NJ: Princeton University Press.

[30]Thanks to Brad Armour-Garb, Patrick Girard, and Zach Weber for useful comments. Special thanks to the anonymous referee(s) for their valuable criticisms of an earlier draft of the chapter.

8. Burgess, A. (2018). Truth in fictionalism. In M. Glanzberg (Ed.), *Oxford handbook of truth*. Oxford University Press.
9. Chalmers, D. (2011). Verbal disputes. *Philosophical Review, 120,* 515–566.
10. Eklund, M. (2002). Inconsistent languages. *Philosophy and Phenomenological Research, 64,* 251–271.
11. Eklund, M. (2015). Replacing truth? In A. Burgess & B. Sherman (Eds.), *Metasemantics: New essays on the foundations of meaning* (pp. 293–310). Oxford: Oxford University Press.
12. Field, H. (1980). *Science without numbers*. Princeton: Princeton University Press.
13. Field, H. (1989). *Realism, mathematics, and modality*. New York: Basil Blackwell.
14. Gendler, T. (2000). The puzzle of imaginative resistance. *The Journal of Philosophy, 97,* 55–81.
15. Gendler, T. (2016) Imagination. In E. N. Zalta (ed.) *The stanford encyclopedia of philosophy* (Winter 2016 Edition). https://plato.stanford.edu/archives/win2016/entries/imagination/.
16. Grim, P. (2004). What is a contradiction? In Priest et al. (2004), 49–72.
17. Johnston, C. (2014). Conflicting rules and paradox. *Philosophy and Phenomenological Research, 88,* 410–433.
18. Joyce, R. (2001). *The myth of morality*. Cambridge: Cambridge University Press.
19. Kripke, S. (1975). Outline of a theory of truth. *The Journal of Philosophy, 72*(19), 690–716.
20. Kroon, F. (2004). Realism and dialetheism. In Priest et al. (pp. 245–263).
21. Kroon, F. (2018). Pushing the boundaries of pretence. *Analysis, 78*(4), 703–712.
22. Kung, P. (2010). Imagining as a guide to possibility. *Philosophy and Phenomenological Research, 81,* 620–663.
23. Kung, P. (2016). You really do imagine it: Against error theories of imagination. *Noûs, 50,* 90–120.
24. Lewis, D. (1978/1983). Truth in fiction. *American Philosophical Quarterly, 15,* 37–46. (Reprinted with Postscripts in his *Philosophical Papers, 1,* Oxford: Oxford University Press, 1983, 261–80).
25. Lewis, D. (1998). Logic for equivocators. In D. Lewis (ed.) *Philosophical papers* (Vol. 1, pp. 97–110). Oxford: Oxford University Press. (First published in *Noûs* 16 (1982): 431-441).
26. Lewis, D. (1986). *On the Plurality of Worlds*. Oxford: Blackwell Publishers.
27. Lewis, D. (2004). Letters to Priest and Beall. In Priest et al. (2004), 176–7.
28. Lewis, D. (2005). Quasi-realism is fictionalism. In M. Kalderon (Ed.), *Fictionalism in metaphysics* (pp. 314–321). Oxford: Oxford University Press.
29. Patterson, D. (2009). Inconsistency theories of semantic paradox. *Philosophy and Phenomenological Research, 79,* 387–422.
30. Priest, G. (1979). The logic of paradox. *Journal of Philosophical Logic, 8,* 219–241.
31. Priest, G. (1987). *In contradiction: A study of the transconsistent*. The Hague: Martinus Nijhoff.
32. Priest, G. (1997). Sylvan's box. *Notre Dame Journal of Formal Logic, 38,* 573–582.
33. Priest, G. (1998). What's So bad about contradictions? *Journal of Philosophy, 95,* 410–26. (Reprinted in Priest et al. (2004), pp. 23–38).
34. Priest, G. (2002). Paraconsistent logic. In D. M. Gabbay & F. Guenthner (Eds.), *Handbook of philosophical logic* (2nd ed., Vol. 6, pp. 287–393). Dordrecht: Kluwer Academic Publishers.
35. Priest, G., Beall, J. C., & Armour-Garb, B. (Eds.). (2004). *The law of non-contradiction: New philosophical essays*. Oxford: Oxford University Press.
36. Scharp, K. (2013). *Replacing truth*. Oxford: Oxford University Press.
37. Scharp, K. (2013). Truth, the liar, and relativism. *Philosophical Review, 122*(3), 427–510.
38. Stock, K. (2003). The tower of goldbach and other impossible tales. In M. Kieran & D. Lopes (Eds.), *Imagination, philosophy and the arts* (pp. 106–124). London: Routledge.
39. Walton, K. L. (1990). *Mimesis as make-believe: On the foundations of the representational arts*. Cambridge, MA: Harvard University Press.

Dialetheism and Impossible Figures

Chris Mortensen

Abstract Dialetheism is the thesis that there are true contradictions. But humans tolerate contradictions without necessarily believing them to be true. Two sources are (a) contradictions in mathematics, which are dialethic only if one is a realist, non-fictionalist about mathematics (b) contradictory images such as the impossible triangle, which are certainly not true in the real world. This paper explores these ideas, and concludes that impossible figures give us examples of non-dialethic contradictory contents.

1 Introduction

We begin with some definitions. An L-*theory* is a set of propositions (of a background language) closed under the deducibility relation ⊢ of some logic L (we often drop the L when it is clear what is meant). An *inconsistent* theory is a theory containing some proposition A and its negation ~A. A theory is *trivial* if it is identical with its language, i.e. if contains every proposition of its language; otherwise, it is *nontrivial*. A *paraconsistent* logic is one which has nontrivial inconsistent theories. Inevitably, a paraconsistent logic lacks the Boolean rule *ex contradictione quodlibet ECQ*, from a contradiction every proposition may be validly deduced, i.e. A, ~A ⊢ B.

Dialetheism is the thesis that there are true contradictions. (An alternative, whose equivalence is not disputed here, is to say that some proposition and its negation are both true.) If dialetheism is true, then natural logic must be paraconsistent, since the world (more exactly, the true theory) is manifestly non-trivial. For that reason, dialetheism has sometimes been called *strong paraconsistency* [1]. But the converse thesis fails, that is, dialetheism does not follow from the thesis that natural logic is paraconsistent. This became apparent in semantical studies which showed that relevant logics and other good candidates for natural logic had to admit nontrivial inconsistent theories independently of dialetheism. For such semantical studies see Routley and Routley [6], or Routley and Meyer [7]. This weaker thesis, namely that

C. Mortensen (✉)
University of Adelaide, Adelaide, Australia
e-mail: chris.mortensen@adelaide.edu.au

A. Rieger and G. Young (eds.), *Dialetheism and its Applications*, Trends in Logic 52,
https://doi.org/10.1007/978-3-030-30221-4_7

natural logic admits nontrivial inconsistent theories, irrespective of dialetheism, has sometimes been called *weak paraconsistency.*

2 Sources of Paraconsistency

What are the sources of paraconsistency (strong or weak)? The literature contains a number of candidates. First there are the *semantical paradoxes,* such as The Liar, Grelling's etc. Now as usually presented (e.g. by Graham Priest) these paradoxes represent a case for strong paraconsistency: e.g. the Liar sentence is both true and false. While the present author has sympathy for this position, especially for its brilliant simplicity, nonetheless it is set aside here, in the interests of reaching a conclusion that should be attractive to all.

A second source of paraconsistency is the foundations of mathematics, such as the *set-theoretic paradoxes*. These include Russell's, according to which the Russell set R both is and is not a member of itself (or, it is both true and false that R is a member of itself). This might initially look like more strong paraconsistency. However, the evident similarity between Russell's paradox and the Barber paradox should not tempt anyone into thinking that the latter, the Barber, similarly gives us that it is both true and false that the Barber shaves themself. The Barber is no paradox, just a fictional story, certainly no reason to think it *true*. In contrast, we derive the Russell set from naïve comprehension which is natural and simple, and thus arguably true.

But this rather gives things away, because to argue for dialetheism from paradoxes in mathematics requires an argument against mathematical fictionalism, and there are plenty of fictionalists these days. For example, the present writer confesses a dislike for set-theoretic realism, particularly the null set. On the other hand, even if such a fictionalist strategy succeeds against strong paraconsistency in mathematics, weak paraconsistency remains untouched: inconsistent mathematical contents require distinctions to be made which ECQ simply cannot tolerate. We return to this below.

One more source of dialetheism is the observation that our best physical theories contain persistent anomalies, such as black holes, quantum gravity, the first instant of time, to the point that one should not feel *entirely* confident that our best theories *long term,* our best long term source of physical truths, might continue to be anomalous. This is hardly a conclusive argument for strong paraconsistency, needless to say. It is more of a caution to be open-minded about the possibility.

And then there is cognitive paraconsistency. In my experience, many paraconsistentists cannot accept dialetheism of any kind, but agree that natural logic is a logic of *information*, or alternatively intentional states, broadly construed as arising in the context of *cognition*. The argument then concludes that since cognition is inconsistency-tolerant, so must natural logic be. This position is sometimes described as the *cognitive justification of paraconsistency.* Now, since beliefs or other cognitive states need not be true, cognitive paraconsistency is not dialetheism, but rather a species of weak paraconsistency. For an example in which the perception of motion is seen as inconsistent, see Mortensen [3].

3 Impossible Figures

We turn then to our main theme, the contribution of impossible figures to the understanding of paraconsistency. The impossible figures movement got underway properly in 1934 when the young Swede Oscar Reutersvärd drew what has come to be called somewhat erroneously in the literature as the Penrose Triangle (see the top left of the five forms below). An important contributor to the movement was the formidable M. C. Escher, but he was not the first. The terminology "impossible figures", IF, apparently due to Teddy Brunius, is fairly settled by now, so we stick with that. Even so, the kind of impossibility which is up here, is not mere physical impossibility, that is contrary to natural law, but something stronger: incompatibility at least with logic or maybe mathematics also (it is not intended here to beg the question either for or against logicism).

It will help if we can look at some IFs (Fig. 1).

Of these five forms, schematized from the originals, the top three were first drawn by Reutersvärd. The top two left, the Penrose Triangle and the Stairway, were rediscovered by Roger Penrose in 1956. The bottom left was drawn by Escher in his masterpiece *Belvedere* (1958), and bottom right the Steps is by Escher's student Bruno Ernst (1984).

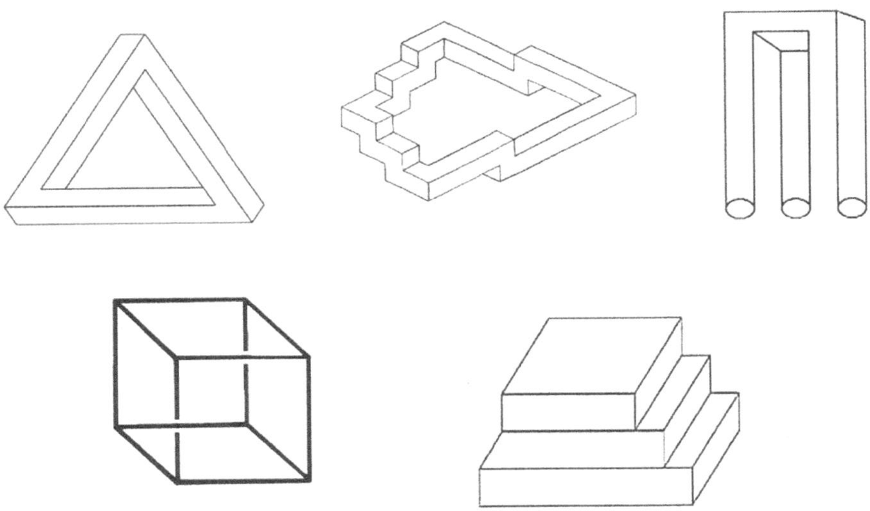

Fig. 1 Five forms

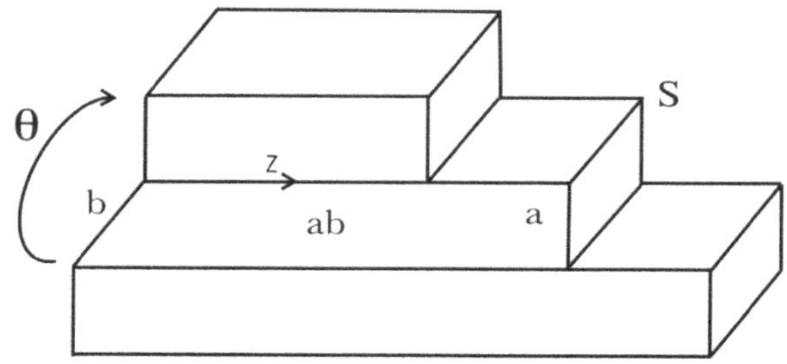

Fig. 2 The stairway lettered

Now the claim has been made that these are impossible. But clearly they are sitting there for all to see. So what is impossible about them?

Many pictures have a content, and among contents there are 2-D contents and 3-D contents. 3-D contents are obtained when we project into the third dimension: good examples include perspective and occlusion. The 2-D aspects of the above images are obviously not impossible; but *something* is impossible about them, which indicates that it is the 3-D contents that are impossible. More exactly, what makes it impossible is that it has contradictory 3-D content: the mind projects a contradictory 3-D theory as part of the content, of how it seems.

We illustrate this with a quick proof for the case of the Ernst Steps. First the figure is lettered (taken from [2, 130]) (Fig. 2).

Now the argument goes: (1) **a** is vertical (2) **b** is horizontal (3) **a** is coplanar with **b**, since both lie in the plane **ab**, and (4) **a** and **b** arbitrarily extended do not meet. Note that the fourth premiss (4) is necessary, since without it (1)–(3) are mutually consistent—a vertical and a horizontal can meet in the one plane, but not if they are skew to each other. Taking (3) and (4) together, then, we have that **a** and **b** are parallel; but then if **a** is vertical, so must be **b**. That contradicts (2), the horizontality of **b.** Note in passing that these premisses are all part of the 3-D contents of the image, that is why simple 2-D won't do.

This argument essentially follows Ernst. It can be seen in more detail in Mortensen [2, 130–3], along with arguments for contradictoriness in the cases of the other Forms (Chaps. 9–15).

One significant matter concerns the premiss that **ab** is a plane. That is right about the way the figure looks. But if the figure derives from a physical object, then it might be that **ab** is not a plane, but twists from horizontal to vertical without that being noticeable. Then there would be a consistent object as a source of the figure, the way the figure looks would be the way a certain consistent object looks. Indeed, one can see why this can arise in a consistent world: there seems to be a default setting in our perceptual apparatus; if a twist is not perceived, our perceiver defaults to *flat*.

The same can be said for several of the other Forms. It is well-known that it is possible to build objects which photograph looking like that. (A possible exception is the Fork, top right.) For example, the Triangle can be built not-joined-up; but photographed from a particular angle gives the illusion that it looks joined up. Our perceptual apparatus obviously employs another default mechanism, whereby items at a small angular distance to one another look like they are also at a small radial distance to one another. It is easy to demonstrate this by lining up fingers from opposite hands so that they look touching even though far apart.

But here is an important point: *Reutersvärd was not drawing illusory aspects of existing physical objects: he made his figures up*. In so doing, he showed that it was possible to have contradictory visual content independently of there being any physical objects that might look that way from a preferred aspect. As Reutersvärd noted himself, *his figures were not illusions.*

And here is another important point. *Figures, drawings and other images do not carry with them any ontological commitment.* It is all too easy to pick up a pencil and draw! The Lady with the Unicorn is just a fiction. That is, the existence of impossible figures does not deliver dialetheism: what it delivers is weak paraconsistency. Mind you, the same is true for text. It is easy to cancel out real-world assertion, you just start with "Once upon a time…" and you are immediately exempt from providing evidential support for truth.

Indeed, this opens the way for us to see that humans like to play around not merely with false content, but even with contradictory content. In his short story "Sylvan's Box", Priest [5] made this point. The story turns on the supposition that Priest and Griffin discovered among Richard Sylvan/Routley's *Nachlass* a box containing an object with contradictory properties. Priest was making the weak paraconsistentist point that different contradictory stories are not generally equivalent to each other, as they would be if they were governed by ECQ. The present writer registers emphatic assent, and notes in passing that this also opens a door for mathematical fictionalism and inconsistent mathematics. But that is a longer story.

4 Conclusions

1. The contents of the human mind and imagination go beyond the merely consistent.
2. Humans like to play around with inconsistent contents; without existential commitment to inconsistent truths i.e. dialetheism.
3. Drawing impossible figures demonstrates the capacity for weak paraconsistency.
4. In any case, the world occasionally affords us inconsistent content deriving from conveniently-viewed objects and utilising our various perceptual default mechanisms.

References

1. Mortensen, C. (1989). Paraconsistency and C1. In G. Priest, R. Routley, & J. Norman (Eds.), *Paraconsistent logic: Essays on the inconsistent* (pp. 289–304). Miinchen: Philosophia Verlag.
2. Mortensen, C. (2010). *Inconsistent geometry*. London: College Publications.

3. Mortensen, C. (2012). Motion perception as inconsistent. *Philosophical Psychology,* online June 2012, 1–12; paper December 2013, *26*(2), 913–924.
4. Mortensen, C., et al. (2013). How many impossible images did Escher produce? *The British Journal of Aesthetics, 53*(4), 425–441.
5. Priest, G. (1997). Sylvan's box. *Notre Dame Journal of Formal Logic, 38*, 573–582.
6. Routley, R., & Routley, V. (1972). Semantics of first degree entailment. *Nous, 6*, 335–358.
7. Routley, R., & Meyer, R. K. (1973). The semantics of entailment I. In H. Leblanc (Ed.), *Truth syntax and modality* (pp. 199–243). Amsterdam: North-Holland.

Making Truth Safe for Intuitionists

Andrew Tedder and Stewart Shapiro

Abstract We consider a handful of solutions to the liar paradox which admit a naive truth predicate and employ a non-classical logic, and which include a proposal for *classical recapture*. Classical recapture is essentially the property that the paradox solvent (in this case, the non-classical interpretation of the connectives) only affects the portion of the language including the truth predicate—so that the connectives can be interpreted classically in sentences in which the truth predicate does not occur. We consider a variation on this theme where the logic to be recaptured is not classical but rather intuitionist logic, and consider the extent to which these handful of solutions to the liar admit of intuitionist recapture by sketching potential ways of altering their various methods for classical recapture to suit an intuitionist framework.

1 Classical Recapture

The Liar paradox has been solved, many times over. It is widely (but not universally) agreed that the principles or rules involving truth lead to contradiction, and thus triviality, using classical or intuitionistic logic. The principles in question are the instances of the truth-scheme:

$$T\langle A\rangle \rightarrow A$$
$$A \rightarrow T\langle A\rangle$$

where $\langle A\rangle$ is a name (or code) of the sentence A. Alternately, one can use the corresponding rules of inference, allowing the derivation of A from $T\langle A\rangle$, and vice versa,

A. Tedder (✉)
Institute of Computer Science, The Czech Academy of Sciences, Prague, Czech Republic
e-mail: ajtedder.at@gmail.com

S. Shapiro
Ohio State University, Columbus, OH, USA
e-mail: shapiro4@osu.edu

A. Rieger and G. Young (eds.), *Dialetheism and its Applications*, Trends in Logic 52,
https://doi.org/10.1007/978-3-030-30221-4_8

or else a prima facie stronger "transparency rule" that allows the substitution of any formula in the form $T\langle A\rangle$ with the corresponding formula A, and vice versa.

Given the problem of the Liar (and the other semantic paradoxes), there are two kinds of solutions. One is to maintain classical (or intuitionistic) logic and to propose restrictions on the truth rules, and the other is to maintain the full truth rules—either the truth-scheme, the corresponding rules, or transparency—and to propose a different, weaker logic, one that blocks the derivation of the contradiction, or at least blocks the triviality.[1] Our concern here is with the latter, with theories that accept the full truth-rules, but operate with a weaker logic.

There are a host of such solutions on the philosophical (and logical) market using a variety of non-classical logics, such as many-valued logics, substructural logics, contraction-free logics, and various combinations thereof.

A standard to which these solutions are usually held is sometimes called *classical recapture*. The idea is that the paradox solvent should affect that portion of the language including the offending vocabulary—the truth predicate in most cases—but not the portion of the language that excludes the offending vocabulary. That is to say, the non-classical logic is supposed to restrict the use of certain classical inferences *where the truth predicate is concerned*, but when dealing with sentences in which the truth predicate does not occur, all classically valid inferences should be admissible.

A picture of what usually goes on is as follows: first, one starts with a non-trivial classical theory. This is typically taken to be Dedekind-Peano arithmetic or Zermelo-Franekel set theory (usually first-order). One then adds the truth predicate, with the appropriate axioms or rules, unrestricted, while weakening the logic. One then argues that, with some well-motivated additions, the initial mathematical theory is still included in the resulting non-classical theory—via classical recapture. So, the argument goes, the weakening of the logic only affects the addition to the theory including the truth predicate, not the original mathematical theory.

Our first observation here is that the original theory is almost always a classical one, not an intuitionistic mathematical theory.

There seems to be a schism among the philosophers, mathematicians, and logicians who study and use non-classical logics. On one hand, there are those who focus exclusively on intuitionistic logic, and perhaps logics intermediate between classical and intuitionistic. One focus of this community is on constructive mathematics, in the tradition of L. E. J. Brouwer and Arend Heyting. A related focus is on anti-realism, and, in particular, on an anti-realist understanding of the logical terminology, through the extensive work of Michael Dummett, Dag Prawitz, Neil Tennant, and others. Of course, intuitionistic logic rejects the validity of excluded middle, but accepts *ex falso quodlibet*, sometimes called "explosion". So it does not seem to resolve the alethic paradoxes.

The other group of non-classical philosophers, mathematicians, and logicians focus on the many-valued logics and relevant logics built on **FDE**, such as **R**. Many

[1] To be sure, one can restrict both the truth rules and propose a different logic, but we do not consider such options here.

of these logics are paraconsistent, rejecting explosion (and related principles like thinning: $A \rightarrow (B \rightarrow A)$), while maintaining excluded middle. Many of those in this camp are interested in theories with an unrestricted truth predicate; others are interested in maintaining a requirement of relevance between the premises and the conclusions of a valid argument.

Others in this camp adopt "paracomplete" logics that reject the validity of excluded middle, as well as other principles accepted by intuitionism, such as $\neg\neg(A \vee \neg A)$. Unlike intuitionistic logic, these paracomplete logics typically accept the equivalence (in some sense) between A and $\neg\neg A$. A prominent focus of this sub-group is theories with an unrestricted truth predicate.

These two groups of logicians—those who focus on intuitionistic logic, and everyone else—are largely disjoint and do not interact with each other very much.[2] Moreover, virtually all of the work on truth that concerns us—those who maintain the truth rules while weakening the logic—are in the second camp, the one that pays little attention to intuitionistic logic. As noted, their focus on *classical* recapture.

It seems that the focus on the logicians and philosophers in the first group—the one that considers (only) intuitionistic logic—are less concerned with truth, or at least with an unrestricted truth predicate. As noted, the focus is on constructive mathematics and/or anti-realism.

The aim of this paper is to assess the prospects for what may be called *intuitionistic recapture*. We start with a more or less standard intuitionistic theory. Heyting Arithmetic (HA) will do, but perhaps a better case study would be Heyting arithmetic together with an intuitionistic version of Church's thesis (stating, in effect, that all number-theoretic functions are recursive). That theory is actually inconsistent with classical logic—and becomes trivial (i.e., explodes) if excluded middle is added. We wish to add an unrestricted truth predicate to the language, while restricting the logic somehow, and then show how to get the original intuitionistic theory back via a kind of recapture.

We will not start from scratch, coming up with a brand new logic for this purpose. Frankly, we think there are enough logics on the market right now (at least for present purposes). We explore the prospects for adapting some of the accounts of truth to this case. The paradox paradigms we'll work with include:

- Neil Tennant's [44] **Core Logic** approach
- David Ripley's [34] *cut-free* approach
- Hartry Field's [9] **K3**-based approach
- Jc Beall's [4] **LP**-based *detachment-free* approach
- Graham Priest's [27] *contraction-free* approach

We'll briefly survey each approach and consider the extent to which it can be modified to fit an intuitionist framework.

---

[2]Notable exceptions are the extensive work by Michael Dunn, Neil Tennant's work on *core logic*, and a forthcoming book by Robles and Mendez [36].

2 Plans A and B

When it comes to bridging the gap between intuitionist logic and those non-classical logics which are our focus here, we shall distinguish two *plans*:

(Plan A) Replace the negations of the logics in question with an appropriate (*sub-*) *intuitionist* negation.
(Plan B) Build an intuitionistically flavoured logic *inside* the logics in question by means of a variation on the Gödel translation from the classical modal logic **S4** to intuitionist logic.

Of these plans, A is clearly the gold standard, since B is somewhat roundabout and involves, at least, adding **S4**-style modal vocabulary to these non-classical logics. However, an issue for plan A arises immediately in considering existing proof systems for **LP**, **K3**, and the family of *relevant* logics employed by Priest and others. These logics all share a core DeMorgan negation, embellished by some additional principles. As the name suggests, a feature in common between the negations of all these logics is that all validate both double negation elimination and the intuitionistically unacceptable DeMorgan law:

$$\neg(A \wedge B) \vdash \neg A \vee \neg B$$

This may not seem a serious problem, *prima facie*, but the standard proof systems for these logics rely heavily on *all* of the DeMorgan laws.[3] The issue concerns how *negations* are handled. For instance, in the standard style of sequent system for **FDE**, **LP**, and **K3**, one employs rules for introducing formulae with various main-connectives, besides negation, on the left and the right of the turnstile. This much is more or less standard. One then has rules for introducing negated atomic formulae, and rules for introducing formulae with negated connectives.

For instance, the following rule introduces negated *conjunctions* on the left:

$$\frac{\Gamma, \neg A \vdash \Delta \qquad \Gamma, \neg B \vdash \Delta}{\Gamma, \neg(A \wedge B) \vdash \Delta}$$

Assuming the standard rules for $\vee$,[4] the rule above provides us the offending DeMorgan law:

$$\frac{\dfrac{\neg A \vdash \neg A, \neg B \qquad \neg B \vdash \neg A, \neg B}{\neg(A \wedge B) \vdash \neg A, \neg B}}{\neg(A \wedge B) \vdash \neg A \vee \neg B}$$

[3]For a taste of these proof systems, see the Tableaux systems of Priest's [28, Chaps. 7 and 8] or [5], or the sequent systems of [1, Sect. 19].

[4]One can insert one's preferred rules for $\vee$ to the same effect (perhaps needing some additional applications of structural rules).

So this feature of logics with DeMorgan negation is a stickier issue than one might think. It has the result that for our project to proceed via plan A, we would need some new proof system for the **FDE**-related logics.

Without digging into the involved logical work of adapting existing proof systems to include a sub-intuitionist negation, or of devising some new kinds of proof system to do this task, the best option is to focus on plan B. To that end, the Sect. 4 lays out a model structure for **S4**-style modalities in **FDE**, **K3**, and **LP**. The language is defined as usual to include the connectives $\wedge$, $\vee$, $\neg$ (and $\rightarrow$, $\leftrightarrow$ to be defined), as well as quantifiers $\forall$, $\exists$. Before digging into this approach, we'll consider Ripley and Tennant's approaches and the Plan A options available to them.

3 Cut and Transitivity: Ripley and Tennant

3.1 Tennant

Neil Tennant is a notable exception to the above schism in the community of non-classical logicians. He has long advocated a logical framework, now dubbed *core logic*, that is both intuitionistic and relevant (although not following the more familiar relevance logicians, based on Anderson et al. [1, 2]; see, for example, Tennant [41, 42]).

All derivations in core logic are in *normal form*, in that there can be no occurrence of a formula that is both the conclusion of an elimination rule for a given connective and the major premise of an introduction rule for that same connective. Consequently, the validity relation of core logic is not transitive: the result of chaining together two core proofs might not be a core proof.

Tennant established a nice meta-theorem that, in effect, restores the benefits of combining proofs. If a set of premises Γ entails a conclusion A via intuitionistic logic, then either there is a core proof of A from a subset of Γ or there is a core proof of absurdity from a subset of Γ. So if one shows that a given argument is valid in intuitionistic logic, then there is a core deduction of either the same argument or a better one—an argument with fewer premises and/or an argument whose conclusion is absurdity. The result is an analogue of the usual algorithm for transforming a proof in intuitionistic logic into a normal proof—typically as part of a cut elimination theorem.

There is a similar result for a system that Tennant calls "classical core logic" which is the result of adding something equivalent to excluded middle to core logic. If a set of premises Γ entails a conclusion A via classical logic, then either there is a classical core proof of A from a subset of Γ or there is a classical core proof of absurdity from a subset of Γ.

Tennant [43] has recently turned his attention to truth. In line with his basic logicism and inferentialism, the truth predicate is introduced via an introduction rule

$$A \vdash T\langle A\rangle$$

and an elimination rule:

$$T\langle A\rangle \vdash A$$

Both rules are unrestricted. If the background language and theory are rich enough, a Liar sentence λ can be formulated, where λ is equivalent to (or even identical with) $\neg T\langle\lambda\rangle$.

In the resulting system, there is indeed a core proof of λ, a derivation of $\neg T\langle\lambda\rangle$ with no undischarged premises [43, p. 586; Lemma 3]. And there is a refutation of λ, a core proof of absurdity whose only undischarged premise is $\neg T\langle\lambda\rangle$ [43, p. 586; Lemma 4]. However, Tennant shows that these two proofs cannot be "chained together". Any attempt to normalize the two derivations results in a loop, and never terminates. So there is no reason to think that the system allows a core proof of absurdity, with no undischarged premises.

Something similar holds in classical core logic as well. The truth rules do allow the derivation and the refutation of a Liar sentence, but there is no global damage. The proof and refutation do not chain together, because the resulting combination cannot be normalized.

Nothing changes when arithmetic or other mathematical terminology is added to the system, along with appropriate axioms or rules of inference. So an unrestricted truth predicate can be added to any intuitionistic or classical theory. However, one can only reason with core logic (or classical core logic) in the resulting system.

The aforementioned meta-theorems allow for a kind of recovery. If a reasoner does not invoke either of the truth rules, then she can proceed with ordinary intuitionistic (or classical) logic. She is assured that if she derives a conclusion that way, then she can either derive the same conclusion or absurdity from the same or fewer premises in core (or classical core) logic.

There are two ways to interpret the resulting framework. One can think of it as a variety of dialetheism. We do indeed have a kind of "true contradiction", as witnessed by the core proof of λ and the core refutation of λ. But the system does not "explode". There is, it seems, no way to combine that proof and that refutation into an outright proof of absurdity from no undischarged premises.

Tennant does not understand the system that way. As he sees things, the refutation of λ serves as a kind of defeater for the original proof of λ. Once the truth predicate is added to the system, one cannot think of a given sentence as established—as proved—unless it has a core proof *and* we are also confident that there is no defeater—no core refutation of the same formula. As Tennant [43, p. 578] puts it:

> When a speaker sincerely asserts a declarative sentence, the listener is given to understand that the speaker has a warrant to back the assertion, or at least possesses an effective method for finding such a warrant, if one is called for. Also—as needs to be emphasized—the

> listener is given to understand that the speaker knows of no *countervailing* or *undercutting* or *defeating* construction or considerations that would diminish or cancel the justificatory force of the proof on offer. In mathematics, for example, the default consensus is that proof is dispositive. For any mathematical sentence, if one has a proof of it, then one cannot also have a disproof of it.

It is crucial to mathematics that one can chain together proofs of previously established results. One can do this only if one is confident that the overall system is consistent, or "dispositive" as Tennant puts it: there is no proof and refutation of the same sentence. Despite the second incompleteness theorem, we can be reasonably confident that ordinary mathematics is "dispositive" in this way. Or, perhaps better, the consistency of the enterprise is a kind of presupposition of it. Mathematicians do not trace their results back to core proofs (or core classical proofs). Once an unrestricted truth predicate is added, however, one must indeed take care to make sure that any proof is not matched by a disproof. It might seem that this is a good reason to not introduce the truth predicate.

3.2 Ripley

Ripley [34, 35] works from a "bilateralist" inferentialist perspective, taking the meanings of the logical terminology to be given by assertability and deniability conditions. The framework is a multiple-conclusion sequent calculus. One driving thought is that if Γ and Δ are sets of sentences, then the sequent:

$$\Gamma \vdash \Delta$$

is valid just in case it is "out of bounds" to simultaneously assert every member of Γ and deny every member of Δ (following Restall [32, 33] and Brandom [7]). The usual left and right rules for the logical terminology in classical logic are motivated in terms of these primitive notions.

The system does not have a rule of cut. But if we stick to only the basic system, one can show that cut is a derived rule. That is, if we have that both:

$$\Gamma, A \vdash \Delta \quad \text{and} \quad \Gamma \vdash A, \Delta$$

then we also have:

$$\Gamma \vdash \Delta$$

Then Ripley shows that one can add a fully transparent truth predicate to the system. And, as usual, we assume that there is a Liar sentence λ, which is equivalent to (and perhaps even identical to) $\neg T\langle\lambda\rangle$.

The resulting system is a conservative extension of classical logic. The same goes if we begin with any formalized mathematical theory. But like Tennant's core logic and classical core logic, the resulting system is not transitive; there is no "cut lemma". Indeed, the truth predicate makes for some counterexamples to cut.

Thomas [46] observed that the treatment readily extends to an intuitionistic background. Begin with any intuitionistic sequent calculus that does not have cut as a primitive rule. The most common approach, by far, is to start with the classical system, but to restrict the sequents to those with at most one sentence on the right (or exactly one, if there is a sentence standing for absurdity). The result is intuitionistic logic.

As in the classical case, one can prove that cut is an derived rule (if no new rules are added to the system). And, as in the classical case, one can add a fully transparent truth predicate. The result, again, is that cut is no longer a derived rule and, indeed, there are counterexamples to it.[5]

So both of our Plan A approaches make truth safe for intuitionistic logic giving up transitivity. In both systems, one cannot, in general, just use previously proved lemmas when proving theorems. That is the price to be paid. Given the importance of cut, or transitivity, to both classical and intuitionistic mathematics, this cost is considerable. We leave it to the reader to decide if this cost for a transparent truth predicate is too high.

But once the cost is paid, recovery is straightforward. If the truth rules do not come into play, then cut is indeed a derived rule, and can be used with abandon.

4 Model Structure for $\mathbf{FDE}_{\mathbf{S4}}$

Much of the work on truth involves the sub-classical logic called *first-degree entailment* (**FDE**), or extensions thereof. To develop our Plan B, we sketch here a way to add the modal logic **S4** to **FDE**. We dub the system $\mathbf{FDE}_{\mathbf{S4}}$.[6]

[5]Thomas develops a Kripke-style model-theoretic semantics for this framework, but those details do not matter here. We might add that it seems to us that the Ripley account of logical consequence, inspired by Restall and Brandom, is not appropriate for intuitionism. The intuitionist *does* take it to be "out of bounds", i.e., incoherent, to assert a sentence in the form $\neg\neg A$ while denying A. But, of course $\neg\neg A \vdash A$ is not intuitionistically valid. Equivalently, it is incoherent to deny an instance of excluded middle, but $\vdash (A \vee \neg A)$ is not valid. To put the point another way, suppose that one starts with an intuitionistic deductive system, and then defines a new technical term as follows: An argument $[\Gamma, \phi]$ is *Ripley-valid* just in case it is incoherent to assert every member of Γ and to deny ϕ. Then Ripley-validity is the same as classical validity.

[6]This sketch isn't new—propositional modal and many-valued logics have been studied extensively, and details are available in [28, Chap. 11].

4.1 The Framework

We assume a standard formal language. Officially, we have only the connectives '$\neg$', '$\wedge$', and '$\vee$'. The material conditional is introduced as an abbreviation: '$A \supset B$' is short for '$\neg A \vee B$'. We have the usual quantifiers '$\forall$' and '$\exists$', and the two usual modal operators '$\Box$' and '$\Diamond$'.

A *frame* $\mathcal{F}$ is a 5-tuple $\langle W, @, R, D, d\rangle$ satisfying the following conditions[7]:

- $W \neq \varnothing$ (a set of *worlds*, *points*, or whatever kind of thing you prefer)
- $@ \in W$
- $R \subseteq W^2$ is reflexive and transitive, and for all $w \in W$, $R@w$

R is the accessibility relation on the worlds, and @ is the 'base world' of the frame. All worlds are accessible from @.

- $D \neq \varnothing$
- $d : W \longrightarrow \wp(D)$, s.t. $d(@) \neq \varnothing$, and for all worlds w, w', $Rww' \Rightarrow d(w) \subseteq d(w')$

D is the *domain* of the frame, the collection of objects in all worlds. For each world w, $d(w)$ is the set of objects that exist in w. We have it that the domains do not shrink along the accessibility relation. So the domain of every world is non-empty.

A model $\mathcal{M}$ of $\mathbf{FDE_{S4}}$ consists of a frame $\mathcal{F} = \langle W, @, R, D, d\rangle$, as above, and three functions I, I^1, I^0, with the following conditions:

- If c is a constant, then $I(c) \in d(@)$

Our first function I gives the denotation of the constants. We insist that each denoted object exists in the base world, and thus in all worlds of the frame, and that the constants are *rigid*, in the sense that each one denotes the same object in all worlds.

- If P is an n-place predicate and w a world, then $I^1(P, w) \subseteq d(w)^n$
- If P is an n-place predicate and w a world, then $I^0(P, w) \subseteq d(w)^n$

$I^1(P, w)$, $I^0(P, w)$ are called the *extension* and *anti-extension* of P in w, respectively. The former gives the n-tuples (among the objects that exist in w) that P is true of, in w, and the latter gives the n-tuples (among the objects that exist in w) that P is false of, in w.[8]

If $\mathcal{M}$ is a model, then define a *variable assignment* on $\mathcal{M}$ to be a function from the set of variables to the domain D of $\mathcal{M}$. If g is a variable assignment on $\mathcal{M}$ and w is a world in the frame of $\mathcal{M}$, say that g is *apt for* w if, for each variable x, $g(x) \in d(w)$.

[7] $\Rightarrow$ and $\Leftrightarrow$ stand in for the metalanguage conditional and biconditional, respectively.

[8] Arguably, mathematical predicates are *stable*, in the sense that if an atomic sentence is true (respectively false) in a given world, then it is true (false) in all accessible worlds (see Linnebo [15] or [16]). So if our attention restricted to mathematical languages, we could just think of our extensions and anti-extensions as functions from predicates to n-tuples of the whole domain D. The translation from the present language into that of intuitionism, in the next sub-section, will have this effect.

In words, a variable assignment is apt for a world just in case it assigns objects in that world to each variable.

As usual, constants and variables are called *terms* in the language. Let $\mathcal{M}$ be a model, g a variable assignment on $\mathcal{M}$, and t a term. If t is a constant, then its denotation on $\mathcal{M}$ under g is $I(t)$, and if t is a variable, then its denotation on $\mathcal{M}$ under g is $g(t)$.

We can now give a recursive definition of a valuation function v. It takes a model, a world, a variable assignment apt for that world, and a formula in the language, and gives a subset of $\{1, 0\}$. If 1 is a member of the valuation function for a world in a model under a variable assignment apt for that world, then we say that the formula is *true* in that world under that variable assignment. And if 0 is a member of the valuation function for a world in a model under a variable assignment apt for that world, then we say that the formula is *false* in that world under that variable assignment. In the general case, we allow that a formula can be true only, false only, both true and false, or neither true nor false—in a given world in a given model under a given variable assignment. This happens when the valuation function yields $\{1\}$, $\{0\}$, $\{1, 0\}$, and $\varnothing$, respectively.

Let $\mathcal{M}$ be a model, w a world in $\mathcal{M}$, g a variable assignment apt for $\mathcal{M}$. The following are the clauses for the valuation function v. We sometimes suppress mention of the model and variable assignment.

For the base clause, let P be an n-place predicate and let $t_1, \ldots, t_n$ be terms.

- $1 \in v(P(t_1, \ldots, t_n, w)$ if and only if the n-tuple consisting of the denotations of $t_1, \ldots, t_n$ is in $I^1(P, w)$
- $0 \in v(P(t_1, \ldots, t_n, w)$ if and only if the n-tuple consisting of the denotations of $t_1, \ldots, t_n$ is in $I^0(P, w)$

This captures the idea that I^1 gives the extension of each predicate and I^0 gives the anti-extension of each predicate.

Now let A, B be any formulas. The clauses for the logical connectives are straightforward:

- $1 \in v(\neg A, w) \Leftrightarrow 0 \in v(A, w)$
- $0 \in v(\neg A, w) \Leftrightarrow 1 \in v(A, w)$
- $1 \in v(A \wedge B, w) \Leftrightarrow 1 \in v(A, w) \cap v(B, w)$
- $0 \in v(A \wedge B, w) \Leftrightarrow 0 \in v(A, w) \cup v(B, w)$
- $1 \in v(A \vee B, w) \Leftrightarrow 1 \in v(A, w) \cup v(B, w)$
- $0 \in v(A \vee B, w) \Leftrightarrow 0 \in v(A, w) \cap v(B, w)$

A negation is true (respectively false)—in a model, at a world, under an apt variable assignment—just in case the negated formula is false (true). A conjunction is true just in case both conjuncts are, false just in case one conjunct is; and a disjunction is true just in case one disjunct is, and false just in case both disjuncts are.

Let g be a variable assignment, x a variable, and $a \in D$. Let $g(x/a)$ be the variable assignment that agrees with g on all variables, except possibly x, and where

$g(x/a)(x) = a$. Notice that if g is apt for w and $a \in d(w)$, then $g(x/a)$ is also apt for w.

As usual, a universally quantified formula is true just in case all instances are, false just in case at least one instance is; and an existentially quantified formula is true just in case at least one instance is, and false just in case all instances are.

- $1 \in v(\forall x A, w) \Leftrightarrow$ for all $c \in d(w)$, 1 is a member of the valuation of A for $\mathcal{M}$ in w under $g(x/c)$
- $0 \in v(\forall x A, w) \Leftrightarrow$ for some $c \in d(w)$, 0 is a member of the valuation of A for $\mathcal{M}$ in w under $g(x/c)$
- $1 \in v(\exists x P, w) \Leftrightarrow$ for some $c \in d(w)$, 1 is a member of the valuation of A for $\mathcal{M}$ in w under $g(x/c)$
- $0 \in v(\exists x P, w) \Leftrightarrow$ for all $c \in d(w)$, 0 is a member of the valuation of A for $\mathcal{M}$ in w under $g(x/c)$

Our final items here, of course, are the modal operators. Here the relevant observation is that if a variable assignment is apt for a given world, in a given model, then it is also apt for any accessible world. The reason is the feature that worlds do not shrink along the accessibility relation.

- $1 \in v(\Box A, w) \Leftrightarrow$ for all $w' \in W$, $(Rww' \Rightarrow 1 \in v(A, x))$
- $0 \in v(\Box A, w) \Leftrightarrow$ for some $w' \in W$, $(Rww' \;\&\; 0 \in v(A, x))$
- $1 \in v(\Diamond A, w) \Leftrightarrow$ for some $w' \in W$, $(Rww' \;\&\; 1 \in v(A, x))$
- $0 \in v(\Diamond A, w) \Leftrightarrow$ for all $w' \in W$, $(Rww' \Rightarrow 0 \in v(A, x))$

We can now define a notion of logical consequence. Let Γ and Δ be sets of sentences. Let $\mathcal{M}$ be a model.

Definition 1 $\Gamma \vdash_{\mathcal{M}} \Delta$ if and only if, for each world w in $\mathcal{M}$, if for each $B \in \Gamma$, $1 \in v(B, w)$ then there is an $A \in \Delta$ such that $1 \in v(A, w)$.

In words, an argument is valid in a model $\mathcal{M}$ just in case every world in which all of the premises are true at least one conclusion is true.

Definition 2 $\Gamma \vdash_{\mathbf{FDE_{S4}}} \Delta$ if and only if, for each model $\mathcal{M}$, $\Gamma \vdash_{\mathcal{M}} \Delta$.

As usual, then, an argument is valid in $\mathbf{FDE_{S4}}$ just in case it is valid in all $\mathbf{FDE_{S4}}$ models.

4.2 Other (Sub-)classical Logics

The foregoing framework is readily adapted to a variety of extensions of **FDE**, including many of those used to deal with the alethic paradoxes. We just restrict the models. Let $\mathcal{M}$ be an model, and let I, I^1, I^0 be the interpretation functions of $\mathcal{M}$. So for each world w and each n-place predicate P, $I^1(P, w)$ is the extension of P in w in $\mathcal{M}$, and $I^0(P, w)$ is the anti-extension of P in w in $\mathcal{M}$.

Definition 3 $\mathcal{M}$ is a **classical model** if and only if for each world w and each n-place predicate of the language, $I^1(P, w) \cup I^0(P, w) = d(w)^n$ and $I^1(P, w) \cap I^0(P, w) = \varnothing$.

In words, in classical models, the extension and anti-extension of each predicate are disjoint in each world, and, together, they exhaust the domain of the world. Classical models are, of course, equivalent to models of ordinary classical **S4**. It is easy to see that, every sentence A is either true only or false only in each classical model: $v(A, w) \in \{\{1\}, \{0\}\}$. There are no gaps and no gluts.

Definition 4 $\Gamma \vdash_{\mathbf{CL_{S4}}} \Delta$ if and only if, for each classical model $\mathcal{M}$, $\Gamma \vdash_{\mathcal{M}} \Delta$.

This, of course, is the ordinary consequence relation of classical **S4**, in a multi-conclusion framework.

A modal variant of Priest's "Logic of paradox" **LP** is easily formulated here. An $\mathbf{LP_{S4}}$ model is one in which, in each world, the extension and anti-extension of each predicate exhaust the domain of the world:

Definition 5 $\mathcal{M}$ is a **LP model** if and only if for each world w and each n-place predicate of the language, $I^1(P, w) \cup I^0(P, w) = d(w)^n$.

It is easy to see that, every sentence A is either true only, false only, or both true and false, in each **LP** model: $v(A, w) \in \{\{1\}, \{0\}\, \{1, 0\}\}$. There are no gaps, but there may be gluts.

Definition 6 $\Gamma \vdash_{\mathbf{LP_{S4}}} \Delta$ if and only if, for each **LP** model $\mathcal{M}$, $\Gamma \vdash_{\mathcal{M}} \Delta$.

Finally, we sketch a modal variant of so-called "strong Kleene" logic (**K3**). A $\mathbf{K3_{S4}}$ model is one in which, at each world, the extension and anti-extension of each predicate are disjoint:

Definition 7 $\mathcal{M}$ is a **K3 model** if and only if for each world w and each n-place predicate of the language, $I^1(P, w) \cap I^0(P, w) = \varnothing$.

It is easy to see that, every sentence A is either true only, false only, or neither, in each **K3** model: $v(A, w) \in \{\{1\}, \{0\}, \varnothing\}$. There are no gluts, but there may be gaps.

Definition 8 $\Gamma \vdash_{\mathbf{LP_{K3}}} \Delta$ if and only if, for each **K3** model $\mathcal{M}$, $\Gamma \vdash_{\mathcal{M}} \Delta$.

5 Translation of Intuitionist Logic

The language of intuitionism has the four connectives: $\wedge, \vee, \neg, \rightarrow$, and the two quantifiers $\forall, \exists$, with the usual recursion rules for defining formulas. Following Gödel [11], we define a translation function—τ from the language of intuitionism to the foregoing modal language.

If A is atomic, then $A^\tau = \Box A$
$(A \wedge B)^\tau = A^\tau \wedge B^\tau$
$(A \vee B)^\tau = A^\tau \vee B^\tau$
$(\neg A)^\tau = \Box\neg A^\tau$
$(A \rightarrow B)^\tau = \Box(A^\tau \supset B^\tau)$
$(\forall x A)^\tau = \Box\forall x A^\tau$
$(\exists x A)^\tau = \exists x A^\tau$

Let $\vdash_{\text{int}}$ be the usual intuitionistic first-order consequence relation. The following is then well-known:

Theorem 1 $A_1, \ldots, A_n \vdash_{int} B$ *if and only if* $A_1^\tau, \ldots, A_n^\tau \vdash_{\mathbf{CL_{S4}}} B^\tau$.

So we have a way to translate an intuitionistic language into a classical modal language, one that preserves the consequence relation. This will be our vehicle for intuitionistic recapture, below. We formulate the given mathematical theories in a modal language, using various logics. For recapture, we make the background modal logic classical. Then we will have intuitionistic logic, via the translation—τ.

It will be helpful to illustrate our conclusions with some theory. One of the most studied intuitionist theories is known as *Heyting Arithmetic* (**HA**). It is essentially the same as ordinary, Dedekind-Peano arithmetic (**PA**), but with an intuitionistic logic in the background. It is captured with the following axioms and induction schema:

1. $\forall x(x' \neq 0)$
2. $\forall x, y((x' = y' \rightarrow x = y) \wedge (x = y \rightarrow x' = y'))$
3. $(A(0) \wedge \forall x(A(x) \rightarrow A(x'))) \rightarrow \forall x A(x)$,

with addition and multiplication defined as usual, using the induction schema.

A great deal is known about this theory. One can formulate a version of Church's thesis (**CT**):

$$\forall x\exists! y A(x, y) \rightarrow \exists u\forall x\exists z(T(u, x, z) \wedge A(x, U(z)),$$

where T is a predicate defining computability, and U produces the output of a computation.[9] On the informal BHK gloss on the quantifiers, the formula in the form $\forall x\exists! y A(x, y)$ says that given a number x, one can effectively find a unique number y such that $A(x, y)$. This instance of **CT** thus says that, given the antecedent, there is a recursive function that computes such a y from each x.

It is well-known that **CT** is consistent with **HA**. This follows from standard realizability techniques [12]. It is also easy to see that **CT** is *in*consistent with classical (**PA**). Just let $A(x, y)$ be a formula that says that either $y = 0$ and the Turing machine with code x halts on input x, or else or $y = 1$ and the Turing machine with code x does not halt on input x. Using excluded middle, it is easy to prove that $\forall x\exists! y A(x, y)$. But, of course, there can be no Turing machine that computes this "self-halting"

[9] $T(u, x, z)$ is read as "z is the code of a complete computation of the Turing machine with code u, given input x. And if z is the code of a complete computation, then $U(z)$ is the output.

problem.[10] So our test theory, for intuitionistic recapture, is **HA** augmented with each instance of **CT**.

6 Interlude on the Conditional

As noted, the conditional in **FDE**, and its extensions, is introduced as an abbreviation. It is just the material conditional:

$$A \supset B \equiv_{\text{df}} \neg A \vee B.$$

Unfortunately, one or both of the usual classical (and intuitionistic) inference rules for the conditional fail in **FDE** and in its non-classical extensions **K3**, **LP** introduced above. First, the elimination rule, sometimes called "modus ponens" or "detachment", fails in **FDE** and **LP**. Suppose that a sentence A is both true and false (i.e., has truth value $\{1, 0\}$), and that B is false only (i.e., has truth value $\{0\}$). Then $(A \supset B)$ (i.e., $\neg A \vee B$) is true (and false), but B is not true. So the inference from $(A \supset B)$ and A, to B is not truth-preserving.

Second, the usual introduction rule, sometimes called "conditional proof'.' fails in **FDE** and **K3**. In those two logics, of course, one can infer A from A, but one cannot conclude that $A \supset A$ is a logical truth. If A is neither true nor false (i.e., has the empty set as truth value) then so is $A \supset A$. Indeed, there are no logical truths in **FDE** and **K3**.

These are sometimes regarded as defects of these logics, making them unsuitable (or at least awkward) as accounts of truth. It is hard to get by without a suitable conditional, one that validates at least some of the usual classical (and intuitionistic) rules in most contexts. To take just one example, the usual way to introduce restricted quantification is with a conditional. To say, for example, that all P's are Q's, is to say

$$(\forall x)(Px \rightarrow Qx).$$

This does not seem to work as intended if the conditional lacks the usual properties.

So a number of theorists have sought to introduce another conditional into the language to supplement the material conditional of **FDE** and its non-classical extensions. To be sure, this conditional cannot have all of the properties of the classical, or intuitionistic conditional, on pain of the Curry paradox.

The goal is to do better than the material conditional. Beall [4] dubs this enterprise *the quest*.[11]

---

[10]Further details and references on the topic of **HA** can be found many places, including [47].

[11]Beall himself was engaged in this endeavour in [3].

7 LP, Detachment Free

Beall himself was originally engaged in the *quest* [3] for a suitable conditional. His more recent work [4], however, eschews it, opting for only the material conditional of **LP** (along with the other logical terminology, of course). According to Beall, detachment (aka arrow-elimination, aka modus ponens) is never logically valid. The key word here is "logically".

Beall argues that the detachment freedom of **LP** is a *feature* of the framework, not a bug. For our purposes, classical recapture in this setting comes in the form that if one can safely ignore **LP**'s third, glutty, truth value $\{1, 0\}$, the result is classical logic, plain and simple.[12]

The recapture plan has a nice formulation in the foregoing framework in which sequents consist of *sets* of premises and conclusions. Recall that the formula "$A \supset B$" is an abbreviation of "$\neg A \vee B$". We have:

$$A \supset B, A \vdash_{\mathbf{LP}} B, A \wedge \neg A.$$

In words, from a material conditional and its antecedent, one can infer that either the antecedent is a true contradiction or that the consequent is true.

This sequent thus displays the fact that the only counterexamples to detachment occur when the antecedent gets the glutty value. So if we can be confident that the antecedent of a material conditional *doesn't* get that value, then the instance of detachment is rationally admissible—even if not logically valid. In such contexts, one can just ignore the contradictions that appear in the succedent.

As it happens, a central theme of Beall's earlier work [3] is that paradoxes arise *only* from uses of the truth predicate, and related semantic notions. So if the truth-rules play no role in a given deduction, then one can safely ignore the contradictions on the right. His later detachment-free framework [4] involves giving up the depth-relevant conditional defended in [3] (in the throes of *the quest*), in favor of the material implication.

We submit that the result of combining these two views is interesting, and allows for a intuitionist treatment via what we call Plan B. To repeat, the views in question are that the truth predicate is the primary culprit in paradox, that a glutty, **LP**-style response is appropriate, *and* the detachment free approach.[13] This elegant resolution readily translates into the Gödel-translated intuitionist system inside **LP**.

First, it is easy to see that the failure of detachment in **LP** lifts naturally to the constructive system definable in $\mathbf{LP_{S4}}$, via the translation τ.

Fact 1 $A^{\tau}, (A \rightarrow B)^{\tau} \nvdash_{LP_{S4}} B^{\tau}$

---

[12]These comments address *propositional* **LP**, but for first order **LP** the recapture of classical logic is obtained by demanding the extensions and anti-extensions of predicates are *exclusive*, which amounts to the same restriction. Also worth noting is that this approach has a predecessor in the discussion in [27, pp. 117–119].

[13]Beall may not want to hold this combination of views at present.

Proof A counterexample is as follows: Let $W = \{x\}$, $R = \{x, x\}$, $v(A, x) = \{1, 0\}$, and $v(B, x) = \{0\}$. Then $v(\Box A, x) = \{1, 0\}$ and $v(\Box B, x) = \{0\}$. Since $v(A, x) = \{1, 0\}$, $v(\neg A, x) = \{1, 0\}$. So $1 \in v(\neg A \vee B, x)$, and so $1 \in v(\Box(\neg A \vee B), x)$. In frames that consist of a single "world", the modal operators are trivial. □

It's also easy to see that Beall's strategy lifts naturally into the constructive interpretation.

Fact 2 $A^\tau, (A \to B)^\tau \vdash_{LP_{S4}} B^\tau, (A \wedge \neg A)^\tau$

Proof Suppose that $\mathcal{M}$ is a $\mathbf{LP_{S4}}$ model, $x \in W^{\mathcal{M}}$, and $1 \in v(\Box A, x) \cap v(\Box(\neg A \vee B), x)$. So, for all x-accessible y, $1 \in (v(A, y) \cap v(\neg A, y)) \cup (v(A, y) \cap v(B, y))$. Hence, either $1 \in v(\Box B, x)$ or $1 \in v(\Box(A \wedge \neg A), x)$ as desired.

So, putting the pieces together, if we are in a context in which contradictions can be ruled out—perhaps on non-logical grounds—then it is safe to infer—again on non-logical grounds—that the translation of the elimination rule (or detachment) is acceptable. So, in such contexts, we have classical **S4** for our modal principles. So the foregoing Gödel translation recapitulates intuitoinistic reasoning, and so we have intuitionistic recovery.

As a simple example of the nature of the recovery, recall some of the axioms and principles of Heyting arithmetic:

1. $\forall x(x' \neq 0)$
2. $\forall x, y((x' = y' \to x = y) \wedge (x = y \to x' = y'))$
3. $(A(0) \wedge \forall x(A(x) \to A(x'))) \to \forall x A(x)$,

and the scheme we call CT, which is inconsistent with classical logic:
$\forall x \exists ! y A(x, y) \to \exists u \forall x \exists z (T(u, x, z) \wedge A(x, U(z))$
Those are rendered as follows (simplifying a bit):

- $\Box \forall x(\Box \neg \Box x' = 0)$
- $\Box \forall x, y(\Box(\Box x' = y' \supset \Box x = y) \wedge \Box(\Box x = y \supset \Box x' = y'))$
- $\Box[(A^\tau(0) \wedge \Box \forall x(\Box(A^\tau(x) \supset A^\tau(x')))) \supset \Box \forall x A^\tau(x)]$

and

- $\Box[\Box \forall x \exists ! y A^\tau(x, y) \supset \exists u \Box \forall x \exists z(\Box T(u, x, z) \wedge A^\tau(x, U(z))]$

This is a mouthful, perhaps, but it formally works given our construction.

8 K3, Savior of Truth

Field's [9] approach involves the use of the paracomplete logic **K3**, which can be obtained from **FDE** as noted in Sect. 4.1. The system is a sort of dual to the paraconsistent, or dialetheic, approaches. For Field, the liar sentence, and related paradoxical sentences, are not *false* (since then they would need to be true as well). Instead, they take an undesignated (gappy) truth value, the "middle" value in **K3**. Paradoxical sentences have the empty set as their truth-value (a truth-value that is not available in **LP**).

Kripke's famous construction [14] also employs **K3**, but Field's approach is more nuanced. In particular, he develops a conditional, which we call '$\rightarrow$'. The details are interesting, but for our purposes, we only need some key features.

Field obtains *classical* recapture by showing that if excluded middle holds of the relevant formulas, then his conditional '$\rightarrow$' is equivalent to the material conditional '$\supset$':

Fact 3 $A \vee \neg A, B \vee \neg B \vdash (A \rightarrow B) \leftrightarrow (\neg A \vee B)$

There are thus two reasons to be pessimistic that anything like our Plan A can be carried out for intuitionistic recovery in Field's system. First, as noted, **K3** extends **FDE** and so is incompatible with intuitionist logic. It has it that a given formula A is equivalent to $\neg\neg A$, and it validates the intuitionistically unacceptable DeMorgan laws. Moreover, the DeMorgan laws are crucial in interpreting the negations of, say, conjunctions.

Second, and perhaps more important, the last thing an intuitionist wants to do, in effecting recovery, is to assume instances of excluded middle for the semantic-free vocabulary. The whole point of constructive mathematics is to not make such an assumption.

Nevertheless, the aforementioned features, and the key "fact" do facilitate what we call Plan B. The class of **K3** models which assign to no sentences the third value is a class of models of classical logic, and this fact above shows that the same is true in Field's extended system. Moreover, if the "middle" truth value is eschewed, then excluded middle holds, and so the above "fact" is that Field's conditional collapses to the classical conditional.

The reader can verify that this also hapens in the modal framework $\mathbf{K3}_{S4}$ and the system resulting from extending $\mathbf{K3}_{S4}$ by the addition of Field's conditional. The reader is welcome to verify that the Gödel translation in question does generate intuitionist logic when the gappy middle value is removed, but the key results are straightforward. So, when it comes to intuitonistic recovery, the result is similar to that of the previous section.

9 Contraction Freedom: Priest and Others

There are wide variety of contraction-free approaches to the semanitic paradoxes. Typically, these reject the structural rules of contraction, stated below in a usual Gentzen-style guise.[14]

---

[14]Using Gentzen's original formulation, the data type of the premise and conclusion are something like multisets or sequences.

$$\frac{\Gamma_1, A, A, \Gamma_2 \vdash \Delta}{\Gamma_1, A, \Gamma_2 \vdash \Delta}$$

The result is a substructural logic.[15] Our focus here is on Hilbert systems and on the conditional in particular. Roughly, the structural rule above corresponds to the provability of the following two axioms[16]:

(1) $(A \rightarrow (A \rightarrow B)) \rightarrow (A \rightarrow B)$ W
(2) $((A \rightarrow B) \wedge A) \rightarrow B$ WI

It has been well known since Meyer et al. [23] that either of these, along with a transparent truth predicate, the inference rule of modus ponens, and the idempotency of conjunction, allows for the derivation of arbitrary sentences from a Curry sentence:

$$C \leftrightarrow (C \rightarrow \bot).$$

This, of course, leads to disaster.

So the key idea behind non-contractive logics is have an unrestricted, or transparent truth predicate, along with the usual rules for conjunction and disjunction (i.e. those corresponding to lattice meet and join), *and* to retain a conditional that satisfies the rule of modus ponens, but rejects both (1) and (2).

This can be effected in a number of ways. In proof theory, one can reject, or somehow restrict the contraction rules by making changes to a Hilbert style system or natural deduction system. In model theory, a standard approach has been to employ a *ternary relation* semantics for relevant logics, as presented in [39], among other places, (or something similar), wherein one has the kind of careful control over the properties of the conditional to reject (1) and (2), while keeping many other desirable features of conditionals.

Priest is a famous defender of this kind of approach to semantic paradoxes, so following him, we'll employ model theoretic tools in discussing the problem. In particular, we'll focus primarily on the logic **B**, which is the basic system generated by the ternary relation semantics. This system is an extension of **FDE** by a conditional obeying only some minimal properties (the rule form of modus ponens, rule forms of transitivity, and identity).[17]

First, we'll introduce the basic ternary relation framework. This is a frame structure, in the same vein as Kripke semantics for modal and intuitionist logic. For present

[15] The classic work on substructural logic is Restall's textbook [31], though additional details are available in [25, 40].

[16] The sans-serif titles refer to a correspondence between these arguments and *combinators*, which are central in proof and model theoretic presentations of substructural logics. Roughly, with the structural rule (along with appropriate rules for $\rightarrow$ and $\wedge$), one can prove each of the following. Conversely, these axioms allow one to prove that the structural rule is admissible in a Hilbert system (when the premise and conclusion sequents of the rule are appropriately translated as sentences). The details here are complex, and [6] provides a good overview of the area. For present purposes it is enough to note the correspondence.

[17] There is another system, occasionally called **BB** which is slightly weaker than **B** which still has **FDE** as it's conditional-free fragment, but our focus is on **B** as this is most natural in the standard ternary relation semantics.

purposes, a pleasing feature of this approach is that it provides fine control along the lines of what in the intuitionist tradition is called *separability*. So one can alter either of the conditional and negation while, in large part, leaving the other unchanged.

The options we'll sketch include two variations of what we call Plan A. The first of these is to construct a common subsystem between **B** and intuitionist logic. One can then use this rather weak logic when the truth predicate or other paradoxical items are present, and *intuitionist logic* when they aren't. This is a weaker logic than Priest and others in the contraction-free camp usually employ, but it is in the neighbourhood of those logics. The other option we discuss relies on a fact about the strong relevant logic **E**, principally investigated in [1, 2]. Namely, one can translate the intuitionist conditional and negation into the language of **E**, when this is extended by propositional quantification. It happens, however, that this does not provide an elegant method for intuitionist recapture.

In addition, we sketch two options for Plan B. The first involves adding frame conditions to the ternary relation semantics to obtain *classical* logic, and then building **S4** modalities on top of this. Second, some work of Mares [17] on embedding classical modal logics into relevant modal logics suggests another way forward. This last option involves carrying out Plan B while simply *ignoring* the relevant conditional for the truth-predicate-free portion of the language and just using material implication. The result is essentially the same as that of Sect. 7.

9.1 Ternary Relation Semantics

A ternary relation frame $\mathcal{F}$ is a 4-tuple $\langle W, N, R, *\rangle$. This includes a set of worlds W, a distinguished set of (normal) worlds $N \subseteq W$, a relation $R \subseteq W^3$, and the Sylvan-Plumwood star $*$.

To state the frame conditions, we define a partial-order $\sqsubseteq$ on W:

$$x \sqsubseteq y \iff \exists z \in N(Rzxy).$$

First, R must be antitone with respect to $\sqsubseteq$ in its first two argument places, and monotone in its third place.[18] That is:

$$(Rxyz \;\&\; x' \sqsubseteq x \;\&\; y' \sqsubseteq y \;\&\; z \sqsubseteq z') \Rightarrow Rx'y'z'.$$

In addition, we add the following two conditions relating to the star:

[18] Strictly speaking not all of these tonicity conditions on R are required for present purposes. All are required to interpret the *fusion* or *multiplicative conjunction* (as it's called in the tradition of Linear Logic) and an additional conditional along with the usual conditional. Details on these connectives in the ternary relation semantics are available in [31] and elsewhere. We keep the full set of tonicity conditions here for the sake of broadest applications to a variety of logics including these additional connectives, but our attention is restricted to $\rightarrow$.

- $x^{**} = x$
- $x \sqsubseteq y \Rightarrow y^* \sqsubseteq x^*$

and require that N is upward closed w.r.t. $\sqsubseteq$, that is:

$$\text{If } x \in N \text{ and } x \sqsubseteq y, \text{ then } y \in N.$$

To build a model from a frame, we introduce a *binary* valuation: $v : \mathcal{L} \times W \longrightarrow \{1, 0\}$.[19] As usual v is arbitrary for atomic formulas, and obeys the following conditions for complex sentences:

$$\begin{aligned}
v(A \wedge B, w) = 1 &\iff v(A, w) = v(B, w) = 1 \\
v(A \vee B, w) = 1 &\iff v(A, w) = 1 \text{ or } v(B, w) = 1 \\
v(A \rightarrow B, w) = 1 &\iff \text{for all } x, y, \text{ if } Rwxy \text{ and } v(A, x) = 1, \text{ then } v(B, y) = 1 \\
v(\neg A, w) = 1 &\iff v(A, w^*) = 0.
\end{aligned}$$

Of course, to interpret mathematics in this language, we will need to interpret the quantifiers. This introduces many additional complications, and while these details are important, we set them aside for this paper. The key works on quantified relevant logic are Fine [8] and Mares and Goldblatt [18].

Finally, we need a *heredity* condition on $\sqsubseteq$, similar to that of the Kripke semantics for intuitionistic logic:

$$\text{If } x \sqsubseteq y, \text{ then for every sentence } A, v(x, A) = 1 \Rightarrow v(y, A) = 1.$$

If this is assumed for atomic sentences, it can be proved for all sentences by an induction on the complexity of sentences. The details are simple, basically just relying on the tonicity conditions placed on the various frame elements, and are available in many places, but the classic presentation is in [37]. The heredity condition in effect tells us that $\sqsubseteq$ is anti-symmetric, and hence is a partial order. That is, if $x \sqsubseteq y$ and $y \sqsubseteq x$, then x and y disagree on no formulae, and we have that the worlds, in essence, collapse into one.

We (re)define "valid argument" as follows, adapting the definition to binary valuations, and replacing $W^{\mathcal{M}}$ with N:

Definition 9 $\Gamma \vdash_{\mathcal{M}} \Delta$ iff for all $x \in N$, if for each $B \in \Gamma$, $v(B, x) = 1$ then there is an $A \in \Delta$, $v(A, x) = 1$.

[19] Unlike the valuation defined in Sect. 4.1, we can get away with a binary valuation rather than a four-valued valuation since the negation here is interpreted by $*$, rather than in terms of additional values.

One of the key features of this semantics is that $A \rightarrow B$ is valid iff for every world x, if $v(x, A) = 1$ then $v(x, B) = 1$. This property captures the intuitive idea that the relevant implication captures the consequence relation *at each world*, and this feature is important in considering the simplified semantics below.

With only the frame conditions listed above, we get a DeMorgan negation like that in **FDE**. It is shown in [38] that the semantics, without R, is sound and complete with respect to **FDE**.

The following Hilbert system is sound and complete with respect to the basic ternary relation semantics:

9.1.1 B Axioms

- $A \rightarrow A$
- $(A \wedge B) \rightarrow A$, $(A \wedge B) \rightarrow B$
- $((A \rightarrow B) \wedge (A \rightarrow C)) \rightarrow (A \rightarrow (B \wedge C))$
- $A \rightarrow (A \vee B)$, $B \rightarrow (A \vee B)$
- $((A \rightarrow C) \wedge (B \rightarrow C)) \rightarrow ((A \vee B) \rightarrow C)$
- $(A \wedge (B \vee C)) \rightarrow ((A \wedge B) \vee (A \wedge C))$
- $\neg(A \wedge B) \leftrightarrow (\neg A \vee \neg B)$
- $\neg(A \vee B) \leftrightarrow (\neg A \wedge \neg B)$
- $\neg\neg A \leftrightarrow A$

9.1.2 B Inference Rules

- From $A \rightarrow B$ and A, infer B
- From $A \rightarrow B$ and $C \rightarrow D$, infer $(B \rightarrow C) \rightarrow (A \rightarrow D)$
- From A and B, infer $A \wedge B$
- From $A \rightarrow B$ infer $\neg B \rightarrow \neg A$

To reiterate, **B** is an extension of **FDE** by a conditional satisfying some fairly minimal properties.[20] This is evidenced by the fact that the axioms listed above express the core facts about deducibility in **FDE**—for instance, in **FDE**, if $A \vdash B$ and $A \vdash C$, then $A \vdash B \wedge C$, and $\neg(A \wedge B) \dashv\vdash \neg A \vee \neg B$. As above, this last "feature" makes the logic unsuitable for expressing constructive mathematics, with or without an unrestricted or transparent truth predicate.

The extension of **B** by the Law of Excluded Middle ($A \vee \neg A$) has been called **BX** by, for instance, Beall [3]. This is the logic which extends **LP** by the **B** conditional, and is the one Priest most often defends as the correct logic for paradox solution (for

[20] As an example of the minimality of **B**'s conditional, one can only prove $(A \rightarrow B) \wedge (B \rightarrow A)$ in **B** when A, B are *syntactically* identical. This is closely related to classic work in relevant logic on the **T**-W problem (or **P**-W problem) as investigated in [21].

instance, in [27]). This can be obtained from the above semantics by the addition of the frame constraint:

$$\text{If } x \in N \text{ then } x^* \sqsubseteq x$$

Finally one can add a Boolean negation to **B** by further enforcing the constraint:

$$\text{If } x \in N \text{ then } x \sqsubseteq x^*$$

which corresponds to explosion in the form $(A \wedge \neg A) \rightarrow B$.

A shared feature of **B** and **BX** is that contraposition is a valid rule of inference, but the axiom associated with this rule, $(A \rightarrow B) \rightarrow (\neg B \rightarrow \neg A)$, is not a **B** validity.[21]

9.2 *Plan A: Building Intuitionistic Logic from a Constructive Subsystem of B*

The two connectives which differ between **B** and intuitionist logic are the conditional and negation. As it stands, neither is appropriate. We'll start with the conditional, by considering the fragment of **B** which doesn't contain negation (i.e. the positive fragment). Then we will turn to negation.

It is fairly straightforward to get a decent conditional for our purposes, since **B**'s conditional is strictly weaker than the intuitionist conditional. In the present framework, intuitionist implication corresponds to the implication characterised by the principles which are often called K and S (again, because of connections to combinatory logic; see [31] for details). K corresponds to *thinning* (or *weakening*): $A \rightarrow (B \rightarrow A)$, and S corresponds to the distribution of the conditional over itself: $(A \rightarrow (B \rightarrow C)) \rightarrow ((A \rightarrow B) \rightarrow (A \rightarrow C))$. These principles correspond to the following frame constraints[22]:

- $Rxyz \Rightarrow x \sqsubseteq z$ K
- $\exists s(Rxys \;\&\; Rszu) \Rightarrow \exists s\exists t(Rxzs \;\&\; Ryzt \;\&\; Rstu)$ S

Adding these two frame conditions will extend **B** by the intuitionistic conditional. So, in circumstances in which there is no threat of paradox, the intuitinionist mathematician can simply adopt these frame conditions.

To be sure, this does not affect **B**'s negation. In particular, no addition of conditions to **B**'s star results in making $\neg\neg A \rightarrow A$ invalid—similarly for the bad direction of the DeMorgan law. Something else has to be done for the negation.

---

[21] The frame condition corresponding to this formula is $Rxyz \Rightarrow Rxz^*y^*$.

[22] Those familiar with combinatory logic or lambda calculus may recognise in these principles the combinators called K and S. In combinatory terms, these have the following reduction behaviours: Kxy reduces to x, and Sxyz reduces to $(xz)(yz)$. Note the resemblance between these reduction schemata and the frame constraints corresponding to the axioms in question.

We suggest that the best option for Plan A is to introduce an entirely new negation, and reinterpret the theory in question with this new negation rather than **B**'s negation—taking advantage of the "separability" feature of the system, mentioned above. A natural way is to proceed by Dunn's *perp* [24], or Restall's *compatibility* relation [31].[23] Following Restall, we would interpret this new negation by means of a binary relation, rather than a unary function, on worlds. For this, add $C \subseteq W^2$ to the frame (antitone in both places), and introduce a negation $\sim$, with the following valuation clause:

$$v(x, \sim A) = 1 \iff \text{for all } y \in W \text{ s.t. } Cxy, v(y, A) = 0.$$

The following frame condition makes the negation intuitionistic:

$$Cxy \iff \exists z(x \sqsubseteq z \;\&\; y \sqsubseteq z)$$

To summarize, then, the program is to start with a subintuitionist theory, a fragment of **B** employing some weaker negation frame conditions which can be extended to the above condition without collapsing into classical logic.

It is, however, quite unclear, at least to us, what this subintuitionist logic looks like. But if one could be presented and developed, it could be used to reason about an intuitionist theory including unrestricted truth, and then strengthened to full intuitionist logic for the portion of the language not including the truth predicate.

9.3 A Second Plan A: Embedding Intuitionist Logic

Another option in the vein of Plan A relies on some important early results in the development of relevant logics. Intuitionist logic, under a translation, is exactly contained the logic **E** (for Entailment) extended by propositional quantification. That is, using propositional quantification in **E**, one can give definitions of connectives $\rightharpoonup$ and $\sim$ such that if A is a sentence built up out of $\rightharpoonup, \sim, \wedge, \vee$, then **E** (with propositional quantification) validates A iff A is intuitionistically valid (interpreting these connectives as the intuitionist conditional, negation, conjunction, and disjunction respectively). The key results here are given in [2, pp. 55–64]. For the interested reader, the definitions are:

- $A \rightharpoonup B \iff \exists C(C \wedge ((C \wedge A) \rightarrow B))$
- $\sim A \iff A \rightharpoonup \forall B(B)$

So another option for obtaining intuitionist logic from **B** is to (1) extend **B** to **E**, (2) add to **E** the resources of propositional quantification, (3) define the connectives

[23] The difference here is largely one of temperament—if one is of a positive mindset, then compatibility seems right, if of a negative mindset, then perp or *incompatibility* seems best.

above, and (4) in the statement of the theory one is attempting to recapture reinterpret the conditional as $\rightharpoonup$ and negation as $\sim$.

Unfortunately, the details here are quite complex—much more so than in the other options we consider in this paper.[24] In addition, the resources of propositional quantification make the resulting logic non-axiomatisable, as shown by [13]. A more elegant way to proceed is available for **R**, if one adds some propositional constants. Details are available in [19]. This is much more elegant, and seems a more reasonable option to introduce some propositional constants than propositional quantification.

9.4 *Plan B: From the Logic B to Classical Logic, to Classical* **S4** *(and then to intuitionistic logic, via the Gödel translation)*

In order to get plan B to work, we need to introduce more structure to the frames in order to interpret some **S4** style modals. With the semantics given in the usual Ternary relation setup, there are many ways to go about this, but some work better than others. We'll employ a framework due to Fuhrmann [10], who adds a relation $S \subseteq W^2$ interpreting modals which interacts with R and $*$ in the following ways:

1. S reflexive and symmetric
2. S is antitone in the first place and monotone in the second.
3. $x \sqsubseteq y \Rightarrow \forall z(Syz \Rightarrow Sxz)$
4. $\exists a(Rxya \,\&\, Saz) \Rightarrow \exists b(Syb \,\&\, \exists c(Sxc \,\&\, Rcyz))$

With the following valuation conditions:

$$v(\Box A, x) = 1 \iff \forall y(Sxy \Rightarrow v(A, y) = 1)$$
$$v(\Diamond A, x) = 1 \iff \exists y(Sx^*y \,\&\, v(A, x^*) = 1)$$

Condition (4) above is somewhat convoluted, and this is required to capture the interaction between the conditional and modals (interpreted by R and S). Condition (3) corresponds to the **K** axiom: $\Box(A \to B) \to (\Box A \to \Box B)$, (2) corresponds to the validity of the rule of necessitation:

From A infer $\Box A$.

And condition (1) provides us with $\Box A \to A$ and $\Box A \to \Box\Box A$ to make $\Box$ and $\Diamond$ **S4** style modalities.

So, with this option, we can take **B**, and define a necessitated conditional in terms of **B**'s conditional, and build a subintuitionist negation by taking the Gödel translation of **B**'s **FDE**-like negation. Then we can introduce the additional frame constraints on

[24] For instance, additions to the model structure are required to obtain **E** [2, pp. 171–172].

the conditional as in the previous Sect. 9.4, and generate the intuitionist conditional by extending **B**'s negation to classical negation using:

$$\text{If } x \in N \text{ then } x^* = x.$$

With this negation, the Gödel translation will generate the intuitionist negation as desired.

The only detail necessary to see that this works, is to show that the conditional you obtain by both making R satisfy both K and S becomes the classical conditional when you introduce $x^* = x$ and $Rxyz \Rightarrow Rxz^*y^*$ (we'll call this the "contraposition condition").

It's well known that you can start with the conditional of the strong relevant logic **R**, and introduce a boolean negation by collapsing worlds and their star-mates, and obtain a logic which is *sub-classical* in a sense. Meyer and Routley [20, 22] showed that the resulting system has $(A \wedge \neg A) \to B$ and $A \to (B \vee \neg B)$ as theorems, but still doesn't validate $A \to (B \to A)$ or $A \to (\neg A \to B)$.[25] This goes to show that the two step process of altering the conditional and negation by altering the frame conditions for R and * is necessary, if we are to use the usual frame conditions.

9.5 Plan B: Classically Complete Relevant Logics

Mares [17] shows that the relevant logic **R** extended by **S4** style modals (as in Sect. 9.4), is an extension of classical **S4**. First, **R** is obtainable from **B** by adding the following frame constraints:

- $Rxxx$
- $Rxyz \Rightarrow Ryxz$
- $\exists s(Rxys \ \& \ Rszu) \Rightarrow \exists t(Rxtu \ \& \ Ryzt)$
- $Rxyz \Rightarrow Rxz^*y^*$

Mares shows that if one adds $x \in N \Rightarrow x^* = x$, then if A is a sentence containing only $\wedge, \neg, \Box$ as connectives, A is valid in $\mathbf{R_{S4}}$ iff it is valid in **S4**. So, one can define one's way to classical **S4** in this way, and then use the Gödel translation as usual to obtain intuitionist connectives.

The major downside here is that we wind up ignoring the relevant conditional, in favor of material implication, but the relevant conditional is crucial to having any deductive strength in these already weak logics. It would be at least bizarre to introduce a relevant conditional in order to have even reasonably powerful mathematical theories in the presence of the truth predicate, and then to simply ignore this conditional when we can safely ignore the truth predicate. It seems to give the game away (if one remembers what the game is). We were driven to introduce a conditional in order to do substantial mathematical reasoning within the base theory, even when

[25] They call the resulting system **CR**.

there is a truth predicate in play. But, it seems, we have to ignore this conditional when doing intuitionistic recapture.[26]

The additional constraint that if $x \in N$, then $x^* = x$ (which Mares calls "normality") amounts to rejecting true contradictions and false instances of LEM, just as dropping the values both and false does from **FDE** as presented earlier. So this proposal is very much like Beall's proposal, just starting from **R** rather than from **LP**. This gives rise to one other option.

9.6 Plan B on the Cheap

In light of the last option, we can go back to the recapture in Sect. 7, where, following Beall, we just eschewed a conditional, beyond the material one. The idea here is to employ **BX** when in the presence of the truth predicate. And when one does not encounter the truth predicate, we ignore the conditional of **BX** in favor of material implication, and simply throw out all true contradictions, as in Beall's proposal. And then get intuitionism via the Gödel translation.

As with the Mares-variation on Plan B, this would involve reinterpreting the intuitionist theory in question, in particular by replacing all instances of $\rightarrow$ with $\supset$. This fact makes this proposal, though elegant against the background of other options for weak relevant logics, inelegant as against Beall's picture, which, again, found no use for the $\rightarrow$.

The notable upshot is that one can give interesting mathematical theories in weak non-contractive logics, for instance in the work of [48], but there seems to be few interesting theories available against the background of **LP**. One can construct non-trivial inconsistent models of set theory and arithmetic in **LP** (for instance as in [26, 30]), but these theories lack a robust set of *theorems*—i.e. sentences true on all models—as discussed by [45]. What makes **LP** interesting in non-classical mathematics (that it allows for more models of the theories in question) is also its major downfall as an inference engine in these theories (it seems to allow far too many *countermodels*, and hence very few theorems).

So while **LP** has an elegant recapture method, by Beall's proposal, it doesn't seem to allow for the development of robust non-classical mathematics, of the kind one would like to develop by taking interesting classical theories and adding a naïve truth predicate or comprehension scheme. On the other hand, with the addition of a relevant conditional one can provide the means for more interesting mathematical theories with these naïve elements, but at the cost of elegance in the recapture.

---

[26] Recall that, in Field's system (Sect. 8), the conditional just became the material conditional upon recovery—upon assuming instances of excluded middle in that case. And that facilitated Plan B. Here we have no such luck, and, it seems, have to simply ignore the conditional that was introduced in order to do the mathematics internally.

9.7 *Upshot*

There are a handful of issues we have come across a few times in this section on intuitionist recapture in weak relevant logics.

1. The fact that the arrow and negation are separable in this semantic framework, since interpreted by different frame elements, has the result that no recapture method is so simple as is available in the other logics we've considered. The process seems to require a multi-stage development first altering the conditional (via R) and then negation (via * or C), unless one goes with the nuclear option and just constrains the frames to have only one world. But then Plan B fails miserably. Moreover, this option has a less intuitive appeal than in Field or Beall's approach. In Beall's case, though adopting material implication as the logical conditional results in losing many intuitive inference principles, it is nice for recapture purposes. As for Field, though he has an additional conditional, the result that this is equivalent to material implication the presence of LEM means that he too has a very intuitive condition for collapsing the logic to classical logic.
2. Plan B on the relevant approach as we have considered it is pretty inelegant. There may be a different way of going about Plan B starting with **B**, but we haven't found it. It is possible that there is a variation on the Gödel translation which might do the work we want, but have no idea what it is. It would seem to have to be a non-trivial departure from the norm to do the job of strengthening either the **FDE** or **LP** negation to intuitionist negation. In either case, Plan B, while straightforward on the other accounts we've seen, is much less obviously appropriate in this setting.
3. Plan A for **B**, on the other hand, does have something nice going for it. All the work can be done within the frame, and the conditions for generating the intuitionist connectives are straightforward and (reasonably) intuitive. The major downside here is that the proponent of **B** would have to either (i) reinterpret the negation of the theory in question for the purposes of recapture or (ii) alter their theory to include a constructive negation from the start. Logics in the area have seen significant development recently, thanks to Robles and Mendez [36] in particular. A logic in this area seems the most natural fit for the kind of project we've discussed here.

10 Conclusion

We have considered a variety of logical frameworks and methods for classical recapture. Ripley and Tennant's systems are natural fits for a direct, Plan A, approach, where as the systems of Beall and Field are natural fits for Plan B. The non-contractive approach favoured by Priest and others in the relevant or substructural logics community have a harder job, with neither a Plan A or a Plan B approach providing a natural fit with the existing logics.

This paper has been a sketch of the territory, and there is much more to be said on the topic. For instance, we have more or less taken for granted that researchers interested in classical recapture should also be interested in intuitionist recapture, but of course this claim brings along a collection of issues relating to logic and its role in our best theories. In addition, while we have more or less completely sketched the approaches available to some of the theories we consider, there is a great deal more to be said and done with regard to the non-contractive approach, for which our discussion did not even broach the topic of the interpretation of quantifiers. This is an area with potential for interesting, fruitful work, but it's a step beyond the scope of the current work. Our hope is that these projects, and others which arise from our work here, can prove fruitful.

References

1. Anderson, A., & Belnap, N. (1975). *Entailment: the logic of relevance and necessity* (Vol. 1). Princeton University Press.
2. Anderson, A., Belnap, N., & Michael Dunn, J. (1992). *Entailment: The logic of relevance and necessity* (Vol. 2). Princeton University Press.
3. Beall, J. C. (2009). *Spandrels of truth*, Oxford.
4. Beall, J. (2015). Free of detachment: logic, rationality, and gluts. *Nous*, *49*(2), 410–423.
5. Beall, J. C. & Allen Logan, S. (2017). *Logic: The basics* (2nd ed.), Routledge.
6. Bimbó, K. (2014). *Proof theory; Sequent calculi and related formalisms*. Boca Ration, FL: CRC Press.
7. Brandom, R. B. (1994). *Brandom, making it explicit*. Havard University Press.
8. Fine, K. (1988). Semantics for quantified relevance logic. *Journal of Philosophical Logic*, *17*(1), 27–59.
9. Field, H. (2008). *Saving truth from paradox*. Oxford University Press.
10. Fuhrmann, A. (1990). Models for relevant modal logics. *Studia Logica*, *49*, 501–514.
11. Gödel, K. (1933). Eine Interpretation des intuitionistischen Aussagenkalküls. *Ergebnisse eines mathematischen Kolloquiums*, *4*, 39–40.
12. Kleene, S. C. (19 45). On the interpretation of intuitionistic number theory. *Journal of Symbolic Logic*, *10*, 109–124.
13. Kremer, P. (1993). Quantifying over propositions in relevance logic: Nonaxiomatisability of primary interpretations of $\forall p$ and $\exists p$. *Journal of Symbolic Logic*, *58*(1), 334–349.
14. Kripke, S. (1975). Outline of a theory of truth. *The Journal of Philosophy*, *72*(19), 690–716.
15. Linnebo, Ø. (2013). The potential hierarchy of sets. *Review of Symbolic Logic*, *6*(2), 205–228.
16. Linnebo, Ø., & Shapiro, S. (Forthcoming). *Actual and Potential Infinity*. Nous.
17. Mares, E. (1993). Classically modal complete relevant logics. *Mathematical Logic Quarterly*, *39*(1), 165–177.
18. Mares, E., & Goldblatt, R. (2006). An alternative semantics for quantified relevant logic. *Journal of Symbolic Logic*, *71*(1), 163–187.
19. Meyer, R. K. (1973). Intuitionism, entailment, negation. In H. Leblanc (Ed.), *Truth, syntax, and modality* (pp. 168–198), North Holland.
20. Meyer, R. K., & Martin, E. P. (1973). Classical relevant logics. *Studia Logica*, *32*, 51–63.
21. Meyer, R. K., & Martin, E. P. (1982). Solution to the P-W problem. *The Journal of Symbolic Logic*, *47*(4), 869–887.
22. Meyer, R. K., & Routley, R. (1974). Classical relevant logics II. *Studia Logica*, *33*, 183–194.
23. Meyer, R. K., Routley, R., & Dunn, J. M. (1979). Curry's paradox. *Analysis*, *39*, 124–128.

24. Michael Dunn, J. (1993). Star and Perp: two treatments of negation. *Philosophical Perspectives, 7*, 331–357.
25. Paoli, F. (2002). *Logics, substructural: A primer*. Springer.
26. Priest, G. (2000). Inconsistent models of arithmetic part II: The general case. *Journal of Symbolic Logic, 65*, 1519–1529.
27. Priest, G. (2006). *In contradiction: A study of the transconsistent* (2nd ed.). Oxford University Press.
28. Priest, G. (2008). *An introduction to non-classical logics* (2nd ed.), Cambridge.
29. Priest, G., & Sylvan, R. (1992). Simplified semantics for basic relevant logics. *Journal of Philosophical Logic, 21*(2), 217–232.
30. Restall, G. (1992). A note on Naïve set theory in LP. *Notre Dame Journal of Formal Logic, 33*(3), 422–432.
31. Restall, G. (2000). *An introduction to substructural logics*, Routledge.
32. Restall, G. (2005). Multiple conclusions. In P. Hajek, L. Valdes-Villanueva, & D. Westerstahl (Eds.), *Logic, Methodology and Philosophy of Science: Proceedings of the Twelfth International Congress* (pp. 189–205). Kings College Publication.
33. Restall, G. (2013). Assertion, denial and non-classical theories. In K. Tanaka (Ed.), *Paraconsistency: Logic and applications* (pp. 81–99). Springer.
34. Ripley, D. (2012). Conservatively extending classical logic with transparent truth. *Review of Symbolic Logic, 5*(2), 354–378.
35. Ripley, D. (2013). Paradoxes and failure of cut. *Australasian Journal of Philosophy, 91*(1), 139–164.
36. Robles, G., & Mendez, J. M. (2018). *Routley-Meyer ternary relational semantics for intuitionistic-type negations*. Academic Press.
37. Routley, R., & Meyer, R. K. (1973). The semantics of entailment. In H. Leblanc (Ed.), *Truth, Syntax, and Modality. Proceedings of the Temple University Conference on Alternative Semantics* (pp. 199–243), North-Holland, Amsterdam, Netherlands.
38. Routley, R., & Routley, V. (1972). The semantics of first degree entailment. *Nous, 6*, 335–358.
39. Routley, R., Plumwood, V., Meyer, R. K., & Brady, R. T. (1982). *Brady, relevant logics and their rivals: The basic philosophical and semantical theory*, Ridgeview.
40. Schroeder-Heister, P., & Dosen, K. (1994). *Substructural logics*. Oxford University Press.
41. Tennant, N. (1987). *Anti-realism and logic: Truth as eternal*. Oxford University Press.
42. Tennant, N. (1997). *The taming of the true*. Oxford University Press.
43. Tennant, N. (2015). A new unified account of truth and paradox. *Mind, 124*, 571–605.
44. Tennant, N. (2017). *Core logic*. Oxford University Press.
45. Thomas, M. (2014). Expressive limitations of naive set theory in LP and minimally inconsistent LP. *Review of Symbolic Logic, 7*(2), 341–350.
46. Thomas, M. (2017). *A kripke-style semantics for paradox-tolerant nontransitive intuitionistic logic*. Retrieved November 6, 2017, from https://drive.google.com/file/d/0Bx_KuRX8hgkZenBSbXZ6YnZNRkk/edit.
47. van Dalen, D. (2001). Intuitionistic logic. In L. Goble (Ed.), *The blackwell guide to philosophical logic* (pp. 224–258). Blackwell.
48. Weber, Z. (2012). Transfinite cardinals in paraconsistent set theory. *Review of Symbolic Logic, 5*, 269–293.

A Substructural Logic for Inconsistent Mathematics

Guillermo Badia and Zach Weber

Abstract A logic for inconsistent mathematics must be strong enough to support reasoning in proofs, while weak enough to avoid paradoxes. We present a substructural logic intended to meet the needs of a working dialetheic mathematician—specifically, by adding a de Morgan negation to light linear logic, and extending the logic with a relevant conditional. The logic satisfies a deduction theorem. Soundness and completeness is established, showing in particular that contraction is invalidated. This opens the way for a robust naive set theory; we conclude by showing how the set theory provides a foundation for induction.

1 Inconsistent Mathematics

Is there such a thing as an *inconsistent structure* in mathematics? History abounds with inconsistent *theories*,[1] of course, but according to standard model theory, none of them correspond to a structure. Further, according to standard logical principles, 'all' models of inconsistent theories are isomorphic, so any inconsistent theory is categorical.[2]

This last claim looks dubious. The *naive set comprehension axiom*, for example, has been known to be inconsistent for over a century, and prevailing wisdom is that it must be abandoned; but from the beginning, many rejected this possibility out of

[1] See for example [5].
[2] As challenged in [18, 19].

G. Badia
School of Historical and Philosophical Inquiry, University of Queensland, Brisbane, Australia
e-mail: guillebadia89@gmail.com

Z. Weber (✉)
Department of Philosophy, University of Otago, Dunedin, New Zealand
e-mail: zach.weber@otago.ac.nz

A. Rieger and G. Young (eds.), *Dialetheism and its Applications*, Trends in Logic 52,
https://doi.org/10.1007/978-3-030-30221-4_9

hand, e.g. retaining certainty that collections like the set of all ordinals must exist [17, p.111, 117].[3] The idea that *every property determines a set*

$$x \in \{z : \varphi\} \leftrightarrow \varphi$$

has struck many as worth some logical adjustments—'solving for $\leftrightarrow$' in the above formula, so to speak; see e.g. [2, 13] for early steps. The claim that all inconsistent theories are categorical looks doubly dubious because it turns on an instance of *ex falso quodlibet*, explosion: if there are no φs, then all φs are ψ.

The claim that the inconsistent has no structure has been challenged, by the *inconsistent mathematics* program. Explosion is challenged by *paraconsistent logic*. Advances in paraconsistent logic in the latter 20th century showed that it is possible to consider inconsistent but non-trivial mathematical theories. Da Costa expresses the spirit of exploration:

> It would be as interesting to study the inconsistent systems as, for instance, the non-euclidean geometries: we would obtain a better idea of the nature of paradoxes, could have a better insight on the connections amongst the various logical principles necessary to obtain determinate results, etc. ... It is not our aim to eliminate the inconsistencies, but to analyze and study them [9].

Initial results were obtained in set theory [4, 25] (in [23]), [8] (in [1]) and arithmetic [11, 16], and especially the model theory for understanding these theories [10]; see also [21, 24].

The program has had two main themes: in any mathematics based in a paraconsistent logic, one wants as far as possible to satisfy:

Recapture	Nothing (too important) is lost
Expansion	New results are gained

The important concern behind 'recapture' is to show that real, everyday arguments are viable—"recovering the bulk of intuitive mathematics" [25, p. 903]. The important concern behind 'expansion' is that all this work will be worthwhile: not only would naive set theory be able to do what ZFC can do, but also more, opening up new vistas on the other side of the horizon. As such, there are two countervailing demands on logic(s) for inconsistent mathematics: it must be

- *weak* enough to support large abstraction schemes, such as the set comprehension axiom, without triviality
- *strong* enough to be able to support actual proofs and carry through to familiar results

Achieving the recapture and expansion goals requires finding a suitable logic, and mainly, a suitable implication connective. The purpose of this paper is to present a logic that has been 'field tested' [30] and can be up to the job.

If there is structure in the inconsistent, we will need a logic to find it.

---

[3] Perhaps by an argument from 'deep familiarity'; the same type of sentiment is invoked in a different context: "The situation would be like having a purported 'obviously true' principle that implied that midtown Manhattan doesn't exist. But I know it exists; I live there. Please come visit!" [14].

2 Adequacy Conditions on a Logic

The logic is designed to house naive set theory; the main focus is on *propositional* logic and in particular finding suitable implication connective(s). In light of paradoxes, we discuss the desiderata we wish our logic to satisfy.

An effective conditional should have the following properties.

- Obey modus ponens
- *Not* be the material conditional (else satisfy disjunctive syllogism, and not be paraconsistent [22, Chap. 1]; cf. [27, Chap. 1])
- *Not* obey contraction (else Curry/Grišin's paradoxes [6])
- Either contrapose, or weaken, but not both (else satisfy explosion)
- Have a reasonable introduction rule/deduction theorem
- Support a theory of identity compatible with the axiom of set extensionality

In the Australasian tradition, vertebrate paraconsistent conditionals come from *relevant logic*. But a conditional that obeys an 'irrelevant' conditional-introduction rule appears to be indispensible in mathematical proofs; some statements of the form 'φ implies ψ' are true even when the implication makes use of background assumptions other than φ, or does not use φ at all. On the other hand, a conditional that respects relevance, where 'φ implies ψ' indicates that nothing but φ was used, is indispensible, too, to handle identity (see Sect. 5). Our solution is that this list can be satisfied, but by two conditionals. The main one, $\Rightarrow$, meets all the criteria except contraposing and compatibility with set extensionality. So the other, a relevant $\rightarrow$, is added that does not weaken but can allow substitution of equivalents.

So far, much of this is compatible with the *linear logic* approach [20, 28], and calls up the logic BCK. But an effective *negation* is needed, too:

- Be exclusive, or exhaustive, but not both
- Not support ex contradictione quodlibet
- Support argument by cases (χ follows from $\varphi \vee \psi$ if it follows from both disjuncts)
- Support reductio ad absurdum arguments (if $\neg\varphi$ is absurd, then φ)
- Have full de Morgan properties (duality between conjunction and disjunction)

One of the most basic reasons for these demands is that the target is *inconsistent* mathematics, which will include some dialetheias (true contradictions). At the structural level, proving contradictions (appears to) require the law of excluded middle, $\varphi \vee \neg\varphi$, which is also needed to make sense of reductio arguments. Moreover, dialetheism correctly says of itself that it is false: all contradictions are false. This follows nicely if excluded middle holds *and* there is duality: $\neg(\varphi \& \neg\varphi)$ ought to be a theorem. Our solution is that this list can be met with two negations. The main one, $\neg$, is dialetheic negation, essentially as given in Priest's logic LP. It is exhaustive and not explosive. Another negation, 'φ implies $\bot$', is explosive and supports reductio ad absurdum arguments.

So far so good. But putting these solutions together leaves us in a bind. To avoid contraction, we cannot have

$$\varphi \Rightarrow \varphi \& \varphi$$

But to have argument by cases, we must have, well, argument by cases, which proves

$$\varphi \vee \varphi \Rightarrow \varphi$$

And yet to do proofs, e.g. to preserve duality between finite unions and finite intersections, we need de Morgan laws connecting & with $\vee$. That is, we want a multiplicative (intensional) conjunction, an additive (extensional) disjunction, and negation bridging principles between the two. Now, the $\Rightarrow$ arrow does not contrapose, so this looks prima facie possible, but as far as we know combining all these ideas has never been attempted substructural logic. Thus a main result in this paper is just that such a logic *exists*.

A final point to make about any satisfactory logic here is that there are problems with substitution. The plausible principle that equivalents[4] are intersubstitutable in any context χ,

$$(\varphi \Leftrightarrow \psi) \Rightarrow (\chi(\varphi) \Rightarrow \chi(\psi))$$

is incoherent, for reasons related to Grišin's paradox (Sect. 5). If substitution is allowed, we will have

$$(\varphi \Leftrightarrow \psi) \Rightarrow (\varphi \Leftrightarrow \psi) \& (\varphi \Leftrightarrow \psi)$$

because $(\varphi \Leftrightarrow \varphi) \& (\varphi \Leftrightarrow \varphi)$ is true. But then given the curry set

$$x \in c \Leftrightarrow (x \in x \Leftrightarrow \bot)$$

we could use formula substitution to reason as follows:

(1) $c \in c \Leftrightarrow (c \in c \Leftrightarrow \bot)$
(2) $(c \in c \Leftrightarrow \bot) \Rightarrow (c \in c \Leftrightarrow \bot) \& (c \in c \Leftrightarrow \bot)$
(3) $c \in c \Rightarrow (c \in c \Leftrightarrow \bot) \& (c \in c \Leftrightarrow \bot)$
(4) $c \in c \Rightarrow c \in c \& c \in c$

the last step by substitution. This would reinstate contraction in the Curry derivation, and trivialize the theory. Hence $\Leftrightarrow$ cannot be a congruence on formulas.

All these needs in view, we turn to the logic.

---

[4]Where $\varphi \Leftrightarrow \psi$ is $(\varphi \Rightarrow \psi)\&(\psi \Rightarrow \varphi)$.

3 Proof Theory for SubDL

With applications in mind, we are not concerned with designing a nice logic *qua* logic. We want a tool to prove theorems in inconsistent mathematics. We call this system subDL, for 'substructural dialetheic logic', with acknowledgment of [26].[5] For mathematical background on the approach taken here, see [12].

3.1 Gentzen System

We start with a language having connectives $\&, \vee, \neg$, an arrow $\Rightarrow$ and $\bot$. A single conclusion sequent is written $\Gamma \succ \varphi$, where Γ, Δ, Σ ..., are finite—possibly empty—multisets of formulas. A Gentzen calculus for the logic can be presented as follows:

Initial sequents:

$\phi \succ \phi \qquad \bot \succ \phi \qquad \succ \neg(\bot) \qquad \succ \phi \vee \neg\phi \qquad \phi, \neg\psi \succ \neg(\phi \Rightarrow \psi)$
$(\phi \,\&\, \psi) \vee (\phi \,\&\, \chi) \succ \phi \,\&\, (\psi \vee \chi) \qquad \phi \,\&\, (\psi \vee \chi) \succ (\phi \,\&\, \psi) \vee (\phi \,\&\, \chi)$
$\neg((\phi \,\&\, \psi) \vee (\phi \,\&\, \chi)) \succ \neg(\phi \,\&\, (\psi \vee \chi)) \qquad \neg(\phi \,\&\, (\psi \vee \chi)) \succ \neg((\phi \,\&\, \psi) \vee (\phi \,\&\, \chi))$
$\neg(\phi \vee \psi) \succ \neg\phi \,\&\, \neg\psi \qquad \neg\phi \,\&\, \neg\psi \succ \neg(\phi \vee \psi)$
$\neg(\phi \,\&\, \psi) \succ \neg\phi \vee \neg\psi \qquad \neg\phi \vee \neg\psi \succ \neg(\phi \,\&\, \psi)$
$\neg\neg\phi \succ \phi \qquad \phi \succ \neg\neg\phi$

Operational rules:

$$\frac{\Gamma, \phi \succ \psi}{\Gamma \succ \phi \Rightarrow \psi}\,(\succ \Rightarrow) \qquad \frac{\Gamma \succ \phi \qquad \Delta, \psi, \Sigma \succ \chi}{\Delta, \Gamma, \phi \Rightarrow \psi, \Sigma \succ \chi}\,(\Rightarrow \succ)$$

$$\frac{\Gamma \succ \phi}{\Gamma \succ \phi \vee \psi}\,(\succ\vee)a \qquad \frac{\Gamma \succ \psi}{\Gamma \succ \phi \vee \psi}\,(\succ\vee)b \qquad \frac{\Gamma, \phi, \Delta \succ \psi \qquad \Gamma, \chi, \Delta \succ \psi}{\Gamma, \phi \vee \chi, \Delta \succ \psi}\,(\vee\succ)$$

$$\frac{\Gamma \succ \psi \qquad \Delta \succ \phi}{\Gamma, \Delta \succ \phi \,\&\, \psi}\,(\succ \&)$$

$$\frac{\Gamma, \chi, \Delta \succ \psi}{\Gamma, \phi \,\&\, \chi, \Delta \succ \psi}\,(\&\succ)a \qquad \frac{\Gamma, \phi, \Delta \succ \psi}{\Gamma, \phi \,\&\, \chi, \Delta \succ \psi}\,(\&\succ)b \qquad \frac{\Gamma, \psi, \chi, \Delta \succ \phi}{\Gamma, \psi \,\&\, \chi, \Delta \succ \phi}$$

$(\&\succ)c$

Structural rules:

$$\frac{\Gamma, \Delta \succ \phi}{\Gamma, \psi, \Delta \succ \phi}\,(\text{Weakening}) \qquad \frac{\Gamma \succ \phi \qquad \Delta, \phi, \Sigma \succ \psi}{\Delta, \Gamma, \Sigma \succ \psi}\,(\text{Cut})$$

[5]To obtain subDL, start with the relevant logic DK from [25]; remove structural contraction, and any axiom that equates one occurrence of a formula with two; then add a conditional that obeys a deduction theorem.

The structural rule of contraction is, of course, absent from the system while exchange (the re-ordering of premises) is just a feature of using multisets.

3.2 *Hilbert System*

Now let us consider the following Hilbert system which, as we will see, is equivalent to the Gentzen system just presented. That is, given a collection S of sequents, a sequent s is provable from S in our Gentzen calculus iff $t(s)$ is deducible from $\{t(q) : q \in S\}$ in our Hilbert calculus, where t is an operation that given a sequent $\phi_0, \ldots, \phi_n \succ \psi$ gives back the formula $\phi_0 \,\&\, \ldots \,\&\, \phi_n \Rightarrow \psi$.

Axioms:

A1 $\vdash \phi \Rightarrow (\psi \Rightarrow \phi)$
A2 $\vdash (\phi \Rightarrow \psi) \Rightarrow ((\chi \Rightarrow \phi) \Rightarrow (\chi \Rightarrow \psi))$
A3 $\vdash (\phi \Rightarrow \psi) \Rightarrow ((\psi \Rightarrow \chi) \Rightarrow (\phi \Rightarrow \chi))$
A4 $\vdash (\phi \Rightarrow (\psi \Rightarrow \chi)) \Rightarrow (\psi \Rightarrow (\phi \Rightarrow \chi))$
A5 $\vdash \phi \Rightarrow \phi \vee \psi \qquad \vdash \psi \Rightarrow \phi \vee \psi$
A6 $\vdash (\phi \Rightarrow \chi) \Rightarrow ((\psi \Rightarrow \chi) \Rightarrow (\phi \vee \psi \Rightarrow \chi))$
A7 $\vdash \phi \Rightarrow (\psi \Rightarrow \phi \,\&\, \psi)$
A8 $\vdash (\phi \Rightarrow (\psi \Rightarrow \chi)) \Rightarrow (\phi \,\&\, \psi \Rightarrow \chi)$
A9 $\vdash \phi \,\&\, (\psi \vee \chi) \Leftrightarrow (\phi \,\&\, \psi) \vee (\phi \,\&\, \chi)$
$\vdash \neg(\phi \,\&\, (\psi \vee \chi)) \Leftrightarrow \neg((\phi \,\&\, \psi) \vee (\phi \,\&\, \chi))$
A10 $\vdash \neg(\phi \vee \psi) \Leftrightarrow \neg\phi \,\&\, \neg\psi$
A11 $\vdash \neg(\phi \,\&\, \psi) \Leftrightarrow \neg\phi \vee \neg\psi$
A12 $\vdash \neg\neg\phi \Leftrightarrow \phi$
A13 $\vdash \phi \vee \neg\phi$
A14 $\vdash (\phi \,\&\, \neg\psi) \Rightarrow \neg(\phi \Rightarrow \psi)$
A15 $\vdash \phi \,\&\, \psi \Rightarrow \phi \qquad \phi \,\&\, \psi \Rightarrow \psi$

Rule:

$$\frac{\Gamma \vdash \phi \qquad \Delta \vdash \phi \Rightarrow \psi}{\Gamma, \Delta \vdash \chi} \text{(Modus Ponens)}$$

3.2.1 Validity [Multiset Consequence]

Given an arbitrary list (possibly with repetitions) of formulas $\delta_0, \ldots, \delta_n$ (or a multiset) we define a *deduction* of ϕ from $\delta_0, \ldots, \delta_n$ as a finite sequence of formulas, with ϕ as its last member such that

(1) Every element of the sequence is

  (1.1) some δ_i, or
  (1.2) an axiom of the Hilbert system, or
  (1.3) it comes by an application of MP from formulas ψ and $\psi \Rightarrow \phi$ appearing before in the sequence.

(2) Every element of the sequence has an associated multiset listing the formulas different from itself immediately involved in its appearance on the sequence ($\varnothing$ if it is either an axiom or a premise, the two formulas required for an application of MP, if it results from such an application)
(3) The submultiset of the union of all the associated multisets of elements in the sequence containg elements from the multiset $\delta_0, \ldots, \delta_n$ is a submultiset of $\delta_0, \ldots, \delta_n$.

A formula ϕ is deducible from a multiset $\delta_0, \ldots, \delta_n$ if there is a deduction as defined above. We write

$$\delta_0, \ldots, \delta_n \vdash \phi$$

in that case, which exaplains the notation in modus ponens. The idea is that the notion of deduction for this system represents the internal consequence relation of the Gentzen system.

How does this block, say, the problematic deduction $\varphi \vdash \varphi\&\varphi$? Let us try to run the usual argument

(1) φ Premise, Associated multiset: $\varnothing$
(2) $\varphi \Rightarrow (\varphi \Rightarrow \varphi\&\varphi)$ Axiom, Associated multiset: $\varnothing$
(2) $\varphi \Rightarrow \varphi\&\varphi$ MP, Associated multiset: $[\varphi, \varphi \Rightarrow (\varphi \Rightarrow \varphi\&\varphi)]$
(2) $\varphi\&\varphi$ MP, Associated multiset: $[\varphi, \varphi \Rightarrow \varphi\&\varphi]$

Now, however the multiset $[\varphi, \varphi]$ is *not* a submultiset of $[\varphi]$, so this is emphatically *not* a deduction in our system.

3.2.2 Derived Formulas

While substitution of equivalent formulas is not generally valid, some instances are provable. Namely, in any positive context (i.e. not under a negation) we can substitute 'one formula at a time' using the transitivity of the conditional (A2 and A3), as can be shown by checking on formula complexity. E.g. if $\phi \Leftrightarrow \psi$ then $\phi\&\chi \Rightarrow \psi\&\chi$.[6]

Already the identity axiom

$$\phi \Rightarrow \phi$$

[6]On the other hand we don't get the paradox from the end of Sect. 2 because the instance

$$(\phi \Leftrightarrow \psi) \Rightarrow ((\phi \Leftrightarrow \phi)\&(\phi \Leftrightarrow \phi) \Rightarrow (\phi \Leftrightarrow \psi)\&(\phi \Leftrightarrow \psi))$$

uses substitution 'two at a time'.

follows: from A1 and A4, we get $\psi \Rightarrow (\phi \Rightarrow \phi)$. Then take any axiom for ψ and apply modus ponens. Using A4, A7 and A8, we know that

$$\phi \,\&\, \psi \Rightarrow \psi \,\&\, \phi$$

is a theorem, that is, & is commutative. Observe that

$$(\phi \,\&\, \psi \Rightarrow \chi) \Rightarrow (\phi \Rightarrow (\psi \Rightarrow \chi))$$

is a theorem, which in what follows we will use interchangeably with the residuation axiom A8. For $(\phi \,\&\, \psi \Rightarrow \chi) \Rightarrow ((\psi \Rightarrow \phi \,\&\, \psi) \Rightarrow (\psi \Rightarrow \chi))$ is an instance of A2. Permuting, $(\psi \Rightarrow \phi \,\&\, \psi) \Rightarrow ((\phi \,\&\, \psi \Rightarrow \chi) \Rightarrow (\psi \Rightarrow \chi))$. Now, $\phi \Rightarrow (\psi \Rightarrow \phi \,\&\, \psi)$ is an instance of A7, so with A3, we get that $(\phi \,\&\, \psi \Rightarrow \chi) \Rightarrow (\phi \Rightarrow (\psi \Rightarrow \chi))$ as desired.

Distribution follows too:

(1) $\phi\&\psi \Rightarrow (\phi\&\psi) \vee (\phi\&\chi)$ (A5)
(2) $\psi \Rightarrow (\phi \Rightarrow (\phi\&\psi) \vee (\phi\&\chi))$ (from (1), A7, A2)
(3) $\phi\&\chi \Rightarrow (\phi\&\psi) \vee (\phi\&\chi)$ (A5)
(4) $\chi \Rightarrow (\phi \Rightarrow (\phi\&\psi) \vee (\phi\&\chi))$ (from (3), A7, A2)
(5) $\psi \vee \chi \Rightarrow (\phi \Rightarrow (\phi\&\psi) \vee (\phi\&\chi))$ (from (2), (4), A6)
(6) $(\psi \vee \chi)\&\phi \Rightarrow (\phi\&\psi) \vee (\phi\&\chi)$ (from (5), A1, A8)

We include it as an axiom because it is better to be explicit than to avoid redundancy.

3.2.3 Deduction and Equivalence

Theorem 1 *The Hilbert system satisfies a deduction theorem.*

Proof By induction on the definition of deducibility we show that for arbitrary multiset Γ, and formulas ψ and ϕ, if $\Gamma, \psi \vdash \phi$, then $\Gamma \vdash \psi \Rightarrow \phi$.

If ϕ appears in Γ, then $\Gamma \vdash \phi$ and $\varnothing \vdash \phi \Rightarrow (\psi \Rightarrow \phi)$, so in fact $\Gamma \vdash \psi \Rightarrow \phi$ because trivially the multiset Γ is just the multiset union of Γ and $\varnothing$.

If ϕ is an axiom we argue as in the previous case.

If $\phi = \psi$, then $\Gamma \vdash \phi \Rightarrow \phi$ is easy to get.

Now, if ψ follows from Γ, ϕ by an application of modus ponens, then there must be disjoint multisets (since no formula in Γ can be used twice in a deduction) $\delta_i, \ldots, \delta_j$ and $\delta_k, \ldots, \delta_m$ which are in turn segments of Γ, ϕ such that $\delta_i, \ldots, \delta_j \vdash \tau$ and $\delta_k, \ldots, \delta_m \vdash \tau \Rightarrow \psi$. If ϕ is in none of $\delta_i, \ldots, \delta_j$ or $\delta_k, \ldots, \delta_m$, then, from weakening, $\delta_k, \ldots, \delta_m \vdash \phi \Rightarrow (\tau \Rightarrow \psi)$, and, by exchange, $\delta_k, \ldots, \delta_m \vdash \tau \Rightarrow (\phi \Rightarrow \psi)$. So from modus ponens, $\delta_i, \ldots, \delta_j, \delta_k, \ldots, \delta_m \vdash \phi \Rightarrow \psi$, so $\Gamma \vdash \phi \Rightarrow \psi$ as desired. If ϕ is in $\delta_k, \ldots, \delta_m$ but not in $\delta_i, \ldots, \delta_j$, then me may assume $\delta_k, \ldots, \delta_m$ to be $\delta_k, \ldots, \delta_{m-1}, \phi$. By inductive hypothesis, $\delta_k, \ldots, \delta_{m-1} \vdash \phi \Rightarrow (\tau \Rightarrow \psi)$, and reasoning as before, we obtain that in fact $\delta_i, \ldots, \delta_j, \delta_k, \ldots, \delta_{m-1} \vdash \phi \Rightarrow \psi$, so $\Gamma \vdash \phi \Rightarrow \psi$. The only other possibility is that ϕ is in $\delta_i, \ldots, \delta_j$ but not in $\delta_k, \ldots, \delta_m$, then me may assume $\delta_i, \ldots, \delta_j$ to be $\delta_i, \ldots, \delta_{j-1}, \phi$. Then by inductive hypothesis, $\delta_i, \ldots, \delta_{j-1} \vdash \phi \Rightarrow \tau$ and since $\delta_k, \ldots, \delta_m \vdash \phi \Rightarrow (\tau \Rightarrow \psi)$, we have with modus ponens and A4, $\delta_i, \ldots, \delta_{j-1}, \delta_k, \ldots, \delta_m \vdash \phi \Rightarrow \psi$, so $\Gamma \vdash \phi \Rightarrow \psi$. □

Theorem 2 *The Hilbert system is theoremhood equivalent to the Gentzen system. That is, ϕ is a theorem of the Hilbert system iff $\succ\phi$ is a provable sequent of the Gentzen system.*

Proof This follows along the lines of the proof of Corollary 2.21 from [12]. Given a collection S of sequents, the first step is to show that a sequent s is provable from S in our Gentzen calculus only if $t(s)$ is deducible from $\{t(q) : q \in S\}$ in our Hilbert calculus, where t is an operation that given a sequent $\phi_0, \ldots, \phi_n \succ \psi$ gives back the formula $\phi_0 \,\&\, \ldots \,\&\, \phi_n \Rightarrow \psi$. This is established by induction on the length of the proof. Let us work through one example.

Consider an arbitrary instance of the rule $(\Rightarrow \succ)$:

$$\frac{\gamma_0, \ldots, \gamma_j \succ \phi \qquad \delta_0, \ldots, \delta_k, \psi, \sigma_0, \ldots, \sigma_l \succ \chi}{\delta_0, \ldots, \delta_k, \gamma_0, \ldots, \gamma_j, \phi \Rightarrow \psi, \sigma_0, \ldots, \sigma_l \succ \chi} (\Rightarrow \succ)$$

We need to show that from

$$\gamma_0 \,\&\, \ldots \,\&\, \gamma_j \Rightarrow \phi$$

and

$$\delta_0 \,\&\, \ldots \,\&\, \delta_k \,\&\, \psi \,\&\, \sigma_0 \,\&\, \ldots \,\&\, \sigma_l \Rightarrow \chi$$

we can deduce that

$$((\delta_0 \,\&\, \ldots \,\&\, \delta_k \,\&\, \gamma_0 \,\&\, \ldots \,\&\, \gamma_j) \,\&\, (\phi \Rightarrow \psi) \,\&\, (\sigma_0 \,\&\, \ldots \,\&\, \sigma_l)) \Rightarrow \chi$$

Now, certainly from A8,

$$\psi \Rightarrow (\delta_0 \,\&\, \ldots \,\&\, \delta_k \,\&\, \sigma_0 \,\&\, \ldots \,\&\, \sigma_l \Rightarrow \chi)$$

Then, using A3 and A4,

$$(\phi \Rightarrow \psi) \Rightarrow (\phi \Rightarrow (\delta_0 \,\&\, \ldots \,\&\, \delta_k \,\&\, \sigma_0 \,\&\, \ldots \,\&\, \sigma_l \Rightarrow \chi))$$

so, by A4,

$$\phi \Rightarrow ((\phi \Rightarrow \psi) \Rightarrow (\delta_0 \,\&\, \ldots \,\&\, \delta_k \,\&\, \sigma_0 \,\&\, \ldots \,\&\, \sigma_l \Rightarrow \chi))$$

and, from A3,

$$\gamma_0 \,\&\, \ldots \,\&\, \gamma_j \Rightarrow ((\phi \Rightarrow \psi) \Rightarrow (\delta_0 \,\&\, \ldots \,\&\, \delta_k \,\&\, \sigma_0 \,\&\, \ldots \,\&\, \sigma_l \Rightarrow \chi)).$$

Hence, we can use A8 and commutativity of & in conjunction with A3 if necessary to get the desired goal.

Finally, it suffices to show that if ϕ is a theorem of the Hilbert system, then $\succ\phi$ is a provable sequent in the Gentzen calculus. We again proceed by induction on the length of the proof. If ϕ is an axiom, it is straightforward that $\succ\phi$ is also provable. For the induction step, let us assume that ϕ follows from ψ and $\psi \Rightarrow \phi$, which must both be theorems by an application of modus ponens (notice we are dealing with a deduction here from the empty multiset of premises). By induction hypothesis, $\succ\psi$ and $\succ\psi \Rightarrow \phi$ are provable, from which we can get $\succ\phi$ using $\Rightarrow\succ$ and Cut. □

4 Semantics for SubDL

Now we introduce a semantics for the Gentzen system. This will not be the intended interpretation of the logic, but merely an instrumental device for the purposes of showing that the logic is not incoherent.

4.1 Gentzen Matrices

The notion of a *Gentzen Matrix* is studied in [12, Chap. 7]. We produce an algebraic object specifically designed to correspond to our Gentzen calculus (and then, by equivalence, to the Hilbert system). Consider a structure M of the form

$$(A, \cap, \cup, \neg, \Rightarrow, <, 0)$$

such that A is a set of objects, $\cap$, $\cup$, $\neg$ and $\Rightarrow$ are operations on A of the obvious arity, $<$ is a binary relation and 0 is a distinguished element (or a zero-place operation). We want to interpret sequents of the form $\Gamma \succ \phi$ as claims about the relation $<$ on a given structure M.

Start with a homomorphism h from the set of formulas of our language **Fmla** into M such that the propositional variables get assigned elements of A and complex formulas get assigned elements in the expected way corresponding to the operations $\cap, \cup, \neg, \Rightarrow$ and 0 (that is, conjunction corresponds to $\cap$, disjunction to $\cup$, implication to $\Rightarrow$, negation to $\neg$ and $\perp$ to 0). A model **M** is a pair (M, h). We then write, where Γ is the multiset $\phi_0, \ldots, \phi_n$ and $\cap(x_0, \ldots, x_n)$ abbreviates the expression $\cap(x_0, \cap(x_1, \cap(\ldots, \cap(x_{n-1}, x_n)\ldots)))$,

$$\mathbf{M} \Vdash \Gamma \succ \psi \quad \text{iff} \quad M \vDash \cap(h(\phi_0), \ldots, h(\phi_n)) < h(\psi)$$

When Γ is the empty multiset, we interpret it as $\neg 0 = h(\neg\bot)$. For instance, when Γ is the empty multiset and Δ is the multiset $\delta_0, \ldots, \delta_n$, we would have that

$$\mathbf{M} \Vdash \Gamma, \Delta \succ \psi \quad \text{iff} \quad M \vDash \cap(\neg 0, \cap(h(\phi_0), \ldots, h(\phi_n))) < h(\psi)$$

and also that

$$\mathbf{M} \Vdash \Gamma \succ \psi \quad \text{iff} \quad \mathbf{M} \Vdash \succ \psi \quad \text{iff} \quad M \vDash \neg 0 < h(\psi)$$

A sequent $\Gamma \succ \psi$ is *valid* in the class of models K in this semantics if for every model $\mathbf{M} \in K, \mathbf{M} \Vdash \Gamma \succ \psi$.

Our aim is to find a class of models that would validate all the sequents provable in our Gentzen calculus. For that purpose consider the models satisfying the following first order conditions, which will guarantee the soundness of the Gentzen system:

$\forall x(x < x)$
$\forall x, y(\cap(x, \neg y) < \neg(x \Rightarrow y))$
$\forall x, y, z, u, v((\cap(x, \cap(y, z)) < u) \supset (\cap(x, \cap(\cap(y, v), z)) < u)$
$\forall x, y, z, u, v((x < y \;\&\; \cap(z, \cap(y, u)) < v) \supset (\cap(z, \cap(x, u)) < v)))$
$\forall x, y(\cap(x, y) < \cap(y, x))\forall x, y(\neg(\cap(x, y)) < \cup(\neg(x), \neg(y)) \;\&\; \cup(\neg(x), \neg(y)) < \neg(\cap(x, y)))$
$\forall x, y(\neg(\cup(x, y)) < \cap(\neg(x), \neg(y)) \;\&\; \cap(\neg(x), \neg(y)) < \neg(\cup(x, y)))$
$\forall x(\neg(0 < \cup(x, \neg(x))))$
$\forall x, y, z, u, v, w((x < y \;\&\; (\cap(z, \cap(u, v)) < w)) \supset ((\cap(z, \cap(\Rightarrow (y, u), \cap(x, v))) < w)))$
$\forall x, y, z, v((x < y \;\&\; z < v) \supset (\cap(x, z) < \cap(y, v)))$
$\forall x(x < \neg\neg x \;\&\; \neg\neg x < x)$
$\forall x, y, z, u(\cap(x, y) < z \supset \cap(x, \cap(u, y)) < z)$
$\forall x, y, z(x < y \supset (x < \cup(y, z)))$
$\forall x, y, z(x < z \supset (x < \cup(y, z)))$
$\forall x, y, z, u, w((< (\cap(x, \cap(y, z)) < u) \;\&\; (\cap(x, \cap(w, z)) < u)) \supset ((\cap(x, \cap(\cup(y, w), z)) < u)))$
$\forall x, y, z, u, v(((\cap(x, \cap(y, z)) < u)) \supset ((\cap(x, \cap(\cap(y, w), z)) < u)))$
$\forall x, y, z, u, v(((\cap(x, \cap(w, z)) < u)) \supset ((\cap(x, \cap(\cap(y, w), z)) < u)))$
$\forall x, y, z(((\cap(x, y) < z)) \supset ((y <\Rightarrow (x, z))))$
$\forall x(0 < x)$
$\forall x, y((x \cap y) \cup (x \cap z) < x \cap (y \cup z)) \;\&\; (x \cap (y \cup z) < (x \cap y) \cup (x \cap z)))$
$\forall x, y(\neg((x \cap y) \cup (x \cap z)) < \neg(x \cap (y \cup z))) \;\&\; (\neg(x \cap (y \cup z)) < \neg((x \cap y) \cup (x \cap z))))$
$\forall x, y, z, u, v(((\cap(x, \cap(y, \cap(z, u))) < v)) \supset ((\cap(x, \cap(\cap(y, z), u)) < v)))$

The class of models so defined make possible the following.

4.2 Completeness

Theorem 3 *The Gentzen system in Sect. 3.1 is sound and complete with respect to the structures described above.*

Proof Because the semantics is designed for this theorem, there is little to argue. The soundness argument proceeds, as usual, by induction on the length of the proof. All the initial sequents are valid as well as the rules.

The completeness of the Gentzen system with respect to this class of structures is by a canonical model argument. Let the domain of the canonical model be the collection of all formulas and the operations be defined in the obvious way—i.e., the operation $\neg$ is negation, 0 is $\bot$, $\cap$ is just & and so on. The canonical relation $<$ is defined as follows:

$$\phi < \psi \quad \text{iff the sequent } \phi \succ \psi \text{ is provable}$$

All the conditions the model needs to satisfy correspond to either some initial sequent or some rule of the Gentzen system, so they trivially come out true. The homomorphism serving as assignment of values in this structure to the formulas of our language is simply the identity. □

4.3 A Four Valued Algebra and the Invalidity of Contraction

Finally the following model (found using PROVER9)[7] shows that $\phi \succ \phi \,\&\, \phi$ is not a provable sequent, so contraction is not derivable in the system. Take the domain to be the set {0, 1, 2, 3}. The operations and relation $<$ are defined as follows:

	$\cap(0,0)=0$	$\cup(0,0)=0$	$\Rightarrow(0,0)=2$	
	$\cap(0,1)=0$	$\cup(0,1)=3$	$\Rightarrow(0,1)=2$	$<(0,0)$
	$\cap(0,2)=0$	$\cup(0,2)=2$	$\Rightarrow(0,2)=2$	$<(0,1)$
	$\cap(0,3)=3$	$\cup(0,3)=3$	$\Rightarrow(0,3)=2$	$<(0,2)$
	$\cap(1,0)=0$	$\cup(1,0)=3$	$\Rightarrow(1,0)=1$	$<(0,3)$
	$\cap(1,1)=0$	$\cup(1,1)=1$	$\Rightarrow(1,1)=2$	$<(1,1)$
$\neg(0)=2$	$\cap(1,2)=1$	$\cup(1,2)=2$	$\Rightarrow(1,2)=2$	$<(1,2)$
$\neg(1)=3$	$\cap(1,3)=3$	$\cup(1,3)=3$	$\Rightarrow(1,3)=3$	$<(1,3)$
$\neg(2)=0$	$\cap(2,0)=0$	$\cup(2,0)=2$	$\Rightarrow(2,0)=0$	$<(2,2)$
$\neg(3)=3$	$\cap(2,1)=1$	$\cup(2,1)=2$	$\Rightarrow(2,1)=1$	$<(2,3)$
	$\cap(2,2)=2$	$\cup(2,2)=2$	$\Rightarrow(2,2)=2$	$<(3,0)$
	$\cap(2,3)=3$	$\cup(2,3)=2$	$\Rightarrow(2,3)=3$	$<(3,1)$
	$\cap(3,0)=3$	$\cup(3,0)=3$	$\Rightarrow(3,0)=3$	$<(3,2)$
	$\cap(3,1)=3$	$\cup(3,1)=3$	$\Rightarrow(3,1)=3$	$<(3,3)$
	$\cap(3,2)=3$	$\cup(3,2)=2$	$\Rightarrow(3,2)=2$	
	$\cap(3,3)=3$	$\cup(3,3)=3$	$\Rightarrow(3,3)=3$	

Note that $\cap(1,1)$ is 0 but we do not have that $1 < 0$ in this model. Hence, the sequent $p \succ p \,\&\, p$ fails in this model when p gets assigned the value 1. Contraction is invalid in `subDL`.

[7] At https://www.cs.unm.edu/~mccune/mace4/.

This is notable because the law of excluded middle is related to contraction, so adding LEM with all the negation properties we've assumed could be thought to bring contraction back. It does not.

5 Extending with Relevant Implication

Set theory is the theory of extensions. It is natural to assume, then, that naive set theory will validate an axiom asserting that extensionally identical sets are really identical: if the φs are exactly the ψs, then $\{x : \varphi\} = \{x : \psi\}$, as Frege famously asserted in his Basic Law V. What happens if we assume $\forall z(z \in x \Leftrightarrow z \in y)$ implies $x = y$?

We presume a lesson of Curry's paradox is that any connective & that residuates the conditional, $(\varphi \,\&\, \psi \Rightarrow \chi) \Leftrightarrow (\varphi \Rightarrow (\psi \Rightarrow \chi))$, must not validate $\varphi \Rightarrow \varphi \,\&\, \varphi$. But *instances* of idempotence in the presence of $=$ are still provable, and this is sufficient for *Grišin's paradox*. Note that

$$x = x \,\&\, x = x$$

always holds, since $x = x$ is true and $x = x$ is true. Then given the instance of Leibniz identity

$$x = y \Rightarrow ((x = x \,\&\, x = x) \Rightarrow (x = y \,\&\, x = y))$$

we have contraction for $=$,

$$x = y \Rightarrow x = y \,\&\, x = y$$

by permutation. This becomes a major problem since comprehension generates a set term for every formula. Contraction for $=$ trickles back to contraction for $\in$. The idea of the proof is just this. Let $\mathcal{V}_0 = \{x : x = x\}$ and $\mathcal{V}_1 = \{x : x = x \,\&\, \varphi\}$. Then

(1) $\varphi \Leftrightarrow (x = x \Leftrightarrow x = x \,\&\, \varphi)$
(2) $\varphi \Leftrightarrow (x \in \mathcal{V}_0 \Leftrightarrow x \in \mathcal{V}_1)$
(3) $\varphi \Leftrightarrow \forall x(x \in \mathcal{V}_0 \Leftrightarrow x \in \mathcal{V}_1)$
(4) $\varphi \Leftrightarrow \mathcal{V}_0 = \mathcal{V}_1$
(5) $\varphi \Leftrightarrow \mathcal{V}_0 = \mathcal{V}_1 \,\&\, \mathcal{V}_0 = \mathcal{V}_1$
(6) $\varphi \Leftrightarrow \varphi \,\&\, \varphi$

proving contraction for arbitrary formulas, which is bad.

There are more extensionality paradoxes, as in [6]. Let $\varnothing = \{x : \bot\}$ and let

$$g = \{x : g = \varnothing\}$$

Then

$$x \in g \Rightarrow g = \varnothing$$

Since identity contracts,

$$g = \varnothing \Rightarrow g = \varnothing \& g = \varnothing$$

But using the definition of g,

$$g = \varnothing \& g = \varnothing \Rightarrow x \in g \& x \in g$$

Now, assuming substitution holds in the form $a = b \ \& \ \varphi(a) \Rightarrow \varphi(b)$, then

$$x \in g \& g = \varnothing \Rightarrow \bot$$

is a valid instance. But $x \in g \Rightarrow g = 0$, so cutting these together we have

$$x \in g \& x \in g \Rightarrow \bot$$

And finally, since

$$x \in g \Rightarrow x \in g \& x \in g$$

we get

$$x \in g \Rightarrow \bot$$

This appears to prove that g is extensionally equivalent to $\varnothing$. If so, though, then the defining condition of g is true, whereby it has everything as a member, and

$$\bot$$

This is sometimes called the Hinnion-Libert paradox; see [15].

What this derivation *cannot* mean is that $g = \varnothing$. It must mean that, while g is in some sense co-extensive with the empty set, it is not identical to the empty set. Being coextensive in the sense of $\Leftrightarrow$ does not imply identity. This solves the problem. The Very Serious Problems posed for any would-be naive set theorist appear to be avoided.

Along with substitution, the triviality proof relies on $\Rightarrow$ obeying weakening, in the instance $\varphi \Rightarrow (x = x \Rightarrow \varphi)$. In *relevant logics*, weakening is invalid and hence naive set theories like Brady's in DKQ [3] can include both extensionality and Leibniz identity, at least in the form

$$x = y \ \& \ z = z \rightarrow (x \in z \rightarrow y \in z)$$

where $\rightarrow$ is a relevant arrow. A relevant set theorist will still have to watch out for sets like $\mathcal{V}_2 = \{x : x \notin \mathcal{V}_2\}$, since for all x, $x \in \mathcal{V}_2$ and $x \notin \mathcal{V}_2$. So we *don't* want

$$\exists z(x \in z \ \& \ y \notin z) \rightarrow x \neq y$$

since then we'd have $\forall x(x \neq x)$. But with Brady's axiom we just get $x \neq x \vee \mathcal{V}_2 \neq \mathcal{V}_2$. It is important too for relevant logics in this vicinity that permutation is invalid: $\varphi \rightarrow (\psi \rightarrow \chi)$ does not always imply $\psi \rightarrow (\varphi \rightarrow \chi)$. Brady's conjunction does not residuate the conditional and the deduction theorem fails for these logics. So relevant logic can preserve the axiom of extensionality, and to our knowledge is the only non-classical logic to be able to do so in the face of full set comprehension.

Hence we are interested in adding conservatively a *relevant* conditional $\rightarrow$ satisfying the following initial sequents:

$$\succ \bot \rightarrow \phi \qquad \succ \phi \rightarrow \phi \qquad \phi \rightarrow \psi \succ \phi \Rightarrow \psi$$
$$\neg(\phi \Rightarrow \psi) \succ \neg(\phi \rightarrow \psi) \qquad (\phi \rightarrow \psi) \succ (\neg\psi \rightarrow \neg\phi)$$
$$\succ (\phi \,\&\, \psi) \vee (\phi \,\&\, \chi) \rightarrow \phi \,\&\, (\psi \vee \chi) \qquad \succ \phi \,\&\, (\psi \vee \chi) \rightarrow (\phi \,\&\, \psi) \vee (\phi \,\&\, \chi)$$
$$\succ \neg(\phi \vee \psi) \rightarrow \neg\phi \,\&\, \neg\psi \qquad \succ \neg\phi \,\&\, \neg\psi \rightarrow \neg(\phi \vee \psi)$$
$$\succ \neg(\phi \,\&\, \psi) \rightarrow \neg\phi \vee \neg\psi \qquad \succ \neg\phi \vee \neg\psi \rightarrow \neg(\phi \,\&\, \psi)$$
$$\succ \neg\neg\phi \rightarrow \phi \qquad \succ \phi \rightarrow \neg\neg\phi$$

Theorem 4 *A conditional $\rightarrow$ as described above can be added conservatively to our Gentzen system: adding a conditional satisfying these properties does not make any new sequent provable in the $\rightarrow$-free fragment of the language.*

Proof It suffices to show that indeed a conditional

$$\phi \rightarrow \psi := (\phi \Rightarrow \psi) \,\&\, (\neg\psi \Rightarrow \neg\phi)$$

satisfies all the desired properties.[8]

$\bot \succ \phi$ is an initial sequent so, using the introduction rule for $\Rightarrow$, we get that $\succ \bot \Rightarrow \phi$. On the other hand $\succ \neg(\bot)$, so using the weakening rule, $\neg(\phi) \succ \neg(\bot)$ and then, by the $\Rightarrow$ introduction rule, we get that $\succ \neg(\phi) \Rightarrow \neg(\bot)$ is a provable sequent. Hence, by the & introduction rule, we see that, indeed, $\succ \bot \rightarrow \phi$.

The fact that $\succ \bot \rightarrow \phi$ is provable follows since $\phi \succ \phi$ is an initial sequent and we can use the $\Rightarrow$ and & introduction rules.

The sequent $\phi \rightarrow \psi \succ \phi \Rightarrow \psi$ is provable by the & elimination rule in conjunction with the fact that $\phi \Rightarrow \psi \succ \phi \Rightarrow \psi$ is an initial sequent.

$\neg(\phi \Rightarrow \psi) \succ \neg(\phi \Rightarrow \psi)$ is an initial sequent, so using the $\vee$ introduction rule, we see that $\neg(\phi \Rightarrow \psi) \succ \neg(\phi \Rightarrow \psi) \vee \neg(\neg\psi \Rightarrow \neg\phi)$ is also provable. Furthermore, $\neg(\phi \Rightarrow \psi) \vee \neg(\neg\psi \Rightarrow \neg\phi) \succ \neg((\neg\psi \Rightarrow \neg\phi) \,\&\, (\phi \Rightarrow \psi))$, so $\neg(\phi \Rightarrow \psi) \vee \neg(\neg\psi \Rightarrow \neg\phi) \succ \neg(\phi \rightarrow \psi)$ is a provable sequent, and thus, by the Cut rule, we get that $\neg(\phi \Rightarrow \psi) \succ \neg(\phi \rightarrow \psi)$ is provable.

The sequent $\phi \Rightarrow \psi, \phi \succ \psi$ is easily seen to be provable given the $\Rightarrow$ elimination rule and the initial sequents $\phi \succ \phi$ and $\psi \succ \psi$. However $\neg\neg\phi \succ \phi$ is an initial sequent and hence, by Cut, $\phi \Rightarrow \psi, \neg\neg\phi \succ \psi$. But $\psi \succ \neg\neg\psi$ is an initial sequent so, by Cut, $\phi \Rightarrow \psi, \neg\neg\phi \succ \neg\neg\psi$. Using the $\Rightarrow$ introduction rule,

[8]This is not a serious definition of the relevant conditional; this would make various irrelevancies true, for example $p \rightarrow q$ if $\neg p$ and q. For a conservativity proof it makes no difference if the so defined conditional also satisfies other properties.

we have that $\phi \Rightarrow \psi \succ \neg\neg\phi \Rightarrow \neg\neg\psi$. Since $\neg\psi \Rightarrow \neg\phi \succ \neg\psi \Rightarrow \neg\phi$ is an initial sequent, by the & introduction rule, we obtain that $\phi \Rightarrow \psi, \neg\psi \Rightarrow \neg\phi \succ (\neg\neg\phi \Rightarrow \neg\neg\psi)\ \&\ (\neg\psi \Rightarrow \neg\phi)$ is provable. By one of the & elimination rules, we see that, in fact, then $\phi \rightarrow \psi \succ \neg\psi \rightarrow \neg\phi$ is a provable sequent.

The sequents $\succ(\phi\ \&\ \psi) \vee (\phi\ \&\ \chi) \rightarrow \phi\ \&\ (\psi \vee \chi)$ and $\succ\phi\ \&\ (\psi \vee \chi) \rightarrow (\phi\ \&\ \psi) \vee (\phi\ \&\ \chi)$ are provable given the initial sequents for distribution, $\Rightarrow$ introduction and & introduction.

The remaining sequents regarding $\rightarrow$ simply follow using the De Morgan and double negation initial sequents. □

Therefore one may extend `subDL` with a relevant conditional, and then have the extensionality principle in the form

$$\forall z(z \in x \leftrightarrow z \in y) \Rightarrow \forall z(x \in z \Rightarrow y \in z)$$

which is good.[9] The two conditionals will satisfy the nice rule: if $\varphi \rightarrow \psi$ then $\varphi \Rightarrow \psi$. Since the latter obeys modus ponens, so will the former.

6 Adding Quantifiers

Take now a structure M of the form

$$(U, \cap, \cup, \neg, \Rightarrow, <, 0, (c_i)_{i<\omega}, f)$$

where U is some non-empty collection, $\cap, \cup, \neg, 0$, and $(c_i)_{i<\omega}$ are all operations on U, $f : U \longrightarrow \wp(\omega)$ is such that $f(x\#y) = f(x) \cup f(y)$ for # in $\{\cap, \cup, \Rightarrow\}$, $f(c_i(x)) = f(x) \setminus \{l\}$ and $f(\neg(x)) = f(x)$, and $<$ is a binary relation on U.

Besides the conditions guaranteeing that $(U, \cap, \cup, \neg, \Rightarrow, <, 0)$ is a Gentzen matrix for our logic, we may demand further that the following infinite list of axioms is satisfied:

$$\forall x, y, z((\cap(x, y) < z) \supset (\cap(x, c_l(y)) < z)) \quad (l < \omega),$$
$$\forall x, y((x < y \wedge \notin f(x)) \supset (x < c_l(y))) \quad (l < \omega).$$

Now given a predicate language L with a countable list $x_0, x_1, \ldots$ of variables and a universal quantifier $\forall$ we might interpret it as follows. Suppose our language has only a binary predicate symbol R to simplify things. We will take as the models of L those structures M such that every two elements subset of ω is the image of some element of U by f. Now for any atomic formula $R(x_i, x_j)$ we let its value on U be some preimage of $\{i, j\}$. Propositional connectives get assigned the obvious interpretation whereas a quantifier $\forall x_i$ is interpreted as c_i. Now we can see that the purpose of f was to keep track of the free variables in a given formula.

---

[9]See [29].

This semantics guarantees that the system obtained by adding the following quantifier rules is sound:

$$\frac{\Gamma, \phi(y) \succ \psi}{\Gamma, \forall x\phi \succ \psi}\ (\forall\succ) \qquad \frac{\Gamma \succ \phi}{\Gamma \succ \forall x\phi}\ (\succ\forall)$$

where in ($\succ\forall$), x does not appear free in Γ. An existential quantifier $\exists$ can then be added similarly.

Completeness again follows. It suffices to consider once more the canonical model where the domain is the collection of all formulas, the operations are defined by the connectives and quantifiers, f is the function that for any formula gives the subset of numbers which are indices of the free variables of the formula, and again $<$ is interpreted by $\succ$.

7 Non-triviality?

We know that contraction is not valid in `subDL`. It does not follow from this, however, that it is coherent to add naive set theory to the logic. Traditionally, one would need to show that, in such a set theory, at least one formula (like $\varnothing \in \varnothing$) is not provable, establishing *non-triviality*. At present we do not have a way to do this. There are two approaches to consider in future work.

- Model theoretically, one builds up a structure in the fixed-point style of Kripke and then Brady, as in [3]. The problem here is that all known Brady constructions that validate the law of excluded middle will *invalidate* weakening; and all the structures that validate weakening do not seem apt for excluded middle.
- Proof theoretically, one gives a cut elimination argument, like in [31]. It is via such proof theoretic techniques that dropping contraction was first identified as a way to maintain naive comprehension. Again though the problem is that all known approaches do not adapt easily to include the law of excluded middle.

Neither approach is ruled out; we only report that, as of writing, we have not found a way.

There is a philosophical reason that proving non-triviality is, in some sense, pointless, as pointed out in [30]. Non-triviality proofs are only meaningful relative to some stronger system. But the object of the exercise is to make inconsistent set theory possible, and one good reason to do this is to have an ultra-general foundation of mathematics. If the envisioned system is itself the most general and, one might think, *correct* logico-mathematical framework for reasoning, then there is none stronger. Any non-triviality proof that mattered to a serious dialetheism, someone who was a revisionist about classical mathematics, would only be an Absolute Non-Triviality proof, not a relative one—naive set theory proving itself non-trivial. Then, of course, if the system is trivial, it will prove its own non-triviality. All of which is to say that there is a sense in which non-triviality proofs matter *only* as reassurance to our comrades still in thrall to classical logic (as Meyer once said).

It is worth pointing out that subDL has removed all of the known incoherence-triggering logical principles, up to the most recent experiments. There are no known paradoxes that will explode the set theory—and there are many known paradoxes. But lest we intone any famous last words, we move on with cautious optimism.

8 Induction and Recursion in Naive Set Theory

What can such a set theory do? In [30] we claimed that it validates proofs by induction. Here is the argument, which takes advantage of some highly impredicative sets.

8.1 Two Cases of Induction

Suppose X is a set and $f_1, \ldots, f_n, n < \omega$, are functions from V to V, where $V = \{x : \top\}$ is the set theoretic universe. We say that a set x is closed under a function f if

$$y \in x \rightarrow f(y) \in x.$$

For two sets x and y,

$$x \subseteq y =_{df} \forall z(z \in x \rightarrow z \in y).$$

$$x \sqsubseteq y =_{df} \forall z(z \in x \Rightarrow z \in y).$$

Now we want to consider the following property P for a set x:

$X \subseteq x$ and x is closed under each of the functions $f_1, \ldots, f_n$.

Let us consider, by unrestricted comprehension, a set

$$i = \{y : y \in X \bigvee_{i=1}^{n} \exists z(z \in i \,\&\, y = f_i(z))\}.$$

Due to its circular definition, this set term may fail to pick a unique element of the universe V. It is easy to see that i satisfies property P. Clearly, $X \subseteq i$. On the other hand, (for $1 \leq k \leq n$),

$$f_k(y) = f_k(y) \,\&\, y \in i \rightarrow \exists v(f_k(v) = f_k(y) \,\&\, v \in i),$$

but,

$$\exists v(f_k(v) = f_k(y) \,\&\, v \in i) \rightarrow f_k(y) \in i$$

almost immediately. Hence, i must be closed under each of the functions $f_1, \ldots, f_n$.

We have then that the set (again given by comprehension) $\{y : P(y)\}$ is not empty since $i \in \{y : P(y)\}$. One can consider next

$$\bigcap\{y : P(y)\} = \{y : \forall z(z \in \{y : P(y)\} \rightarrow y \in z)\}.$$

We see immediately that this set satisfies an induction principle with respect to property P, that is: if $X \subseteq x$ and x is closed under each of the functions $f_1, \ldots, f_n$, we can deduce that $\bigcap\{y : P(y)\} \sqsubseteq x$ for then $x \in \{y : P(y)\}$ would be the case.

However, we usually need information about the form of the elements of a given inductive set. For instance, in the case of the natural numbers we want to be able to claim that every natural number is either 0 or the successor of some natural number. Since we know that some circular sets satisfy P and the information about membership in this sort of collection is only partial we cannot be sure about the form of the elements of $\bigcap\{y : P(y)\}$. However, we can use circularity again to remedy this issue to a reasonable extent.

Consider a set

$$I = \{y \in \bigcap\{y : P(y)\} : y \in X \bigvee_{i=1}^{n} \exists z(z \in I \,\&\, y = f_i(z))\}.$$

Once more, a set like this exists by unrestricted comprehension but it may not be unique. Note that I satisfies property P and we also know that every element of I has either the form of some element of X or $f_m(y)$ $(1 \leq m \leq n)$ and $y \in I$. Notice that trivially $I \subseteq \bigcap\{y : P(y)\}$.

Then we have the following (cf. [31, Theorem 2.1]).

Theorem 5 *Let Φ be a formula, X a set term and $f_1, \ldots, f_n$ functions. Then, there is a set I such that*

$$(*)\ I = \{y \in \bigcap\{y : P'(y)\} : y \in X \bigvee_{i=1}^{n} \exists z(z \in I \,\&\, y = f_i(z))\},$$

and

$$(i)\&(ii) \Rightarrow I \sqsubseteq \{y : \Phi(y)\},$$

for

(i) $\forall x(x \in X \to \Phi(x))$
(ii) $\forall x[\Phi(x) \to \Phi(f_m(x))]$ *for* $1 \leq m \leq n$.

Proof Existence comes by comprehension, which gives us sets with the form (*). If we have facts (i) and (ii), it is immediate that $\{y : \Phi(y)\}$ satisfies property P. From (i), $X \subseteq \{y : \Phi(y)\}$, while from (ii),

$$x \in \{y : \Phi(y)\} \to f_m(x) \in \{y : \Phi(y)\},$$

and since we know that

$$f_m(x) = f_m(x) \,\&\, x \in \{y : \Phi(y)\} \to x \in \{y : \Phi(y)\},$$

$\{y : \Phi(y)\}$ is closed under f_m. But then $\{y : \Phi(y)\} \in \{y : P(y)\}$, so

$$I \subseteq \bigcap\{y : P(y)\} \sqsubseteq \{y : \Phi(y)\},$$

which is what we wanted. □

Now, suppose we consider instead of P, the following property P':

$$X \in x \text{ and } x \text{ is closed under each of the functions } f_1, ..., f_n.$$

Theorem 6 *Let* Φ *be a formula,* X *a set term and* $f_1, ..., f_n$ *functions. Then, there is a set* I *such that*

$$I = \{y \in \bigcap\{y : P'(y)\} : y = X \bigvee_{i=1}^{n} \exists z(z \in I \,\&\, y = f_i(z))\},$$

and

$$(i)\&(ii) \Rightarrow I \sqsubseteq \{y : \Phi(y)\},$$

for

(i) $\Phi(X)$
(ii) $\forall x[\Phi(x) \rightarrow \Phi(f_m(x))]$ *for* $1 \leq m \leq n$.

Proof A trivial modification of the proof of the previous result. □

The two induction principles above are theorem schemata rather than proper theorems, that is, they give a theorem for every set term X, formula Φ and functions $f_1, ..., f_n$.

8.2 Natural Numbers

As an application of the above we can show that there are sets of all natural numbers (since in general we cannot get a unique set of all natural numbers because of circularity) satisfying inductive principles.

We denote $S(x) = x \cup \{x\}$ following von Neumann. Let P^N be the following property (for a set x):

$$0 \in x \text{ and } x \text{ is closed under } S.$$

Corollary 1 *Let* Φ *be a formula. Then, there is a set* ω *such that*

$$\omega = \{y \in \bigcap\{y : P^N(y)\} : y = 0 \vee \exists z(z \in \omega \,\&\, y = S(z))\},$$

and

$$\Phi(0) \,\&\, \forall x(\Phi(x) \rightarrow \Phi(S(x))) \Rightarrow \omega \sqsubseteq \{y : \Phi(y)\}$$

9 Conclusion

As a logic, the one presented here is of interest from the perspective of multiset consequence, especially as it satisfies a deduction theorem (cf. [7]). The connection between intensional conjunction and extensional disjunction via de Morgan negation is novel. One immediate open task is to give an interpretation of this logic beyond the formula-driven Gentzen matrices, or at least to make a detailed investigation of the algebra presented here.

As a base for inconsistent mathematics, this logic combines strength with flexibility right up to the limit of what is possible for paraconsistent set theory (and, we hope, not beyond). Work in modal logic, arithmetic, and topology using this logic are underway. There is structure in the inconsistent. To study it is the most delicate of balancing acts, requiring some very finely tuned equipment. The logic presented here is progress. Let us see what is next.[10]

References

1. Batens, D., Mortensen, C., Priest, G., & van Bendegem, J.-P. (Eds.). (2000). *Frontiers of paraconsistent logic*. Philadelphia: Research Studies Press.
2. Brady, R. (1971). The consistency of the axioms of the axioms of abstraction and extensionality in a three valued logic. *Notre Dame Journal of Formal Logic*, *12*, 447–453.
3. Brady, R. (2006). *Universal logic*. Stanford: CSLI.
4. Brady, R. T. (1989). The non-triviality of dialectical set theory. In [23], pp. 437–470.
5. Brown, B., & Priest, G. (2004). Chunk and permeate i: The infinitesimal calculus. *Journal of Philosophical Logic*, *33*, 379–88.
6. Cantini, A. (2003). The undecidability of grisin's set theory. *Studia Logica*, *74*(3), 345–368.
7. Chvalovský, K., & Cintula, P. (2012). Note on deduction theorems in contraction-free logics. *Mathematical Logic Quarterly*, *58*(3), 236–243.
8. da Costa, N. (2000). Paraconsistent mathematics. In [1], pp. 165–180.
9. da Costa, N. C. A. (1974). On the theory of inconsistent formal systems. *Notre Dame Journal of Formal Logic*, *15*, 497–510.
10. Dunn, J. M. (1979). A theorem in 3-valued model theory with connections to number theory, type theory, and relevant logic. *Studia Logica*, *38*, 149–169.
11. Dunn, J. M. (1980). Relevant robinson's arithmetic. *Studia Logica*, *38*, 407–418.
12. Galatos, N., Jipsen, P., Kowalski, T., & Ono, H. (2007). *Residuated lattices*. Elsevier.
13. Grišin, V., (1982). Predicate and set theoretic calculi based on logic without contraction rules. *Mathematics of the USSR Izvestija*, *18*(1), 41–59.
14. Hamkins, J. D. (2012). The set-theoretic multiverse. *Review of Symbolic Logic*, *5*, 416–449.
15. Libert, T. (2005). Models for paraconsistent set theory. *Journal of Applied Logic*, *3*.
16. Meyer, R. K., & Mortensen, C. (1984). Inconsistent models for relevant arithmetics. *Journal of Symbolic Logic*, *49*, 917–929.
17. Moore, G. H. (1982). *Zermelo's axiom of choice*. Berlin: Springer.
18. Mortensen, C. (1995). *Inconsistent mathematics*. New York: Kluwer Academic Publishers, Dordrecht.
19. Mortensen, C. (2011). *Inconsistent geometry*. College Publications.
20. Ono, H., & Komori, Y. (1985). Logic without the contraction rule. *Journal of Symbolic Logic*, *50*, 169–201.
21. Priest, G. (2000). Inconsistent models of arithmetic, ii: The general case. *Journal of Symbolic Logic*, *65*, 1519–29.
22. Priest, G. (2008). *An introduction to non-classical logic* (2nd ed.). Cambridge.
23. Priest, G., Routley, R., & Norman, J. (Eds.). (1989). *Paraconsistent logic: essays on the inconsistent*. Munich: Philosophia Verlag.

[10] Thanks to Petr Cintula, Patrick Girard, Dave Ripley, Lloyd Humberstone, members of the Melbourne Logic Group, and the group at the Second Kyoto Workshop on Non-Classical Logic. This work was supported by a Marsden Grant from the Royal Society of New Zealand.

24. Restall, G. (1992). A note on naïve set theory in *LP*. *Notre Dame Journal of Formal Logic, 33*, 422–432.
25. Routley, R. (1977). Ultralogic as universal?, in *Relevance logic newsletter* (Vol. 2, pp. 51–89). (New edition as Volume 4 of *Exploring Meinong's Jungle and Beyond* , edited by Zach Weber, with contributions. Synthese Library, forthcoming 2018).
26. Routley, R., & Meyer, R. K. (1976). Dialectical logic, classical logic and the consistency of the world. *Studies in Soviet Thought, 16*, 1–25.
27. Routley, R., Plumwood, V., Meyer, R. K., & Brady, R. T. (1982). *Relevant logics and their rivals*. Ridgeview.
28. Terui, K. (2004). Light affine set theory: A naive set theory of polynomial time. *Studia Logica, 77*, 9–40.
29. Weber, Z. (2010). Extensionality and restriction in naive set theory. *Studia Logica, 94*(1), 87–104.
30. Weber, Z., Badia, G., & Girard, P. (2016). What is an inconsistent truth table? *Australasian Journal of Philosophy, 94*(3).
31. White, R. B. (1993). A consistent theory of attributes in a logic without contraction. *Studia Logica, 52*, 113–142.

The manufacturer's authorised representative in the EU is Springer Nature Customer Service Centre GmbH, Europaplatz 3, 69115 Heidelberg, Germany. If you have any concerns regarding our products, please contact ProductSafety@springernature.com

Printed and bound by CPI Group (UK) Ltd, Croydon, CR0 4YY
22/09/2026
02221058-0004